GUILÁ NAQUITZ

Archaic Foraging and
Early Agriculture in Oaxaca, Mexico

This is a volume in
STUDIES IN ARCHAEOLOGY

A complete list of titles in this series is available from the publisher.

GUILÁ NAQUITZ

Archaic Foraging and Early Agriculture in Oaxaca, Mexico

Edited by

Kent V. Flannery

Museum of Anthropology
University of Michigan
Ann Arbor, Michigan

1986

ACADEMIC PRESS, INC.

(Harcourt Brace Jovanovich, Publishers)

Orlando San Diego New York London
Toronto Montreal Sydney Tokyo

Frontispiece. Autumn in the upper piedmont near Guilá Naquitz, ca. 7000 B.C. A Naquitz phase forager removes a cottontail (*Sylvilagus* sp.) from a snare while his son collects prickly pear fruit (*Opuntia* sp.) with a knotted net bag. To the boy's right is an oak tree (*Quercus* sp.); to his left is a nanche (*Malpighia* sp.) heavy with fruit. The adult forager is careful to avoid the large *mala mujer* (*Jatropha urens*) in the lower right foreground, a plant which can produce a nettle-like sting. Behind him we see a *garambullo* or organ cactus (*Myrtillocactus* sp.), its trunk partly hidden by the fruit-laden boughs of another nanche. In the lower left foreground we see a flowering prickly pear, partly hidden by a *susí* bush (*Jatropha neodioica*) with its edible nuts borne in paired seed cases. Still further back is a saw-toothed *lechuguilla* (*Agave karwinskii*), sending its *quiote* or inflorescence 3 meters into the air. More shrub oaks and *susí* bushes appear behind the *lechuguilla*. In the center foreground, between the *susí* and *mala mujer,* we see a single *Hechtia* plant.

The rabbit, acorns, susí nuts, nanches, and prickly pear fruits will all be eaten this autumn; the *garambullo* will bear most of its fruit later in the winter, when Guilá Naquitz has been abandoned. The agave inflorescence will be used as a fire-drill hearth, and the *Hechtia* will be used for fiber. (Drawing by Nancy Hansen.)

NH

For
SCOTTY MACNEISH
who introduced me to the joys
of rummaging through other people's 10,000-year-old trash

ACADEMIC PRESS, INC.
Orlando, Florida 32887

United Kingdom Edition published by
ACADEMIC PRESS INC. (LONDON) LTD.
24–28 Oval Road, London NW1 7DX

LIBRARY OF CONGRESS CATALOGING IN PUBLICATION DATA

Main entry under title:

Guilá Naquitz.

Bibliography: p.
Includes index.
1. Guilá Naquitz Cave (Mexico) 2. Indians of Mexico—Agriculture. 3. Agriculture—Origin. 4. Excavations—(Archaeology)—(Mexico) 5. Indians of Mexico—Antiquities. 6. Mexico—Antiquities. I. Flannery, Kent V. II. Title.
F1219.1.O11G85 1985 972'.74 85-4051
ISBN 0-12-259830-X (alk. paper)

PRINTED IN THE UNITED STATES OF AMERICA

86 87 88 89 9 8 7 6 5 4 3 2 1

Contents

II THE CULTURAL AND ENVIRONMENTAL SETTING 29

Contributors

Numbers in parentheses indicate the pages on which the authors' contributions begin.

Eric O. Callen[1] (173), Department of Plant Pathology, McGill University, Montreal, Canada PQ H3A 2T7

Hugh C. Cutler (275), Missouri Botanical Garden, and Washington University, St. Louis, Missouri 63130

J. N. Elias (297), Department of Medicine, Indiana University Medical Center, Indianapolis, Indiana 46202

Kent V. Flannery (3, 19, 31, 43, 65, 147, 163, 169, 171, 175, 239, 249, 255, 285, 303, 321, 331, 425, 435, 501, 511), Museum of Anthropology, University of Michigan, Ann Arbor, Michigan 48109

Frank Hole (97), Department of Anthropology, Yale University, New Haven, Connecticut 06520

Lawrence Kaplan (281), Department of Biology, University of Massachusetts, Boston, Massachusetts 02125

Mary Elizabeth King (157), Laboratory of Anthropology, Museum of Indian Arts and Culture, Santa Fe, New Mexico 87504

Michael J. Kirkby (43), School of Geography, University of Leeds, Leeds LS2 9JT, England

Silvia Maranca (65), Museu Paulista da USP, 04263 São Paulo, Brazil

Chris L. Moser (65), Riverside Municipal Museum, Riverside, California 92501

Robert G. Reynolds (385, 439), Department of Computer Science, Wayne State University, Detroit, Michigan 48202

J. R. K. Robson (297), Department of Family Medicine, Medical University of South Carolina, Charleston, South Carolina 29425

James Schoenwetter (179), Department of Anthropology, Arizona State University, Tempe, Arizona 85287

C. Earle Smith, Jr. (265), Department of Anthropology, University of Alabama, University, Alabama 35486

Landon Douglas Smith (179), USDA Forest Service, Southwestern Region, Computer Systems, Albuquerque, New Mexico 87102

Charles S. Spencer (331), Department of Anthropology, University of Connecticut, Storrs, Connecticut 06268

Michael E. Whalen (141), Department of Anthropology, University of Tulsa, Tulsa, Oklahoma 74104

[1]Deceased.

Robert Whallon (369), Museum of Anthropology, University of Michigan, Ann Arbor, Michigan 48109
Jane C. Wheeler (239, 285), Department of Anthropology, University of Colorado, Boulder, Colorado 80309
Thomas W. Whitaker (275), U. S. Department of Agriculture, and Department of Biology, University of California, San Diego, La Jolla, California 92038
Anne V. Whyte (43), Division of Ecological Sciences, UNESCO, 75700 Paris, France
David J. Wilson[2] (521), Museum of Anthropology, University of Michigan, Ann Arbor, Michigan 48109

[2]Present address: Department of Anthropology, Southern Methodist University, Dallas, Texas 75275

Preface

This is the report of an interdisciplinary study of the origins of agriculture in the southern Mexican highlands. It begins in the preagricultural era and ends with the earliest evidence for cultivated plants in the eastern Valley of Oaxaca. The disciplines employed include archaeology, botany, zoology, palynology, and computer modeling.

There are a number of precedents for this kind of study. I have been lucky enough to participate in some earlier projects on the origins of agriculture, in Mesoamerica with Richard S. MacNeish (1960–1964) and in the Near East with Robert J. Braidwood (1960) and Frank Hole (1961–1963). The Guilá Naquitz report builds on their earlier work, and would probably never have been undertaken had I not been able to tag along on their projects.

Guilá Naquitz was a small cave, not in a class with larger sites such as Coxcatlán Cave in Mexico's Tehuacán Valley, or Hayonim Cave in Israel's Galilee Hills. Its small size, however, meant that it could be excavated in its entirety, giving us a very good look at what MacNeish has called a "microband camp." Microband camps are only one type of site in a complex settlement system, and Guilá Naquitz is only one of four preceramic sites on which we ultimately intend to report. Therefore, while this is our final report on the preceramic levels at Guilá Naquitz, it is certainly not our final word on the Oaxacan preceramic sequence. Eventually, when other sites such as Cueva Blanca and Gheo-Shih have been reported, we will be able to see the Guilá Naquitz data in clearer perspective.

Since Guilá Naquitz was first tested in 1966, it will be evident that this was a long-term study. Indeed, the excavation itself was the most rapid phase of the project. Analyzing the artifacts took several years, and the pollen study took longer than that. The censuses of wild vegetation described in Chapter 18 took most of a decade (1966–1976). Only after those censuses were complete could computer modeling of subsistence strategies begin, and that modeling consumed another 4 or 5 years. Since we needed previously uncollected data on annual variation in rainfall and wild productivity, there was no way the ecological phases of the study could have been done any faster. We now see the value of such long-term analysis and have more or less resigned ourselves to it, but many of our Mesoamerican colleagues remain puzzled as to how it could take 15 years to analyze such a small site.

I can sympathize with their position, because I used to feel that way myself. As a graduate student 20 years ago, I used to kid Bob Braidwood about how long it was taking him to complete the interdisciplinary study of Jarmo. First tested in 1948, Jarmo was finally published in 1983, after a lapse of time that no longer seems astonishing to me at all. In fact, as I told Braidwood on the occasion of his *Festschrift* presentation in 1982, my experience with Guilá Naquitz has converted me to what might be called the "Paul Masson" approach to archaeology: "I will publish no site. . . before its time."

Acknowledgments

A number of institutions contributed to the support of the research described in this volume, which was part of a project entitled "The Prehistory and Human Ecology of the Valley of Oaxaca." Smithsonian Research Foundation Grant 019 funded the excavation of Guilá Naquitz in 1966; our analysis of artifacts, plant remains, pollen, and animal bones continued into the decade of the 1970s and was supported by National Science Foundation Grants GS-1616, GS-2121, and GS-42568. The NSF also funded additional studies, including several years of wild plant censuses (1967–1976). Finally, the University of Michigan supported an extensive series of computer analyses and simulations between 1976 and 1984. In this age of padded budgets and inflated grants, however, I should probably point out that the entire excavation of Guilá Naquitz cost only $5000, and the subsequent analyses less than $7000.

The John Simon Guggenheim Foundation made the writing and editing of this volume possible by providing me with a fellowship during the 1981–1982 academic year.

I am grateful to Mexico's Instituto Nacional de Antropología e Historia for permission to excavate Guilá Naquitz and other preceramic sites. In Mexico City, José Luis Lorenzo (Jefe, Departamento de Prehistoria) and Ignacio Bernal (Director of I.N.A.H., 1968–1971) were instrumental in arranging our permits. In Oaxaca, Lorenzo Gamio (1966–1972) and Manuel Esparza (1973–1981) were the I.N.A.H. representatives who facilitated our work.

Our colleagues from the University of the Americas and its regional headquarters, the Museo Frissell de Arte Zapoteca in Mitla, provided us with a laboratory and a home away from home during the early years of the Oaxaca Project. In particular, we could always depend on John Paddock for his help and archaeological advice, and on Darío Quero for his hospitality. Cecil Welte, of the Oficina de Estudios de la Humanidad del Valle de Oaxaca, gave us the benefit of his extensive knowledge of the area. Humberto and Liza Arredondo (then of Mexico City) loaned me a vehicle for the initial survey of Oaxaca caves. Richard S. MacNeish, who had done a preliminary study of projectile points in the Frissell Museum collections, was the first to suggest that I begin my preceramic surveys in the Mitla area. My intellectual debt to MacNeish, to whom this volume is dedicated, is obviously very great indeed.

Marcia Johnson and Camie Moody typed the manuscript of the book, and a series of talented artists provided the illustrations. Three who should be singled out for special mention are Katherine Clahassey, who did the charts and diagrams in Chapters 28–32; Nancy Hansen, who did the frontispiece, the artifact drawings in Chapters 6–8, and the plant illustrations in Chapter 4; and Lisa Klofkorn, who did the contour maps in Chapter 26. Virtually all the excavation and artifact photographs are by Chris L. Moser.

Finally, we cannot close our acknowledgments without a word about the kindness and generosity of the people of Mitla. Our Mitla workmen were some of the best excavators and nicest people we have ever met. Indeed, we found the Mitleños to be just as Elsie Clews Parsons (1936:x) found them 30 years before, "*buena gente, gente muy cariñosa.*"

I

THE PROBLEM AND THE MODEL

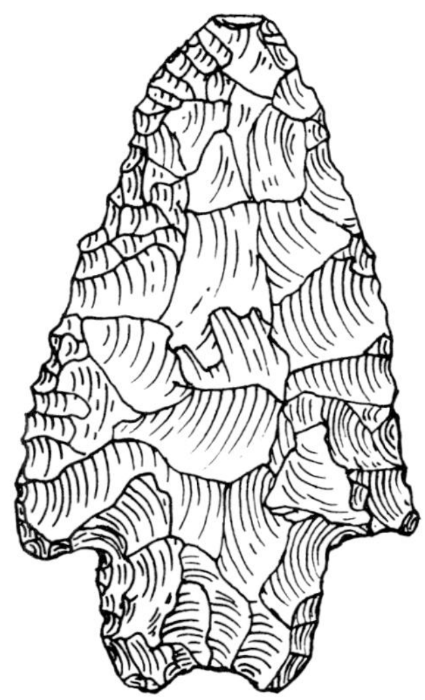

1

The Research Problem

Kent V. Flannery

We save the *chipil* seeds to plant for the time when there are none in the fields so we will have greens to eat.

Zapotec informant,
speaking to ethnobotanist
Ellen Messer, ca. 1972
(Messer 1978:88).

INTRODUCTION

During the late Pleistocene period, at a time roughly equivalent to the Wisconsin glacial advance in North America, the climate of central Mexico seems to have been colder and drier than today's (Bradbury 1971). Areas such as the Tehuacán Valley of southern Puebla were inhabited by a late Pleistocene complex of animals that today would be found on the cooler and more arid plains of Coahuila and southern Texas (Flannery 1967). As the Wisconsin glaciation drew to a close, temperatures in central Mexico rose until the mountain valleys of Puebla and Oaxaca had been taken over by a rich forest of cacti and legume trees, giving way at higher elevations to oak and pine. This transition to a semiarid, semitropical complex of vegetation, not unlike today's, appears to have been well along by 8000 B.C.

Sometime between the establishment of this post-Pleistocene vegetational regime and the start of the fifth millennium B.C., the Indians of this part of Mexico began the cultivation of a series of plants, some of them native members of the Holocene thorn forest. For centuries the prehistoric inhabitants of Puebla, Oaxaca, Morelos, Guerrero and the Basin of Mexico had lived off the land, learning the secrets of the wild vegetation. They had learned how to roast the maguey, or century plant, to make it edible, how to make wooden tongs to pick off the spiny fruit of the organ cactus, how to cook up a syrup from the pod of the mesquite, how to leach tannic acid from bitter acorns, and how to find wild onion and runner bean flowers in the piedmont underbrush and to predict when they would be ready to harvest. They had developed a technology that included protective maguey-fiber sandals, the atlatl, or spear-thrower, the fire-making drill, the coiled basket, the net carrying-bag, a variety of snares and traps, the one-hand mano for seed grinding, and the boulder mortar. Camping sometimes in caves, sometimes in the open, for hundreds of years they followed a strategy in which they moved to the wild resources as the latter came ripe, dispersing into smaller groups or coming together in larger ones as the seasons of abundance permitted. By 7000 B.C. they were carrying their drinking water with them by means of the bottle gourd, a domestic plant whose early history is still mysterious and which is not thought to be native to Mexico.

Finally, the Indians of central Mexico added another strategy—the manipulation of certain edible plants by selection and planting. Three of the most important of these were squashes (*Cucurbita* spp.), beans (*Phaseolus* spp.), and maize, or Indian corn (*Zea mays*), all of which are discussed in this volume. These cultivated plants made their first appearance as relatively minor components of a diet still dominated by wild species, but their importance was to increase with time as favorable genetic changes made their cultivation an even more attractive activity.

AGRICULTURE: "HOW" OR "WHY"?

The origin of agriculture was an important moment in cultural evolution, and the question most frequently asked about it by anthropologists is, "Why did it happen?" I would like to reflect on that question for a moment, because it is so different from the one usually asked by paleontologists about biological evolution.

For example, the weight of fossil evidence has convinced paleontologists that birds evolved from reptiles, but one rarely hears a paleontologist ask, "*Why* did birds evolve from reptiles?" Rather, one usually hears the question, "*How* did birds evolve from reptiles?" To a paleontologist, "why?" is a teleological question; it implies that something was done for a purpose. Evolution, on the other hand, is generally considered to be opportunistic: Changes take place not because evolution has a goal to reach, but because opportunity is present and a favorable mutation takes place in the right place at the right time.

Paleontologists will tell you that there are two major competing theories about *how* birds evolved from reptiles (Lewin 1983). According to the "arboreal" theory, the intermediate stage was a reptile who became adapted to gliding at the end of a leap from a high place. According to the "cursorial" theory, the intermediate stage was a reptile who developed "lift" that allowed it to increase its foraging range while running and leaping. However, the question, "*Why* did birds evolve?" is unlikely to elicit the answer, "Because reptiles wanted to catch more insects." Some paleontologists would probably say, "We don't yet know." Still others might answer, "Because certain mutations took place at a time when conditions were right, and natural selection favored those individuals." This is not a specific explanation for the origin of birds but a reference to an underlying principle of evolution.

I have always been struck by the fact that while paleontologists most frequently ask *how*, anthropologists most frequently ask *why*, since it underlines a basic difference between the two fields. Paleontologists are concerned with evolutionary mechanisms, and they try to answer the *how* question by studying the intervening fossil forms from an evolutionary transition. Anthropologists are concerned with human aspirations, with what Rindos (1980) has called "intentionality," and they try to answer the *why* question by studying culturally patterned behavior. Paleontologists do not picture reptiles saying, "Let's turn into birds," but many anthropologists do picture hunter–gatherers saying, "Let's plant these seeds on the talus slope below our cave."

Nor is this an outrageously unscientific thing for anthropologists to picture. Let us consider, for a moment, the wild herb *Crotalaria pumila*, a member of the legume family, which is known as *chipil* in the Valley of Oaxaca. While not considered a cultivated plant, the *chipil* has a symbiotic relationship with man; it is one of the first second-growth weeds to invade fallow cornfields in the region, and its leaves are gathered, along with wild *epazote* (*Chenopodium* sp.) and *quintonil* (*Amaranthus* sp.), to be used as flavoring in tamales or stews. Ethnobotanist Ellen Messer (1978) found some Zapotec-speaking villagers saving the seeds of *Crotalaria*, and when she asked why, she was given the answer seen at the start of this chapter: "We save the *chipil* seeds to plant for the time when there are none in the fields so we will have greens to eat."

There are several points that can be drawn from this example. First of all, human cultures do display intentionality, and it may be completely rational and practical. They have perfectly good reasons for planting *chipil*, although those reasons would not necessarily be recoverable by an archaeologist 5000 years from today. The Zapotec certainly do not see themselves as occupying a stage of "incipient *chipil* cultivation", but if favorable genetic changes were to appear in certain *Crotalaria* plants and the Zapotec were to select for those traits when saving seeds, there could one day be a domestic strain of *chipil*. Botanical colleagues might be able to explain to future archaeologists how the change had taken place, but not necessarily why. As in the bird evolution example above, they could refer it to an underlying principle of genetic change and selection, leaving the problem of intentionality aside.

Anthropological archaeologists, however, would continue to kick intentionality around, and there is a good chance they would try to explain *chipil* domestication in terms far more deterministic than Messer's informants. Based on some of the models we examine later in this chapter, they might argue that the Zapotec were forced to augment their diet with *chipil* cultivation because of (1) climatic changes that reduced their maize crop, (2) population increases that made the maize crop inadequate, (3) greater production demands made on them by a local government, or (4) some nutrient missing in maize but present in *Crotalaria*.

While all these explanations are plausible, the quotation from Messer's informant suggests a much simpler and less frequently cited reason for planting *chipil*. The Valley of Oaxaca is a semiarid, semitropical environment in which many variables—most notably rainfall—show great fluctuation and unpredictability from year to year. While annual rainfall usually ranges between 450 and 650 mm, over the period from 1930 to 1970 there were 4 years with less than 400 mm, 5 years with over 700 mm, 2 years with less than 300 mm, and 1 year with over 1000 mm (Kirkby 1973:Fig. 58). In very dry years, the *chipil* may not survive at all, while in very wet years it may be flooded out by standing water on the fields. The Zapotec deal with this unpredictable annual variation by saving *chipil* seeds to plant whenever nature does not provide enough greens. It is not a response to overpopulation, crop failure, or political pressure, but to uncertainty.

This is perhaps the most general point to be drawn from our *Crotalaria* example. The origins of agriculture involve both human intentionality and a set of underlying ecological and evolutionary principles. When we, like paleontologists, ask *how* a particular case of domestication happened, we are probably going to be confronted with a biological principle such as mutation, natural selection, or (as we shall see below) coevolution. This is probably the level on which the most universal aspects of domestication lie. When we, like anthropologists, ask *why*, we may find ourselves confronted with reasons that are local and can only be modeled after we have done our best to reconstruct the local cultural pattern.

Small wonder, then, that anthropologists have had so little success producing universal explanations for the origins of agriculture. Concerned as they are with human aspirations and intentionality, they have focused on what is probably the most regionalized and least universal aspect of agricultural origins. No matter how elegant a hypothesis of population pressure or climatic stress one could propose for preceramic Mexico, if we could go back to 8000 B.C. and interview the first cultivators in that region, we might simply be told, "We save the bottle gourd seeds to plant for the time when there are none, so that we will have something to carry our water in."

Of course, we can ignore human intentionality and concentrate on the underlying biological aspects of domestication, producing a universal model as Rindos (1980) has recently done (see below). The trouble is that while such a model may satisfy biologists, it often does not satisfy anthropologists (see Ceci 1980; Cohen 1980). The reasons for this lack of satisfaction are not hard to discover: Anthropologists want cultural explanations for cultural behavior and are not usually satisfied when culture is reduced to biology. Anthropologists know that human hunter–gatherers are mammals, primates, and predators, but that is not what anthropologists find most interesting about them. It is no accident that so much of recent writing on hunter–gatherers deals with "decision making," "strategy," "consensus," "information flow," and the application of models drawn from economics (Keene 1981a; Moore 1981; Reynolds 1978; Reynolds and Ziegler 1979; Winterhalder and Smith 1981).

Recently, Wilson (1979) presented a diagram showing the different ways in which biologists, primatologists, anthropologists, and sociologists view human social behavior; a modified version is reproduced here as Fig. 1.1. As one moves down the diagram toward biology, one gains perspective and loses resolution; as one moves up the diagram toward sociology, one gains resolution and loses perspective. Reducing the origins of agriculture to a case of plant–animal symbiosis or coevolution achieves biological perspective but lacks resolution for Mexico or the Near East. Focusing on human decision making in Mexico may provide high resolution for that area yet be inapplicable for the Near East. Perhaps no model will achieve much acceptance until it effectively includes both the universality of the biological principles and the specificity to tie it into the human cultural pattern for a given region.

The research problem we have chosen to attack in this volume is essentially that: to develop a model for the origins of agriculture in Oaxaca, Mexico, that deals not only with some of the underlying and more universal processes of domestication but also with the specific events of the Oaxaca cultural sequence. Our goals in this first chapter, however, are more limited. We deal first with the empirical data on early agriculture in Mexico as they are currently understood. Following that, we examine recent theoretical frameworks for the origins of agriculture as they relate to the Oaxaca–Puebla region.

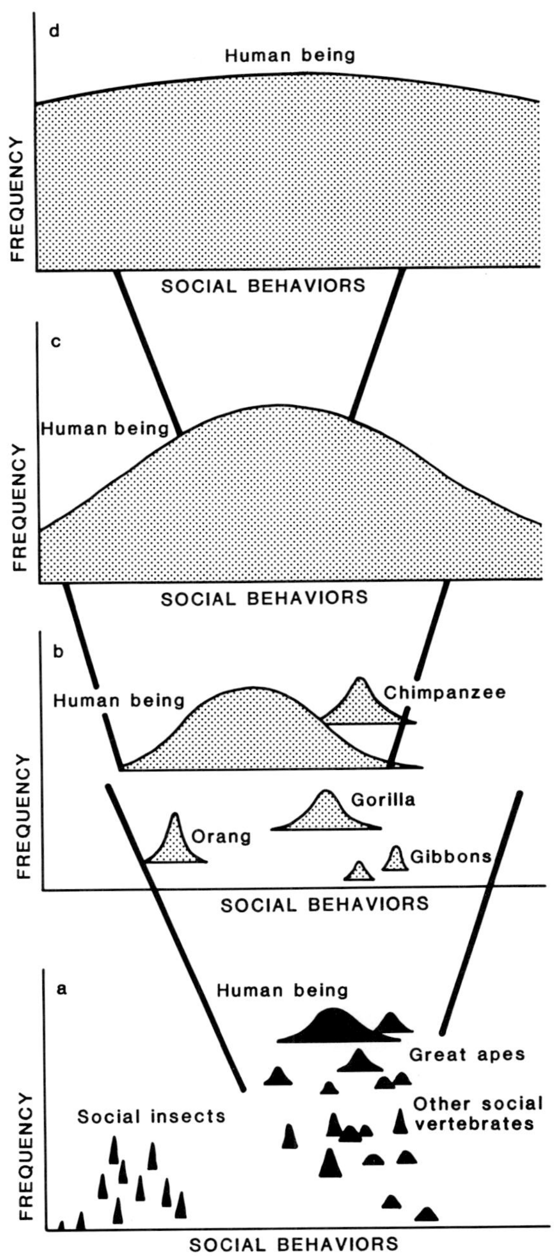

FIGURE 1.1. Wilson's (1979) diagram comparing the ways social behavior is focused on by (a) biologists, (b) primatologists, (c) anthropologists, and (d) sociologists. In going from a through d, we gain greater resolution but lose perspective, and vice versa. (Redrawn, with modifications, from Barash 1982:Fig. 7.2)

THE PALEOBOTANICAL EVIDENCE

It is unlikely that we will ever know the precise chronology of early domestication in Mexico; preceramic sites are too scattered and too infrequently excavated, conditions for plant preservation too unpredictable, and radiocarbon dates too frequently ambiguous. In Table 1.1, I have listed the earliest known occurrences of gourds, squashes, runner beans, maize, and teosinte, but the table only underscores how many more excavations we will need before we have an adequate sample. All dates are given using the original 5570-year half-life of ^{14}C as recommended by the journal *Radiocarbon*.

TABLE 1.1
The Oldest Archaeological Occurrences of Some Early Mexican Domesticates and Their Wild Ancestors[a]

Name of plant	Oldest archaeological occurrence	Comment	Subsequent archaeological occurrences
Lagenaria siceraria (bottle gourd)	Rind fragments in Zones C–B1 of Guilá Naquitz, Oaxaca (7400–6700 B.C.); 7000 B.C. levels at Ocampo, Tamaulipas.	Rinds are good evidence, but seeds are better.	One rind in Zone XIII of Coxcatlán Cave, Puebla (5050 B.C.).
Cucurbita pepo (summer squash)	One "pepo-like" seed in Zone D of Guilá Naquitz (8750–7840 B.C.); larger sample at 7000 B.C. at Ocampo, Tamaulipas.	Seeds are fairly good evidence, but peduncles are even better.	14 seeds and peduncles in Zones C–B1 at Guilá Naquitz (7400–6700 B.C.).
Wild runner beans	100+ in 8750–6700 B.C. levels at Guilá Naquitz; 14 between 7000 and 5500 B.C. at Ocampo, Tamaulipas.	Tamaulipas specimens are a wild ancestor of domestic runner bean; Guilá Naquitz specimens are not.	One runner bean from roughly 200 B.C. at Coxcatlán Cave.
Zea mexicana (teosinte)	Pollen grains in Zones C–B1 of Guilá Naquitz (7400–6700 B.C.).	It is still disputed whether teosinte pollen can be distinguished from that of maize.	Waterlogged seeds from 5090 B.C. level at Zohapilco, Basin of Mexico.
Zea mays (maize)	18 cobs in Zone XIII of Coxcatlán Cave, Puebla (5050 B.C.).	Mangelsdorf says "wild"; Beadle would say "domestic" (see text).	74 more cobs in 5000–3000 B.C. levels of Tehuacán (Puebla) caves.

[a]Source: Flannery 1973: Table 2.

At the time I began work in Oaxaca, virtually all our paleobotanical evidence for early domestication came from three areas studied by MacNeish: (1) dry caves in the Sierra de Tamaulipas (MacNeish 1958), (2) dry caves in the Sierra Madre near Ocampo, Tamaulipas (Mangelsdorf *et al.* 1964), and (3) dry caves in the Valley of Tehuacán, Puebla (Byers 1967). Since then, additional data have come from (4) Niederberger's carbonized or waterlogged material from the open-air site of Zohapilco (Tlapacoya) in the Basin of Mexico (Niederberger 1979), and (5) my own excavations at Guilá Naquitz and other sites in Oaxaca, reported in this volume. Bearing in mind how limited an area these five excavation projects cover (and the fact that in some cases the earliest evidence consists of one seed, one pod, or some equally small sample), let us now briefly consider three of the most important kinds of early cultivated plants.

GOURDS AND SQUASHES

Perhaps the first plants domesticated in the New World were members of the family Cucurbitaceae, a group that includes not only the squashes and pumpkins but also the bottle gourd *Lageneria siceraria*. This gourd, as botanist Charles B. Heiser has pointed out, "must have been one of man's most important plants before the invention of pottery," and may even have been one of the world's most widely distributed plants in the era of preceramic nomadism (Heiser 1979:71). Especially in arid and semiarid environments, foragers needed some way to transport drinking water. The !Kung Bushmen of the Kalahari and the Natufian hunters of the Negev used ostrich egg shells, some Near Eastern hunters may have used goatskin bags, and the Indians of Mesoamerica and the Andes had the bottle gourd.

Since this gourd is one of the very first cultivated plants to appear in the New World archaeological record, many researchers suspect that it may have been the first plant domesticated there. Lathrap (1977) has suggested that in addition to its use as a water bottle, the gourd's suitability as a net float led to its worldwide dissemination along coasts. One of the intriguing aspects of seeing the gourd as the New World's first domesticate is that it removes the origins of agriculture from the realm of food stress and population pressure, making it a technological breakthrough. Hunter–gatherers moving into an environment where bottle gourds did not occur naturally needed only to carry along some seeds to provide themselves and their descendants with a supply of water bottles. Like the origin of ground-stone tools, this is a technological change that could have happened at any population density and under a wide range of environmental conditions. Moreover, once cultivation of gourds had begun, it could have acted as a preadaptation for the growing of other plants of the same family: Wild squashes, whose seeds are edible (though their flesh is bitter to nonexistent), would have been instantly recognizable as something easy to grow, store, and transport.

A problem in any reconstruction of gourd domestication, however, is the uncertainty about its wild ancestor. The bottle gourd is supposed to be native to Africa, but as Heiser (1979) points out, the evidence for its homeland is far from conclusive, and the plant was widespread in Asia and America in prehistoric times. It has been shown that gourds can float in seawater for at least 347 days—"ample time to have crossed the Atlantic," according to Heiser—and still have viable seeds. In fact Heiser (1979:114) suspects that the bottle gourd floated to America many times, perhaps as much as 50,000–100,000 years ago, and may have washed up anywhere from Brazil to Mexico.

Once established on the New World coast and used by man, the bottle gourd could have become a commensal plant whose seeds would sprout naturally in human campsites. Thus, "there is no real need to postulate the existence of a wild gourd being found outside of man's habitations at any time in the Americas" (Heiser 1979:116). Heiser could picture the gourd originally spreading as a weed rather than as a domestic plant, achieving a truly symbiotic relationship with humans—the latter dependent on gourds for water bottles and net floats, the gourd dependent on human settlements for its survival—which was followed eventually by domestication. This model of mutualism is not unlike the approach taken by Rindos (1980).

If the bottle gourd was Mexico's earliest domesticate and served as a preadaptive model for cultivating the other cucurbits, this might eventually be confirmed by the archaeological record. At the moment, there are simply too few archaeological occurrences to show this, but we can say that both gourds and squashes are early in the Archaic record (Table 1.1). Rind fragments of bottle gourd occur in MacNeish's caves near Ocampo, Tamaulipas by 7000 B.C. (Whitaker *et al.* 1957) and in Zones C–B1 of Guilá Naquitz by 7400–6700 B.C. (Chapter 20). Fourteen seeds and peduncles of *Cucurbita pepo* occur in those same levels at Guilá Naquitz, and one *pepo*-like seed was found in zone D (8750–7840 B.C.). Guilá Naquitz also produced specimens of *Apodanthera*, the coyote melon, showing that local wild cucurbits were being collected as well (Chapter 20). Finally, pollen studies by Schoenwetter and Smith (Chapter 15) produced one cucurbit pollen grain from Zone B3 of Guilá Naquitz, identified as *Cucurbita* cf. *moschata*. This squash, not represented by macroscopic remains from Guilá Naquitz, was previously documented by five specimens from Coxcatlán Cave, Puebla, in the 4500–3000 B.C. time range (Cutler and Whitaker 1967). Perhaps future excavations will more clearly establish a temporal priority for bottle gourds over edible cucurbits in Mexico.

BEANS

Three species of *Phaseolus*—common beans (*P. vulgaris*), runner beans (*P. coccineus*), and tepary beans (*P. acutifolius*)—have wild ancestors in Mexico. The oldest archaeologically documented specimens are wild runner beans from MacNeish's caves in Ocampo, Tamaulipas (Kaplan 1965) and from Guilá Naquitz (Chapter 21). The 14 specimens from Tamaulipas are wild *Phaseolus coccineus*, the ancestor of the domesticate, and date to 7000–5500 B.C. The 100-odd specimens from Guilá Naquitz, dating to 8750–6700 B.C., are small black runner beans belonging to a wild species that has left no domesticated descendants. Their unusually high frequency, relative to other plants, makes it seem possible that the occupants of Guilá Naquitz were experimenting with growing them near the cave (Chapter 24), but if this is so, it was a strategy that was ultimately abandoned. The history of incipient cultivation may one day show many attempts to grow plants that were eventually given up in favor of more responsive species; if Callen (1967b) was correct, *Setaria*, or foxtail grass, may be another example of this phenomenon.

TEOSINTE AND MAIZE

Maize (*Zea mays*) has by far the most enigmatic and controversial origin of any major cultivated plant. This can be directly attributed to the fact that no wild race of maize currently exists. This raises two possibilities: Either "wild maize" has become extinct or maize descended from a different wild plant.

The theory perhaps best known to archaeologists is that of Paul C. Mangelsdorf, who along with R. G. Reeves proposed in 1938 that cultivated maize descended from a now-extinct wild pod-popcorn in which individual kernels were enclosed in and protected by chaff rather than by a cupulate fruit case (Mangelsdorf 1947, 1958; Mangelsdorf and Reeves 1938, 1939). Most archaeologists are familiar with this view because Mangelsdorf, together with Walton Galinat, has been involved in analyzing virtually every important collection of early archaeological maize from Mexico, including the famous collections from the Tehuacán caves (Mangelsdorf *et al.* 1967).

A different theory, originally proposed in 1877 and vigorously revived in the 1970s by Beadle (1972, 1977), Galinat (1970, 1971), and de Wet and Harlan (1972), holds that maize may be descended from the widespread Mexican grass, teosinte (*Zea mexicana*). If this theory is correct, no "wild maize" ever existed, and maize and teosinte may in fact be the same species. This theory is supported by anatomical studies that claim to show a "clear-cut connecting link between the maize cob and the cupulate fruit case of teosinte" (Galinat 1970) and by scanning electron microscope studies that show no significant differences between maize and teosinte pollen (Banerjee and Barghoorn 1972). However, despite Beadle's suspicion that no more than 6–10 independently inherited gene differences separate maize and teosinte, Mangelsdorf (1974) can still argue that no genetic manipulation has yet shown how one plant could have given rise to the other.

I summarized the evidence on both sides of the controversy a decade ago (Flannery 1973:290–296), and there is no point in repeating it here. Since that time, however, Iltis *et al.* (1979) have discovered a perennial teosinte, *Zea diploperennis*, growing in Jalisco. Mangelsdorf (1983:94) has now argued that "the ancestry of corn is biparental, with *Zea mays* and *Zea diploperennis* playing coequal roles: the former contributing the botanical features that corn has had in all stages of its evolution under domestication; the latter providing many modern races with vigorous root systems and resistance to diseases." From this perspective, the annual teosintes would be recently evolved weeds that played no role in corn's ancestry. Unfortunately, Mangelsdorf goes further than this, stating that "the entire array of archaeological evidence throughout Middle America clearly suggests that teosinte itself has probably never been used either as a cultivated or gathered fruit crop. Teosinte fruit cases are distinctive, durable, highly archaeologically preservable, and yet virtually never found prior to or during incipient agriculture" (Mangelsdorf 1983:89).

This statement overlooks the fact that Chalco-type teosinte has been found at the archaeological site of Zohapilco (Tlapacoya, Basin of Mexico) in levels dating to 5090 B.C. (Lorenzo and Gonzales 1970; Niederberger 1979). While this antedates by only 40 years the radiocarbon date for the earliest known corncobs from Coxcatlán cave in the Tehuacán Valley, it is a discovery that can hardly be ignored. Small *Zea* pollen grains, believed by Schoenwetter to belong to teosinte, occurred in a sample from Zone C at Guilá Naquitz, dated to 7450–7280 B.C., well before any macroscopic evidence for actual maize (Chapter 15). Of course, the mere presence of teosinte pollen does not prove that teosinte was being eaten. Indisputable use of teosinte, however, went on in Formative Oaxaca (1500–500 B.C.), where it occurred intermixed with maize kernels in carbonized seed samples from Tomaltepec (J. E. Smith 1981) and San José Mogote (C. E. Smith, Jr., unpublished data, 1980). At Fábrica San José, Ford's ethnobotanical study of the corncobs showed introgression with teosinte "to such an extent that one must conclude that this annual wild grass was growing in the corn fields themselves" (Ford 1976:265). In other words, Mangelsdorf may have underestimated the use of teosinte in early Mesoamerica, especially since we have so few excavated sites from the most relevant periods.

Mangelsdorf is quite right, however, in pointing out that no site yet excavated documents a gradual genetic change from teosinte to maize. Whether this means (1) that such a change did not take place or (2) that we have too few sites in too few areas depends on which of the two theories you like. Most recently Iltis, one of the discoverers of *Zea diploperennis*, has proposed a new and controversial reconstruction of the origins of maize. According to Iltis (1983), a "catastrophic" sexual mutation converted the male teosinte inflorescence (tassel) into the female ear of maize. By "modifying Beadle's teosinte hypothesis to accommodate some of Mangelsdorf's criticisms," Iltis believes, the catastrophic sexual transmutation theory can "lay to rest over 40 years of debate" (Iltis 1983:893). If Iltis is correct, the transition from teosinte to maize may have happened so fast that it will be very hard to detect in the archaeological record.

Unfortunately, Guilá Naquitz contributes no new information to this debate, presumably because its preceramic levels are too early to catch the domestication of *Zea*. The youngest preceramic level at Guilá Naquitz dates somewhere near 6700 B.C., while the first maize cobs from Tehuacán date to 5050 B.C. (Table 1.1). Hence, cultivation of *Zea* may not have taken place until some two millennia after bottle gourds and squash had come under domestication.

We did recover four small, primitive-looking maize cobs at Guilá Naquitz, but their provenience was such that little can be concluded from them. All occurred in small lenses of ash that lay stratigraphically above Zone B1 (the youngest preceramic level) and below the isolated scatters of Formative sherds that preceded the deposition of Zone A (early Postclassic). Since these lenses of ash had no artifacts, all we can do is guess that they date somewhere between 6700 B.C. and the period of the earliest Formative sherd scatters, 1000 B.C.(?). The four specimens (Fig. 1.2) have been examined by Richard I. Ford and George Beadle, and both agree that, depending on which side of the maize–teosinte argument you prefer, they could be considered either (1) maize–teosinte hybrids or (2) primitive maize that shows strong teosinte influence in its ancestry. In view of the specimens' undated stratigraphic position, there is little more to say at this time.

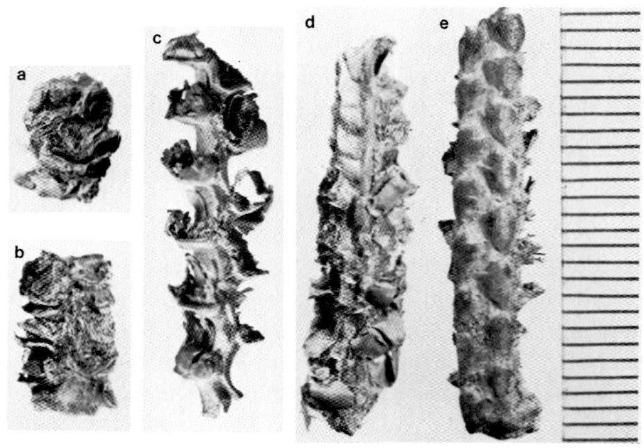

FIGURE 1.2. Primitive-looking maize specimens or maize–teosinte hybrids from undated ash lenses resting on the upper surface of Zone B1 at Guilá Naquitz (see text). a, b, two fragments of the same cob, Square C9; c, two-rowed "cob" from Square D10; d, e, two specimens from Square D7. (Scale in millimeters.) **Specimens are now in Departamento de Prehistoria, Instituto Nacional de Antropología, Mexico City.**

THE MAIZE-BEAN-SQUASH ASSOCIATION

Every textbook on Mesoamerica points out that corn, beans, and squash were an "agricultural triumvirate" for the Indians of that area. Indeed, it is sometimes suggested that one of the Indians' major contributions was the discovery that all three plants could be grown together in the same field. Not only do beans contribute nitrogen to the field as corn extracts it, the combination of the two crops in the diet produces a more complete plant protein. This is because beans are rich in lysine, an amino acid deficient in corn (Kaplan 1965).

It would be wonderful if we could attribute this triumvirate to the genius of the Mesoamerican Indian; but, in fact, nature probably provided the model. The annual teosintes are weedy, pioneer plants that colonize natural scars in the landscape, such as fallow cornfields, scree slopes, or areas opened by fire. In thorn-forest areas of Guerrero and the Michoacán–State of Mexico border today, abandoned milpas are rapidly taken over by massive stands of teosinte up to 2 m high (Flannery and Ford 1972). Wild runner beans and wild squash occur naturally in such fields, with the bean twining

FIGURE 1.3. Wild runner beans twining around stalk of mature teosinte near Chilpancingo, Guerrero, 1971.

around the stalks of the teosinte (Fig. 1.3) and the cucurbits growing among the bases of the same stalks.

The Indians could hardly have failed to notice this naturally occurring weedy association whenever they cleared the thorn forest in such a region, and they eventually duplicated it in their cultivated fields. This may be yet another case of the kind of mutualism pictured by Heiser (1979) and Rindos (1980), with man providing the environmental disturbance and *Zea, Phaseolus,* and *Cucurbita* providing the economically useful plant association. However, in terms of the approach I outline in Chapter 2 of this volume, it is also an example of information flow: Nature, in effect, demonstrated to incipient cultivators the format of the multispecies milpa.

MODELS AND THEORIES FOR EARLY DOMESTICATION

Years ago it was assumed that early domestication in Mexico had resulted from a "discovery"—humans, after some prolonged period of living off the wild vegetation, discovered that seeds thrown away from one of their meals had sprouted, thereby providing them with the knowledge necessary for cultivation. Frequently, it was suggested that this had taken place on the talus slope below an occupied cave or rockshelter.

It is true that some plants are "weedy camp followers" that grow readily on the talus slope of a cave, or any other area disturbed by humans; as we have seen, teosinte and the wild curcurbits are noted for this. However, it is unlikely that domestication resulted from such a discovery. We know of no group of hunter–gatherers so ignorant that they are unaware that plants grow from seeds. Those who rely heavily on wild plants as food usually have extremely detailed knowledge of the habits, life cycle, and environmental requirements of a wide variety of plants, and in spite of this they do not cultivate them.

Today one almost never hears domestication proposed as a discovery or an invention. Rather, the last few decades have seen the emergence of a whole new series of models and theories, far more ecologically based and carefully thought out. Each of these models has something to recommend it, yet each runs into problems with the empirical archaeological data, and the latter—although far from ideal—provide one test for any model of agricultural origins. In this section we see what useful insights we can derive from each of these models.

Climatic Change

We have already mentioned the late Pleistocene and early Holocene environmental changes that took place in Mexico not long before cultivation began. It is no surprise that as the evidence for these changes has increased, so has archaeologists' curiosity about the role they may have played.

According to one model, the late Pleistocene occupants of Mexico lived primarily by hunting large game such as mammoth, mastodon, or native American horse. With the extinction (or northward migration) of many large species at the end of the Pleistocene, hunter–gatherers were forced to rely increasingly on small game and plant foods; and their increased dependence on the latter led to agriculture. According to another model, the climate of Mexico was thought to have gone through a hot, dry period between 5000 and 3000 B.C. (called the "Altithermal" in North America) that might have influenced agriculture. However, archaeologists were not in agreement as to whether this hot, dry period had (1) forced people to turn to agriculture by depleting their other food resources or (2) delayed the success of early agriculture until after 3000 B.C. For example, in a 1966 synthesis, Willey (1966:83) considered both dry and wet periods as possible stimuli:

> With the increasing heat and dryness of the Altithermal climatic era, after 5000 B.C. this casual, minimal cultivation seems to have been taken more seriously. . . . Then, with the return to the cooler and more moist conditions of the Medithermal, after 2500 B.C., plant cultivation apparently "exploded."

Let us now bring up some complications with these models. First, the late Pleistocene stratigraphic levels from Coxcatlán Cave make it clear that the Indians of that era did not rely solely on large fauna but hunted dozens of small species and even engaged in communal drives of jackrabbits (Flannery

1966, 1967). Moreover, the late Pleistocene climate was, if anything, drier than the early Holocene, and probably significantly cooler. Once the Holocene climate had been established in regions such as Oaxaca, pollen studies do not reveal a period so hot and dry that the post-Pleistocene vegetational communities would have been eliminated, although there were climatic fluctuations during the Archaic that might have caused those communities to migrate upslope or downslope (Schoenwetter and Smith, this volume:Chapter 15). In short, I see no evidence for climatic changes, either at the end of the Pleistocene or between 5000 and 3000 B.C., that would have forced the adoption of agriculture.

We cannot rule out, however, the possibility that climatic changes slowed down or speeded up incipient agriculture. This possibility is examined in Chapter 31 by Reynolds during the course of a computer simulation of incipient agriculture in the Valley of Oaxaca. Having simulated adaptation under today's climatic regime, Reynolds runs the program again—first under conditions wetter than today's, then under conditions drier than today's. This allows us not only to see whether such changed conditions would speed up or slow down incipient agriculture but also to examine the program output to find out how they did so.

What is the most important insight we might retain from these various climatic change models? In my opinion, it can be found in an article by H. E. Wright (1977) entitled, "Environmental Change and the Origin of Agriculture in the Old and New Worlds." Wright discusses the fact that the late Pleistocene climate of the Zagros–Taurus Mountains in the Near East was so cold and dry as to leave the area treeless and severely reduce the potential of the mountains for supporting wild cereals. He then adds, "With the end-Pleistocene climatic change about 11,000 years ago, a new suite of plants immigrated into the region, including the wild cereal grains. These and other wild plants were added to the diet of man" (Wright 1977:297). In other words, emmer wheat cultivation in the mountains of Iraq was a very unlikely occurrence prior to 9000 B.C.

This does not mean that end-Pleistocene climatic change *caused* agriculture, or that agriculture might never have begun without it. After all, even during the peak cold of late Pleistocene times, there were more sheltered localities, such as the Ghab Basin of Syria, that probably provided a refuge for some wild cereals (Niklewski and Van Zeist 1970). However, the range of the wild cereals expanded enormously during the early Holocene, making them accessible to thousands more people and increasing the likelihood of their use as food.

Much the same could be said of Mexico. End-Pleistocene changes did not cause agriculture, but the early Holocene saw an enormous expansion of the thorn–scrub–cactus forest in which so many of the wild ancestors of the early domesticates grew. Theoretically, agriculture might have begun without these changes, but it would probably not have taken place as early in Tehuacán and Oaxaca, to name only two regions.

Population Pressure

Another model that has risen in the past, and then fallen, and lately risen again is one in which humans were forced to adopt agriculture when their population reached the point where wild plants and animals were an insufficient source of food. There are two variants to this model. In one, the pressure for agriculture comes from local imbalances in population "density equilibrium" (Binford 1968); in the other, "overpopulation" is seen as a worldwide phenomenon that underlies all early cases of food production (Cohen 1977).

In a now-classic paper on post-Pleistocene adaptations, Binford (1968) proposed a density equilibrium model for the adoption of agriculture, which I later applied to the Near East (Flannery 1969). Binford argued that adaptation would change only in the face of some disturbance in the equilibrium between population and environment, such as (1) a change in the physical environment that would bring about a reduction in the density of chosen plant or animal foods or (2) a change in population density or distribution that would raise human populations too close to the carrying capacity of the immediate area. Following the end-Pleistocene climatic changes described by Wright (above), the Near East would have become a mosaic of "favorable" habitats (e.g., oak–pistachio woodland) and "marginal" habitats (e.g., gravel desert or arid steppe). The wild cereals also have definite "optimum" zones, in which they grow densely, and "marginal zones," in which they do poorly. Following Binford's model, the optimum habitats should have been the centers for population growth ("donor areas"), with the marginal habitats ("recipient areas") receiving the emigrant overflow. It is in these marginal zones that man–land disequilibrium would have been felt first. Thus, in 1969 I suggested that farming might have begun not in the optimum areas of wild cereal growth (where wild wheat and barley already grow as densely as they would in a cultivated field), but around the margins, where it would be necessary to cultivate cereals in order to achieve the same densities that occur naturally in the optimum zone.

There may be parts of the world where this process actually took place. Particularly in the Levant, where sedentary life based on the collection of wild cereals may actually have preceded agriculture, there may have been a mosaic of regions with great disparities in population density. Unfortunately for the model, however, available archaeological data (such as they are) do not show strong population increases in optimum areas, such as the southern Zagros or the Lebanese woodland, but the very opposite—some of the most striking end-Pleistocene increases in numbers and sizes of sites are in marginal habitats, such as the Negev (Marks 1971). While I am not prepared to discard this model out of hand, I also feel it needs further refinement because it comes too close to making climatic change and population growth into prime movers. Let us suppose that population growth in the end-Pleistocene Levant was a product of life in sedentary communities, which can shorten birth spacing and reduce infant mortality (Binford and Chasko 1976). Let us further suppose that sedentary life began not as a response to optimal habitats but as part of a shift from foraging to collecting in the sense that Binford (1980) has recently defined those terms (see below). Suddenly we have a model that is substantially more complicated, and with causality not nearly as clear-cut. (In other words, our nice, neat, density equilibrium model has become a typical anthropological problem.)

Before discussing the difficulties of applying the local density model to Mexico, let us consider the far more ambitious and global model of Cohen (1977), who actually uses the term "overpopulation" in his subtitle. Cohen begins by asking two questions: (1) Why would human populations successfully adapted to a hunting–gathering life-style (thought by some to be mankind's most stable adaptation) abandon it in favor of farming? and (2) Why would so many of the world's people adopt agriculture at roughly the same time? Cohen concludes that agriculture would only have been adopted if the demand for calories was out of balance with the productive potential of wild foods and that "only one possible explanation—actual population growth—could account for an ecological imbalance that affected so many parts of the world." Cohen (1977:279) argues that hunting and gathering populations "had saturated the world approximately 10,000 years ago and had exhausted all possible (or palatable) strategies for increasing their food supply within the constraints of the hunting–gathering life style," thus leaving themselves no alternative but agriculture. Cohen realizes that this implies that human populations, unlike those of any other mammal, grow continuously without stabilizing at some carrying capacity; and he accepts this implication.

I think the two questions asked by Cohen are good ones, particularly the second. It does seem to be the case that agriculture began independently in the Near East, Southeast Asia, Mesoamerica, and the Andes sometime during the period 10,000–5000 B.C. Given the whole of human prehistory, that is a relatively brief period of time. To be sure, there are occasional claims for agriculture in earlier times, but most rest on shaky stratigraphic or radiocarbon evidence. The most recent of these to be brought back in line with the 10,000–5000 B.C. period is barley, which had been tentatively radiocarbon dated to 19,060 B.P. at Wadi Kubbaniya in Egypt (Wendorf *et al.* 1979). New analyses of the actual barley grains themselves give a date of 4850 B.P., suggesting they were not actually associated with the charcoal that yielded the earlier dates (Wendorf *et al.* 1983).[1] I expect similar future readjustments for most other cases of cultivated plants that are currently dated before 10,000 B.C.

Cohen is therefore quite right when he says that any universal theory for the origins of agriculture must explain not only how it began, but also why it began at that particular time. The end-Pleistocene climatic changes mentioned previously could have influenced this, but surely there had been warm interglacial periods earlier in human prehistory. Many authors have speculated on why agriculture did not begin in one of those earlier periods, several of them suggesting that human populations "had not reached a cultural stage of a nature appropriate to take the necessary steps to domestication" (Wright 1977:296). Braidwood and Howe (1960:180) have pointed to a "regionalism" in tool assemblages that was not seen before the late Paleolithic; perhaps, they once argued, this reflects a degree of "settling into" the regional environment that was necessary before agriculture could begin. Now Cohen has suggested, in effect, that world population would have been too sparse during earlier interglacials to exhaust the wild food resources.

The main problems in applying either Cohen's overpopulation model or Binford's density equilibrium model to Mexico are presented by the empirical archaeological data. Populations in highland Mexico were so small when agriculture began that phrases such as "overpopulation," "food crisis," and "exhaust all possible strategies" appear to be exaggerations. Elsewhere (Flannery 1983:35) I have estimated that early Archaic populations in the Tehuacán–Oaxaca valleys consisted of no more than 1 person per 9–29 km^2, and "this figure includes only the valley areas, not the surrounding mountains" where so much foraging was done. In Chapter 24, I calculate that a catchment circle with a 5-km radius would contain more plant food than the occupants of Guilá Naquitz could eat in their lifetimes. As if this were not enough, it is even possible that agriculture in Mexico began with a plant that was not even domesticated for food. If the bottle guard was the first domesticate, the causes had nothing to do with food shortage; it was a technological breakthrough, equivalent to assuring oneself a reliable supply of flint.

On a deeper and more theoretical level, I share with Hassan (1981) a certain reluctance to see human population growth as an independent variable, unrestrained by the factors that govern other mammalian populations. In zoology, population increase is generally taken as the sign of a successful adaptation or as evidence that a favorable evolutionary advance has already been made, rather than as a cause for further evolution. In paleontology, I cannot think of a single new species believed to have arisen because its fossil ancestors became too numerous; becoming numerous is a mark of fitness. What increasing population growth often does do is to bring about *density-dependent* changes in behavior, but these changes are usually reversible when population declines.

In his book on demographic archaeology, Hassan (1981:Fig. 12.3) combines all available "best estimates" on Pleistocene world population increase into a single diagram, which is reproduced here as Fig. 1.4. Note that world population first began to climb steeply about 100,000 years ago, at the start of the Würm glacial advance, when the transition from handaxe users to Middle Paleolithic flake-tool users was virtually complete. These flake-tool users ("Neanderthals" *sensu latu*) gave way to Upper Paleolithic *Homo sapiens sapiens* around

[1]Wendorf, Close, and Schild are to be commended for the dedication with which they pursued the true date of this barley, subjecting it to Electron Spin Resonance Spectroscopy and Linear Accelerator Radiocarbon Counting. Most archaeologists would have let the joy of the early dates overrule their own scientific curiosity.

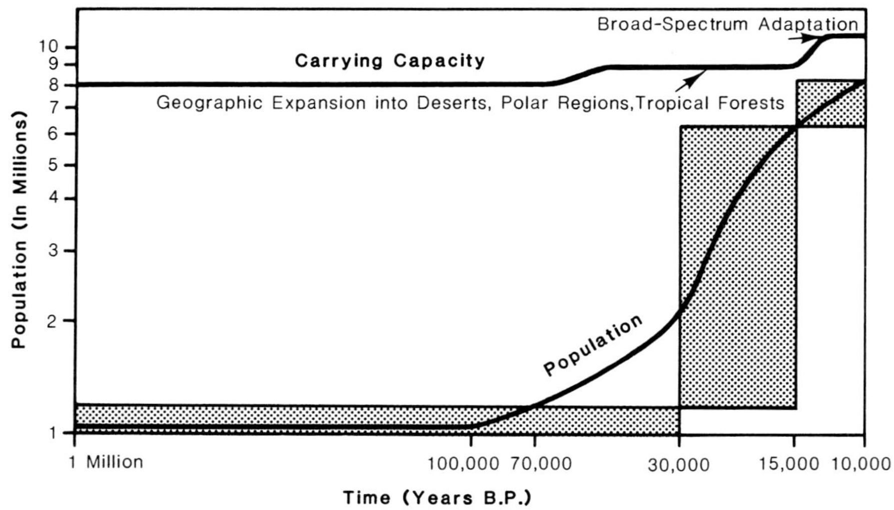

FIGURE 1.4. Reconstructed increases in world population and carrying capacity for humans during the Pleistocene. (Redrawn from Hassan 1981:Fig. 12.3.)

30,000 years ago, setting off a precipitous population increase that saw "fully modern" humans continuing an expansion into deserts, polar regions, and tropical forests of the Old World, as well as colonizing the Americas.

Hassan's graph shows the carrying capacity for end-Pleistocene hunter–gatherers rising above 10 million worldwide as the result of the "broad spectrum adaptation." The latter is an empirically detectable phenomenon for which I suggested a name in 1969 and about which I have more to say below. The most significant feature of Hassan's chart is that it shows carrying capacity moving ahead of population as the result of changes in foraging efficiency that preceded (and made possible) population growth. This is, I believe, more in line with the way zoologists view evolution and population growth. The steep rise of Hassan's population curve after 100,000 B.P. is almost certainly associated with a major evolutionary change, perhaps of an irreversible (genetic) type. The approach of the population curve to carrying capacity after 15,000 B.P., on the other hand, might have set off behavioral changes in response to increased competition and decreased mobility (see below), but most of these changes would be reversible should population density fall for any reason.

Hassan's chart ends at 10,000 B.P., near the start of the period singled out by Cohen as crucial for incipient agriculture. What happened next, of course, was that successful agriculture was in turn followed by an enormous rise in population in those areas of the world where it developed. Significantly, however, this happened because agriculture raised the carrying capacity of those areas to the point where environmental resistance to human population growth was substantially lower, not because of an irreversible (genetic) change such as the one we have proposed for 100,000 B.P. This is an important point to remember when we consider Rindos' (1980) coevolution model below.

What are the most important insights we might retain from the various population pressure models that have been proposed? One of them is surely Cohen's insistence that we account for the events of the period 10,000–5000 B.C. What I would like to do, however, is drop the terms "overpopulation" and "food crisis" from his hypothesis and rephrase it this way: By 10,000 B.P., all the world's major landmasses had been populated to the point where (even though the actual densities might not have been that high) emigration had been considerably reduced as a strategy for adjusting man–land relationships.

For most of the Pleistocene, emigration had undoubtedly been one of the easiest and most frequently used options open to early humans, whether to correct for unpredictable shortages or to resolve conflict (Lee and deVore 1968). By 10,000 B.C., however, most human groups would have had neighbors on all sides, making it increasingly necessary for them to solve their problems on a local basis. The late Pleistocene regionalism of which Braidwood and Howe speak is one line of evidence for this, as is the broad spectrum adaptation discussed below. Complicating this picture still further is the possibility that the end-Pleistocene climatic changes described by Wright may have increased, rather than decreased, the environmental variation with which hunter–gatherers had to deal. As Hassan (1981:206–207) has expressed it:

> By the end of the Pleistocene, the climatic fluctuations accompanying the transition from glacial to postglacial conditions were great. These conditions would have enhanced group mobility. However, because the overall population density during the Terminal Pleistocene was much higher than it had been (especially in favorable habitats) and because contiguous groups provided a barrier against territorial expansion or long-range migration, localization of population and regional differentiation was intensified. Localization also enhanced the exploitation of a wide range of resources which served as further protection against unpredictable fluctuations.

This passage by Hassan effectively synthesizes the most defensible parts of Wright's climatic change model, Cohen's population growth model, Binford's optimal habitat model, and Braidwood and Howe's notions of regionalization. What Hassan does *not* do is accept Cohen's notion that end-Pleistocene hunter–gatherers had "exhausted all possible strategies" short of agriculture. Post-Pleistocene hunter–gatherers in many parts of the world opted for sea mammal hunting, salmon fishing, intensive seed collecting, or some other option instead of agriculture. Not all these strategies had as their goal "an increase in calories per unit space." What many seem to show—perhaps on the global scale that Cohen insists we should seek—is a penchant for dealing with uncertainty, or unpredictable variation, at a level of competence shown by no previous society.

Broad-Spectrum Adaptation

Since Hassan's model incorporates the notion of a broad-spectrum adaptation, this would seem to be the place to bring that concept up to date. In the same paper in which I tried to apply Binford's density equilibrium model to the Near East (Flannery 1969), I pointed out that early agriculture in that region was preceded by a trend from exploiting a more "narrow spectrum" of environmental resources to a more "broad spectrum" of edible wild products. In addition to the ungulates that had so long been a dietary staple, Near Eastern peoples now ate progressively larger amounts of fish, crabs, water turtles, molluscs, land snails, partridges, migratory waterfowl, and perhaps even wild cereal grasses. The same period was characterized by several preadaptations for agriculture, such as the development of ground-stone technology and storage facilities. While ground-stone tools appear to have originated as tools for milling red ocher, the technology was available when humans turned to extensive use of cereal grains. As for storage pits, their use expanded so rapidly that one site in Israel dating to 10,000–8000 B.C. contained more such facilities than all previously known sites in the Near East put together (Flannery 1969:78, 1973:281).

The use of storage facilities by hunter–gatherers is a topic that promises to shed considerable light on their strategies for coping with uncertainty. Binford (1980) has recently suggested that most hunting–gathering societies occupy a position along a strategic continuum that reflects differences in mobility and subsistence patterns. At one end of the continuum are *foragers*, the most mobile, and at the other end *collectors*, the most nearly sedentary. To use Binford's phrase, foragers "map onto" resources through residential moves and adjustments in group size. As resource opportunities change throughout the year, foragers travel to where the food is, and their settlement pattern becomes dispersed or aggregated according to whether resources are dispersed or aggregated. Collectors, on the other hand, supply themselves with specific resources through logistically organized task groups. That is, the band as a whole tends to remain in one spot while smaller groups go out and bring back resources; there may be men's deer-hunting trips or women's fruit-collecting trips, but these differ from whole-scale residential moves like those of foragers. As Binford (1980) and Thomas (1983) have stressed, most groups fall somewhere between the two extremes, and there are foragers such as the !Kung San Bushmen who "map onto" *mongongo* nut groves while organizing their big-game hunting logistically through all-male hunting parties.

While both foragers and collectors may practice storage, the extent to which it is used can affect settlement and subsistence strategies. Storage, as Binford (1980:15) points out, is a way of extending the time of use of a specific resource beyond its period of natural availability in the habitat. Yet storage reduces the mobility of a group, so that "with increases in storage dependence there will be an expected increase in the logistical component of a settlement system." If, as Hassan suspects, group mobility was reduced during the terminal Pleistocene because contiguous groups provided a barrier to unrestrained emigration, we might expect the period 10,000–5000 B.C. to show both (1) an increase in storage features and (2) an increasing number of shifts from mobile foraging to more sedentary collecting. And Binford (1980:18) adds that still further increases in logistically organized procurement strategies should be expected with shifts toward agricultural production.

Binford also argues that since the purpose of storage is to extend the season of availability, then the shorter the growing season, the greater the reliance on storage. In an interesting graph (Binford 1980:Fig. 4), he compares the storage dependence of various contemporary hunter–gatherers with the "effective temperatures" of their regions, showing that groups in colder climates rely more on storage. Given the regions where Binford has done fieldwork, it is understandable that he sees temperature as the critical limiting variable; however, this would not work for us in Tehuacán and Oaxaca where growth is limited more by deficient moisture. It is not so much frost that ends the growing season at Guilá Naquitz but rather a 5-month drought with mean monthly rainfalls below 10 mm (Kirkby 1973:16–17). Hence, the key to sedentary collecting in the Tehuacán–Oaxaca region was to find something that could be stored from November until the rains began again in April. In some higher-altitude localities, such as Guilá Naquitz, it is clear that an effort was made to store acorns (Chapter 5). In lower-altitude localities, such as Coxcatlán, one sees few candidates for storage until maize came along.

We have also seen, in our discussion of *Crotalaria* at the start of this chapter, that the November-to-May dry season is not the only problem in the Oaxaca region. Harsh as it is, the dry season is at least a predictable problem, since the Indians know it is coming every year. On the other hand, the annual variation in rainfall—which can range from drought to flood—follows no observable cycle and is therefore unpredictable. Kirkby's (1974) work on individual and community responses to rainfall variability in Oaxaca suggests, at least to me, that the Zapotec response is aimed at reducing the differences between wet and dry years. In this regard, I am struck by the analogous behavior of the Tewa Indians studied by Ford (1968). Like highland Mexico, the Pueblo region of the Southwest is an area with great contrasts in productivity between rainy and dry years. Like the Zapotec, the Tewa are not so much interested in controlling nature as in making it more predictable; both in their technology and in their ritual, they seek not to maximize but to reduce the extremes and thereby lessen variation.

In light of all these additional lines of data, I would like to rephrase the broad-spectrum adaptation in the following terms. The period 10,000–5000 B.C. witnessed annual and seasonal climatic variation at least as great as today's, perhaps even exacerbated by the fact that in many parts of the world the late Pleistocene vegetation was being replaced by Holocene floral communities that included the wild ancestors of many eventual domesticates. In the past, emigration had been one way of dealing with unpredictable variation, but now mobility had been reduced by the societal barriers of an expanding hunter–gatherer population. During 10,000–5000 B.C., therefore, humans had to learn how to cope with unpredictable environmental variation without leaving their own region, and they did so on a scale never before seen. Diversification of the subsistence base to include many more species on lower trophic levels was only one strategy in a milieu that included the escalation of storage, the development of groundstone tools, improved fishing technology, and an increasing trend from foraging toward logistically based collection. What all these alternatives seemingly shared was a goal of resiliency, risk reduction, amelioration of environmental extremes, and an increase in resource predictability. From the standpoint of human intentionality, the cultivation of a series of weedy annual plants that like to pioneer disturbed habitats may simply have been one of a series of strategies that, like storage, were designed to help even out the differences between good and bad years. And if collectors "move goods to consumers with generally fewer residential moves" (Binford 1980:15), then agriculture was the ultimate collecting strategy.

Human societies are particularly well suited to dealing with long-term seasonal, annual, predictable, and unpredictable variation because they have a multigenerational "memory" and can share information on the success of past subsistence-settlement strategies in ways no other creature can. In Chapter 31 of this volume, Reynolds explores the use of a long-term, memory-using, efficiency-improving model for simulating the activities of incipient cultivators in the Valley of Oaxaca. I do not want to get too far ahead of our story here, but Reynolds' results suggest that a great deal of the prehistoric record can be understood as an effort to cope with uncertainty, even at relatively low population densities and in relatively benign environments.

Agriculture: Coevolution and Symbiosis?

As end-Pleistocene populations grew and diversified the plant species they were using, they undoubtedly disturbed their habitats more and more. Human modification of the environment, even such a simple act as clearing a campsite in the thorn forest, increased the prevalence of a second-growth species at the expense of primary vegetation. Even before agriculture began, as Heiser (1979) has stressed in the case of the bottle gourd, certain plants had achieved a symbiotic relationship with human hunter–gatherers.

In a recent article and an even more recent book, Rindos (1980, 1984) has developed this concept of plant–human symbiosis into "a new model for the origin and development of domesticated plants and agricultural systems" (Rindos 1980:751). I hope that my brief summary does not do too much violence to the model, for it is too lengthy to consider here in all its details. Basically, Rindos sees domestication as a symbiotic interaction between man and the plants on which he feeds, the product of long coevolution between the two. Citing numerous cases from the animal world, he discusses the symbiotic relationships between ants and acacias, termites and their fungus gardens, and squirrels and oak trees. Squirrels, viewed from Rindos' perspective, are one of the oak trees' mechanisms for dispersing their acorns; thus man becomes a mechanism by which maize disperses its seeds. As Aschmann (1980) puts it, Rindos "has taken the capability of choice or decision making in evolving agricultural systems away from humans and given it to the plants." Indeed, Rindos makes it clear that human intentionality is unnecessary to his model and may parsimoniously be set aside. Consistent with this view, Rindos see the morphological changes characteristic of domestic strains as having been favored by the symbiotic relationship, not selected by man, and he cites examples of plants unable to survive in the wild without their animal partner.

Agriculture, in Rindos' model, *follows* domestication; it is "a set of integrated activities that affects the environment inhabited by the domesticated plant throughout its life cycle" (Rindos 1980:752). Even here he insists that "agriculture cannot, in any meaningful way, be restricted to the human–plant interaction. If we try to find any activity or collection of activities to define the man–plant relationship, we are at the same time defining other animal–plant relationships" (Rindos 1980:754–755). Small mammals, he argues, accidentally "plant" seeds in nitrogen-rich soils near their nests. True, but how many small mammals deliberately hand-sort and plant large seeds, as the Mesoamerican Indians do, or make decisions about which species to plant "after a subtle evaluation of costs, yields, and markets related to the particular environment at the farmer's disposal" (Aschmann 1980)?

Obviously, I am sympathetic to many aspects of Rindos' model. After all, in 1968 I argued for a relationship between preceramic foragers and prickly pear cactus (Flannery 1968) that sounds like the one Rindos is proposing, and in Chapter 2 I provide some quantification for that relationship. My main worry about Rindos' model is that, in the process of finding some analogies for agriculture elsewhere in the animal kingdom, he has concluded that human domestication of plants can be filed away as just one more case of a well-known biological process. In 1957, when I was a zoology student at the University of Chicago, it might have seemed like a good idea, but now that I am an anthropologist I have reservations.

Is agriculture really the same kind of plant–animal coevolution displayed by ants and acacia trees? Let us consider the interaction between early Archaic foragers and *Phaseolus* beans. Wild beans have two genetic characters that aid in propagation: (1) Their pods twist like a corkscrew when ripe, shattering on drying in such a way that the seeds are well dispersed, and (2) their seed coats are so impermeable that they will sprout only after the rainy season has begun in earnest, rather than opening prematurely after an aberrant early thundershower. Both characters were favored by natural selection, but both reduce the beans' attractiveness to humans; the former character makes them harder to harvest, while the latter requires extensive soaking in water to make them edible. In spite of these characters, it is clear that phenotypically "wild" beans were extensively used by Archaic foragers in both Tamaulipas and Oaxaca (Table 1.1). There is even some circumstantial evidence, summarized in Chapter 24, to suggest that the phenotypically "wild" runner beans at Guilá Naquitz were having their density artificially increased by cultivation.

Ultimately, a domestic strain of bean developed in which (1) the pod became limp, straight, and nonshattering, and (2) the permeability of the seed to water became greater (Kaplan 1965). Both these changes benefited humans but were deleterious to the bean's seed dispersal mechanism. If I read Rindos correctly, he would argue that these changes were produced by natural selection as the bean entered into a coevolutionary relationship with humans; and because they prevented the bean from reproducing adequately, became the reason humans were later drawn to cultivate *Phaseolus*. This is an interesting thought, but I doubt that many botanists would believe that a strain of bean with such a deficient seed dispersal mechanism would be able to survive unless man was already cultivating it, in defiance of natural selection.

Moreover, all the archaeological data at our disposal suggest that cultivation began before such morphological changes. In Mexico, Callen (1967b) found statistical evidence for incipient cultivation of *Setaria*, including the selection and planting of larger seeds, in the absence of any genetic changes that would make the plant dependent on human intervention. At Guilá Naquitz, Whitaker and Cutler (Chapter 20) have identified cultivated squash in levels with rind fragments that show that edible flesh had not yet developed. On the fringes of Mesopotamia, Helbaek (1969) spoke of "cultivated wild barley," a crop that could be shown to differ from true wild barley by its average seed size but that had not yet undergone genetic change in the direction of the nonshattering rachis.

Yet another archaeological example comes from the village of Tell Mureybit on the Euphrates River in northern Syria (van Loon *et al.* 1968, 1970). Dating to 8200–7500 B.C., Mureybit produced a series of earth ovens, used for the parching of grain, and some two dozen flotation samples totalling several thousand carbonized seeds. Van Zeist and Casparie (1968) describe the collection as dominated by einkorn wheat (*Triticum boeoticum*) of the large-seeded variety *thaoudar.* This is a cold-tolerant wheat that thrives in the Turkish mountains 100–150 km north of Mureybit but cannot be found in the lower and warmer environment of the preceramic site. The inescapable conclusion is that the Mureybit villagers had removed einkorn from its native habitat and were growing it on the Euphrates floodplain, where they harvested it with flint sickles and ground it in limestone mortars after parching. Two-row barley (*Hordeum spontaneum*), lentils (*Lens* sp.), and bitter vetch (*Ervum* sp.) were also present, giving Mureybit an early version of the same complex of cereals and legumes that characterized later Near Eastern villages.

Most significantly, none of the cereals studied by van Zeist and Casparie were phenotypically domestic; they lacked the tough rachis and fragile glumes characteristic of domestic einkorn (*Triticum monococcum*) and barley (*Hordeum vulgare*). Obviously, therefore, their cultivation began before the kind of genetic change that ultimately made them dependent on humans for survival. The alternative possibility—that Mureybit's main food was brought in by the ton from wild cereal stands located 100–150 km away—seems remote on the basis of ethnographic data. Nor do the sickles, grinding stones, and parching pits make it seem that intentionality played no role in the cereal use of the Mureybit villagers.

In other words, while I like many aspects of Rindos' model, the archaeological data strongly suggest that the crucial genetic changes resulted from human selection *after* cultivation of phenotypically wild plants had begun. And the latter is not true coevolution; albeit less sophisticated and intentional, it is more analogous to the genetic engineering that now goes on in laboratories all over the country. Features such as the corncob are monstrosities, artifacts of human intervention that could not survive in nature. When man selected for limp bean pods or pendulous chile pepper fruits, he was working against natural selection, not with it; to borrow Dawkins' (1976) phrase, the "selfish gene" met its match in "selfish man." And in contrast to biological coevolution, humans do not seem to have undergone *genetic* change—whether morphological or behavioral—as a result of their interaction with corn, beans, or squash. There is no gene for agriculture; any significant change the incipient cultivators underwent was cultural, and it was reversible. The ethnographic record provides us with cases of previously full-time agriculturalists who have largely reverted to hunting and gathering under changed conditions (see Lathrap 1968), just as a number of mammalian density-dependent behaviors may be given up when densities change.

Not surprisingly, Rindos' 1980 article ran into some strenuous objections from anthropologists. Cohen (1980) was quick to point out that if domestication was simply a case of symbiosis, we might expect it to be strung out through all of human prehistory rather than clustered so significantly between 10,000 and 5000 B.C. Ceci (1980), like Aschmann, was disturbed that "deliberation is taken away from man or, conversely, granted to all living things." Shaffer (1980) argued that "any model ignoring the different human responses to plant and animal domestication has limited value for assessing domestication's impact on the history of world flora and fauna or on the history of human behavior."

Rindos seems puzzled at this response. While acknowledging that his model "is not subject to empirical verification" and that he has "identified no clear causes for agriculture" (which he believes "has no single cause"), he is "forced to wonder" why such an unnecessary variable as human intentionality "should cause such a fuss" (Rindos 1980:769–770). "Human genetics," he points out, "proceeds well without recourse to intentionality." Unfortunately, Rindos could not have picked a worse example. While it may tell us a great deal about blood types and eye color, human genetics has not told us much about the differences between hunter–gatherers and agriculturalists; nor is it likely to.

I suggest that what we have here is a classic case of scholars from two different disciplines talking past each other. We have already discussed Wilson's (1979) diagram showing the differences in resolution and generality between anthropologists and biologists (Fig. 1.1). Rindos, trained as a botanist, is content with reducing domestication to a biological model that is admirably general but (by his own admission) lacks the resolution to explain the origins of agriculture in any area. He apparently does not realize that anthropologists, trained to seek explanations in terms of cultural, social, political, and economic decisions, are most interested precisely in the testable, high-resolution explanations that his model fails to provide. The fact that there are biological analogies for human behavior is as big a bore for anthropologists as human intentionality is for Rindos. This is especially true since archaeological data suggest that, contrary to his expectations, intentional cultivation seems to have preceded the genetic changes leading to domestication.

So far, every model or theory we have considered has provided us with some insights we can defensibly maintain. In Rindos' case, I would like to preserve the notion of plant–animal symbiosis, but I feel it applies more accurately to the period of hunting and gathering than to later times. The mutualism of man and wild bottle gourds; the affinity of teosinte, wild runner beans, and squashes to areas of human disturbance; and the dispersal of prickly pear seeds by man (see Chapter 2) all seem to me acceptable cases of symbiosis. What happened next was a series of changes in cultural behavior, which produced genetic monstrosities in opposition to natural selection. This was not coevolution in the biological sense for it did not involve genetically programmed behavioral changes in the animal, nor were the genetic changes in the plant induced by natural selection. As a cultural phenomenon, it must be attacked by a science that focuses on culture rather than genetics.

Multivariate Models

It will be clear by now that if we save the defensible insights from each of the theories and models discussed above, we have a multiplicity of factors to deal with rather than a single, neat, prime mover explanation for the origins of agriculture. This is the situation Hassan (1981) found himself in when discussing incipient agriculture in the Near East, and his solution was a multivariate model that considers technocultural preadaptation, population density, sedentariness, climatic fluctuation, and many other factors (Hassan 1981:Fig. 13.3; Fig. 1.5, this chapter). Without question, many scholars would prefer to have a parsimonious algebraic formula—to be able to say simply, "Agriculture will begin if and only if x has reached a certain level y and z takes place." Every such attempt I have seen, however, is either contradicted by the archaeological evidence or leaves a great many important questions unanswered.

During the course of this volume we develop our own multivariate model for the origins of agriculture in Oaxaca, testing it not against some abstract logic but against the archaeological record. The biggest danger in such an approach is that the closer one comes to providing an appropriate model for incipient agriculture in Oaxaca, the less he has to say about the universal aspects of agricultural origins. For this reason, we incorporate into the model as many of the universal processes discussed above as we find convincing.

UNIVERSAL ASPECTS OF THE MODEL

In our model, the end-Pleistocene climatic changes described by Wright and the growth of world population described by Cohen and Hassan combined between 10,000 and 5000 B.C. to bring about a density-dependent shift in human cultural behavior over much of the world. Emigration and high mobility declined in importance, and strategies for dealing with predictable (seasonal) and unpredictable (annual) variation on a local basis began to emerge. The regionalism in tool kits mentioned by Braidwood and Howe and the broad spectrum expansion of lower trophic level foods I once described were but two aspects of this shift. Storage became more important, and the diversity of foraging and logistic collecting strategies proposed by Binford become more apparent in the archaeological record. So also do Binford's differences in population among the mosaic patches of the post-Pleistocene environment.

Humans are well suited to the working out of resilient strategies for environmental variability, since they possess multigenerational memory and have unique methods for exchanging information and establishing cooperative relationships over long periods. Their successful adaptation to uncertainty during the period 10,000–5000 B.C. was not the result of a change in genetically programmed behavior, and no anthropologist is likely to be satisfied with a model that treats it as barely different from ant–acacia coevolution.

Humans did, as Rindos suggests, establish mutually beneficial relationships with a number of plant and animal species in the post-Pleistocene environment, but these relationships did not all lead to domestication. Human environmental disturbance certainly favored second-growth species over others, but as late as A.D. 1800 there were still many areas of the world where successful human adaptation did not include agriculture. In Mexico, agriculture may have begun as one of many strategies aimed at reducing the differences between wet and dry years, at making nature "more predictable" in Ford's terms. And it may have begun with a plant that, although symbiotic with man according to Heiser, was used as an artifact rather than a source of food. Having

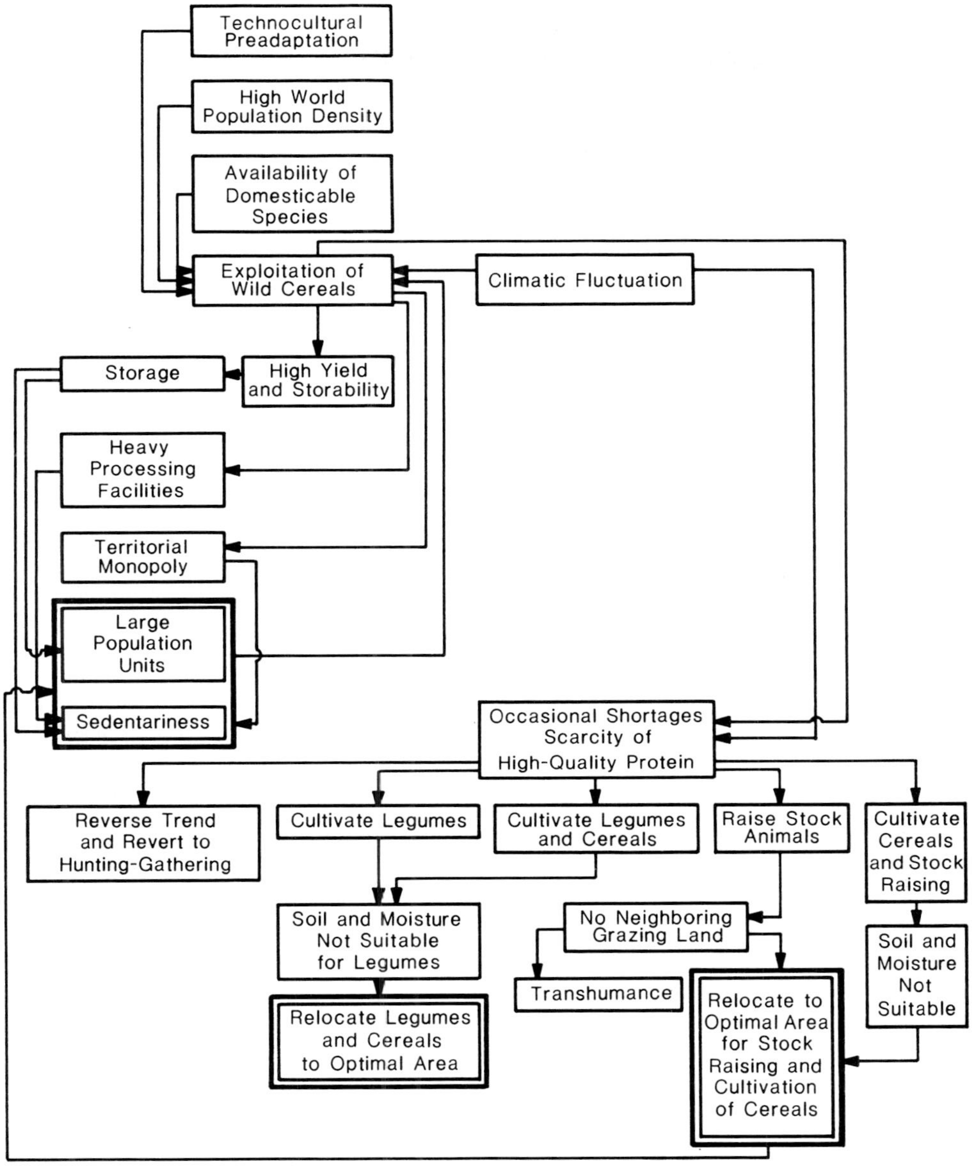

FIGURE 1.5. A multivariate model for agricultural origins in the Near East. (Redrawn from Hassan 1981:Fig. 13.3.)

disturbed the environment unintentionally for centuries, man now began to disturb it intentionally so he would have enough bottle gourds. And if bottle gourds, why not their relatives, the squashes? And why not store them, as Binford suggests, to shorten the dry season? In other words, it is not necessary to use terms such as "food crisis" or "overpopulation" to explain early agriculture in Mexico. By increasing the density of storable plants, you shorten the dry season, lessen the necessity to move, favor the shift from foraging to collecting, and make it easier to deal with uncertainty without being forced to leave your region. Thus, the specific steps taken in Oaxaca may be examples of the more universal hunter–gatherer behaviors recently codified by Binford (1980).

Why did agriculture "take off"? It did not always, and many possible early attempts, such as those proposed for *Setaria* and the "wild" Guilá Naquitz runner beans, must have ended in failure. In 1968, I suggested that most of the attempts that "took off" did so because of genetic changes that reinforced the payoff for the cultivator. (Rindos discusses many of those same changes, except that in his model they *precede* cultivation.) While occasional mutations favorable to human collectors must have existed in preagricultural times, I follow the archaeological record in suggesting that not until cultivation had begun was there a mechanism—namely, human selection and planting—that would increase the frequency of such variants in spite of natural selection's tendency to eliminate them.

In the case of the squash mentioned above, Whitaker and Bemis (1975) suggest that only the seeds were edible at first. After cultivation had begun, humans selected for (and planted) those that had the largest numbers of seeds. Selection for greater numbers of seeds led to the emergence of flesh. Further selection for less-bitter flesh led eventually to the domestic varieties we see today, whose large size and palatable flesh make them even more attractive to cultivate than the *Cucurbita pepo* at Guilá Naquitz—which had phenotypically domestic seeds but still had no flesh. I have argued that such favorable changes kicked off a positive feedback loop between cultivator and plant, which increased the investment of time in cultivation (Flannery 1968) but would not have been as difficult to reverse as a true case of coevolution (see Gilbert and Raven 1975).

EPILOGUE

We have indicated that the research problem we chose for ourselves at Guilá Naquitz was to develop a model that would not only deal with some of the underlying and more universal aspects of early domestication but also tie that process into the specific cultural pattern for the Valley of Oaxaca. By far the bulk of this volume is devoted to a case study that deals with what happened in Oaxaca between 10,000 and 5000 B.C. It should be clear from this chapter, however, that we do not see events in Oaxaca as independent of the more universal factors we have discussed. These included end-Pleistocene climatic changes; end-Pleistocene world population growth; plant–animal symbioses prior to agriculture; a widespread, probably density-dependent adaptation to uncertainty and variation on the local level; worldwide broad-spectrum dietary expansion; and a different kind of coevolution, following the adoption of agriculture, that was genetic for plants but cultural for humans.

When we turn to the specific Oaxaca case, however, we want the "resolution" anthropologists like to get from their data. In Chapter 2, therefore, we discuss a model that is specific to Oaxaca, and in Chapters 3–29 we present the "fossil record" from Guilá Naquitz. Over the course of those chapters, we attempt to reconstruct the preagricultural lifeway of the eastern Valley of Oaxaca, seeking to understand the extent to which incipient agriculture was an outgrowth of that lifeway. Using the archaeological and environmental data presented in Chapters 3–29, Reynolds (Chapter 31) has designed a computer simulation to see whether we can model (and hence understand better) the way the Guilá Naquitz foragers added agriculture to their subsistence pattern. This simulation also allows Reynolds to test the effects of climatic change and population growth on the rate at which agriculture was adopted, allowing us to evaluate those factors in our final summation. In other words, we try in this volume to achieve some balance between perspective and resolution, between biology and culture, between universal process and regional intentionality. To be sure, the question "How?" will more often lead to a universal; but for anthropologists, who study rational, cultural, decision-making beings, the question "Why?" will just not go away.

Of course, anthropologists must be careful not to overlook an answer as simple as the one Ellen Messer got from her Zapotec informant.

2

Ecosystem Models and Information Flow in the Tehuacán–Oaxaca Region

Kent V. Flannery

> In short, we are describing a system in which everything affects everything else, and the complexity of the system of interlocking cause–effect pathways confronts us with a superficially baffling problem in scientific analysis. It is precisely this interlocking feature which is the most characteristic identifying aspect of a system.
>
> (Watt 1966b:2, reprinted by permission of Academic Press)

INTRODUCTION

It should be clear from our discussion in Chapter 1 that the origin of agriculture is a complex problem involving symbiosis, population growth, genetic change, technological change, human strategy, and cultural adaptation. It is, like the system described by ecologist Kenneth Watt in the quotation above, a complexity of interlocking cause–effect pathways in which every variable affects several others. My feeling is that a problem that involves so many mutual-causal processes is not profitably attacked by "linear-causal" or "prime mover" models; it requires a systems approach.

In 1968, in a volume published by the Anthropological Society of Washington (ASW), I presented the skeleton of a systemic, mutual-causal model for the origins and development of agriculture in the Tehuacán–Oaxaca region (Flannery 1968). This book attempts to put a little more flesh on that skeleton. It responds to one of the first comments on my ASW article by Ezra Zubrow, who asked me simply, "Why didn't you quantify your model?" I had to explain that since my own Oaxaca project had barely started, the model was based almost entirely on MacNeish's Tehuacán data, and we had only recently begun to collect the information necessary to quantify and test it.

Perhaps the biggest mistake I made was to call the ASW paper "Archeological Systems Theory and Early Mesoamerica." This confused some critics, who assumed I was referring to Bertalanffy's (1962) "General System Theory" or Miller's (1965) "cross-level hypotheses of living systems." While I am sympathetic to the approaches of those investigators, the term "General System Theory" appears nowhere in my 1968 article. In fact, I explicitly stated that in the course of that paper I would attempt to apply, "on a prehistoric time level, the kind of *ecosystem analysis* advocated most recently by Vayda and Rappaport, with modifications imposed by the nature of the archeological data" (Flannery 1968:68, italics added).

As Watt's (1966a) volume suggests, systems models have long been part of the biologist's standard arsenal for the analysis of living communities of organisms. Had I called my ASW paper "Ecosystem Models and Mesoamerican Prehistory," many readers would probably have been less confused about what I was trying to do. As it is, Doran's (1970) and Salmon's (1978) critiques—which suggest that systems approaches have relatively little to offer archaeology—do not really deal with the analysis of ecosystems. Rather, they concern themselves with whether General System Theory is really a "theory," with whether systems analyses are really "explanations," and with more formal and mathematical systems approaches, such as those used in engineering. Archaeologists have not really stampeded to such approaches.

The fact is that the kind of systems analysis outlined in Watt's volume is something much less ivory-tower: the modeling of "recurrence relationships" and "mutual-causal processes" in ecological systems. Watt (1966b) uses the example of a simplified, abstract forest ecosystem, including weather, trees, humans, insect defoliators, warblers, spiders, hymenopterous parasites, and insect diseases. In his example, an unusually warm spring season sets the trees to flowering faster and earlier. Insect defoliators respond by rapid growth, utilizing the unusual amount of available food. The warm spring brings warblers back in force to consume the insects, but it also depresses the rates of insect diseases, which do best in cool, damp weather. The more insects eaten by warblers, the fewer are available to spiders and to insect parasites. All these variables affect the population density of insect defoliators, who in turn regulate the probability of tree survival, and hence the economic life of the human lumbering community. If that human community chooses to use pesticides, they affect not only insect defoliators but also the warblers and spiders who consume them.

Watt points out that such a system can only be understood by viewing it as a whole, yet the whole is too complex for normal analysis. Hence, the analyst must break it down into subsystems, analyzing each without losing sight of the necessity to reintegrate the whole at the end of the analysis (Watt 1966b:3). Two basic concepts of Watt's analysis are (1) the notion of recurrence relationships, such as the relations between warbler populations and insect populations above, and (2) the concept of optimization, "the notion that the central aim of systems analysis is that of making the optimal choice from among an array of alternative strategies at each of a sequence of times" (Watt 1966b:4). For reasons that are more fully discussed in Chapter 30, in this volume I retain the concept of recurrence relationships but substitute the notion of "steady improvement over time" for "optimization."

Finally, Watt points out that computer simulation "figures prominently in the arsenal of tricks used by systems analysts." This is because "most systems optimization problems are so complex that they cannot be worked out in any straightforward fashion using paper and pencil" (Watt 1966b:5). Reynolds' simulation of the transition to agriculture in the Valley of Oaxaca (Chapter 31) is such a computer program, using the recurrence relationships discovered by our project and the notion of "steady improvement" mentioned above.

MATTER, ENERGY, AND INFORMATION

All ecosystems are characterized by exchanges of matter, energy, and information among their components (Flannery 1972:400). Most ecological studies monitor the transfer of matter or energy, but leave aside the information exchanges that regulate such transfers. This is particularly unfortunate in the case of human ecology, where the role of information is undoubtedly greater than for any other species. One stumbling block has been the widespread impression that while matter can be measured in grams, and energy in calories, there is no workable unit of measurement for information. Some archaeologists, however, have begun to consider the use of the *bit*, defined as the smallest unit of information that can resolve the ambiguity between two alternatives. For example, the question, "shall we forage in that particular piñon nut grove?" can be reduced to the binary opposition Yes/No, possibly resolvable by one crucial bit of information passed on by someone who has recently walked through the grove.

Moore (1981) acknowledges that while cultural systems function through flows of matter, energy, and information, researchers have only done a good job modeling the first two. "The ephemeral quality of information flows . . . has limited their study. Most theories have no explicit role for information sharing and processing as a variable" (Moore 1981:194). Moore's paper is one of the few that attempt to build a model in which human foragers' demonstrated need and tendency to exchange information about resources are given "equal time" with matter and energy. Significantly, he was forced to design his own computer simulation because he found no acceptable set of computational formulae for information flow. These are the same constraints that led Reynolds to design his own program for this volume; at present, if you want your model to "learn" and "improve" through time, you will probably have to design it yourself. Reynolds was ideally suited for this project because he had previously tried an information-sharing, decision-making approach to foraging societies (Reynolds 1978; Reynolds and Ziegler 1979).

In addition to sharing information and making consensus-based decisions, human foragers can store information beyond the lifetime of an individual. For example, in a later section of this chapter we discuss the fact that the presence or absence of purple bean flowers in the thorn–scrub–cactus forest was a crucial bit of information that could set up a wild bean harvest months later. Consider the following information transfers that could be based on the sighting of such flowers:

1. "I saw bean flowers there last May."
2. "My cousin says he saw bean flowers there last May."
3. "A microband from across the river says they saw bean flowers there last May."
4. "During a dry year in my childhood, we found beans there."
5. "My grandfather said that in case of drought or emergency, in his childhood you could usually find beans there."

Here, then, is one of the ways human foragers differ from the ungulates studied by optimal foraging theorists (see Chapter 30): Not only can they make use of information based on the experiences of distant relatives and acquaintances, they can store information on the availability of certain resources under specific climatic conditions and use it years later. This is how information on plant availability and foraging performance is used in the model that Reynolds presents in Chapter 31.

THE CHOICE OF AN ECOSYSTEM

The choice of the Valley of Oaxaca as an "open-air laboratory" for this kind of study was a relatively easy one. In 1964, MacNeish's Tehuacán Archaeological–Botanical Project (on which I had served as faunal analyst) was drawing to a close, and it was time to select a project of my own. I looked around for a valley (1) close enough to Tehuacán so that the artifact complexes, vegetation, and fauna would all be familiar to me, yet (2) not so arid and agriculturally marginal as Tehuacán. MacNeish had chosen Tehuacán for the dryness of its caves, and I realized that by opting for a less arid valley I might be sacrificing archaeological preservation. However, I was willing to risk this because I wanted to see what differences incipient cultivation might display in a valley that was less dependent on irrigation than Tehuacán.

The Valley of Oaxaca looked promising because its large early urban center, Monte Albán, had presumably been supported by successful agriculture. I also knew the valley had caves and rockshelters, because Lorenzo and Messmacher (1963) had recently located some near Mitla. In addition, MacNeish had examined projectile points in the surface collections of the Frissell Museum of Zapotec Art in Mitla and was convinced many of them were from preceramic periods. Several of Lorenzo and Messmacher's shelters lay only 150 km from Coxcatlán Cave, making it likely that any preceramic complexes I found could be directly compared with Tehuacán's. By Christmas of 1964, my preliminary survey had convinced me that at least some Oaxaca caves had plant preservation, although it seemed unlikely that there would be as many superb dry caves as Tehuacán had.

Hindsight, as a lovable old philosopher once said, is always 20–20; and with the wisdom of hindsight, I realized there were many things left undone in Tehuacán that I wished we had accomplished. Many of these things had to do with the quantification of subsistence data, and, as my old ASW model began to take shape, our need for quantification became even clearer. By the time the Oaxaca Project began, I had some of the following research goals in mind:

1. We needed to locate a series of preceramic sites occupied at different seasons of the year, keeping our fingers crossed that some would have good plant preservation. Luckily, Guilá Naquitz fulfilled this hope (see Smith, Chapter 19; Whitaker and Cutler, Chapter 20; Kaplan, Chapter 21).
2. Having learned from our dry caves which plant and animal species were important in the Oaxaca preceramic, we needed to undertake censuses of them in their native habitats in order to determine how many kilograms of edible portion were available per hectare. This would put us in a better position to quantify the flow of matter. Eventually, we realized that we needed to understand the differences in productivity between wet, dry, and average years, so the censuses took a decade to complete (see Chapter 18).
3. In order to monitor the flow of energy, we needed to have professional nutritionists convert our kilograms of plant and animal foods into calories and protein. (This is done by Robson and Elias in Chapter 23.)
4. Having quantified the present-day environment, we needed to find out whether it was sufficiently similar to the preceramic environment to serve as a guide for interpreting the past. This was accomplished by studies of pollen (Schoenwetter and Smith, Chapter 15) and microfauna (Flannery and Wheeler, Chapter 16).
5. Having assembled all these data, we wanted someone to "operationalize" them, to design a computer model that would simulate the foraging strategies of the Oaxaca Archaic. We wanted to model the flow of matter and energy, to be sure, but without forgetting the role of information and long-term memory in producing the "steady improvement [in adaptation] over time" of which we spoke earlier. The ultimate test of this model, of course, would be the extent to which it matched the archaeological record (see Chapters 30–32).
6. Having modeled the preagricultural pattern, we then wanted to simulate the adoption of incipient agriculture to see if it would provide insights into the way early agricultural strategies grew out of preexisting foraging strategies. I also hoped we could use the model to examine the importance of variables such as population growth and climatic change, which had been proposed as causal factors by previous authors.
7. Finally, since we knew that many aspects of our model would be open to debate, disagreement, and modification by our colleagues, we wanted the centerpiece of the volume to be a straightforward site report on Guilá Naquitz, which could stand alone and be useful even to archaeologists who did not share our point of view. In this way, we would at least succeed in accounting for the money the National Science Foundation and the Smithsonian Institution had spent on our research, although of course we hoped to do more.

CONSTRUCTING A MULTIPLE-LOOP MODEL

In the remainder of this chapter, we consider the form that an anthropocentric model for the preceramic ecosystem in the Tehuacán–Oaxaca highlands might take. As Watt's (1966b) hypothetical example suggests, we need a model composed of numerous subsystems, or "loops," each of which can be lifted out, studied, and eventually put back into the whole. Depending on the policies at a given moment, the recurrence relationships in such loops can be positive or negative, and they can either counteract deviations from the previous state of the system or reinforce them. In the ASW paper I drew the framework for my model from "The Second Cybernetics," a seminal article by Maruyama (1963) that contains many more insights than I could use at the time. In my opinion, Maruyama's example of a multiloop system is still one of the clearest and most instructive in the literature.

In our Fig. 2.1 we reproduce Maruyama's Fig. 3, his simplified model for the operation of a hypothetical urban system. The seven variables used by Maruyama are linked to each other by a series of feedback loops, some positive and some negative. A plus sign on the arrow between two variables indicates that they change in the same way; for example, as *P* increases, so does *G*. A minus sign, on the other hand, indicates change in the opposite way—as *S* increases, *D* decreases. These pluses and minuses reflect the policies of the system, and a change in policy from plus to minus can have dramatic effects.

Negative feedback loops are ones that help to counteract deviations from the preexisting pattern and return the system toward its previous state. For example, consider the loop *P–G–B–D–P*. As the number of people in the city goes up, so does the amount of garbage per area, raising the bacteria per area, therefore raising the number of diseases and eventually lowering the number of people—returning the system toward the original population level, although not necessarily reaching it.

Positive feedback loops are ones that help the system diverge farther from the previous pattern, preventing its return toward the previous state. Consider the loop *P–M–C–P*. As

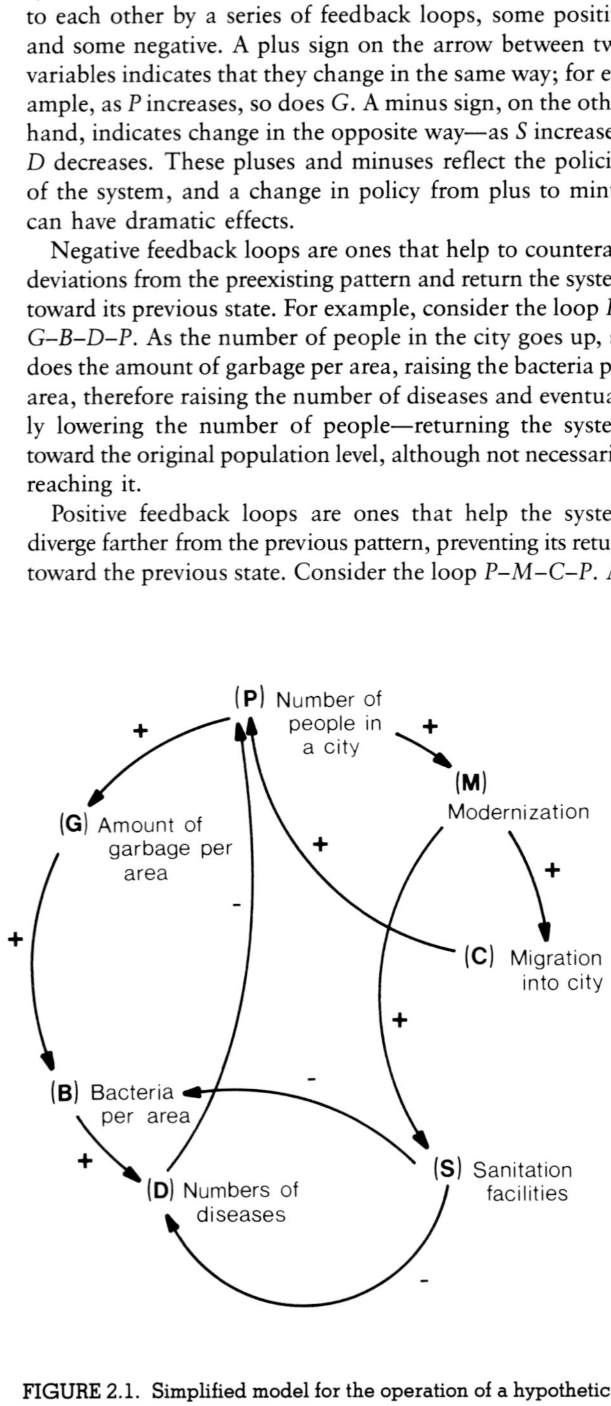

FIGURE 2.1. Simplified model for the operation of a hypothetical urban system. (Redrawn from Maruyama 1963:Fig. 3.)

the number of people in the city goes up, it stimulates modernization, which encourages migration into the city, further raising population and stimulating more modernization and more immigration. As the system diverges farther from its original condition, new loops can appear that cause major changes elsewhere. As an example of this, consider the arrow *M–S* that sets off the loop *M–S–D–P–M*. An increase in modernization can lead to better sanitation facilities, reducing bacteria per area, lowering diseases, and thereby partially canceling out the effect of negative feedback loop *P–G–B–D–P*.

What are the merits of such a model? First of all, even the simple act of putting in the pluses and minuses forces you to be explicit about the relationships among variables, rather than just listing them. A more sophisticated model would even quantify the rates at which variables change in response to others—for example, the number of kilograms of garbage per hectare per year that would be produced by each new person entering the city or the number of bacteria per hectare produced by each kilogram of garbage. Such quantification turns your work into the model of a real system, not just a "heap" (which is what systems analysts call a mass of variables with no explicit relationships). It also puts you in position to make your system "run," to simulate its ongoing progress as Reynolds does later in this volume.

Second, I believe that such a model deals with causality in a way that is more realistic than the linear-causality models sometimes used in archaeology. For example, which is the independent variable in Fig. 2.1? Many archaeologists would like it to be *P*, the growth of population in the city over time. As we have seen, however, the value of *P* is partly a function of immigration (*C*) and disease (*D*). Indeed, no variable in the system is wholly independent; while they are certainly not all of equal importance, each has contexts in which it is a cause and others in which it is an effect. Such a model may lack parsimony, but it more closely approximates the mutual-causal relationships archaeologists encounter in the real world.

Third, there is another advantage I see in the way systems models deal with causality. In any ecosystem, there are variables that are demonstrably associated. In the more deterministic models frequently used by archaeologists, these associations are often seized upon and proclaimed as "laws." For example, in Watt's model, "The higher the temperature, the lower the incidence of insect diseases"; in our model, "The greater the increase in seed permeability, the greater the likelihood that bean cultivation will be intensified." In a systems model these are not laws but simply the "recurrence relationships" of which Watt (1966b) speaks. They are important, but they are no more than components of a greater whole.

DERIVING A SYSTEMS MODEL FROM THE TEHUACAN AND OAXACA DATA

In 1966, when I began to write my ASW paper, the Oaxaca project had barely begun. We had as yet no quantitative

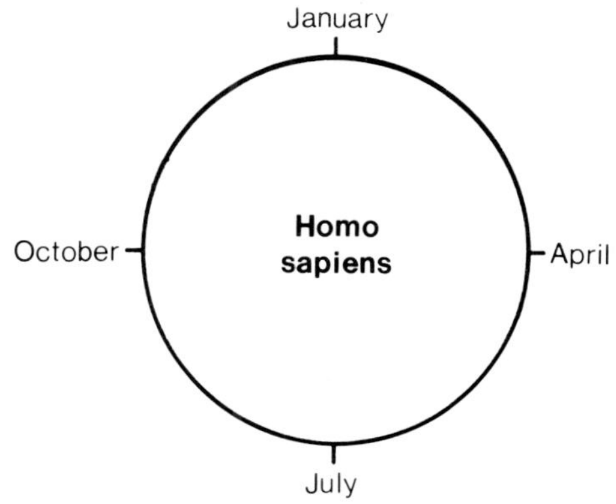

FIGURE 2.2. The annual cycle of activities of a microband of human foragers, to be used as the core of an anthropocentric ecosystem model. (Design by Richard I. Ford, unpublished, 1970).

analyses of cave remains, no decade of plant censuses, no nutritional studies of aboriginal food. Hence, most of my data came from MacNeish's Tehuacán project or from a study of the characteristics of its flora and fauna. In a very real sense, therefore, my ASW paper can be considered a preliminary model generated largely by the Tehuacán data—one that can now be tested against a whole new body of data derived from Oaxaca. For example, the plant census data we subsequently collected in Oaxaca now enable us to put some quantification into the model, which was impossible in 1966.

Since the model is a complex one, it cannot really be displayed in one single drawing. Rather, we must build it subsystem by subsystem, or feedback loop by feedback loop. We begin with the core of the model, which is shown in Fig. 2.2. In this diagram, *Homo sapiens* represents a microband[1] of human foragers in the Tehuacán–Oaxaca highlands, and the encircling wheel represents their annual round of foraging activities, with January, April, July, and October serving as landmarks. To this core we can now attach some of the procurement subsystems I described in my ASW paper.

[1]For the definition of microband, see Chapter 3.

Maguey Procurement

As I suggested then, the preceramic forager "was not simply extracting energy from his environment, but participating in it; and his use of each genus was part of a system that allowed the latter to survive, even flourish, in spite of heavy utilization" (Flannery 1968:69). Figure 2.3 shows how this happened with the maguey, or century plant (*Agave* spp.), a tough and primitive member of the Amaryllis family that is available even at the leanest times of the year.

The wheel surrounding *Agave* is its life cycle. New magueys sprout and gradually grow to maturity over a 6–8-year period, at which point the inflorescence, or *quiote*, appears. The plant sends out its pollen and disperses its seeds, after which it begins to die. The seeds are not its main means of propagation, however, for mature agaves reproduce vegetatively and may be surrounded by suckers, or *hijos*, whose development is slowed by competition with the parent plant.

Today's Indians know that the best time to eat the agave is after it has bloomed, and the pod and *quiote* fragments in preceramic cave deposits indicate that the preceramic foragers knew that as well. The appearance of the inflorescence, frequently toward the end of the rainy season, provided foragers with the information that that particular agave was ready. They might cut off the *quiote* at this point, after which a natural fermentation took place in the moribund plant that softened it and increased its sugar content. Months later—even in the heart of the dry season, when few other plants were available—they could return to dig up that

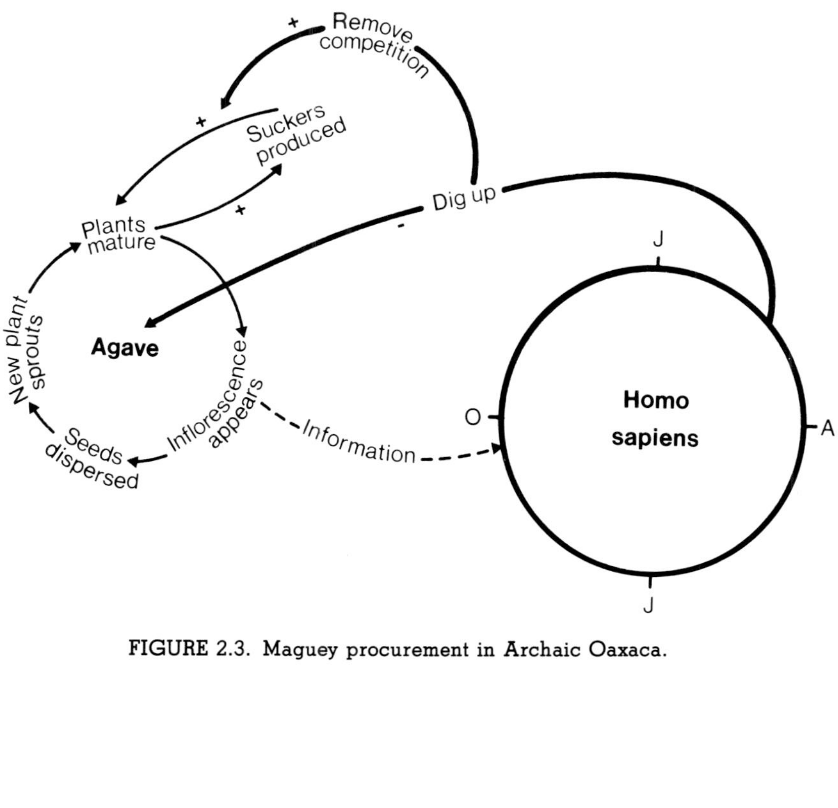

FIGURE 2.3. Maguey procurement in Archaic Oaxaca.

maguey and roast it in an earth oven for 24 to 72 hours, depending on its size. The digging up of the parent plant removed competition from the suckers or vegetative offspring, allowing them to develop into mature plants themselves and thereby keeping agave populations at roughly the same level.

The pluses and minuses in Fig. 2.3 indicate the variables that change in the same way (digging up eliminates one mature agave but may promote the development of one or more *hijos*). The diagram shows the critical point of information flow but not matter and energy. The latter variables, however, can now be partially modeled on the basis of data given in Chapters 18 and 24. For example, we can now suggest that there are an average of 122 inflorescence-bearing *Agave potatorum* per hectare near Guilá Naquitz; these agaves represent 341 kg of edible portion, providing 429,660 kcal/ha; and each occupant of the cave ate an estimated 140 g of agave per day during his or her stay.

Cactus Fruit Procurement

Organ cacti of at least four species were eaten in the Tehuacán caves. The variety was less in the somewhat higher Valley of Oaxaca, and at least one genus—*Myrtillocactus*—has no archaeological record of use because its fruits appear during February, when no one was at Guilá Naquitz.

Perhaps the most useful cactus in both regions was the prickly pear (*Opuntia* spp.), which was used both for its seasonal fruits (tunas) and its tender young stem segments (nopales). Nopales are available throughout much of the year and were despined and roasted over hot coals at Guilá Naquitz (see Fig. 10.5). They provided the preceramic foragers with an abundant (if less than delicious) vegetable.

The fruits, or tunas, are available seasonally, and the way they were used by the Indians may provide an example of the kind of symbiotic relationship proposed by Rindos (see Chapter 1). I described this relationship briefly in 1968 on the basis of the Tehuacán data, and it is diagramed here in Fig. 2.4. On the average, prickly pear fruits from the eastern Valley of Oaxaca weigh 50 g and contain 50 seeds (Chapter 24), each of which has a tough outer layer that actually impedes the seed's chances of sprouting until it has been weakened by stomach acids. *Opuntia* remains in preceramic coprolites show that the seeds "almost invariably survive the human digestive tract and escape in the feces" (Flannery 1968:72). In other words, no matter how intensively tunas were harvested, the Indians were doing the plant a favor by weakening the protective outer layer and propagating the seeds.

As Fig. 2.4 suggests, the crucial information for the human population was the appearance of the flowers (and then the immature fruits) with the advent of the spring rains. This let the Indians know that they should return to the *Opuntia* groves in July to harvest tunas. This harvest had to be done in competition with fruit bats, birds, rodents, and collared peccaries, all of whom also like cactus fruits. Some of the Tehuacán caves contained wooden tongs for use in picking the spiny fruits off the stem. The spines could be singed off, and the fruits could be split open and sun-dried to keep them from spoiling for several weeks. Ironically, as Fig. 2.4 shows, it is the fruits that the Indians *ate* that were more likely to have their seeds dispersed; fruits that were uneaten often went to waste (as, for example, when they were left behind in the cave to desiccate and be recovered by archaeologists). Thus the procurement of cactus fruits was an activity that was beneficial to both participants. It did not, however, produce a genetically different strain of *Opuntia* that was dependent on man.

Wild Bean Procurement

Wild beans (*Phaseolus* sp.) and wild onions (*Allium* sp.) were two plants whose presence in the Holocene thorn forest

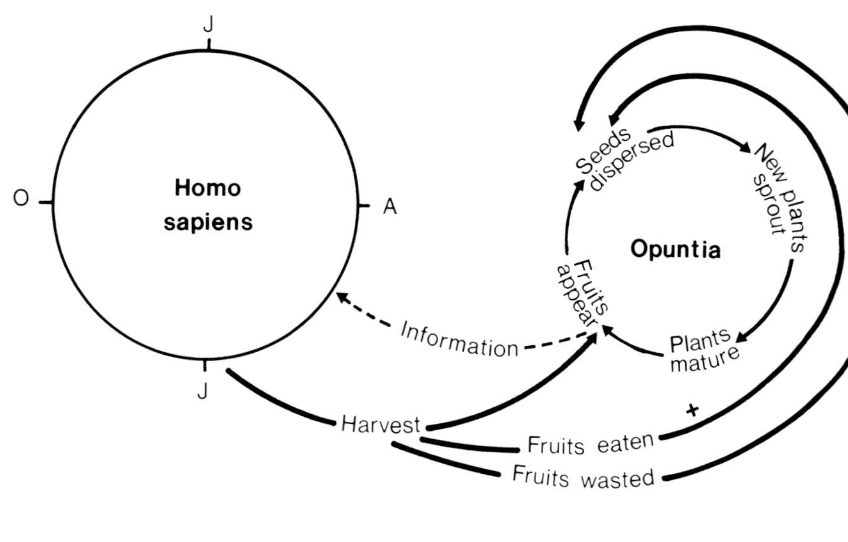

FIGURE 2.4. Procurement of prickly pear fruit in Archaic Oaxaca.

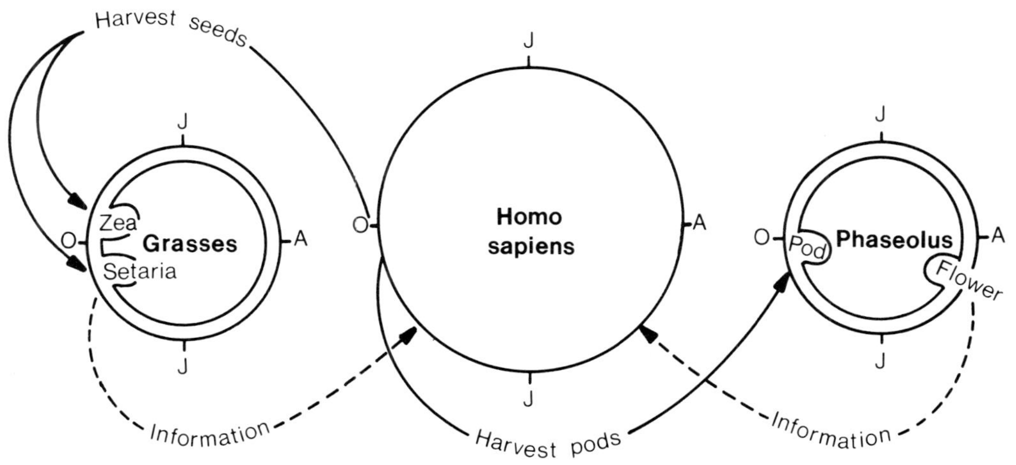

FIGURE 2.5. Procurement of wild beans (right) and two genera of wild grasses (left) in the central Mexican highlands during the Archaic.

could be detected by their flowers. In the case of the wild beans of the Guilá Naquitz region (Fig. 2.5), the purple flowers appear in May during a normal year. These flowers provide the necessary locational cues so that foragers could return to the same area in September or October to harvest the plant, either for the beans in its pods or for its tuber (*jicamita*). In the case of the local wild onions, the flowers are white and tend to appear in damp microhabitats.

Beans are one of the plants that do not show strong evidence for a symbiotic relationship with humans. Their seed dispersal depends on a corkscrew-twisting pod that shatters easily, a genetic character that hampers harvesting. Archaeological data suggest that this character did not disappear until it was selectively bred out by humans following domestication; indeed, some beans may initially have been cultivated for the tuber rather than the seeds.

Wild Grass Procurement

The harvesting of wild grasses was one of the autumn activities of preceramic foragers in Tamaulipas, Tehuacán, and Oaxaca. Much of this activity took place in the tributary barrancas of the valley piedmont, a more humid habitat in an otherwise semiarid environment. Two of the important genera involved were *Setaria,* or foxtail grass, and *Zea*, which includes teosinte.

Setaria and teosinte share a number of biotopes, such as the Guerrero piedmont shown in Fig. 2.6. In such a habitat, *Setaria* comes ripe earlier, often in late September; teosinte comes ripe some weeks later, often in mid-October. As shown in Fig. 2.5, the gradually ripening heads of foxtail grass would provide the information necessary to bring foragers to the piedmont at the right time. By camping in the area for a month or two, they could harvest both grasses in succession, as well as any wild cucurbits and/or runner beans that were twining around the teosinte (see Chapter 1). No productivity figures are available for *Setaria*, but in 1971 Ford and I collected data to show that teosinte could yield anywhere from 152.5 to 627 kg of edible portion per hectare (Flannery and Ford 1972).

Tree Legume Procurement

Several leguminous trees, including huizache (*Acacia* sp.) and *guaje* (*Leucaena* sp.), produce pods or seeds that could be used as food by preceramic foragers. By far the most important legume tree in the Tehuacán–Oaxaca environment, however, was the mesquite (*Prosopis juliflora* and related species). In their most favorable habitats in the eastern Valley of Oaxaca, mesquite groves can produce 109,960 pods (or about 183.6 kg of edible portion) per hectare. Such a harvest would yield 545,292 kcal and 10,465 g of protein (Chapter 18). Mesquite pods can be handpicked and probably were transported back to the Tehuacán and Oaxaca caves in the

FIGURE 2.6. Mature *Setaria* growing on the margins of a still-immature stand of teosinte near Chilpancingo, Guerrero, 1971.

many types of baskets and net carrying bags recovered from preceramic levels. Competing with deer, jackrabbits, coatimundis, ring-tailed cats, and other animals, the Indians harvested mesquite pods to cook up into a kind of *miel*, or syrup. Although the seeds can be toasted and eaten, archaeological data indicate they were often discarded by the thousands; this presumably aided in the dispersal of the plant by spreading the seeds over a much larger area than would have resulted from normal July–August tree fall.

ZEA VERSUS *PROSOPIS*: AN EXAMPLE OF TWO GENERA LINKED THROUGH THEIR RELATIONSHIP TO HUMANS

One of the characteristics of a multiple-loop system, as we saw in Maruyama's example, is that changes in one loop can have an impact on a distant loop in another part of the system. Frequently these loops are so widely separated in time and/or space that one's intuition would never lead one to expect their relationship. In this section I consider a relationship between the *Zea* and *Prosopis* loops in the Oaxaca model we are building. At the end of the Tehuacán project I did not have enough data to appreciate this relationship, and it was still undeveloped at the time of my ASW paper. By the time of my 1973 article in the *Annual Review of Anthropology*, some of the outlines of the relationship were emerging, and I elaborate on these here.

Under wild conditions, *Zea* and *Prosopis* do not generally occupy the same space, because the canopy of a good mesquite stand would shade out the teosinte. Mature mesquite groves occur with greatest frequency along the alluvium near streams on the valley floor, and the backslopes of the natural levees of suitable rivers. Teosinte is a pioneer weed of natural scars in the landscape, particularly clearings in the thorn-forested piedmont. Even the growth of agriculture during the period 8000–5000 B.C. did not initially move cultivation out of the piedmont barrancas and into the major river floodplains. Early agriculture of that period remained "barranca horticulture," as MacNeish has called it, possibly because the piedmont barrancas received higher rainfall than the valley floor and because the clearing of forested levees along the major rivers would have involved a great deal more work. Besides, some of those forested groves were producing 183.6 kg of mesquite pods per hectare.

For the purposes of this chapter it is not necessary for us to specify whether early maize was a domestic form of teosinte (as Beadle (1977), Galinat (1971), and Iltis (1983) believe) or came from a now-extinct wild corn (as Mangelsdorf (1974) believes). At 5000 B.C. it was a primitive cultivar with tiny cobs, easier to harvest than wild teosinte but probably no more productive.

> In many areas it would not have been worthwhile to remove the mesquite cover of the main river floodplains to cultivate such a plant. A more reasonable strategy would have been to leave the mesquite on the main valley floor to produce its 180 kg per hectare each year, while growing [*Zea*] in the piedmont barrancas where it was at home. . . . That such a strategy was used is suggested by the fact that for thousands of years after *Zea* had been domesticated, no valley-floor villages appeared in Mesoamerica, although MacNeish (1964) suspects pit-house settlements in the tributary barrancas. (Flannery 1973:299)

This strategy was eventually altered, however, by gradual genetic changes that increased the length of the maize cob, its number of kernel rows, and the size of its kernels. We have some estimates of the related increase in productivity of maize, thanks to a linear-regression relationship between mean corncob length and yield in kilograms per hectare that A. Kirkby (1973:Fig. 48) recorded in Zapotec Indian cornfields in Oaxaca. Based on this relationship, and using figures on the mean length of corncobs recovered by MacNeish's Tehuacán excavations, Kirkby then calculated an "estimated yield" for various periods of prehistory.

The earliest cobs from Coxcatlán Cave (5050 B.C.) suggest a yield of only 60–80 kg/ha; later preceramic cobs from Tehuacán (ca. 3000 B.C.) suggest yields of 90–120 kg/ha. By 2000 B.C., maize reached an estimated yield of 200 kg/ha, surpassing mesquite in crude weight of edible portion per unit area (for the moment, we are ignoring all other differences between the two foods). By 1000 B.C., maize yields should have been up to 300 kg/ha or more.

A later study by Kirkby (1974) shows that Zapotec Indian farmers do not usually consider cultivation and land clearance to be worthwhile unless a yield of at least 200–250 kg/ha (shelled maize) can be expected. Taking all of the above observations into account, we can therefore propose the following possibility: Permanent, agriculturally based villages did not appear in the natural levee zone of the major river floodplains until maize productivity was high enough to make it worthwhile removing the mature mesquite cover of that zone. A possible corollary to this statement would be that the threshold for mesquite removal might have come (1) when maize surpassed mesquite in productivity (> 184 kg/ha), (2) when maize productivity reached the 200–250 kg/ha favored by today's Zapotec Indians, or (3) at some third value that can only be determined by future archaeological research.

In Fig. 2.7 we present a diagram of the situation prior to the reaching of the threshold described above. On the right we see the life cycle of the mesquite tree *Prosopis juliflora*, from the growth of new saplings to the maturing of the tree and the development of green pods in June. The appearance of these unripe pods was the crucial information transmitted to the human population, letting them know that mesquite would be ready to harvest by August. As mentioned above, seeds discarded when the pods were processed by the Indians probably had a better chance of sprouting (and certainly got wider dispersal) than those that remained in their pods on the tree to dry out.

On the left of Fig. 2.7, we see the annual cycle of early cultivated maize, gradually increasing from 50–100 kg/ha in its piedmont barranca clearings as the Archaic period progressed. Its yield was yet another source of information for the human population, who would certainly have had a set of policies concerning the amount of labor they were willing to invest in any given plant. At the time of Fig. 2.7 (perhaps 3500 B.C.), their policy was to cultivate *Zea* in the

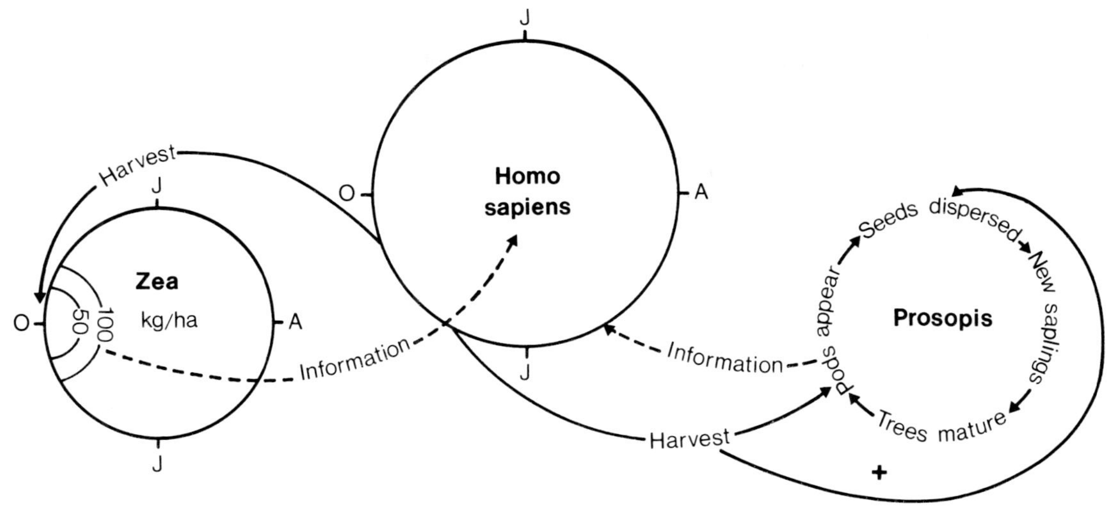

FIGURE 2.7. Human interaction with wild mesquite (right) and cultivated *Zea* (left) in Archaic Oaxaca.

piedmont and harvest *Prosopis* on the alluvium. A substantial increase in the productivity of maize, however, had the potential of providing new information that could change the policies for both maize and mesquite for years to come. In Chapter 32 of this volume, after examining both the data from Guilá Naquitz and the computer simulation by Reynolds, we present a second *Zea–Prosopis* diagram (Fig. 32.1) that we believe models the threshold of this policy change.

In Maruyama's terms, the favorable genetic changes leading to more productive maize may have been the "initial kick" that converted the *Zea* subsystem into a positive feedback loop. The increasing yield of maize, along with the greater ease of harvest of its tough-rachis, soft-glume cob, may have encouraged the Indians to spend more time cultivating corn and less time harvesting other, less responsive plants. If this is the case, the archaeological record should show not merely (1) an increase in commitment to maize but also (2) significant changes in the "mix" of wild species used as the human population changed its priorities. In other words, policy changes in one loop could have had profound changes in other loops.

The genetic mutations that contributed to these policy changes could not be farther removed from human intentionality; this is the biological component of the transformation. However, without human intentionality, mutations that so damage a plant's seed dispersal mechanism could not have survived, let alone had their frequency increased. Archaeological data suggest that humans had already made the decision to cultivate cucurbits and beans in Mexico, and wheat and barley in the Near East, *before* the crucial genetic changes that made the domestic varieties dependent on man. Our model must therefore take into account not only matter, energy, and information but also genetics and human intentionality. Systems models are well suited to this because, as Reynolds demonstrates in Chapter 31, they can be designed to handle variables as different as kilocalories, search area, and decision making.

MODELING ENVIRONMENTAL VARIATION

We have already suggested in Chapter 1 that Archaic foragers had to adapt to two major sources of environmental variation. One of these sources, seasonal variation, was rhythmic and relatively predictable; in Oaxaca, the summer rains normally begin in May and end in November, with peaks in June and September. This means there will usually be an abundance of plant resources by the end of the rainy season and scarcity by the end of the dry season every year.

The second source, annual variation, is random and unpredictable. As indicated in Chapter 1, annual rainfall in Oaxaca averages 450–650 mm but may be as low as 300 mm or as high as 1000 mm. In rainy years, certain genera will produce a bumper crop; in dry years, yields of almost all genera will be reduced. Moreover, some trees, such as the piñon pine, may be so "exhausted" by a bumper crop that they produce substandard yields for the next few years regardless of rainfall (Richard I. Ford, personal communication, 1980).

In this volume, we do little to model seasonal variation since Guilá Naquitz seems to have been occupied only during the August–December period, and we therefore have inadequate data on activities of the January–July period. Future analyses of neighboring sites, such as Cueva Blanca and Gheo-Shih, may help to complete our understanding of the annual cycle of activities.

The modeling of annual variation, however, will be one important focus of Reynolds' simulation. In that simulation, an unpredictable sequence of wet, dry, and average years will be fed into the model in an effort to see how the Archaic foragers might have responded to such variation.

SCHEDULING

As I suggested in the ASW paper, so many possibilities for resource procurement were open to these ancient foragers that it would have been impossible to engage in all of them, even seasonally. "It happens that there are times of the year when a number of resources are available simultaneously, producing a situation in which there is some conflict for the time and labor of the group. Division of labor along the lines of sex, with men hunting and women collecting, is one common solution to these conflicts, but not all conflicts are so easily resolved" (Flannery 1968:75). The solution for more complex situations may be called *scheduling*.

In this volume, we investigate these solutions as follows. In Chapters 25–29, the pattern of debris on living floors at Guilá Naquitz is analyzed by Spencer, Whallon, and Reynolds, with an eye to determining (among other things) the extent to which sexual division of labor allowed the simultaneous performance of tasks that would otherwise have competed for the foragers' time.

In Chapter 31, on the other hand, Reynolds attempts to model the scheduling of plant procurement activities by the Guilá Naquitz population. A hypothetical band of foragers is presented with the 11 most common plants used in preagricultural levels of the cave, along with data on the environmental zone where each occurs, its density, and its nutritional makeup. It is up to our foragers to schedule the order in which these plant resources are procured, the environmental zone(s) visited, the area searched, and the amount collected. The efficiency is evaluated as they improve steadily over time while enduring their unpredictable succession of wet, dry, and average years. Eventually, our foragers are given cultivated squash, beans, and primitive maize to fit into their system of priorities, and we see whether their schedule of activities changes along the lines we have suggested.

SUMMARY

In other words, while the centerpiece of this volume is a site report on Guilá Naquitz Cave, we also use the volume to expand on, and test some aspects of, a model for wild plant gathering and incipient cultivation in the Valley of Oaxaca. At the core of the model is a small human population that moves through its annual wild plant harvests year after year for centuries. It is linked to a series of plant populations through exchanges of matter, energy, and information like those we saw in Figs. 2.3–2.7. Because there are so many of these linkages, we cannot represent them all in one diagram; we must imagine an *n*-dimensional space in which acorns, piñon nuts, prickly pear, mesquite, agaves, wild beans, squashes, and bottle gourds all intersect with the human population. Our human foragers cannot establish a static, unvarying relationship with these plant genera, however, because they are buffeted by a series of random variables, including a succession of wet, dry, and average years in which the productivity of the wild plants changes.

One of our suspicions—to be tested and clarified in this volume—is that our foragers dealt with both seasonal and annual variation in this ecosystem by working out a diversified strategy of resource scheduling, which was resilient enough to survive such perturbations over long periods of time. We hope that Reynolds' operationalized version of our model will tell us how this *could* have been done, but we also recognize that the archaeological record is the ultimate test of whether it *was* done in the way our model suggests.

Another of our suspicions is that incipient cultivation, as in the maize–mesquite example given earlier in this chapter, changed the information flows in the system in such a way as to alter its policies. To accommodate early agriculture, rescheduling should have taken place, with some wild plant genera rising or falling in priority according to their relationships with each other and with man. We hope that our model predicts some of the directions these changes might have taken, and we hope the archaeological record will allow us to see which of our predictions (if any) are correct. We also hope to be able to use the model to examine the extent to which variables such as population growth and climatic change would have affected the adoption of agriculture.

Incompleteness of the Model

Despite our best efforts, many aspects of my old ASW model remain unquantified even after this volume. Some of this can be corrected in future publications of the Oaxaca project, and some probably cannot.

For example, hunting and trapping were important activities for our preceramic foragers, and in 1968 I included the procurement of cottontail rabbits and white-tailed deer in my model. While animal procurement is discussed in Chapter 22 and elsewhere in this volume, it is not included in Reynolds' simulation. This is because we have decided to defer any major treatment of hunting until the publication of Cueva Blanca, a neighboring site that produced more abundant evidence for that activity than did Guilá Naquitz. Without hunting and trapping included, our model in this volume obviously falls short of completion.

In addition, we still cannot fully quantify the diagrams shown in Figs. 2.3–2.7 without additional data that would take years to collect. For example, to know exactly how human eating of prickly pear fruit would affect *Opuntia* densities in the Oaxaca piedmont, we would have to follow a sample of seeds all the way through their life cycle, from their deposition in coprolites until the point where we had figures on just how many of them would grow to be mature cacti. To collect the same data for *Agave*, we would have to dig up a sample of mature magueys and follow to adulthood those vegetative *hijos* that had flourished as a result. Until such data are collected, our model will continue to have some loose ends.

This should not be too discouraging, for it is likely that any model attempting to integrate as many variables as this one does will have areas of missing data. Building the model of any ecosystem is a slow process, and we have to expect that modeling a prehistoric ecosystem will be doubly hard. In the remainder of this volume, therefore, we try to think positively and concentrate on the data that *are* there, the insights that *are* forthcoming, and the flesh that *can* be added to our 1968 skeleton.

II

THE CULTURAL AND ENVIRONMENTAL SETTING

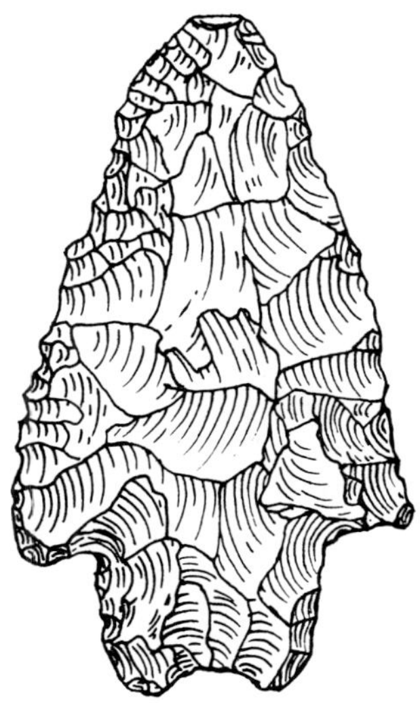

3

Guilá Naquitz in Spatial, Temporal, and Cultural Context

Kent V. Flannery

INTRODUCTION

Two previous chapters have discussed (1) the origins of agriculture as a universal theoretical issue in anthropology, and (2) the problems of building an ecosystem model for plant collecting and incipient cultivation in the Tehuacán–Oaxaca region. In this chapter we concentrate on the specific archaeological sites used to reconstruct the Archaic lifeway in the Valley of Oaxaca, most importantly Guilá Naquitz Cave. We begin with an overview of the valley, gradually focusing down on its eastern, or Tlacolula arm, and the 16-km^2 area surrounding the Precolumbian "Fortress of Mitla." Here it becomes apparent that there are at least two environmental contexts for Guilá Naquitz. The first is the "original primary vegetation" of the area as reconstructed by C. E. Smith (1978); the second is the present-day vegetation, obviously disturbed by human activity, in which the "original" vegetation survives only in relict patches.

Space and environment, however, are only two of the contexts in which Guilá Naquitz can be discussed. It also was occupied during a specific time period, and by a group of people whose cultural pattern extended over a wider area of Mexico. These temporal and cultural contexts are considered later in the chapter.

THE VALLEY OF OAXACA

The Valley of Oaxaca lies in the southern highlands of Mexico, between 16°40′ to 17°20′N and 96°15′ to 96°55′W. It is drained by two rivers: the upper Río Atoyac, which flows from north to south, and its tributary, the Río Salado, or Tlacolula, which flows westward to join the Atoyac near the present city of Oaxaca. The valley is shaped like an irregular Y or three-pointed star, whose center is Oaxaca City and whose southern limit is defined by the Ayoquesco gorge, where the Atoyac River leaves the valley on its way to the Pacific coast. The climate is semiarid, with 500 to 700 mm of annual rainfall confined largely to the summer months. The valley floor lies at altitudes between 1420 and 1740 m, the lower elevations being semitropical and the higher elevations warm-temperate. It is surrounded by cooler and wetter mountains that rise to more than 3000 m (Fig. 3.1).

Bedrock Geology

The Ríos Atoyac and Salado and their tributaries have created the valley by erosion and dissection of three major geological formations (Lorenzo 1960). The oldest of these is a basal complex of Precambrian gneisses and schists, so thoroughly metamorphosed that their original petrographic composition is unknown. The second formation is a series of Mesozoic limestones, thought to be mainly Cretaceous and perhaps distantly related to the limestones of the Chiapas highlands. The most recent of the three major formations is a complex of Middle Tertiary (perhaps Miocene) volcanic tuffs known as ignimbrites. Ignimbrites are ash flow tuffs laid down not by explosive eruptions but by glowing avalanches that may smoulder for years; in Oaxaca they include such familiar variants as rhyolite and rhyodacite (Williams and Heizer 1965).

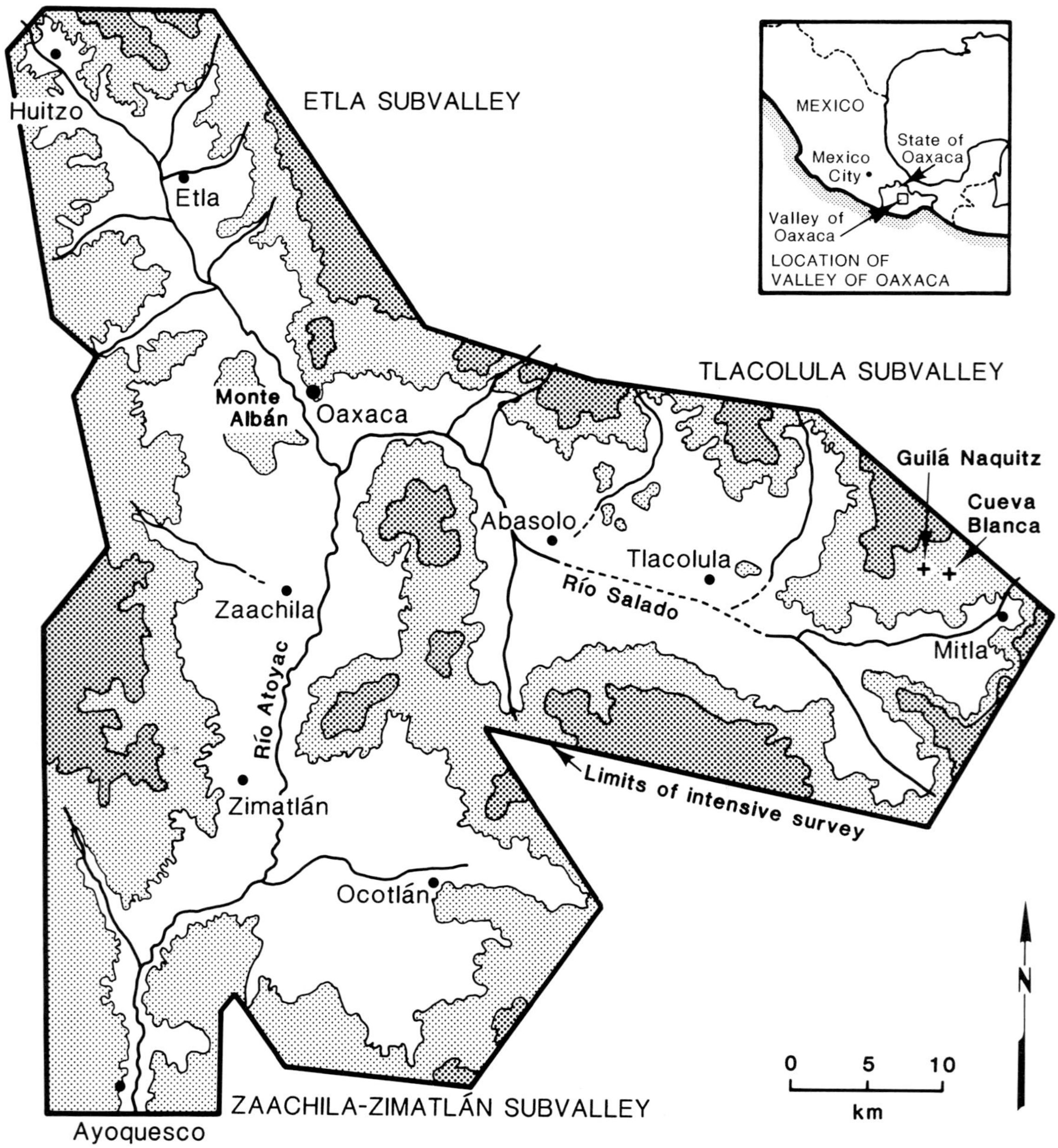

FIGURE 3.1. The intensively surveyed portion of the Valley of Oaxaca, showing the Etla, Tlacolula, and Zaachila-Zimatlán subvalleys. Light stipple indicates piedmont, heavy stipple the steeper mountains.

Physiographic Zones

Processes of Quaternary erosion and alluviation have sculptured these parent materials into four main physiographic zones: mountains, piedmont, high alluvium, and low alluvium (Kirkby 1973:9). The mountain zone, generally above 2000 m, is characterized by steeper slopes, cooler temperatures, and higher precipitation than the other zones. The extensive piedmont formed originally as a series of coalescing fans along the base of the mountains, but is now dissected into a line of ridges and arroyos delimiting the outer edge of the valley floor. Its average slope is 1–2° except where gullied, and it constitutes a zone of transition (in both physiographic and vegetational terms) between the mountains and the valley floor.

Of the two zones of alluvium, it is the high alluvium that is the most extensive and important. Representing the Pleistocene floodplain of the Ríos Atoyac and Salado, it has a gradient of less than 1° and forms the main part of the valley floor. The low alluvium, or present floodplain, occurs as a distinct geomorphic unit along less than 50% of the courses of the Atoyac and Salado, and may not have existed

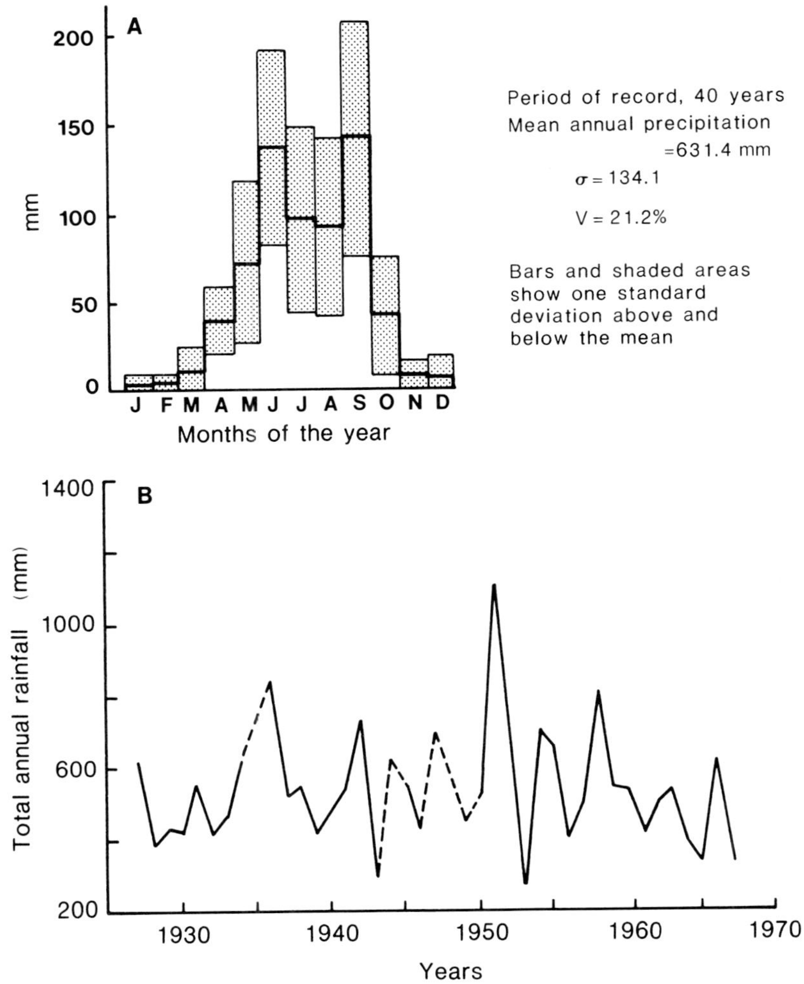

FIGURE 3.2. Rainfall in the Valley of Oaxaca. A, mean monthly precipitation at Oaxaca City; B, annual precipitation at Tlacolula between 1926 and 1968. (Redrawn from Kirkby 1973:Figs. 6 and 58.)

prior to A.D. 1500 (Kirkby 1973:15). It is a zone of lower rainfall and higher temperatures than the mountains or piedmont, although its subsurface water table may be high.

The native vegetation of the Valley of Oaxaca is greatly influenced by these physiographic zones and by the gradients of temperature, rainfall, evapotranspiration, and length of growing season that exist between the low alluvium and the high mountains. In simplest terms, the season of frost increases in length as one ascends higher into the mountains, while the likelihood of drought increases as one descends to the floodplain. The result is that the mountains have vegetation that can tolerate lower mean annual temperatures (8° at 3100 m) but needs higher mean annual precipitation (1000 mm); the floodplain vegetation is adapted to higher mean annual temperatures (21°C at 1540 m) and survives the lower mean annual precipitation (630 mm at Oaxaca City) by drawing on the subsurface water table. The piedmont, lying in between, has a tremendous variety of plants because of its transitional position.

The rainfall pattern "shows a sharp distinction between the dry winter months from November to March whose mean monthly rainfalls are less than 10 mm, and the summer rainy season. This characteristically begins in late April to May but does not become well established until June with its mean rainfall of 137 mm. The rainy season has two peaks, one in June and a second in September (mean rainfall = 144 mm). By October mean rainfall has fallen to 44 mm, and the rainy season usually ends as abruptly as it began" (Kirkby 1973:17; see also Fig. 3.2, this chapter).

Native Vegetation

It is not easy to reconstruct the original wild vegetation of a valley whose floor has been disturbed by agriculture for more than 7000 years, whose piedmont has been grazed by goats for at least 400 years, and whose mountains are filled with woodcutters and charcoal burners. However, botanist C. Earle Smith has attempted such a reconstruction based on (1) surviving remnants of the original vegetation, (2) principles of plant growth as they apply to the physiographic features described above, (3) knowledge of vegetation in similar environments in neighboring valleys, and (4) plant remains recovered by archaeological excavation (Smith 1978:17–22, Map 2). All evidence suggests that the Valley of Oaxaca was originally covered by forest, brush, or low monte—unlike the open, grassy valley we see today, which is a product of agricultural land clearance.

Along the Río Atoyac and its major tributaries, wherever there were clay soils and water table within 3 m of the surface, Smith sees a mesophytic (riverine) forest composed of willow (*Salix*), alder (*Alnus*), baldcypress or ahuehuete (*Taxodium*), fig (*Ficus*), *Annona, Cedrela*, and possibly wild avocado (*Persea*). There would have been many evergreen species in this forest; and it would have been tall and shady, with a canopy perhaps 15 m to the lowest branch. In areas where the water table was similar but soils were sandy, such a mesophytic forest would have been lighter. Constant demands for firewood and building materials have reduced this forest to occasional patches of alder and willow, or isolated ahuehuetes like the group at Santa María del Tule.

On the high alluvium beyond the natural levees of the major rivers, where subsurface water table falls to between 3 and 6 m, Smith sees a forest dominated by mesquite (*Prosopis*) but thickly intergrown with other legume trees such as huizache (*Acacia*) and members of the families Burseraceae, Malvaceae, and Euphorbiaceae. While this formation would have included a few evergreen taxa, most species were deciduous, and the canopy would have been closed only during the summer rainy season. This is a zone whose wild plant resources would have been richest between May and late August.

Beyond the mesquite forest, where the water table drops below 6 m, began a zone of thorn–scrub–cactus forest that once covered the margins of the high alluvium and the lower slopes of the piedmont. This vegetation zone, dependent on seasonal rainfall rather than groundwater, would have been relatively bleak in the late dry season but relatively rich after the summer rains. Smith (1978:20) reconstructs this thorn–scrub forest as "similar to that described for the northeastern side of the Tehuacán Valley (Smith 1965)" and sharing many of the same species. While many of the elements of the mesquite forest struggle on into this zone, it is also characterized by trees like the *guaje* (*Leucaena*) and *tepeguaje* (*Cassia*), and shrubs like the *mala mujer* (*Jatropha urens*). Maguey (*Agave* spp.) and yucca are common, as is the prickly pear (*Opuntia* spp.). Organ cacti include *Lemaireocereus, Myrtillocactus,* and *Cephalocereus*. "At its finest development, this association must have had a canopy ranging between 5 and 10 m. tall, depending on exposure, soil moisture, etc." (Smith 1978:24). This canopy, however, would have remained open because the rigors of the dry season prevented plants from concentrating densely.

As one ascended to the upper piedmont and lower mountain zone, the thorn–scrub–cactus forest would gradually have thickened and eventually come to include oak trees. Today this transition takes place between 1700 and 1900 m elevation, but, since pollen data show climatic fluctuations in prehistory (Chapter 15), we cannot specify exactly where it was at any one time. Smith's (1978) reconstruction shows an oak–pine forest covering the mountains around the valley of Oaxaca. While oaks (*Quercus* spp.) would have dominated the lower reaches of this forest, the balance would gradually have shifted in favor of pines (*Pinus* spp.) as one moved higher in elevation. One of the pines documented in the archaeological record—the piñon pine—has apparently disappeared from this forest, perhaps because all pine species are selectively overharvested by charcoal burners. Pine has apparently been the preferred genus for hearth charcoal since at least 1000 B.C.

In Smith's reconstruction, the original oak–pine forest would have consisted of "much larger trees spaced much farther apart," with a canopy "completely closed in areas of oak dominance" (1978:25). Even today the oak zone is one of magnificent beauty, with trees such as black walnut (*Juglans*), black zapote (*Diospyros*), manzanita (*Arctostaphylos*), and madroño (*Amelanchier*) overlooking shady canyons carpeted with oak leaves. One of the areas richest in edible wild plants is the transition from upper thorn forest to oak forest; its resources are at their peak during the terminal rainy season and early dry season (September–December).

The Tlacolula District

Of the three subvalleys forming the irregular Y of the Valley of Oaxaca, the driest is the Tlacolula arm (Kirkby 1973:15–25; Smith 1978:9–11). Lying at 1620 m elevation, Tlacolula, the district capital, has a mean annual precipitation of only 550 mm and a potential evapotranspiration of 2020 mm. By contrast, Oaxaca City has an annual precipitation of 630 mm and a potential evapotranspiration of only 1890 mm; comparable figures for Etla, the capital of the rainiest arm of the valley, are 650 mm precipitation and 1670 mm evapotranspiration.

Accompanying Tlacolula's overall aridity is a longer-lasting dry season. While Oaxaca City generally has no recorded precipitation between December and February, Tlacolula "frequently records no rainfall for the entire November through March period" (Smith 1978:9). The length and severity of this dry season strongly affects the seasonality of vegetation in the Tlacolula subvalley, especially the deciduous species.

Mitla and the Eastern Tlacolula Valley

This volume is mainly concerned with the eastern third of the Tlacolula valley, a region whose landscape is dominated by mountains and mesas of Tertiary volcanic tuffs (Fig. 3.3). In a sense, this arm of the valley begins at Mitla, where the upper Río Salado (known locally as the Río Grande de Mitla) enters the valley by way of a steep-sided canyon carved out of ignimbrite. This canyon is lined with caves and rockshelters, the most famous of which is Cueva del Diablo (Parsons 1936:295).

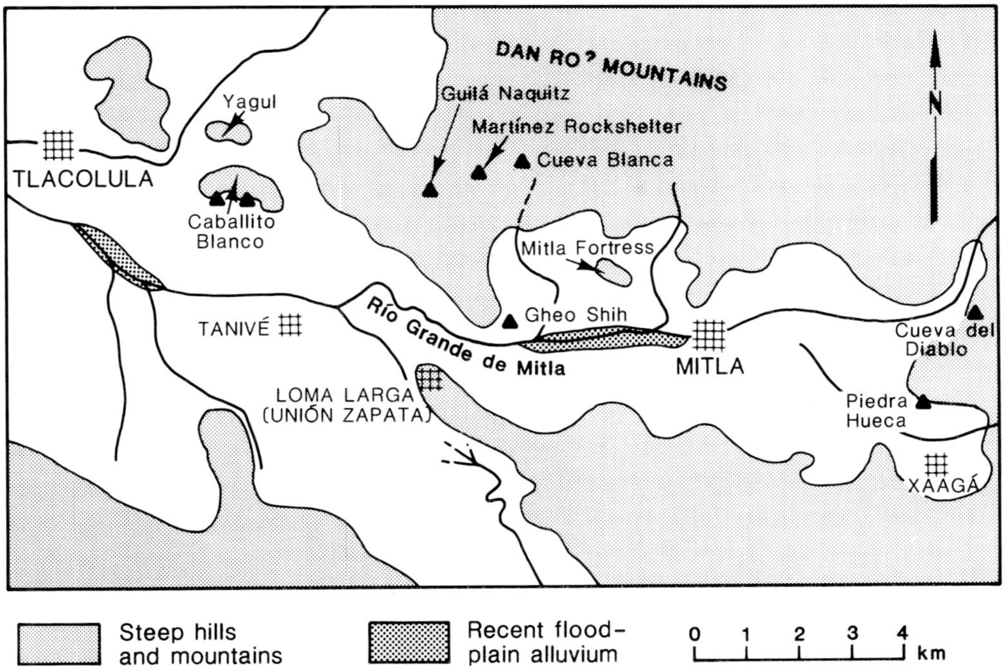

FIGURE 3.3. The eastern end of the Tlacolula subvalley, showing Guilá Naquitz and other places mentioned in the text.

Running southwest out of the canyon, the river turns and flows due west for about 4 km until it passes under the bridge at Mitla. Here it is deeply incised in a ribbon of silty high alluvium only a kilometer wide, running east–west at an elevation of 1688 m (5537 feet). To either side of this alluvial strip, an apron of piedmont rises gently toward the mountains that overlook the valley from a distance of roughly 2 km.

Of all these mountains, the most impressive is the range to the north that the Mitla Zapotec call *Dan Ro?*, "big mountain, the mountain which 'does not shake during earthquake' " (Parsons 1936:295). Dan Ro? is more than 15 km long east–west and 12 km wide north–south, and its pine-forested ridgetop lies at 3000 m on the continental divide. Both Schmieder (1930:Map 4) and Parsons (1936:295) report a ritual cave, *Biliyär Calaver*, near the top of the ridge a day's journey from Mitla. It is the canyons and lower slopes of the Dan Ro?, however, that have the greatest number of occupied caves and rockshelters.

Beyond the bridge, the Río Grande de Mitla continues westward for another 4 km until the valley narrows to 1000 m between rocky spurs. Here, near the hamlet of Loma Larga (Unión Zapata), an *agencia* of Mitla, the lands of Mitla end and the central Tlacolula valley begins. Along the north bank of the river, just east of the narrows, lies the preceramic open-air site of Gheo-Shih. Beyond it, the river continues westward into the Tlacolula plain and becomes the Río Salado.

From the standpoint of the preceramic era, one of the most interesting areas of the Valley of Oaxaca is that bounded on the west by Unión Zapata, on the east by Mitla, on the south by the Río Grande, and on the north by the higher reaches of the Dan Ro?. The area is shaped like a vast amphitheater whose curving back wall is formed by the retreat of the mountains to a point some 2.5 km north of the river. The floor of the amphitheater is rock, covered thinly in places with a poor, sandy soil derived from the weathering of volcanic tuff. At center stage is the Mitla Fortress, a steep ignimbrite mesa whose crest was fortified in Postclassic times. This mesa was a source of silicified ignimbrite for Precolumbian tools (Holmes 1897:287; Williams and Heizer 1965:47; and Whalen, Chapter 7, this volume), and chipped-stone artifacts lie scattered on the surface for many kilometers in every direction. Because of the shallow soils, erosion, and wind deflation in this area, preceramic tools are easier to find on the surface here than elsewhere in the Valley of Oaxaca. Furthermore, along the north bank of the Río Grande de Mitla there are places where the indurated Pleistocene alluvium—almost always deeply buried elsewhere in the valley—lies exposed near the surface. Such exposures sometimes reveal preceramic open-air sites such as Gheo-Shih.

We have no reason to believe that this natural amphitheater of mountains, piedmont, and narrow alluvium was used more intensely by hunters and gatherers than any other part of the valley. It simply has, for a variety of reasons, better environmental conditions for the preservation of preceramic sites. Here the base of the Dan Ro? is dissected by steep-sided canyons of strongly welded or sillar-type ignimbrites. The cliffs are suitable for cave formation, and the canyons themselves served as routes for travel between the valley floor and the pine-forested ridgetops. Most canyon streams are ephemeral at best, but a few originate at springs in the Dan Ro?. The area is so dry that open-air sites normally buried by alluviation have instead been deflated or uncovered until they are readily found on survey. Because of these conditions, we concentrated our preceramic surveys and excavations in the Mitla region and must rely on that area heavily because preceramic sites are so difficult to find elsewhere.

The 16-km^2 region west of Mitla and north of the river had other advantages from the standpoint of our research goals: Now largely uninhabited ejido land, its flora comes as close to resembling Smith's original "primary vegetation zones" as any area left in the valley. Once incorporated by a large ranch or hacienda named for the Mitla Fortress, part of it was still known as "El Rancho del Fuerte" when Parsons (1936:18) visited Mitla between 1929 and 1933. In 1966, the canyons and slopes on the back wall of the amphitheater were forested communal lands that the *agencia* of Loma Larga used sporadically for goat grazing. In the years since the overthrow of the haciendas and the establishment of the ejido system, the vegetation of this area was given a chance to return to the point where Smith (1978:22) was able to say, "only in the Mitla arm of the valley, toward Guilá Naquitz Cave, are there some remnants of the original vegetational cover."

Best represented of these remnants is Smith's "thorn–scrub–cactus forest," still recognizable from the talus of Guilá Naquitz in spite of centuries of human exploitation. It occurs as relict patches in the present-day vegetation zone we have called "Thorn Forest A" (Chapter 4). Less recognizable, but still struggling bravely to return whenever it is given a chance, is Smith's "mesquite forest," which once occurred downslope from the thorn–cactus–scrub. It has been largely replaced by the present-day zone we now call "Mesquite Grassland B"; but even within this floral community, there are still groves of large mesquite trees, and should the land stand fallow for 10 years or so, it eventually reverts to *Prosopis* and *Acacia*. We were indeed fortunate that these remnants of the original vegetational cover of the eastern Valley of Oaxaca coincided with the area of our best preceramic sites, for without them a great many of the analyses in the second half of this volume would not have been possible.

THE DISCOVERY OF GUILA NAQUITZ

Guilá Naquitz Cave is a small shelter at the base of a large ignimbrite canyon wall located high above the valley floor some 5 km northwest of Mitla (Fig. 3.4). The cliff occurs at an elevation of 1926 m (6319 feet) above sea level (nearly 300 m above the Río Grande de Mitla) and faces southeast toward the Mitla Fortress. The cave, 16°57′ N and 96°22′ W, lies near one of the main footpaths through the Dan Roʔ mountain range between Mitla and Díaz Ordaz (Sto. Domingo del Valle). The area is ejido land belonging to Loma Larga (Unión Zapata).

Actually, the cave has no name; we have borrowed the name of the rock formation, which the Mitla Zapotec refer to simply as "the white cliff." *Guilá*, "cliff" (from *gui*, "rock"), and *naquitz*, "white," are the Mitleño pronunciations of words that can be traced back at least as far as sixteenth-century Valley Zapotec. In that era, such a cliff would have been referred to as *Quieláa Naquichi* (Córdova 1578).

In December of 1964, armed with a copy of Lorenzo and Messmacher's (1963) "Hallazgo de Horizontes Culturales Precerámicos en el Valle de Oaxaca," I flew to Mexico City to speak with José Luis Lorenzo and Ignacio Bernal. Lorenzo showed me the projectile points Messmacher had collected near Mitla and Xaagá. Bernal provided me with a card catalogue of the first 275 archaeological sites found by his surveys in the Valleys of Oaxaca, Ejutla, and Miahuatlán. Both encouraged me to continue on to Oaxaca to examine Lorenzo and Messmacher's rockshelters and look for others.

At that time my sister, Liza Flannery Arredondo, and her husband, Humberto Arredondo, lived in Mexico City where Humberto was a business executive. Liza and Hue generously loaned me their Mercedes Benz sedan to take to Oaxaca for cave survey. Had they seen some of the places I eventually took it, they might well have had second thoughts; it was akin to loaning Bonnie and Clyde your passenger car for a bank robbery and subsequent getaway. Miraculously, the Mercedes climbed canyons and negotiated donkey trails where few 4-wheel-drive vehicles could have gone, getting high-centered only twice and wedged between boulders only once.

On December 16, 1964, I visited the first of the small rockshelters in the Caballito Blanco Mesa near Tlacolula. This became OC-1 ("Oaxaca Cave No. 1") in the survey numbering system I had borrowed from MacNeish. It soon became apparent that the Caballito Blanco shelters had mainly shallow deposits related to the Monte Albán I–V occupations of the Tlacolula region. The small shelters at Piedra Hueca near Xaagá also proved disappointing, for although preceramic flint tool types could be found on the talus slope below, the shelters themselves were so shallow that most amounted to little more than surface scatters.

At the Museo Frissell de Arte Zapoteca in Mitla I was greeted enthusiastically by John Paddock, whose excavations at Yagul in 1958 had provided my first field experience in Oaxaca. Paddock accompanied me on the Xaagá survey and introduced me to several Mitla villagers who knew the surrounding mountains well because they had herded goats there as young boys. Darío Quero, who had been a child when Elsie Clews Parsons did her fieldwork in Mitla,[1] directed me to Cueva del Diablo where Parsons had done the first "testing" of a Mitla cave (Parsons 1936:295). Parsons' description make it appear that she found Postclassic miniature offering vessels, which confirmed my suspicion that this important ritual cave had mostly very late deposits; it was also too wet to have any preservation of perishables.

One day as we surface-collected the dissected plains near the Mitla Fortress, one of the Zapotec employees of the Frissell Museum pointed off toward an unsurveyed area of mountains and piedmont some kilometers to the northwest. He remembered a cave somewhere up there, he said, that he

[1]In her Preface, Parsons (1936:x) thanks "papacito, tiny Darío with the clothes and manners of a vivacious *anciano*" for being one of the people who made her stay enjoyable. His hospitality during our stay in Mitla some 30 years later was greatly appreciated as well.

FIGURE 3.4. Guilá Naquitz Cave.

had taken refuge in when he and his goats were caught in a spring rainstorm. The Mitleños called it *Biliyär Keyh* ("Cueva Oscura") because it appeared as a lozenge-shaped black dot from a distance. On December 23 I parked the Mercedes near the Mitla–Matatlán crossroads and struck out on foot for the area my informant had pointed to. After about an hour I had isolated a suspicious-looking black dot in a volcanic tuff cliff and climbed toward it through groves of organ cactus and stinging *mala mujer.* As I reached the talus slope a Trinidad point, an Abasolo point, and a Kent point practically leaped up at me, and inside the cave I found crude blades and steep denticulate scrapers lying on a bed of Precolumbian corncobs, maguey quids, and pot sherds. This site became OC-27, known as *Biliyär Keyh* to Zapotec speakers but *Cueva Redonda* to the Spanish speakers of nearby Loma Larga.

Overjoyed with the surface collection, I was interrupted briefly by a group of rabbit hunters from Loma Larga who wondered what a gringo was doing on their ejido land. "Looking for caves," I explained, and one of them smiled and said, "There's a much bigger one around the bend of the next canyon. We call it *la Cueva Blanca.*" I decided to spend the rest of the afternoon surface-collecting the Cueva Redonda talus and return the next day.

On December 24, 1964, I climbed the next canyon and found OC-29, OC-30, and OC-31, the "Cueva Blanca group." More than any other site, OC-30 launched the Oaxaca project, for it seemed clear that Cueva Blanca would have preceramic occupation over a considerable area and to substantial depths. The next day I spent a Christmas far from home and family writing the first draft of a grant proposal.

A whole year was to pass before I was back in Mitla again, supplied with a 4-wheel drive pickup truck and a grant from the Smithsonian Research Foundation. On January 22, 1966, I began the survey for preceramic sites again, assisted this time by Eligio Martínez of the Frissell Museum, Chris L. Moser of the University of the Americas, and Silvia Maranca, a Brazilian student who had recently spent a year at the Smithsonian.

Fanning out from Cueva Blanca, our survey reached the canyon below a long cliff Eligio knew only as *Guilá Naquitz,* "La Peña Blanca." There were small caves in it everywhere, and while Eligio and Silvia surface-collected the middle part of the cliff, I climbed up to Gui Gohtz Pass where the canyon began. On January 26, 1966, my survey notebook records the following entry:

> OC-43. Small cave (8 × 10 m with 3m-high ceiling) just NE of Gui Gohtz Pass, along W. Cliff of canyon running S–N toward Cueva Blanca. Cave faces E or SE onto canyon. Estimated time range, preceramic and Postclassic. Very dry, with good preservation of corn cobs, maguey quids, grass and other plant remains. Could be 1.5–2.0 m deep, and first 50 cm will be practically solid plant remains. Surface collection includes one-half of a projectile point with a concave-based stem; lots of steep scrapers; manos; metates; Postclassic sherds; and lots of chipped stone debitage.

This small cave, OC-43, arbitrarily named Guilá Naquitz after the cliff in which it occurs, is the focus of this volume. Survey continued for another week or so, after which time it was decided to test several of the caves and rockshelters that had been found. To make this process somewhat easier, we constructed a road into the area so that we could drive our workmen and equipment to within a few hundred meters of the caves. Once this had been done, we revisited each cave with a crew of four Zapotec workmen from Mitla and had them spend all afternoon intensifying our surface collection in the thorn-forest cover of the talus. In the case of Guilá Naquitz, this more intensive surface collection produced a second projectile point with a concave-based stem. This complete specimen, which we now know belongs to the Pedernales type, was different from any found at Cueva Blanca or Cueva Redonda and hence suggested that we were dealing with at least two preceramic assemblages.

During February of 1966, we tested several caves out of our sample of more than 60, and decided that Guilá Naquitz and Cueva Blanca were the best. Once this had been determined, we expanded the excavation at those two sites in April and completed them in May. Survey for preceramic sites did not end there, however; renewed searches in 1967 produced the open-air preceramic sites of Gheo-Shih (OS-70) and OS-71, also near Mitla. Even at this writing, surveys of the Tlacolula–Mitla region directed by Stephen A. Kowalewski are continuing to locate preceramic open-air sites, and it seems likely that more such sites will turn up in the future as land clearance and canal digging expose buried Archaic horizons.

THE CHRONOLOGICAL AND CULTURAL PLACEMENT OF GUILA NAQUITZ

The chronological placement of Guilá Naquitz is discussed in detail in Chapter 14. The radiocarbon dates from its preceramic living floors run from 8750 B.C. (Zone D) to 6670 B.C. (Zone B2), and the artifacts from these levels give us no reason to disbelieve the dates. In general, the preceramic assemblage from Zones E-B1 at Guilá Naquitz are internally consistent enough to allow us to assign them all to a single phase, called "Naquitz."

There is one living floor at another site that we would also assign to the Naquitz phase. This is Zone E at Cueva Blanca, which has provided four radiocarbon dates ranging between 9050 and 8100 B.C. (see Table 3.1). On the basis of these dates, it is possible that Zone E at Cueva Blanca is slightly earlier than the lowermost zones at Guilá Naquitz.

The Naquitz phase occupies a span of time that is interesting for two reasons. First, its earliest dates fall very near the supposed transition from the Paleoindian period, with its now-extinct Pleistocene fauna, to the early Archaic period, with its Holocene fauna. Second, by the end of the Naquitz phase we have our first evidence for local plant domestication.

Obviously, we had hopes of correlating our preceramic sequence in Oaxaca with the chronological phases worked out for the Tehuacán Valley by Johnson and MacNeish (1972), but it proved impossible to correlate the Naquitz phase precisely with Tehuacán. Naquitz contains only "modern" fauna and should therefore be more recent than the Early Ajuereado phase in the Tehuacán Valley, which yielded animals now extinct in both Tehuacán and Oaxaca. Despite its Holocene fauna, however, Naquitz cannot be correlated with the El Riego phase in the Tehuacán Valley, since the El Riego phase runs from 7000 to 5000 B.C. and includes numerous projectile point types so far unknown from Naquitz. Johnson and MacNeish (1972:39–40, Fig. 4) regard Naquitz as being contemporaneous with Late Ajuereado in the Tehuacán Valley, while having an artifact complex more like that of the Santa Marta phase in central Chiapas, which

TABLE 3.1
Tentative Correlation of Preceramic Cultural Phases in the Valleys of Oaxaca and Tehuacán

Oaxaca phase[a]	Site and level	Tehuacán phase
Martínez (± 2000 B.C.)	Yuzanú site Martínez Shelter, 10–15 cm	Abejas (3400–2000 B.C.)
Blanca (3300–2800 B.C.)	Cueva Blanca C Cueva Blanca D	
Jícaras (5000–4000 B.C.)	Gheo-Shih A, B	Coxcatlán (5000–3400 B.C.)
Naquitz (8900–6700 B.C.)	Guilá Naquitz B1 Guilá Naquitz B2 Guilá Naquitz B3	El Riego (7000–5000 B.C.)
	Guilá Naquitz C Guilá Naquitz D Guilá Naquitz E Cueva Blanca E	Late Ajuereado
Late Pleistocene (> 10,000 B.C.?)	Cueva Blanca F	Early Ajuereado (> 10,000 B.C.?)

[a]Exact beginning and ending dates are not given for the Oaxaca phases because the sample of radiocarbon dates is too small for such precision. The same is true for the Early and Late Ajuereado phases in the Tehuacán Valley. Dashed line indicates the transition from Pleistocene to Holocene.

also runs from 7000 to 5000 B.C. (MacNeish and Peterson 1962). Actually, Naquitz shares some projectile point types with late Ajuereado, El Riego, and Santa Marta, but is not identical to any one of those.

In the Valley of Oaxaca, the Naquitz phase was followed by the Jícaras phase, whose type site is Gheo-Shih. Gheo-Shih is easier to correlate with the Tehuacán sequence, since it produced a much larger sample of projectile points. The most common point is the Pedernales type, with its short, broad stem and concave base. MacNeish *et al.* (1967:78 and Fig. 67) report a single Pedernales point from Level 4 of El Riego Cave West in the Tehuacán Valley, a component of the Coxcatlán phase. This Pedernales point from El Riego Cave, considered unique and "aberrant" in the Tehuacán collections, looks so similar to Gheo-Shih specimens that we would not be surprised to learn that it came from Oaxaca. Other projectile point types, such as La Mina, Trinidad, and San Nicolás, link the Jícaras phase in the Valley of Oaxaca to the Coxcatlán phase in Tehuacán.

In cultural terms, we could summarize Guilá Naquitz's position in the following way. During the Early Ajuereado phase, under the cooler and drier conditions of the late Pleistocene period, the occupants of the southern Mexican highlands hunted native wild horse, pronghorn antelope, and jackrabbit, sometimes by communal drives (Flannery 1966, 1967). With warming temperatures at the onset of the Holocene, the thorn–scrub–cactus forest reconstructed by Smith (1978) spread over Oaxaca and Tehuacán, and many Pleistocene animals disappeared; so, apparently, did communal game drives. The early Archaic witnessed the establishment of the microband–macroband foraging pattern originally defined by MacNeish (1964, 1972) and discussed elsewhere in this chapter. The Late Ajuereado foragers in Tehuacán and the Naquitz phase foragers in Oaxaca may have been among the first to display this early Archaic settlement-subsistence pattern. What followed in both areas was incipient agriculture.

GUILA NAQUITZ IN ITS DEMOGRAPHIC CONTEXT

To understand the kind of sample of prehistory that Guilá Naquitz might represent, we must consider the population density of preceramic hunter–gatherers in the southern Mexican highlands and the numbers of camps they might have left behind. I have already done this in a recent paper (Flannery 1983), which I only summarize briefly here.

Out of a sample of more than 60 caves we visited, no more than half a dozen appeared to have preceramic deposits we could actually excavate; a few others had Archaic projectile points on the talus but no depth of deposit inside. Similarly, stray preceramic artifacts were not uncommon on the surface in the Mitla area, but Gheo-Shih was our only open-air site with enough deposit to excavate. If we assume that Gheo-Shih was occupied by a macroband of 25 persons whose "territory" was the eastern half of the Tlacolula Valley, there might have been no more than 50 persons in the entire Tlacolula arm of the Valley of Oaxaca during the Jícaras phase. Extrapolation from this would give us 150 persons in the entire 700 km^2 of flat valley floor in the Valley of Oaxaca, or 1 person per 4.7 km^2 of valley floor. However, if we assume that the Gheo-Shih macroband's territory was the entire Tlacolula Valley, there might have been only 3 such bands in the Valley of Oaxaca, for a total population of 75 and a density of 1 person per 9 km^2.

These figures can be compared with our estimates for the Tehuacán Valley during the El Riego phase, where MacNeish's (1972:Fig. 3) map shows 7 macroband camps and 13 microband camps from the entire 2000-year span of the phase, over a surveyed area of 1400 square miles.[2] Remembering that microband camps were simply the lean-season settlements made by the same people who lived at macroband camps during high-resource seasons, we included only macroband camps in our calculations and can give the following estimates for the El Riego phase: (1) if all macroband camps were occupied simultaneously (unlikely), 1 person per 10 square miles; (2) if they were occupied sequentially (more likely), 1 person per 70 square miles. In other words, judged by the archaeological sites recovered, early Archaic populations in the Tehuacán–Oaxaca region are estimated to have been within the range of other hunter–gatherers listed by Steward (1955:125), "ranging from a maximum which seldom exceeds one person per 5 square miles to one person per 50 or more square miles" (Fried 1967:55). These low population figures help to explain why I have always been skeptical of population-pressure theories of the origins of agriculture, at least for Mesoamerica.

Of course, the group occupying the eastern Valley of Oaxaca did not exist in isolation. Based on estimates by Wobst (1974) and Birdsell (1968), I have suggested that the preceramic hunter–gatherers of the Valley of Oaxaca were "organized into *families* of 2 to 5 persons who harvested wild foods together at places like Guilá Naquitz, and in turn belonged to *local groups* averaging 25 persons, all of whom might camp together at places like Gheo-Shih when resources permitted. These local groups would have been linked by marriage into an *effective breeding population* of 175 or more persons occupying neighboring valleys, and a still larger *dialect tribe* of at least 500 persons," all speaking a mutually intelligible dialect of Proto-Otomanguean (Flannery 1983:36). Such a dialect tribe would almost certainly have included the occupants of the Tehuacán Valley, with whom the Mitla foragers may occasionally have exchanged artifacts—Pedernales points going to Tehuacán, Coxcatlán

[2]I am happy to have the opportunity here to correct a typographical error in the text of Topic 8 of *The Cloud People* (Flannery 1983). I gave the area of the Tehuacán Valley as "70 by 20 miles, for a total of 140 square miles" (p. 35). The latter figure should, of course, be 1400 square miles. This means that even if all 7 macroband camps were occupied simultaneously (a very unlikely situation), the El Riego population density would be 105–175 persons in 1400 square miles, or about 1 person per 10 square miles (not "1 person per square mile" as I stated on p. 35). I thank Norman Hammond for being the first person to bring this error to my attention.

points to Oaxaca (Flannery *et al.* 1981:59–60; and Hole, Chapter 6, this volume).

Given all these population estimates, what kind of a sample was Guilá Naquitz? I have already estimated that (assuming macrobands changed their base camp every year) MacNeish's 7 El Riego phase macroband sites might represent no more than 0.4% of the macroband camps made during the 2000 years of the El Riego phase (Flannery 1983:35). In the case of the eastern Valley of Oaxaca, let us assume that the Naquitz phase lasted for 2200 years, from 8900 to 6700 B.C. (Chapter 14). Gheo-Shih could, on the basis of our current evidence, have been occupied at least twice; it might therefore represent one one-thousandth (0.1%) of all the macroband camps made in the Tlacolula arm of the valley. The situation is even worse for Guilá Naquitz; if we assume every macroband of 25 persons broke up annually into 5 microbands of 5 persons each, there could have been more than 10,000 microband camps made during the Naquitz phase. The 6 preceramic living floors at Guilá Naquitz might therefore represent less than 0.06% of the microband camps made during the Naquitz phase in the eastern Valley of Oaxaca.

GUILA NAQUITZ IN THE CONTEXT OF BINFORD'S FORAGER–COLLECTOR DICHOTOMY

In Chapter 1 we examined Binford's (1980) proposal that most hunting–gathering societies occupy a position along a strategic continuum from foraging to collecting. Foragers, the most mobile, travel to where the food is, and their settlement pattern becomes dispersed or aggregated according to whether resources are dispersed or aggregated; collectors, the most nearly sedentary, tend to remain in one favored locality while logistically organized task groups go out and bring back resources. Binford also discussed the role of food storage in prolonging the season of abundance, and we argued that in Oaxaca, such storage would be a way of adapting to drought rather than cold. We also acknowledged that most hunters and gatherers fall somewhere between Binford's two extremes, and that some groups forage for plant foods while organizing their ungulate hunting logistically through all-male hunting parties.

This last statement is similar to the way I would describe the group that occupied Guilá Naquitz. I would place them near the foraging end of the continuum because they appear to have changed residence several times a year as different resources became available; they were at Guilá Naquitz in the autumn when acorns were abundant, moved elsewhere during the January–May dry season, and may have camped in the valley-floor mesquite groves during the June–August rains. However, at various times of the year their deer hunting may have been done by small groups of men who went out from the main camp and later returned with the meat. Occasional pits, apparently for acorns, also make it clear that the Guilá Naquitz group engaged in at least small-scale storage.

GUILA NAQUITZ IN THE CONTEXT OF MACNEISH'S MACROBAND–MICROBAND DICHOTOMY

In his work in the Tehuacán Valley, MacNeish (1964, 1972) distinguished two types of preceramic settlements. These were *macroband camps,* occupied for a season or more by 15–25 persons, and *microband camps,* occupied by 2–5 individuals for any period from a day or two to most of a season. Using Steward's (1938, 1955) model for Great Basin hunter–gatherer settlement patterns as his major framework, MacNeish reasoned that microband camps were produced by family collecting bands like those of the Paiute or Shoshone. Macroband camps would then represent places where several families coalesced during a time of abundant resources. MacNeish's reasoning was supported by the fact that archaeological microband camps in Tehuacán often contained both "men's tools" (e.g., projectile points) and "women's tools" (e.g., metates) as well as foods presumably collected by both men (e.g., deer) and women (e.g., wild fruits). This suggested that many microbands in Tehuacán were family groups (men, women, children) rather than the kinds of task-specific all-male or all-female groups associated with Binford's "collecting" strategy.

At least for the early Archaic periods in Tehuacán, such as El Riego and Coxcatlán, MacNeish's reconstruction of the settlement pattern would fit Binford's description of foragers—people who made residential moves in search of resources and adjusted their group size between microband and macroband. MacNeish's original suggestion was that macrobands probably coalesced in the rainy season when plant foods were abundant, then dispersed as microbands during the dry season.

Now the sites we have found in Oaxaca can be added to those from Tehuacán (see Fig. 3.5), providing a larger sample and making it clear that there are even more settlement types than we had originally thought. We would still see the Naquitz phase as a time of foragers who "mapped onto resources through residential moves and adjustments in group size," but we now believe it would be a mistake "to assume a simple correlation between macrobands and the rainy season, microbands and the dry season. In a bumper year for mesquite, the macroband coalescence might take place on the alluvium in July; in a bumper year for acorns, it might take place in the upper piedmont during November. During a drought year, there might never be an opportunity for macroband settlement" (Flannery 1983:33).

GUILA NAQUITZ IN THE CONTEXT OF OAXACA–TEHUACAN SETTLEMENT TYPES

In 1983 I presented a five-part settlement typology for the Oaxaca–Tehuacán area (Flannery 1983), which is reprinted here as Fig. 3.6. As more sites are found and excavated it is likely that this typology will need modification, but I will recapitulate it here in order to place Guilá Naquitz in perspective.

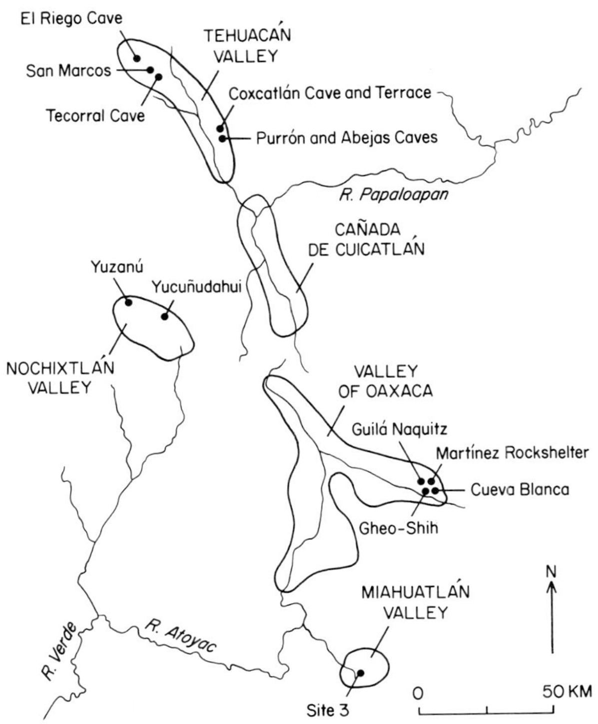

FIGURE 3.5. The Valleys of Tehuacán, Cuicatlán, Nochixtlán, Oaxaca, and Miahuatlán, showing excavated preceramic sites and important preceramic surface finds.

1. The site of Gheo-Shih would be an example of a macroband camp in the Mitla region. Gheo-Shih covers 1.5 ha and might have been occupied by 25 persons, probably during June–July–August when the valley-floor alluvium was experiencing peak mesquite pod production, and when rainy-season cultivation of early domestic plants could take place. There are oval concentrations of rocks and tools that may indicate residence in small shelters or windbreaks, and a boulder-lined area that may have been used for ritual or public activity. Gheo-Shih will be the subject of a future site report.

In the Tehuacán Valley, Coxcatlán Cave and Coxcatlán Terrace (MacNeish *et al.* 1972) would be examples of macroband camps.

2. Zones E–B1 at Guilá Naquitz (the focus of this volume) represent a series of microband camps in the upper piedmont; they contain both men's and women's tools. We reconstruct these levels as living floors left by families of 4–5 persons who arrived toward the end of the mesquite harvest season (late August or early September?) and stayed until the end of the acorn harvest (late December?). While the group did a little hunting, their most important activity was probably the harvesting of many species of wild plants. The microbands who camped in Guilá Naquitz may have done so after the breakup of a macroband camp somewhere else or may have done so in years when no macroband camps were possible. They were foragers in Binford's sense, and family collecting bands in Steward's and MacNeish's sense. Thus, they tell us only about autumn activities in the Mitla region and only about one of many adjustments in local group size.

In the Tehuacán Valley, levels 4 and 5 of El Riego Cave (MacNeish *et al.* 1972) were probably left by the same kind of family microbands whose major focus was wild plant harvesting.

3. Zones D and C at Cueva Blanca also represent microband camps in the Mitla region, but they seem to have been oriented more heavily toward hunting than were Zones E–B1 at Guilá Naquitz. Projectile points and deer bones were relatively well represented, while grinding implements were rare. At this writing, we have not yet decided whether these Cueva Blanca occupations represent all-male deer-hunting camps indicative of "logistically organized hunting" in the San Bushman sense (or Binford's "collector" sense) or whether they are family microband camps with a strong hunting focus. Cueva Blanca will be the subject of a future site report.

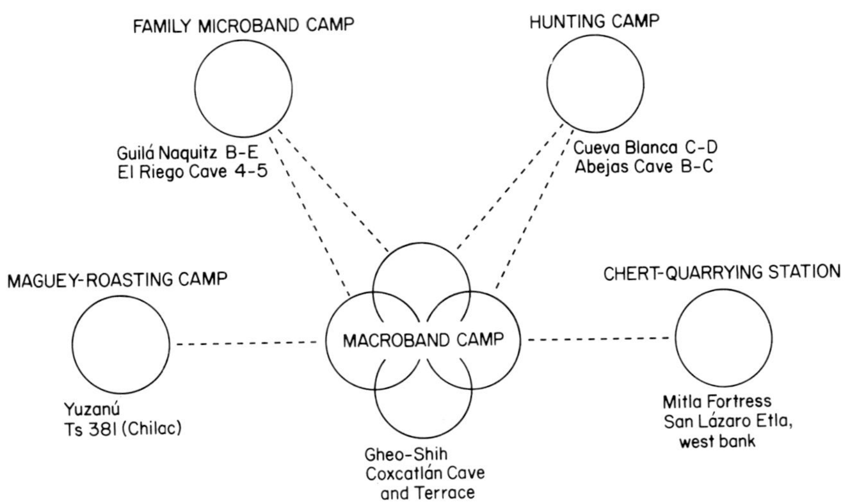

FIGURE 3.6. Model for the hypothetical integration of five preceramic settlement types in the Oaxaca–Tehuacán area. Each circle represents a microband in MacNeish's terminology. Two archaeological examples of each settlement type are given.

In the Tehuacán Valley, Zones C–B of Abejas Cave (MacNeish *et al.* 1972) also represent this kind of hunting-oriented microband camp.

4. In the heart of the dry season (e.g., February) there are too few plants available in the Guilá Naquitz area to keep a family alive, and often the deer have moved up to the higher mountains where the lower evapotranspiration provides them with more to browse on. One of the activities of this lean season is the roasting of maguey, some mature specimens of which are available at all times of the year. The heart of the maguey was roasted in large rock-lined pits or earth ovens, which show up at numerous maguey-roasting camps. I believe that the large roasting pit found by Lorenzo (1958) at Yuzanú near Yanhuitlán, Oaxaca, is an example of such a feature. In the Tehuacán Valley, the west half of Site Ts381 near Chilac (MacNeish *et al.* 1972) has a maguey-roasting pit assigned to the El Riego phase. Similar features are in evidence along the cliffs near our excavations at the Martínez Rockshelter, a site that will be reported in a future publication. I expect most (but not all) maguey-roasting camps to date to the dry season and to be found at elevations above the valley floor.

5. A fifth type of site, not considered a true camp, was the flint- or chert-quarrying station. In the Mitla region, perhaps the most prominent quarry occurred on the slopes of the so-called Mitla Fortress, described by Whalen in Chapter 7. Here surface finds of preforms and unfinished projectile points indicate that preceramic knappers frequently visited the veins of silicified volcanic tuff exposed on the Fortress. In the western Valley of Oaxaca, similar evidence can be found around a chert outcrop on the west bank of the Río Atoyac near San Lázaro Etla. Farther to the north, limestone outcrops on Yucuñudahui mountain in the Nochixtlán Valley have also produced preceramic projectile points (Ronald Spores, personal communication, 1975).

This brief discussion makes it clear that Guilá Naquitz belongs to only one of several preceramic settlement types in the Mitla region. Further complicating this situation is a still-unresolved question that arises from the chronological differences among the sites. Gheo-Shih, Cueva Blanca, and Abejas Cave B-C are all more recent than the Naquitz and El Riego phases. This raises the possibility (discussed in Chapter 32) that the foraging strategy of the early Archaic, with its family microbands, was gradually transformed into a collecting strategy with logistic microbands in the late Archaic. If this happened, I would expect to find more permanent-looking macroband camps and more all-male or all-female microband camps in the late Archaic. To date, our sample of excavated sites is probably too small to show this conclusively, but it is worth thinking about. It would certainly do no violence to the sequence of sites that have been excavated in Oaxaca and Tehuacán.

SUMMARY

One of the research goals of the Oaxaca Project was to better understand the transition from foraging to agriculture. Guilá Naquitz contributed to this goal by providing us with six living floors of the Naquitz phase (8900–6700 B.C.). This falls squarely within the period 10,000–5000 B.C. that we identified in Chapter 1 as crucial for the origins of agriculture worldwide.

To understand how the occupants of Guilá Naquitz used their prehistoric environment, we have considered the cultural context of that period. This was a time when the inhabitants of the Oaxaca–Puebla highlands were making the transition from a Paleoindian to an Archaic way of life. They lived mainly on wild plants and animals, although minor evidence for plant domestication appears before the end of the Naquitz phase. They lived in temporary camps, moving from place to place as the harvest seasons changed. Sometimes, when resources permitted, they coalesced into groups of 25–30 individuals; at other times, they fragmented into family groups of 3–5 persons. The Mitla area contains examples of several types of preceramic sites—caves, open-air sites, microband and macroband camps, summer and winter occupations—and we cannot understand Guilá Naquitz without considering the larger settlement system of which it was only one component. It would appear to have been an autumn camp, occupied for perhaps 4 months by a family of 4–5 persons.

One way to visualize the situation is this: Imagine a 2200-year period during which the climate of the Mitla region goes through phases when it is somewhat warmer or somewhat cooler, somewhat drier or somewhat moister. Imagine the primary vegetation zones shifting upslope or downslope in response to these phases; sometimes the oaks descend below Guilá Naquitz, sometimes they withdraw to higher slopes. Outside, a small population of Indians moves through the area as the seasons change, sometimes nucleating to take advantage of a localized bumper crop and sometimes dispersing to exploit more widely scattered resources. On six occasions during that 2200-year period, you notice people in Guilá Naquitz Cave and you press the shutter of your camera: six snapshots, one for each preceramic living floor. They would be interesting photos, but each would represent only an instant frozen in time. None would really do justice to the dynamic processes of long-term cultural and environmental fluctuation, made to look so stable and eternal by your pressing of the shutter.

4

The Physical Environment of the Guilá Naquitz Cave Group

Michael J. Kirkby

Anne V. Whyte

Kent V. Flannery

INTRODUCTION

During 1964 and 1966, the Oaxaca project visited more than 60 caves and rockshelters in a search for preceramic occupations in the eastern Valley of Oaxaca. About 10 sites were picked for testing, and on the basis of the test results, 4 were selected for more extensive excavation. All the best preceramic caves occurred in the amphitheater-shaped area mentioned in the previous chapter—a region bounded on the east by Mitla, on the west by Loma Larga, on the north by the Dan Roʔ mountain range, and on the south by the Río Grande de Mitla.

Many of these caves are in the lower cliffs of the Dan Roʔ, on ejido lands once owned by the ex-Hacienda El Fuerte, some 4–5 km northwest of Mitla and 2–3 km from the Postclassic Mitla Fortress (Fig. 4.1). The widespread distribution of ignimbrite, a volcanic rock that readily forms caves, has allowed the formation of several cave groups in this arm of the Valley of Oaxaca. The physical environment of the El Fuerte cave group, however, offers definite advantages over other caves for human occupance. The caves are natural and have formed in low cliffs at the base of the mountains forming the continental divide. Above the caves, the land rises steeply to the rich wild food resources of the oak–pine forest, and below them the steep slopes give way to the lower gradients of the alluvial fans and pediment surfaces that comprise the piedmont zone. Lower still, the piedmont grades into the alluvial deposits of the Río Grande de Mitla, whose course is now incised between 3 and 7 m into its own deposits at a distance of 3 km from the cave area. The caves are therefore located at the important boundary between the mountain and piedmont zones, so that people using the caves had direct access to the resources of several physiographic and ecological zones.

The caves that have been tested or excavated occur in two groups. The larger group, including Guilá Naquitz (OC-43), Martínez Rockshelter (OC-48), Silvia's Cave (OC-47), and Cueva de los Afligidos (OC-45), is situated along a continuous cliff at an elevation of ca. 1900 m. The smaller group, including Cueva Blanca (OC-30) and Cueva Redonda (OC-27), is found 1.5 km to the northeast along another cliff at an elevation of ca. 1800 m (Fig. 4.2).

The Guilá Naquitz cave group has formed at the base of a cliff that varies from south to east in aspect and from 20 to 40 m in height. It forms the rim of a small valley, only 0.4 km^2 in area, which is drained by a small ephemeral stream flowing 50 m below the base of the cliff. It is below the cliff on the steep slopes of this valley, and along the stream course, that some of the richest wild plant resources are to be found. Immediately above the cliff, the land flattens to form a plateau 1.5 km^2 in area, which provides another, though less important, source of wild plant foods. The plateau provides a flat area that can be used as marginal

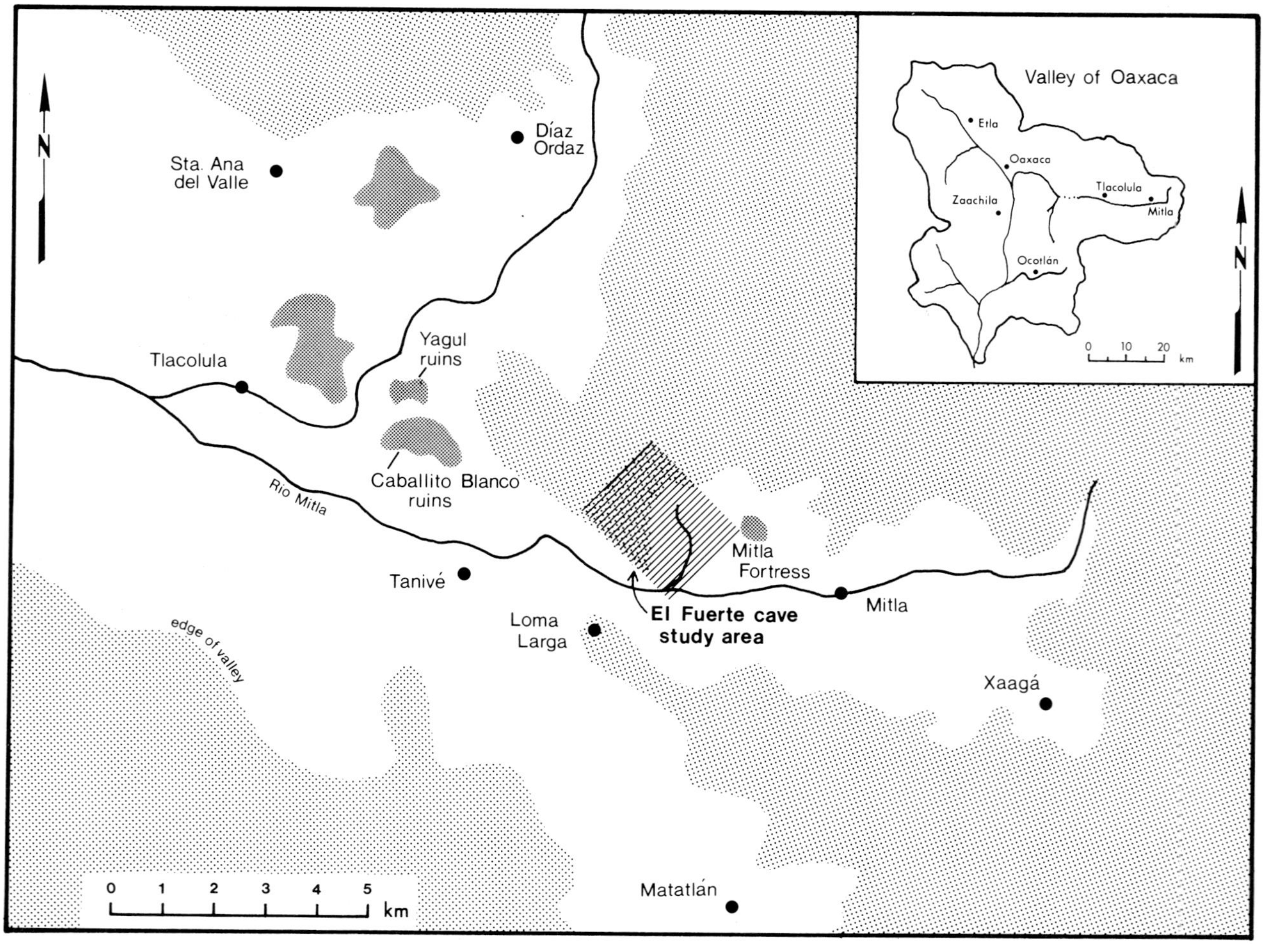

FIGURE 4.1. The eastern Tlacolula Valley, showing the El Fuerte cave area.

agricultural land, and across it passes an important route between the valley of Díaz Ordaz and the Mitla area (Fig. 4.3).

The Cueva Blanca group is situated at the base of a low cliff, only 10 m high, and the various caves face in different directions between west and south as the cliff curves around, following the direction of a stream that flows from the mountains. This ephemeral stream drains an area of 4.3 km^2, and in the vicinity of the caves its course changes from a 50-m-deep , V-shaped valley, characteristic of the mountain zone, to a shallow (2–5-m-incised) piedmont valley cutting into the surface of its own alluvial fan. This group of caves is located closer to piedmont land flat enough for cultivation, and with terracing even the 20° talus slope below the caves has been tilled at one time or another.

GEOLOGY AND SOILS

Most of the area is underlain by ignimbrite, but on the piedmont surface below the caves are thin deposits of consolidated sands derived from the ignimbrite (Fig. 4.4). Ignimbrite is a layered volcanic tuff. Each layer of debris is released in a single flow and becomes welded together as it cools. Earth movements since deposition have extensively uplifted and faulted the ignimbrites. Dips are generally less than 10° but are variable in inclination and direction. The layers are of variable thickness and resistance to erosion, and each layer tends to be most resistant in the middle (Williams and Heizer 1965). Caves form in the more resistant, cliff-forming layers, and appear to have been formed by weathering and collapse along alcove-shaped stress surfaces. Progressive deepening of alcoves leads to the formation of caves without the agency of water erosion, and this type of cave formation is most common in relatively arid regions such as the Valley of Oaxaca.

The dryness of the cave is important both for its selection as a prehistoric living and storage site and for the preservation of archaeological remains, particularly vegetable materials. The existence of dry caves is always to some extent a fortunate chance but is encouraged by a number of factors:

1. Potential evapotranspiration is greater than rainfall for most months of the year.

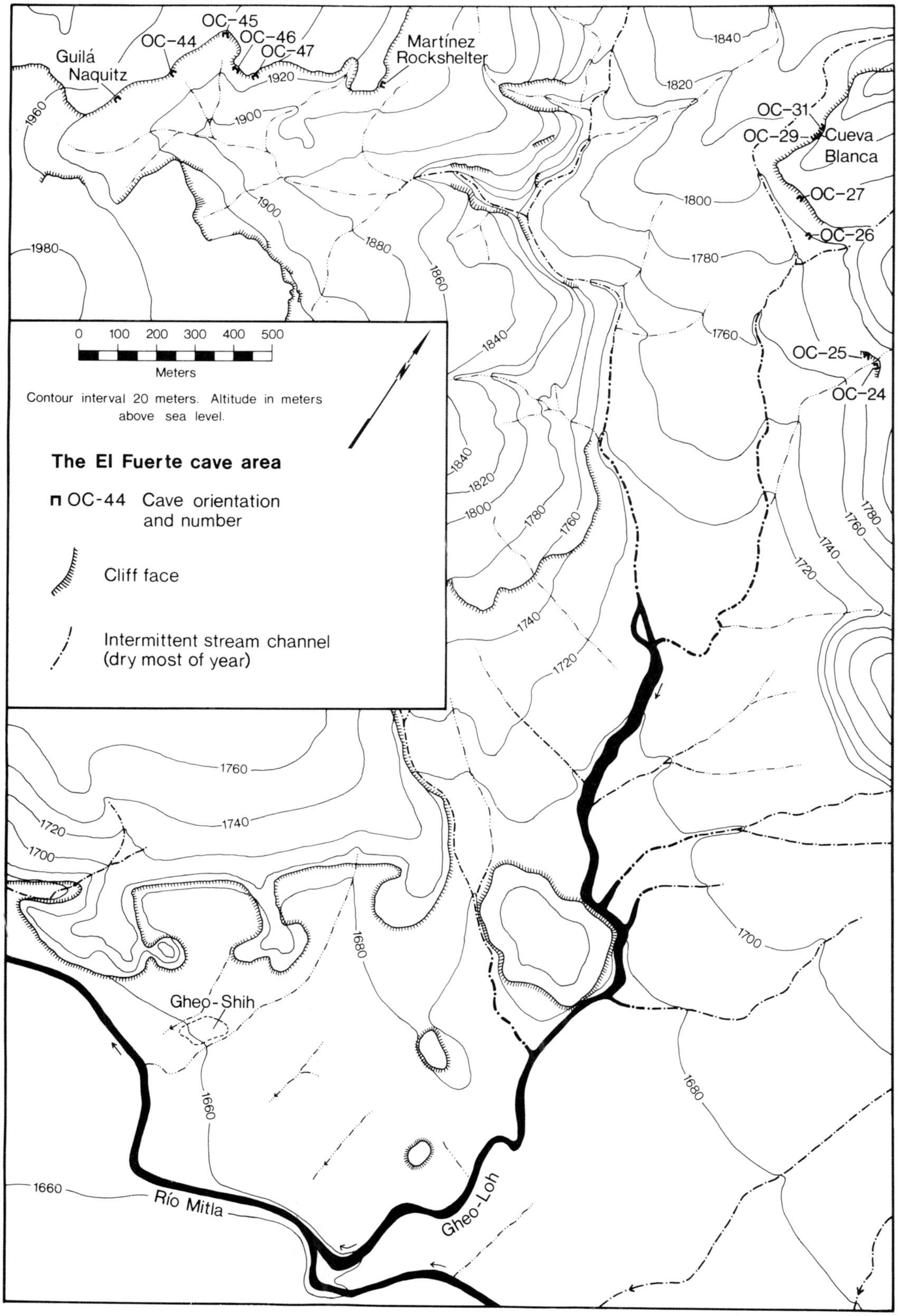

FIGURE 4.2. Topographic map of the El Fuerte cave area, showing stream channels, cliffs, and caves.

FIGURE 4.3. Photo mosaic of the Guilá Naquitz cliff, with locations of OC-43 (Guilá Naquitz), OC-44, OC-45 (Cueva de los Afligidos), OC-47 (Silvia's Cave), and OC-48 (Martínez Rockshelter) given by vertical lines. In the foreground are numerous examples of *Agave potatorum*, an edible maguey.

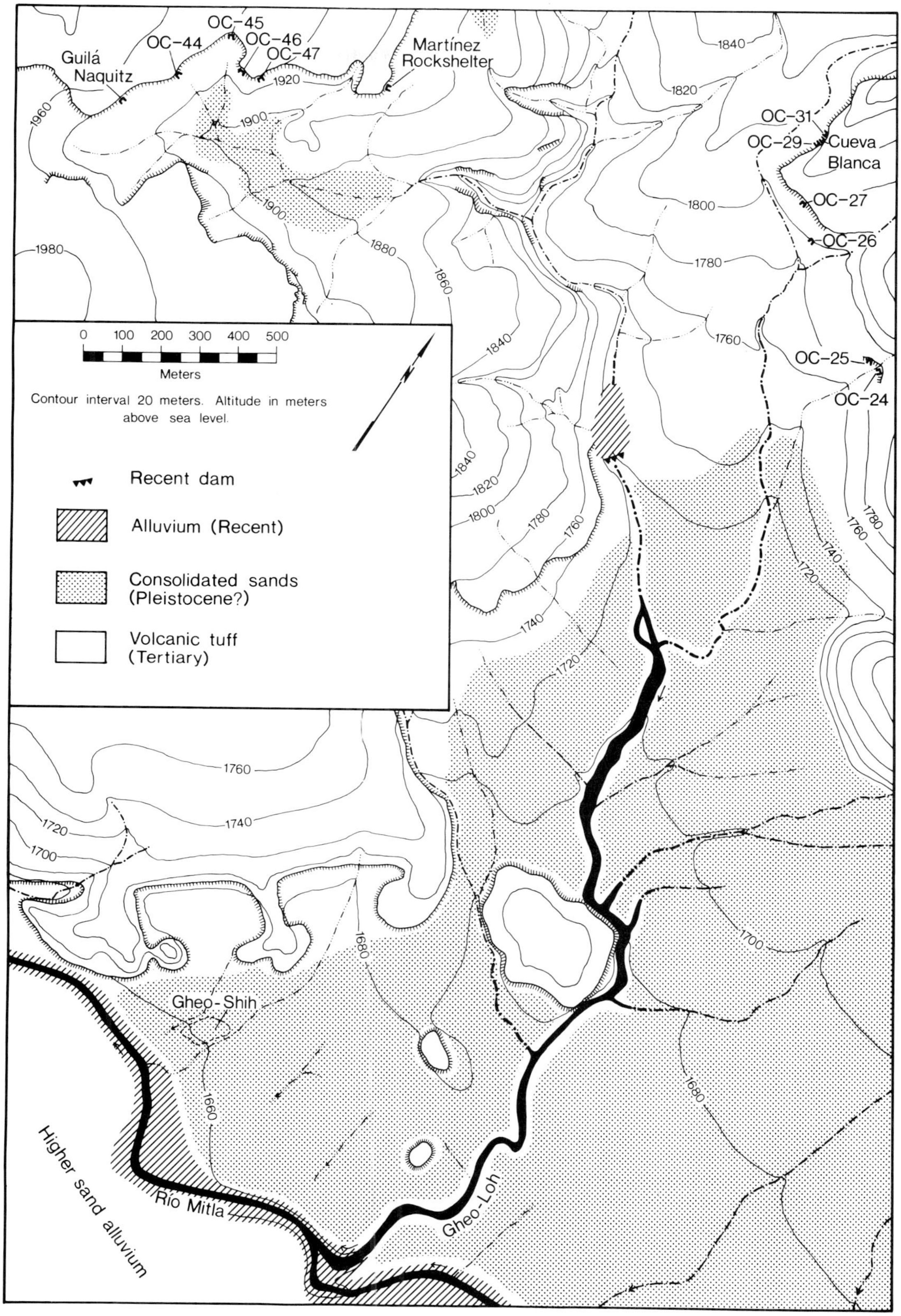

FIGURE 4.4. Geology and soils of the El Fuerte cave area.

2. The cliff overhang is sufficient to shield the cave from most of the direct rainfall.
3. The cave was not initially formed by running water.

The Mitla caves are particularly suitable for dry living and storage conditions and for good archaeological preservation because

1. They are located in the driest part of the Valley of Oaxaca.
2. The ignimbrite layers are less resistant at the top and bottom of each layer so that these weather out, and the cliff formed by the resistant middle part of each layer provides a protective overhang for the caves.
3. The ignimbrite is not a readily soluble rock, so that caves tend to be formed by arching rather than through solution by running water.

Below the caves, the piedmont surface is mainly covered by a thin (ca. 5-m) layer of consolidated sands, probably Pleistocene in age, derived from the ignimbrite. They contain no datable material and weather to soils very similar to those developed directly on the ignimbrite. Throughout the area, therefore, soils are generally stony sands (more than 15% of particles are coarser than 2 mm diameter) with low clay content (less than 5%). Except on the narrow stream floodplains, these soils are less than 25 cm deep, have low water-holding capacity, and are low in nutrients.

CLIMATE AND HYDROLOGY

The Mitla area is the driest part of the Valley of Oaxaca. Potential evapotranspiration exceeds mean rainfall for every month of the year, so that only specialized semiarid vegetation is able to grow; exceptions occur along watercourses and in bottomland areas with a water table at less than a 5-m depth. Estimated climatic figures for the Mitla valley bottom and the higher cave area are given in Table 4.1. The low ratios of rainfall to potential evapotranspiration for the Mitla valley bottom indicate that, under dry farming conditions, winter crops cannot be grown, perennial crops can only be grown if they are xerophytic (such as nopal or maguey), and the Mitla area is marginal for summer annual crops such as maize or beans. The cave area has higher ratios of rainfall to potential evapotranspiration than the bottom of the Mitla Valley (1.10 for the summer as opposed to 0.48), so that on climatic grounds alone an annual maize crop in the summer months near the caves has a good chance of survival.

There are no appreciable areas near the caves where the water table is at less than a 5-m depth, and wells are not used in the area. However, the local streams have only traveled short distances from their source areas in the mountains, so that perennial pools and seeps are common in the beds of the larger streams. The Cueva Blanca group of caves is situated only 60 m from the nearest suitable stream, whereas occupants of the Guilá Naquitz group had to travel over 1 km to reach perennial pools (Fig. 4.5). Although frosts occur in the Mitla cave area, it is the absence of available water that limits plant growth in the winter, rather than the occasional frosts.

The combination of (1) relatively steep slopes with poor, thin soils and (2) a marginal rainfall area without the benefit of available irrigation water severely limits the agricultural potential of the cave area for maize. Only the piedmont surface is flat enough for maize cultivation without terracing; and even here, the soils developed on the Quaternary sands are thin and low in nutrients. After a 3–5-year period of fallow, the productivity today from this type of land is about 0.2 metric tons/ha in an average year. Even today (and without considering the alternative of collecting wild plants), a farmer can barely work enough land to feed himself and a family of five with such low maize yields. Before the Postclassic period, corncob sizes (and hence yields) would have been impossible to subsist on maize grown in this area as the main food supply (A. Kirkby 1973: Fig. 48). However, cultivation of primitive varieties of cucurbits, such as seen in Guilá Naquitz, should have been possible in the upper Thorn Forest based on rainfall alone.

VEGETATION ZONES AND THEIR FACIES

The general pattern of natural vegetation (Fig. 4.6) probably provides a more sensitive indicator of variations in available moisture than the climatic estimates. Vegetational

TABLE 4.1
Estimated Climate for the Mitla Region[a]

Area	Elevation (m)	Mean annual temperature (°C)	No. of frosts per yr.	Mean annual rainfall (mm)	Mean annual potential evapotranspiration (mm)	Rainfall ÷ potential evapotranspiration whole year	June – Sept.
Valley bottom	1650	19.9	2	480	2000	0.24	0.48
Cave area	1900	18.6	5?	600	1300	0.46	1.10

[a]Values are based on the meteorological station data for Tlacolula and Oaxaca, with supplementary rainfall data collected by the Oaxaca project for Mitla village. Additional information provided by valley-wide correlations between potential evapotranspiration and monthly temperature and by the calculated environmental lapse rate.

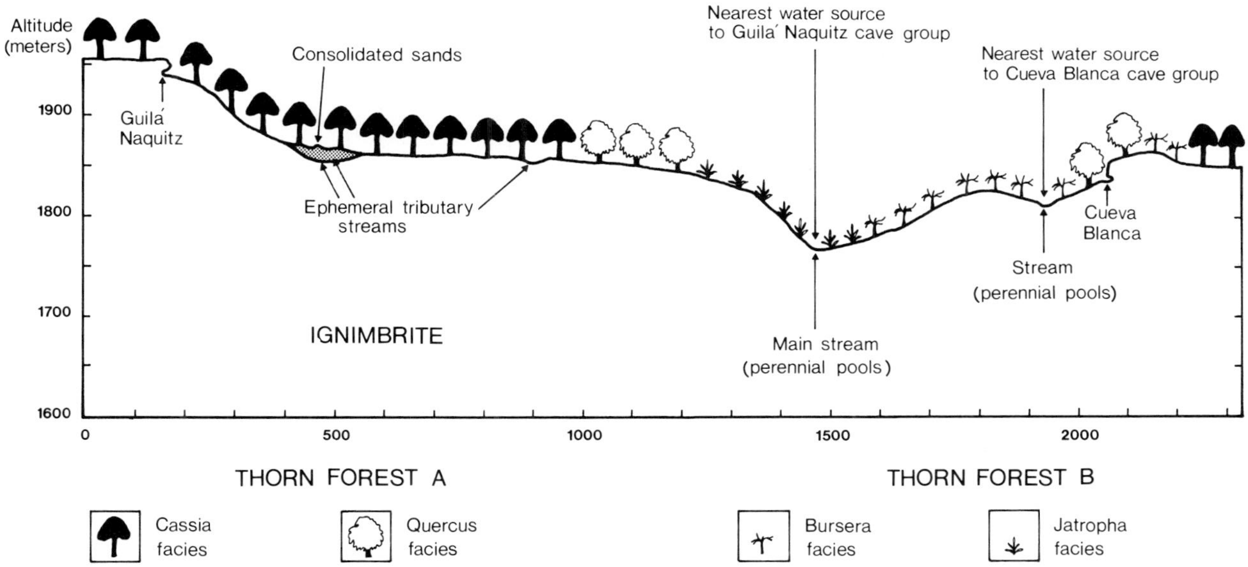

FIGURE 4.5. Cross section from Guilá Naquitz to Cueva Blanca showing geology, water sources, and vegetation zones.

zones show an overall altitudinal zonation, with an oak-dominated zone coming down to about 2000 m elevation, our Thorn Forest A zone (see below) descending to about 1800 m, and our Thorn Forest B zone extending to about 1750 m. The oak facies of our Thorn Forest A zone is mainly confined to steep north-facing slopes where there is less moisture loss through evapotranspiration. Along stream courses are concentrated the more moisture-loving plants such as *Baccaris* ("seep willow," or Zapotec *yak šeh*), *Diospyros* (wild black zapote), *Salix bonplandiana* (willow), and *Alnus* (alder), here presented in the order in which they first occur as one goes upstream. However, there appears to be little extension of the hillside vegetation zones down the valleys, showing that only the channels themselves are rich in moisture.

Lower Oak Zone

The Lower Oak zone, which today occurs some 100 m above Guilá Naquitz, probably represents an impoverished vestige of the "original" oak–pine zone reconstructed by Smith (1978). Pine trees have literally been eliminated from the vicinity of the cave, including the piñon pine whose nuts appear in the prehistoric refuse. According to Smith (1978), farther upslope one can find *Pinus michoacana*. The only oak left near Guilá Naquitz is *Quercus impressa*, which extends down into Thorn Forest A. Other useful trees of this zone are the manzanita (*Arctostaphylos polifolia*), whose wood was used as fuel for maguey-roasting pits (Fig. 4.7); the madroño (*Amelanchier denticulata*); the *tejocote*, or hawthorn (*Crataegus mexicana*), whose fruits were eaten at Cueva Blanca; and a wild black zapote (*Diospyros* sp.; Fig. 4.8).

Thorn Forest A

Thorn Forest A is the vegetation zone in which, under today's climatic conditions, Guilá Naquitz is located. It has at least two facies, (1) one in which oaks are present (*Quercus* facies) and (2) a second in which *tepeguaje* is common (*Cassia* facies). While present, oaks are never common in Thorn Forest A; *Agave* is common; and *Ferrocactus* is uncommon.

Thorn Forest A is characterized by a dense growth of thorny shrubs, trees of varying heights, and tall columnar cacti. The zone is extraordinarily rich in edible plants, and near the cave are relatively undisturbed patches that give us a glimpse of the "original" thorn–scrub–cactus vegetation reconstructed by Smith (1978) for the upper piedmont.

Among the obvious food resources of Thorn Forest A are the acorns supplied by the scattered oaks on north-facing slopes (Figs. 4.9 and 4.10). Legume trees are common; most preferred of these is the *guaje* (*Leucaena esculenta*), whose abundant pods contain edible seeds (Fig. 4.11), but there are also its less-favored relatives *tepeguajes* (*Lysiloma divaricata* and *Cassia polyantha*) or huizaches (*Acacia farnesiana*). Copal trees (*Bursera* sp.) grow right outside the cave (Fig. 4.12).

Distinctive features of the zone are monumental columnar cacti such as *Lemaireocereus treleasi* (Spanish *cardón* or *tunillo*, Zapotec *bidsz-lats*) and *Myrtillocactus schenkii* (Spanish *garambullo,* Zapotec *bidsz-žob*) (Figs. 4.13 and 4.14), both of which have edible fruit. Another common cactus is the prickly pear (*Opuntia* spp.), whose fruits (tunas) are edible raw and whose tender young stem sections (nopales) are edible if cooked (Fig. 4.15). Its relative the cholla, or jumping cactus (also *Opuntia* sp.), is equally abundant.

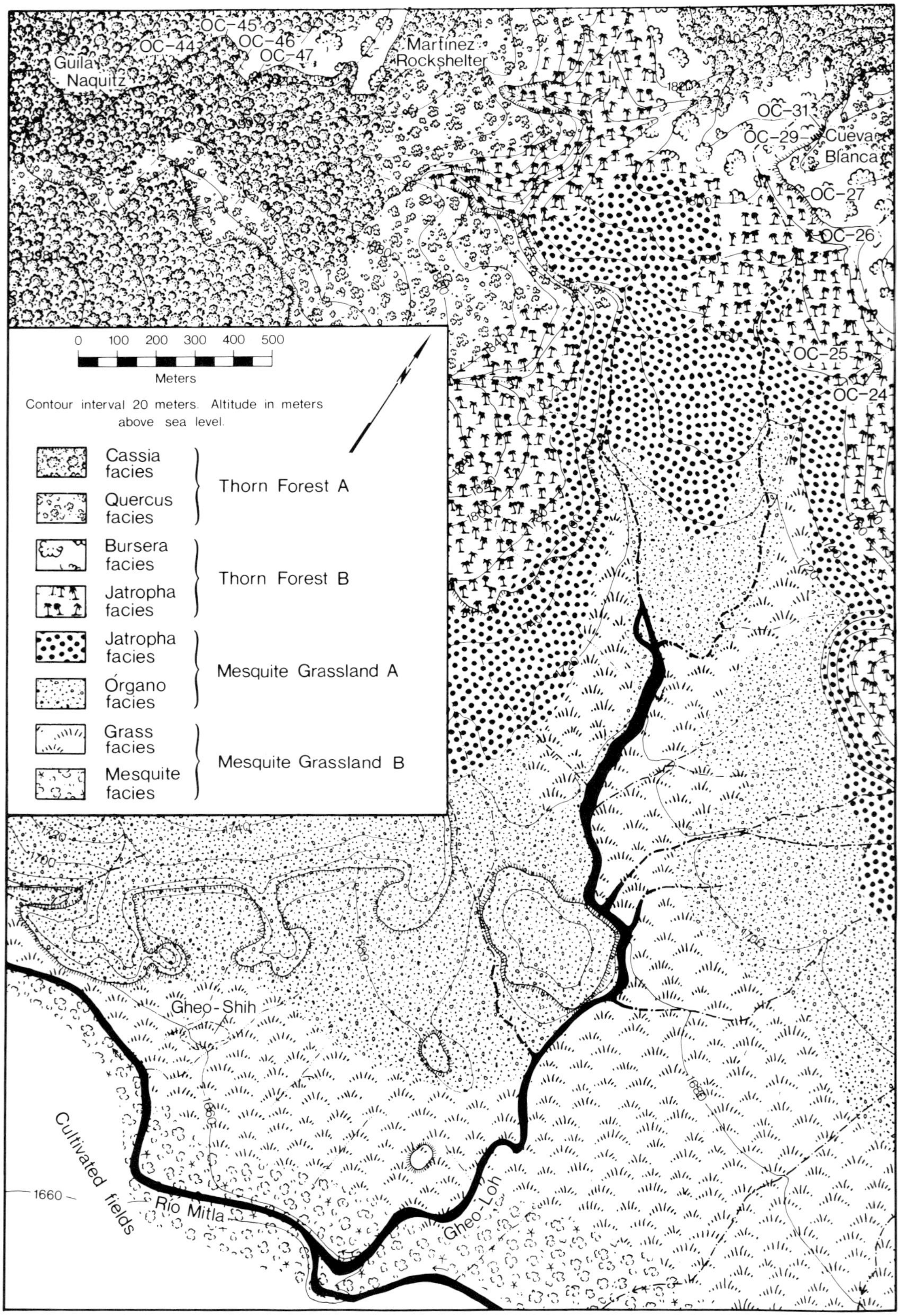

FIGURE 4.6. Distribution of present-day vegetation zones and their facies in the El Fuerte cave area. The important mesquite facies of Mesquite Grassland B can be found 1 km to the southeast of the area shown here, along the Río Mitla.

FIGURE 4.7. Manzanita (*Arctostaphylos polifolia*).

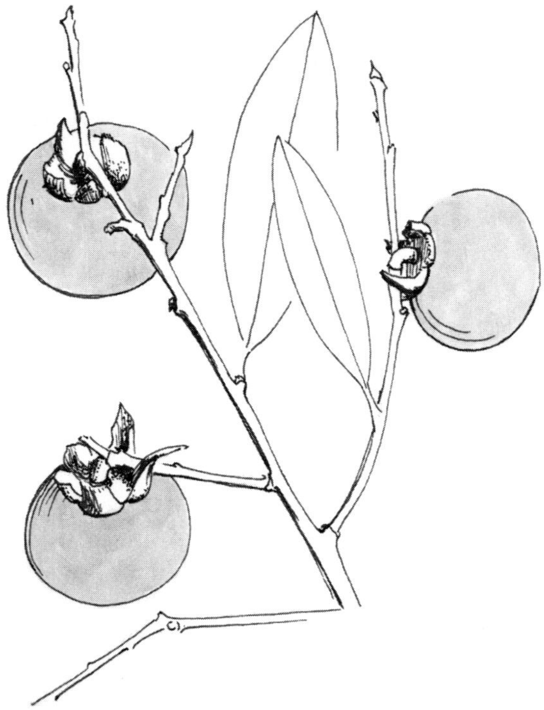

FIGURE 4.8. Wild black zapote (*Diospyros* sp.).

Thorn Forest A is characterized by many species of *Agave*, from the larger magueys *A. potatorum* (Zapotec, *dobh-gih*) and *A. marmorata* to the much smaller *lechuguilla* (*A. karwinskii*), whose inflorescence was sometimes used as a fire drill hearth in ancient times. *Agave potatorum* is edible, but only after the heart has been baked for 24 to 72 hours in a rock-lined earth oven or roasting pit (Flannery 1968: 70–71).

In addition to the wild black zapotes in stream canyons, this zone has trees and bushes of the nanche, or West Indian cherry (*Malpighia* sp.), whose fruits can be eaten raw. There are also a number of food plants for which we have only Zapotec names. Chief among these is the *yak susí* (*susí* tree), a low shrub with edible nuts encased in spherical seed coats that are borne in pairs (Fig. 4.16). The *susí* (*Jatropha neodioica*) is apparently the only edible member of its genus in the region; its relative the *mala mujer* (*Jatropha urens*) is one of the most avoided plants in the thorn forest since it produces a stinging rash that can last for days.

Other plants of this zone include the *chapuliztle* (*Dodonaea viscosa*), a bush that makes the whole region aromatic after rains; *Randia watsonii*, a thorny shrub with bittersweet fruits; the saw-edged sotol (*Dasylirion* sp.), whose prickly colonies cluster along blowouts and deflated areas; and *Hechtia*, a terrestrial bromeliad that can be used for fiber.

Thorn Forest B

Thorn Forest B has a somewhat less dense growth than Thorn Forest A, and a greater number of woody plants are shrubs rather than trees (Fig. 4.17). Oaks are absent; *Cassia* and *Agave* are less common; *Ferrocactus* is more common. The zone has *Bursera* and *Jatropha* facies.

As might be expected, Thorn Forest A and B share many species. Most of those already discussed grow more densely in A, but there are some that do very well in B also. These include the *pájaro bobo*, or morning glory tree (*Ipomoea pauciflora*), ocotillo (*Fouquieria* cf. *formosa*), and the pincushion cactus known as *chilillo* (*Mammillaria* sp.). The *chilillo's* name is derived from its red, chile pepper–shaped fruits, which are sweet (rather than spicy) and can be eaten raw (Fig. 4.18).

Other plants shared by Thorn Forest A and B include wild cucurbits such as the coyote melon (*Apodanthera aspera*) and wild runner beans (*Phaseolus heterophyllus* and *P. atropurpureus*). There are also special humid areas near springs and seeps where one can find clusters of wild onions (*Allium* sp.).

FIGURE 4.9. Thorn Forest A, *Quercus* facies. The oak tree in the right foreground bore more than 2000 acorns.

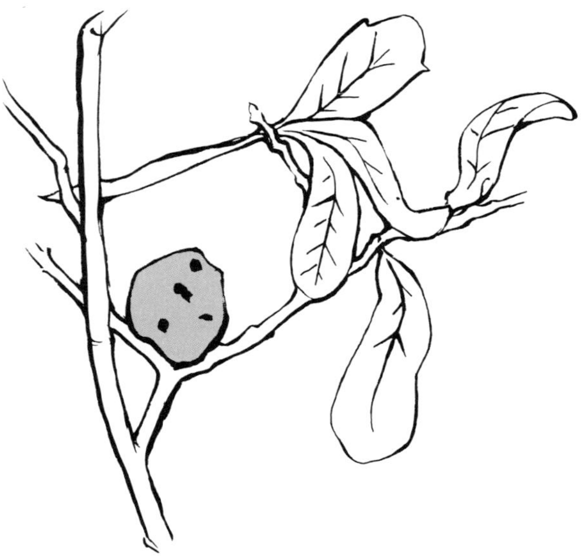

FIGURE 4.10. Oak gall from Thorn Forest A. These edible galls were sometimes collected by the occupants of Guilá Naquitz (see Chapter 19).

Mesquite Grassland A

Mesquite Grassland A has at least two facies: (1) In its *mala mujer* facies, *Jatropha urens* and/or *Dodonaea viscosa* are common, *Bursera* is absent, and the landscape is dominated by grass, with a relatively low density of shrubs and no trees at all. (2) In its organ cactus facies, *Jatropha urens* is rare and *Dodonaea* absent, columnar cacti are common, and mesquite (*Prosopis juliflora*) is present though not abundant. This facies is characterized by its preponderance of organ cactus and its relative paucity of shrubs.

While Mesquite Grassland A is a zone of transition, its relative impoverishment has undoubtedly been increased by overgrazing since the Spanish Conquest. It is perhaps the zone poorest in wild plant foods, except for such things as cactus fruits.

Mesquite Grassland B

Mesquite Grassland B is all that remains today of the "original" primary vegetation zone reconstructed by Smith (1978) as mesquite forest (Fig. 4.19). It has two facies: (1) Where most disturbed by man, it has a grass facies, with grass dominant, mesquite present but not abundant, and *Jatropha* and *Dodonaea* absent. The open-air site of Gheo-Shih lies

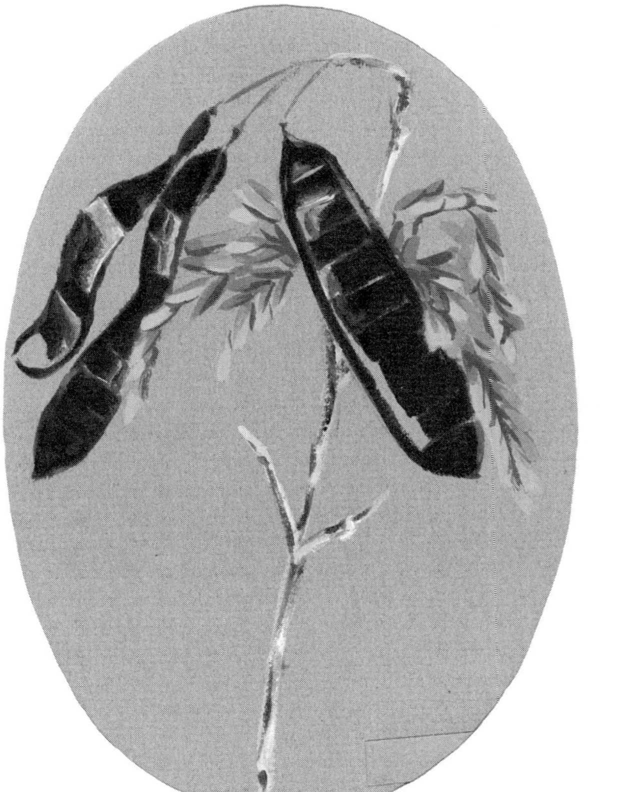

FIGURE 4.11. *Guaje* (*Leucaena esculenta*).

FIGURE 4.12. Copal tree (*Bursera* sp.) on the talus of Guilá Naquitz. Note epiphytes.

in this barren facies, where human activity will not allow the native vegetation to return (Fig. 4.20). (2) Where left undisturbed for a reasonable length of time, a mesquite facies develops that is dominated by *Prosopis juliflora*. The facies characteristically seeks out areas of relatively shallow subsurface water table (less than 5 m), where phreatophytes such as mesquite flourish and form a rather dense growth of trees up to 6 m tall (Fig. 4.21). The pods of these mesquite trees, producing up to 180–200 kg of edible portion per hectare, are one of the main resources of this facies (Fig. 4.22). Other important plants include the *rompecapa* or desert hackberry (*Celtis pallida*), whose fruits can be eaten raw (Fig. 4.23); the so-called crucifixion thorn (*Dalea* sp.); the milkweed-like *binya'a,* a member of the Asclepidaceae; the isote (*Yucca* sp.), whose fiber was used in ancient times; and *Acacia farnesiana* (Fig. 4.24).

NATIVE ANIMALS

The native animals of the Guilá Naquitz area are similar in species composition to the fauna of thorn forest areas in the Tehuacán Valley (Flannery 1967). In both areas, of course, land clearance and overhunting have caused some animals to seek refuge in the higher mountains, while the more tolerant species flourish even in second growth.

Surely the most important local animal was the white-tailed deer (*Odocoileus virginianus*), still present in the upper reaches of the Dan Roʔ range but long since eliminated from the Guilá Naquitz area. The collared peccary (*Dicotyles tajacu*) tolerates human disturbance even more poorly and probably abandoned the area even sooner.

Smaller animals of the region include the coyote (*Canis latrans*), the gray fox (*Urocyon cinereoargenteus*), the opossum (*Didelphis marsupialis*), the raccoon (*Procyon lotor*), the cacomixtle (*Bassariscus astutus*), the coatimundi (*Nasua narica*), and three genera of skunks (*Mephitis, Conepatus,* and *Spilogale*). Of all the small mammals, however, the most important in the prehistoric diet appear to have been rabbits. Still common in the region are the jackrabbit (*Lepus mexicanus*), the *tepetoztle*, or large Mexican cottontail (*Sylvilagus cunicularius*), and the *conejito*, or small Eastern cottontail (*Sylvilagus floridanus connectens*).[1] The cottontails, in particular, are still abundant in the Guilá Naquitz area (see Chapter 18).

Present in virtually every level at Guilá Naquitz was the mud turtle (*Kinosternon integrum*), which inhabits pools of water near seeps and springs, as well as mudholes along the Río Grande de Mitla. Two small lizards, *Cnemidophorus* and

[1]*Sylvilagus floridanus connectens* is one of the smallest races of cottontail, as small as an Audubon cottontail and easily confused with the latter until one examines the tympanic bullae. Further work with the smaller cottontails Flannery collected in Tehuacán has convinced him they are also *S. floridanus connectens,* rather than *S. audubonii* as he originally reported (Flannery 1967).

FIGURE 4.13. Thorn Forest A, *Cassia* facies. The organ cacti in the left foreground are *Myrtillocactus schenkii.*

Sceloporus, are among the reptiles common in the area. Small rodents are numerous and include *Liomys irroratus, Oryzomys couesi, Reithrodontomys* spp., *Peromyscus* spp., *Baiomys musculus, Sigmodon hispidus,* and *Neotoma mexicana* (see Chapter 16 for discussion).

The birdlife of the region is rich, and seems to flourish in spite of human modification of the landscape. Most important to the prehistoric foragers were the doves, pigeons, and quail, which still inhabit the thorn forest. Included are the band-tailed pigeon (*Columba fasciata*), the *huilota* or mourning dove (*Zenaidura macroura*), the *torcaz* or white-winged dove (*Zenaida asiatica*), the *paloma barranqueña* or white-fronted dove (*Leptotila verreauxi*), the *tortolita* or ground-dove (*Columbigallina passerina*), the bobwhite (*Colinus virginianus*), and the Montezuma quail (*Cyrtonyx montezumae*). The call of the chachalaca (*Ortalis poliocephala*) could be heard from the talus of Guilá Naquitz, and the barn owl (*Tyto alba*) roosted inside. Hawks, including *Buteo jamaicensis*, are common in the area and their feathers may have been used for fletching atlatl darts.

One animal that does seem to have vanished entirely from the Valley of Oaxaca fauna is the pocket gopher *Orthogeomys grandis.* Common in Formative sites, it seems to have been eliminated so thoroughly that our workmen could not recall ever having seen one.

MORE DISTANT ENVIRONMENTS

A word should be said about the environment zones that might have been encountered by hunter–gatherers entering or leaving the eastern Valley of Oaxaca. Travelers ascending further into the Dan Roʔ mountains today eventually reach a cloud forest with tall pines, mosses, orchids, and numerous epiphytes (Fig. 4.25). This high montane forest remains humid even when the lower valley is in the grip of the dry season, and it may have provided crucial forage for deer, peccary, and other animals between January and March. This is an important point to bear in mind, because the resources of the valley and piedmont are extremely meager during this period. We do not know where the occupants of Guilá Naquitz went during the late dry season, but one possibility would be this mountain zone at nearly 3000 m. There are

FIGURE 4.14. *Myrtillocactus* (Zapotec *bidsz-žob*), showing characteristic tiny fruits.

FIGURE 4.15. Prickly pear (*Opuntia* sp.), showing flowers, mature fruits (tunas), and stem section (nopal).

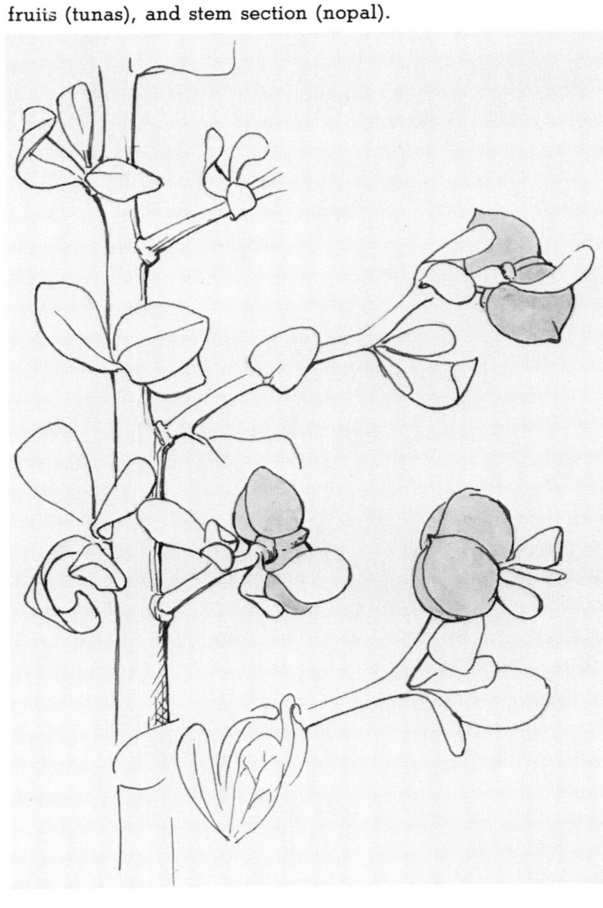

FIGURE 4.16. *Yak susí* (*Jatropha neodioica*), showing characteristic pairs of seeds in their spherical seed coats.

caves at the top of the Dan Ro?, such as *Biliyär Gusi?* ("Lightning's Cave") and *Biliyär Calaver* ("Skull Cave"), formerly used for ritual by the Mitla Zapotec (Parsons 1936:295).

Leaving Mitla to the south, enroute to the Isthmus of Tehuantepec, travelers today reach a zone of arid tropical thorn forest not unlike that surrounding Purrón Cave in the southern Tehuacán Valley. This hot, dry canyon area, below 1000 m elevation, features a magnificent complex of giant organ cactus (Fig. 4.26), including the pitahayo (*Lemaireocereus* cf. *pruinosus*) with its large red fruits (Fig. 4.27). Another native species is the pochote or silk-cotton tree (*Ceiba parvifolia*), whose tubers are edible and whose seeds (borne in pods) are available during the February dry season (Fig. 4.28). This zone also has tropical animals such as the black iguana (*Ctenosaura* sp.). Such are the resources that might have been available to the occupants of Guilá Naquitz should they have chosen to descend below the Valley of Oaxaca into the headwaters of the Tehuantepec River, a region our 1966 surveys showed to be well supplied with caves.

FIGURE 4.17. Thorn forest B, *Jatropha* facies, below caves OC-27 and OC-28.

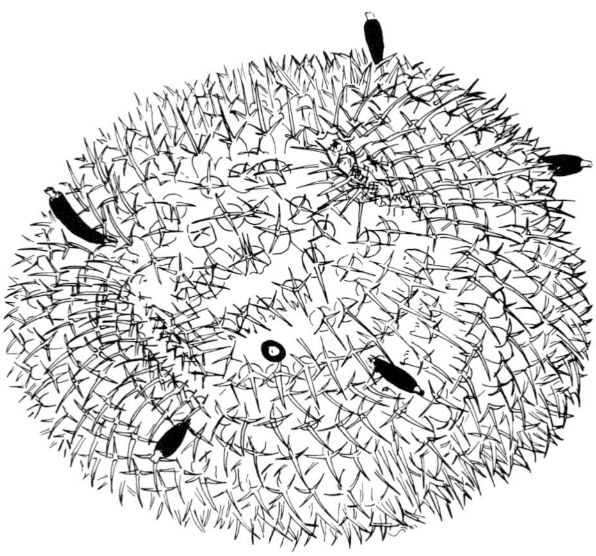

FIGURE 4.18. The *chilillo* (*Mammillaria* sp.), showing chile pepper-shaped fruits.

CONCLUSION: FACTORS IN PRECERAMIC SITE LOCATION

The distribution of caves in the ex-Hacienda El Fuerte region, especially those dry enough for habitation, is strictly limited by the physical environment. Considering the choice of locations available to preceramic hunter–gatherers, the caves they selected were *resource oriented*; that is, access to important natural resources was more important than access to other settlements with a similar subsistence economy.

The El Fuerte caves offer several advantages as seasonal camps.

1. They are close to permanent (although small) supplies of drinking water.
2. They are located with regard to a major natural resource: the wild plant foods of the Thorn Forest A zone. This is one of the two most important resource zones in the area today.
3. From the caves there is unusually easy access to other major resource zones: (a) Access to the important high-altitude hunting and collecting zones of the Dan Ro? mountains is afforded by a route following the course of the stream below Cueva Blanca; this route is still used

FIGURE 4.19. Expanse of stunted mesquite grassland near San Juan Guelavía, Oaxaca. This area, not farmed today because of its salinity, provides clues to what the aboriginal mesquite forest might have been like.

by hunters today. (b) The important food resources of the Mesquite Grassland B zone (mesquite facies) near the Río Grande de Mitla is only 3 km away over gently sloping land. This is the second of the two most important resource zones in the area today.

4. There are probably many caves close to important resources and dry enough for temporary occupation, especially higher in the mountains, but the El Fuerte caves have the added attraction of being dry enough for year-to-year storage. This undoubtedly enhanced their value to wild food collectors, who could bring food from the higher and wetter elevations and store it in the caves. Access to storage areas may have been equal in importance to access to collecting areas, and the El Fuerte caves are perhaps uniquely well located for this. This unusual dryness, combined with their access to the resources of several ecological zones, made the caves prime locations for early seasonal camps.

Thus geologic, topographic, climatic, and vegetational factors of the local physical environment have combined to provide caves well suited as sites for early storehouses and temporary camps, but their unique archaeological significance depends not only on their fitness as early sites but also on the unusual dryness of the caves that has allowed long-term preservation of perishable materials.

FIGURE 4.20. Mesquite Grassland B, grass facies. The preceramic site of Gheo-Shih lies in the foreground, the Río Mitla alluvium in the background.

FIGURE 4.21. Mesquite Grassland B, mesquite facies. Groves of *Prosopis juliflora* lie in the foreground, the Dan Roʔ range in the background.

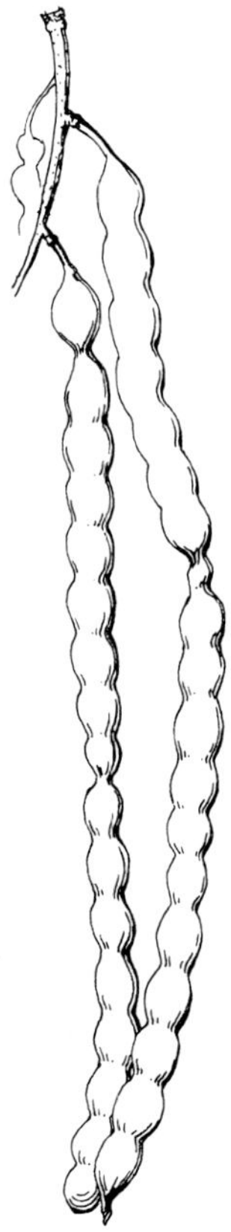

FIGURE 4.22. Pods of mesquite (*Prosopis juliflora*).

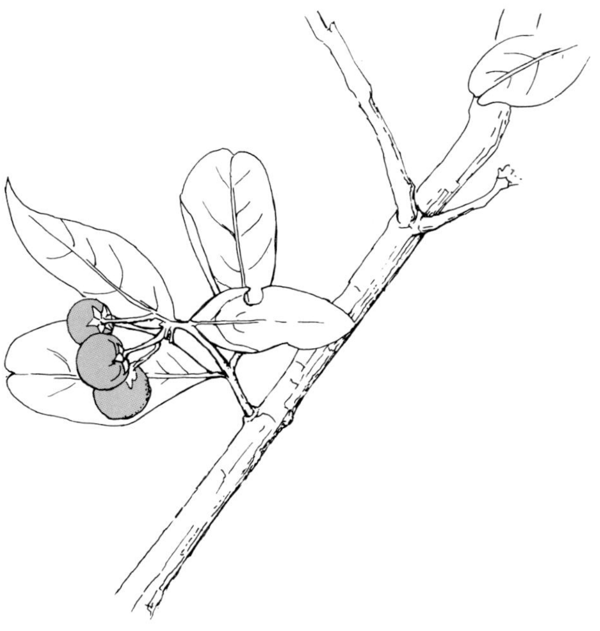

FIGURE 4.23. The *rompecapa* or hackberry (*Celtis pallida*), showing characteristic fruits.

FIGURE 4.24. Huizache (*Acacia farnesiana*), showing pods, seeds, and thorns.

FIGURE 4.25. Dense pine forest above the Valley of Oaxaca (ca. 2500 m elevation).

FIGURE 4.26. Forest of columnar cacti in arid tropical canyon below San Lorenzo Albarradas (ca. 1000 m elevation).

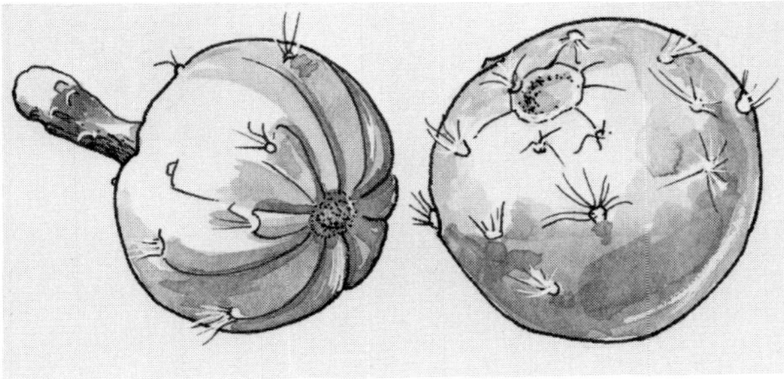

FIGURE 4.27. Fruits of the wild pitahayo (*Lemaireocereus* sp.).

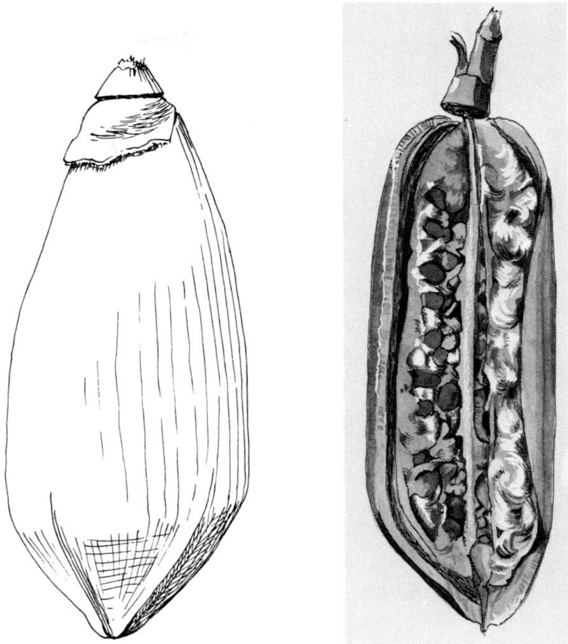

FIGURE 4.28. Pods and seeds of the pochote (*Ceiba parvifolia*).

III

EXCAVATION AND ARTIFACT ANALYSIS

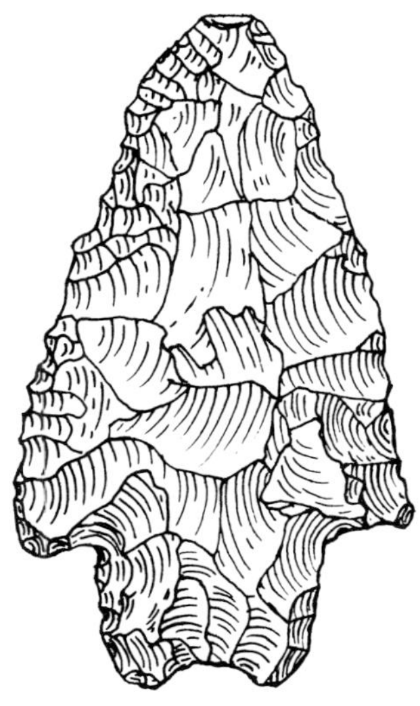

5

The Excavation of Guilá Naquitz

Kent V. Flannery

Chris L. Moser

Silvia Maranca

INTRODUCTION

Guilá Naquitz (Site OC-43) was discovered on January 26, 1966. Part of the following month was spent converting the old ox cart trail from Mitla to Díaz Ordaz into a road that would allow us access to the Rancho El Fuerte caves by means of our 4-wheel-drive pickup truck. Between February 21 and 25, we tested Guilá Naquitz in order to determine its stratigraphic sequence, establish whether *in situ* preceramic levels were present, ascertain how far back in the sequence plant remains were preserved, and evaluate its potential for further excavation (Fig. 5.1).

While MacNeish was excavating at Purrón Cave in the Tehuacán Valley, he was frequently asked what he was finding. "Well," ran his favorite reply, "at least the botanist had a good day." This response reflected the fact that Purrón Cave had superb plant preservation but not many highly diagnostic artifacts per square meter. Testing at Guilá Naquitz revealed a similar situation: a preceramic occupation with outstanding preservation but low artifact densities. Clearly, if we were to establish the nature of the tool assemblage, we would have to excavate all or most of this small cave.

The month of March was spent in testing other caves and extending our survey. Finally we returned to Guilá Naquitz on April 14 with a larger crew and excavated there until May 10, 1966. At that point we were convinced that, while there might still be preceramic features somewhere on the talus, we had removed the entire area of preceramic occupation protected by the overhang of the cave. In total, our excavations involved all or part of 64 1 × 1-m squares.

Flannery and Chris L. Moser directed the February testing, with Eligio Martínez serving as foreman for 6 workmen from Mitla. In April, Flannery and Moser were joined by Silvia Maranca, and the crew of Mitla workmen was increased to 10. This group of Zapotec-speaking Mitleños was the heart and soul of our project and should be considered the real excavators of Guilá Naquitz (Fig. 5.2). Included were Don Juan Martínez, his brother Ambrosio, and his son Ernesto; Felix and Angel Sosa; Pablo García, Carlos Pérez, and Genaro Luis; and the late, and greatly missed, Alfredo Sosa. Their patience, intelligence, and careful attention to detail was indispensable to our success, and their unflagging high spirits and sense of humor kept us continuously entertained.

Somehow our Ford pickup, "El Caballito Blanco", made it up the boulder-strewn slopes of the mountain every morning, and we dismounted in a wind-deflated gully below the cave. Somehow all the equipment made it up to the cliff, including the 10-gallon *garrafón* of drinking water that Don Juan would entrust to no one else. Once at the cave, Moser would tether his faithful watchdog on the talus to keep out any stray goats that wandered away from their herders on the slopes below us.

February mornings were cool, the sky clear and blue, and the cave so shady that we had to use Tri-X film. Over the sound of trowels and paintbrushes, an interminable shaggy-dog story would be told in Zapotec by one of the excavators. Inevitably, the punch line would be in Spanish, but this was little consolation to those of us who had not understood the preceding 1500 words. Gradually we accumulated a list of punch lines for which we had no stories: "Mátalo, porque ya está comprado;" "Con dos cobijas, una nueva encima y una vieja abajo." The frequent inclusion of lines such as, "Confiésame, Padre, porque yo he pecado," made it clear that many stories involved priests and confessionals, but the sins

FIGURE 5.1. View of Guilá Naquitz from the east during excavation. Organ cacti (*Myrtillocactus sp.*) dominate the foreground, with numerous low bushes of *susí* (*Jatropha neodioica*) hidden in the underbrush.

were confessed in Zapotec to the accompaniment of high-pitched laughter. Though we gradually picked up the minimum vocabulary necessary to describe the archaeological deposits—*giht* for the cucurbit rinds, *dohb* for the maguey quids, *gui bidžúi* for the obsidian—we never really deciphered any of the stories.

Nor did the entertainment cease at lunchtime. Out of the hamper came tortas, fruit, and other goodies prepared by Lupe back at the *Posada La Sorpresa*, along with a dozen hard-boiled eggs. No, make that eleven hard-boiled eggs. The twelfth was "the egg of the day," a raw one indistinguishable from the others, which had been slipped into the basket by Eligio Martínez. Tension mounted as we all began to crack and peel our eggs, and we waited to see who would get the egg of the day. Finally there would come a crack, a sickening squish, and an epithet in English or Zapotec. Eventually, Moser began cracking his egg on Eligio's head as a precautionary measure.

One memorable lunchtime there came an ear-splitting whoop, and seconds later one workman emerged from the shadow of the cliff holding the tattered remnants of what had once been his shirt. Everyone knew what had happened: One of the screen operators had captured a scorpion and cut off the last segment of its tail, leaving it harmless but alive and very angry. Approaching one of the trowelmen as he hunched over his lunch, he had pulled his comrade's collar away from his neck and suspended the creature over the resulting opening by holding its tail between his fingers. As the trowelman looked up, the last thing he saw was an obviously live scorpion plummeting down between his shoulder blades and his shirt. Now most of the workmen were on the ground, helpless with laughter. "It's the only way," one of them giggled, "we can get him to change that old shirt."

No one ever lost his temper at such times, not even the victim. He simply waited; he knew his turn would come, and probably sooner than his tormenter expected.

Through the long afternoons, our crew of Mitleños became better and better at following the cultural stratigraphy of the cave. They knew each living floor intimately, remembered what had been found in it and in which square, and had their

FIGURE 5.2. The real excavators of Guilá Naquitz Cave. Clockwise from lower left, Ambrosio Martínez, the late Alfredo Sosa, Genaro Luis, Pablo García, Felix Sosa, Don Juan Martínez, Eligio Martínez, Angel Sosa, Carlos Pérez, and Ernesto Martínez. An unidentified urban cowboy serves as scale.

own code name for each. Zone C, with its soft white ash, was referred to as *leche en polvo,* "powdered milk". Zone D, with its granular brown matrix of burned oak leaves, was referred to as *el Nescafé*, "instant coffee." Zone E was simply *yuš* (Zapotec for "sand"). We were constantly amazed at the near-photographic memory our trowelmen displayed for the squares they had dug and the retrospective comments they made that were confirmed by field notes from weeks before.

At quitting time the shaggy-dog stories and practical jokes began again. An orange left over from lunch would be offered to one workman, and he would eagerly begin removing the peel with his teeth—only to discover that a companion had already removed the fruit through an imperceptible razor cut, refilled the peel with bat guano from a nearby cave, and sealed the cut with *Ipomoea* gum. And woe to anyone who, like little Alfredo, fell asleep in the truck on the way home. How anyone could fall asleep in the back of a pickup coming down the rocky slopes of a mountain in 4-wheel drive is a secret Alfredo never shared with us. Nor did he ever learn that if he fell asleep his sandals would be surreptitiously tied together with twine. As we rolled into the posada in Mitla, Felix and Eligio would leap out, drop the tailgate, and shout for Alfredo. Usually he was in the air, somewhere between the tailgate and the earth, when he realized his huaraches were tied.

It was with some sadness that we closed the excavation on May 10, 1966. Guilá Naquitz had been a great site, a joy to excavate, with remarkably clear stratigraphy and excellent preservation throughout. Beyond that, it had been a colorful experience shared with a group of people none of us would ever forget.

EXCAVATION TECHNIQUES

We began our work with the knowledge that most Mexican dry caves so far excavated had had complex stratigraphy, involving living floors of varying sizes perforated at intervals by intrusive pits. We wanted to adhere as closely to the natural, or cultural, stratigraphy as we could, while maintaining standardized horizontal units within which remains could be plotted. Our techniques combined the approach used by MacNeish (1961) at Tehuacán with that used in the

Near East by Hole *et al.* (1969). The horizontal control came from a grid of 1 × 1-m squares laid out over the entire area to be excavated. Vertical control was established by exposing the cultural stratigraphy in the wall profiles of a few initial test squares and then, working against those drawn and recorded vertical faces, removing each living floor separately as the next square was dug. An effort was made to isolate a square on two or more sides before excavation so that the slope and variation of each living floor, as well as any intrusive features, could be seen from the outset. Guilá Naquitz "cooperated" by having some of the clearest and easiest-to-follow color changes between strata we have ever seen, as the photographs in this chapter should show.

The orientation of Guilá Naquitz lent itself to a magnetic north–south grid, with letters (B, C, D, etc.) running west to east from the rear of the cave and numbers (3, 4, 5, etc.) running south to north across the chamber (Fig. 5.3). We laid out 64 1 × 1-m squares, excluding an area of exposed bedrock in the northeast corner of the cave. Our major datum point was a nail driven flush into the wall of the cave where it was touched by the southwest corner of Square C7; this remained the primary referent for all depths and grid squares.

Each square was assigned three workmen: a trowelman, who did the actual excavation; a bucketman, who carried the trowelman's backdirt to the screen; and a screen operator, who processed that backdirt. These three habitually worked together as a team so there could be no confusion as to which bucket of dirt had produced which object in the screen. When possible, the members of a team included individuals who were used to working together, such as a father and son, two brothers, or two brothers-in-law, a practice Flannery had observed Robert Braidwood using in the Near East.

Digging at Guilá Naquitz was so soft that it was never necessary to use a tool larger than a trowel. In addition to a trowel, however, each excavator was provided with a screwdriver (for loosening the indurated sand at the base of the deposits), an ice pick and paintbrush for exposing *in situ* artifacts, a sugar scoop for passing backdirt to the bucket, and a whisk broom for keeping the square neat. Each worker was also provided with a small petate, or reed mat, about 30 × 40 cm in size, on which to organize any objects that were found—another of Braidwood's procedures (Fig. 5.4).

The excavation of a typical square would begin with an examination of the profile and a discussion of exactly how deep the trowel would have to go to loosen the first living floor. Were a given stratum thicker than 10 cm, we followed MacNeish's procedure of removing it by arbitrary substrata, even if it appeared homogeneous. The trowelmen began at one profile and worked toward the center of the square, loosening only the uppermost living floor (Fig. 5.5). As they worked, they would organize the material they found on their petates, separating it into different piles such as acorns (Fig. 5.6), maguey quids, hackberry seeds, animal bones, flint chips, and so on. When they encountered objects such as projectile points, scrapers, bifaces, grinding stones, wooden tools, or textiles, these were piece-plotted by recording the depth, distance from north, and distance from west within each square. All material from each living floor was saved separately by 1-m square so that unretouched flakes (and flakes that were so dirty that the retouch was not detectable when first encountered) could be added to the plan and counted by square.

No trowelman, no matter how careful, could possibly find all the small items in a living floor. For this we had two sizes of screens. Using a sugar scoop, each trowelman would place the cave matrix (from which all visible items had been removed) into a bucket. Once full, this matrix was carried by the bucketman to a worker who operated two sizes of screens (Figs. 5.7 and 5.8). First the matrix was passed through the larger screen (6-mm mesh), which caught all objects the size

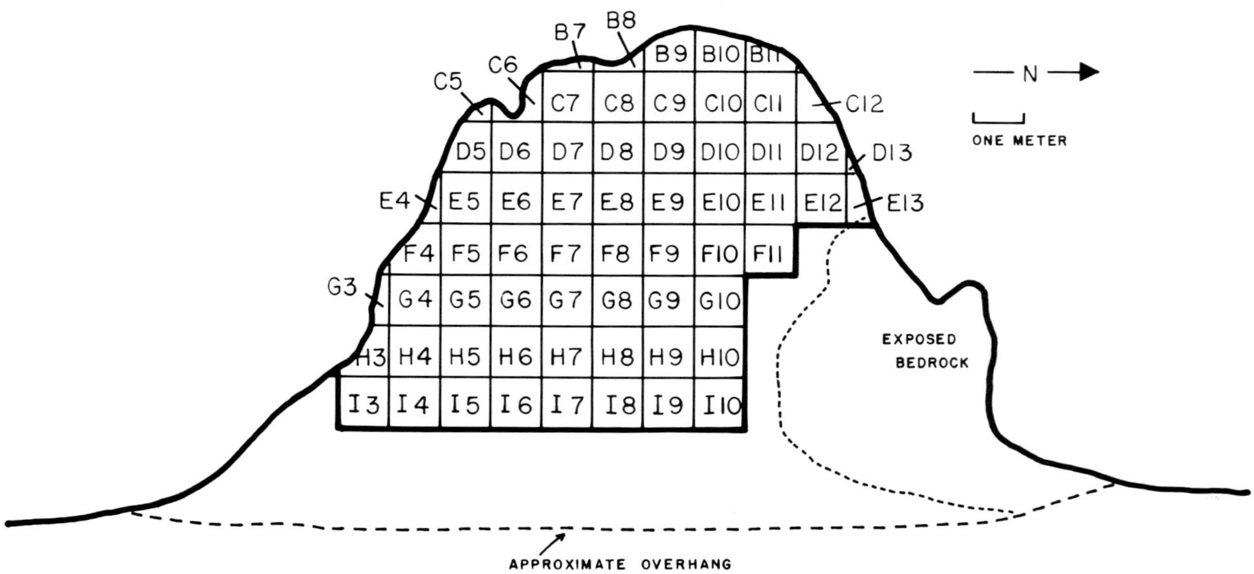

FIGURE 5.3. Plan of excavated 1 × 1-m squares at Guilá Naquitz.

FIGURE 5.4. Trowelman's tool kit for the excavation of Guilá Naquitz: small petate, or reed mat, trowel, ice pick, screwdriver, paintbrush, whisk broom, and sugar scoop.

of a cucurbit seed that had not been noticed by the excavator. Dirt passing through the larger screen fell directly into the smaller screen (2-mm mesh), which easily caught seeds as small as those of a chile or prickly pear. Together, the bucketman and the screen operator perused each screen, recovering tiny seeds, rodent vertebrae, and other items (Fig. 5.9). In addition, samples of matrix for flotation in water were taken to provide a check on the screens in case even smaller remains had escaped. Fortunately, passing these flotation samples through carburetor mesh suggested that only slender grass seeds (such as *Heteropogon* sp.) were escaping, and even these were rare.

When the trowelman had completed a given living floor within a 1-m square, the material from his screen was brought back to be added to the material on his petate. At this point, we assembled each category of plant (*guajes,* mesquite, onions, acorns, and so forth) and placed each in a separate twist of aluminum foil with its own jewelry tag (Fig. 5.10). Later, in our laboratory in Mitla, these plant categories were cleaned and placed in clear plastic envelopes with the same tag. All animal bones were placed in strong coin envelopes and labeled by 1-m square and living floor; artifacts were

FIGURE 5.5. Working from the east profile, Don Juan Martínez begins Square F10; in the background, Pablo García works in Square G9.

FIGURE 5.6. Mass of acorns recovered from Zone E, Square H6.

given similar treatment, with those that had been piece-plotted given their own jewelry tags with the three-dimensional coordinates. Finally, if the square had been free of intrusive features, samples of wood charcoal for radiocarbon dating and samples of matrix for pollen analysis were taken from each living floor.

Guilá Naquitz was a shallow site; deposits reached about a 120-cm depth in the rear and about a 40-cm depth near the talus. Deepest squares might have as many as six or seven living floors, while peripheral squares might have only one or two. Intrusive features were numerous but so clear that their separation from the earlier deposits was relatively easy (see Figs. 5.11 and 5.12); even rodent burrows could easily be isolated (see Chapter 16).

Both horizontal and vertical control were maintained by drawing and photographing profiles as squares were completed. Considerations of cost have prevented us from publishing every profile, but we do present four major east–west cross sections and two major north–south cross sections in this chapter (Figs. 5.15, 5.16, 5.20, 5.21, 5.27).

STAGE 1: INITIAL TEST

We decided that the best place to make our initial test was the column of Squares C7, D7, E7, F7, and G7. We suspected that this column would give us a look at the deepest deposits and provide us with a cross section from the back wall of the cave to near its mouth. Accordingly, these were the first five squares staked out (Fig. 5.13).

We began by digging alternate squares D7 and F7. Because these were the first squares tested, we had no idea what

FIGURE 5.7. Felix Sosa watches Genaro Luis process matrix through 6-mm-mesh screen into 2-mm-mesh screen below.

FIGURE 5.8. Bucketman Felix Sosa watches Pablo García sift material in 2-mm-mesh screen.

stratigraphic levels to expect. We therefore proceeded slowly, removing what Old World archaeologists call "spits" of no more than 10 cm at a time. Artifacts, plants, and animal bones discovered were cataloged according to their depth and to a detailed description of the stratum in which they occurred. When we reached a new cultural stratum that differed in color and matrix from the one above it, we stopped troweling downward until the whole square had been excavated to that natural (actually "cultural") break. We reasoned that once we had determined the stratigraphy of the cave and assigned names to each stratum, the materials from these initial test squares could be assigned to those strata by their depth and the detailed descriptions of the stratigraphic contexts in which they occurred. This proved easy because of the clarity of the color changes.

The fact that the stratigraphy was clear, however, did not mean it was not complex: our two test squares, only 1 m apart, turned out to have such different sequences that zone designations could not be made until the intervening square, E7, had been dug.[1]

[1]We have followed MacNeish in using the term *stratigraphic zone* for the cultural strata in our Mitla caves. In retrospect, the term *level* or *layer* seems more appropriate than *zone*, which sounds more like a horizontal division than a vertical one. However, since *zone* is well established in the literature it seemed pointless to change it.

FIGURE 5.9. In foreground, Ernesto Martínez, Genaro Luis, and Alfredo Sosa pick small seeds and rodent remains out of 2-mm-mesh screen. In background, Carlos Pérez and Eligio Martínez do the same.

FIGURE 5.10. Eligio Martínez and Chris Moser sort and package remains from Square D9.

FIGURE 5.11. Square E10 seen from the east, with Feature 12 (an intrusive Postclassic bell-shaped pit containing two large stones and 3 m of rope) shown interrupting the preceramic strata. Such intrusive features were usually as clear as the one shown here.

Square D7 began with a 20-cm-thick layer of grass, gray ash, oak leaves, and Early Postclassic (Monte Albán IIIb–IV) potsherds. This layer was the uppermost true living floor in every square in the cave, so we eventually designated it Zone A. It contained maize, beans, squash, avocados, chiles, zapotes, cotton, and other cultivated plants, and was a zone rich in storage pits and other features. Zone A appears to have been laid down by a group of people engaged in farming and maguey roasting in the piedmont hinterland of Postclassic Mitla. Radiocarbon samples dated this occupation to A.D. 620–740. Zone A has been discussed in a preliminary fashion (Flannery 1970:14–15; Flannery and

FIGURE 5.12. Silvia Maranca removes Feature 18, a Postclassic ceramic vessel in a maguey fiber harness, intrusive through Zone A and into the top of Zone B1 in Square I4.

Smith 1983), and will be reported on fully in a future volume on the Postclassic levels of the Mitla caves.

While Zone A was the most recent actual stratum in the cave, there were four features present from the later Monte Albán V period (radiocarbon-dated to A.D. 1270). Included were a maguey-roasting pit, a storage pit with an unused maguey heart, and two beds of grass and oak leaves in which the Monte Albán V occupants had apparently slept. One of these beds (Feature 6) overlay Zone A in our F7 test, and provided the first difference we detected between Squares F7 and D7. While this Monte Albán V occupation had not produced an actual living floor, we eventually came to use the term *Zone Super-A* to refer to the complex of four features. Zone Super-A will also be reported on in the future volume mentioned above.

In Square F7, the Feature 6 bed of grass and oak leaves extended from the surface to a 10-cm depth, while Zone A extended from a 10- to 20-cm depth. Below this was a slender layer of white-and-gray ash that was readily distinguishable from Zone A and contained no ceramics. This superficially unimpressive layer, we were later to realize, was part of Zone B, a stratum that in other squares would become the thickest and most complex in the cave.

As Don Juan Martínez continued downward in F7, he came to a thicker layer, this time of gray ash and coffee-colored decomposing oak leaves, extending down to a depth of 40 cm below the surface. This important level, which we initially referred to as "the basal gray ash layer," eventually became Zone D, and from its artifact assemblage it definitely appeared to be preceramic. Only in the south half of Square F7 was Zone D overlain by a wispy film of white ash that we now realize was an attenuated patch of Zone C. In Square F7, this white film contained no artifacts and could literally not even be separated from Zone D.

Below Zone D, Don Juan came to a layer of soft, reddish-yellow sand that appeared to be the product of natural weathering of the volcanic tuff floor of the cave. Largely sterile, it contained a few isolated pockets of flints and plant remains at a depth of roughly 40–60 cm. This occupation, the oldest in the cave, we designated Zone E. Below it, the soft sand turned gradually to hard, sterile sand and eventually to solid ignimbrite.

Meanwhile, in Square D7, Alfredo was finding a different situation. At least part of his test was taken up by Feature 1, a very large Monte Albán IIIb–IV storage pit running diagonally across the square and filled with maize and other plants. To the north of this pit, Zone A resembled its counterpart in Square F7, but below Zone A came a long series of alternating layers of gray ash and white ash that did not correspond to anything seen in F7 (see Fig. 5.14). First came a white ash lens without any artifacts. Next came a layer of floury gray ash about 20 cm below the surface. This was

FIGURE 5.13. Looking west into Guilá Naquitz after the first five squares (C7–G7) had been staked out.

followed by a second layer of white ash, followed in turn by a second layer of floury gray ash at a depth of 50 cm below the surface. Then came a third layer of white ash, present in only part of the square. Suddenly, below this third white ash layer, at a depth of 60–70 cm, the deposit changed to gray ash and coffee-colored decomposing oak leaves identical to our basal gray ash layer in F7. Sure enough, below this basal gray ash we found the same soft reddish-yellow sand we had encountered in the bottom of F7. Our two test squares, therefore, were linked by Zones A, D, and E; what remained to be explained were all the intervening layers, and to do this we would clearly have to dig E7.

In the lower levels of Square D7 we found our first preceramic features. A circular pit 46 cm in diameter had been dug down from Zone D into the soft sand of Zone E. Designated Feature 2, it contained acorns, a fire-cracked rock, and a fragment of discoidal one-hand mano. Nearby we encountered another acorn-filled pit, Feature 3, whose east half extended into Square E7. This pit, also from Zone D, actually continued downward into bedrock.

Felix Sosa was the trowelman for Square E7. Since the stratigraphy of D7 had been so much more complex than that of F7, we decided to start with the west profile of E7 and trace its complicated gray and white ash layers east toward the simpler E7–F7 profile. Almost immediately, Felix ran into Feature 5, a *tecomate*, or neckless jar, 50 cm in diameter, that had been buried in the cave by the Monte Albán IIIb–IV occupants of Zone A. It may once have held the water supply for that zone, but its rim and upper walls had collapsed with the passage of time. Feature 5 had disturbed the preceramic stratigraphy to a depth of 40 cm below the surface, but the gray and white ash layers could be traced around it on all sides. In addition, the base of this large *tecomate* rested on Zone D, that characteristic layer of gray ash and coffee-colored decomposing oak leaves that ran continuously from D7 through E7 to F7 (see Fig. 5.15).

Once Feature 5 had been removed, the sequence in E7 became clearer. The uppermost white ash lens of D7 faded out before it reached E7. The uppermost floury ash level thinned out as it reached E7 and went on to become the

FIGURE 5.14. View of Square D7 immediately after excavation, looking from the southeast toward the intersection of the west and north walls. The north profile (right) can be compared to the drawing in Fig. 5.15.

slender layer of white-and-gray ash we had found in Square F7 just below Zone A (see above). This level we now recognize as Zone B1.

Below Zone B1 came the second of the white ash layers seen in Square D7. As it ran east into E7, it sloped upward and became thinner and wispier until it disappeared near the surface. This level we now recognize as Zone B2.

Below Zone B2 came the second of the floury gray ash layers seen in Square D7. It also sloped upward and grew thinner as it ran east, but it spanned all of Square E7 and only disappeared when it was truncated by Feature 6. This level we now recognize as Zone B3, and in Square E7 it produced one of our most interesting preceramic artifacts: a fragment of net displaying lark's head knots where its horizontal and vertical cords met (see Chapter 9). This net was piece-plotted as resting against the south profile of the square at a depth of 25 cm below the surface, 10 cm in from the west profile, where it was accompanied by flint debitage, mesquite seeds, *guaje* pods, and maguey quids.

Below Zone B3, Felix encountered the third (and thickest) of the white ash layers seen in Square D7. It ran east through E7, becoming more and more tenuous as it went, until it had shrunk to the wispy film of white ash that we had found overlying (and inseparable from) Zone D in Square F7. This level we now recognize as Zone C.

Finally, Square E7 ended with the excavation of Zones D and E. We recovered the east half of Feature 3, the acorn-filled pit dug down into bedrock from Zone D, whose west half had been discovered in Square D7 (see Figs. 5.16 and 5.17). The presence of two such pits—Features 2 and 3—in the small area we had so far excavated served to alert us to the importance of acorns in the subsistence strategy of preceramic Guilá Naquitz.

At this point, we began to draw the profile of Squares D7, E7, and F7. How crucial the excavation of E7 was to our understanding of the cave stratigraphy can been seen by covering that square with a piece of paper while looking at Fig. 5.15 and trying to imagine how we would have explained the difference between D7 and F7 without it.

The more squares we dug, the clearer the stratigraphic picture of the cave became. The removal of Square C7 by Felix on February 24 showed how Zones A and B1 had been truncated by Feature 1, while Zone B3 passed below that feature. The excavation of Square G7 by Don Juan during February 22 and 23 revealed the increasing simplification of the stratigraphy as we moved east toward the mouth of the cave: Below Feature 6 we found a few patches of Zone A and some good exposures of Zones D and E but no traces of Zone B. Finally, the excavation of Square H7 brought to light Feature 7, a large maguey-roasting pit associated with Zone Super-A, which had essentially destroyed all earlier levels in that square.

STAGE 2: SQUARES D8 AND E8

The completion of Squares C7–H7 gave us a 6-m trench whose east–west profile could be used as a guide to excavation of the rest of the cave. We drew the north profile of these squares (Fig. 5.15) and photographed the south profile (Fig. 5.16). In addition, we drew and photographed the west profile of Square D7 so that, if work continued, we would eventually be able to draft a north–south section through Squares D5–D12 (Fig. 5.18). At this point we were nearly convinced that full excavation of the cave would be worthwhile, but we decided to take out two more squares, D8 and E8, just to be sure. In the excavation of these squares, we would have the advantage of being able to work from the north profile of D7 and E7, which was now drawn and labeled.

Squares D8 and E8 were excavated by Don Juan Martínez and Felix Sosa during February 24 and 25 (Fig. 5.19). The sequence in D8 was not unlike that in D7 except that there were no intrusive features to disturb the stratigraphy. Zone A was a thick layer of ash and plant remains overlying what appeared to be a bed of grass. Below it, Zone B1 had expanded to include two components—an upper white ash with few artifacts and a lower gray ash with flint debitage, oak leaves, acorns, and nanche seeds. We took a pollen sample from the

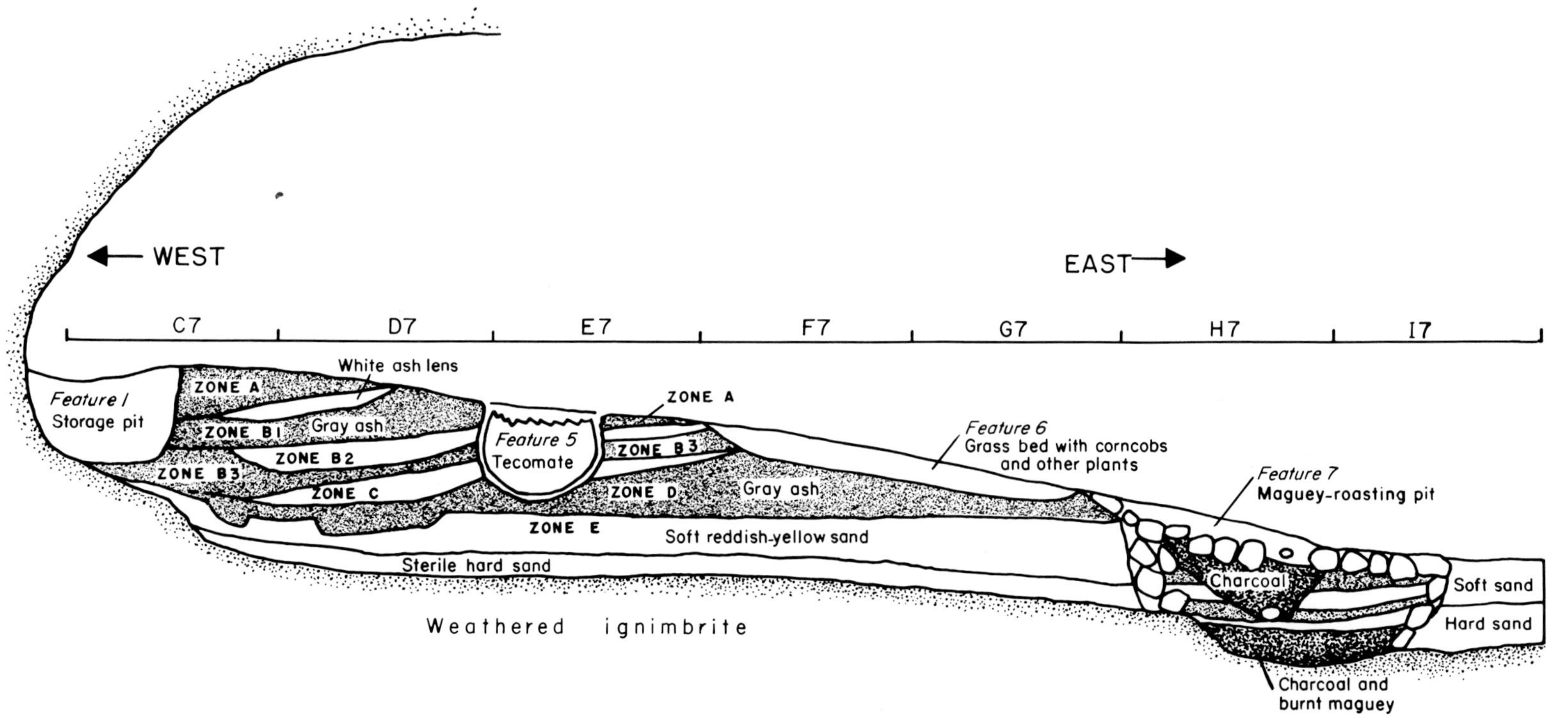

FIGURE 5.15. North profile of Squares C7–I7. (Feature 5 did not actually penetrate the north profile; it is shown for the purpose of establishing its position relative to Square E7 and Zone B1–D).

lower component. Still farther down, Zone B2 was an irregular white ash layer that expanded in places and became thin and wispy in others. It contained flint, acorns, mesquite seeds, and abundant chunks of wood charcoal. These chunks of charcoal became radiocarbon sample SI-515, which yielded a date of 6670 B.C. ± 160 (see Chapter 14).

Below this, at a depth of 45–50 cm below the surface, Zone B3 was a layer of floury gray ash spanning the whole square and providing us with good context for a pollen sample. Zone C, at a depth of 50–60 cm, was an irregular layer of white ash that expanded in places and became wispy in others, sometimes consisting of two components. Zone D was the usual layer of gray ash and coffee-colored decomposing leaves, and in this square it contained a notched flake and a drill in addition to the usual unretouched flakes. Zone E, the soft reddish-yellow sand, was not particularly rich. Taken all in all, however, we were very happy with Square D8 because its lack of intrusive features had given us a chance to examine an undisturbed sequence of preceramic occupation floors.

We were not so lucky in E8, for there we hit Feature 4, a grass-lined Monte Albán IIIb–IV pit that had disturbed Zones B1 and B2 in various parts of the square. Zone B3 remained intact, however, and Zone C was present although irregular in thickness. While poor in artifacts, Zone C produced some cucurbit seeds that excited us a great deal because of their undisturbed preceramic context. Moreover, Zone D was rich in material. Two fragments of maguey fiber cordage were piece-plotted in this zone, and the Zone D refuse included flints, *guaje* pods, acorns, wild bean pods, nanches, mesquite seeds, prickly pear seeds (from the 2-mm-mesh screen), and cottontail rabbit bones. Zone E was present but not as rich in material.

At this point we ended our testing of Guilá Naquitz, convinced that the cave was worthy of much fuller excavation. While there were many intrusive Postclassic features, these were easy to recognize and remove; and despite them, there were extensive areas of preceramic occupation with well-preserved plant remains. We therefore planned to return to Guilá Naquitz after a period of testing other caves.

STAGES 3–5: BEGINNING THE MAIN EXCAVATION

We returned to Guilá Naquitz on April 14, 1966 and completed the excavation in five stages. Stage 3 consisted of removing Squares C8, B8, F8, G8, and H8, extending the column of squares begun when D8 and E8 were excavated in late April. This allowed us to continue our north–south section drawing (Fig. 5.20) as well as drawing a new east–west profile along the north wall of Squares B8–H8 (Fig. 5.21). The squares dug in Stage 3 were good ones, relatively free of intrusive pits from Zone A, and usually contained long stretches where Zones B1, B2, and B3 were superimposed. However, Zone B2 gradually turned to gray ash in this part of the cave, and in Square C8 we found an area where it could not be separated from Zone B3. Such areas we came to refer to as "Zone B2+3."

Square F8 was an important one for Zone D. In the southwest corner of that square we found a white ash–filled hearth some 30 cm in diameter and 8 cm deep and designated it Feature 22 (Fig. 5.22). Nearby lay a collared peccary jaw and a flint flake. A chipped-stone core fragment was piece-plotted, and acorns, nanches, maguey quids, and prickly pear appeared in the ashy matrix.

Stage 4 saw the removal of Squares F9–H9, F10–H10, and F11, beginning with the north profile of Squares F8–H8 and tracing the living floors north until they ended against the dome of exposed bedrock under the northeast corner of the cave overhang. At this point, a photograph could be taken of the west profile of Squares F9–F11 in preparation for tracing the living floors west.

Square F9 was an important one for Zone C. Here the layer was thick enough to have two components, an upper white ash and a lower dark brown ash and compost. In the upper component, we were able to piece-plot a broken discoidal one-hand mano, a limestone pebble pigment grinder, and a piece of cordage. In the lower component, we were able to piece-plot a complete, subrectangular one-hand mano of veined quartzite (Fig. 5.23).

Square H10, on the other hand, had no Zone C but was an important square for Zone E. In the soft reddish-yellow sand of that zone, we piece-plotted an unfinished Lerma projectile point.

Excavation Stage 5 consisted of removing Squares B9–E9, E10–E13, and D10–D13, an L-shaped block flanked on its south side by the B8–E8 profile drawing and on its east side by the F9–F11 profile photograph. While tracing the various levels north and west from these profiles (Fig. 5.24), we encountered more preceramic features. Pablo García, troweling in Zone B1 in Square E9, located a hearth in the center of the square. Designated Feature 11, this hearth was roughly 45 cm in diameter and 15 cm deep, filled with white ash and charcoal and flanked on at least two sides with stones (Figs. 5.25 and 5.26). This level also produced a good sample of chipped-stone tools and a bit of cordage, along with acorns, *guajes*, nanches, hackberry, mesquite, and maguey quids. Zone C in the same square, a thick layer of white ash over brown compost, could be dated to the autumn season by its abundant acorns and its hardened deer antler. Also present were *susí* hulls, mesquite, cucurbits, piñon nuts, hackberry, nanche, prickly pear, and maguey.

Felix Sosa, troweling in Zone D in Square D9, found yet another acorn storage (or processing) pit. Called Feature 10, this was a pit 40 cm in diameter and 35 cm deep in the northwest corner of the square, dug down from the level of Zone D. While it still had a layer of acorns in the bottom, the pit had filled after abandonment with the same coffee-colored ash and leaves seen in the rest of Zone D. Acorns and chipped-stone debitage were common in the debris around it.

In fact, Square D9 turned out to be one of the more interesting in this part of the cave. In Zone B2 we isolated a rodent burrow that turned out to have a mummified pygmy mouse (*Baiomys* sp.) inside. In Zone B3, we were able to piece-plot a fire drill fragment. Square C9 was equally interesting, especially for Zones B1–B3. Zone B1 produced a

FIGURE 5.16. Photo mosaic of the south wall of the initial test trench (Squares C7–H7). Squares C7 and D7 are filled mainly by Feature 1, an intrusive Postclassic storage pit. The base of Feature 3, a pit dug down from Zone D, is visible in bedrock at the juncture of Squares D7 and E7. In E7, white ash lenses separating Zone B1, B2, and B3 can be seen sloping upward from right to left; they almost reach the surface before being truncated by Zone A, a 10-cm-thick layer of grass and ash capping the entire profile. The lowermost white ash—beginning near bedrock at the D7–E7 boundary and sloping upward to near the surface in the middle of F7—is Zone C. The gray ash below it is Zone D, and the light-colored layer of soft sand visible below D in Square G7 is Zone E. All of Square H7 (and part of G7) is occupied by Feature 7, a Postclassic maguey-roasting pit. Holes in the profile in C7 and D7 are places where loose stones had to be removed before they fell.

one-hand mano as well as a ball of organ cactus gum that may have been collected for tool hafting. Zone B2 produced a one-hand mano and a metate fragment. Finally, in Zone B3 we piece-plotted the distal half of a Pedernales point; miraculously, this turned out to be the missing portion of the same Pedernales point whose proximal half we had found on the talus of the cave during our January 26 survey (see Chapter 3).

A number of interesting discoveries were made in Squares D10 and D11 as well. In Zone B2 of Square D11, we piece-plotted a plaque of coiled basketry 34 cm from the east edge and 70 cm from the south edge of the square. In Zone C of the same square we recovered another fire drill. Square D10 had such clear stratigraphy and lack of rodent burrows in its lower levels that we elected to take radiocarbon samples from Zones C, D, and E.

STAGES 6–7: COMPLETING THE MAIN EXCAVATION

Stage 6 of our excavation involved tracing the levels already discovered in Squares C7–H7 southward into Squares D6–H6. At this point, the south profile of D6–H6 was drawn (Fig. 5.27); and working south from this new profile, we removed Squares G3–G5 and H3–H5. The L-shaped block of squares thus removed from the south part of the cave was analogous to that removed from the north part of the cave during Stage 5.

FIGURE 5.16 (*continued*).

It was the column of Squares D6–H6 that produced the most interesting material from this stage. We excavated H6 first because we suspected it would be relatively uncomplicated and its removal would give us two profiles to work from as we moved west through the remaining squares of this column. As we had suspected, the only preceramic level in H6 was Zone E, which was loaded with acorns (see Fig. 5.6). By the time we reached Square F6 (Fig. 5.28), Zones B1, C, and D had all reappeared. In Zone B1 of F6 we were able to piece-plot a broken projectile point and a fragment of slab palette stained with red ochre; Zone C produced a fragment of mortar. In Zone D we isolated the burrow of a spiny pocket mouse (*Liomys* sp.) who had brought in some intrusive maize from Zone A, then curled up in his burrow, died, and was mummified (see Chapter 16).

The removal of Square F6 left E6 with an exposed profile on its west and north sides (Fig. 5.29). Figures 5.30 and 5.31 show how easy it was to remove one stratum at a time from such a square, owing to the remarkably clear color changes that characterized Guilá Naquitz. Zone B2+3—the one Moser is shown removing in the photographs—produced considerable flint, a few maguey quids, acorns, prickly pear, hackberry, wild beans, and pollen and flotation samples. Upper Zone C, the undisturbed white ash layer in the same photographs, produced deer and cottontail bone. Lower Zone C, the brown compost layer below it, yielded pollen and flotation samples, deer bone, and a variety of plant remains. The pollen sample was particularly interesting because it included a few grains of *Zea* sp., in the size range of teosinte pollen.

Square D6 had a number of Postclassic features, but once Felix and Pablo had worked these out the remaining preceramic deposits were productive. We piece-plotted two slab metate fragments in Zone B1, then two mano fragments in Zone B2.

By now we had removed all or part of 44 squares, and all that remained during Stage 7 was to excavate the periphery of the excavation. This involved completing the northwest corner (Squares B10–B11 and C10–C12), the southwest corner (Squares C5–F5 and E4–F4), and the extreme eastern margin (Squares I3–I10). In fact, I3–I10 proved to lie beyond the preceramic occupation, but interesting discoveries were made in both corners of the cave.

FIGURE 5.17. Looking west down the initial test trench toward the west profile of Square D7. Note Feature 1 occupying the upper left corner of the profile and interrupting the characteristic white ash layers of Zones B and C. In the foreground, the lower portion of Feature 3 appears as a circular pit into bedrock. Several rodent burrows are also visible in the profile.

We came upon another hearth (Feature 14) in Zone B1 of Square C10. Measuring 33 cm in diameter and 16 cm deep, it was filled with white ash but did not have the lining of stones seen in Feature 11 (Fig. 5.26). In Zone B2+3 of the same square, we piece-plotted a mortar fragment. In Square C11, Zone B1 produced another mortar fragment, while Zone B3 yielded the stem from what may have been a Trinidad point.

Square E5 had excellent preservation in all preceramic levels, although Zone C had shrunk to a 10-cm layer of brown compost without the usual cap of white ash. Zone D in this square produced a flint biface 45 cm from the north edge and 4 cm from the west edge of E5. Charcoal from nearby yielded a radiocarbon date of 8750 B.C. Zone D in the adjacent square, F5, was equally rich, supplying us with a radiocarbon date of 7840 B.C. This charcoal sample came from the vicinity of our oldest *Cucurbita pepo* seed and was surrounded by fragments of deer bone, acorns and piñon nuts, hackberries, agave quids, *susí* and prickly pear. All of this material was nicely sealed under an undisturbed expanse of Zone C, which included both its upper (white ash) and lower (brown compost) components, although these were compressed into a layer only 10 cm thick.

THE PRECERAMIC LIVING FLOORS: SUMMARY DESCRIPTIONS

Having described the stratigraphy of the grid squares at Guilá Naquitz, noting the changes displayed by each stratum as we traced it from one part of the cave to another, we can now summarize the characteristics of those strata.

Bedrock at the base of the cave was whitish volcanic tuff. This gradually weathers to reddish-yellow sand, and an indurated layer of such sand was found above bedrock. This sterile layer, presumed to be Pleistocene, yielded pollen samples showing a climate cooler than today's. Above the indurated stratum was a layer of softer reddish-yellow sand that is presumed to be the original surface found by the first occupants of the cave. These first occupants evidently lived

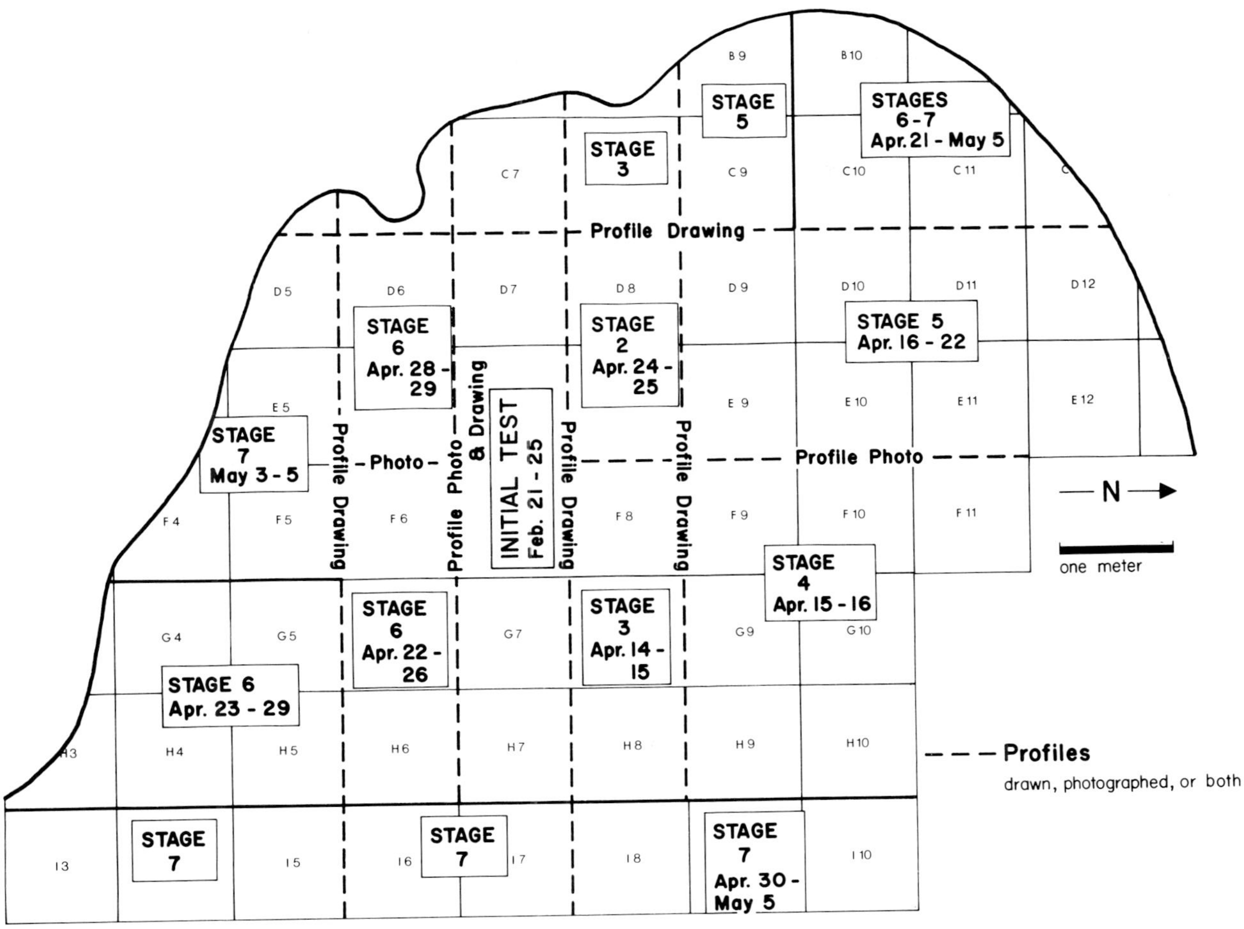

FIGURE 5.18. Stages of excavation at Guilá Naquitz, showing the order in which groups of squares were excavated.

in the cave long enough to leave pockets of artifacts, plant remains, animal bones, and charcoal in the upper part of the sand, but not long enough to produce a white or gray ash layer like those of the later occupations.

Zone E

Zone E may be summarized as a layer of reddish-yellow sand that extended throughout the northern half of the cave, occasionally reaching 20 cm in thickness, but present in the southern half only in places. Most of the matrix was weathered ignimbrite, but it also contained chipped-stone tools and plant remains left by a small number of people during a short stay or a succession of short stays (Fig. 5.32). Deer and rabbit bones were present. Acorns were very abundant and piñon nuts, mesquite, hackberry, *Opuntia* seeds, and maguey quids were also common, suggesting an occupation in the August–December period. A single unfinished projectile point of Paleoindian or very early Archaic type was present, along with other chipped-stone tools. Although no hearths had survived as features, there were enough fragments of charcoal to indicate that some fires had been set. Despite its partial derivation from the weathered ignimbrite of the cave, we consider Zone E to be a living floor because it appears to be a surface on which people camped, plants and animals were eaten, and artifacts were chipped, with many of those materials sinking into the sand in the course of being walked on.[2]

Zone D

Zone D was a layer of soft, dark gray ash and coffee-colored decomposing leaves present in virtually all squares near the back of the cave but absent or very reduced near the front. In contrast to Zone E, virtually the entire matrix of Zone D consisted of material introduced into the cave by humans. In some places, the zone was chocolate brown because of large numbers of burned oak leaves; these may

[2]In the case of our Mitla caves, we have used the term *living floor* for any layer that consists at least partly of material brought into the cave by humans and that seems to have been used at least partly as a surface on which people camped, walked, worked, and/or constructed features. It is not intended as a universal definition (it does not correspond, for example, with units called "living floors" at some open-air Lower Paleolithic sites). We use it primarily to contrast with midden layers on which people did not actually live or with levels produced by the forces of nature, such as layers of wind-blown sand or dust.

FIGURE 5.19. Guilá Naquitz as it looked on February 24, 1966. The initial test trench has been removed and workmen are excavating two extra squares, D8 and E8. Moser's faithful watchdog keeps stray goats out of the cave.

originally have been brought into the cave as bedding, but at some point they had caught fire. In other places the color was gray from wood ash, presumably resulting from the kicking around of hearth materials. One such hearth, Feature 22, was found in Square F8. Three other features (Features 2, 3, and 10) were pits, presumed to be for the storage or processing of acorns because of the large quantities of that plant food found in and around them.

The average thickness of debris in Zone D was 20 cm, and it might present an autumn encampment (August–December) by a small group of people. The presence of four features leaves no doubt that it was a true living floor. Bifacial and unifacial retouched tools, steep denticulate scrapers, utilized flakes, and one-hand manos were among the artifacts found (Fig. 5.33). Deer, rabbit, and mud turtle contributed the bulk of the animal bone. Beside the abundant acorns, plant foods included mesquite, *susí* nuts, nanches, piñon nuts, hackberry, *guaje*, prickly pear stems and fruits, maguey quids, and wild onions. Three of the more exciting ethnobotanical discoveries were bottle gourds, *Cucurbita pepo* squash, and wild runner beans, all present on a living floor with radiocarbon dates of 7840 and 8750 B.C.

Zone C

Zone C was a layer of variable thickness (usually 5–20 cm), consisting of a layer of almost pure white ash underlain in many squares by an adherent strip of gritty, brown-to-black compost. This zone was found throughout the cave, from Squares F9–F11 on the one hand to E6, F6, and D7 on the other. In some squares it became wispy and hard to trace, but in others it was quite prominent. The gritty black compost appeared to be mainly decomposed oak leaves; it looked as if the group had begun their occupation by covering parts of their living surface with oak leaves and grass, perhaps as bedding. Later, this bedding (as well as the accumulated plant material above it) had caught fire and smouldered. The white ash appeared to be hearth material that had become widely spread around. No hearths had survived as features

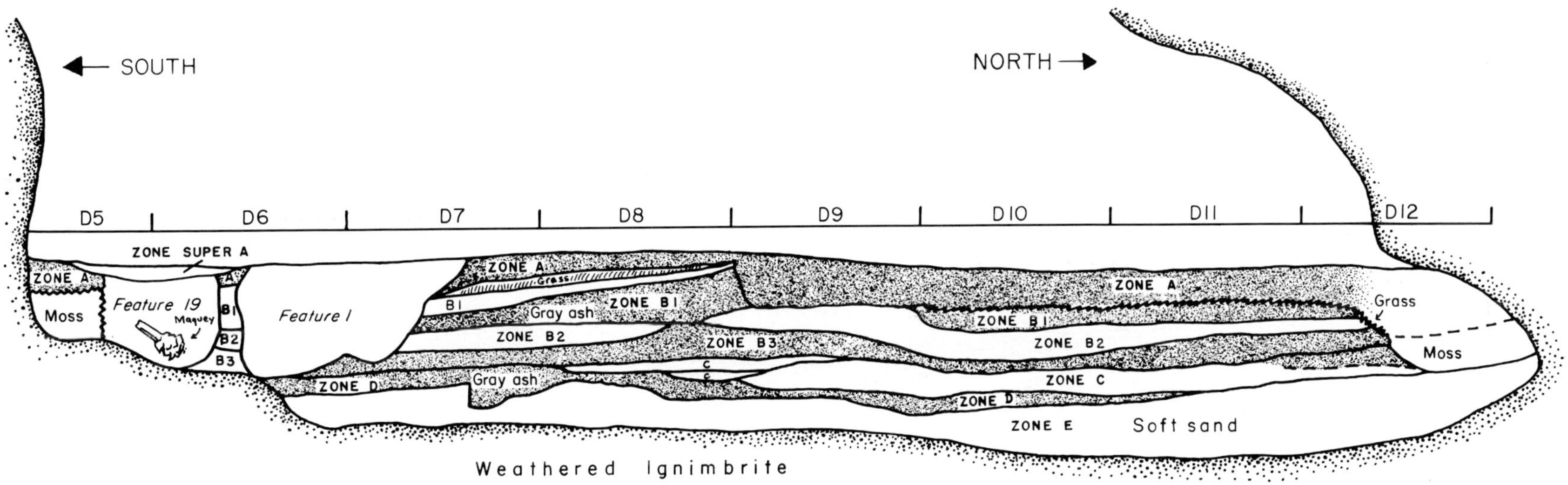

FIGURE 5.20. West profile of Squares D5–D12. The profile of Square D7 can be compared with the photograph in Fig. 5.17.

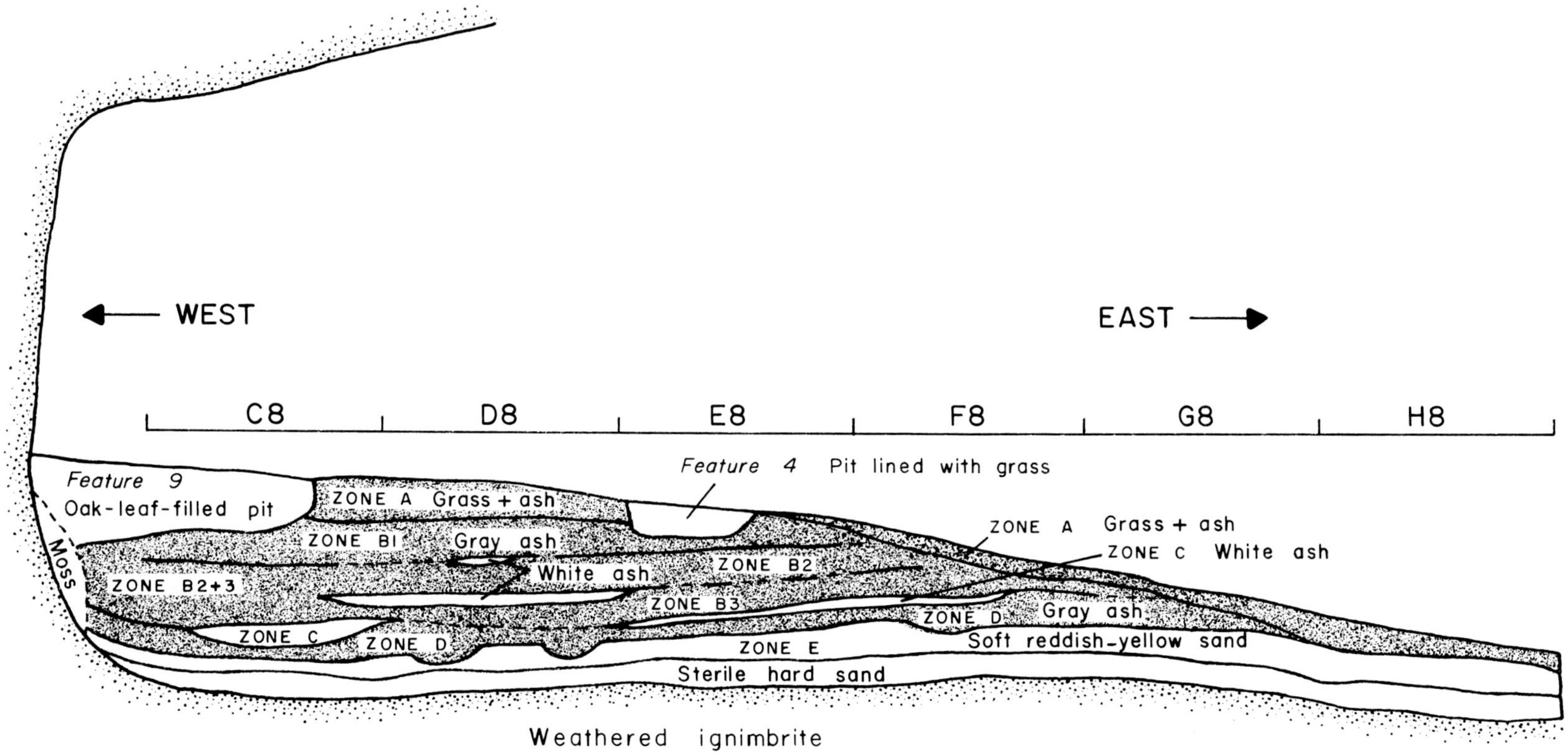

FIGURE 5.21. North profile of Squares C8–H8.

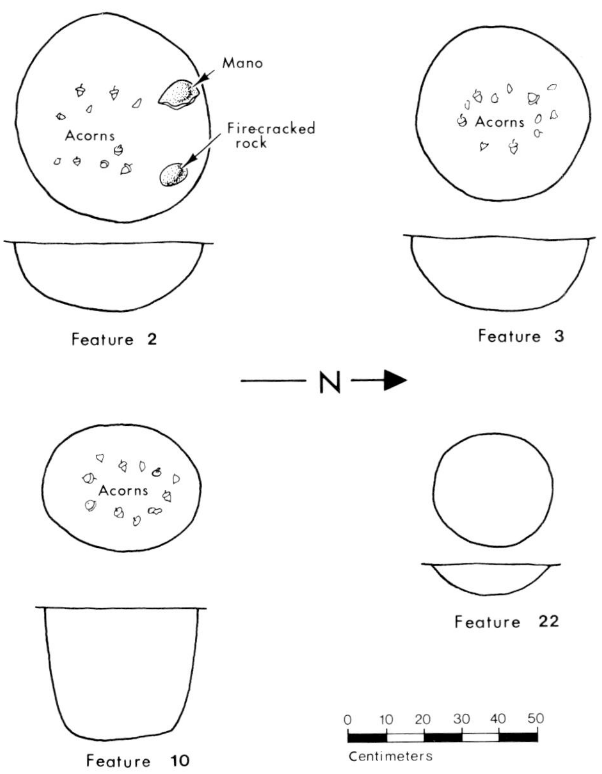

FIGURE 5.22. Features associated with Zone D.

FIGURE 5.23. Veined quartzite(?) one-hand mano found *in situ* in Square F9, lower Zone C. (Scale in centimeters; arrow points north.)

in this zone, but in Square D9 we found an unusual concentration of charcoal and burned twigs that may represent the remains of a surface campfire. Such a campfire, later kicked around underfoot until it was almost unrecognizable, would account for the large expanse of white ash.

Zone C may represent the single encampment of a small group of people made during the August–December period when mesquite pods, hackberry fruits, nanches, *susí* nuts, acorns, and piñon nuts were available. Once again, wild runner beans and early domesticated squashes were present, this time associated with charcoal yielding radiocarbon dates of 7280 and 7450 B.C. Despite its lack of features, we considered Zone C to be a true living floor: It consists entirely of material brought into the cave, including campfire ashes presumably spread around by being walked on (Fig. 5.34).

Superficially, Zone C resembled the white ash components of our more complex Zone B. However, it differed from the latter in that it appeared to constitute a single occupation that ran through almost the entire cave, never merging with any of the B zones and consistently underlying them, just as it consistently overlay Zone D. Furthermore, its radiocarbon dates confirm its stratigraphic position, earlier than B and more recent than D. To be sure, there were a few squares (such as E6 and F9) where Zone C was so thick that it could be divided into upper (white ash) and lower (compost) components, with each component therefore excavated separately. Never, however, did we find any stratigraphic evidence that the white ash component of Zone C was in any way related to Zone B.

Zone B

Zone B was so complicated that most of the cave had to be excavated before the relationships of all its subzones became clear. Indeed, it can only be understood by means of an idealized diagram (Fig. 5.35) and a short account of how we worked out its complexity.

Briefly, Zone B appears to be the remains of a very thick and intensive series of repeated occupations by a fairly small group during the August–December season. Each occupation had produced a floury layer of gray or white ash, and so little time had elapsed between occupations that the beds coalesced in places. For example, in most of the F squares, as well as in squares such as B9 and B10, there was only one stratum, which we originally called "Zone B." In other squares, such as D11, E9, and E10, this stratum thickened and bifurcated into a gray ash, Zone B1, and a white ash, Zone B2. Eventually, we found that our old Zone B and our new Zone B1 contained bones from the very same animals and chipped-stone flakes from the same block of raw material; at that point it became clear we were dealing with a single layer that could simply be called B1 everywhere.

In still other squares, such as C8, D10, and E6, Zone B2 thickened to a white-and-gray ash layer we now refer to as Zone B2+3. Finally, in squares such as C9, D9, and E5, Zone B2+3 bifurcated into a white ash, Zone B2, and a gray ash, Zone B3.

FIGURE 5.24. Excavation of Square E9: A, Squares E9 and E10 isolated on two sides by the removal of Squares E8 and F9–F11 (layers shown were considered the west profile of F9–F10 at the time the photograph was taken). B, Pablo García excavates the south half of Zone B1 in Square E9. Zone A has been removed from E9 but remains in E10. The south half of Zone B1 has been removed, and Pablo is reaching Feature 11. The thin gray layer below, still unexcavated, is Zone B2; the thick white ash below that is Zone C.

Eventually, it became evident we were dealing with an occupation that had begun in Squares C9–C11 and D8–D10 on the north and Squares E5–E6, D5–D6, and C6–C7 on the south. These two patches of Zone B3 may represent two separate work areas within the cave. Gradually, the debris expanded to include Squares F4–F5, D11, and E7–E10, with Zone B2 forming above B3. While Zones B2 and B3 remained clear and distinct in Squares C7, C9, C11, D5, D6, D9, and E5, they coalesced to form a Zone B2+3 in Squares C8, C10, D10, E6, F4, and F5. For convenience, we will sometimes refer to this complex of subzones B2, B2+3, and B3 as "Lower Zone B" (Figs. 5.36 and 5.37).

FIGURE 5.25. Feature 11, a hearth flanked by stones, found in Zone B1, Square E9 (Scale in centimeters; arrow points north.)

With the passage of time, the occupation expanded to Squares B8–B11, C12, D12, E11, F6–F8, and G4 and G5, with Zone B1 forming above B2. Around the periphery of the occupation, Zone B1 spilled over into areas that had no previous Zone B debris, becoming the most widespread of all preceramic components. We will sometimes refer to this extensive subzone as "Upper Zone B" (Fig. 5.38).

What we have, therefore, is a mounded, multilayered occupation that is thickest in the original areas of Zone B3 debris and thins out gradually as it radiates out from those areas. Its tracing was made difficult because the B zones were the layers most often perforated by intrusive features from Zones A and Super-A. With patience, however, they could be worked out. For example, we know that Zones B2 and B3 in the northwest part of the cave are the same B2 and B3 found in the southeast part because we could trace them continuously from Squares D8 and D9 through C8 and C7 to C6 and D6 and on to E5.

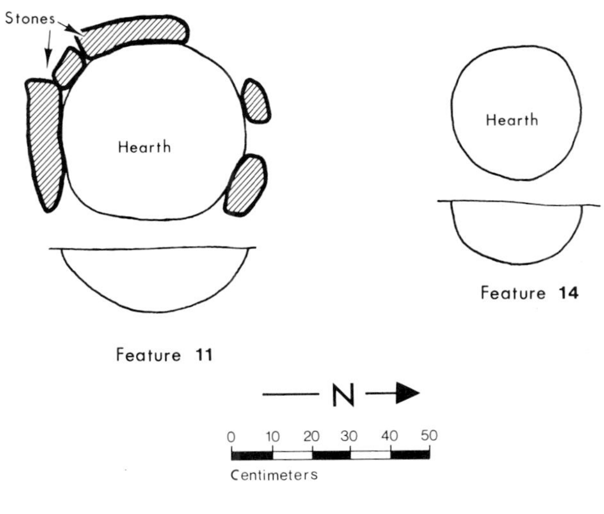

FIGURE 5.26. Features associated with Zone B1.

There could be little doubt that Zone B1 was a living floor, complete with two hearths (Features 11 and 14), one of them lined with rocks. But what about Lower Zone B? It appeared that Zone B3 might represent the debris lying on two separate activity areas and that B2 might represent a second layer of debris that covered them after no great interval of time. No hearth had survived as a feature in either subzone, but given the amount of ash and charcoal it was clear that fires had been set. Perhaps any campfires had been superficial, and their contents had simply been scattered during the laying down of the next subzone. Eventually we decided that B2 and B3 were small living floors and that B2+3 was simply an area where they had become so attenuated and stepped on that they could not be stratigraphically separated.

All the B zones were rich in plant remains and artifacts. Included were one-hand manos, projectile point fragments, scrapers, bifaces, cores, utilized flakes, wooden artifacts, cordage, maguey fiber knots, and fragments of netting and basketry. Acorns, piñon nuts, maguey quids, nanche and hackberry seeds, *susí* nuts, mesquite and *guaje* seeds, prickly pear fruits and stem sections were all present. Wild runner beans and domestic squashes were well preserved. The major animals were white-tailed deer, cottontail rabbit, and mud turtle. Charcoal from Zone B2 yielded radiocarbon dates of 6670 and 6910 B.C.

THE IMPLICATIONS OF ZONE B

The most obvious question to be answered about Zone B concerns the time intervals separating B3, B2, and B1. Do these represent three separate occupations made at different seasons of the same year? Are they three separate occupations made during the same season? Are they three separate occupations made during three different years? Finally, why is Zone B so much more complicated than the earlier zones?

Because of the plant remains from Zone B, we can eliminate some of these possibilities. If B3, B2, and B1 had been separate occupations from different seasons of the same year, we would expect the species of plants in each to be different; instead, like the plants from Zones C and D, all suggest the August–December period. If we were dealing with three separate occupations during this same autumn period, we would expect B3 to have more of the plants available in August (mesquite, hackberries), while B1 had more of the plants available in December (acorns, piñon nuts). Instead, although the sample sizes are different, all three subzones produced roughly the same plants. While they differ somewhat from Zones E, D, and C (for example, they tend to have more mesquite and fewer acorns), Zones B3, B2, and B1 are all very much alike.

Could we be dealing with three separate autumn occupations in rapid succession—for example, over a 3-year period?

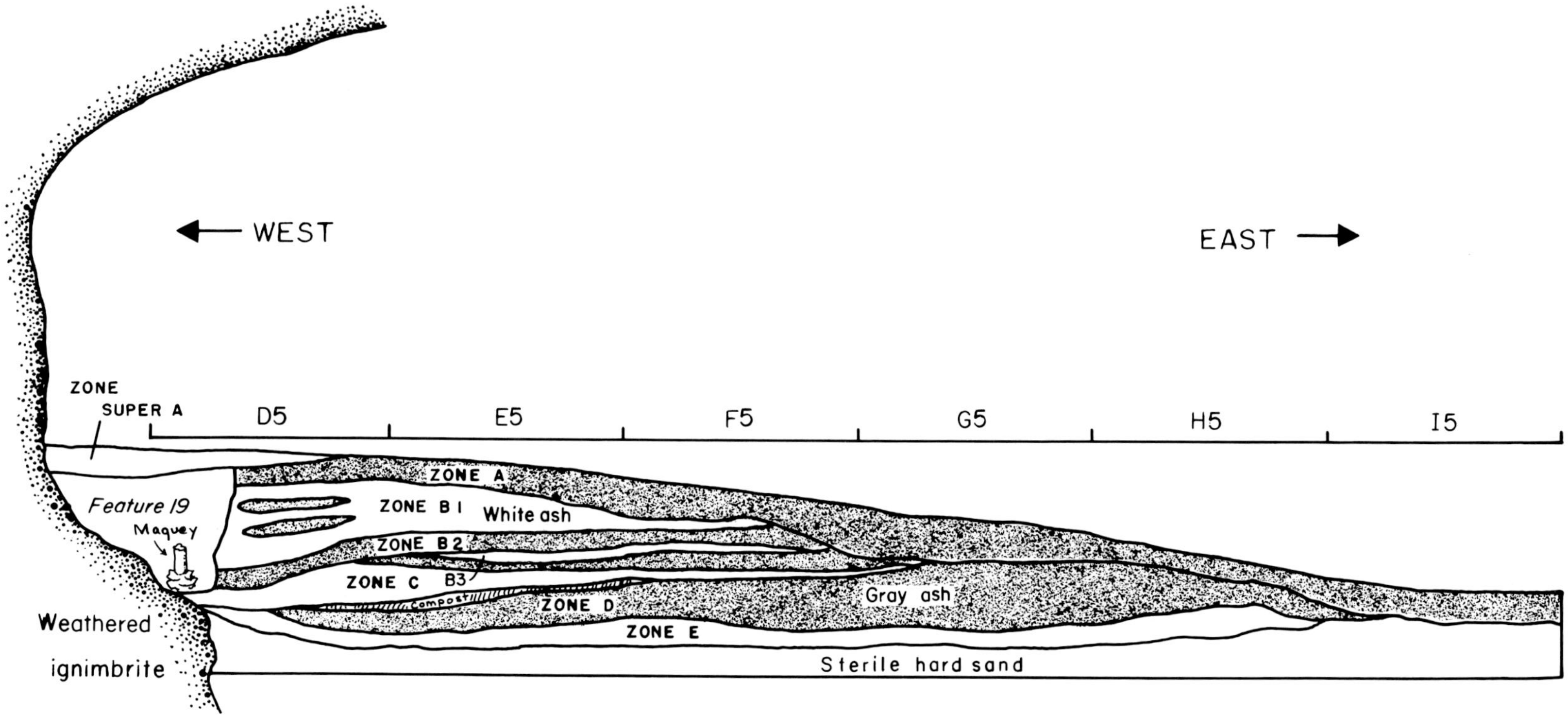

FIGURE 5.27. North profile of Squares D5–I5. (This section was actually drawn as the south profile of Squares D6–I6 but has been reversed so that it conforms to the format for Figures.

FIGURE 5.28. Don Juan Martínez completes the troweling of Zone E in Square F6. In this view from the east, Feature 3 is visible in bedrock behind his bucket. The removal of F6 left Square E6 (behind Don Juan) isolated on two sides for the excavation shown in Figs. 5.30 and 5.31.

We feel that this is the most likely possibility. However, we cannot totally rule out the possibility that Lower Zone B represents one two-stage occupation, while Upper Zone B represents a single-stage occupation made not long after (for example, the next year).

As for the question of complexity, we would like to consider the following possibility: Most of our preceramic zones may once have been more complex but have become compacted and "simplified" through the passage of time and the weight of the cave occupants walking on them. Zone B may indeed have been a longer occupation than the others, involving several consecutive years of ash deposition. However, since it was the final preceramic occupation, it may also have been subjected to less compaction and weathering.

Hints that some earlier occupations may once have been more complex can be found in Zones C and D. For example, in Squares E6 and F9 it was possible to isolate an "upper C" (white ash) and a "lower C" (brown compost). In other squares, these two components were still visible, but had been so compacted they could not be separated. Is it possible they may once have been as clearly distinguishable as B3 and B2? As for Zone D, it could be separated into an upper component (gray ash) and a lower component (burned oak leaves) in Square F8 near Feature 22. Even in this square, the two components were so compacted they were hard to separate. The implication of this, we feel, is that the apparent simplicity of the lower zones relative to Zone B may be partly (though not wholly) an artifact or their greater age and postdepositional history.

THE POSTCLASSIC LEVELS

The Postclassic levels at Guilá Naquitz, as we have indicated, will be reported in a future volume along with contemporary occupations from other caves. However, we do provide brief descriptions of them here inasmuch as their occupants were responsible for most of the disturbance of the preceramic strata.

FIGURE 5.29. Square E6 seen from the north on April 28, 1966. Square F6 and E7 have already been removed, so the stratigraphy of E6 can be seen on two sides before excavation.

Zone A

Zone A was a thick layer of grass and soft gray ash that overlay the lower deposits in all squares. It was by far the most extensive living floor in the cave and had the best preservation, presumably because it was much younger than Zones B–E. Associated ceramics (which were not abundant) indicated a placement in the Monte Albán IIIb–IV period, and the radiocarbon dates were A.D. 620–740. There were effigy "claw" vessels, *sahumadores* (*incensarios*), and pottery like that from the nearby sites of Mitla and Lambityeco. Because some Zone A features intrude into other Zone A features, the deposits may reflect fairly steady use of the cave as a repository for collected wild plants and harvested domestic plants during several seasons and/or over a period of several years by a small- to medium-sized group of people, possibly coming from the rural sustaining area for Mitla.

The people who used the cave during Zone A times made a number of permanent facilities for themselves. A series of very large storage pits were dug into the older deposits near the back wall of the shelter (Features 1a, 1b, 9, 15, and 13). Other storage pits were scattered throughout the site, about a dozen in all. Later, many of these pits had been used for

FIGURE 5.30. Chris Moser excavates Square E6 (view from the east). The gray ash layer that has been partly removed is Zone B2 + 3. The layer of thick white ash resting on dark brown compost (at the level of the sugar scoop) is Zone C. This was one of the squares where upper and lower Zone C could be excavated separately.

FIGURE 5.31. Another view of Square E6, this time from the northeast. Moser is removing the last of Zone B2 + 3, a gray ash layer. Below is a pure white ash layer, upper Zone C, resting on a thinner layer of dark compost, lower Zone C. Still lower down is Zone D, another gray ash layer; below this is Zone E, a layer of soft sand that appears light in the photograph.

trash. One of them (Feature 13) contained what may have been the debris from a "new maize" ceremony, including abundant harvested maize, "pulque cups," effigy vessels and claw vessels, *incensarios*, and a possible ball of copal. Harvested wild products were abundant in the zone and included acorns, *susí*, hackberry, maguey, prickly pear, mesquite, *guaje*, and wild onion. Piñon does not seem to be as well represented as it was in the preceramic layers, and may have been in the process of disappearing from the area (it no longer is present near the cave). In addition, there were many domestic plants that presumably had been grown not too far away: maize, large black twining beans, small black bush beans, *Canavalia*, avocados, pumpkins, bottle gourds, chile peppers, cotton, and white zapotes (*Casimiroa*). The bush bean plants had been pulled up by the roots, just as is the custom today at Mitla. The cotton is interesting because (1) it confirms a statement in the *Relación de Tlacolula y Mitla* (Canseco 1580) that indicates that Mitla grew its own cotton before the Conquest (it no longer does so today) and (2) one boll contained a boll weevil, the oldest known from an archaeological context (Warner and Smith 1968). Mangelsdorf's preliminary study of the maize indicates it is a highly drought-adapted form of *chapalote* (Mangelsdorf n.d.).

Zone Super-A

Zone Super-A was not a living floor; rather, it was a set of four features left by a small group (perhaps no more than two people) who occupied the cave for too short a time to lay down a living floor. These features either rested on or penetrated into Zone A. Their associated ceramics belonged to the Monte Albán V period, and one feature was radiocarbon-dated to A.D. 1270.

These features included a large maguey-roasting pit (Feature 7) near the cave mouth, a storage pit near the rear wall containing an unused maguey heart (Feature 19), and two beds of grass and oak leaves (Features 6 and 16) that presumably had been used by two persons who occupied the cave while waiting for the maguey-roasting process to be completed (this usually takes 24–72 hours). It is possible that at this period the cave was used merely as a station for roasting one of the local maguey species (*Agave potatorum*) and that this activity was undertaken by a couple of people who came out from the city of Mitla and later returned to Mitla with the baked maguey. Other caves in the same cliff also have maguey-roasting pits, most dating to the period A.D. 1250–1500.

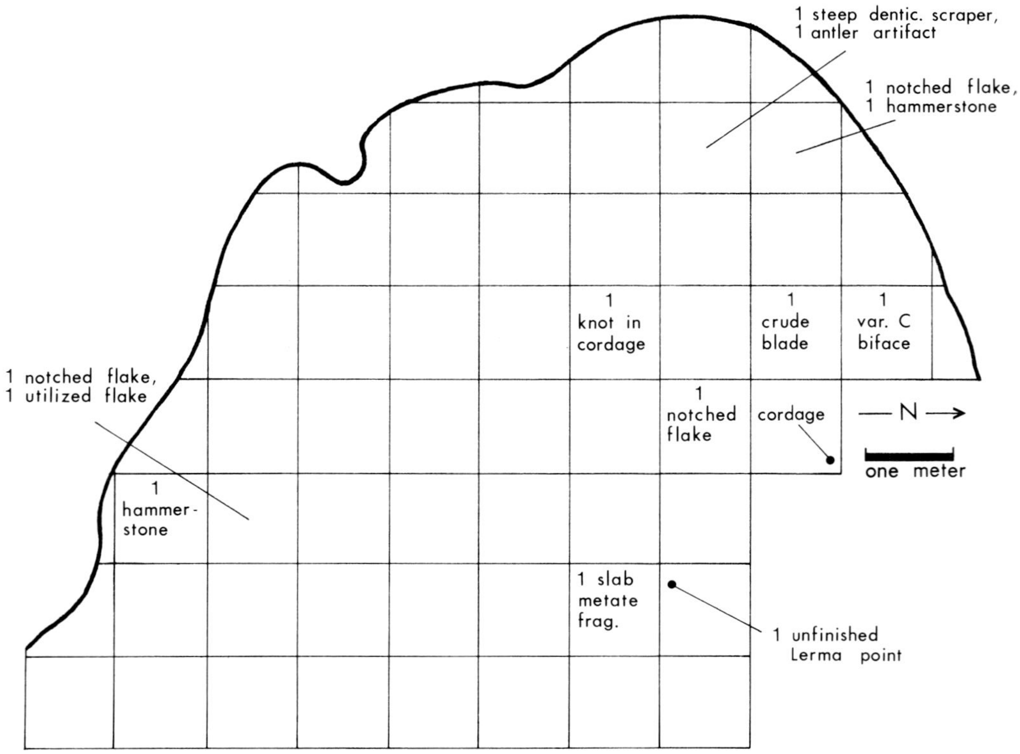

FIGURE 5.32. Zone E, showing location of some significant artifacts and features.

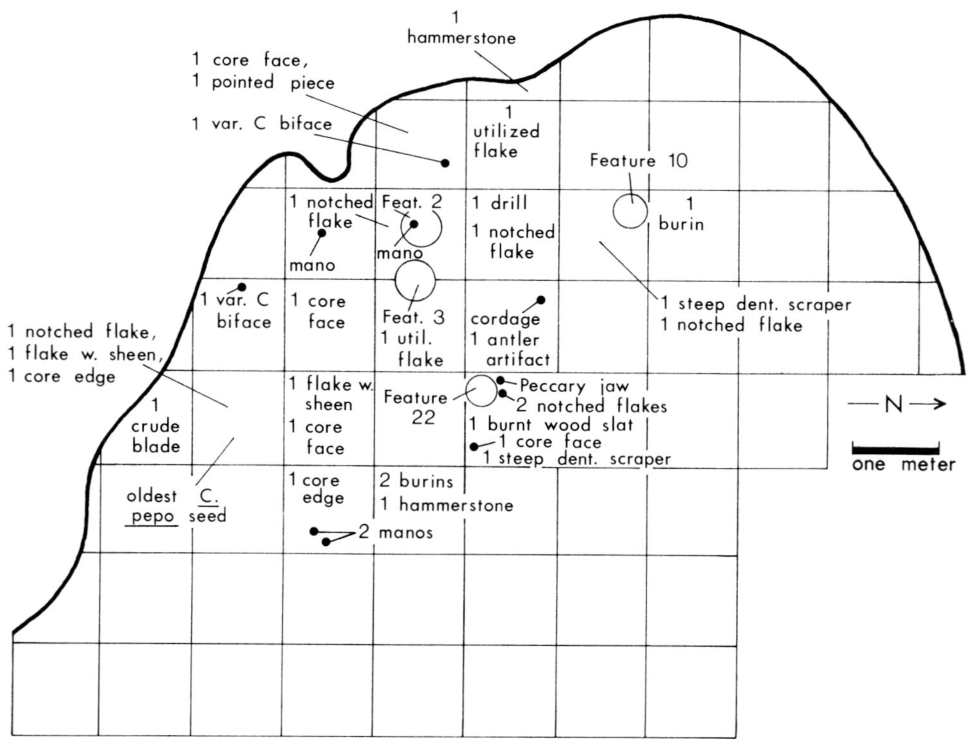

FIGURE 5.33. Zone D, showing location of some significant artifacts and features.

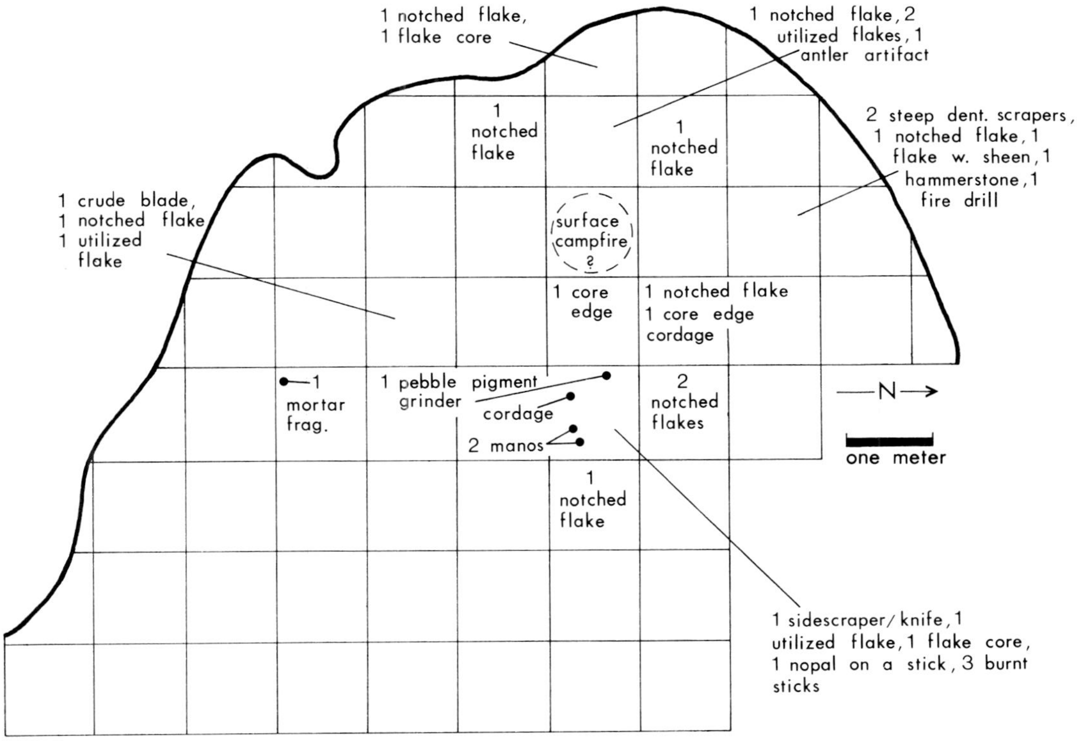

FIGURE 5.34. Zone C, showing location of some significant artifacts.

Artifacts from Zone Super-A included sherds of thin black ollas and dishes of light gray clay with a dark-fired rim, which are identical to Monte Albán V pottery found at the site of Mitla itself. In addition, there were a number of bifaces and heavy denticulate scraper planes probably used in maguey shredding or processing. Many of these scraper planes are like the ones that Holmes (1897) found at the ruins of Mitla and near the Mitla quarries. They superficially resemble preceramic scraper planes, but our evidence indicates they are in fact Postclassic, something already pointed out by Williams and Heizer (1965).

Because of this Postclassic occupation, the preceramic levels of Guilá Naquitz had been perforated by more than a dozen intrusive features. This had two major effects. First, some preceramic tools, including Archaic projectile points, were dug up and redeposited in Postclassic features. Second, some areas of preceramic occupation were so badly disturbed that we decided not to consider any of the material in the immediate vicinity to be reliably dated. In addition, some rodent burrows had introduced Postclassic material to preceramic depths (see Chapter 16). Such burrows were generally easy to recognize and isolate. Nevertheless, their presence made us reluctant to accept any domestic plants from preceramic levels as reliably dated unless they came from a safe distance from any burrow or Postclassic feature. Chile peppers, avocados, and *Phaseolus vulgaris* beans were plants rejected as possible preceramic domesticates on such grounds. As a matter of policy, any plant we list as being present in Zones B–E of Guilá Naquitz comes from a context in which we are confident there is no intrusion from Postclassic levels.

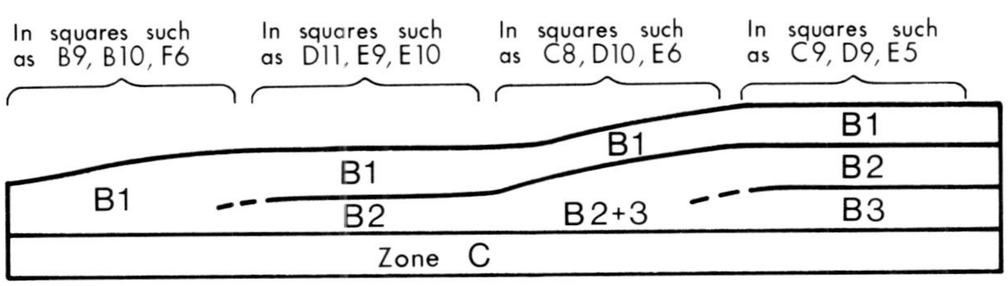

FIGURE 5.35. Idealized diagram of the internal stratigraphy of Zone B.

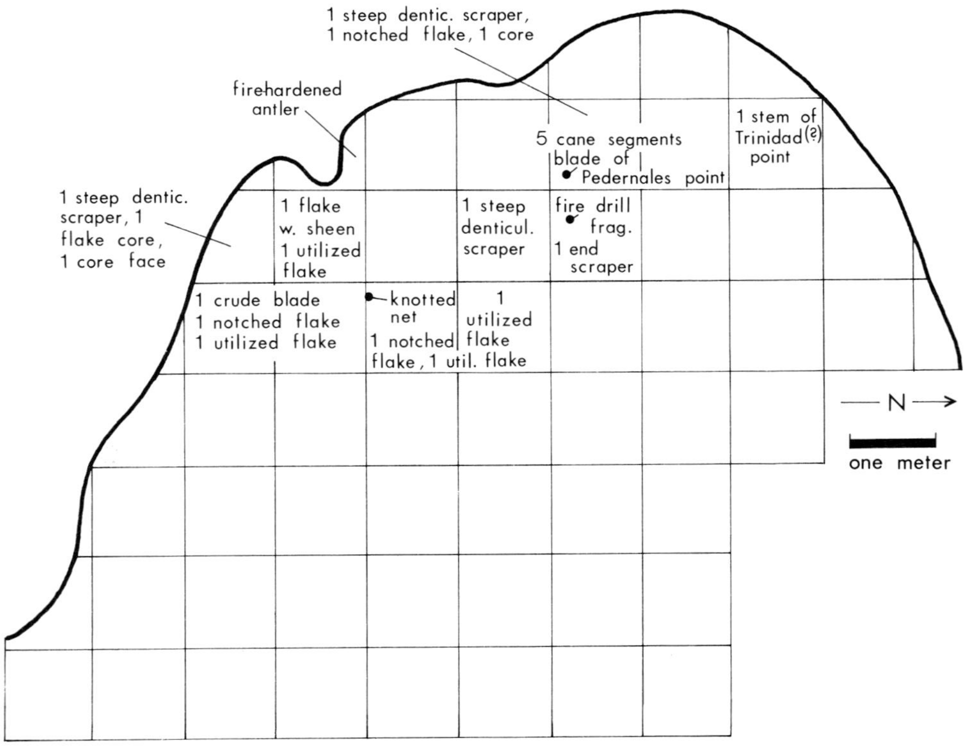

FIGURE 5.36. Zone B3, showing location of some significant artifacts.

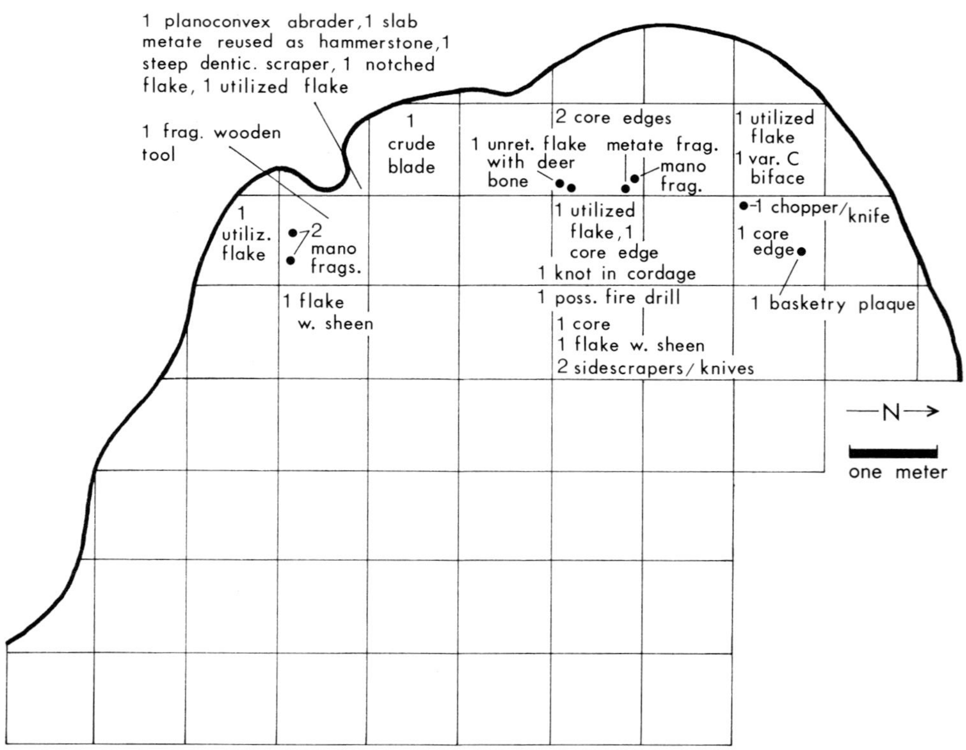

FIGURE 5.37. Zone B2, showing location of some significant artifacts.

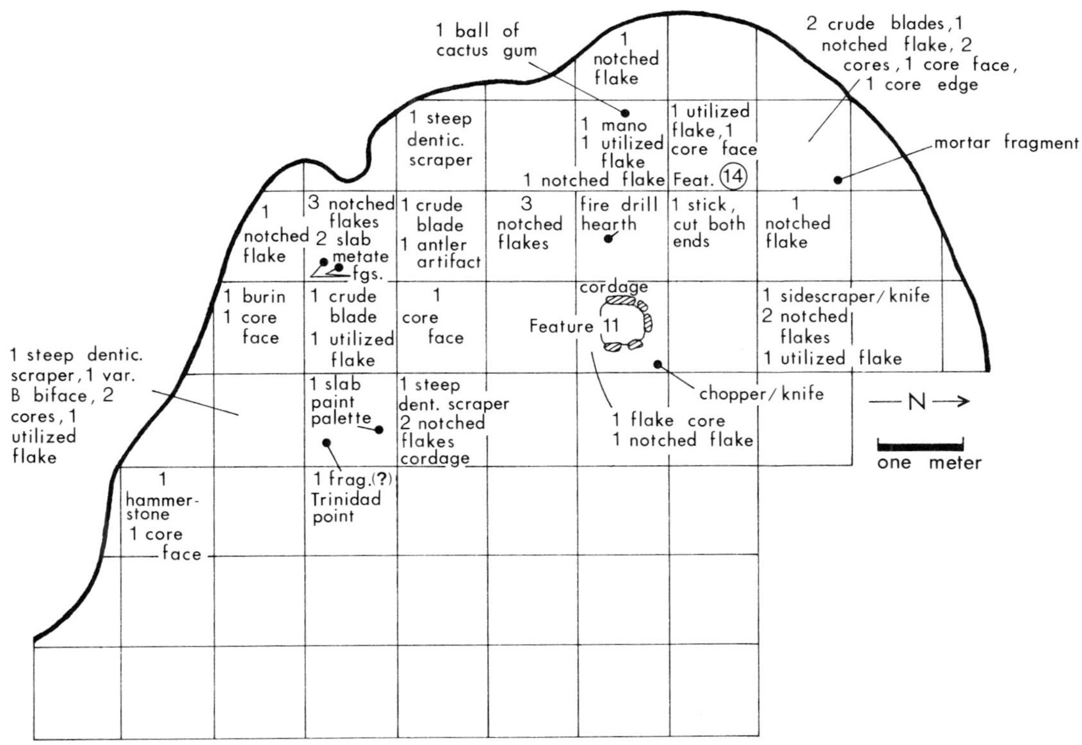

FIGURE 5.38. Zone B1, showing location of some significant artifacts and features.

EPHEMERAL FORMATIVE OCCUPATIONS

Finally, we should mention some evidence for occasional visits to Guilá Naquitz during the Formative period. These visits were evidently so ephemeral that they produced no living floors, no activity areas, and no features. The evidence consists of a handful of scattered Formative potsherds that turned up in one of two contexts: (1) redeposited in Zone A or one of its features or (2) right at the disconformity between Zones A and B1. The periods represented by these sherds were the San José phase (1150–850 B.C.), Monte Albán I (500–200 B.C.), and Monte Albán II (100 B.C.–A.D. 100).

The most likely explanation for these sherds is that during the long hiatus between the abandonment of Zone B1 and the deposition of Zone A, Guilá Naquitz was occasionally visited by people who did not stay long enough to produce a living floor. Assuming such people brought along their water in pottery vessels, some of these vessels may occasionally have been broken or abandoned at the cave. By the time the Zone A occupants arrived in A.D. 700, there may even have been a light scatter of Formative sherds on the surface of the cave. The same situation applies to Cueva Blanca, where ephemeral visits went back at least to the Tierras Largas phase (1400–1150 B.C.). Such rare Formative ceramics, like the Postclassic materials from these caves, will be discussed in a future volume.

6

Chipped-Stone Tools

Frank Hole

INTRODUCTION

This is the first of a series of studies on the chipped-stone tools of preceramic Oaxaca. It deals only with the 1716 pieces of chipped stone from Zones B–E at Guilá Naquitz, but it deals with them in the context of a typology worked out for Cueva Blanca, Martínez Rockshelter, Gheo-Shih, and other preceramic sites in the eastern arm of the Valley of Oaxaca. Guilá Naquitz, a small microband camp, did not contain the full range of tools known for the preceramic; no one site of that period did. Therefore, any overall synthesis of the preceramic chipped-stone industry will have to be deferred until the other sites mentioned above have been reported on as well.

One advantage of dealing with each site separately is that it allows us to focus on seasonality and function. We know that Guilá Naquitz was occupied mainly between the end of the mesquite harvest season (late August–early September) and the end of the acorn harvest season (December–early January). We know from the organic debris that wild plant collecting was a relatively major activity and hunting a less dominant activity when compared with other sites. We have reason to believe that the group occupying the cave was no more than 4–5 persons, possibly a single family. Apart from whatever stylistic–chronological information the tools contain, therefore, they should also reflect the activities of a single family during the autumn season. Moreover, in view of the activity areas Reynolds has reconstructed for each living floor in Chapter 28, various groups of tools may even represent men's butchering activities or women's plant processing activities. Frankly, the possibility of one day being able to identify tool kits based on seasonality and/or division of labor seems more interesting than simply using the tool types to establish a chronology of preceramic phases. Once future studies of nearby sites dating to different seasons have been carried out, we will be in a better position to specify which intersite differences are due to chronology and which are due to site function and seasonality.

TECHNOLOGY AND RAW MATERIAL

Coming to the Mexican preceramic from a background in Old World prehistory, I was impressed first by the crude appearance of the artifacts. In the case of Oaxaca, this impression is heightened by the fact that most tools are made of coarse raw material, mostly secondarily silicified ignimbrite (volcanic tuff). This is the most abundantly available raw material in the Mitla area, occurring in several forms. First, veins of silicified material occur in the local ignimbrite bedrock, and since they weather more slowly than the latter, they may project out in tabular form from the bedrock so that chunks can easily be broken off. Second, as local streams and barrancas erode headward into the ignimbrite mountains, their beds become paved with cobbles and pebbles. The softer ignimbrite disintegrates, while the silicified pebbles survive to be picked up and used as cores.

There is also at least one low ignimbrite mesa in the area that seems to have been used as a quarry for silicified material. This is the so-called Mitla Fortress, identified originally by Holmes (1897) and later by Williams and Heizer (1965) as a source of stone for artifacts. A still more recent survey for chert and silicified ignimbrite sources in the Valley of Oaxaca by Whalen (Chapter 7, this volume) also pinpoints the Fortress as a major Precolumbian quarry. However, the occupants of Guilá Naquitz could have picked up all the silicified ignimbrite they wanted within 2 km of the cave, providing they were not fussy about the quality.

If they did want higher-quality raw material, the occupants of Guilá Naquitz had to go farther afield. Perhaps the finest single chert or chalcedony source in the area is a limestone cliff at Rancho Matadamas in the Etla region of the valley, some 50 km to the northwest (Whalen, Chapter 7). Smaller sources of chert are available only 25 km away (e.g., at Rojas de Cuauhtemoc), but they are of a distinctly lower quality than the Matadamas source. The occupants of Guilá Naquitz apparently used chert both from the inferior sources 25 km away and from the superior source 50 km away but not nearly as frequently as they used the local silicified ignimbrite. In general, the waxy Matadamas chert or chalcedony was preferred for well-made tools such as projectile points, while the vast majority of such tools as notched flakes and utilized flakes were on silicified ignimbrite. In some cases, the cave occupants seem to have found sources of silicified ignimbrite so fine grained that it fractures almost as well as flint and is hard to distinguish from the latter.

Another factor that contributes to the crude appearance of the Oaxaca tools is the coarseness of the chipping technique. The percussion chipping technology is about on a par with that used during the later Acheulean in Old World archaeology. In those instances where there was bifacial chipping, it seems to have served as much to correct for inadequate flakes as it did to produce different sorts of tools. In other words, the local flint knappers seem to have been unable to produce good large flakes of proper shape and thickness without secondary bifacial retouch. One is also impressed by the lack of standardization in the preceramic tools. This goes along with the crude technique generally and makes precise typology difficult. For this reason I have preferred to "lump" tools rather than to "split" them, preferring to wait for more data before establishing too many subdivisions.

The foregoing remarks are descriptive of the bulk of the chipped-stone tools. The only major exceptions are the projectile points and bifaces, which show a care in manufacture and a conformity to type that are not found with the other tools. Taking this into account, however, it is still evident that many "types" of points, encompassing a wide range of variation, were in use at any one time.

One of my concerns has been to compare the Oaxaca preceramic tools with those from the nearest group of comparable sites, the Tehuacán caves studied by MacNeish *et al.* (1967).[1] However, in view of the generally unstandardized nature of the tools, plus the fact that I have not had the opportunity to examine the Tehuacán material firsthand, it seems unwise to push these comparisons too far. When different workers sort tools, they often arrive at different types, and it is not easy to standardize the results if we have only published photographs rather than the actual tools to work from. Moreover, when the tools are so crude, it is inevitable that the nature of the raw material itself will determine some of the types. For example, the difference between "flat" and "domed" scrapers might lie in the availability of certain types of raw material and have nothing whatsoever to do with cultural, chronological, or functional distinctions.

In Table 6.1, I have given the rough equivalents between my chipped-stone tool categories for Oaxaca and those of MacNeish *et al.* for Tehuacán. It should be stressed that these equivalents are only tentative because they are based solely on published descriptions and photos of the Tehuacán material. The table also makes it clear where MacNeish *et al.* have "split," while I have "lumped" for the reasons given above.

[1]Perhaps the second nearest comparable group of sites are the caves of the Valsequillo Basin near Puebla. One of these, Cueva del Texcal, has been published since I did my study of Guilá Naquitz (García Moll 1977). There are many similarities between García Moll's Texcal I phase (7000–5000 B.C.) and MacNeish's El Riego phase.

TABLE 6.1
Approximate Equivalence between Preceramic Tool Categories in Oaxaca and Tehuacán[a]

Oaxaca	Tehuacán
Hammerstones	Pebble hammerstones; spherical battered pebbles
Flake cores	Crude conical and hemiconical nuclei
Core faces } Core edges }	Core fragments
Choppers/Knives	Flake choppers
Steep denticulate scrapers	Scraper planes (various types); crude discoidal scrapers; ovoid planoconvex scrapers
Sidescrapers/knives	Thick flakes, 1 and 2 edges retouched
End scrapers	End scrapers (various types)
Burins	Burins
Drills	No stylistic equivalent
Pointed pieces	Flake gravers(?)
Crude blades, plain	Crude blades, unprepared striking platforms
Notched flakes	Some utilized flakes and some "spokeshave-like tools"
Flakes with sheen	No equivalent
Utilized flakes	Thin flakes, 1 and 2 edges utilized; thick flakes, 1 and 2 edges utilized
Projectile points	Projectile points
Bifaces, Variety B	Thin, crude, ovoid bifaces
Bifaces, Variety C	Large bifacial disks(?)
Tanged pick	No equivalent
Debitage	Chipping debris

[a]The Oaxaca categories are from the present text and the Tehuacán categories are from MacNeish *et al.* 1967.

TYPOLOGY

The tool types distinguished here represent a preliminary attempt to order a large body of new material. The method of analysis is similar to that employed in the author's study of chipped-stone tools from the Near East (see Hole *et al.* 1969) and is therefore somewhat at variance with the techniques and nomenclature used by American archaeologists.

Basically, I attempted to determine the purpose of the chipping and hence focused on the nature of the retouch and its position on the tool. In such a typology, the size and form of the piece is given less weight (e.g., whether a flake is thick or thin is less important than whether it has retouch on one edge and the retouch is steep).

Invariably in a group of specimens so crudely worked—where there is a minimum of standardization of finished pieces, where one tool may have been used for several tasks, and where several types are made on the same kind of flake—it is sometimes hard to decide into which of two related types a given tool should be fitted. To solve this problem, I have weighted the degree of modification, and when one piece has chipping in two different locations (which would put the piece into two different types), I have placed it in the type having the "greatest weight." For example, if a piece has steep, denticulate retouch on one edge and a notch on another, I have recorded it as a "steep denticulate scraper" rather than as a "notched flake." However, one should bear in mind that many Oaxaca preceramic types grade from one to another. Notched flakes can be regarded as one point in a continuum that has steep denticulate scrapers at one of its extremes.

One uses types to help compare material from different sites, to help establish relative chronologies, and to help determine what the people who made them were doing. These three purposes are not inherently mutually exclusive, but types created to help answer one question may not help with another. In the case of the Guilá Naquitz artifacts, I have stressed what I perceive to be the functional implications of the various types and the relations in this regard of one type to another. My suggestions are based on a set of fairly widespread assumptions about the relationship between the form and retouch of a tool and the tasks for which it is intended. The plant and animal remains from Guilá Naquitz provide independent evidence that many of those tasks were carried out at the cave. The presence of certain tools in the same activity area with certain plants or animals provides further evidence that the tools were used to perform those tasks, but it is still no more than circumstantial evidence. We hope that in the future, as we examine nearby preceramic sites with different tool assemblages, our expanded sample of activity areas will help strengthen the bridging arguments between these types and their functions.

THE BASICS OF THE GUILA NAQUITZ INDUSTRY

The chipped-stone industry from Guilá Naquitz B–E is basically a flake industry. It began with the striking of flakes from polyhedral cores with unprepared striking platforms; a total of 11 cores and 19 core fragments were found. At least some of the 8 hammerstones found in the cave may have been used for the detachment of flakes and for further percussion flaking. In a clear example of recycling, some discarded cores were later reused as hammerstones.

Most flakes struck from these cores were nondescript, needing to be retouched to correct for their asymmetry. A few came off fortuitously in the form of crude blades (10 were found), but these were exceptions. Many flakes were probably used in their naturally sharp form, and if this use continued for some time, various patterns of wear might develop. Some 21 flakes showed retouch from use, while 6 had developed sheen along an edge.

Some flakes were deliberately retouched: 40 of them showed a notch or notches on at least one edge. More careful retouch had turned at least 4 flakes into sidescrapers/knives. Steeper retouch was used to produce other types of tools: choppers/knives (2), steep denticulate scrapers (13), and an end scraper (1). At least 1 flake had been retouched at one end to form a drill.

Four flakes, struck with burin blows, had been transformed into tools for slotting or grooving. Still another piece had been given a small, sharp point.

Six of the larger flakes from Guilá Naquitz had been percussion flaked on both surfaces, producing a series of reasonably attractive and well-made ovoid bifacial tools. By far the best-made tools, however, were a series of flakes carefully bifacially chipped and pressure retouched into points for atlatl darts. Four of these were found in Zones B–E.

It should be clear from the above that many of the tool types at Guilá Naquitz represent stages in the progression of a flake from unused, to used but not deliberately retouched, to casually retouched, to extensively retouched. Figure 6.1 is an attempt to represent this situation in a diagrammatic form, showing the stages of manufacture and the interrelationships of various chipped-stone categories. We present those categories in roughly the sequence given above, beginning with basic flake production and ending with the best-made bifacial tools.

POUNDING AND FLAKING TOOLS

Hammerstones

Figs. 6.2 and 6.3.

Number of specimens: 8.

Dimensions: diameter, estimated at 4–7 cm (when unbroken).

Description: These are chunks of silicified ignimbrite, chert, or flint with evidence of battering from use as a hammer. Several of the specimens appear to be flake cores that were reused in this manner. Many hammerstones were discarded in the early stages of battering, while some (including the two illustrated in Fig. 6.2) evidently split in the process. When subjected to more prolonged pounding, the specimens were virtually worn down to the shape of a round ball. Some hammerstones could have been used to pound up

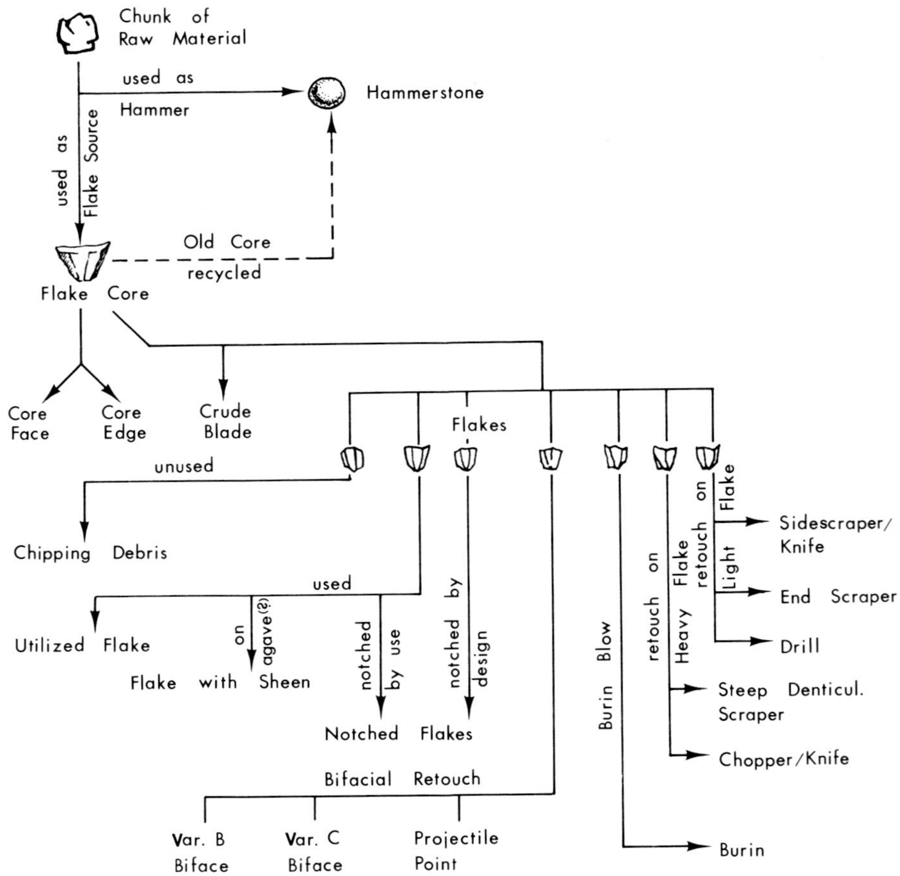

FIGURE 6.1. Diagram showing relationships among chipped-stone tool types from preceramic Guilá Naquitz.

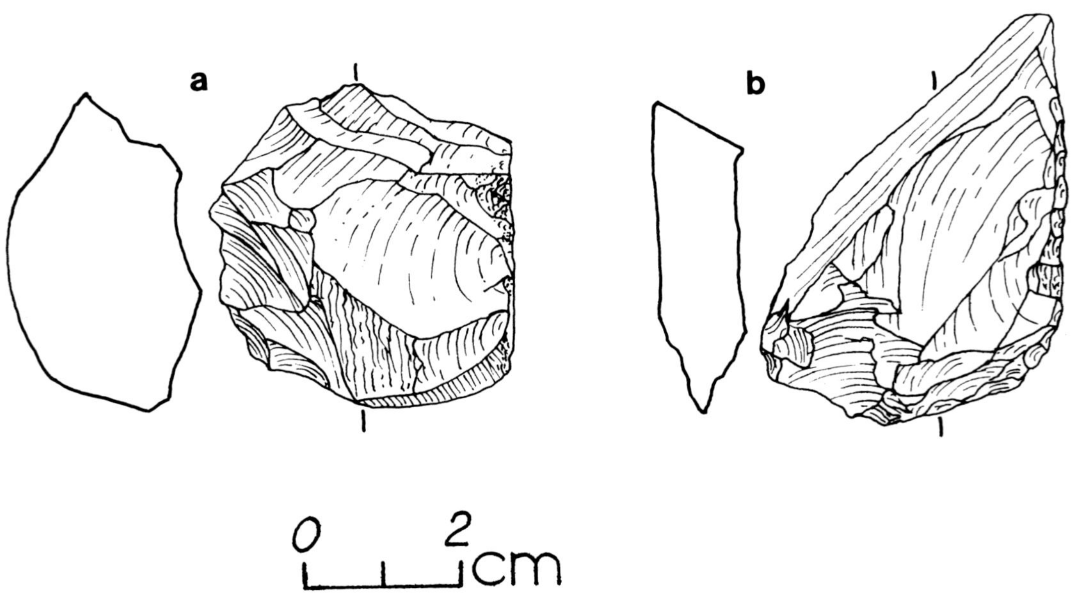

FIGURE 6.2. Fragments of hammerstones made on battered flake cores: a, Zone D, Square B9; b, Zone C, Square D11.

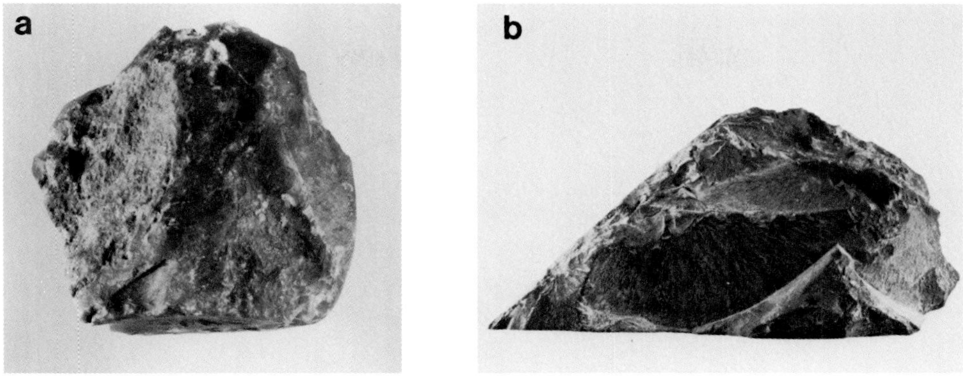

FIGURE 6.3. Fragments of hammerstones made on silicified tuff flake cores: a, Zone D, Square B9; and b, Zone C, Square D11 (maximum width, 62 mm).

acorns or mesquite pods in a mortar, while the battering on others suggests they were used for percussion flaking or for striking flakes off cores.

Rough equivalents in the Tehuacán sequence include "pebble hammerstones" (MacNeish *et al.* 1967:126–127) and "spherical battered pebbles" (1967:85). Some of the latter do appear to be flake cores reused as hammers (1967:Fig. 71) and would thus be similar to many of our specimens.

CORES AND CORE FRAGMENTS

Prepared cores are rare in the Oaxaca preceramic; most flakes were struck from unprepared nodules of stone. One type of core seen in later preceramic phases, the discoidal core, was not present in Levels B–E at Guilá Naquitz. It will be discussed in a later report.

Flake Cores

Figs. 6.4–6.6.
Number of specimens: 11.
Dimensions: height, 3.5–8 cm; diameter of striking platform, 4–10 cm.
Description: These are chunks of silicified ignimbrite, chert, or flint that have one or more platforms from which flakes of a usable size were struck. It will be clear from the illustrations that there is often a superficial resemblance between some flake cores and some steep denticulate scrapers. In such cases, the criterion used to distinguish the two categories was flake size: The flakes struck from these cores were large enough to be used as typical Naquitz phase tools. Within the Tehuacán sequence, it is MacNeish *et al.'s* "crude conical and hemiconical nuclei" with unprepared platforms (1967:26, Fig. 11) that most resemble our cores.

Core Faces

Fig. 6.7a, b.
Number of specimens: 10.
Description: These are fragments from the sides of flake cores removed by a blow directed down from the platform. The blow occurred far enough back from the edge of the platform to remove the whole face rather than just one flake.

Core Edges

Fig. 6.7c, d.
Number of specimens: 9.
Description: A *core edge* is defined as the edge of a striking platform, removed (in almost all cases) by a blow directed from the side of the core rather than from above.

CHOPPERS, SCRAPERS, AND KNIVES

Choppers/Knives

Figs. 6.8–6.10.
Number of specimens: 2.
Dimensions: length, 8–11 cm; width, 5.5–8.5 cm; thickness, 1–3 cm.
Description: Most of these are large, heavy flakes, chunks, or nodules of silicified ignimbrite or chert that have been chipped to give a sharp cutting edge or edges. The retouch is usually coarse and bifacial, giving the edge a sinuous line; such pieces grade into Variety C bifaces, but choppers/knives are larger and cruder than the former. Moreover, some choppers/knives from other Oaxacan preceramic sites are not bifacially chipped. In general, the edges of choppers/knives are more battered and less symmetrical than those of Variety C bifaces, and the retouch did not shape the pieces as it did in the case of the bifaces.

As their name implies, these tools could have been used for both cutting and chopping. Some may have been used for the heavy cutting and chopping of plants, but their frequency and distribution at Cueva Blanca (unpublished data) suggest they might often have been used for the butchering of deer.

In the Tehuacán area, the tool type most resembling our choppers/knives is MacNeish *et al.'s* "flake choppers," which

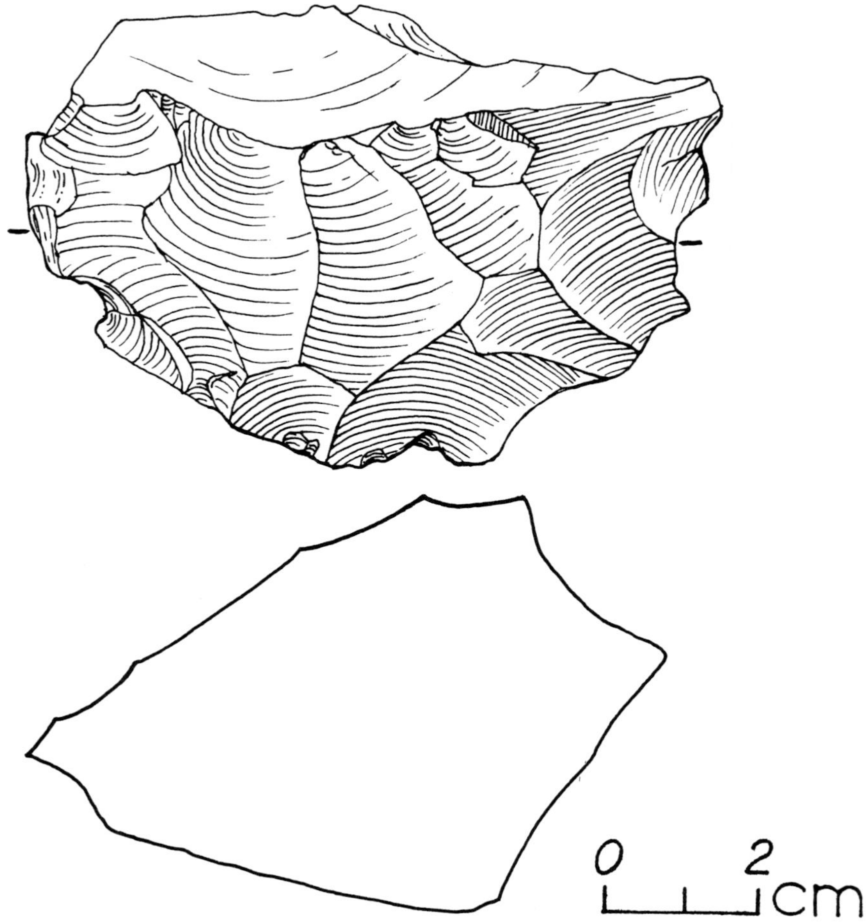

FIGURE 6.4. Flake core, Zone C, Square B9.

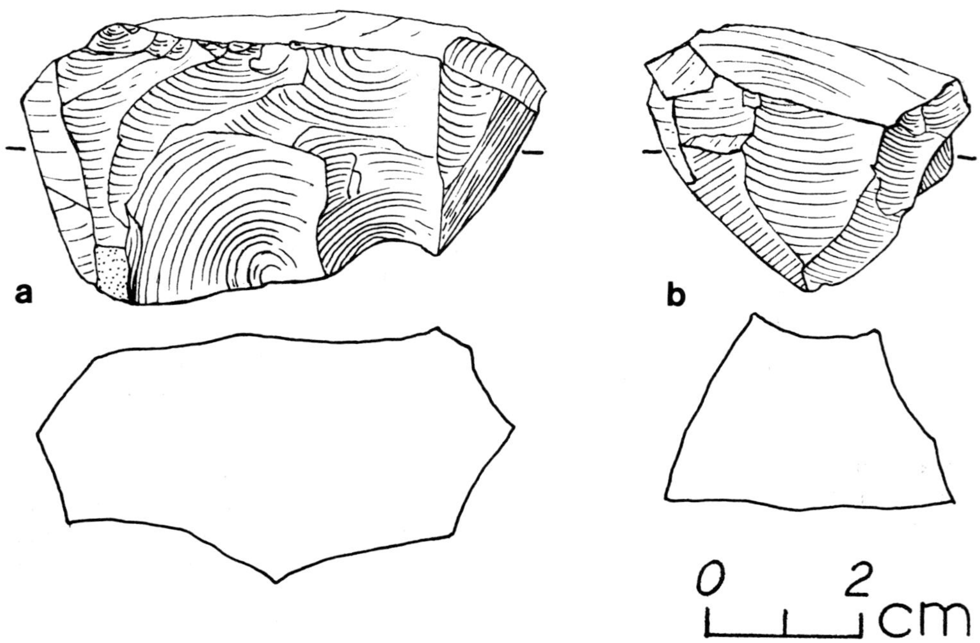

FIGURE 6.5. Flake cores: a, Zone C, Square F9; and b, Zone B1, Square C11.

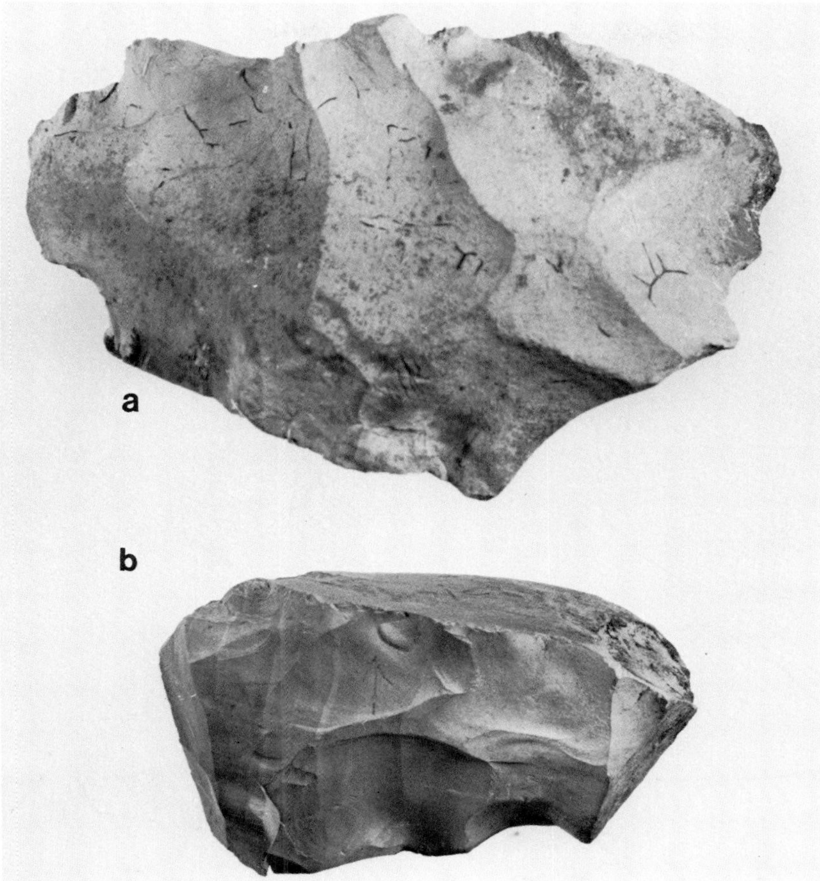

FIGURE 6.6. Flake cores, silicified tuff: a, Zone C, Square B9 (maximum length, 100 mm); and b, Zone C, Square F9.

are "most popular during the Ajuereado phase" (1967:85, and Fig. 70, top).

Steep Denticulate Scrapers

Figs. 6.11 and 6.12.

Number of specimens: 13.

Dimensions: length, 4–9 cm; width, 4–9 cm; thickness, 1–5 cm.

Description: These heavy flakes or chunks of silicified ignimbrite or chert have one or more edges chipped by coarse percussion retouch that is relatively steep to the plane of the chipping surface. The retouched edges are denticulated or notched rather than smoothed by the chipping. The flakes removed during retouch were generally no more than 1 cm long and therefore could not have been used as typical Naquitz phase tools. Had the removed flakes been significantly larger, one might question whether some of these scrapers might not in fact be cores. As it is, cores that subsequently received steep retouch would be classified in this type.

The larger examples of this type probably most closely resemble the "scraper planes" of MacNeish *et al.* (1967:36–39, and Figs. 17 and 18). Most of the Oaxaca examples, however, are made on flakes that are probably too small to have been employed conveniently as planes. These smaller specimens are probably more equivalent to the "crude discoidal scrapers" and "ovoid plano-convex scrapers" from Tehuacán (1967:40–43, and Figs. 21 and 24). Some of our steep denticulate scrapers were probably used in the processing of plant material or the shaping of wood; some of the larger specimens may indeed have been used in the shredding or pulping of fibrous plants such as maguey and nopales.

The steep denticulate scrapers from Guilá Naquitz are similar in size to those from other caves in the area (Cueva Blanca, Martínez Rockshelter), averaging about 110 g in weight. In this respect, the cave specimens differ markedly from the steep denticulate scrapers found at the open-air preceramic site of Gheo-Shih, where such scrapers weighed only about half as much on the average (Table 6.2). A higher proportion of the steep denticulate scrapers from Guilá Naquitz and the other caves are large enough to have been used comfortably as a hand-held scraper plane, and the average weights reflect this. Although the figures are subject to errors in measurement and problems caused by the necessity of dealing with broken as well as whole specimens, the agreement among the cave sites is impressive and suggests that steep denticulate scrapers at Gheo-Shih were notably smaller. We

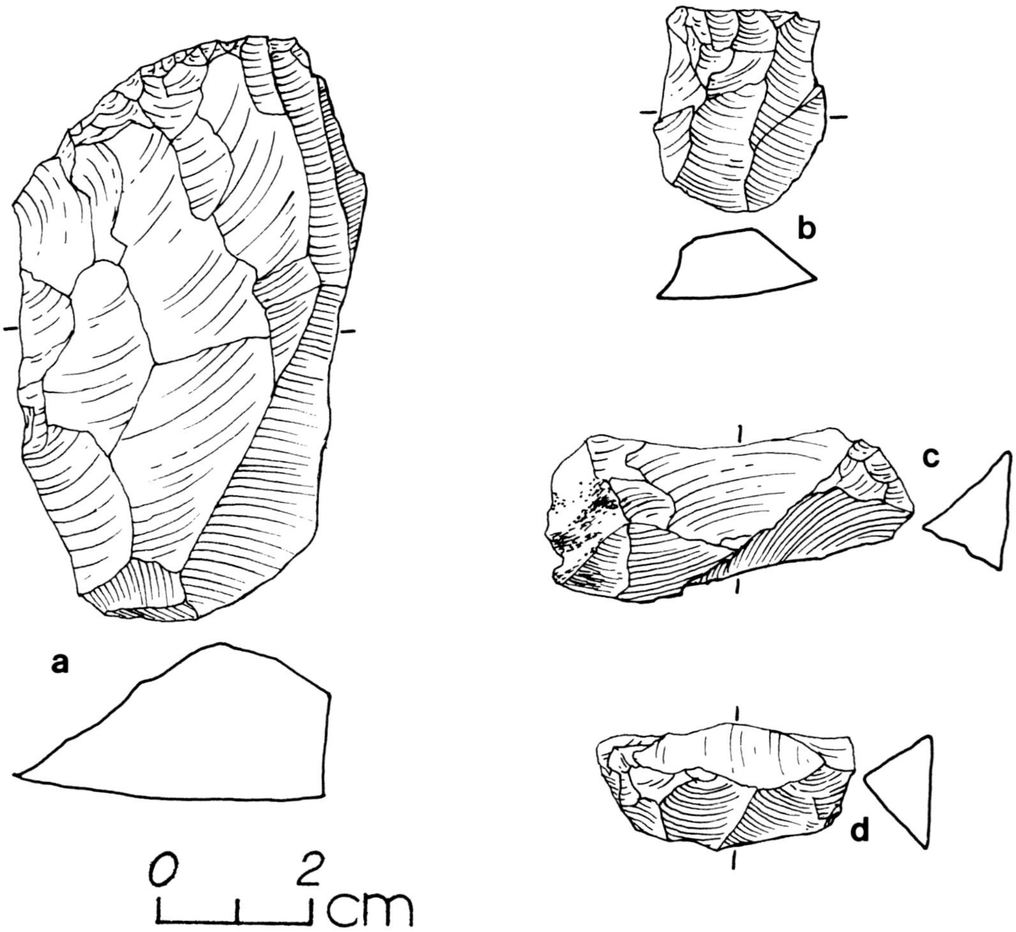

FIGURE 6.7. Flake core fragments: a, core face, Zone D, Square F8; b, core face, Zone D, Square C7; c, core edge, Zone B2, Square D11; and d, core edge, Zone D, Square F5.

TABLE 6.2
Comparison of the Weights of Steep Denticulate Scrapers from Guilá Naquitz, Cueva Blanca, Martínez Rockshelter, and Gheo-Shih

Site	Zone	Number of scrapers	Number whole	Weight (kg)	Average weight (kg)
Guilá Naquitz	B	8	5	.74	.09
	C	2	1	.13	.07
	D	2	2	.10	.05
	E	1	1	.40	.40
	(Avg.)				.11
Cueva Blanca	B	127	93	14.00	.11
	C	9	5	.85	.08
	D	32	23	3.50	.19
	E	8	3	.38	.05
	(Avg.)				.11
Martínez Rockshelter	A	2	2	.39	.19
	B	8	5	.79	.09
	C	4	3	.41	.10
	(Avg.)				.11
Gheo-Shih	Surface collected	199	173	14.75	.07
	Excavated	144	76	6.75	.05
	(Avg.)				.062

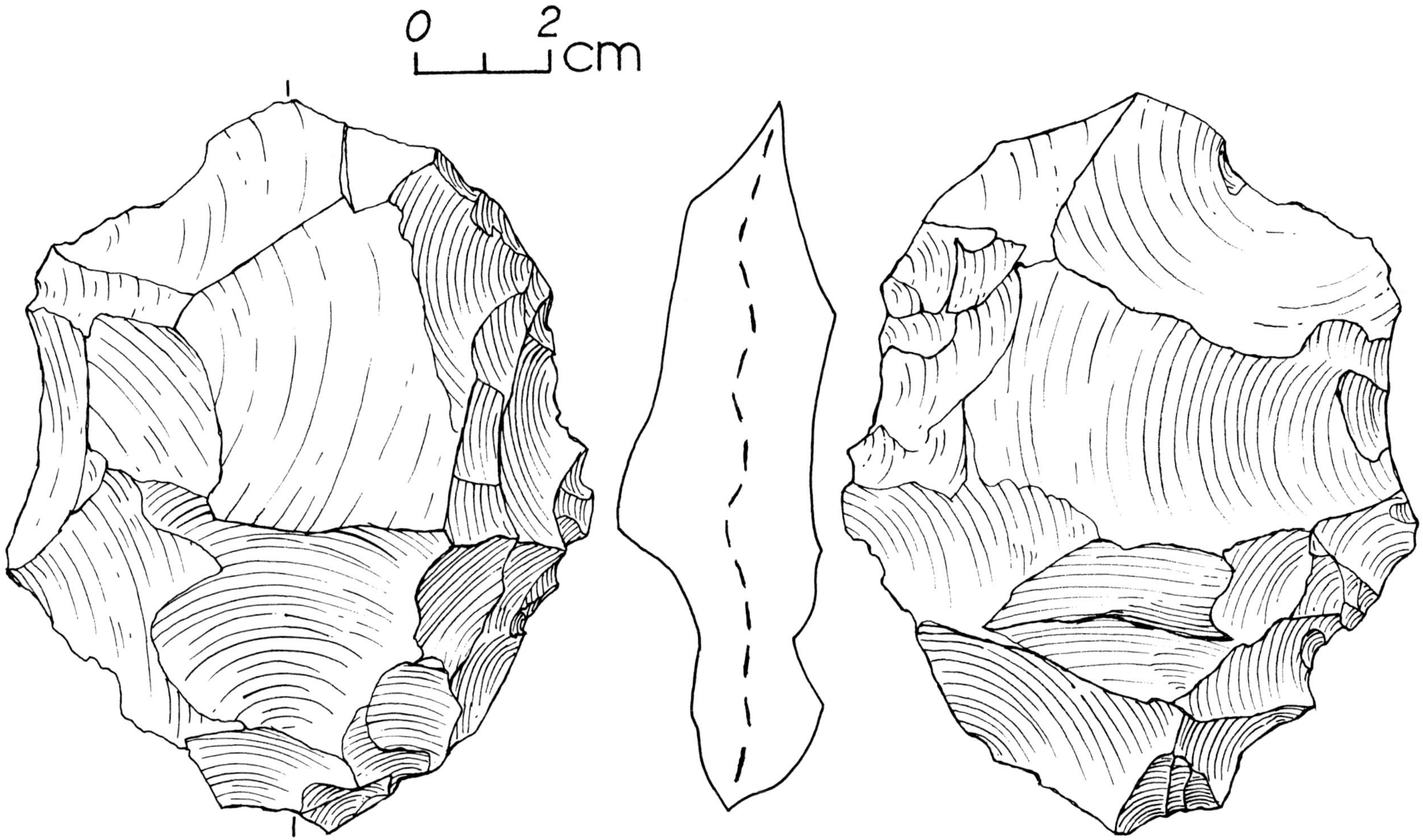

FIGURE 6.8. Chopper/knife, Zone B2, Square D11.

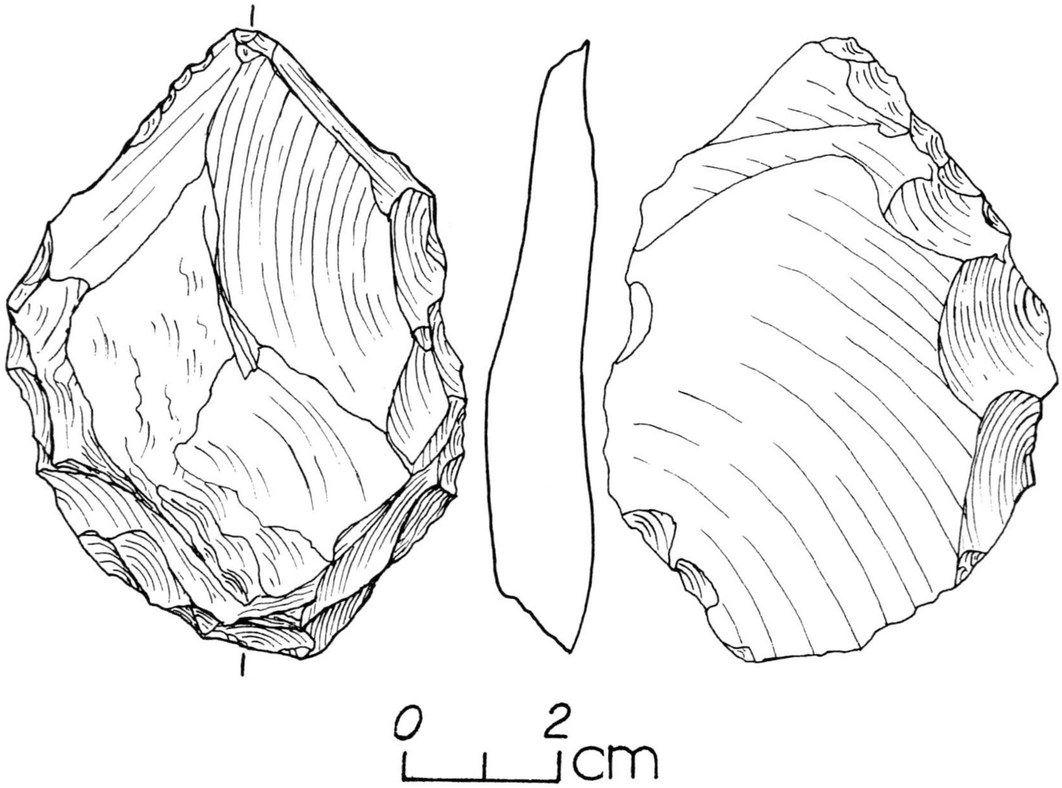

FIGURE 6.9. Chopper/knife, Zone B1, Square E9.

should consider the possibility that, in spite of their morphological similarity, the scrapers from Gheo-Shih were put to different purposes from those used in the caves. This point will be pursued further in the Gheo-Shih report.

Steep denticulate scrapers were also present at the Cueva del Texcal near Puebla (García Moll 1977: Lám. 26).

Sidescrapers/Knives

Figs. 6.13a–c and 6.14a–c.
Number of specimens: 4.
Dimensions: length, 5–8 cm; width, 3–6 cm; thickness, 5–15 mm.

Description: These are flakes with one edge completely chipped, with shallow step or scaling retouch. The edge is smooth (rather than denticulate) and may be either straight or curved in outline. These tools were probably used as scrapers and/or knives in the skinning of animals and the preparation of hides.

In terms of the Tehuacán sequence, our sidescrapers/knives are probably most like some of MacNeish *et al.'s* "thick flakes, one and two edges retouched" (1967:48–50, Figs. 29 and 31).

End Scrapers

Figs. 6.13d and 6.14d
Number of specimens: 1.
Dimensions: length, 3.5 cm; width, 3 cm; thickness, 5–7 mm.

Description: The single example from Guilá Naquitz is a flake with one end blunted by steep retouch. The edge is convex in outline. Traditionally, these have been considered tools for hide working. Since the other preceramic sites have a larger sample of end scrapers, they will be discussed at greater length in future reports.

End scrapers were common in the Tehuacán preceramic, and as a result MacNeish *et al.* (1967:30–43) have devoted an entire chapter to them.

TOOLS FOR SLOTTING AND PERFORATING

Burins

Figs. 6.15a, b and 6.16a.
Number of specimens: 4.
Dimensions: height, 3–4 cm; width, 2–3 cm; thickness, 5–10 mm.

Description: The Guilá Naquitz specimens are simple burins that have an acute angle formed by the removal of flakes roughly perpendicular to one another or from the intersection of a transverse break and a perpendicular flake. A few burins from other preceramic sites in the area have more than one flake removed from each direction and are thus technically polyhedric. However, there are so few burins

FIGURE 6.10. Choppers/knives: a, silicified tuff specimen from Zone B1, Square E9; and b, chert specimen from Zone B2, Square D11 (maximum length, 110 mm).

in general that it was considered not worthwhile to subdivide the category.

Burins are generally considered to have been used for slotting, gouging, and cutting wood or bone. In this case, they might have been used for preparing wooden hafts, such as atlatl foreshafts, or for cutting wooden tools such as those described in Chapter 10. Our Oaxaca burins are most like the simplest types from Tehuacán (MacNeish *et al.* 1967:45–47).

Drills

Figs. 6.15c and 6.16b.
Number of specimens: 1.
Dimensions: length, 5.2 cm; width, 2.5 cm; thickness, ca. 6 mm.

Description: The single specimen from Guilá Naquitz is an elongated flake with retouch on one end, producing a point. The retouch occurs on both edges of the tip and is directed from both the dorsal face and the bulbar face. In general, drills from preceramic Oaxaca show little apparent standardization. At Gheo-Shih, some drills show wear that suggests they were used to convert sandstone discs into possible beads, but no such discs were found at Guilá Naquitz. One possibility for the Guilá Naquitz specimen is that it was used to drill the sockets in wooden fire drill hearths such as those described in Chapter 10.

No drills found in the Tehuacán area appear stylistically similar to our specimen.

Pointed Pieces

Number of specimens: 1.

Description: Pointed pieces are defined as flakes with a tip or point created by retouch or use. They are distinguished from drills by the fact that they show very few signs of deliberate retouch; rather, the chipping usually seems to have resulted from light use. At their best, some specimens may resemble MacNeish *et al.'s* "flake gravers" (1967:47, Fig. 26). At their simplest, as in the case of the lone Guilá Naquitz specimen, they may be no more than a use-retouched perforator for starting or enlarging holes in wood or hide.

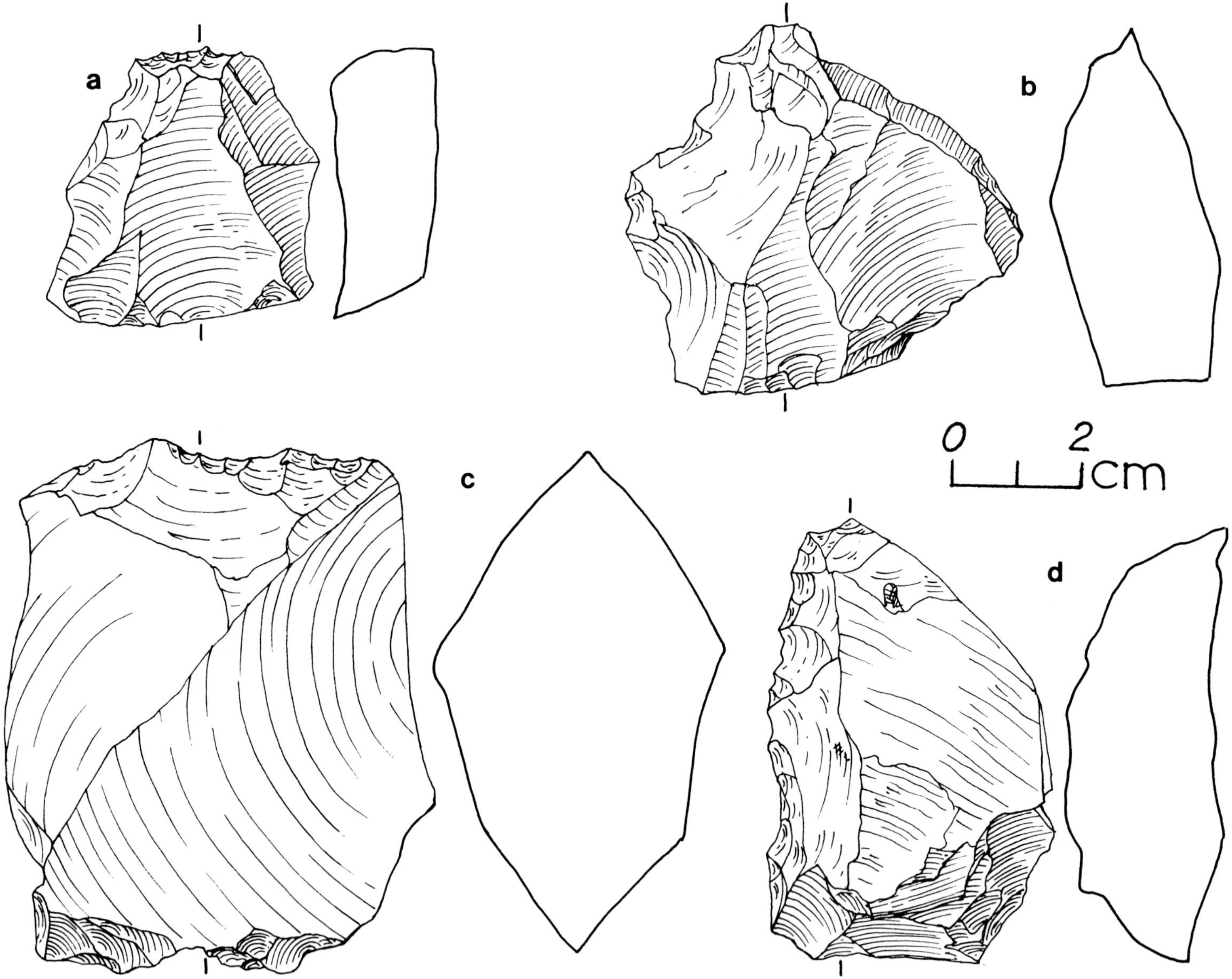

FIGURE 6.11. Steep denticulate scrapers: a, Zone E, Square C10; b, Zone B1, Square F5; c, Zone B3, Square C9; and d, Zone C, Square D11.

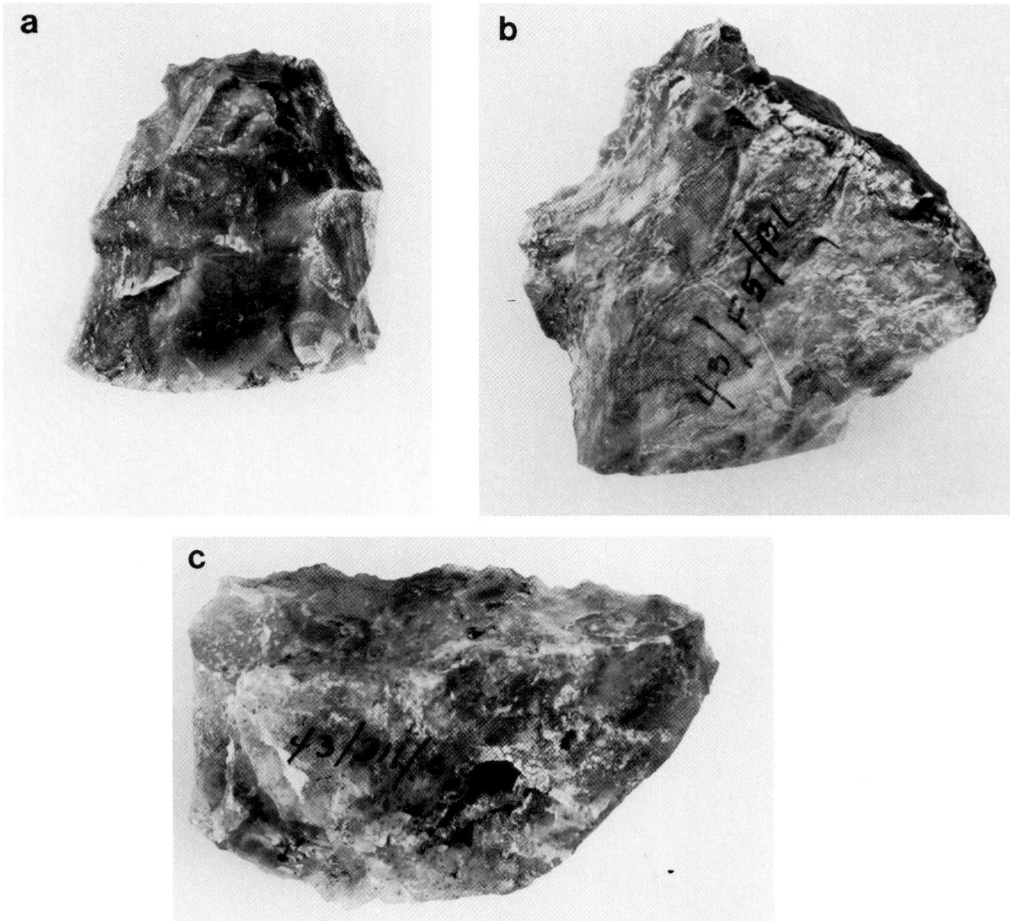

FIGURE 6.12. Steep denticulate scrapers: a, Zone E, Square C10; b, Zone B1, Square F5; and c, Zone C, Square D11 (maximum length, 70 mm).

OTHER FLAKE AND BLADE TOOLS

Crude Blades, Plain

Figs. 6.17 and 6.18.
Number of specimens: 10.
Dimensions: length 3.5–6.5 cm; width, 0.8–3 cm; thickness 3–8 mm.

Description: *Blades* are defined as thin flakes at least twice as long as they are wide and with relatively parallel sides. Most of the Guilá Naquitz specimens appeared to be fortuitous. They were not struck from prepared pyramidal cores; rather, they appear to be flakes, struck from typical flake cores, that just happened to come out long and narrow. As will be discussed in later reports, however, that is not necessarily true of all the examples from other preceramic sites. None of the Guilá Naquitz blades were retouched, and relatively few seem to have been used for any task other than light-duty cutting.

It is significant that most blades were made of local chert, with only a few being of silicified ignimbrite. The difference in raw material is such that it is much less likely that long, parallel-sided flakes would come off a silicified ignimbrite core. In regional terms, the blades from Guilá Naquitz and the Martínez Rockshelter are the "crudest" specimens, while the smallest blades (and the largest number) came from Gheo-Shih. However, given the small sample from Guilá Naquitz, this may not mean much.

"Crude blades with unprepared striking platforms" were also present in the Tehuacán preceramic as far back as the Ajuereado phase (MacNeish *et al.* 1967:20, Fig. 3).

Notched Flakes

Figs. 6.19 and 6.20.
Number of specimens: 40.
Dimensions: virtually coextensive with those of the population of flakes in general. The best examples are: length, 3–8 cm; width, 1.5–5 cm; thickness, 6–15 mm.

Description: These are flakes of any size or shape that have a notch or notches chipped into an edge. The chipping that

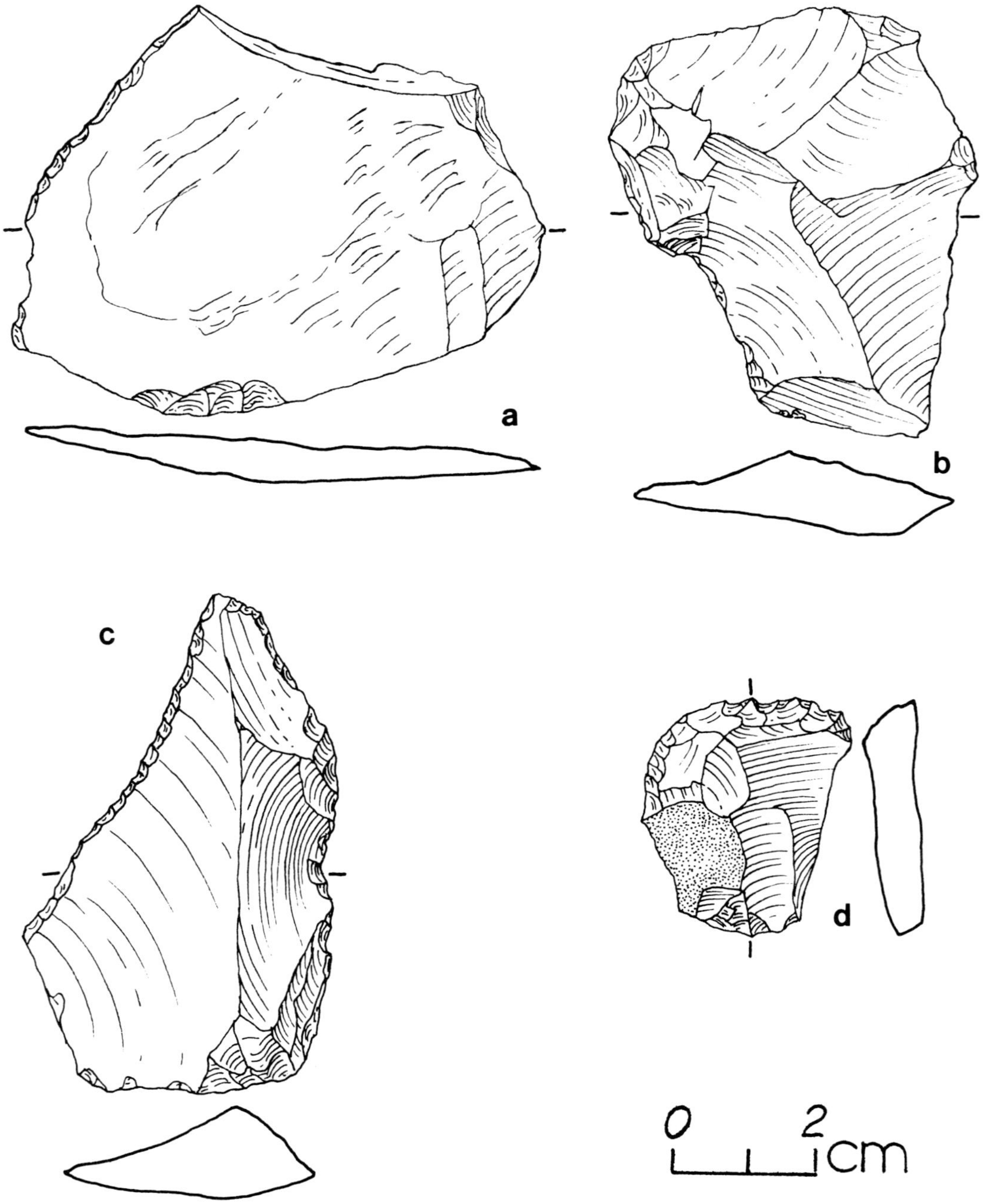

FIGURE 6.13. Scrapers: a, b, two sidescrapers, Zone B2, Square E9; c, sidescraper, Zone B1, Square E11; and d, end scraper, Zone B3, Square D9.

produced the notch may have been deliberate or caused through use on a hard material. Some flakes, in fact, may have both purposeful and use-retouched notches. These flakes are ordinarily less chunky and have thinner edges than the pieces classified as steep denticulate scrapers, but at one end of the continuum they approach the latter.

Notched flakes were the largest single category of artifact at Guilá Naquitz (and most preceramic sites in the area). A few of the deliberately retouched examples may have been used as spokeshaves for working wooden or cane tools, but the tremendous variety and abundance of notched flakes at Oaxaca sites suggest they were used for many activities. Probably, the occupants of Guilá Naquitz picked up any conveniently shaped flake when they had a cutting task to perform, modifying it further as they saw fit.

Some of MacNeish *et al.'s* "spokeshave-like tools" from Tehuacán (1967:47, Fig. 27) overlap with this category, but they do not match the full range of our notched flakes.

FIGURE 6.14. Scrapers: a, silicified tuff sidescraper, Zone B2, Square E9 (maximum width, 75 mm); b, chert sidescraper, Zone B2, Square E9; c, silicified tuff sidescraper, Zone B1, Square E11; and d, chalcedony end scraper, Zone B3, Square D9.

Flakes with Sheen

Figs. 6.21 and 6.22.

Number of specimens: 6.

Dimensions: length, 3–7 cm; width, 2–5 cm; thickness, 5–15 mm.

Description: These are flakes that have one or more edges that display the kind of glassy sheen, or "corn gloss," usually seen on flint sickle blades from Near Eastern sites. Flakes with such sheen occur at the other preceramic sites in the area, so the specimens from Guilá Naquitz are in no way unique. Most examples have thin, sharp edges, but similar sheen was noted even on one steep denticulate scraper. These tools were probably held in the hand; there is no evidence for hafting, and, indeed, the shape of many of the flakes would make hafting difficult.

There have been numerous articles written on the possible causes of sickle sheen, with some authors suggesting that it is caused by friction between the silica or opal deposits in grass stems (Curwen 1930; Witthoft 1967). However, a series of unpublished experiments that my students and I conducted at Rice University suggested that sheen can also form when the grass stems (or other material that is being cut) are dirty. In other words, some sheen may in fact be a kind of

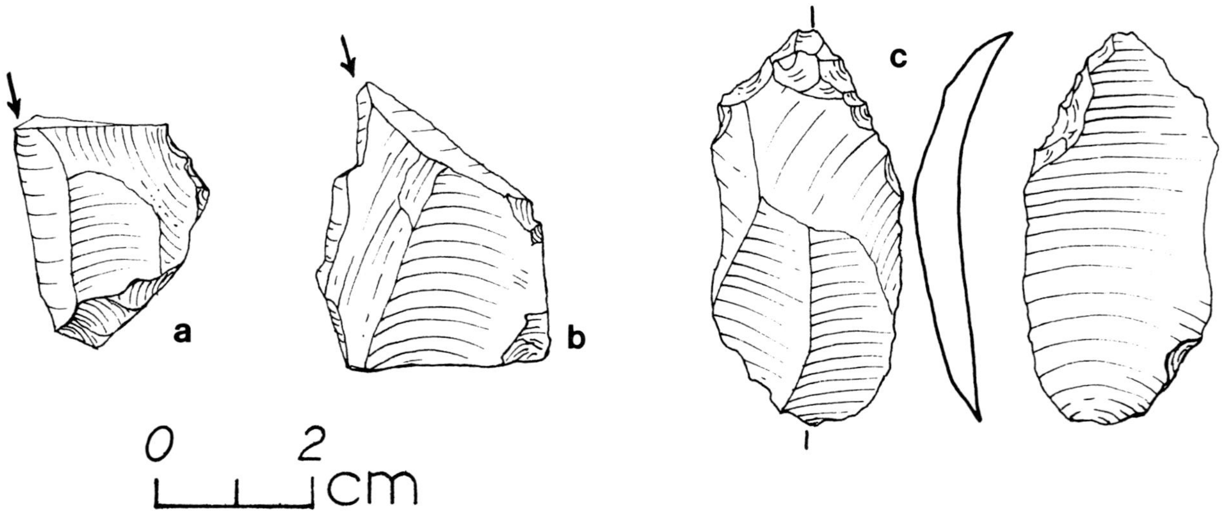

FIGURE 6.15. Burins and drill: a, burin, Zone D, Square G7; b, burin, Zone B1, Square E5; and c, drill with tip partially broken, Zone D, Square D8.

"soil polish" analogous to the luster on some Neolithic or Mississippian hoes, caused more by the fine-grained dirt on the plants than by the plants themselves.

What might have caused the sheen on these flakes from the Oaxaca preceramic? One obvious suggestion, based on the remains from Guilá Naquitz, would be the trimming of the leaves from maguey hearts prior to roasting. Maguey hearts were dug up from underground, and the fragments of leaf base found at Guilá Naquitz suggest that they were left dirty rather than washed with water. In the process of sawing through the tough, fibrous leaves in order to trim the heart for roasting, any flint would be subjected to a great deal of abrasion. There are also many leaves on each maguey heart. Tentatively, therefore, we could suggest that the sheen on these flakes may result from the combination of dirt, sand, and tough fiber encountered over a long period of maguey processing.

No equivalent tool category is mentioned for preceramic Tehuacán, but we would expect these flakes with sheen to appear elsewhere if we have interpreted them correctly.

FIGURE 6.16. Burin and drill: a, silicified tuff burin, Zone B1, Square E5; and b, chalcedony drill with broken tip, Zone D, Square D8 (maximum length, 50 mm).

Utilized Flakes

Figs. 6.23 and 6.24.

Number of specimens: 21.

Dimensions: virtually coextensive with those of the population of flakes in general; best examples—length, 5–7 cm; width, 3–7 cm; thickness, 5–15 mm.

Description: These are flakes or chunks of silicified ignimbrite, chert, or flint that have light, shallow chipping limited to part of an edge. In view of the irregular occurrence of the chipping, it appears to have resulted from light use. These flakes are the second most common type of artifact at Guilá Naquitz, and once again it looks as if the occupants simply picked up whatever conveniently shaped flake might have been available to perform a cutting task. Had they been unused, we would have lumped these with the chipping debris. Had they been used for a longer period of time, they might well have turned into one of our notched flakes or flakes with sheen.

In terms of comparisons with the Tehuacán preceramic, our utilized flakes probably correspond to MacNeish *et al's.* "thin flakes [with] one or two edges utilized" and "thick flakes [with] one or two edges utilized" (1967:47–51, Figs. 28–31).

PROJECTILE POINTS

All projectile points from Guilá Naquitz were bifacially worked, and all are relatively large, heavy points that were

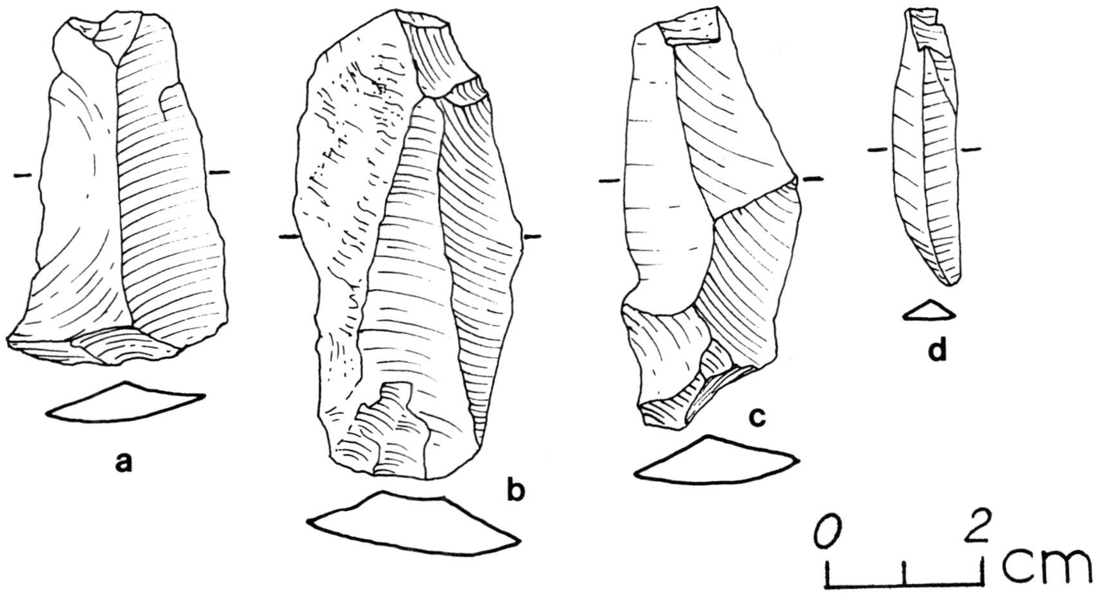

FIGURE 6.17. Crude blades: a, Zone E, Square E11; b, Zone B2, Square C7; c, Zone B2 + 3, Square E6; and d, Zone D, Square F4.

presumably intended for use either with a thrusting spear or with the atlatl, or spear-thrower. Most specimens from Guilá Naquitz were chipped from relatively good chert or chalcedony, although both silicified ignimbrite and low-quality chert or flint were widely used for projectile points in the Oaxaca preceramic.

Projectile points are traditionally considered the most useful artifact for identifying preceramic phases in Mexico and for comparing one preceramic site with another. Their value lies in the fact that they are relatively better made and more standardized than other chipped-stone tools, they tend to contain more "information" in the form of stylistic attributes, and they show some relatively distinctive changes through time.

Unfortunately, the sample of in situ projectile points from Guilá Naquitz is so small it would be foolish to attempt to describe them individually. Only four whole or partial points were found in Zones B–E. Two more whole or partial points

FIGURE 6.18. Crude blades, all silicified tuff: a, Zone E, Square E11; b, Zone B2, Square C7 (maximum length, 62 mm); c, Zone B2 + 3, Square E6; and d, Zone D, Square F4.

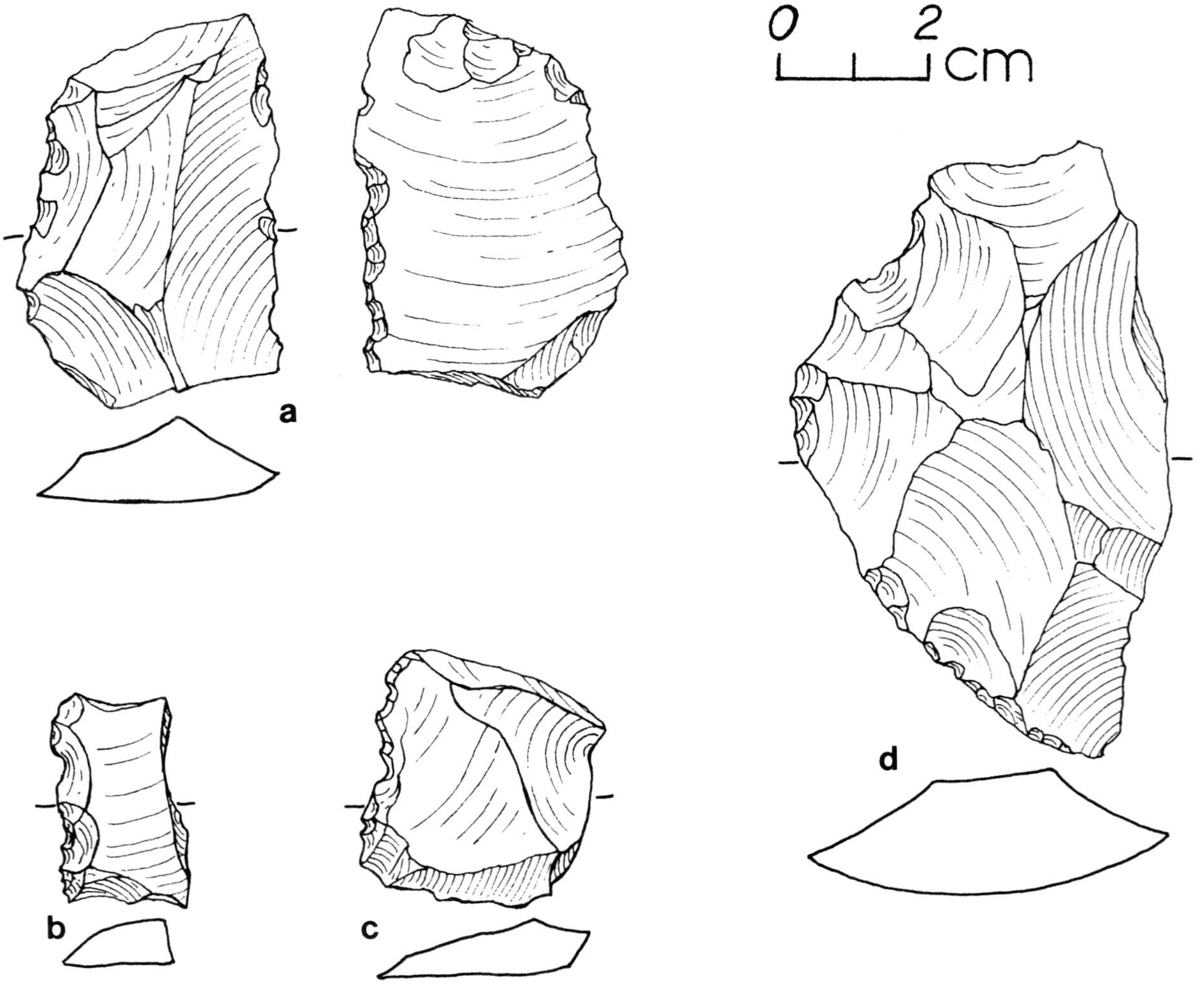

FIGURE 6.19. Notched flakes: a, Zone B1, Square F7; b, Zone B1, Square D8; c, Zone B1, Square C11; and d, Zone B1, Square D6.

were found in storage pits associated with Zone A, a Monte Albán IIIb–IV level; on typological grounds, they are believed to be points redeposited from preceramic levels, perhaps at the time the pits were dug down from Zone A into the underlying deposits. In addition, two more whole or partial points were found on the surface when the cave was originally located on survey—one near the base of the cliff nearby and one on the talus many meters below the cave. Through a great stroke of luck, the fragment from the talus turned out to be the proximal (or basal) half of a point whose distal half was found in situ in Square C9 of Zone B3, and the two could actually be glued back together. However, that still gives us only 8 points or fragments thereof, representing no more than 7 artifacts. Clearly, either hunting was not as important an activity at Guilá Naquitz as it was at some other Oaxaca sites, or loss of points was less common than at other sites.

Because of this small sample, I shall wait until a future report to discuss the entire projectile point sequence from preceramic Oaxaca. It makes more sense to do so after sites like Cueva Blanca and Gheo-Shih, with their much larger samples, have brought our number of points to above 50. In anticipation of that later report, however, I present some measurements for the Guilá Naquitz points according to the standard format developed for all Oaxaca projectile points (Table 6.3). Figure 6.25 illustrates the way these measurements were taken. This procedure was developed after several trials and considerable discussion among a variety of people concerning nomenclature, feasibility of taking consistent measurements, and significance of variance measurements. It is hoped that ultimately, when our full sample from all the Oaxaca sites can be presented, statistical procedures will enable us to determine which attributes of the points can best serve our needs.

In terms of projectile point typology, we have also chosen to be conservative, partly as the result of our small sample. We had no particular desire to create new type names, and wherever possible we have followed the typology of MacNeish *et al.* (1967:56–79), which in turn is based partly on the preexisting Texas nomenclature (Suhm and Jelks 1962). We already know that this will not work for the entire population of Oaxaca projectile points, but for the small collection from Guilá Naquitz it is a good place to start.

TABLE 6.3
Measurements of Projectile Points and Point Fragments from Guilá Naquitz

Identification		Maximum dimensions[a] (mm)					Ratios			Blade width[b]				Stem width[c]			Notch depth		Notch width	Blade angles[d]			
type	provenience	L	W	TH	LB	LS	$\frac{L}{W}$	$\frac{L}{TH}$	$\frac{LB}{L}$	A	B	C	$D–D^1$	A	B	C	base	stem	right	A^1	B^1	C^1	D^1
Lerma	Zone E, Square H10	76	35	13.7	—	—	2	5.5	—	21	34	16	35	—	—	—	—	—	—	30	17	—	28
Trinidad(?)	Zone B3, Square C11	—	—	7.8	—	—	—	—	—	—	—	—	—	—	24(?)	18(?)	—	—	—	—	—	—	—
Trinidad(?)	Zone B1, Square F6	—	40	6.4	50	—	—	—	—	23	37	39	40	—	—	—	—	—	—	38	33	24	31
Pedernales[e]	Blade, Zone B3; base, surface	65*	46	9.8	52*	13	1.4	6.6	.8	25	38	45	46	25	21	19	3	6	18	38*	30*	25*	29*
Pedernales	Surface	68	39	8.4	54	16	1.7	8	.8	22	34	33	39	24	22	21	3	6	15	34	26	20	22
Almagre	Feature 13	—	—	9.0	—	9	—	—	—	—	35(?)	45	45	29	20	12	—	—	—	—	—	—	—
Corner notched	Feature 12	—	—	6.1	—	13	—	—	—	—	—	—	—	—	—	—	—	5	18	—	—	—	—

[a]Dimensions listed are length (L), width (W), thickness (TH), length of blade (LB), and length of stem (LS).
[b]When possible, four blade-width measurements (A, B, C [see Figure 6.25b], and $D–D^1$ [see Figure 6.25a]) were taken.
[c]When possible, three stem-width measurements (A, B, and C; see Figure 6.25c) were taken.
[d]When possible, four blade-angle measurements ($X–A^1$, $X–B^1$, $X–C^1$, and $X–D^1$; see Figure 6.25b) were taken.
[e]Asterisked (*) measurements were estimated, allowing for broken tip.

FIGURE 6.20. Notched flakes, all silicified tuff: a, Zone B1, Square D8; b, Zone B1, Square C11; and c, Zone B1, Square D6 (maximum length, 80 mm).

Zones B–E at Guilá Naquitz have been radiocarbon-dated to 8750–6670 B.C., and most of the projectile point types we have identified from those levels would not be out of place at other sites of that time period. Included are the Lerma, Pedernales, and Trinidad types; an Almagre point redeposited in a later level fits comfortably in that epoch as well.

Unfinished Lerma(?) Point

Figs. 6.26 and 6.27d.
Number of specimens: 1 in situ, Zone E, Square H10.
Dimensions: see Table 6.3.
Description: These lanceolate points are described by MacNeish *et al.* (1967:56–57, and Fig. 35) as resembling a laurel leaf in outline, with "a slight tendency to be widest just below the midpoint." Most are diamond-shaped in cross section and fashioned almost entirely by percussion flaking. Our single example seems to be an unfinished specimen, one of whose ends had not yet been completely chipped before it was discarded. The raw material is waxy, brownish-gray chert or chalcedony.

Lerma points "seem to have been in existence in Mexico before 7000 B.C." and have been found in the Ajuereado phase of the Tehuacán Valley as well as at mammoth kill sites in the Basin of Mexico (Aveleyra 1956; MacNeish *et al.* 1967:56–57).

Trinidad(?) Points

Figs. 6.26b, 6.27a, and 6.28a.
Number of specimens: 2 (1 possible blade fragment from Zone B1; 1 possible stem fragment from Zone B3).
Dimensions: see Table 6.3.
Description: Trinidad points are broad bladed and have "relatively short, wide, contracting stems with convex bases. Their bodies, which are longer than the stems, are roughly the shape of isosceles triangles and have slightly convex edges. The shoulders are usually marked with fairly prominent short barbs" (MacNeish *et al.* 1967:62, and Fig. 43). They may show retouch on alternate sides of opposite edges. The two incomplete specimens from Guilá Naquitz resemble Trinidad points from Cueva Blanca and other Oaxaca sites more closely than they do any other projectile points in our sample, but because of their fragmentary nature our identification should be considered tentative. The Zone B1 specimen is of reddish-gray silicified tuff.

Trinidad points occur in the El Riego, Coxcatlán, and Abejas phases at Tehuacán and are traced back to at least 7000 B.C. This would fit with our dates of 6910–6670 B.C. from Zone B.

Pedernales Points

Figs. 6.29–6.31.
Number of specimens: 1 found in situ, Zone B3, Square C9; 2 more from the surface near the cave.
Dimensions: see Table 6.3.
Description: Pedernales points have a broad blade with slightly convex edges, and their shoulders are usually marked by fairly prominent short barbs. However, their most distinctive feature is the stem, which is short and straight and

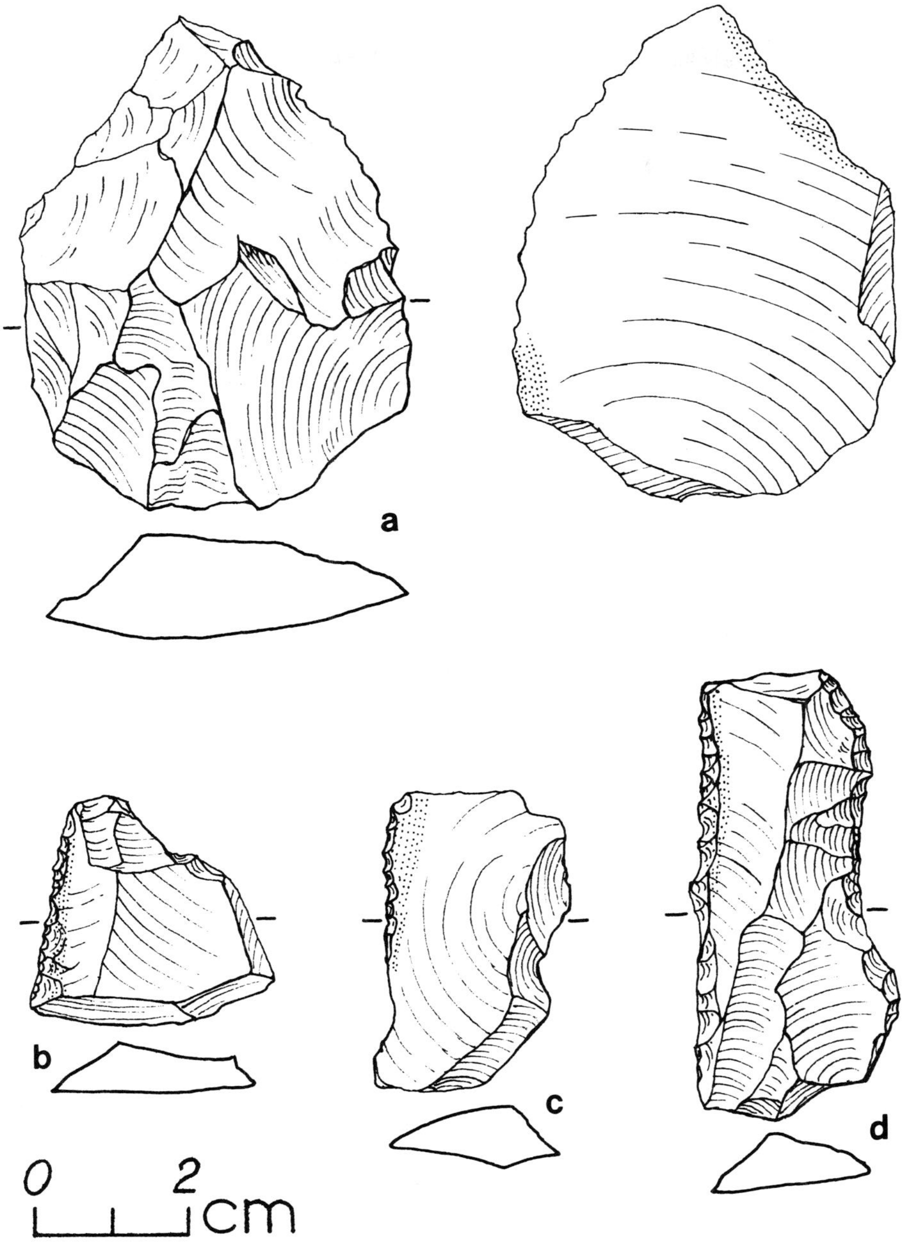

FIGURE 6.21. Flakes with sheen: a, flake with sheen on ventral surface (left, dorsal surface and cross section; right, ventral surface), Zone B2, Square E6; b, Zone B2, Square E9; c, Zone D, Square F6; and d, Zone D, Square F5. Areas of sheen are indicated by stippling.

has a concave base. Pedernales points from Oaxaca, while largely percussion flaked, may have careful secondary retouch along the blade edges and the basal concavity of the stem. Sometimes (as in the case of Trinidad points) this retouch is on alternate sides of opposite edges. It could have been done with the tip of a fire-hardened, antler tine pressure flaker like those described in Chapter 12.

All Guilá Naquitz specimens were made of high-quality, waxy chalcedony, although some Pedernales points from other Oaxaca sites were of silicified ignimbrite. The specimen whose distal half was found in Zone B3, Square C9, was made from waxy white and yellow chalcedony that almost certainly comes from the quarries at Peña de Matadamas in the Etla Valley, some 50 km northwest of Guilá Naquitz. The proximal half of this same point had been tossed out of the cave and was found on the surface of the talus below. It was unburned, while the distal half that remained in the cave had been burned until it was opaque. Perhaps the proximal end had remained attached to the foreshaft of the atlatl dart and had simply been discarded as useless, while the distal end remained in a successfully hunted deer that was eventually cooked by the occupants of Zone B3.

Pedernales points were common at later Oaxaca sites such as Gheo-Shih but rare in the Tehuacán Valley. A single Pedernales point from Levels 3–4 at El Riego Cave West, a component of the Coxcatlán phase (MacNeish *et al.* 1967:78,

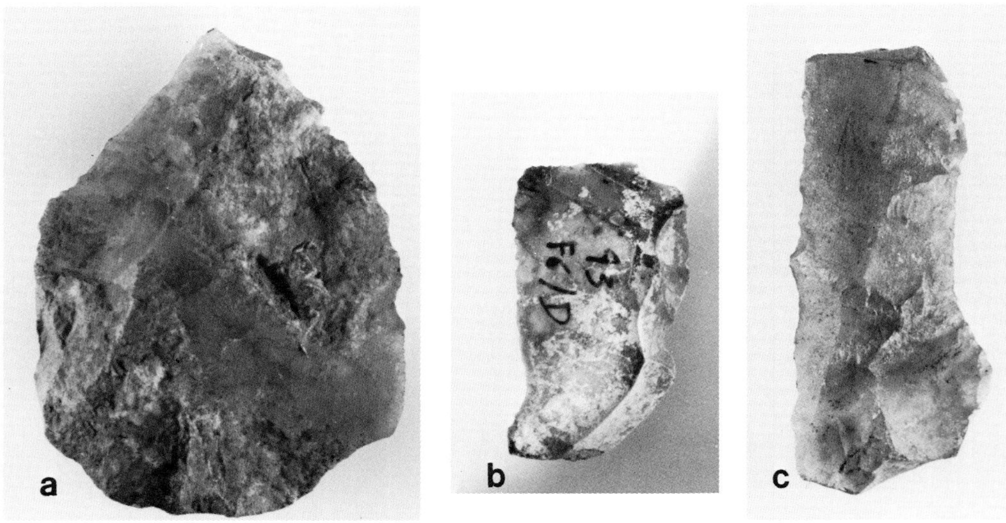

FIGURE 6.22. Flakes with sheen, mostly chert: a, Zone B2, Square E6; b, Zone D, Square F6; and c, Zone D, Square F5 (maximum length, 60 mm). In each case, area of sheen is upper part of left edge.

Fig. 67), was the only specimen found at Tehuacán. It probably dates to ca. 4000 B.C., and resembles Pedernales points from Gheo-Shih more than it does the specimens from Guilá Naquitz; the latter may be among the oldest examples of Pedernales points thus far discovered. Of the Pedernales points illustrated by Suhm and Jelks (1962), ours most closely resemble their Specimen E in Plate 119.

Only two Pedernales points were found at Cueva del Texcal, and the one illustrated (García Moll 1977: Lám. 15) has a deeper basal notch than our specimens.

Almagre Point

Figs. 6.27c and 6.28c.

Number of specimens: none in preceramic levels; 1 from Feature 13, Zone A, presumably redeposited.

Dimensions: see Table 6.3.

Description: Almagre points "have wide, triangular bodies and poorly defined, short, wide, convex to pointed stems. In many cases the stem is little more than a 'bump' in the middle of the base" (MacNeish *et al.* 1967:66). They are made of thin, flat flakes that have percussion scars on both surfaces as well as more delicate secondary chipping along the edges. Almagre points were not particularly common in Tehuacán and the series illustrated from the Sierra de Tamaulipas (MacNeish 1958) is actually more typical than the limited number from Tehuacán (MacNeish *et al.* 1967:Fig. 50) or the Santa Marta Rockshelter (MacNeish and Peterson 1962).

The single Almagre point from Guilá Naquitz is one of those whose stem is little more than a bump. It is made on flinty material that appears to be black, silicified siltstone. Since we have never found Almagre points in Monte Albán IIIb–IV context elsewhere, we assume this specimen is redeposited from a preceramic level. The type is known to occur before 7000 B.C. in Chiapas and western Mexico (MacNeish *et al.* 1967:66), so its presence in the Naquitz phase would not be chronologically troublesome.

Fragment of Corner-Notched Point

Figs. 6.27b and 6.28b.

Number of specimens: none in preceramic levels, 1 from Feature 12 of Zone A, possibly redeposited.

Dimensions: see Table 6.3.

Description: What this fragment most resembles is the basal portion of a large, corner-notched point with a carefully retouched expanding base. Since we have not found such points in Monte Albán IIIb–IV context elsewhere, we suspect this specimen to have been redeposited from a preceramic level. However, it is so fragmentary that we hesitate to assign it to a type.

In other preceramic sites in Oaxaca, there are large corner-notched points that resemble the type called "Marcos" by Suhm and Jelks (1962: Pl. 105). Others look like a larger (and earlier) version of the Palmillas point (1962: Pl. 115). These corner-notched specimens will be discussed in a future report.

BIFACES

Bifaces are large, generally oval or almond-shaped flakes that are percussion flaked on both surfaces. They present a continuum from very finely chipped, thin specimens to coarse, sinuous-edged, large, heavy flakes. The function of these tools probably changes from primarily cutting to primarily chopping as weight and coarseness increase. Some bifaces might even be "preforms" for making projectile points, though many

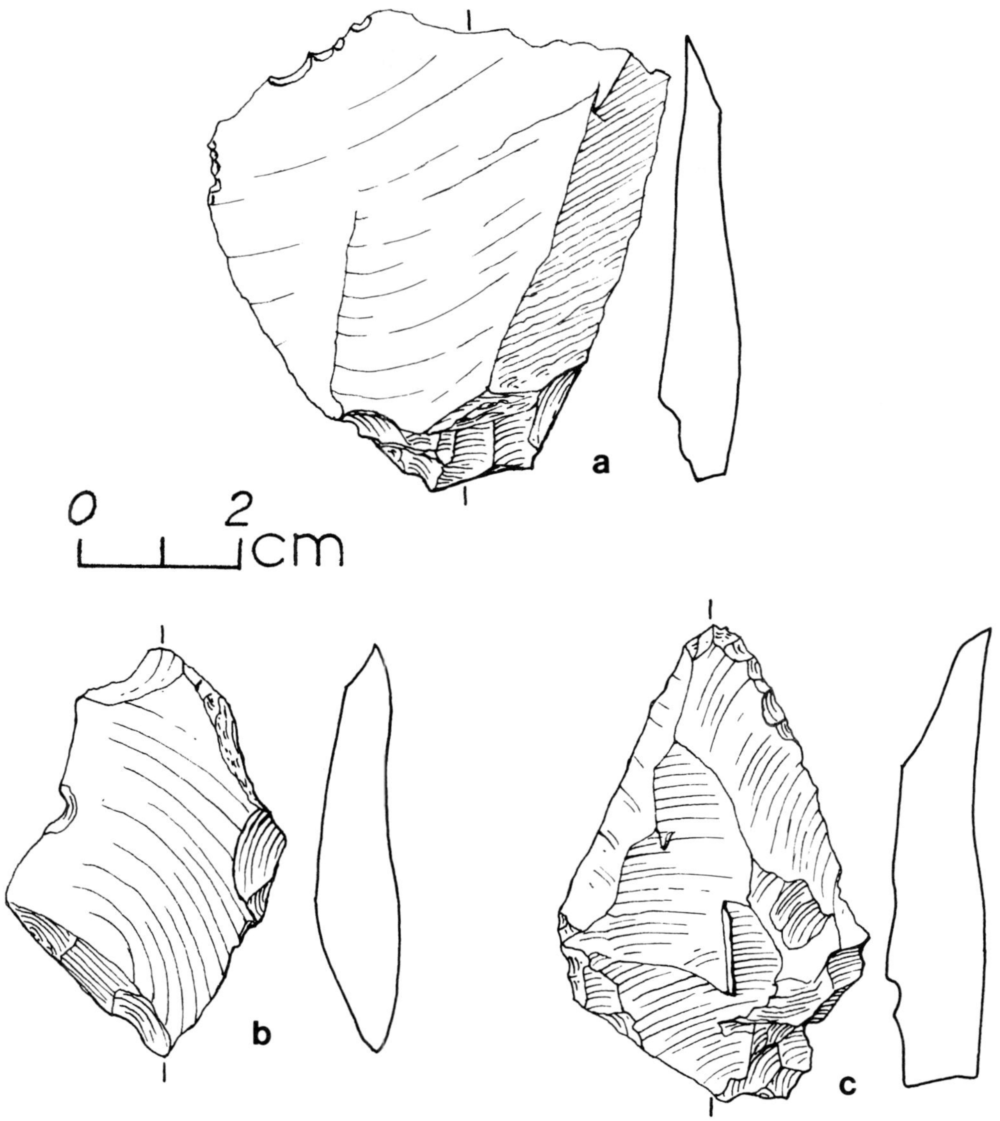

FIGURE 6.23. Utilized flakes: a, Zone C, Square C9; b, Zone B2, Square D9; and c, Zone C, Square F9.

are clearly not. But as with many kinds of tools in the Oaxaca preceramic, it is hard to draw an absolute line between types. For that reason, I have divided the biface series into three varieties, only two of which are present at Guilá Naquitz.

Bifaces, Variety A

Number of specimens: 0.

Description: This variety of biface was absent at Guilá Naquitz but occurred at later preceramic sites in the area. It consists of our best-made bifaces, formed on flakes thinned with careful bifacial retouch on the planar surfaces. The edges are smooth and straight. These tools will be dealt with more extensively in later reports.

Bifaces comparable to our Variety A were found at Cueva del Texcal, Puebla (García Moll 1977: Lám. 11).

Bifaces, Variety B

Figs. 6.32 and 6.33.

Number of specimens: 1.

Dimensions: length, 7 cm; width, 6 cm; maximum thickness, 10 mm.

Description: These are pieces with good bifacial chipping, mostly confined to the edges, leaving the flake relatively thick (unless it happened to be thin as struck from the core). The bases are usually unfinished when present and show either a clean break or a platform. There is a tendency toward an ogival outline, with the ends sometimes blunt. A few examples have battered edges. Because the single in situ example from Zone B1 (Fig. 6.32a) gives no indication of the range of variation in Variety B bifaces, we have illustrated another specimen from Zone A (Fig. 6.32b). While this specimen was found in a Monte Albán IIIb–IV level, it is presumed to be redeposited from a preceramic level because such bifaces are not known to occur in Monte Albán IIIb–IV.

FIGURE 6.24. Utilized flakes: a, silicified tuff specimen, Zone C, Square C9; b, chert specimen, Zone B2, Square D9; and c, chert specimen, Zone C, Square F9 (maximum length, 62 mm).

Bifaces, Variety C

Figs. 6.34–6.36.
Number of specimens: 5.
Dimensions: length, 6–10 cm; width, 4.5–8 cm; thickness, 12–30 mm.

Description: This variety, the crudest kind of biface in terms of manufacture, was also the most common one found at Guilá Naquitz. These are thick flakes with one or more edges chipped by coarse bifacial retouch. The bodies of the flakes are not thinned. The bifaces tend toward round or oval in outline, but there is a great range of variation. The edges are sinuous as a result of the coarse primary chipping; often there is no secondary retouch of the edges. The butts are also rarely chipped. A few examples have only part of an edge that shows bifacial chipping. Although these are far from Variety A bifaces in quality, the nature of the chipped edge is the same, and both varieties could have been used for the same purpose.

Our Variety C bifaces are at least superficially similar to MacNeish *et al.'s* "large bifacial disks" from the Tehuacán Valley (1967:91, Fig. 77), one example of which occurred in El Riego phase deposits.

Tanged Pick(?)

Figs. 6.37 and 6.38.
Number of specimens: 1.
Dimensions: length, 5 cm; width, 4 cm (tang, 3) cm; thickness, 14 mm.

Description: This object, found broken, appears to be a hafted biface of some kind. The dorsal surface is completely percussion flaked, while the ventral surface is chipped primarily around the edges. Evidently the tool had a blade some 4 cm wide and a tang 3 cm wide and about 3 cm long.

Its great thickness and its asymmetry make it unlikely that this object was a projectile point. In other regions, including some Old World areas, such an artifact would be interpreted as a hafted pick of some kind. It could, for example, have been used for excavating maguey-roasting pits or other features in the cave talus or platform. However, in this case its function should be considered as only tentatively interpreted.

Nothing resembling this tool is illustrated in the Tehuacán reports.

UNRETOUCHED PIECES

Debitage

Number of specimens: 1564

Description: This category includes all those fragments of chipped stone that do not show any sign of retouch (whether deliberate or through use) and that cannot be identified as being a specific part of a core. Of course, lack of retouch does not necessarily mean lack of use. Some of the naturally sharp flakes could have been used to cut soft tissue (such as deer meat) without showing wear, except possibly through careful microscopic examination.

Still other flakes might have been used to cut plant material, but for so short a time that they did not accumulate noticeable use retouch or begin to display sheen. As an example of this, we can point to one large flake that was found while cleaning the profile of Square C8 at a depth that would put it either in Zone B or in Zone C (Fig. 6.39). This flake had evidently been used, for it had a long strand of maguey fiber still attached to it; yet it would otherwise have been classified as an unmodified flake. We would assume, therefore, that a certain number of the unretouched flakes were used in some way, but we have no grounds for creating types within them.

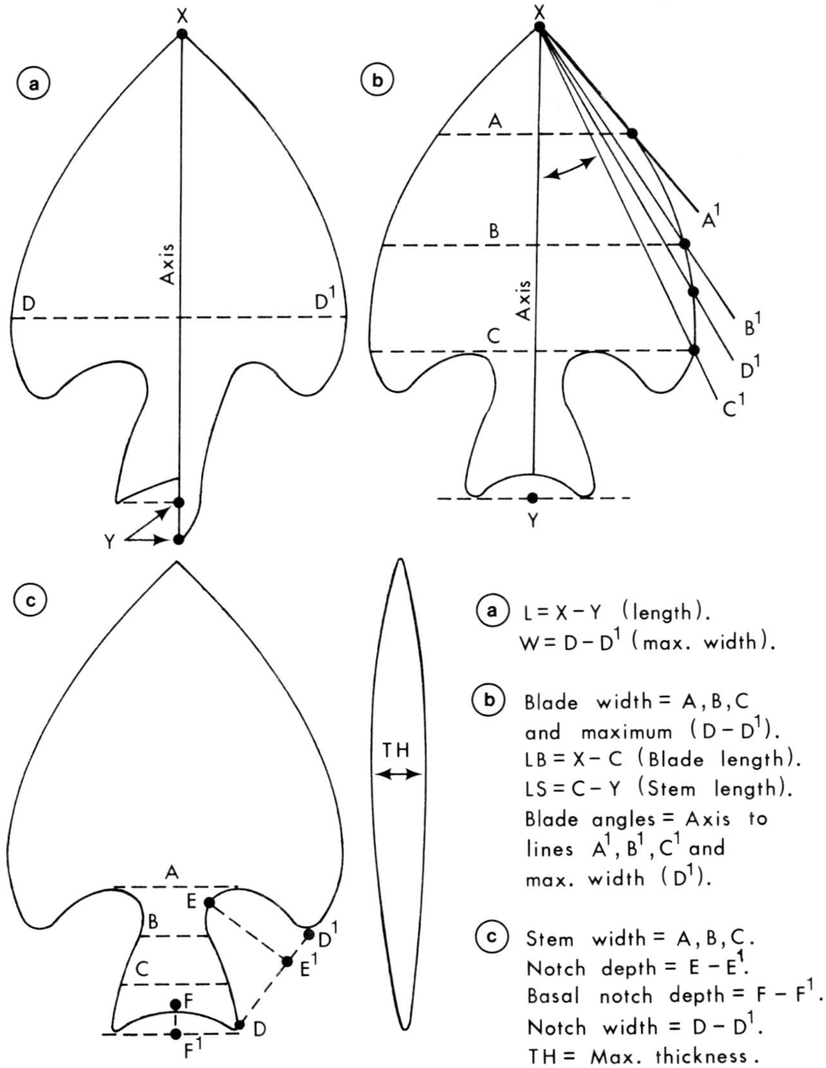

FIGURE 6.25. Diagrams showing how projectile point measurements were taken.

SUMMARY

The entire chipped-stone assemblage from preceramic levels numbered 1716 pieces, of which 1564 show no deliberate or use-induced retouch (Table 6.4). The latter have been referred to as "debitage" for convenience, although many may have been used as naturally sharp flakes. Debitage was the only chipped-stone category considered large enough to be used in the associational analyses in Chapters 26, 27, and 28.

The second largest category in the assemblage was notched flakes. There were 40 of these, but never more than 16 per living floor. Like utilized flakes, which numbered 21 (never more than 5 per level), notched flakes were probably used for a great variety of tasks, including both men's and women's activities. Notched flakes seem to have increased in frequency through time.

Steep denticulate scrapers were present in every level, for a total of 13. These tools may have been used in the processing of plant material, the shaping of wood, or the shredding or pulping of fibrous plants such as maguey and nopales. Roasting of maguey hearts was one of the important activities at Guilá Naquitz, and there are two other tools that could have served in related tasks: a possible tanged pick (for digging roasting pits or other features) and 6 flakes with sheen (perhaps produced by the trimming of tough, fibrous, dirty leaves from the maguey heart). Unfortunately, these tools are not abundant enough for their distribution to be compared with that of agave leaves.

As for hunting activities, they are suggested by four projectile points found in situ and three others found on the surface or in redeposited context. Some sidescrapers, knives, and other items may have been used in butchering or hide preparation, but these tools are not particularly numerous. Whether bifaces (one of the most interesting and distinctive tools in the Oaxaca preceramic) were used in butchering is unknown.

TABLE 6.4
Total Chipped-Stone Assemblage by Category from All Preceramic Zones at Guilá Naquitz

Categories	Stratigraphic zones E	D	C	B3	B2 + 3	B2	B1	Totals
Choppers/knives	—	—	—	—	—	1	1	2
Steep denticulate scrapers	1	2	2	3	1	1	3	13
Sidescrapers/knives	—	—	1	—	—	2	1	4
End scrapers	—	—	—	1	—	—	—	1
Burins	—	3	—	—	—	—	1	4
Drills	—	1	—	—	—	—	—	1
Pointed pieces	—	1	—	—	—	—	—	1
Crude blades	1	1	1	1	1	1	4	10
Notched flakes	3	6	10	3	1	1	16	40
Flakes with sheen	—	2	1	1	—	2	—	6
Utilized flakes	2	2	4	4	—	4	5	21
Projectile points	1	—	—	2	—	—	1	4
Bifaces, Variety B	—	—	—	—	—	—	1	1
Bifaces, Variety C	1	2	—	—	1	1	—	5
Tanged picks	—	—	—	—	1	—	—	1
Hammerstones	2	2	1	—	2	—	1	8
	11	22	20	15	7	13	34	122
Flake cores	—	—	2	2	—	2	5	11
Core faces	—	4	—	1	—	—	5	10
Core edges	—	2	2	—	—	4	1	9
Debitage	60	280	375	69	113	201	466	1564
	71	308	399	87	120	220	511	1716

Zone D produced three burins, a drill, and a pointed piece. While their specific uses are unknown, such tools could have been used to make fire drills, to slot wooden hafts, or to cut and drill other artifacts of wood and bone, such as those described in Chapters 10 and 12.

Almost every living floor produced evidence for flake production in the form of cores or core fragments; only Zone E was without any evidence for nuclei. (Their absence in Zone B2+3 is not considered significant, given their presence in both B2 and B3.) Raw material came from within a 50-km radius, with a considerable amount probably coming from within a few kilometers of the cave. A few flakes (10) seem to have come off the core in the form of crude blades, probably fortuitously in most cases.

The chipped-stone assemblages for Zones B–E are given in detail in Tables 6.5–6.11.

TABLE 6.5
Chipped-Stone Tools from Zone E of Guilá Naquitz

Categories	1 × 1-m squares															Totals
	C10	C11	D5	D9	D10	D11	E10	E11	E12	F9	F10	F11	G4	G5	H10	
Choppers/knives																
Steep denticulate scrapers	1															1
Sidescrapers/knives																
End scrapers																
Burins																
Drills																
Pointed pieces																
Crude blades								1								1
Notched flakes		1									1			1		3
Flakes with sheen																
Utilized flakes							1							1		2
Projectile points															1	1
Bifaces, Variety B																
Bifaces, Variety C									1							1
Tanged picks																
Hammerstones		1											1			2
	1	2					1	1	1		1		1	2	1	11
Flake cores																
Core faces																
Core edges																
Debitage	3	23	2	4	1	6	3	4		1	1	3	9			60
	4	25	2	4	1	6	4	5	1	1	2	3	10	2	1	71

TABLE 6.6
Chipped-Stone Tools from Zone D of Guilá Naquitz

Categories	1 × 1-m squares																						Totals
	B9	C7	C8	C9	C10	D5	D7	D8	D9	D10	E5	E6	E7	E8	F4	F5	F6	F8	G5	G6	G7	G8	
Choppers/knives																							
Steep denticulate scrapers									1									1					2
Sidescrapers/knives																							
End scrapers																							
Burins										1											2		3
Drills								1															1
Pointed pieces		1																					1
Crude blades															1								1
Notched flakes							1	1	1							1		2					6
Flakes with sheen																1	1						2
Utilized flakes			1										1										2
Projectile points																							
Bifaces, Variety B																							
Bifaces, Variety C		1									1												2
Tanged picks																							
Hammerstones	1																				1		2
	1	2	1				1	2	2	1	1		1		1	2	1	3			3		22
Flake cores																							
Core faces		1										1					1	1					4
Core edges																1				1			2
Debitage	3	18	4	5	1	9	67	17	28	9	1	2	10	19	5	10	18	9	3	20	17	5	280
	4	21	5	5	1	9	68	19	30	10	2	3	11	19	6	13	20	13	3	21	20	5	308

TABLE 6.7
Chipped-Stone Tools from Zone C of Guilá Naquitz

	1 × 1-m squares																			
Categories	B9	B10	C8	C9	C10	D8	D10	D11	E5	E6	E7	E8	E9	E10	F5	F6	F9	F10	G9	Totals
Choppers/knives																				
Steep denticulate scrapers								2												2
Sidescrapers/knives																	1			1
End scrapers																				
Burins																				
Drills																				
Pointed pieces																				
Crude blades											1									1
Notched flakes	1		1	1	1			1			1			1				2	1	10
Flakes with sheen								1												1
Utilized flakes				2							1						1			4
Projectile points																				
Bifaces, Variety B																				
Bifaces, Variety C																				
Tanged picks																				
Hammerstones								1												1
	1		1	3	1			5			3			1			2	2	1	20
Flake cores	1																1			2
Core faces																				
Core edges													1	1						2
Debitage	11	7	5	24	16	28	15	38	10	19	37	8	41	23	2	16	43	27	5	375
	13	7	6	27	17	28	15	43	10	19	40	8	42	25	2	16	46	29	6	399

TABLE 6.8
Chipped-Stone Tools from Zone B3 of Guilá Naquitz

	1 × 1-m squares											
Categories	C6	C7	C9	C11	D5	D6	D8	D9	E5	E7	E8	Totals
Choppers/knives												
Steep denticulate scrapers			1		1		1					3
Sidescrapers/knives												
End scrapers								1				1
Burins												
Drills												
Pointed pieces												
Crude blades									1			1
Notched flakes			1						1	1		3
Flakes with sheen						1						1
Utilized flakes						1			1	1	1	4
Projectile points			1	1								2
Bifaces, Variety B												
Bifaces, Variety C												
Tanged picks												
Hammerstones												
			3	1	1	2	1	1	3	2	1	15
Flake cores			1		1							2
Core faces					1							1
Core edges												
Debitage	8	2	15		8	8	6	5	7	8	2	69
	8	2	19	1	11	10	7	6	10	10	3	87

TABLE 6.9
Chipped-Stone Tools from Zone B2 + 3 of Guilá Naquitz

	1 × 1-m squares							
Categories	C8	C10	D10	E6	E7	F4	F5	Totals
Choppers/knives								
Steep denticulate scrapers					1			1
Sidescrapers/knives								
End scrapers								
Burins								
Drills								
Pointed pieces								
Crude blades				1				1
Notched flakes	1							1
Flakes with sheen								
Utilized flakes								
Projectile points								
Bifaces, Variety B								
Bifaces, Variety C				1				1
Tanged picks							1	1
Hammerstones	1	1						2
	2	1		2	1		1	7
Flake cores								
Core faces								
Core edges								
Debitage	16	11	30	38		9	11	113
	18	12	30	38	1	9	12	120

TABLE 6.10
Chipped-Stone Tools from Zone B2 of Guilá Naquitz

Categories	1 × 1-m squares													Totals
	C6	C7	C9	C11	D5	D8	D9	D11	E5	E6	E9	E10	E11	
Choppers/knives								1						1
Steep denticulate scrapers	1													1
Sidescrapers/knives											2			2
End scrapers														
Burins														
Drills														
Pointed pieces														
Crude blades		1												1
Notched flakes	1													1
Flakes with sheen										1	1			2
Utilized flakes	1			1	1		1							4
Projectile points														
Bifaces, Variety B														
Bifaces, Variety C				1										1
Tanged picks														
Hammerstones														
	3	1		2	1		1	1		1	3			13
Flake cores	1										1			2
Core faces														
Core edges			2				1	1						4
Debitage	33	2	19	32	12	31	25	12	5		21	7	2	201
	37	3	21	34	13	31	27	14	5	1	25	7	2	220

TABLE 6.11
Chipped-Stone Tools from Zone B1 of Guilá Naquitz

Categories	1 × 1-m squares																											Totals
	B9	B10	B11	C7	C8	C9	C10	C11	C12	D5	D6	D7	D8	D10	D11	E5	E6	E7	E9	E10	E11	F4	F5	F6	F7	G4	G5	
Choppers/knives																			1									1
Steep denticulate scrapers				1																			1		1			3
Sidescrapers/knives																					1							1
End scrapers																												
Burins																1												1
Drills																												
Pointed pieces																												
Crude blades								2				1					1											4
Notched flakes	1					1		1		1	3		3		1				1		2				2			16
Flakes with sheen																												
Utilized flakes						1	1										1				1		1					5
Projectile points																								1				1
Bifaces, Variety B																							1					1
Bifaces, Variety C																												
Tanged picks																												
Hammerstones																										1		1
	1			1		2	1	3		1	3	1	3		1	1	2		2		4		3	1	3	1		34
Flake cores								2											1				2					5
Core faces							1	1								1		1								1		5
Core edges								1																				1
Debitage	37	3	12	2	8		3		30	11	17	28	37	10	5	17	30	35	19	10	26	10	60	13	11	26	6	466
	38	3	12	3	8	2	5	7	30	12	20	29	40	10	6	19	32	36	22	10	30	10	65	14	14	28	6	511

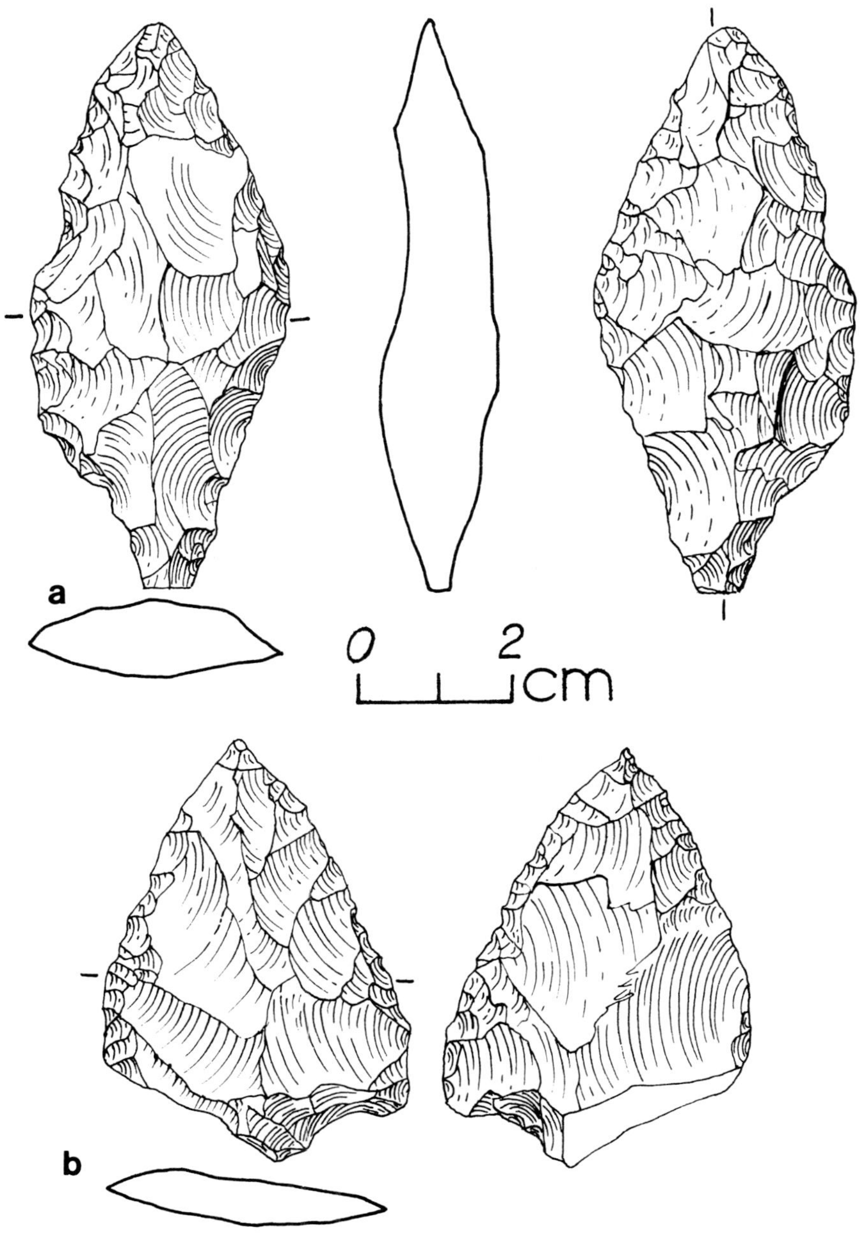

FIGURE 6.26. Projectile points: a, unfinished Lerma(?) point, Zone E, Square H10 (dorsal surface, ventral surface, and two cross sections); and b, possible Trinidad point, Zone B1, Square F6 (dorsal surface, ventral surface, and cross section).

FIGURE 6.27. Projectile points or fragments: a, possible Trinidad point, reddish-gray silicified tuff, Zone B1, Square F6; b, possible fragment of corner-notched point, Feature 12, presumably redeposited; c, Almagre point, black silicified siltstone, Feature 13 (Zone A), presumably redeposited; and d, unfinished Lerma(?) point, waxy brownish-gray chert or chalcedony, Zone E, Square H10 (length, 76 mm).

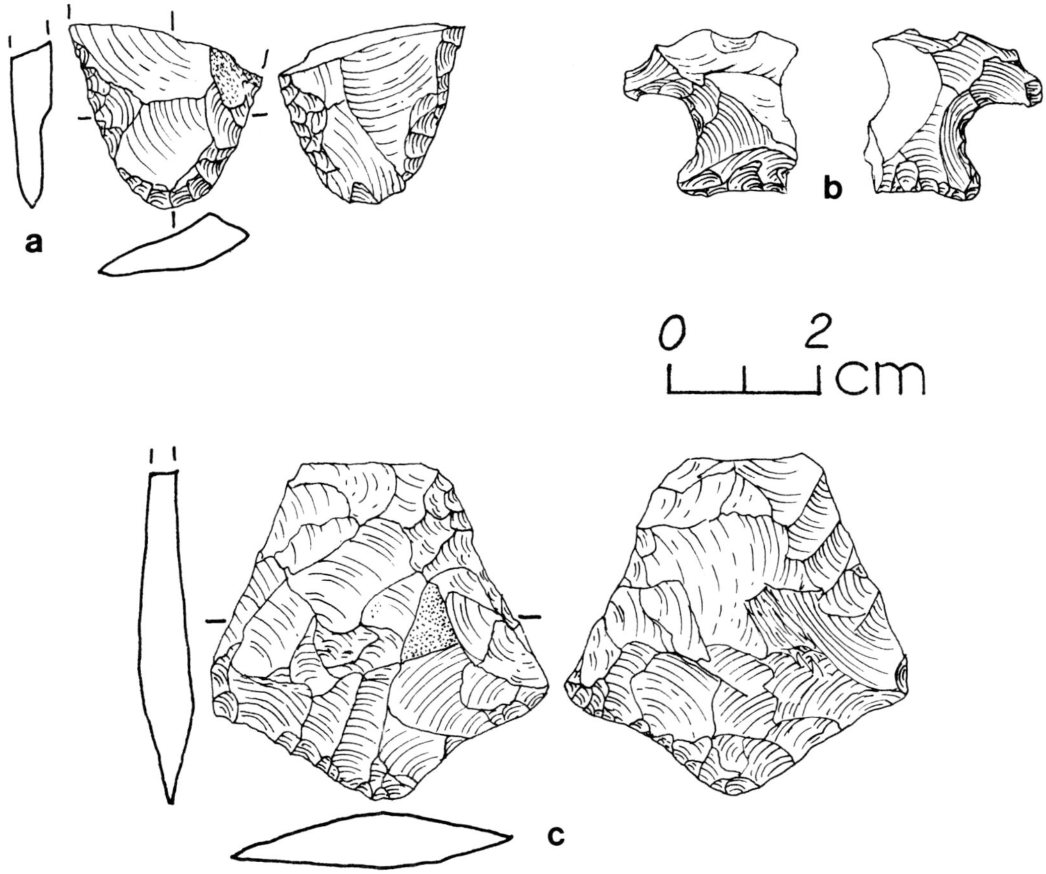

FIGURE 6.28. Projectile points: a, possible contracting stem of Trinidad(?) point, Zone B3, Square C11; b, possible fragment of corner-notched point, Feature 12 (Zone A), presumably redeposited from a preceramic level; and c, Almagre point with broken tip (dorsal surface, ventral surface, and two cross sections), Feature 13 (Zone A), presumably redeposited from a preceramic level.

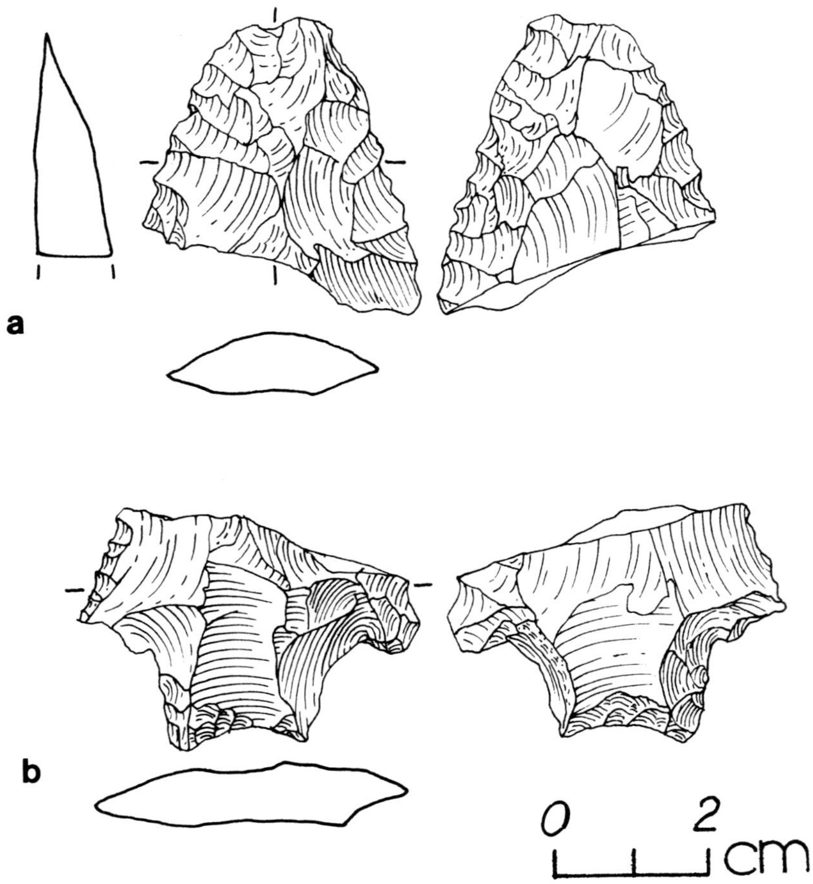

FIGURE 6.29. Broken Pedernales point: a, distal half of point, found *in situ* in Zone B3, Square C9; and b, proximal half of point, found on talus slope below cave during survey. For a view of the two halves joined, see Fig. 6.31b.

FIGURE 6.30. Complete Pedernales point, found on surface near cave.

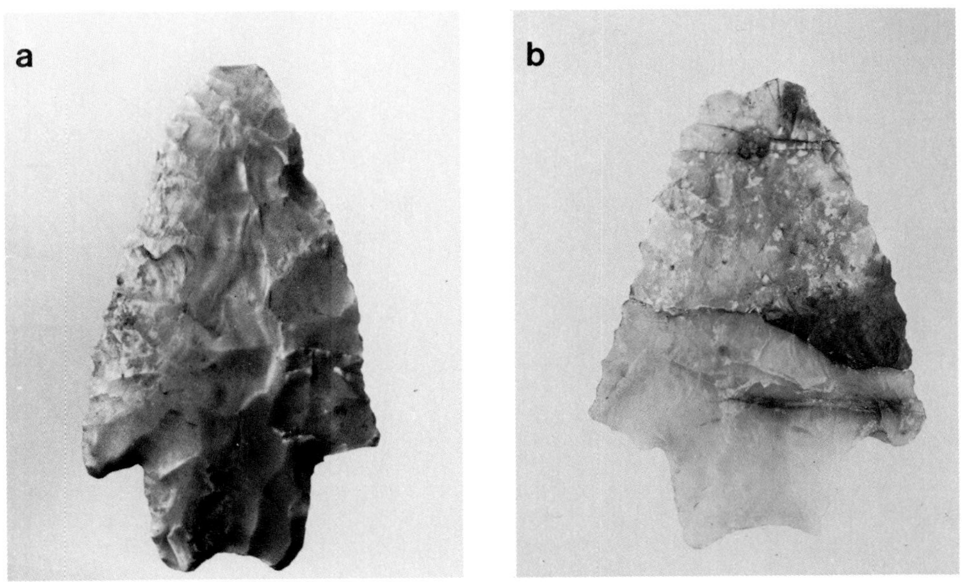

FIGURE 6.31. Pedernales points, Matadamas chalcedony: a, complete waxy yellowish specimen, found on surface (length, 68 mm); and b, waxy white and yellow specimen whose distal half was found in Zone B3, Square C9, and whose proximal half was found on the talus below the cave (the distal half was burned).

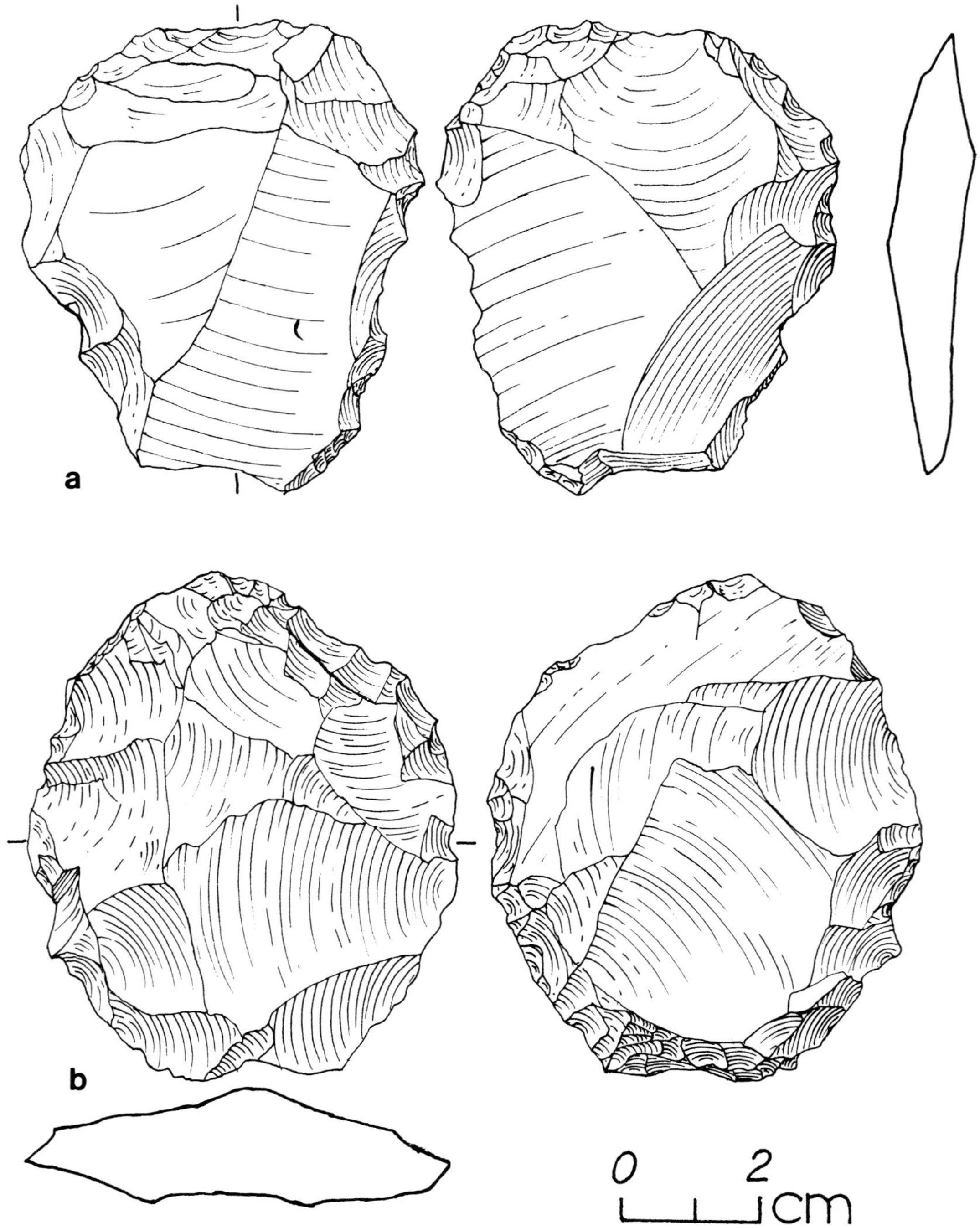

FIGURE 6.32. Bifaces, Variety B: a, Zone B1, Square F5 (both sides and cross section); and b, Zone A, presumably redeposited from a preceramic level (both sides and cross section).

FIGURE 6.33. Bifaces, Variety B: a, silicified tuff specimen, Zone B1, Square F5; and b, chert specimen, Zone A, presumably redeposited (dimensions, 73 × 63 mm).

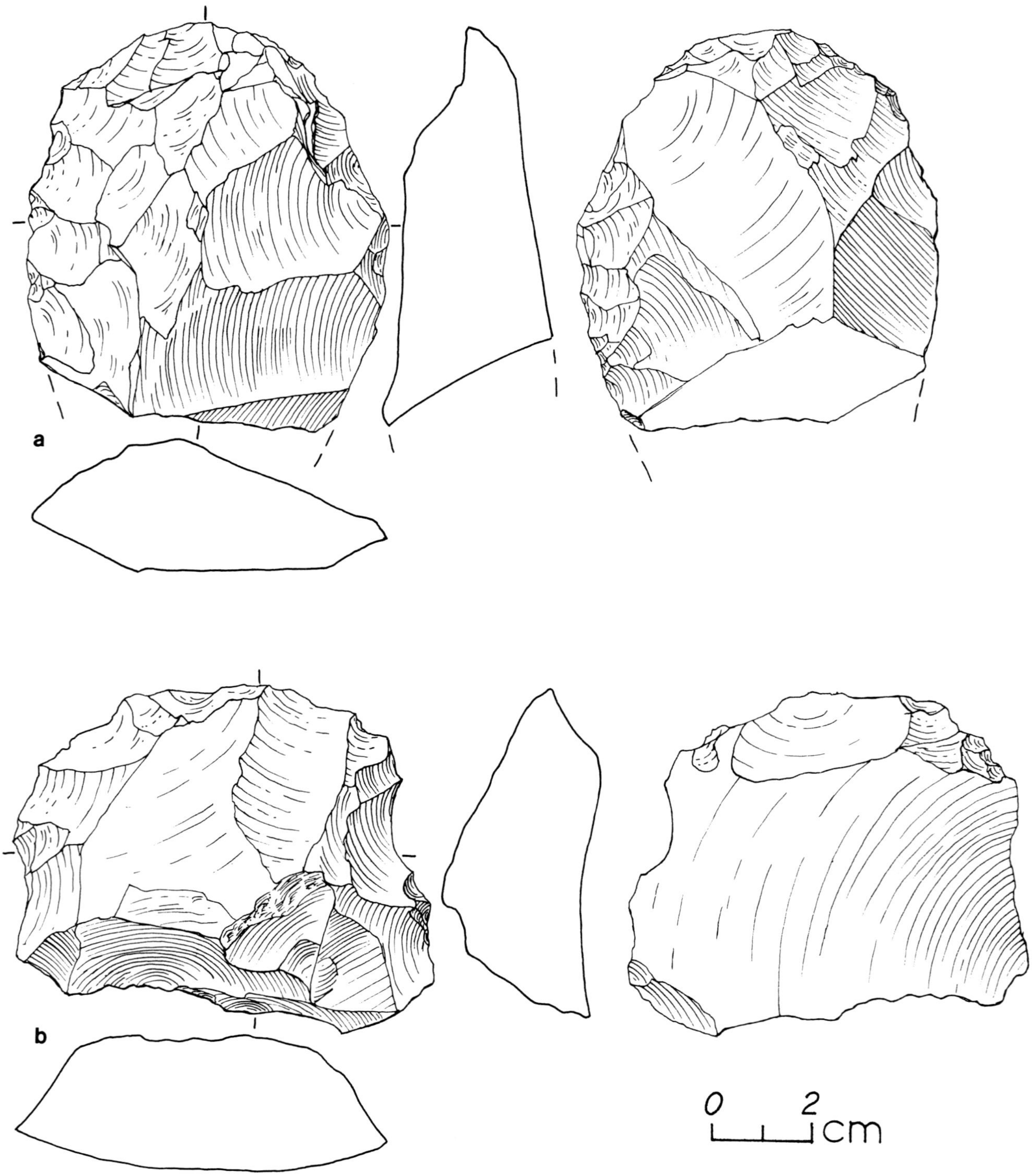

FIGURE 6.34. Bifaces, Variety C: a, Zone D, Square E5 (both sides and two cross sections); and b, Zone D, Square C7 (both sides and two cross sections).

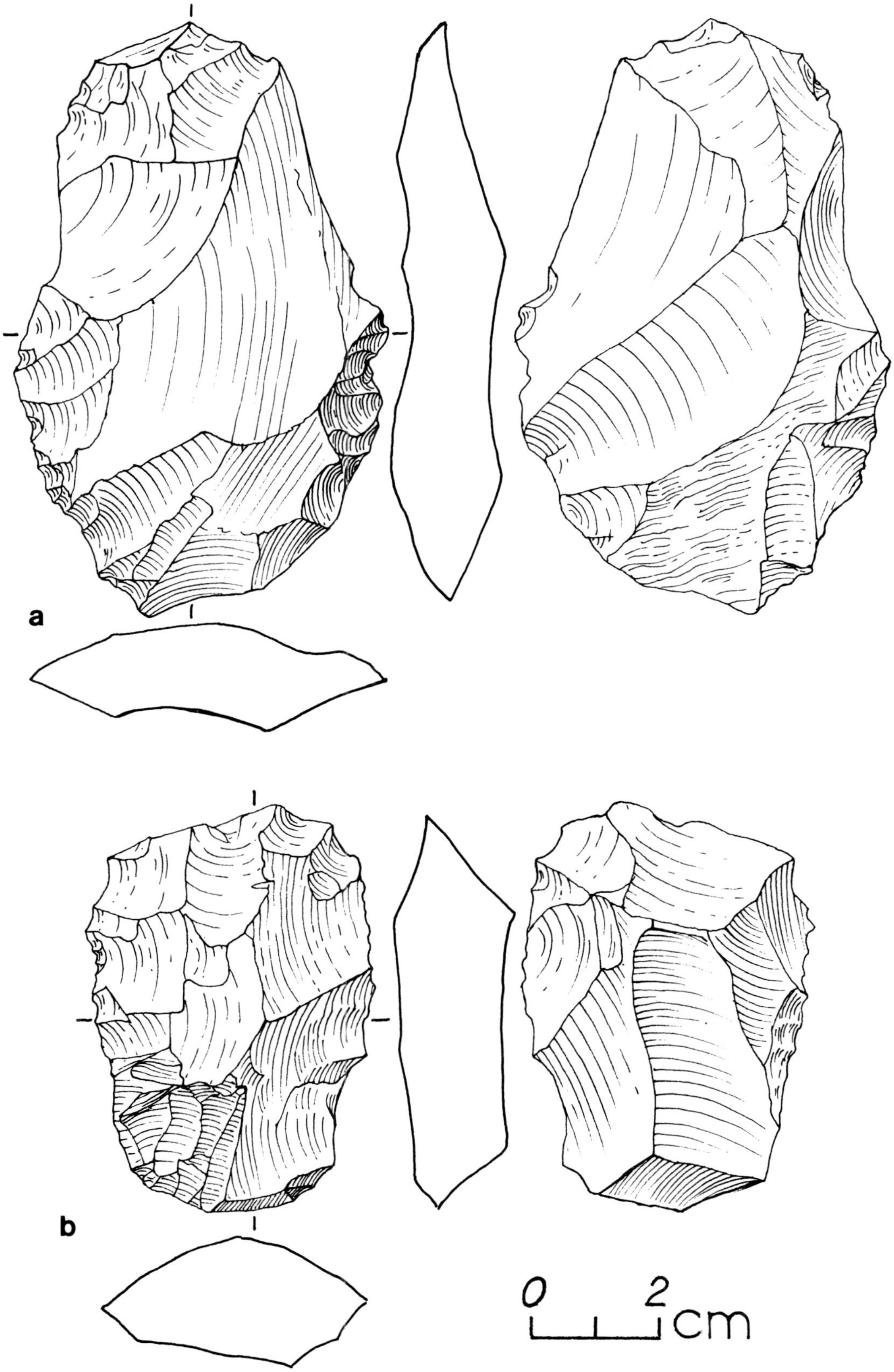

FIGURE 6.35. Bifaces, Variety C: a, Zone B2 + 3, Square E6 (both sides and two cross sections); and b, Zone B2, Square C11 (both sides and two cross sections).

FIGURE 6.36. Bifaces, Variety C: a, Zone D, Square E5; b, Zone B2 + 3, Square E6 (maximum length, 92 mm); c, Zone B2, Square C11; and d, Zone D, Square C7 (a and b are silicified tuff, and c and d are chert).

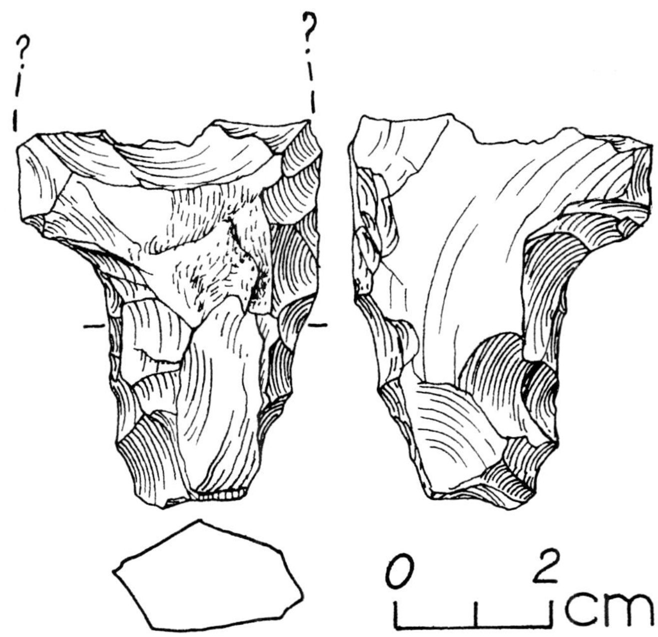

FIGURE 6.37. Possible tanged pick, Zone B2 + 3, Square F5.

FIGURE 6.39. Unretouched flake (chert) with maguey fiber still clinging to the left edge (dimensions, 65 × 55 mm), found while cleaning profile of Square C8, Zone B or C.

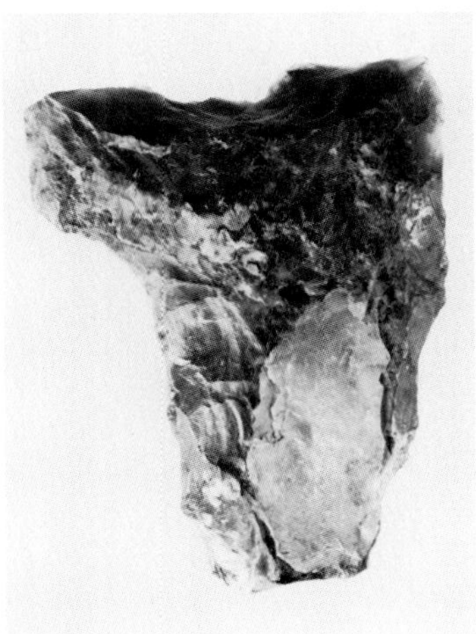

FIGURE 6.38. Possible tanged pick (chert), Zone B2 + 3, Square F5.

7

Sources of the Guilá Naquitz Chipped Stone

Michael E. Whalen

As Hole has pointed out in Chapter 6, abundant raw material for chipped-stone tools was available within a few kilometers of Guilá Naquitz, providing the occupants were not particularly concerned with quality. In cases where they *were* concerned with quality, much better material was available at distances of 50 km, which probably involved no more than a 2-day round trip. In this chapter I consider some of the available sources of chipped stone in order of increasing distance from the cave (Table 7.1). The discussion is based on my own 1971 survey of Oaxacan chert sources (Whalen n.d.), augmented by data from Holmes (1897), Williams and Heizer (1965), and unpublished geological studies by Michael J. Kirkby during the 1960s.

WITHIN 10 KILOMETERS

Guilá Naquitz, like all the Mitla caves, occurs in a massive range of Tertiary volcanics, mainly ignimbrites (Fig. 7.1). While much of this material is unsuitable for chipping, there are localized strata within it that have been silicified to the point of becoming flinty. Since these flinty layers of silicified ignimbrite weather more slowly than the softer volcanic tuff, the material can frequently be found as pebbles in arroyo beds or sticking out as elevated veins on the surface of cliffs and bedrock outcrops. The occupants of Guilá Naquitz could have picked up some silicified raw material in almost any arroyo bed within a few kilometers of the site.

There are also larger sources of chipped stone in the area, the most notable being the Fortress of Mitla some 1.5 km southeast of Guilá Naquitz (Fig. 7.2). This rocky hill, fortified in the Postclassic period, was already identified as a chipped-stone quarry by W. H. Holmes in the past century (Holmes 1897:287). The fortress is described by Williams and Heizer (1965:47) as composed of a series of massive, cliff-forming volcanic tuffs underlain by a greenish, altered ignimbrite that is interbedded with much thicker layers of well-stratified whitish tuffs and tuffaceous muds and silts. Over the millennia, moving groundwaters—enriched in silica through alteration of the overlying ignimbrites—have brought about silicification of the lower tuffaceous sediments.

> These very fine-grained, airborne and fluviatile deposits are extensively silicified and in many places are veined and replaced by chalcedony. Close to the spring near the western base of the hill a few steeply dipping and vertical veins of chalcedony, up to about 30 cm. thick, cut the thinly bedded tuffs; not far away thin lenses of chalcedony conform to the bedding of the tuffs. . . . The gently undulating slopes adjoining the Fortified Hill are thickly strewn with angular fragments of chalcedonic flint. To the east of the spring, as Holmes noted, the flints are almost all pale to dark gray in color; to the west, however, there are abundant yellow, brown, and orange colored flints. (Williams and Heizer 1965:47)

Thus the occupants of the Mitla caves had available to them an extensive flint source to either side of an important local spring. Nor is the Mitla Fortress the only good source of silicified ignimbrite. In 1971, I sampled several other sources northwest of Mitla and found them to be characterized by the same kinds of light brown, dark brown, white, black, and gray flint seen at the Fortress, although the predominant color of all beds is medium-to-dark brown. A great deal of the chipped stone from Guilá Naquitz appears to have come from these local Mitla sources.

TABLE 7.1
Sources of Chipped Stone in the Northern and Eastern Valley of Oaxaca[a]

Source	Nature of material	Distance from Guilá Naquitz (km)
Local arroyos	Silicified ignimbrite	1–2 (E)
Mitla Fortress	Silicified ignimbrite	3.5 (SE)
Caballito Blanco	Silicified ignimbrite	3.5 (W)
Loma Larga	Silicified ignimbrite	4 (S)
Tlacolula	Silicified ignimbrite	6 (W)
Díaz Ordaz	Silicified ignimbrite	7 (N)
Rojas de Cuauhtemoc and related sources	Low-quality chert	25 (NW)
Loma del Trapiche	Chert, chalcedony	47 (NW)
Matadamas	Chert, chalcedony	50 (NW)
Rancho Alemán	Flint	55 (NW)
San Lázaro Etla	Chert	55 (NW)

[a]Data from Whalen (n.d.).

Silicified ignimbrite is also available at localities such as Caballito Blanco (3.5 km west of the cave), Loma Larga (4 km to the south), Tlacolula (6 km to the west), and Díaz Ordaz (7 km to the north). The Caballito Blanco source is spatially well defined, although material was not available in any quantity. Colors seen at this source included black, tan, amber, and grayish white, while the predominant color was dull red, occasionally mottled with yellow or tan.

The Loma Larga source can best be characterized as brownish, badly fractured and impure, although occasional small pieces of high quality can be found. Other colors included dark brown, dark gray, brownish-gray, tan, tan with opaque mottling, light brown, amber, and white.

Silicified materials from the Tlacolula source are commonly dark brown in color. This source was particularly badly dispersed, meager, and hard to characterize. The Díaz Ordaz source is a major outcrop containing flinty material that is almost all black, dark gray, or gray with black striations. Most of it is badly fractured and impure, although occasional small pieces of high quality can be found.

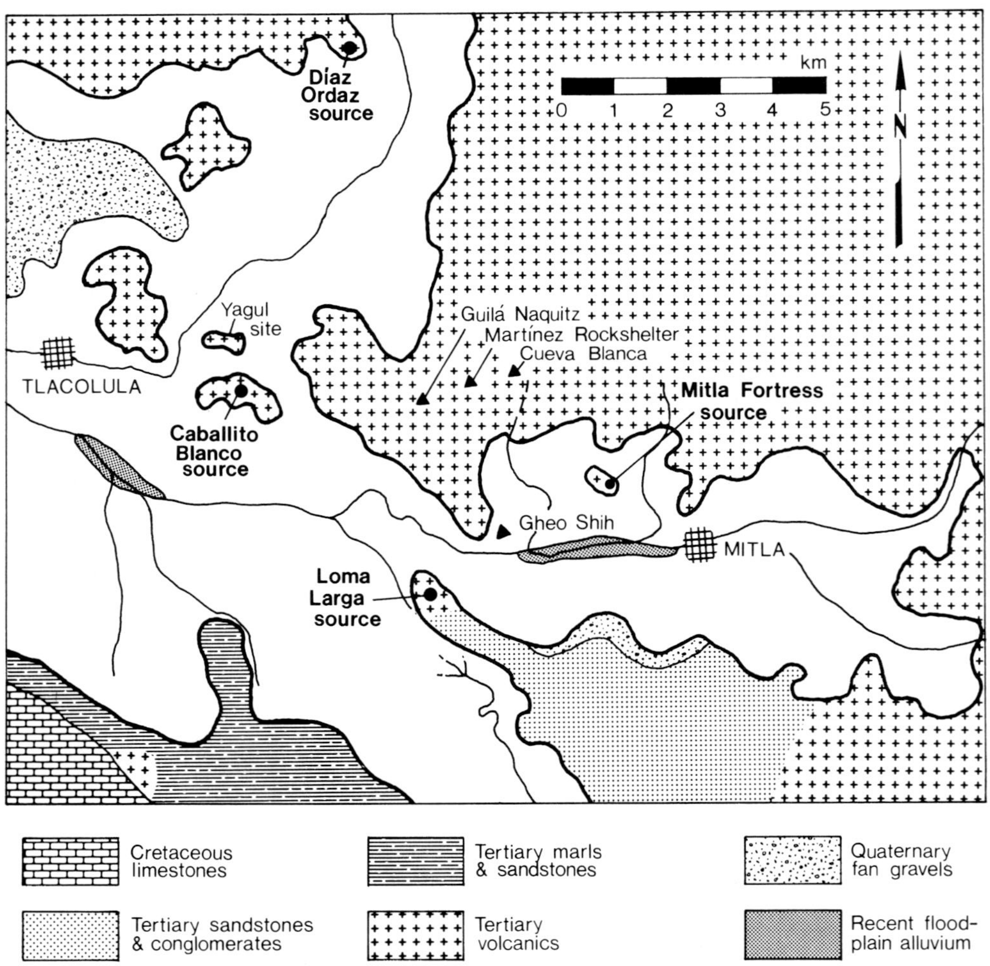

FIGURE 7.1. Sources of chipped stone in the Tlacolula subvalley (geological base map by Michael J. Kirkby).

FIGURE 7.2. The Fortress of Mitla was a major source of silicified ignimbrite.

In summary, the sources available within 10 km of Guilá Naquitz are generally silicified ignimbrites, rhyolites, or tuffaceous sediments in which brown, black, and gray colors predominate. Each source contains such a range of colors that it would be difficult to specify the origin of an individual piece of chipped stone beyond assigning it to the Tlacolula–Mitla region of Tertiary volcanics. On logistic grounds, the Mitla Fortress should have been one of the major sources for Guilá Naquitz because of its proximity.

AT DISTANCES OF 25 KILOMETERS

Traveling west from Guilá Naquitz, one eventually leaves the area of ignimbrite cliffs and enters a zone where the Tertiary volcanics are interspersed with Cretaceous limestones. This central area of the Valley of Oaxaca has some relatively mediocre chert and flint sources, mostly in the limestones. One such source occurs at Rojas de Cuauhtemoc, some 25 km northwest of Guilá Naquitz, and there are others in the limestone formations in the northern piedmont of the western Tlacolula valley. Low-quality siltstone is also available in this central region.

The Rojas source consists mainly of brownish-white chert, occasionally with darker tinges, often highly fractured and of limited use for tool manufacture. Given its location, however—roughly a day's round trip from the Mitla caves—it must be considered a potential source of raw material for Guilá Naquitz.

AT DISTANCES OF 45–55 KILOMETERS

As one continues west into the Etla arm of the Valley of Oaxaca, one reaches a zone of Precambrian metamorphics overlain by localized areas of Cretaceous limestone (Fig. 7.3). These limestones contain by far the highest-quality cherts and chalcedonies available to the ancient knapper, and they were extensively used in Precolumbian times. The best sources occur in a north–south line west of the Atoyac River, from San Lázaro Etla to San Lorenzo Cacaotepec, especially at the boundary between the limestone and the adjacent Precambrian gneiss.

Largest and most important of these sources is the one that runs for more than a kilometer between the *agencias* of Matadamas and Guadalupe Hidalgo, south of Soledad Etla. Known variously as "La Peña de Matadamas" or "Loma Soledad," the setting is a slender ridge on whose western side a series of limestone blocks form low cliff faces (Fig. 7.4). Running through these blocks are veins of waxy, high-quality chalcedony (Fig. 7.5). On the lower, eastern slope of the ridge, chert or chalcedony pebbles are strewn over an area dozens of hectares in extent (Fig. 7.6). Color of the material varies from white to bluish-white, grayish-white, or brownish-white, sometimes mottled. A related source, presumably part of the same basic geological formation, occurs a kilometer to the south at Loma del Trapiche.

The Matadamas ridge is perhaps the largest and highest-quality chert/chalcedony source in the entire Valley of Oaxaca and appears to have been utilized by the occupants of the Mitla caves when they needed good raw material. For example, both Pedernales projectile points found at Guilá Naquitz appear to be made from yellowish-white Matadamas chalcedony. The travel distance of this source from Guilá Naquitz would be approximately 50 km.

Two other chipped-stone sources occur in a limestone formation to the north, some 55 km from the Mitla caves. One

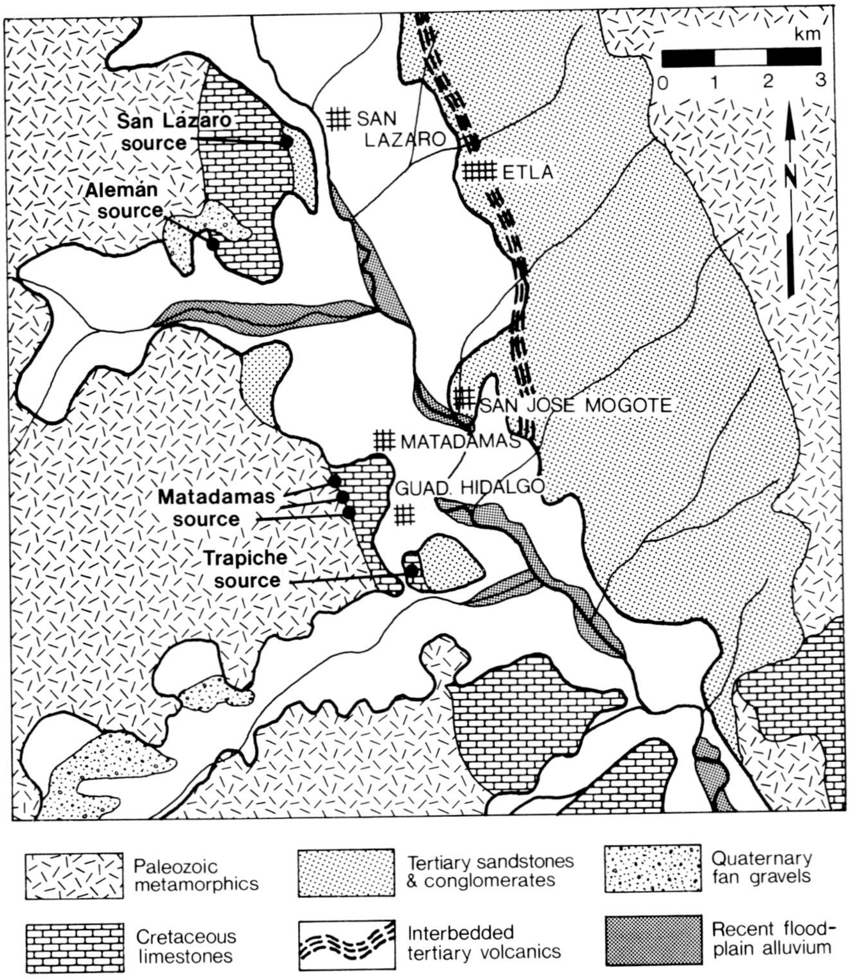

FIGURE 7.3. Sources of chipped stone in the Etla subvalley (geological base map by Michael J. Kirkby).

source lies just across the Atoyac River from San Lázaro Etla, near the eastern edge of the limestones. The other, near the contact between the piedmont gneiss and the western border of the limestones, is Rancho Alemán. The Rancho Alemán source consists of black or very dark brown flint which outcrops for several kilometers along the margin of the limestone, but is sparser and of lower quality than the Matadamas source. The San Lázaro source is small, but more closely resembles the Matadamas cherts and chalcedonies in its whitish color and quality.

In summary, the finest raw material available within the Valley of Oaxaca would seem to be the white and off-white chalcedonies and cherts of the Matadamas ridge in the Etla subvalley. While they did not provide the bulk of the material at Guilá Naquitz, such cherts showed up there in significant quantities, especially for use in bifacial tools such as projectile points. Whether these cherts were procured directly (by means of 50-km trips to the source) or were obtained from neighboring bands of hunter–gatherers in the Etla subvalley is unknown.

Finally, this brief discussion of chipped-stone sources should not be considered definitive in any sense. It deals only with some of the best-known sources and should be regarded as a preliminary statement, pending more intensive geological survey of the Valley of Oaxaca.

FIGURE 7.4. Limestone blocks form low cliff faces on the western brow of the Peña de Matadamas chert and chalcedony source.

FIGURE 7.5. Chunks of waxy bluish-white chalcedony on the surface of the Matadamas source.

FIGURE 7.6. Acres of chert pebbles fill an arroyo bed on the eastern slope of the Peña de Matadamas.

8

Ground-Stone Artifacts

Kent V. Flannery

Twenty-one artifacts of ground stone appeared in Zones B–E of Guilá Naquitz. Included were objects assumed to be for food preparation, tool manufacture, and pigment grinding. In writing this chapter, I would like to acknowledge the help of Marcus C. Winter and Susan H. Lees, who measured the tools and divided them into trial types in our Oaxaca lab. Thanks are also due Michael J. Kirkby, who identified the raw material used. I have revised the trial types after a comparison of our material to the ground-stone tools of the Tehuacán caves, published by MacNeish *et al.* (1967).

PRELIMINARY DISCUSSION

One of the interesting aspects of ground-stone technology in Mesoamerica is how early it appears. For example, the Late Ajuereado phase occupants of the Tehuacán Valley were using mortars, pestles, and manos or mullers at a time when "Paleoindian" projectile point types were still in style. Guilá Naquitz also had a variety of mortars, manos, and metates in levels radiocarbon dated between 9000 and 7000 B.C.

A second point that might be made, however, is that these early ground-stone tools are so generalized and reflect the locally available raw materials so strongly, that a very cautious typology is called for. As is evident from the type descriptions below, our material often resembles that from Tehuacán on the level of general categories, but it less often resembles it on the level of specific types. For one thing, the stones available to be picked up by the occupants of Guilá Naquitz were somewhat different from those available near Coxcatlán or El Riego caves. For another thing, the shape of a ground-stone tool is strongly affected by length of use. After 1 day, a boulder used as a metate will have a very shallow depression. After more use, it may be "trough shaped". After a whole season of use, it may be worn to a "basin." Thus, any shape terminology applied by an archaeologist will be strongly influenced by the point in the use cycle at which a metate is discarded. It is one thing when archaeologists devise categories to distinguish between activities such as rotary or back-and-forth grinding (see below); it is quite another thing when they devise categories that really only reflect the length of time a grinding stone was used. For all these reasons, as well as the small size of our sample, we have kept our categories simple.

Most of the ground-stone tools from Guilá Naquitz were made of ignimbrite, or volcanic tuff,[1] from the Miocene volcanics that constitute bedrock in the eastern Valley of Oaxaca. It appears that the occupants of the cave picked up ignimbrite cobbles from stream beds in the Mitla area to use as one-hand manos. They also carried in ignimbrite slabs to use as metates and larger cobbles or small boulders to use as mortars. Occasionally they brought in raw material such as limestone or quartzite from areas 25–50 km distant, most likely in the Central or Etla regions of the valley. Some of these "nonlocal" raw materials were used for grinding red pigment.

Judging by the plants found in the same levels, the metates and mortars at Guilá Naquitz may have been used for grinding or pounding up such things as acorns, piñon nuts, *susí* nuts, and mesquite beans.

ONE-HAND MANOS

Perhaps the most common grinding implement used at Guilá Naquitz was the tool known in the southwestern United States as a "one-hand mano." For the most part, these were

[1]Apparently volcanic tuff was also preferred for ground-stone tools in preceramic Tehuacán, although the term "volcanic tufa" is used throughout that report (MacNeish *et al.* 1967:101–121). We believe the authors *meant* to say "tuff," since "tufa" is a fossil spring deposit.

rough-and-ready tools made from ignimbrite cobbles picked up in a stream bed and shaped as much by stream action and use as by design. Any of them could have been picked up at the beginning of a cave occupation, used conveniently in one hand by either a man or woman, and discarded at the end of the occupation with the knowledge that a replacement would be easy to find.

These are very generalized tools, and we decided it would be a mistake to be overly specific in our classification of them. Several are fragmentary, and the most complete may give evidence of several kinds of use—rotary grinding, back-and-forth grinding, and pestle-like pounding on an edge. Our categories are essentially morphological, and we should keep in mind our own warning that some morphologies may simply reflect the length of time the object was used.

With their much larger sample, MacNeish *et al.* have been very precise about whether they are talking about mullers, manos, or pestles. "*Manos* are stones used in metates with a back-and-forth motion, whereas *mullers* were used in milling stones with a rotary motion"; *pestles* are "elongated stone objects with one end flattened by a rotary grinding or pecking motion, either intentionally or through use" (1967:101, emphasis added). They correctly point out that the form of grinding affects the morphology of both the hand-held cobble and the milling stone against which it was rubbed. Thus, rotary action produced both ovoid mullers and basin-shaped milling stones, while back-and-forth motion produced both oblong manos and trough metates. However, they also concede that "it is often difficult to tell" whether a rotary or back-and-forth motion was used (1967:101); moreover, in the case of Tehuacán types that most resemble ours, they state that "ovoid manos are extremely similar to ovoid mullers in size and form" (1967:108).

With our much smaller sample of tools from Guilá Naquitz, we decided it would be most practical to refer to all these utilized cobbles simply as "one-hand manos" and to classify them by (1) overall shape and (2) the number and morphology of the grinding surfaces. The two overall shapes were "discoidal" and "subrectangular." Within the discoidal category, there were three types: Cobbles with one convex grinding surface, cobbles with two convex grinding surfaces, and cobbles with one concave and one convex grinding surface. These were distinctions we felt we could make consistently. We did not feel we could consistently distinguish between mullers and manos, especially since some cobbles appeared to have been used in a rotary fashion on one side and back-and-forth on the other. It is also the case that not all archaeologists agree that rotary motion disqualifies a specimen as a mano.

Discoidal One-Hand Manos, One Convex Grinding Surface

Fig. 8.1a, b.
Number of specimens: 2.
Dimensions: diameters, 6.5 cm and 11.2 cm; thickness, 3.9 cm and 5.1 cm.
Raw material: ignimbrite (Zone D) or limestone (Zone B2).
Description: These are disc-shaped stream cobbles that are smoothed on one surface from use in grinding. It is the smoothed surface that is convex; the unutilized surface is irregular. One specimen is slightly battered at one end, as if used briefly as a pounder or pestle. In terms of the Tehuacán typology, these specimens are probably most like MacNeish *et al.'s* "ovoid mullers" (1967:106).

Discoidal One-Hand Manos, Two Convex Grinding Surfaces

Fig. 8.1c.
Number of specimens: 4.
Dimensions: diameter, 9.3 cm (only complete specimen); thickness, 9.3 cm, 4.7 cm, 4.6 cm, and 4.1 cm.
Raw material: ignimbrite.
Description: These disc-shaped stream cobbles are deliberately pecked and ground on all surfaces to achieve a more regular shape. Both upper and lower surfaces are convex and smoothed from use in grinding. The cross section is subrectangular. Striations from use appear on both convex surfaces. Most of these were probably used in rotary fashion, but in several cases a back-and-forth motion cannot be ruled out. Because they are pecked and ground to shape, these artifacts are probably most closely related to the "ovoid manos" of the Tehuacán caves (MacNeish *et al.* 1967:108).

Discoidal One-Hand Mano, One Convex Grinding Surface, One Concave Grinding Surface

Fig. 8.1d.
Number of specimens: 1.
Dimensions: diameter; 9.3 cm; thickness, 4.0–5.7 cm.
Raw material: ignimbrite.
Description: This stream cobble is roughly oval in outline and concave–convex in cross section. The primary grinding surface is convex, appears well smoothed from use in grinding, and shows some parallel striations. One end is battered from pounding, as if used briefly as a pounder or pestle. This artifact illustrates the fact that one-hand manos could be used for more than one purpose and with more than one grinding motion.

Subrectangular One-Hand Mano, Two Convex Grinding Surfaces

Figs. 8.1e and 8.2a.
Number of specimens: 1.
Dimensions: length, 9.6 cm; width, 6.9 cm; thickness, 4.5 cm.
Raw material: veined quartzite(?).
Description: This is an unusually well-finished, symmetrical specimen, perhaps because the excellent raw material encouraged and permitted better craftsmanship. If it was

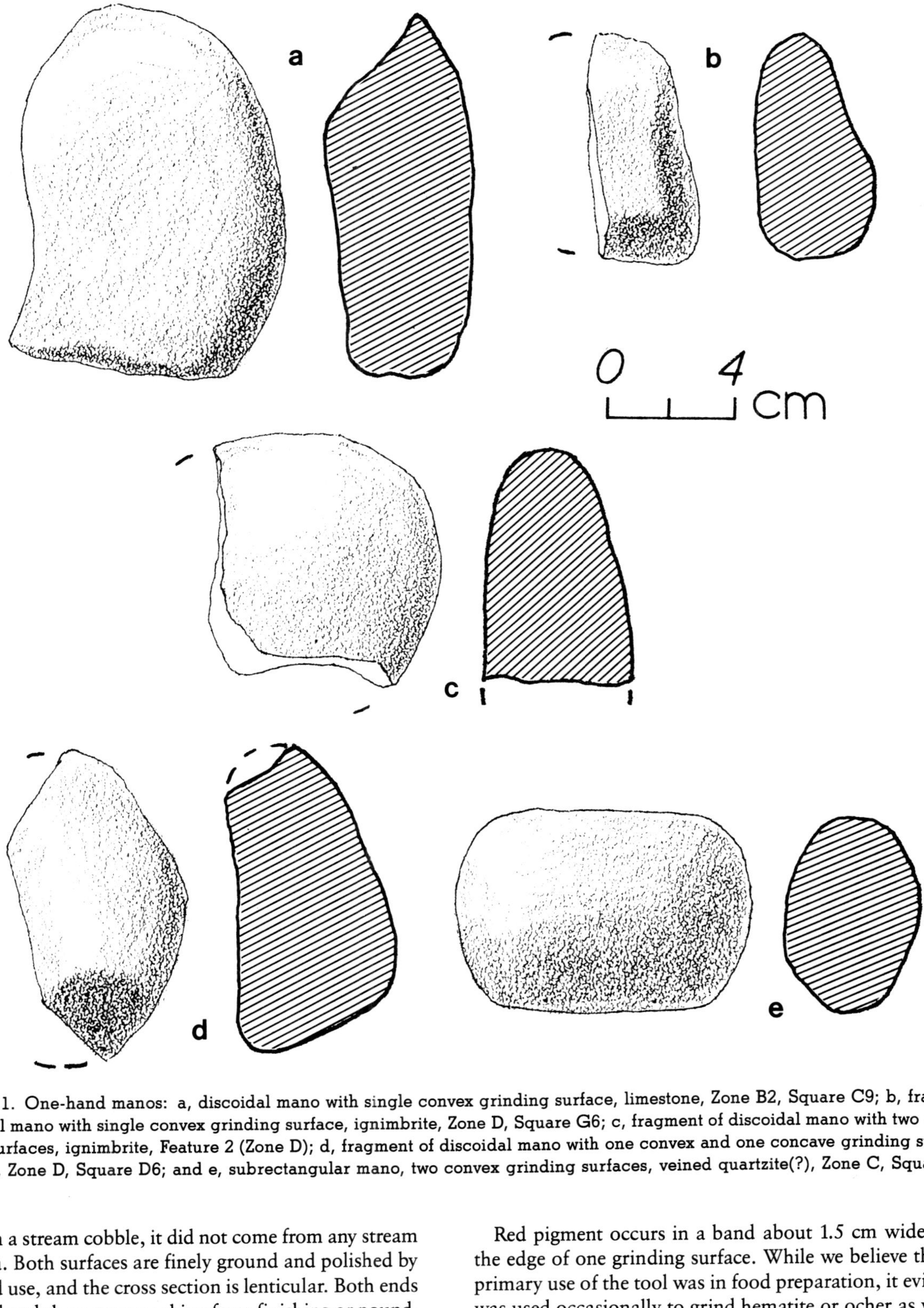

FIGURE 8.1. One-hand manos: a, discoidal mano with single convex grinding surface, limestone, Zone B2, Square C9; b, fragment of discoidal mano with single convex grinding surface, ignimbrite, Zone D, Square G6; c, fragment of discoidal mano with two convex grinding surfaces, ignimbrite, Feature 2 (Zone D); d, fragment of discoidal mano with one convex and one concave grinding surface, ignimbrite, Zone D, Square D6; and e, subrectangular mano, two convex grinding surfaces, veined quartzite(?), Zone C, Square F9.

made from a stream cobble, it did not come from any stream near Mitla. Both surfaces are finely ground and polished by design and use, and the cross section is lenticular. Both ends are ground and show some pecking from finishing or pounding. Indeed, this specimen could have been used both as a pestle and as a back-and-forth mano.

Red pigment occurs in a band about 1.5 cm wide along the edge of one grinding surface. While we believe that the primary use of the tool was in food preparation, it evidently was used occasionally to grind hematite or ocher as well. It would have been far more suitable for this than one of the coarser ignimbrite manos.

FIGURE 8.2. Ground-stone tools: a, subrectangular one-hand mano with two convex grinding surfaces, veined quartzite(?), Zone C, Square F9; b, fragment of slab paint palette with red pigment, ignimbrite, Zone B1, Square F6; and c, fragment of mortar, ignimbrite, Zone B2 + 3, Square C10.

In terms of the Tehuacán artifacts, this mano resembles both the "oblong manos" with two ground surfaces and two pecked ends (MacNeish *et al.* 1967: Fig. 90, bottom) and the "cuboid pestles" (1967:105, and Fig. 84, lower right). As indicated above, it may have performed the functions of both.

Miscellaneous Mano Fragments

Number of specimens: 2.

Dimensions: no measurements possible.

Raw material: ignimbrite.

Description: These are fragments of river cobbles with evidence for at least one convex surface smoothed by use in grinding. They are probably pieces of broken one-hand manos, but it cannot be determined whether they had one or two grinding surfaces.

METATES

Fragments of metates were much less frequent than manos, and no complete specimens were recovered. We assume that the metates involved were used with one-hand manos of the types described above. Based on the Tehuacán cave material, we would therefore expect to find MacNeish *et al.'s* "boulder metate-milling stones" (1967:118, Fig. 98), which were used with their "ovoid mullers", or "basin-shaped metates" (p. 120, Fig. 99), which were used with "one-handed manos."

However, there are some differences between our material and that from Tehuacán. In the latter area, MacNeish *et al.* found that the distinction made in the southwestern United States between slab and block metates was not helpful to them, since "only fourteen excavated specimens appeared to have been made from slabs of rock or from cave rock-fall, and four of these are over two inches thick, which would exclude them from the slab category of the Southwest" (1967:118). At Guilá Naquitz, on the other hand, three of four metates were on slabs under 2 inches thick, so we have found the category useful. This reinforces our suspicion that locally available raw materials may play more of a role in determining which "types" are present than stylistic preferences.

Slab Metate Fragments

Fig. 8.3a.
Number of specimens: 3.
Dimensions: thickness, 1.8 cm, 3.6 cm, and 1.9–2.3 cm.
Raw material: ignimbrite.
Description: These roughly shaped thin slabs of volcanic tuff have a single flat or slightly concave grinding surface. The bottom of the slab is pecked to form an uneven but generally flat surface. None of the three fragments show any evidence of deliberately finished edges. All specimens show parallel striations on the grinding surface. Because of the small size of the fragments, no overall dimensions of the metates can be given.

Most of these metates were probably made on slabs of rockfall or scree that had scaled off the nearby cliff. They may constitute the oldest type of metate used in the area, one fragment appearing in Zone E.

Basin-Shaped Metate Fragment

Fig. 8.3b.
Number of specimens: 1.
Dimensions: thickness, 4.8 cm.
Raw material: ignimbrite.
Description: This boulder has a concave upper surface that is smoothed from use in grinding. The undersurface has been pecked to provide a more even and regular base and in one area has a flat, smooth facet that may indicate a second grinding surface. No overall dimensions of the broken metate could be taken.

Our small fragment cannot be matched with any specific Tehuacán type; it superficially resembles MacNeish *et al.'s* "boulder metate-milling stones," "boulder trough metates," and "basin-shaped metates" (1967:118–121, Figs. 98 and 99). Since "boulder metate-milling stones" show "no evidence of intentional shaping," ours presumably does not belong to that type.

MORTARS

Mortars were one of the most ancient ground-stone tools in the Tehuacán sequence, occurring as far back as the Late Ajuereado phase (MacNeish *et al.* 1967: Table 15). Only three fragments of mortars occurred at Guilá Naquitz; but judging by our radiocarbon dates, they are broadly contemporaneous with the late Ajuereado specimens. Our earliest mortar fragment appeared in Zone C, a level that also produced a subrectangular one-hand mano whose pecked ends suggest it may also have been used as a pestle (Table 8.1).

Mortar Fragments

Figs. 8.2c and 8.4.
Number of specimens: 3.
Dimensions: wall thickness at rim, 2.0 cm (1 specimen); thickness at base, 3.0 cm, 3.3 cm, and 3.3 cm.
Raw material: ignimbrite.
Description: These are large volcanic tuff cobbles or blocks shaped on all surfaces and particularly smooth on the grinding surface. The grinding surface resembles a deeply concave bowl that is hemispherical in cross section, and this attribute has allowed us to distinguish the mortars from metates in spite of their fragmentary nature. In all cases, our metate grinding surfaces are flatter and broader. In terms of the Tehuacán sequence, the Guilá Naquitz mortars are probably most like the "hemispherical mortars" of MacNeish *et al.*, with their simple, blunt rims (1967:116, Fig. 96, upper right).

Our mortars were probably used with the pestle-like pecked and rounded ends of some one-hand manos, but they might also have been used with some of the pebble hammerstones found at Guilá Naquitz. Because these hammerstones were usually made of chert or silicified ignimbrite they have been included with the chipped-stone tools (Chapter 6), but the possibility that some were occasionally used with mortars cannot be ruled out.

As to what the Guilá Naquitz mortars were used for, one task that comes immediately to mind is pounding up acorns. However, the mortar fragments we found all showed up in the later zones (C and B), when the frequency of acorns was declining and other plants, such as mesquite beans, were on the increase. We therefore presume they were used on a variety of plants.

MISCELLANEOUS CATEGORIES

Slab Metate Fragment, Reused as Hammerstone

Fig. 8.3c.
Number of specimens: 1.
Dimensions: length, 8.5 cm; width, 5.6 cm; maximum thickness, 3.1 cm.
Raw material: ignimbrite.
Description: This corner of a slab metate of uniform thickness was deliberately shaped on both sides and had one flat grinding surface. The base is generally flat but does not appear to have been utilized for grinding. After the metate broke, this triangular piece from the corner was used as a hammerstone; three edges show chipping and two show peck marks from hammering. While there are no exact equivalents in the Tehuacán collections, "pebble hammerstones" were

TABLE 8.1
Occurrence of Ground-Stone Tools at Guilá Naquitz Cave[a]

Tool categories	Stratigraphic zones					
	E	D	C	B2 + 3	B2	B1
One-hand Manos						
discoidal, 1 convex grinding surface		1(G6)			1(C9)	
discoidal, 2 convex grinding surfaces		2(D6, Feature 2)	1(F9)			1(C9)
discoidal, 1 convex grinding surface, 1 concave grinding surface		1(G6)				
subrectangular, 2 convex grinding surfaces			1(F9)			
miscellaneous fragments					2(D6)	
Metates						
slab metate fragments	1(H9)					2(D6)
basin-shaped metate fragment					1(C9)	
Mortars						
mortar fragments			1(F6)	1(C10)		1(C11)
Miscellaneous categories						
slab metate fragment, reused as hammerstone					1(C6)	
planoconvex abrader					1(C6)	
slab paint palette fragment						1(F6)
pebble pigment grinder			1(F9)			

[a]Squares in which tools were found are given in parentheses.

common from the Late Ajuereado phase onward (MacNeish *et al.* 1967:126–127).

Planoconvex Abrader

Fig. 8.5a.
Number of specimens: 1.
Dimensions: width, 6.8 cm; maximum thickness, 1.3 cm.
Raw material: ignimbrite.
Description: This is a piece from the midsection of a thin, elongated stone that has been used on both surfaces and one edge. It is plano-convex in cross section. Both surfaces are smooth, and the flat one shows fine scratches parallel to what had originally been the long axis of the tool. Both the convex surface, the flat surface, and one of the narrow edges appear to have been used for abrading or smoothing other tools, possibly artifacts of wood or bone. It occurred in Zone B2, a living floor that also yielded a couple of wooden tools (Chapter 10).

The Tehuacán caves produced a number of "abrader saws" (MacNeish *et al.* 1967:125–126, Fig. 105) that are reminiscent of, but not identical to, this artifact. The "abrader saws" were apparently used only on their narrow edges, while our abrader was used on both side and edge.

Slab Paint Palette Fragment

Figs. 8.2b and 8.5c.
Number of specimens: 1.
Dimensions: thickness, 1.3–1.6 cm.
Raw material: ignimbrite(?).
Description: This is a fragment of very thin slab metate with one flat, smooth grinding surface covered with red pigment. Since the edges of the fragment are broken and irregular, we cannot estimate the original size of the metate or determine whether the edges had been pecked to shape. The base is uneven but generally parallel to the grinding surface.

Slab paint palettes were not uncommon in the Tehuacán caves, where they were used to grind hematite (MacNeish *et al.* 1967:125, Fig. 104).

Pebble Pigment Grinder

Fig. 8.5b.
Number of specimens: 1.
Dimensions: none taken (specimen broken).
Raw material: limestone.
Description: This is a broken stream pebble, smooth and shiny in the portion of cortex that remains. Traces of red pigment remain on the cortex. The pebble was apparently used for the grinding of red pigment, possibly on a slab paint palette; its polish may result from this grinding, as well as from other grinding or polishing tasks. Significantly, limestone pebbles are not available in stream beds near Mitla, so this specimen may have been procured from a stream as much as 25–50 km away (see Fig. 7.6).

Preceramic levels in the Tehuacán caves also produced "polishing pebbles," some with red pigment (MacNeish *et al.* 1967:124, Fig. 102).

SUMMARIES OF INDIVIDUAL LIVING FLOORS

Before proceeding to our summaries of the ground-stone tools from each living floor, there is one general observation that could be made: There was an apparent tendency for grinding activities (or the discard of ground-stone tools) to

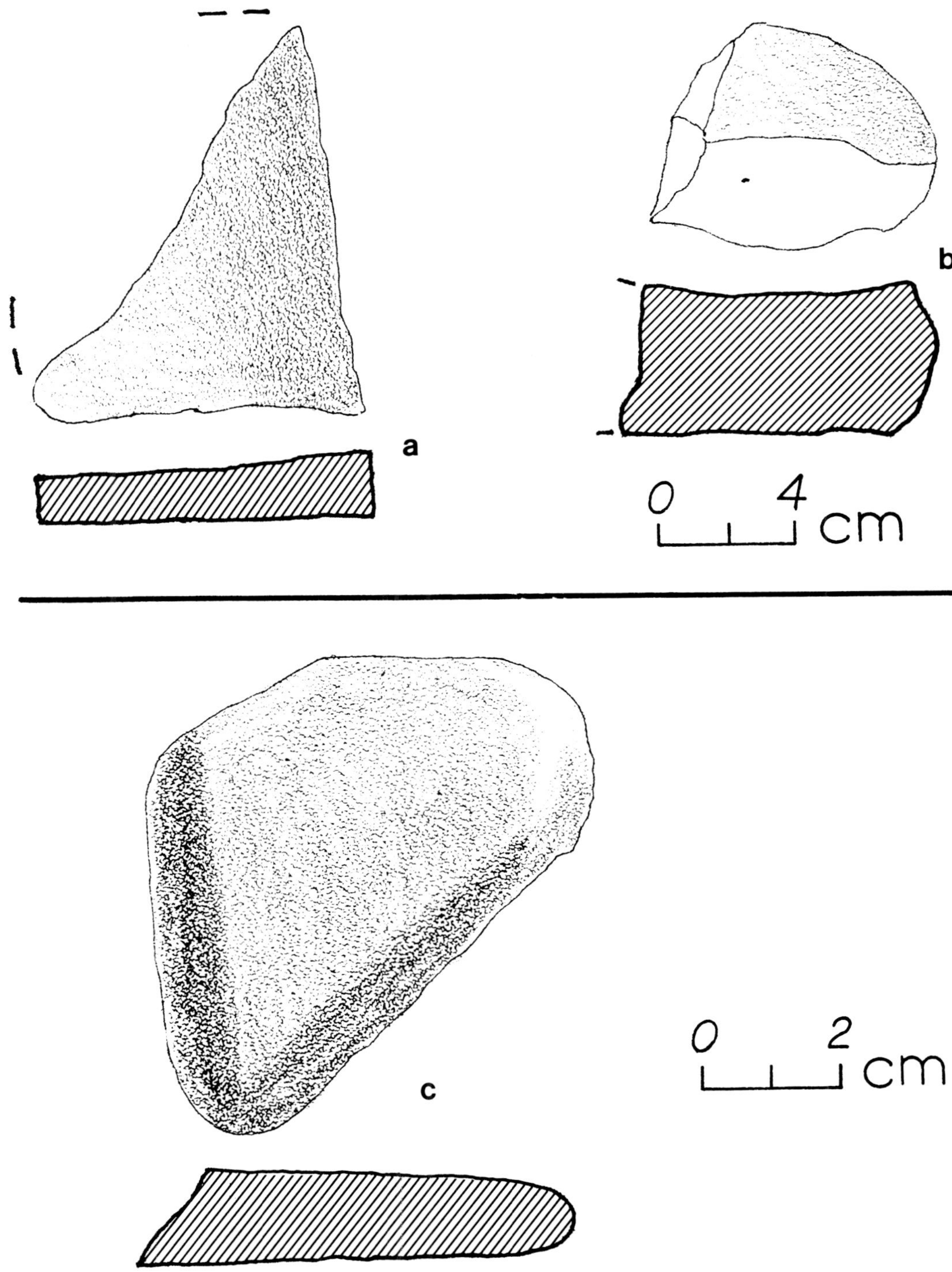

FIGURE 8.3. Metate fragments: a, fragment of slab metate, ignimbrite, Zone B1, Square D6; b, fragment of basin-shaped metate, ignimbrite, Zone B2, Square C9; and c, slab metate fragment reused as hammerstone, ignimbrite, Zone B2, Square C6.

take place in the same general areas of the cave over time. The evidence for this is the fact that a small number of squares seem to have produced most of the ground-stone tools, regardless of living floor. For example, out of a total of 21 items, 3 occurred in Square D6 (and 2 more in nearby C6), 3 occurred in C9 (and 2 more in C10–C11), 3 occurred in F9, and 4 occurred in F6–G6. This suggests that certain areas of the cave floor were considered more appropriate for grinding activity, or for the storing or discard of ground-stone tools, than others.

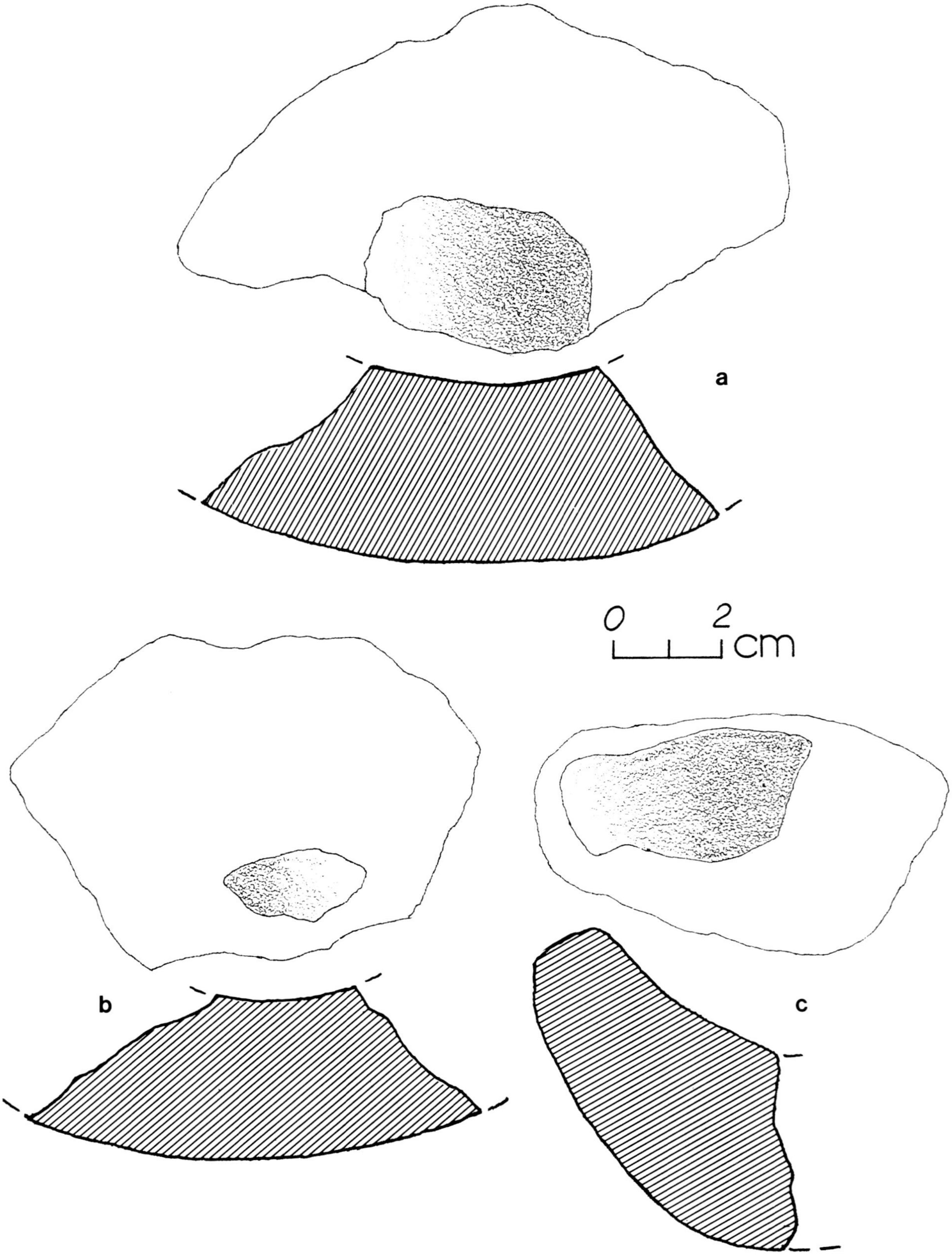

FIGURE 8.4. Mortar fragments: a, mortar fragment, ignimbrite, Zone B1, Square C11; b, mortar fragment, Zone C, Square F6; and c, fragment of mortar rim, ignimbrite, Zone B2 + 3, Square C10.

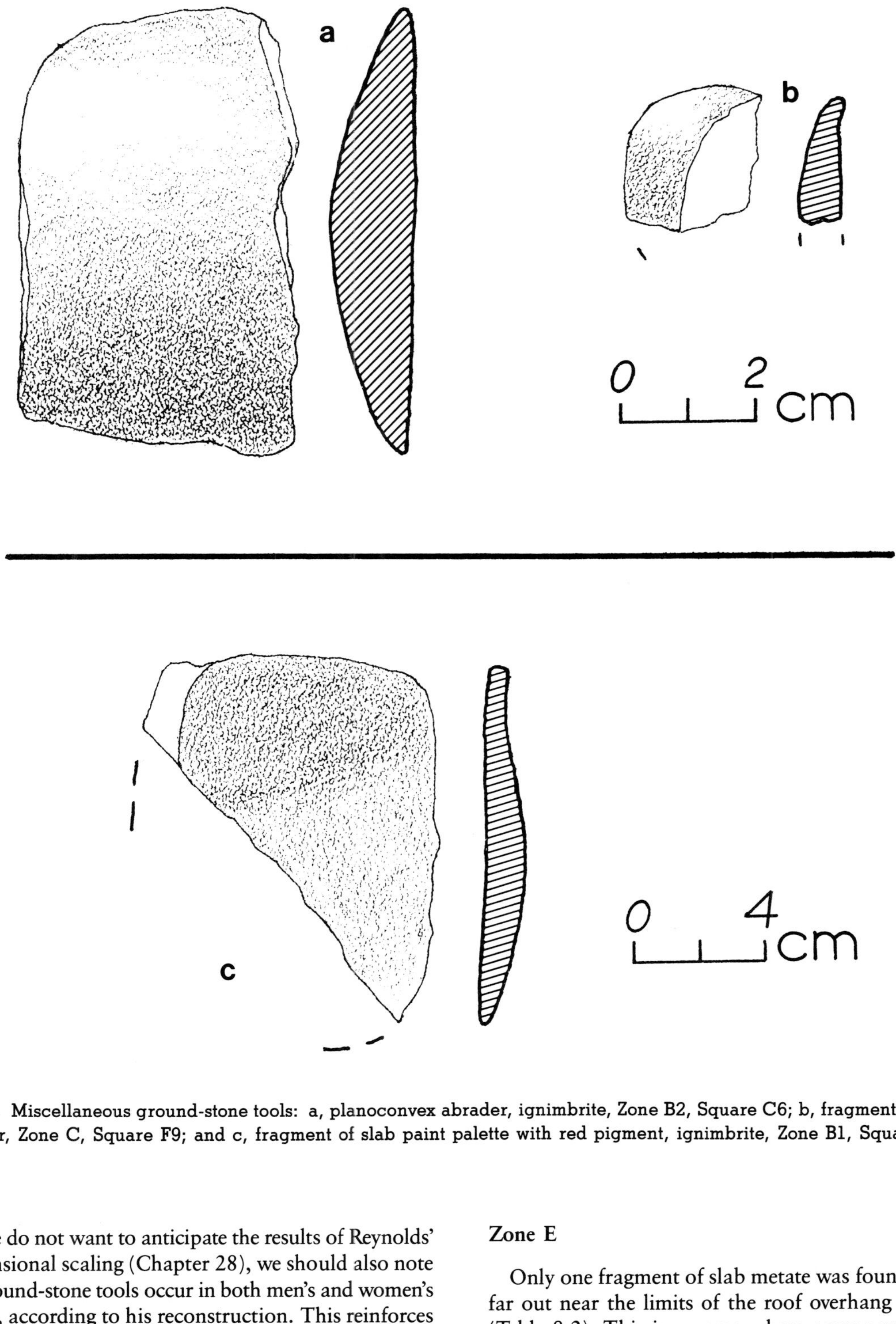

FIGURE 8.5. Miscellaneous ground-stone tools: a, planoconvex abrader, ignimbrite, Zone B2, Square C6; b, fragment of pebble pigment grinder, Zone C, Square F9; and c, fragment of slab paint palette with red pigment, ignimbrite, Zone B1, Square F6.

While we do not want to anticipate the results of Reynolds' multidimensional scaling (Chapter 28), we should also note that our ground-stone tools occur in both men's and women's work areas, according to his reconstruction. This reinforces our suspicion that such things as one-hand manos might as easily be used for men's tasks as women's. Interestingly, our slab paint palette from Zone B1 was found within a men's work area as reconstructed by Reynolds. Of course, we realize that not all stone tools were discarded where they were last used, but their tendency to appear in certain squares is intriguing.

Zone E

Only one fragment of slab metate was found in this zone, far out near the limits of the roof overhang in Square H9 (Table 8.2). This is an area where acorns were one of the most common foods requiring grinding.

Zone D

During the occupation of this zone, two one-hand manos were discarded near the center of the cave, in Feature 2 and

TABLE 8.2
Ground-Stone Tool Assemblages from Zones B–E, Guilá Naquitz Cave

Zone	Square	Tool category	Raw material
B1	C9	Fragment of one-hand mano, discoidal, 2 convex grinding surfaces	Ignimbrite(?)
B1	C11	Fragment of mortar	Ignimbrite
B1	D6	2 fragments of slab metate	Ignimbrite
B1	F6	Fragment of slab paint palette (with red pigment)	Ignimbrite
B2	C6	Slab metate fragment, reused as hammerstone	Ignimbrite
B2	C6	Planoconvex abrader	Ignimbrite
B2	C9	Fragment of one-hand mano, discoidal, 1 convex grinding surface	Limestone
B2	C9	Fragment of basin-shaped metate	Ignimbrite
B2	D6	2 fragments of one-hand manos	Ignimbrite(?)
B2 + 3	C10	Fragment of mortar	Ignimbrite
C	F6	Fragment of mortar	Ignimbrite
C	F9	Fragment of one-hand mano, discoidal, 2 convex grinding surfaces	Ignimbrite(?)
C	F9	One-hand mano, subrectangular, 2 convex grinding surfaces (with pigment)	Quartzite(?)
C	F9	Fragment of pebble pigment grinder (with red pigment)	Limestone
D	Feature 2	Fragment of one-hand mano, discoidal, 2 convex grinding surfaces	Ignimbrite
D	D6	Fragment of one-hand mano, discoidal, 2 convex grinding surfaces	Ignimbrite(?)
D	G6	Fragment of one-hand mano, discoidal, 1 convex grinding surface	Ignimbrite
D	G6	Fragment of one-hand mano, discoidal, 1 convex and 1 concave grinding surface	Ignimbrite
E	H9	Fragment of slab metate	Ignimbrite(?)

Square D2. Feature 2 was a pit that appears to have been for acorn storage. Two more one-hand manos were discarded farther out near the cave entrance, in Square G6. Both the manos found near the center of the cave were similar (with two convex grinding surfaces) and could have been used in the same manner by the same person.

Zone C

All four of the ground-stone tools from this living floor were discarded near the cave entrance, in the row of F squares. Two were one-hand manos, and one of these was a subrectangular quartzite specimen that had apparently also been used as a pestle. Both this subrectangular mano and a nearby limestone pebble grinder had traces of red pigment on them.

Several meters away we found a fragment of mortar, possibly one used with the pestle-like end of the mano already mentioned.

Lower Zone B

No ground-stone tools at all appeared in Zone B3, a small occupation. A single mortar rim fragment was found in Zone B2+3. On the other hand, Zone B2 produced six items, clustered in three squares.

Zone B2 was a living floor in which most of the grinding implements were left toward the rear of the cave, in contrast to Zone C. Included were a one-hand mano and a fragment of a basin-shaped metate in Square C9, an abrader and a metate fragment reused as a hammerstone in Square C6, and two fragments of one-hand manos in Square D6. Acorns, *susí* nuts, and mesquite beans all occurred in this part of the cave and may have been among the food items ground or pounded up.

Upper Zone B

During the deposition of Zone B1, the occupants discarded four ground-stone items. Near the rear of the cave, we found a one-hand mano and a fragment of mortar. Near the front of the cave, in a different work area, a slab palette with red pigment adhering had been left behind. Two additional fragments of slab metate showed up in Square D6. Once again, associated foodstuffs that may have been subjected to grinding or pounding up included acorns, *susí* nuts, and mesquite beans.

9

Preceramic Cordage and Basketry from Guilá Naquitz

Mary Elizabeth King

INTRODUCTION

The preceramic levels at Guilá Naquitz did not produce large quantities of textile materials. The two largest specimens were a fragment of knotted netting and a miniature plaque of coiled basketry. Also recovered were 12 examples of cordage, ranging in length from less than 1 cm to 25 cm. Because of the small sample size, it is unlikely that the full range of local techniques is represented.

All the fabric structures recovered at Guilá Naquitz were also found in the El Riego phase at Tehuacán (MacNeish *et al.* 1967). The El Riego phase produced more kinds of structures, but these might well have also appeared in the Oaxaca caves had the sample been larger. In addition to cordage, knotted netting, and coiled basketry with interlocked stitches, El Riego phase structures included two varieties of looping, weft-twining, warp-faced plain weave and two varieties of coiled basketry, one with noninterlocked and one with twined stitches.

In spite of its limited range of techniques, the Guilá Naquitz sample includes the oldest radiocarbon-dated cordage and basketry from Mesoamerica proper. The cordage fragments from Zones C, D, and E all antedate 7000 B.C. The knotted netting and basketry plaque are from Zone B, whose radiocarbon dates are 6910–6670 B.C. The earliest El Riego phase samples from Tehuacán, with their somewhat greater range of structures, come from Zones XIV and XV of Coxcatlán Cave and probably date in the neighborhood of 5050–4787 B.C. (Johnson and MacNeish 1972). To the north of Mesoamerica proper, samples overlapping in age those from Guilá Naquitz come from the Infiernillo phase of the Sierra de Tamaulipas, dating to 7000–5000 B.C. (MacNeish 1958). Here there are some differences in structures: Knotted nets are absent in the sample, while looped nets are present.

Because of the small size of the sample from preceramic Guilá Naquitz, I have simply divided the material into three basic categories: knotted netting, cordage, and basketry.

KNOTTED NETTING

The only textile from the preceramic levels at Guilá Naquitz is a fragment of knotted netting about 5.5 × 5 cm in size (Fig. 9.1). In all cases but one, the knot used is the cow hitch, or lark's head knot (Fig. 9.2), and the net is constructed so that the same face of the knot always appears on the same side of the fabric (more commonly the face would be reversed in adjacent knotting rows). One knot, possibly a mistake, is a clove hitch. The mesh bar is 0.5 cm in length, making meshes with 2.5-cm square measurement.

The cordage used in the net is of unidentified, noncotton plant fiber with a yarn diameter of slightly less than 1 mm, tightly S-spun and Z-plied.

Some of the El Riego phase knotted netting examples from Tehuacán appear to be almost identical in all respects, even to mesh size; others utilize the overhand knot and opposite spins and plys (Johnson 1967:196). Although Johnson does

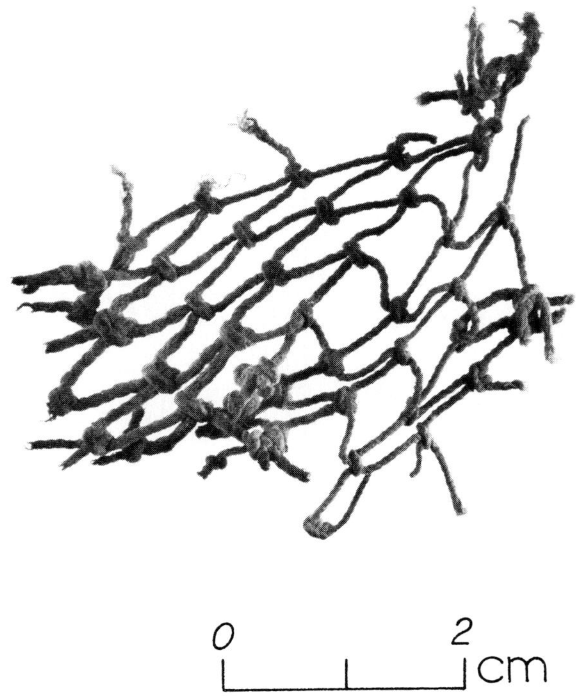

FIGURE 9.1. Fragment of knotted netting, Zone B3, Square E7 (Specimen 60).

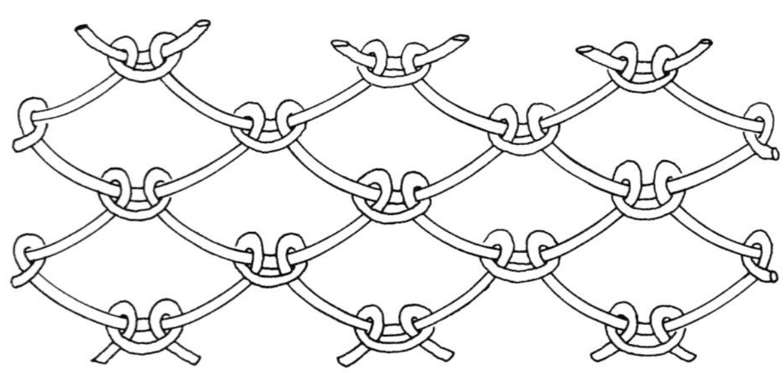

FIGURE 9.2. Diagram of knotted netting using the lark's head knot, with the same face of the knot showing in adjacent rows.

not describe the use of the same face of the knot on one side of the net, her diagram (Johnson 1967: Fig. 157) shows this feature.

Overhand knots, found in some Tehuacán nets, occur at Guilá Naquitz in the form of isolated knots and in one fragment of knotted cordage (Specimen 71) that might conceivably be from a net, though it has a far more randomly knotted appearance than normal netting (Fig. 9.3).

CORDAGE

The cordage from Guilá Naquitz is best summarized in table form (Table 9.1). It is relatively consistent in makeup and quality.

MacNeish *et al.* note that "the ancient Tehuacaneros seemed to care little whether they twisted [cordage] clockwise or counterclockwise" (MacNeish *et al.* 1967:173–174). Their cordage sample for the El Riego phase was much larger than ours—a total of 58 specimens—all 2 and 4 ply. Their figures (1967:172, Table 30) show that 75% of the cordage had a final **Z** twist (ply), while 25% had a final **S** twist. Our smaller sample of 11 plied cords shows that 82% has a **Z** final twist and 18% an **S** final twist. While our sample shows a slightly higher preference for **Z** twist than does that of MacNeish *et al.*'s, the figures are not significantly different when sample sizes are considered.

MacNeish *et al.* feel that this "contrasts with the Southwest of the United States, where these factors show definite trends in time and are connected with whether or not the rolling is accomplished with the right or left hand, with an up or down motion, and on the right or left thigh." MacNeish *et al.*'s comments refer specifically to the findings of Martin

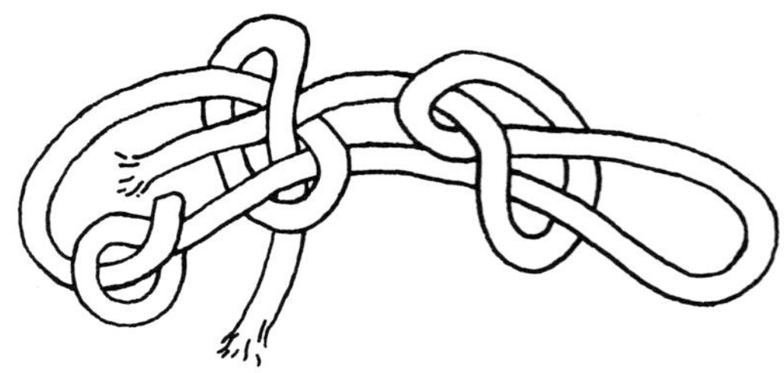

FIGURE 9.3. Diagram of Specimen 71, a fragment of knotted cordage from Zone E, Square E9.

TABLE 9.1
Attributes of Cordage Fragments from Zones B–E, Guilá Naquitz Cave

Specimen[a]	Fiber	Color	Twist	Ply	Diameter (mm)	Length (cm)	Tightness[b] (final twist)	Knots
62	(?)	Dark brown	S	S–Z	3	5	M	—
63	Agave(?)	Dark brown	S	Z	4	14.5	M	—
64	Agave(?)	Tan	S	Z	1.5–2 +	11	M–T	—
65	Agave(?)	Natural tan	S	S–Z	2.5	8	M–T	—
66	Agave(?)	Dark brown	S	Z	4	4	M	—
67	Agave(?)	Tan	Z	S	2–4	25	M	—
68a	Agave(?)	Dark brown	S	Z	6	7	L–M	—
68b	Agave(?)	Dark brown	S	Z	3	2.5	M	—
69	(?)	Dark brown	S	Z	3–4	0.8–0.9	M	(?)
70	(?)	Reddish brown	Z	S	4	4	L	—
71	(?)	Brown	S	Z	—	—	M	Half hitch
72	(?)	Tan	S	—	7	1.7	(?)	Overhand

[a]Specimens 63 and 66 could be parts of the same cord.
[b]L, loose; M, medium; T, tight.

et al. (1952:206–210) for Tularosa Cave and their comparisons to other sites. Although the data for some other sites are perhaps inadequate to show change through time, the usual picture in the southwestern United States[1] is that cotton is initially spun in the Z direction and plied S, while noncotton plant fibers are ordinarily spun or twisted S initially and plied Z. This picture is obscured by a tendency on the part of some scholars to note only the final twist, whether it be a spin, a ply, or a re-ply, perhaps because the initial spin in some plant-fiber cordage is almost nonexistent. Normally, but not always, each successive stage of twisting is opposite to the preceding stage. It is essential to note the initial spin; differences here probably can be laid to differences in spinning method, in tradition, and even occasionally in handedness (though modern left-handed informants who spin do not seem to differ in preferential direction from the norm). Subsequent plying directions should also be noted.

To illustrate why this may be important, two examples of re-plied Guilá Naquitz cordage are S-spun, S-plied (contrary to usual practice), and re-plied Z (Specimens 62 and 65). These are the only re-plied examples, and this seems to indicate preferences for both initial and final twists, or at least an attempt to cling to customary practices, since most cordage is S-spun and Z-plied.

Two examples (both from Zone C) are Z-spun and S-plied, but without fiber analysis there is no way to tell whether this is a consequence of a different fiber being spun with a different method or a consequence of handedness, training, or preference in work habits.

One thing seems clear. Regardless of the preferred direction of spin and ply, I know of no geographic or temporal instances where yarns are spun (and plied) in the preferred direction *only.*

Spin and ply of the Guilá Naquitz specimens are generally medium tight and fairly even. A color range restricted to tans and browns suggests that the yarns are undyed.

Yarn diameters suggest that these yarns were used as cordage, rather than for manufacture into fabrics. None is *as* fine, for example, as the yarns used in the netting fragment, and only a few of the specimens even approach that fineness. Coarse yarns can, of course, be used for both nets and cloth, so this is not conclusive. One yarn (Specimen 72) has an overhand knot (Fig. 9.4), while another small, curved piece of cordage (Specimen 69) may itself be from a knot; but in neither case does this clearly indicate function.

FIBER

Analysis is not yet complete on the fibers utilized in the preceramic levels at Guilá Naquitz. None, however, is cotton. It is likely that the cordage and netting are constructed of yucca, agave, or a similar plant.

BASKETRY

The only basketry specimen from the preceramic levels is a miniature plaque with a diameter of 5.5 cm and a thickness of about 3 mm (Fig. 9.5). It has a bundle foundation with 2–3 rows per cm. The stitches are interlocking (Fig. 9.6), slanted down to the left, about 1–3.5 mm in width, and somewhat spaced. The center starting point is plain but oval

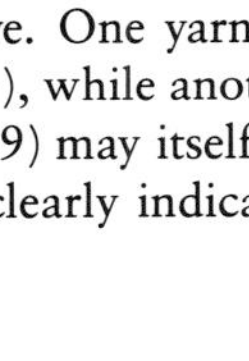

FIGURE 9.4. Specimen 72, a fragment of cordage with an overhand knot, Zone B2, Square D9 (actual length, 1.7 cm).

[1]In Peru, differences in spin follow a geographic distribution to some extent.

FIGURE 9.5. Miniature basketry plaque, Zone B2, Square D11 (Specimen 74).

in shape; the rim is also plain. Stitches are spliced by simply carrying the new stitch fiber under several previously worked stitches.

Materials utilized in the basket have not been identified, but the stitches are of a reddish-brown, relatively hard-surfaced material, possibly strips of yucca or agave leaves, and the foundation is probably of grass stems.

Although coiled baskets with bundle foundations and interlocked stitches are reported from the late El Riego phase at Tehuacán (along with other stitch types), no miniature baskets were noted.

Coiled, bundle foundation basketry is usual for Mexico, and MacNeish suggests that it originated in Mesoamerica and spread northward (MacNeish *et al.* 1967:164).

Miniature baskets, like other miniature artifacts, have several possible functions: as children's toys, as trial pieces, or for ceremonial or other special purposes. Alternatively,

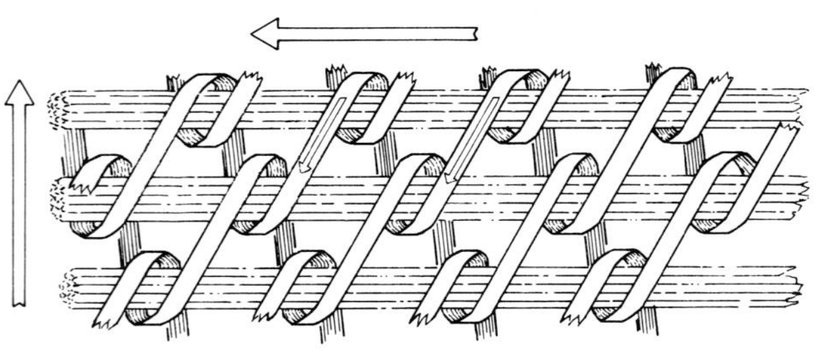

FIGURE 9.6. Diagram of coiled basketry with interlocked stitches on a bundle foundation.

a plaque-like piece of the sort found at Guilá Naquitz might have served as a lid or cover for a small-mouthed container. There were bottle gourds in the preceramic levels (see Chapter 19), and some of these might have been small-mouthed containers.

SUMMARIES OF INDIVIDUAL LIVING FLOORS

In a later section of this volume, Reynolds (Chapter 28) has reconstructed men's work areas, women's work areas, and travel pathways in Guilá Naquitz based on R-mode and Q-mode analysis of the debris in each square. When one plots the occurrence of cordage and basketry specimens on the various activity areas he has proposed, an interesting pattern emerges. Four of 14 specimens occur in what Reynolds has proposed to be either food preparation or women's work areas; 3 more specimens fall on his reconstructed travel pathways. No specimens of cordage or basketry were found in any of Reynolds' proposed men's work areas.

Zone E

Specimen 71, a half hitch knot on S–Z cordage, was discarded in Square E9 (Table 9.2). This falls in Reynolds' Area II, interpreted as the main travel pathway in Zone E. Specimen 62, a 5-cm section of S–S/Z cordage, was found in Square F11. This square was part of Reynolds' Area I, a proposed women's work area.

Zone D

Specimens 68a and 68b are two fragments of S–Z cordage totaling 9.5 cm. Both were found in Square E8, which is part of Reynolds' Area II, a proposed area of food preparation.

Zone C

Interestingly enough, both cordage fragments from this living floor are Z twist and S ply, the only such cordage in the preceramic levels of the cave. Specimen 70, a 4-cm section of Z–S cordage, was found in Square E10; this square has been assigned to Reynolds' Area II, a plant processing area. Specimen 67, consisting of a rather long (25-cm) segment of Z–S cordage, had been discarded in Square F9. This falls within Reynolds' Area V, proposed to be the main travel pathway entering the cave.

Lower Zone B

Because of their small size, Zones B3–B2 were not included in Reynolds' analyses. Zone B3 produced a fragment of knotted netting, perhaps part of a net bag for collecting plants, which had been discarded in Square E7. In Zone B2+3, Specimens 63 and 66 (totaling 18.5 cm of S–Z cordage) appeared in Square E6. Specimen 69, a tiny fragment of S–Z cordage, had been discarded in Square C10.

TABLE 9.2
Specimens of Cordage and Basketry from Guilá Naquitz Cave

Stratigraphic zone	Square (1 × 1 m)	Specimen	Category
B1	E9	64	11 cm of S–Z cordage
B1	F7	65	8 cm of S–S/Z cordage
B2	D9	72	Overhand knot on S-twist cordage
B2	D11	74	Miniature plaque, coiled basketry
B2 + 3	C10	69	0.8–0.9 cm of S–Z cordage
B2+3	E6	63	14.5 cm of S–Z cordage } same(?)
B2+3	E6	66	4 cm of S–Z cordage } same(?)
B3	E7	60	Fragment of knotted net
C	E10	70	4 cm of Z–S cordage
C	F9	67	25 cm of Z–S cordage
D	E8	68a	7 cm of S–Z cordage
D	E8	68b	2.5 cm of S–Z cordage
E	E9	71	Half hitch knot on S–Z cordage
E	F11	62	5 cm of S–S/Z cordage

In Zone B2, a miniature coiled basketry plaque that may have been the lid for a container of some kind was discarded in Square D11. Not far away, in Square D9, appeared an overhand knot on S-twist cordage (Specimen 72). Most Lower Zone B specimens, therefore, occurred in the center or the northwest part of the cave.

Upper Zone B

Zone B1 was a living floor for which Reynolds has proposed a number of activity areas. Specimen 64, an 11-cm segment of S–Z cordage, was found near a hearth (Feature 11). Specimen 65, an 8-cm fragment of S–S/Z cordage, was discarded in Square F7. This square falls in Reynolds' Area V, proposed to be the main travel path entering the cave.

Summary

Obviously, any conclusions to be drawn from the distribution of cordage specimens on the living floors would have to be based on the assumptions that (1) the specimens were discarded approximately where they were found and that (2) Reynolds has correctly interpreted the nature of the activity areas and pathways. If these assumptions are correct, it would not be unreasonable to suggest that one of the major purposes of the Naquitz phase cordage was to tie up bundles of plant material for transport to the cave. The location of knots and cordage fragments would then suggest that some bundles broke or were untied on the major pathway entering the cave, while other bundles broke or were untied in Reynolds' plant processing or food preparation areas, which usually lie in the northwest part of the living floor (Chapter 28).

10

Artifacts of Wood and Related Materials

Kent V. Flannery[1]

The preceramic levels at Guilá Naquitz produced 19 artifacts or modified fragments of wood, reed, cane, cactus stem, or cactus gum. Included were materials for fire making, tool hafting, prickly pear roasting, and other activities. Because of the small sample size and the functional ambiguity of some items, our descriptions of tool categories are no more than a first step toward understanding what some of these materials may have been.

FIRE-MAKING EQUIPMENT

Every Boy Scout is taught that the Indians made fire "by rubbing two sticks together." For Mesoamerica, at least, this conventional wisdom is fairly accurate. One of the most common ways of kindling a fire was by friction between two sticks, one known as a "drill" and the other as a "hearth." The hearth, usually made of a softer and more easily ignited wood, had a socket to receive the drill. In this socket the bit of the drill, usually made of harder wood, was rotated by a person holding the drill shaft between the palms of his hands and rubbing the latter back and forth.

In the Mixtec codices of northern Oaxaca there are numerous depictions of persons kindling a fire with this kind of drill (Fig. 10.1). The dry caves of the Tehuacán Valley produced at least three drills and three fragments of wooden hearths. According to MacNeish *et al.* 1967, the drills were sticks 1 cm in diameter and more than 25 cm long, with "a blunted, burned and polished, or well-worn end." The hearths were pieces of softer wood, flat on one side, with "several charred ground holes" about 1.5 cm in diameter in the flat side (1967:155, and Fig. 131). These Classic and Postclassic specimens from Tehuacán closely resemble our preceramic fire-making equipment from Guilá Naquitz, suggesting more than 9000 years of continuity in this aspect of material culture.

A favored raw material for hearths was the *quiote*, or inflorescence, of the lechuguilla (*Agave karwinskii*), a small agave that grows abundantly on slopes and washouts in Thorn Forests A and B. The dried inflorescence is soft, punky, and highly inflammable. Fire drills are of harder material, as yet unidentified, but possibly including the inflorescence of a different species of agave or yucca.

Fire Drill Hearth

Fig. 10.2.
Number of specimens: 1.
Dimensions: diameter, 1.3 cm; length, 20.5 cm.
Raw material: inflorescence of small agave(?).
Description: This is a section of inflorescence, perhaps of lechuguilla, split lengthwise so as to present one curved and one flat surface. The flat surface is additionally smoothed, perhaps with an abrader like the one found in Zone B2 (see Chapter 8). Three holes have been drilled into the flat surface at intervals of 1.5–4 cm. Each hole is about 1 cm in diameter, smooth, shallow, and slightly charred.

Fire Drills

Fig. 10.3.
Number of specimens: 3.
Dimensions: see below.
Raw material: sticks, not identified to species, possibly including agave inflorescence.

[1] I wish to acknowledge the help of Suzanne K. Fish, who measured the artifacts of wood, reed, and cane in our Oaxaca lab.

FIGURE 10.1. Making fire with a wooden drill and a fire drill hearth with multiple drill sockets (redrawn from Codex Vindobonensis 18-I).

Description: Our first specimen, from Zone C, measures 18.7 cm in length and 1 cm in diameter (Fig. 10.3c). One end has simply been burned off to a stump. The other end has been whittled and smoothed to a point and displays rings of abrasion that encircle it, apparently caused by rotating the point in a small circular aperture, such as the hole of a fire drill hearth. There is a small cut on the shaft approximately 3.8 cm back from the point, possibly made at the time the pointed end was whittled. We would interpret this as a fire drill used until one end burned away then reversed so that the other end could be used briefly.

A second specimen, from Zone B3, consists of several crumbly pieces from which no overall dimensions could be taken (Fig. 10.3b). One fragment does, however, show the well-worn, rounded, scorched end characteristic of most fire drills.

A third possible fire drill fragment, from Zone B2, measures 9.8 cm in length and 1 cm in diameter (Fig. 10.3a). Both ends show burning, which has considerably reduced the length of the fragment. Near one end is a series of fine cut marks and at least one shallowly cut groove, perhaps an effort to remove the burned end and produce a new bit. However, this specimen is so fragmentary we cannot rule out other interpretations for it.

ATLATL-HAFTING EQUIPMENT(?)

We know from the projectile points in our Mitla caves that early hunters in the area used the atlatl or dart-thrower. In his Sierra de Tamaulipas caves, MacNeish (1958) found numerous fragments of atlatl dart mainshafts, foreshafts, and hafting material. The Tehuacán caves were less rich in such

FIGURE 10.2. Fire drill hearth with multiple drill sockets, Zone B1, Square D9 (length, 20.5 cm).

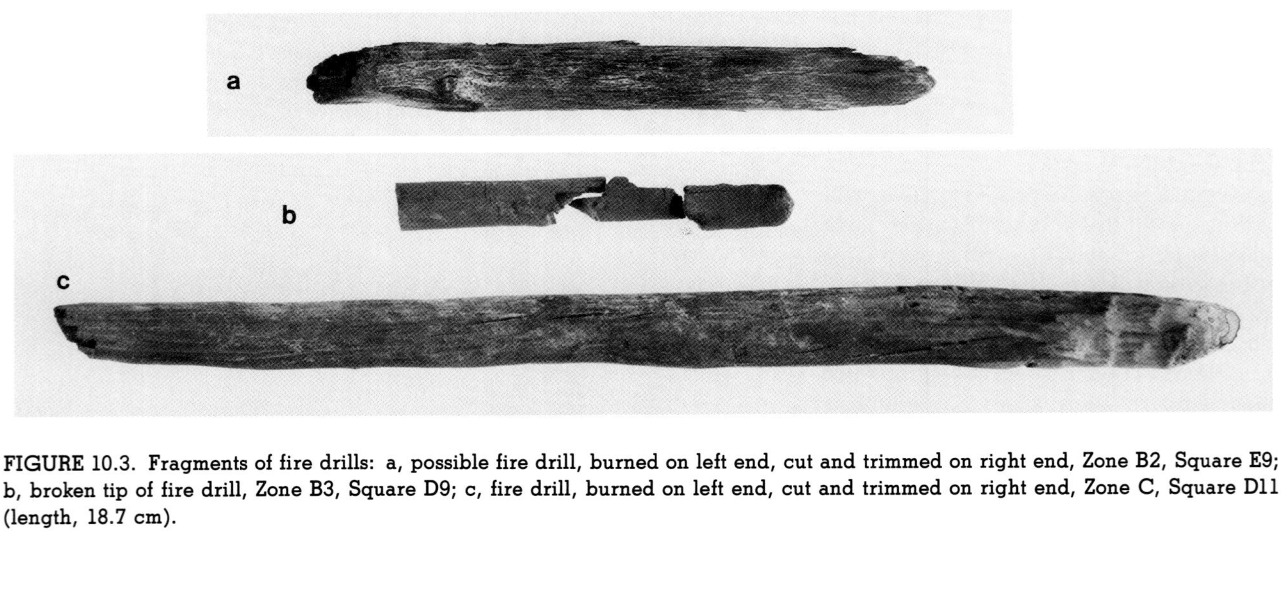

FIGURE 10.3. Fragments of fire drills: a, possible fire drill, burned on left end, cut and trimmed on right end, Zone B2, Square E9; b, broken tip of fire drill, Zone B3, Square D9; c, fire drill, burned on left end, cut and trimmed on right end, Zone C, Square D11 (length, 18.7 cm).

materials (MacNeish *et al.* 1967:Table 27), and Guilá Naquitz produced only a few examples of possible hafting material.

Segments of Reed or Cane

Fig. 10.4a, b.
Number of specimens: 5, all from Square C9, Zone B3.
Dimensions: see below.
Description: In Square C9 of Zone B3 we discovered five fragments of reed or cane, at least some of which had been cut. The lengths of the least deteriorated specimens were 6.5 cm, 6.3 cm, and 4.3 cm. While their actual function is unknown, we tentatively consider these segments to be pieces trimmed from larger cane artifacts rather than artifacts in their own right. One possibility is that they represent the refuse left from trimming the reed or cane mainshafts into which dart foreshafts were inserted (see MacNeish *et al.* 1967:160–161).

Ball of *Lemaireocereus* Gum

Fig. 10.4c.
Number of specimens: 1.
Dimensions: diameter, 7 cm.
Description: This spherical, well-preserved ball of gum is from the large columnar cactus *Lemaireocereus* sp., known to Spanish speakers as the *tunillo* and to the Mitla Zapotec as *bitsz lats* or *bidz lats* (C. E. Smith 1978; Messer 1978).[2] Someone had obviously collected a great deal of gum from this locally available organ cactus, carefully squeezed it into a ball, and kept it in the cave during the occupation of Zone B1. We suggest that one use for the gum might have been in hafting tools, which could have included the gluing of atlatl points to their foreshafts and the attaching of foreshafts to mainshafts. For example, in the Tehuacán Valley, MacNeish *et al.* found "both gum and string fibers" adhering to the cut ends of cane atlatl mainshafts (1967:160).

NOPAL-ROASTING EQUIPMENT

One of the common plants eaten at Guilá Naquitz was the prickly pear cactus, *Opuntia* spp. Its tender young stem segments, called "nopales," are still an important vegetable in Oaxaca. Today they are boiled to render them edible, but we know of no container in which the preceramic occupants of Guilá Naquitz could have boiled anything. Apparently, roasting was an alternative way of cooking nopales. For this purpose, the young stem segments of *Opuntia* were evidently skewered on slender sticks and held over a hearth. Burnt skewers, burnt stems, and even a nopal on a stick were found at Guilá Naquitz.

Opuntia Stem on a Stick

Fig. 10.5.
Number of specimens: 1.
Description: This is a stem segment, or nopal, skewered on a fragment of stick. All the spines had been removed prior to cooking, as is the case with virtually all nopales in the cave. This specimen was found in Square F9 of Zone C toward the mouth of the cave.

Burned Sticks

Fig. 10.6.
Number of specimens: 3.
Dimensions: length, 8.5 cm, 12 cm, and 23 cm.
Description: These three fragments of slender sticks, heavily burned, were found with the skewered *Opuntia* stem already described for Square F9, Zone C. Superficially, these fragments appear to be *Opuntia* skewers that became so burned that they were eventually discarded. Of course, there are alternative possibilities, and our interpretation is influenced by the proximity of these burned sticks to the skewered nopal (Table 10.1).

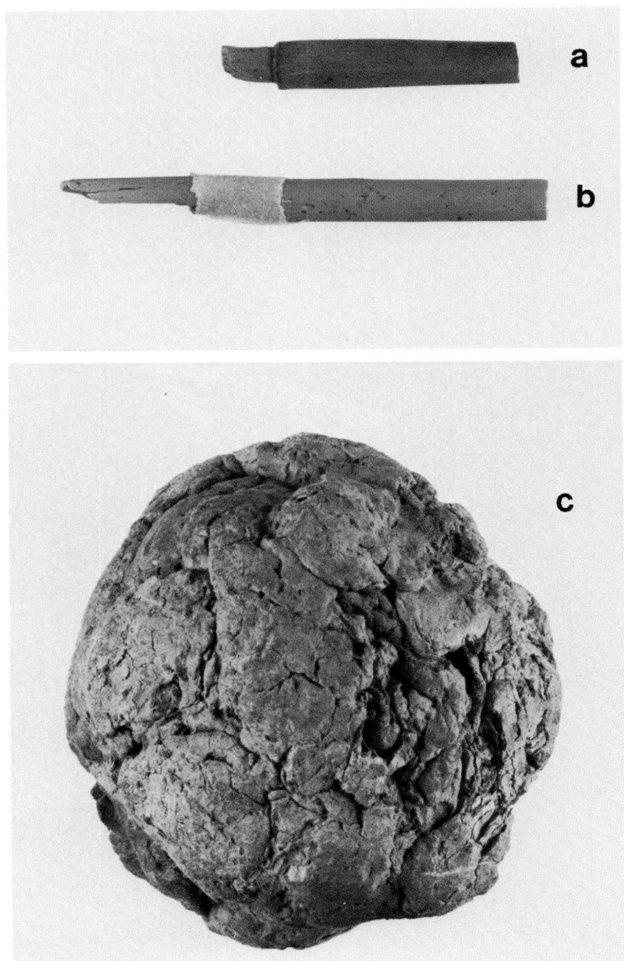

FIGURE 10.4. Atlatl hafting equipment(?): a, b, segments of reed or cane (possibly trimmed from atlatl mainshafts), Zone B3, Square C9; c, ball of *Lemaireocereus* gum, Zone B1, Square C9 (diameter, 7 cm).

[2]According to C. E. Smith (1978), the common species in the Mitla thorn forest is *Lemaireocereus treleasei*.

FIGURE 10.5. *Opuntia* stem section impaled on a stick for roasting, Zone C, Square F9 (stem section, 12.5 × 13.5 cm).

FIGURE 10.6. Burned sticks (possible skewers for *Opuntia* roasting), Zone C, Square F9 (longest specimen, 23 cm).

TABLE 10.1
Occurrence of Artifacts of Wood and Related Materials from Guilá Naquitz Cave[a]

Category	D	C	B3	B2	B1
Fire-making equipment					
fire drill hearth, 20.5 cm long					1(D9)
fire drill, 18.7 cm long		1(D11)			
broken fire drill			1(D9)		
possible fire drill, 9.8 cm long				1(E9)	
Atlatl hafting equipment(?)					
segments of reed or cane			5(C9)		
ball of *Lemaireocereus* gum					1(C9)
Nopal roasting equipment					
Opuntia stem on a stick		1(F9)			
burned sticks		3(F9)			
carbonized *Opuntia* fragments	2(F6)				
Miscellaneous					
stick cut on both ends, 1 cm long					1(D10)
wood fragment, use unknown				1(D6)	
burned wooden slat, use unknown	1(F8)				

[a]Squares in which artifacts were found are given in parentheses.

Carbonized *Opuntia* Fragments

Number of specimens: 2.

Description: These two segments of nopal are heavily carbonized, possibly by the kind of roasting described above. They are too fragmentary to allow us to demonstrate conclusively that they had been skewered like the specimen from Zone C.

MISCELLANEOUS ITEMS

Stick Cut on Both Ends

Fig. 10.7a.

Number of specimens: 1.

Dimensions: length, 1 cm; diameter, 1 cm.

Description: This small section of stick has been cut off or trimmed on both ends. Its use is unknown, and it may, in fact, just be a waste product from the manufacture of something larger.

Wood Fragment, Use Unknown

Fig. 10.7b.

Number of specimens: 1.

Dimensions: length, 3 cm; width, 1.8 cm.

Description: This fragment of wood has clearly been modified, but its use remains enigmatic. Two of its sides have been smoothed and come together in a right angle. There

FIGURE 10.7. Miscellaneous wooden artifacts: a, stick cut on both ends, Zone B1, Square D10; b, wood fragment, use unknown, Zone B2, Square D6; c, burned wooden slat, use unknown, Zone D, Square F8 (length, 12 cm).

are also cutting marks at various places all over the piece. Obviously, it may be a fragment of something larger.

Burned Wooden Slat, Use Unknown

Fig. 10.7c.
Number of specimens: 1.
Dimensions: length, 12 cm; width, 1.9 cm.
Description: This is a burned remnant of a long, narrow slat of wood, virtually flat on its lower surface. Parts of it appear to have been artificially smoothed or shaped. Superficially, its shape reminds one of a primitive atlatl board, but it is too badly destroyed to permit any functional interpretation. Whatever its original use, it ended up as fuel.

SUMMARIES OF INDIVIDUAL LIVING FLOORS

Before discussing each individual living floor at Guilá Naquitz (Table 10.2), there are a few general observations that can be made. First, we are struck by the fact that all presumed fire-making equipment, regardless of zone, tended to be discarded in the same general part of the cave. Two fragments came from Square D9, one from E9, and one from D11. What this might indicate is that there was a particular spot in the cave that, because of its centrality or its relationship to prevailing air currents, was considered appropriate for hearths.

Second, all items we have assigned to nopal roasting came from Squares F6 or F9. This might suggest that the roasting of nopal was usually conducted nearer to the mouth of the cave than to the back wall. Alternately, it might have been done over a centrally located hearth, but with the resulting debris thrown toward the mouth of the cave.

TABLE 10.2
Artifacts of Wood and Related Materials from Zones B–E, Guilá Naquitz Cave

Zone	Square	Category
B1	C9	Ball of *Lemaireocereus* gum
B1	D9	Fire drill hearth, 20.5 cm long
B1	D10	Stick cut on both ends, 1 cm long
B2	D6	Wood fragment, use unknown
B2	E9	Possible fire drill, 9.8 cm long
B3	C9	5 segments of reed or cane
B3	D9	Broken fire drill
C	D11	Fire drill, 18.7 cm long
C	F9	*Opuntia* stem on a stick
C	F9	3 burned sticks
D	F6	2 fragments carbonized *Opuntia* stem
D	F8	1 burned wooden slat, use unknown

Zone E

There is no evidence available for this zone.

Zone D

Two fragments of carbonized *Opuntia* stem suggest that nopal roasting was one of the activities of this zone.

Zone C

During the deposition of Zone C, one of the occupants may have made a fire in Square D9 (see Chapter 5), discarding the fire drill nearby in D11. Closer to the mouth of the cave, someone discarded the charred debris from nopal roasting, including one *Opuntia* stem still skewered on a stick.

Lower Zone B

During the occupation of Zone B3, the occupants may have set a fire somewhere, discarding the fire drill in Square D9. Not far away in C9, someone trimmed the ends off a series of reeds or canes. During the Zone B2 occupation, another fire was evidently kindled, with the drill eventually discarded in Square E9 (not more than a meter from the one left in Zone B3). A wooden artifact of unknown use was also made at this time.

Upper Zone B

During the deposition of Zone B1, the occupants set several fires (including Feature 11 in Square E9), discarding the wooden drill hearth in Square D9. Nearby they also left a stick of unknown function, cut on both ends. A ball of organ cactus gum, possibly for tool hafting, was left toward the rear of the cave when this living floor was abandoned.

11

Woods Used as Fuels

Kent V. Flannery[1]

Occasional fragments of wood charcoal were scattered through the preceramic refuse at Guilá Naquitz. Two kinds of samples were drawn from these fragments: (1) specimens for radiocarbon dating and (2) specimens used to determine the woods preferred as fuels by the cave occupants. In terms of good context and quantity, first priority went to selecting radiocarbon samples; but some squares had enough charcoal so that both purposes could be served (Table 11.1).

The occupants of the cave were eclectic in their use of firewood: Pine, oak, black walnut, mesquite, acacia and other legume trees all went into the hearth, and it was not uncommon for a charcoal scatter to include from two to five genera of wood. Most fragments were small (a centimeter or less in diameter), suggesting that the preceramic foragers simply picked up whatever fallen branches were available within easy walking distance of the cave. Such behavior is not unexpected for hunter–gatherers; but it stands in striking contrast to that of the Formative villagers of the Valley of Oaxaca, who showed a decided preference for prepared pine charcoal. That selective dependence on pine, which continued into the Colonial and modern eras, may help account for the eventual disappearance of piñon pine from the mountains around Mitla (see Chapter 19).

[1]The information presented in this chapter was compiled by Flannery from identifications by R. C. Koeppen, Wilma Wetterstrom, and Suzanne Harris.

TABLE 11.1
Wood Charcoal Identified from Guilá Naquitz Cave[a]

Sample	Stratigraphic zone	Square	Used for ^{14}C sample?	Nature of sample	Woods identified
1	C	D10	yes (M-2097)	Sweepings from campfire(?)	Unidentified, possibly a legume tree
2	D	F6	no	Stray fragments	*Quercus* *Acacia*
3	D	F6	no	Scattered fragments from campfire	*Quercus* *Acacia* *Juglans* *Prosopis(?)* Conifer Unidentified, possibly a legume tree
4	D	D10	yes (M-2098)	Stray fragment	*Pinus*
5	E	F8	yes (M-2101)	Small fragments from campfire(?)	*Quercus* *Acacia* complex

[a]Samples 1 and 5 were identified by R. C. Koeppen, Forest Products Laboratory, U.S. Department of Agriculture; Samples 2 and 3 were identified by Wilma Wetterstrom, Laboratory of Ethnobotany, University of Michigan; and Sample 4 was identified by Suzanne Harris, Laboratory of Ethnobotany, University of Michigan.

Because the native categories of firewood used by Mesoamerican Indians have been a topic of scholarly investigation (see, for example, Metzger and Williams 1966), it is worth noting this possible change in ethnobotanical classification. Between 8000 and 6000 B.C., the Indians of the Valley of Oaxaca evidently considered a wide variety of woody genera to be acceptable firewood. By 1000 B.C. they had settled on pine as the strongly preferred culinary fuel and, in retrospect, this change had serious environmental consequences. Of the five genera represented in the Naquitz phase debris, four are still present near the cave; only *Pinus* has disappeared. We mention this for the benefit of those archaeologists who may still believe that native ethnobotanical classification is an effete topic without implications for human ecology.

Of all the preceramic living floors, Zone D provided the richest data. Square F6 produced a series of charcoal fragments that we suspected to be the scattered debris from a campfire. Included were oak (*Quercus*), acacia, walnut (*Juglans*), a conifer (possibly pine), and a possible fragment of mesquite (*Prosopis*). Square D10, in an area that Reynolds (Chapter 28) has interpreted as having been devoted to food preparation, produced a sample of pine large enough to use for radiocarbon dating. We suspect that had the samples from other living floors been larger, they would have shown roughly the same range of genera.

12

Artifacts of Deer Antler

Kent V. Flannery

In the course of cleaning the remains of white-tailed deer from preceramic Guilá Naquitz, we discovered five fragments of antler tine that appeared to have been deliberately modified. Most were so badly broken that it was difficult to estimate the original size of the artifact, but the longest had probably once been in the 7–10-cm range. In view of the small sample size, we have simply divided these specimens into two general categories.

ARTIFACT CATEGORIES

Possible Antler-Tine Pressure Flaker

Fig. 12.1a.
Number of specimens: 1.
Description: This is a fragment of antler tine with small chips broken off some 10–25 mm below the end as the result of pressure against a hard material, such as chert or silicified ignimbrite. This is the kind of wear we would expect if the tool had been used for doing the secondary chipping on a projectile point or sidescraper. Antler tine flakers were present in the Tehuacán sequence as early as the El Riego phase (MacNeish *et al.* 1967:142, Fig. 119).

Fire-Hardened and/or Scratched Tines

Figs. 12.1b–d.
Number of specimens: 4.
Description: These fragments of antler tine (1) appear to have been deliberately burned to harden them, (2) have been scratched by use as an artifact, or (3) show both attributes.

Since many antler fragments from Guilá Naquitz may well have been charred accidentally after discard, it is not always easy to detect those tines that have been deliberately burned. At least three specimens, however, appear to have been purposefully fire hardened (Table 12.1). One of these (from Zone B3) is burned only at the tip, while the others are scorched farther down the tine. MacNeish *et al.* (1967:142) report that four of their antler tine flakers "have been charred near the tip, probably to make them harder." This suggests that some of our specimens may have been intended for use as pressure flakers but were discarded before they had begun to show the kind of wear seen on the flaker from Zone B1.

Three of our specimens show extensive striations near the tip that presumably resulted from their use as artifacts. To be sure, bucks frequently rub their antlers against trees to get the last of the "velvet" off and sometimes spar with other bucks in ways that result in damage to the antlers. While such a source would be hard to rule out for some scratches, these striations look like the result of human modification rather than natural rubbing or sparring. However, the use to which the tines have been put remains unclear. Equally suggestive "cut and split antler tips" appeared in the Tehuacán caves (MacNeish *et al.* 1967:144–145, Fig. 121).

SUMMARIES OF INDIVIDUAL LIVING FLOORS

Each major living floor at Guilá Naquitz had at least one antler artifact. Perhaps the most interesting occurrence was that of a piece of worked tine in Zone E, a level that produced no unmodified antler and hence no solid evidence that a male deer had been killed during the occupation (see Chapter 22). This fragment of worked tine may therefore have been brought to the cave as an artifact. In all other zones, there is evidence that a buck was killed, and the antler artifacts could therefore have been made on the spot.

Zones D, C, and B3 produced fragments of fire-hardened tines. In Zone B1, an apparent well-worn pressure flaker was discarded near the center of the living floor.

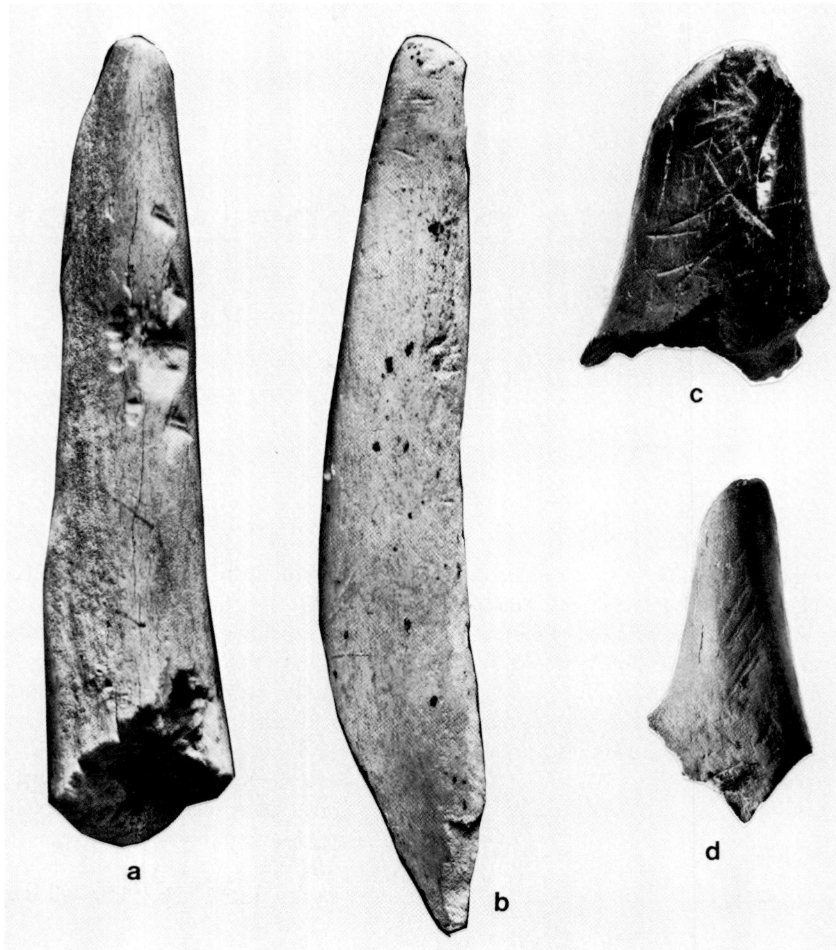

FIGURE 12.1. Modified antler tines from Guilá Naquitz: a, possible antler tine pressure flaker, Zone B1, Square D7; b, scratched and worked tine (maximum length, 60 mm), Zone E, Square C10; c, fire-hardened and scratched tine, Zone C, Square C9; and d, fire-hardened and scratched tine, Zone D, Square E8.

TABLE 12.1
Modified Antler Tines from Guilá Naquitz

Stratigraphic zone	Square	Description
B1	D7	Possible pressure flaker
B3	C6	Tine with fire-hardened tip
C	C9	Fire-hardened and scratched tine
D	E8	Fire-hardened and scratched tine
E	C10	Scratched and worked tine

13

Coprolites and Animal Fur

Eric O. Callen*

*Editor's note: Eric Callen's untimely death in Ayacucho, Peru prevented him from proofreading this chapter or making any final revisions. This version of the chapter has been prepared from Callen's preliminary report and his correspondence with Flannery.

Of the 24 coprolites recovered from Guilá Naquitz, only 6 (Coprolites 1, 4, 8, 9a, 9b, and 10) were from preceramic levels (Table 13.1). All were diagnosed as being of animal, rather than human, origin. Most were probably left by coyotes or gray foxes who scavenged the cave area whenever the human occupants moved out. The Mexican gray fox, in particular, is an omnivore who eats many of the same animal and vegetable foods as man.

The impression that many cave coprolites were nonhuman had grown with time. We started the Tehuacán project expecting that most coprolites would be human. As we began to find more and more coprolites containing large bone fragments, abundant hair, insect parts, beetle larvae, and other such items, we began to realize that domestic dogs, coyotes, foxes, and even skunks had contributed to the samples in the caves (Callen 1967a). By the time analysis of the Guilá Naquitz coprolites began, we had concluded that, in fact, many dry caves have far more animal coprolites than human specimens. In our opinion, there are a number of earlier cave samples that should perhaps be reanalyzed with this possibility in mind.

The six coprolites from preceramic Guilá Naquitz were examined visually before being subjected to detailed phosphate analysis. Two of them had a relatively smooth exterior, were heavy and hard in texture, and had a dirty, putty-like appearance, with embedded hairs and bone in one case and hairs only in the other. This proclaimed them to be "meat coprolites" left by carnivores. The other four specimens did not show their origin as definitely, either by shape, size, weight, or appearance.

All six coprolites were subjected to soaking for a week in trisodium phosphate (Callen 1967a) and given a thorough analysis. Of the two coprolites believed to have been left by carnivores, one (Coprolite 1, Zone D) contained mostly the remains of meat and hair but also some prickly pear (*Opuntia*) tissue and the remains of some grass. The absence of bone suggests that the victim had been an animal larger than a mouse or rat, but on the other hand it had definitely not been either deer or peccary, both of which have very distinctive hairs. Until the hairs in the coprolite can be identified, it is impossible to say whether they come from the victim or from the carnivore's grooming itself.

The other suspected carnivore coprolite (Coprolite 8, Zone D) contained no plant material whatsoever, though there were plentiful remains of meat, mouse hair, jaw and teeth, vertebrae, ribs, bones, and claws. The size of the coprolite and its contents suggest a coyote as the donor. Neither of these carnivore coprolites contained bile, and the trisodium phosphate was a pale transparent yellow, even after the coprolites had been soaked in it for a week. Man and many mammals produce coprolites containing bile, which stains the trisodium phosphate some shade between light and dark brown or even an amber/black. This latter occurs chiefly in man, though it is true that in certain pathological conditions (such as a blocked bile duct), the coprolites can be a pale ivory shade. These two coprolites definitely appear to have had carnivore donors however.

Zone D produced another coprolite that had a different texture and color at each end. Like Coprolite 8, it was from Square F6, and the two halves were designated Coprolites 9a and 9b. Coprolite 9a contained meat remains, bone fragments, and what might have been chips or flakes from a scapula, but no hairs. Present also were a leaf skeleton and grass fibers and the remains of insect mouth appendages, all of this suggesting an animal donor. Coprolite 9b contained meat protein remains, with bone fragments and chewed hair fragments (an unusual occurrence) as well as grass leaf tissue, possible bean pod tissue, and *Opuntia* epidermis. Quite a

TABLE 13.1
Coprolites and Animal Fur from Preceramic Guilá Naquitz

	Zone B2
Coprolite 4	(Zone B2, Square C7). Mainly a meat coprolite with small vertebrae, as well as the remains of a monocotyledon fiber (grass?) and a single seed. Diagnosis: animal origin.
	Zone D
Coprolite 1	(Zone D, Square E7). In spite of the presence of a little bit of prickly pear (*Opuntia*) and grass remains, this was chiefly a meat coprolite. Diagnosis: animal origin, carnivore.
Coprolite 8	(Zone D, Square F6). No plant material. Mouse hair, bones, teeth, ribs and claws, plus meat debris indicating a meat coprolite. Diagnosis: animal origin, carnivore, possibly coyote.
Coprolites 9a and 9b	(Zone D, Square F6). a) Contains bone chips and fragments, insect mouth appendages, the skeleton of a leaflet, and grass fibers. b) Contains hair and bone fragments, but also grass tissue, possible bean pod tissue, and *Opuntia* epidermis. Diagnosis: animal origin.
Rabbit fur	(Zone D, Square E8). Definitely identifiable as *Sylvilagus* sp. (cottontail), but cannot be identified to species because it is fragile and decomposing.
	Zone E
Coprolite 10	(Zone E, Square F9). A little grass fiber, some *Agave* fiber and roasted (?) epidermis is overshadowed by bone and insect remains as well as meat debris. Diagnosis: animal origin.

number of charred fragments could not be definitely identified but appeared to be the remains of roasted *Opuntia* epidermis. This all suggests an animal scavenging around human habitation since the trisodium phosphate color, plus the presence of bones, definitely indicates animal origin.

One specimen (Coprolite 10) came from Zone E and contained mainly bone, meat, and insect remains, with some grass and *Agave* fibers and a little roasted *Agave* epidermis. Here all the evidence, including the color of the trisodium phosphate, points to animal origin, in fact to a carnivore donor.

The last of the preceramic coprolites came from Zone B2. Again, this was mainly a meat coprolite with a number of small vertebrae (probably mouse), as well as what appeared to be grass leaf fibers and a single small seed that has not been identified. The color of the trisodium phosphate and all other evidence points to an animal donor for this coprolite also.

The presence of both *Opuntia* and *Agave* roasted epidermis is of great interest. In the preceramic levels of the Tehuacán caves of southern Puebla, these two plants, both raw and roasted, constituted quite important items in the diet of their inhabitants. Both genera were also roasted at Guilá Naquitz, so it would be logical to have scavenging animals seizing some of these roasted (or even charred) plant remains that had been cast aside by man. Indirectly, therefore, we get some indication of the diet of the Guilá Naquitz cave dwellers.

Finally, Zone D also produced a sample of fur that was definitely identifiable as cottontail rabbit (*Sylvilagus* sp.). This same square (E8) also yielded bones of both Eastern and Mexican cottontail (Flannery and Wheeler, Chapter 22), so the fur sample may suggest the animals were skinned inside the cave.

14

Radiocarbon Dates

Kent V. Flannery

Ten samples of wood charcoal from preceramic levels at Guilá Naquitz were submitted for radiocarbon dating. These samples were selected with great care from areas that appeared to be undisturbed by intrusive pits or rodent burrows. Many other samples of charcoal were rejected because of their proximity to later features or areas where the stratigraphy was less than crystal clear. In order to minimize the effects of methodological differences among analysts, the samples were submitted to three different laboratories: Geochron (GX), the Smithsonian Institution (SI), and the University of Michigan (M).

The results from these 10 radiocarbon assays are given in Table 14.1. Following the policy of the journal *Radiocarbon*, we have calculated our dates from A.D. 1950 using the old 5570-year half-life, resisting any effort to convert to a new half-life or to "calibrate" our dates based on the bristlecone pine curve. In our opinion, the current practice of "calibrating" one's dates only increases the risk of introducing new errors and making dates from different sites less comparable.

No matter how carefully you select your radiocarbon samples, in any group of 10 there are almost sure to be some that come out looking aberrant; and our sample is no exception. Six of these dates look to be internally consistent, increase in age as one proceeds downward through the stratigraphic sequence, and are associated with artifact types for which comparable dates are available from Tehuacán. If this can be taken as evidence for their reliability, then the other four dates are too young.

To begin with the upper preceramic living floors, the dates from Zones B2 and B2+3 are mutually reinforcing. The B2 date, run by the Smithsonian, came out 6670 B.C. The B2+3 date, run by Geochron, came out 6910 B.C. The difference between the two dates, 240 years, is less than two standard deviations. One could split the difference and assign a mean date of 6790 B.C. to Lower Zone B.

Of the three Zone C dates, two are consistent, while one would appear to be too young. The Geochron date of 7280 B.C. and the Michigan date of 7450 B.C. are, for all practical purposes, within one standard deviation of each other. One could split the difference and assign a mean date of 7365 B.C. to Zone C. On this basis, the Michigan date of 4030 B.C. is presumed to be incorrect.

Of the three Zone D dates, once again, two are reasonably consistent while one would appear too young. In addition, the Geochron date of 7840 B.C. is consistent with two of the dates from Zone C. The Michigan date of 8750 B.C., while perhaps too early in absolute terms, has such a large one-sigma range that it ends up being not much more than two standard deviations from the Geochron date. One could split the difference and assign a mean date of 8295 B.C. to Zone D. On that basis, the Michigan date of 2350 B.C. clearly seems incorrect.

Neither date from Zone E looks acceptable in the light of the six internally consistent dates from Zones B–D. About all that can be said for Zone E, therefore, is that it should be at least as old as Zone D and perhaps a little older.

We have no idea why four of these dates came out too recent despite our efforts to select them with care. We note that three of the aberrant dates come from the Michigan lab, and two of these were consecutive samples (M-2100 and M-2101). We also note that two of the aberrant dates came from the same square (D10), while the other two came from a pair of coterminous squares (F8 and F9). However, we have no evidence that this had anything to do with the divergence of these dates from the six that seem acceptable.

COMPARISONS WITH TEHUACAN

The six consistent radiocarbon dates from Guilá Naquitz make it appear that the Naquitz phase extended roughly through the period between 8750 and 6670 B.C. How much

TABLE 14.1
Radiocarbon Samples from Guilá Naquitz Cave

Zone	Square	Lab and no.[a]	Date[b] (yrs. B.C.)	Nature of sample	Comments
B2	D8	SI-515	6670 ± 160	Fragments of charcoal, not identified	Acceptable
B2 + 3	E6	GX-0784	6910 ± 180	Fragments of charcoal, not identified	Acceptable
C	F6	GX-0873	7280 ± 120	Wood charcoal, not identified	Acceptable
	D10	M-2097	7450 ± 300	Fragments of charcoal, legume tree	Acceptable
	F9	M-2100	4030 ± 220	Fragments of charcoal, too small to identify	Too young
D	F5	GX-0783	7840 ± 240	Wood charcoal, not identified	Acceptable
	E5	M-2099	8750 ± 350	Fragments of charcoal, not identified	Acceptable
	D10	M-2098	2350 ± 180	Charcoal, *Pinus* sp.	Too young
E	D10	GX-0872	5545 ± 90	Fragments of charcoal, too small to identify	Too young
	F8	M-2101	4350 ± 220	Fragments of charcoal, *Quercus* and *Acacia*	Too young

[a]Samples were submitted to three laboratories: Geochron (GX), the Smithsonian Institution (SI), and the University of Michigan (M).

[b]All dates are calculated from 1950, using the 5570 half-life.

earlier or later it might have run is unknown. In terms of the Tehuacán sequence, this period would correspond to the last 2000 years of the Ajuereado phase (Johnson and MacNeish 1972).

The Ajuereado phase in the Tehuacán Valley appears to have been a long period, known from only six living floors and four surface collections, and divisible into two parts. During the earlier part of the phase, the Tehuacán environment included such exotic fauna as the pronghorn antelope, native American horse, and Texas gopher tortoise, none of which occur in the state of Puebla today. During the later part of the phase, these exotic forms seem to have been absent, and the faunal remains include only species still found in the Tehuacán Valley (Flannery 1967). Hunting and trapping patterns of the Late Ajuereado phase resembled those of the early El Riego phase, which began about 6500 B.C. While many of the chipped-stone types of the Early Ajuereado phase continued in use during the Late Ajuereado phase, such groundstone tools as cylindrical pestles, *tecomate*-shaped mortars, and muller or mano fragments made their first appearance in Late Ajuereado times (MacNeish *et al.* 1967: Tables 15–17). In very general terms, therefore, the Late Ajuereado phase could be viewed as a time of transition from a Paleoindian to an early Archaic way of life.

The Naquitz phase would also seem to lie at or near that transition. Zone E of Guilá Naquitz produced what appears to be an unfinished Lerma point, one of the tool types linking the Paleoindian and early Archaic periods. A fragment of slab metate was the only ground-stone tool from this living floor. Later levels of the cave produced one-hand manos and mortar fragments like those from Tehuacán. They also produced examples of two projectile point types, Almagre and Trinidad, known to occur before 7000 B.C. elsewhere in Mexico. Obviously, the preceramic phases in the Tehuacán Valley are rather long, and many of their diagnostic artifact types are somewhat generalized; but we see nothing in the Naquitz phase assemblage that makes us uneasy about assigning it to the period 8750–6670 B.C.

IV

COMPARING THE ARCHAIC AND MODERN ENVIRONMENTS

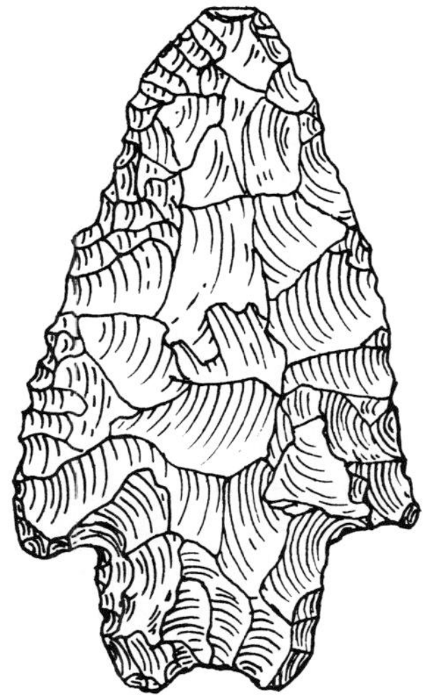

In Chapter 3 of this volume, we discussed C. E. Smith's 1978 reconstruction of the "original" primary vegetation of the Holocene Valley of Oaxaca. Three of his zones—oak–pine forest, thorn–scrub–cactus forest, and mesquite forest—are particularly relevant to the prehistory of the Guilá Naquitz region.

In Chapter 4 we examined the present-day vegetation of the Guilá Naquitz region in some detail. The four basic vegetational zones in the area today—Thorn Forest A and B and Mesquite Grassland A and B—are all that remains of Smith's "original" vegetation after centuries of grazing and agriculture. Within those zones, however, are scattered patches that, through decades of neglect following the overthrow of the hacienda system, have reverted to the point where Smith could use them in his reconstruction of the ancient vegetation.

In Chapter 18 we present the results of a 7-year census, spread over 10 years (1966–1976), in which we quantified the edible plants in those relict patches of vegetation in terms of kilograms of edible portion, grams of protein, and kilocalories per hectare during wet, dry, and average years. These data are used in Chapter 24 to reconstruct the September–December diet of the Guilá Naquitz foragers as well as the relationship between food supply and human population. They are further used by Reynolds in Chapter 31 to model the foraging strategies of the Naquitz phase.

Obviously, all such reconstruction and modeling would be to no avail if the present-day vegetation were so drastically different from that of the Naquitz phase as to shed no light on that preceramic period. We therefore needed to establish just how different the environment was at the time the cave was occupied. We knew that, with the exception of the piñon pine (which may have been selectively removed by charcoal burners), every plant and animal species found in the cave is still native to the area today. However, we wanted other lines of evidence to go with this "macrofossil" data. Those other lines were provided by pollen and microfauna.

Part IV gives our analyses of the latter specimens. In Chapter 15, Schoenwetter and L. D. Smith present their interpretation of the Archaic pollen sequence in light of today's pollen rain in the Valley of Oaxaca, utilizing not only Guilá Naquitz but also neighboring sites such as Cueva Blanca and Martínez Rockshelter. Schoenwetter and Smith's comprehensive treatment could well have stood as a monograph in its own right, and we are very happy to have it. In Chapter 16, Jane Wheeler and I present a briefer comparison of the Archaic microfauna with its present-day counterpart.

There are really two questions to be put to all this "microfossil" data, one more easily answered than the other. The easier question, "Is the present-day vegetation identical to that of the past?" can quickly be answered, "No." The other question, "Are the least-disturbed patches of Oaxaca vegetation a realistic guide to the past?" is more difficult, but based on Part IV we believe the answer should be, "Yes, if used sensibly." As Schoenwetter expressed it in a preliminary study of the Oaxaca pollen, the 8000–6000 B.C. environment near Guilá Naquitz was probably somewhat cooler and drier than today's, but remained "neither as frigid as occurs in the valley today at extreme elevation, nor as xeric as occurs today in the valley's arid sections. It was not sufficiently different to drastically change the character of ecological relations among plants nor to introduce plant types that are not palynologically represented there today" (Schoenwetter 1974:298). In other words, while there are clear fluctuations in the record for the Guilá Naquitz environment, it generally remained within the parameters of the larger-scale Valley of Oaxaca ecosystem. This is probably all we could reasonably ask for in terms of good luck, since it means that the studies in Chapters 18, 24, and 31 were worth doing.

By way of example, let us suppose that the present-day Valley of Oaxaca were to enter a century of moister conditions. Under such conditions, the oak woodland might well expand downslope at the expense of Thorn Forest A; however, products of Thorn Forest A would still be available, albeit at greater walking distance from Guilá Naquitz. In the alternative case of a century of drier conditions, Thorn Forest A might well expand upslope at the expense of the oak woodland; products of the latter zone, however, would still be available, albeit at a greater distance from the cave. In other words, in spite of significant fluctuations, our studies of the 11 plant genera most heavily used by the occupants of Guilá Naquitz would still be useful. Moreover, Reynolds' simulation in Chapter 31 is designed so as to let us model the effects of wetter or drier periods, not unlike the ones seen in the Archaic pollen record.

Many of the economically important plants of the Naquitz phase are represented both in Schoenwetter and L. D. Smith's pollen records and in C. E. Smith's macroscopic plant remains. The pine, oak, *Celtis, Agave,* and *Prosopis* pollen, for example, is fully to be expected, given the number of acorns, piñon nuts, hackberries, maguey quids, and mesquite seeds left behind in the cave. Both bottle gourds and *Cucurbita pepo* were cultivated in the Naquitz phase, and occasional grains of cucurbit pollen show up at Guilá Naquitz and Cueva Blanca. However, the pollen charts also record plants that did not show up in the macroscopic remains, the most interesting being members of the tribe Maydeae. Both *Tripsacum* and teosinte (*Zea mexicana*) are wild relatives of maize whose distribution covers much of southern Mexico (Wilkes 1972), including the state of Oaxaca. Of course, macroscopic maize itself (*Zea mays*) is still unknown before 5050 B.C. (Mangelsdorf *et al.* 1967). However, if it turns out that maize is a domesticated form of teosinte—as Beadle (1977), Iltis (1983), and others believe—one might expect that a long period of teosinte use preceded 5050 B.C. Since no macroscopic remains of the Maydeae occurred in our Naquitz phase deposits, archaeological data cannot clarify which member(s) of that tribe were growing in the region. In such a case, my own inclination is to take the most conservative position: some grass of the *Zea–Tripsacum* group occurred in the Naquitz phase environment, but I am not prepared to specify what it was or how it was used.

15

Pollen Analysis of the Oaxaca Archaic

James Schoenwetter
Landon Douglas Smith

PROJECT DESCRIPTION

Palynological investigations of the Oaxacan preceramic period were designed to address three specific problems: (1) the relative chronological relationships existing among archaeological records of the horizon; (2) the characteristics of changes that may have occurred in ecological conditions through this period in the Valley of Oaxaca; and (3) the nature of palynological evidence bearing on the character, availability, and distributions of plant resources during this period. None of these problems were considered or examined in vacuo, nor was it ever anticipated that palynological evidence would be sufficient for their resolution. The pollen record was conceived as one link in a chain of evidence forged through study of botanical macrofossils, paleontological remains, and studies of the botany, geomorphology, sedimentology, archaeology, and ethnography of the valley that could serve as a source of hypotheses and models for analysis of the past record of human activity and its varied interrelationships with the biophysical world. The palynological investigations were not established to demonstrate facts about prior cultural patterns or prior ecological conditions or chronological relationships that were not demonstrable on the basis of other forms of evidence. They were conceived as a means of developing supportive evidence that expressed the same facts, or related facts, in an independent fashion.

The method employed here is comparative and presumes the need for statistical control. Essentially, the method consists of systematic comparison of pollen values representing known conditions of modern ecology with pollen values ostensibly representing conditions of the past. Where the similarities between the two are identifiable as patterned and unlikely to be an effect of chance, we have accepted the uniformitarian assumption that they are palynological responses to the same environmental or cultural cause. The statistical procedures we have employed are varied and tend not to be those traditionally utilized by pollen analysts. The decision to use nontraditional approaches is based on our recognition of the distinctive problems presented by the contexts sampled to produce the fossil pollen record. Normally, pollen analysts deal with pollen derived from deposits that have been unaffected, or minimally affected, by the vagaries of human behavior. In such cases the pollen analyst may assume—unless evidence exists to the contrary—that changes in the pollen record reflect changes in the biophysical environment. Our investigations, however, rely almost exclusively on pollen samples collected in intimate association with evidence of human cultural activity. In some cases, the sediments from which the samples derive are themselves artifacts—such as ash lenses and midden. We have adopted the assumption that the pollen records obtained were—unless evidence was adduced to the contrary—reflections of human interference with the natural processes of pollen dispersal, deposition, and preservation, following Schoenwetter (1981).

To determine whether a given fossil pollen record represents cultural interference or natural conditions, we compared it with modern pollen rain values considered reasonably likely to represent specific ecological conditions. If it compared favorably in relation to one statistical standard, we tried a second standard as a means of establishing a redundant test. Only in situations where the fossil pollen record is classified by both standards as comparable to a definable ecological pattern have we accepted it as representing conditions of the biophysical environment and progressed to interpretations of paleoenvironment.

This method has required us to undertake intensive study of modern pollen rain variations in the Valley of Oaxaca. In fact, to our knowledge, only one area of similar geographic size has been as extensively sampled or so thoroughly studied in respect to modern pollen rain (West 1978). It has also required us to explore the use of statistical procedures that have been rarely applied in the analysis of pollen records. In order to explain and justify these departures from traditional practice, we must devote a large proportion of this report to issues that bear only indirectly on the conclusions that address the objectives of the investigation. We are well aware that the details of our diversions into methodological and theoretical aspects of archaeological pollen analysis are of little concern or interest to the majority of our readers. But they are crucial to the development of our conclusions, and they cannot be ignored in evaluation of our reconstructions.

These studies were initiated in 1966, at a time when some of the methodological and theoretical problems of archaeological pollen analysis loomed larger than they do today. Many aspects of pollen sampling strategies were also less well comprehended at that time. The former situation encouraged our preoccupation with controls, and the latter led to a tendency to rely on individual samples we presumed would be characteristic of particular spatial or temporal conditions.

The fossil pollen samples described in this report were collected from archaeological sites excavated in 1966 and 1967. During 1966, 1967, and 1968 the pollen they contained was extracted, and pollen counts made, by Schoenwetter and Suzanne K. Fish. Comparable work was also undertaken during that period on a large series of samples collected from archaeological contexts representing the entirety of the Formative, Classic, and Postclassic occupations of the Valley of Oaxaca and on a series of 25 modern surface pollen rain samples. During the 1969 excavation season, Fish undertook a wide-ranging modern surface sampling program in the area. Her subsequent work with these samples resulted in 174 pollen counts reflecting the variations in pollen frequencies occurring under modern conditions in all the major natural environments of the valley.

Fish originally planned, as her doctoral dissertation, to undertake a multiple-regression analysis of the surface pollen rain to identify its value as controls for the interpretation of the fossil pollen records. This program was interrupted in 1971, however. In 1972 Schoenwetter attempted a computer-assisted analysis of the modern pollen rain using homogeneity chi-square calculations as a means of identifying patterned relationships within the data set. That procedure proved unworkable because it demanded the imposition of unwarranted assumptions. The effort was not totally wasted, however, as it provided indications that certain statistically valid patterns were discernible in the surface pollen rain data base. Schoenwetter subsequently approached the problem from a different perspective and utilized an analysis based on confidence interval statistics to identify paleoenvironmental conditions evidenced in the record of samples from Guilá Naquitz (Schoenwetter 1974).

In 1974, Smith—working from a broader background in statistics—began analysis of the surface pollen rain based on multivariate statistical techniques. His initial results demonstrated the applicability of an analysis of variance for the identification of a set of pollen types that was particularly useful for the diagnosis of ecological conditions existing today. Specifically, he was able to show that certain sets of pollen types varied systematically in samples collected from distinctive ecological situations and other pollen types did not. It was not the proportions of pollen in the overall spectrum that were necessarily the best indicators of a given ecological condition but the mathematical interrelationships of the proportions of particular pollen types one to another. These interrelationships can be expressed in a number of ways (e.g., as factors, as correlations, or as principal components).

Smith continued his investigations (Smith 1975) utilizing the expression known as discriminant functions, as this approach required fewest assumptions about the nature of "real world" conditions presumably represented by the pollen record. Discriminant functions, basically, are mathematical formulae that identify distinctions between the pollen records of different sets or groups. After determining the formulae that represent, for example, the distinctions between modern pollen rains reflecting Desert, Thorn Forest, and Pine Forest ecology, those formulae may be applied to the fossil pollen records. This allows one to classify the fossil pollen records in terms of their degree of similarity with the groups of modern pollen records.

Smith's studies along these lines were particularly successful in suggesting unsuspected chronological relationships among the pollen records of the preceramic horizon. Thus encouraged, Schoenwetter and Smith undertook further analyses in 1977 that were designed to serve as redundant tests to deal with the issue of paleoenvironmental reconstructions.

The palynological research reported here, then, represents the integrated and cooperative labor of three workers who each provided skills, knowledge, and orientations none could offer singly. The result is a work of more intensive methodological and technical study than has previously been reported in archaeological pollen analysis and, we believe, provides a more secure and trustworthy reconstruction of paleoecological conditions than was thought possible when the investigations began.[1]

THE MODERN POLLEN RAIN

The reproductive and life cycles of plants and plant association assemblages are conditioned and controlled by a large number of genetic, climatic, edaphic, and biotic factors—and their interrelationships–collectively referred to as the

[1]Yet this research must also be recognized as a product of its times. With the exception of this and three other paragraphs, the text of this chapter was written in 1979. Since then, significant source materials have been published that present the details of evidence and argument for many of our undocumented statements about pollen rain/ecology relationships that might have been referenced (e.g., Birks and Birks 1981:156—283; Bryant and Halloway 1983). Further, discriminant functions analysis has become more regularly employed in archaeological research, an archaeological application of discriminant functions analysis of pollen records has been published (Rankin 1980), and the use of multivariate methods in pollen analysis has become more advanced and sophisticated (e.g., Turner and Hodgson 1979, 1983; Delcourt *et al.* 1984). Updating this chapter was seriously considered, but we were unable to assess or predict the effect of such revision on the integration of information represented by the book as a whole. We have chosen to allow this report to stand on whatever merits it had when it was produced.

ecology of the geographic space in which the plants exist. The most evident effect of the ecology of a given location is observed as the survival of certain plants and the absence of others: the outcome of selection for and against an organism's genetic potential for reproductive success under the specific biophysical conditions occurring at the locality. Study of the presence or absence of taxa, however, is neither the only nor the most effective means of discerning variations in ecological conditions from place to place. The selective pressure exerted upon the genomes of organisms by the biophysical characteristics of a location is pervasive and may be expressed in a great number of less than obvious ways. It may be evidenced, for example, as adaptive variability in the morphology or physiology of organisms within a species (ecotaxa) by adaptive variations in the relationships of one organism to another indicated by their density or sociability (e.g., clines), or by adaptive variations in the timing or quality of portions of the reproductive or life cycles that reduce competitive interactions.

Theoretically, no organism can survive to reproductive maturity unless its genetic characteristics allow physiological tolerance of the unique geographic space that it inhabits. The distinctions in the ecology of loci, then, are discerned and evaluated through analysis of the differences that occur in the organisms that inhabit them. But analysis of similarity and difference is necessarily a matter of degree. One may appraise two locations as similar in ecology insofar as they both provide for the survival of an assemblage of organisms mutually classified as a "forest," or one may appraise the same localities as distinctive in ecology because one supports a forest characterized by the presence of pine trees while the other supports a forest characterized by the presence of oak trees. Identification of the ecology of a geographic location, then, has little to do with the particular biophysical characteristics of the location. The biophysical characteristics of each and every geographic space on the earth—irrespective of size—are unique in some way or another that could have a selective effect on its biota. What is truly at issue is the system of classification one utilizes to discriminate degrees, or levels, of similarity and difference among the biophysical characteristics of the loci. All loci are unique; the question is which are more similar in the biophysical characteristics one is concerned to assess.

The pollen rain preserved at any geographic locus is as affected by the ecological character of that location as the organisms that exist there. It must be remembered that each pollen grain houses a set of living cells that, on dispersal from the male reproductive organ, fulfills a necessary role in the life cycle of the species. Further, this role is undertaken in relative isolation once dispersal is achieved. To accomplish its function the pollen grain must be physically deposited on the stigma of a plant of the same species; it must germinate, thereby exuding unprotected protoplasm into the environment of the stigma; its cellular contents must penetrate and grow through the stigma, style, and ovarian tissues of the receptor female reproductive organ; and, finally, fertilization must be effected. Throughout this complex process, the pollen grain and the living tissue it contains is exposed to the environment of the geographic space in which the parent and the receptor plants exist. Though the period of exposure is not normally long, the genetic characteristics of the disseminated pollen must be such that it is physiologically tolerant of biophysical conditions that occur in that location and capable of fulfilling its function in that place prior to the death of the living cells.

The production and dispersal of pollen grains is also conditioned by the ecological conditions of the location of the male parent. On the one hand, those conditions affect the capacity of the parent to generate given quantities of healthy sperm cells and to release them enclosed within the cell wall of the pollen grain. On the other hand, ecological conditions affect the directions and the distances the pollen grains may travel without interception, destruction, or death.

It is extremely unlikely that any given pollen grain will providentially travel to a resting place on a receptive stigma of the appropriate plant species. To overcome such odds, plants produce enormously greater quantities of pollen grains than eggs, and many species have developed intricate adaptive mechanisms that assure a greater prospect that at least some pollen will inevitably be transferred to appropriate receptor tissue. But very little of the pollen produced and dispersed into the environment fulfills the reproductive function for which it is genetically designed. It falls, ultimately, to a resting place nearer to or further from its source in response to its particular morphological characteristics and the vagaries of environmental phenomena that condition its dispersal and journey. This pollen—the pollen that has been produced and dispersed but that is not reproductively effective—is the pollen rain.

Ecological factors thus condition the characteristic taxa, proportions of taxa, and dispersal of the pollen rain. They also condition the characteristics of the pollen rain that may be preserved and sampled for scientific study. Little of the pollen rain comes to rest upon surfaces where it may be trapped for analysis. The vast majority of pollen of zoogamous plants is dispersed clinging to insects or other animal vectors and is consumed or destroyed long before it can be sampled by devices that strain pollen grains from the air or intercept their progressive fall to the earth's surface. Most pollen grains dispersed directly to the atmosphere undertake relatively short journeys, to judge by controlled observations. Though pollen is not infrequently recovered hundreds or even thousands of miles from its source, the proportion of pollen grains that achieve such long-distance transportation is almost infinitesimal. Most dispersed pollen travels less than a few tens of meters from the parent plant and comes to rest on the stem and foliar surfaces of local vegetation. There the pollen grains lie, for the most part, as an exposed and easily harvested food source for the bewildering variety of herbivorous and saprophytic organisms that inhabit almost all ecological niches. Even that relatively small proportion of the pollen rain that falls or is washed to the ground surface is almost entirely consumed or mechanically destroyed, and much of what remains is subject to chemical corrosion and physical erosion in its ultimate depositional environment.

Thus the pollen rain observed by the palynologist constitutes only a small and biased sample of the pollen that is produced by the spermatophyte taxa of any particular locality. Ecological conditions obtaining at that location affect the production, the dispersal, and the preservation of pollen both directly and indirectly. They also affect the pollen rain differentially at different times of the day, in different days of the season, in different seasons of the year, and in different years.

Given this state of affairs, it is almost incredible that the preserved pollen rain a palynologist may sample has any sort of uniform characteristics. It would be quite comprehensible if each and every one of the small and biased samples we examine was absolutely unique—dissimilar from any other sample that was, or could be, ever collected. But, thanks to the enormous profligacy of nature in pollen production, this turns out not to be the actual case. There are such unbelievable quantities of pollen dispersed that the laws of chance operate effectively in reducing the proportions of the pollen of different taxa in fairly direct relationships. The pollen rain samples of different geographic locations, when considered as populations, are therefore neither individually unique nor randomly organized. They appear to be patterned responses to the overall complex of ecological conditions that occur at the location. Locales in which similar ecological conditions prevail tend to yield similar preserved pollen rains, just as locales in which similar ecological conditions affecting plant growth and maturation prevail tend to produce similar floristic assemblages.

It is generally thought that the predominant ecological factors that condition the character of the pollen rain of a location are those that affect pollen production. Therefore, it is widely believed that if more plants of a given taxon occur at a location, there will be more pollen of that taxon recovered in the preserved pollen rain. The assumption that this belief is valid has prevalence among both palynologists and nonpalynologists; its common sense applicability has been of extraordinary significance in the interpretation of fossil pollen records. There is no doubt that the assumption is valid in many cases, but empirical evidence and an appreciation of the reality of ecosystem relationships document that it is not a reliable assumption. Obviously, since most of the pollen rain of a locus is derived from parent plants of that location, it is very much more likely that pollen from such parents will be preserved at that spot than pollen from more distant plants. But the most frequent plants at a location may not be the the most prolific pollen producers, or they may not be adapted to dispersal of pollen widely in the environment, or they may disperse pollen that is not likely to be sampled by the form of trap used to collect pollen rain for study, or they may disperse pollen that has a greater chance of destruction in the particular depositional environment sampled. There are thus a host of ecological factors that condition the characteristics of a preserved pollen rain sample. The number of local parent plants is in some cases a prevailing factor; in other cases it is not. Only empirical observation may be presently used as a positive means of identification of the palynological expression of different sets of ecological conditions. Palynologists have many theories about the ways in which different ecological factors differentially affect modern and fossil pollen rains, and some use these theories to justify their analyses and interpretations. But we are continually surprised by the conflict between such theories and the results of empirical observation.

To establish statements that meaningfully characterize the nature of a pollen rain, then, one must consider two matters that have nothing whatever to do with the nature of the flora producing the pollen. First, one must recognize that the sample of observed pollen is drawn from a statistical population and it must be defined and identified in terms of its statistical properties. The presence, absence, or proportion of a pollen type in the pollen rain, then, is established statistically for purposes of definition and characterization; it is not established empirically. If, for example, one sample of 100 pollen grains produces 1 grain of oak pollen, and another sample produces 3 grains of oak pollen, and a third sample yields no oak pollen, the samples are empirically distinctive. But, given the size of each sample, they are not statistically differentiable in the proportions of oak pollen; the distinctions observed may be produced through the operation of chance in observation of 100-grain samples drawn from the same statistical population. Similarly, if we observe among these same three samples that the proportion of grass pollen is 35, 38, and 29%, respectively, we may assess the three as members of the *same* population if we utilize the frequency of grass pollen as a characterizing variable. We may make this judgment irrespective of the nature of the flora that exists at the locations from which the samples derive. One sample may come from a desert locale, another from a pine forest, and the third from an oak forest; that makes no difference whatsoever. If grass pollen frequency is established as the characterizing variable, the three samples evidence the occurrence of the same controlling parameter because they are, statistically, samples drawn from the same population. Presumably, since ecological factors condition pollen rain characteristics, the controlling parameter on the grass pollen frequency is some biophysical phenomenon. Apparently, the grass pollen frequency reflects some environmental condition, or set of conditions, that affects grass pollen production, dispersal, and/or preservation equally at the three geographic locations but affects the flora differentially.

Our point is twofold: First, recognition that pollen rains are populations in the statistical sense means that to utilize the surface pollen records as controls for the interpretation of paleoecological or other conditions, we must use appropriate statistical formats of analysis. However, recognition that pollen rains are statistical populations means that they offer potential for both meaningful and nonmeaningful statistical manipulations. Some statistical properties of a pollen rain have nothing to do with the objectives of our analysis; we cannot presume that a given pollen statistic (e.g., a high frequency value for a taxon) reflects the ecological parameters common sense would indicate. It is therefore crucially important that we seek means of demonstrating that any particular statistical property of our data in fact is likely to identify a particular biophysical condition occurring at the locus of sampling.

Second, we must recognize that the identification of ecological conditions occurring at a given geographic locus is, essentially, a matter of classification since each location is similar to all others in some ways and absolutely unique in others. Classification schemes have an empirical basis but they are only approximations of reality. They are designs, mechanisms (artifacts, if one likes) through which we segregate differentiable aspects of the continuum of natural phenomena under investigation *for a purpose.* The significance and value of a classification scheme, then, depends as much on the purpose for which it is employed as it depends on the verity of its approximations of reality.

In order to relate pollen rains to ecological factors, it is necessary to group such factors into sets to develop one or more classifications of the ecology of the geographic area from which the samples of pollen rains are recovered. There is no necessity that such classifications be identical to those that may be developed by others for other purposes. Nor, if we can identify different purposes for them, is there any necessity to restrict ourselves in the number of classification schemes we utilize to determine the relationship of pollen rain to ecological conditions. We may use one classification of biophysical conditions, for example, as a means of discerning the relationship of pollen rain to variations in annual precipitation, another classification as a means of discerning the relationship of pollen rain to variations in annual temperature, and yet a third classification as a means of discerning the relationship of pollen rain to variations in the flora.

Essentially, this rather long statement has been composed as a way of advising the reader of some of the complexities and issues that attend our investigation of the modern pollen rain of the Valley of Oaxaca. The objective of our study is to identify those aspects of the pollen rain that vary directly with variations in ecological conditions. Because we understand that the production, dispersal, and preservation of pollen rain is conditioned by the ecology of the locus of collection, we grant that differences and similarities in the pollen rain of different loci are reflections of differences and similarities in the ecology of those loci. We find, however, that empirical comparison of pollen rains from different locations is not an adequate method for identifying or characterizing variations in ecological conditions. We are limited to the observation of *samples* of pollen rain, and we recognize that they must be treated as such. Therefore, we must use their statistical properties for purposes of comparison rather than their empirical properties. There is a specific disadvantage in this that we must constantly take into consideration: It is possible to recognize and specify statistical properties of a pollen rain that have no necessary relationship to ecological conditions we comprehend or are concerned with. Thus it is not satisfactory to our purpose simply to determine statistically valid variations in the pollen rains of different locations, for we cannot presume that those variations are reflections of any particular ecological factor or set of factors. We must identify those statistically valid variations in pollen rain that vary directly with *known* distinctions in ecological conditions.

This brings us face to face with another problem. Ecological conditions are comprehended, and evidenced, through their effects on the biota of a locality. We approach the analysis of these effects by comparing the similarities and differences we observe and classifying the continuum of nature into units. The purposes of our classification scheme have as much to do with their character as the observations that are made. "Known" distinctions in ecological conditions, then, and "known" similarities, are assessments (or judgments) made by ourselves about nature. They are not necessarily natural in themselves but are mechanisms we use to understand the integrated complexity of nature by dividing it up into less-integrated parts. We are logically free to establish any mechanism we find useful and suitable to the purpose of our investigations. Therefore, we are free to identify "known" distinctions and/or similarities in ecological conditions in terms of multiple standards of reference. There is no practical limit to the number of classification schemes we may use to identify "known" ecological conditions; there is only a limit to the number of classification schemes that will function—or that we can invent to function—to fulfill specified purposes.

Once again, the objective of our study is to identify those aspects of the pollen rain that vary directly with ecological conditions. We must specify the ecological conditions we desire to relate to aspects of the pollen rain, since the statistical properties of the pollen rain do not offer internal evidence of the ecological factor or factors to which they are related. But, we find, we can only specify ecological conditions through classification schemes: and the variety and number of classification schemes we may use depends in last resort on ourselves and not on nature.

At this juncture many readers may be inquiring into the relevance of the foregoing to our study of the modern pollen rain of the Valley of Oaxaca. Simply, it is that these complex—partly philosophical—issues demanded our respect when we set about the problem of designing a procedure that would fulfill the objectives of our overall program of palynological investigation. For one thing, they required us to pay attention to the question of the statistical adequacy of the modern pollen rain sampling program. Clearly, if the number of samples or the numbers of pollen grains observed were not sufficient for analysis of the statistical properties of the pollen rain, we could not expect to use those samples to control interpretations and evaluations of the fossil pollen record. Further, these issues required us to pay attention to the multitude of fashions through which ecological conditions are expressed in the Valley of Oaxaca today and to consider the variety of classification schemes we might use to fulfill the primary objectives of our study. Clearly, if we were unable to specify the ecological characteristics of a sampling locus, we would be unable to determine if there was an evident relationship between ecology and pollen rain at that location.

Thus our study of the modern pollen rain involved us deeply in the study of the particulars of plant ecology in Oaxaca—a task for which our training as pollen analysts and archaeologists was not particularly suited. We did not undertake this research using the sophisticated quantitative procedures that an ecologist would have utilized. But we did not ignore those procedures because we were not qualified to employ or interpret them; we ignored them because we considered them unnecessary. We needed classification schemes that would allow us to assess biotic responses to major ecological variables, and we needed classification schemes that would allow us to assess distinctions among different patterns of vegetation. The former would allow us to organize the surface sample pollen rain information to provide control for reconstruction of paleoecological distinctions among the fossil pollen records. For that purpose we did not desire or need detailed information, since we realized that any paleoecological reconstruction we offered in quantitative terms was likely to lead to fallacies of misplaced concreteness. We also realized that there was a limit to the significance and value of detail we could hope to apply as controls or reconstructions of prior vegetation patterns. The kinds of reconstructions we hoped to evidence palynologically were supposed to be conformable with reconstructions evidenced on the basis of analysis of macrofossil floral and faunal remains, geomorphological and sedimentological information, etc. There was nothing to be gained by analyses of the modern vegetation patterns that were more detailed than those we anticipated would be meaningful and useful for reconstructions of prior patterns.

PLANT ECOLOGY OF THE VALLEY

There were a number of cooperating scientists studying the ecology of the Valley of Oaxaca simultaneously from different perspectives and with different objectives in the 1966–1969 period. The late Wallace Ernst, of the Smithsonian Institution, was specifically interested in the floristic variability of the area. Michael Kirkby pursued investigations of the physical geography and was therefore concerned with the effects of the climate, topography, sedimentology, geomorphology, and hydrology on the biota. Anne Whyte investigated cultural geography, with particular attention to the interrelationship of such variables as seasonality and harvest potential to modern farming and food collecting practices. C. E. Smith undertook investigations that would provide information regarding the characteristic associations of different plant taxa, particularly those that could be identified as resource plants significant to human populations, and the temperature and moisture parameters that affected their distributions. Our investigations centered on identifying plant associations that were, on the one hand, likely to have provided significant food resource potential for prehistoric populations not dependent on farming and also were, on the other hand, easily characterized in terms of their pollen rain.

Because the Mitla arm of the Valley of Oaxaca evidenced the least disturbance of floras, our first ecological studies were restricted to that district. In 1966 Schoenwetter undertook a north–south and an east–west transect analysis with the object of identifying such plant associations. This resulted in the recognition of 16 different plant associations not inclusive of cultivated or obviously fallow fields. A number of these associations, however, appeared likely to be successional floras on previously cultivated or lumbered plots or artifacts created by grazing. Nineteen surface pollen samples were collected during the course of the transect study, representing 12 of the associations identified.

When the pollen values of these samples were evaluated (Schoenwetter *et al.* 1967), it became clear that each plant association recognized was not easily characterizable palynologically. On the basis of pollen frequency values, the samples segregated into six clusters. Certain plant associations clearly were associated with each of these palynological units. These results documented two conclusions: (1) Clear variation does occur in the modern pollen rain of the Valley of Oaxaca, and (2) such variations are more evidently related to ecological variables controlling floristic variability than to the floras themselves. That is, pollen rain characteristics of a sampling location were not predictable from observation of the adult plants growing at the locus. But ecological factors of the location (e.g., relative effective moisture value) could be identified palynologically (Flannery and Schoenwetter 1970:146).

This result was highly significant in that it indicated that the ecological parameters we wished to reconstruct for the past might be reflected in the pollen rain of a sampling locus even if the vegetation growing at the locale had been highly affected by human activity. If the conclusion was valid, surface sampling did not have to be restricted to locations in which relatively undisturbed floras existed; a surface sampling program could yield significant control data even if extended to cultivated fields and portions of the valley where centuries of grazing and millennia of intensive human occupation had drastically altered natural vegetation patterns.

Before such an extension of the sampling program could be initiated, however, it was necessary to establish a clear conception of the nature of the ecological factors we wished the sampling program to be able to identify palynologically. In 1967, in consultation with Kirkby, Whyte, K. Flannery, and C. E. Smith, a number of observations regarding the ecology of the valley were pooled and assessed. It was our consensus that the present range of ecological variability in the valley was most conveniently recognized in terms of three classification schemes. The most general of these was the phytogeographic classification established by Smith (1978). This scheme identifies the major biotic responses to such overriding ecological variables as water table level, annual rainfall modifications created by altitudinal changes and rain shadow effects, and annual temperature values modified by lapse rate variation with altitude. The most particularized classification scheme was that Schoenwetter had utilized. It reflected the additional ecological variables of substrate variations and successional and human disturbance effects, and it provided means of assessing the effects of topographic and altitudinal variation in more specific fashions. But not all of the ecological factors that might be identified using Schoenwetter's classification scheme were easily characterizable palynologically. Thus it seemed more detailed than we required and more detailed than would prove useful to control interpretation and evaluation of the fossil pollen record.

Kirkby and Whyte had developed a classification scheme involving nine units that reflected ecological variations in substrate geophysics, agricultural potential, and plant community in addition to those identified in Smith's classification. Considered together the three schemes allowed recognition of those vegetation patterns that might be identified at the present time in the valley and the ecological factors they reflected that were likely to be easily characterized palynologically. Schoenwetter, Kirkby, and Whyte set out a classification scheme of vegetation zones, plant communities, and—where significant—plant associations for the Mitla arm of the Valley of Oaxaca. This was established as a basic guide for the collection and evaluation of surface pollen samples and was designed to allow reflection of particular ecological conditions: annual temperature as modified by lapse rate, physiologically effective moisture for plant growth and maturation as a function of substrate characteristics and topographic position, canopy character, and dominant floristic elements (see Table 15.1).

Further ecological study was undertaken in 1968 by Fish, with the assistance of C. E. Smith. During a 3-week period, Fish traveled extensively in the valley, observing vegetation and collecting surface sediment samples she and Smith considered representative of the range of variation in modern ecological conditions. Field records at sampling locations noted the geomorphological zone, water table depth, soil type, vegetation pattern, elevation, and crop type and irrigation practice at cultivated locales (Appendix 15.1). Fish recognized six vegetation patterns in the valley that had not been identified in Schoenwetter, Kirkby, and Whyte's study of the Mitla arm of the valley:

1. a plant association in which pine and oak are codominant;
2. a plant association in which *Pithecolobium* (huamuchil) is dominant;
3. a plant association in which *Prosopis, Acacia,* or *Ipomoea* (morning glory tree) is dominant and *Pithecolobium* is subdominant or common;
4. a plant association in which *Acacia* (huizache) is dominant with *Ipomoea* subdominant or common;
5. a plant association in which *Prosopis* and *Acacia* are codominants; and
6. the plant associations limited to riparian habitats along the Río Atoyac and Río Mitla.

Association 1, above, would be classified as a segregate association of the Pine B plant community, reflecting an ecological situation transitional between the two pine zone communities represented in the Mitla arm of the valley. Association 2 would be classified as representing a distinctive plant zone: Mesophytic Forest. It is identifiable as Plant Community A of this zone, reflecting high effective moisture values responsive to a water table 1–3 m below the surface and a dense canopy.

Associations 3, 4, and 5 would be classified as remnant, modified, associations of the Mesquite Forest plant zone. C. E. Smith (1978) has pointed out that this vegetation zone has been markedly changed by the removal of tree species for firewood, grazing, and field preparation, leaving only the most persistent taxa, such as prickly pear and huizache, which may not have been abundant originally. The three associations would be lumped as one plant community since conservatism dictates acceptance of the assumption that their modern distinctions are the result of human practices rather than biophysical environmental conditions. This community reflects the lowest effective moisture values of the valley as a result of its low altitudinal position and low water table (3–6 m). Association 6 would be classified as a second plant community of the Mesophytic Forest plant zone. It reflects conditions of maximal effective moisture occurring in the valley today but high annual temperature and annual rainfall values no greater than those of the other Mesophytic Forest community or the Mesquite Forest zone.

Table 15.2 summarizes the classification schemes established for the vegetational patterns and the ecological conditions they express for purposes of the palynological investigations.

TABLE 15.1
Vegetation Units of the Mitla Arm, Valley of Oaxaca

Plant zone	Plant community	Plant association	Characterizing phytogeography
Pine Forest	A		Pine and/or broad-leafed oak dominant; narrow-leafed oak absent.
	B		Oak dominant, with pine and narrow-leafed oaks both present; manzanita and other Ericaceae often common.
Oak Forest			Pine absent; narrow-leafed oak dominant or codominant with broad-leafed oaks.
Thorn Forest	A		Small-stature trees with spreading crowns characteristic; oaks present or *Cassia (guaje)* common; oaks never dominant; *Agave* common; barrel cacti common.
	B		Oaks absent; *Cassia* and *Agave* uncommon; barrel cacti common; *Bursera* (copal) present.
Mesquite Grassland	A	(i)	*Jatropha* and/or *Dodonaea* common; *Bursera* absent.
		(ii)	*Jatropha* rare; *Dodonaea* absent; *Prosopis* (mesquite) present; columnar cacti common.
	B	(i)	*Jatropha* and *Dodonaea* absent; *Prosopis* present; grass dominant.
		(ii)	*Prosopis* dominant.

TABLE 15.2
Vegetation Patterns and Ecological Conditions, Valley of Oaxaca

<table>
<tr><th>Plant zone</th><th>Plant community</th><th>Plant association</th><th>Characterizing phytogeography</th><th>Annual temperature</th><th>Annual precipitation</th><th>Effective moisture</th><th>Canopy</th></tr>
<tr><td>Pine Forest</td><td>A</td><td></td><td>Pine or broad-leafed oak dominant.</td><td rowspan="3">low</td><td rowspan="3">high</td><td rowspan="3">high</td><td rowspan="3">closed, high</td></tr>
<tr><td></td><td>B</td><td>(i)</td><td>Pine and oak codominant.</td></tr>
<tr><td></td><td></td><td>(ii)</td><td>Oak dominant; pine present.</td></tr>
<tr><td>Oak Forest</td><td></td><td></td><td>Narrow-leafed oak dominant or codominant; pine absent.</td><td>low moderate</td><td rowspan="2">high moderate</td><td>moderate</td><td>loose, low</td></tr>
<tr><td>Thorn Forest</td><td>A</td><td></td><td>Oak present or Cassia common.</td><td rowspan="6">moderate</td><td rowspan="2">low moderate</td><td rowspan="2">layered scattered thickets</td></tr>
<tr><td></td><td>B</td><td></td><td>Leguminous trees characteristic; Bursera present; barrel cacti common.</td><td rowspan="2">moderate</td></tr>
<tr><td>Mesquite Grassland</td><td>A</td><td>(i)</td><td>Bursera absent; Jatropha and/or Dodonaea common.</td><td rowspan="4">low</td><td rowspan="3">none</td></tr>
<tr><td></td><td></td><td>(ii)</td><td>Columnar cacti common; Dodonaea absent, Prosopis present.</td><td rowspan="3">low moderate</td></tr>
<tr><td></td><td>B</td><td>(i)</td><td>Jatropha and Dodonaea absent; Prosopis present.</td></tr>
<tr><td></td><td></td><td>(ii)</td><td>Prosopis common.</td><td>open savanna</td></tr>
<tr><td>Mesquite Forest</td><td></td><td></td><td>Acacia or Prosopis dominant or codominant; Pithecolobium or Ipomoea may be subdominant.</td><td rowspan="3">high</td><td rowspan="3">low</td><td>very low</td><td>low—originally very dense, now modified</td></tr>
<tr><td>Mesophytic Forest</td><td>A</td><td></td><td>Pithecolobium or Taxodium dominant.</td><td>high</td><td rowspan="2">high, layered—originally dense, now radically modified</td></tr>
<tr><td></td><td>B</td><td></td><td>Salix dominant or codominant with Alnus.</td><td>very high</td></tr>
</table>

SURFACE POLLEN RECORDS

Analysis of Frequency Variations

The distribution of surface sample collections of modern pollen rain is shown on Fig. 15.1. The samples were collected from the existing soil surface by the pinch, or scatter, technique recommended by Hevly (1968) at plots averaging 100 m^2. Surficial sediment samples were used, rather than pollen traps, in order to obtain pollen rain records from a depositional environment approximating that sampled for fossil pollen in archaeological site contexts. Of the 191 surface samples collected, 183 yielded sufficient pollen to allow observation of statistically adequate numbers of pollen grains. These observations, along with field observations and the vegetation pattern classification attending each sample, are provided in Appendix 15.1. The frequency values of the surface sample pollen rains displayed on Figs. 15.1–15.4 are based on a pollen sum that excludes maize and squash pollen.

Cultivated Lands. From a floristic perspective, the pollen records fall into two major groups: those collected from plots of cultivated land and those collected from uncultivated plots.

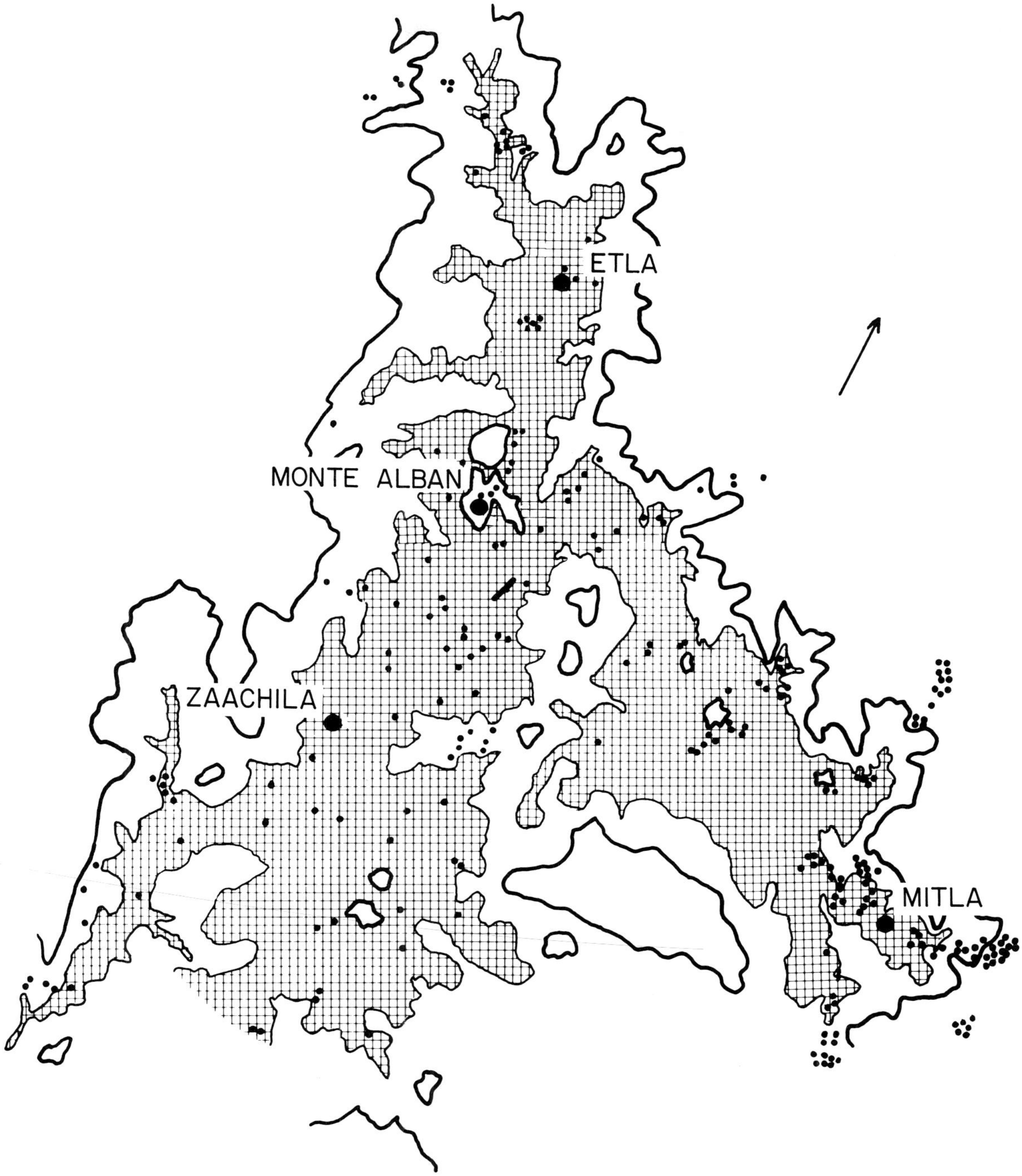

Fig. 15.1. Distribution of surface sample collections. The hatched area is the alluviated district; the contour line defining valley margins is roughly the division of mountain and piedmont topography.

None of the latter collections, however, may be assumed to reference wholly natural floristic conditions. Almost all such collections from the Mesophytic Forest and the Mesquite Forest derive from hedgerow plots or grazed tracts; and the impact of grazing, lumbering, and introduced species on the modern flora of all plant zones in the valley is abundantly evident. The principal pollen frequency distinctions between the two groups lie in the proportions of pollen of cultigens. The 93 samples from uncultivated plots (Figs. 15.2 and 15.3) yielded 56 pollen grains of *Zea* and *Cucurbita,* while the 81 samples from cultivated plots (Figs. 15.2 and 15.4) yielded 598 such grains. The difference is even greater, however, when

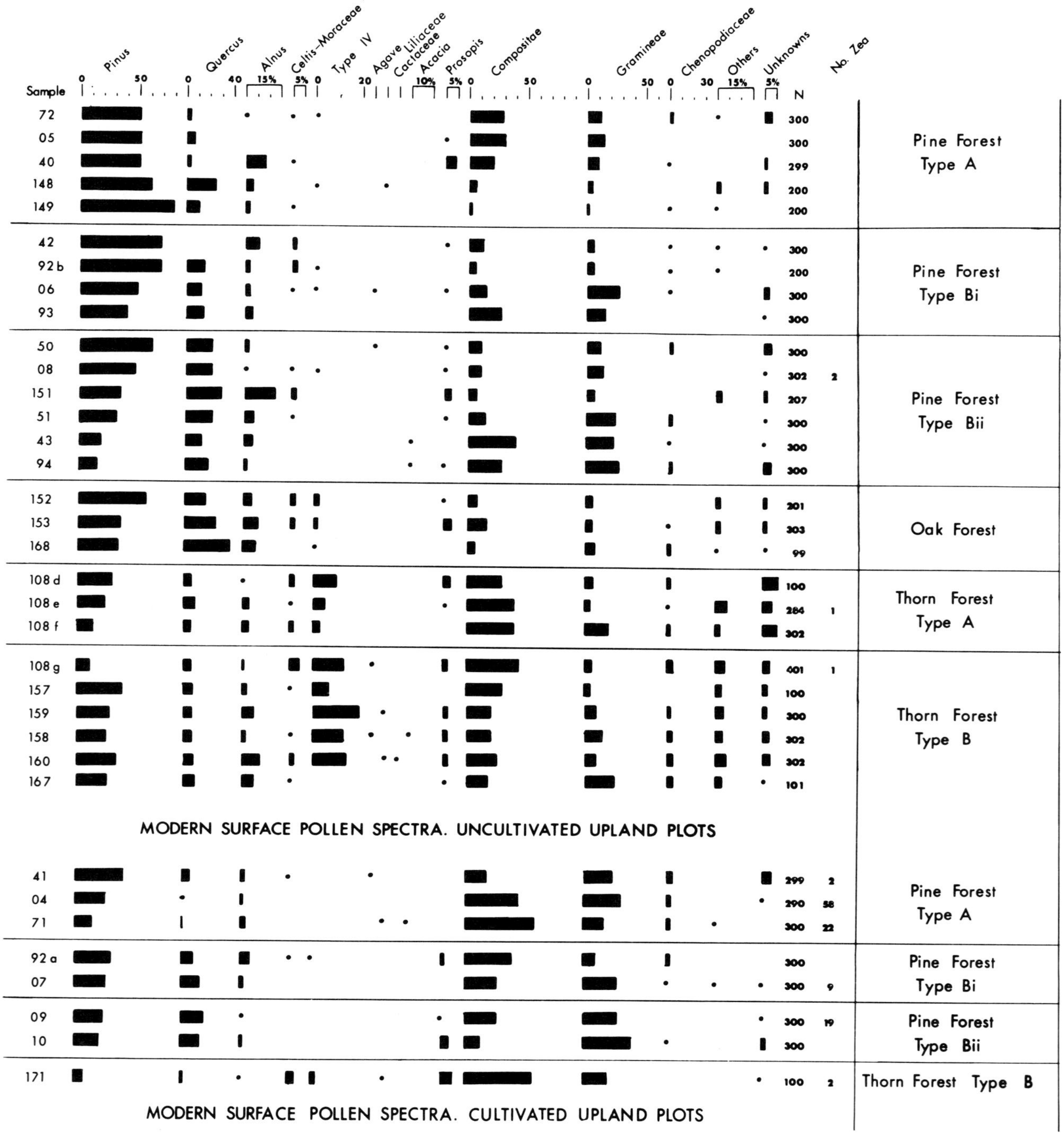

Fig. 15.2. Pollen spectra (cultigens excluded) of modern pollen rains of valley upland plots.

assessed in terms of constancy of occurrence, or ubiquity. Only 26.9% of the samples from uncultivated plots produced any pollen of *Zea* or *Cucurbita*, while 76.5% of the samples from cultivated plots yielded such pollen.

The *proportion* of cultigen pollen occurring in a given sample, though, is not as clear an indication of the occurrence of cultivation at the locus as common sense would indicate. If one utilizes a confidence interval test to identify those samples in which the true proportion of cultigen pollen is statistically greater than zero (at the 0.95 level of confidence), it turns out that only 56.8% of the samples from cultivated plots contain more cultigen pollen than might occur as the result of chance. Thus we may conclude that if the frequency of cultigen pollen in a sample is statistically greater than zero, that sample may be reliably assessed as deriving from the population of samples reflecting ecological conditions that

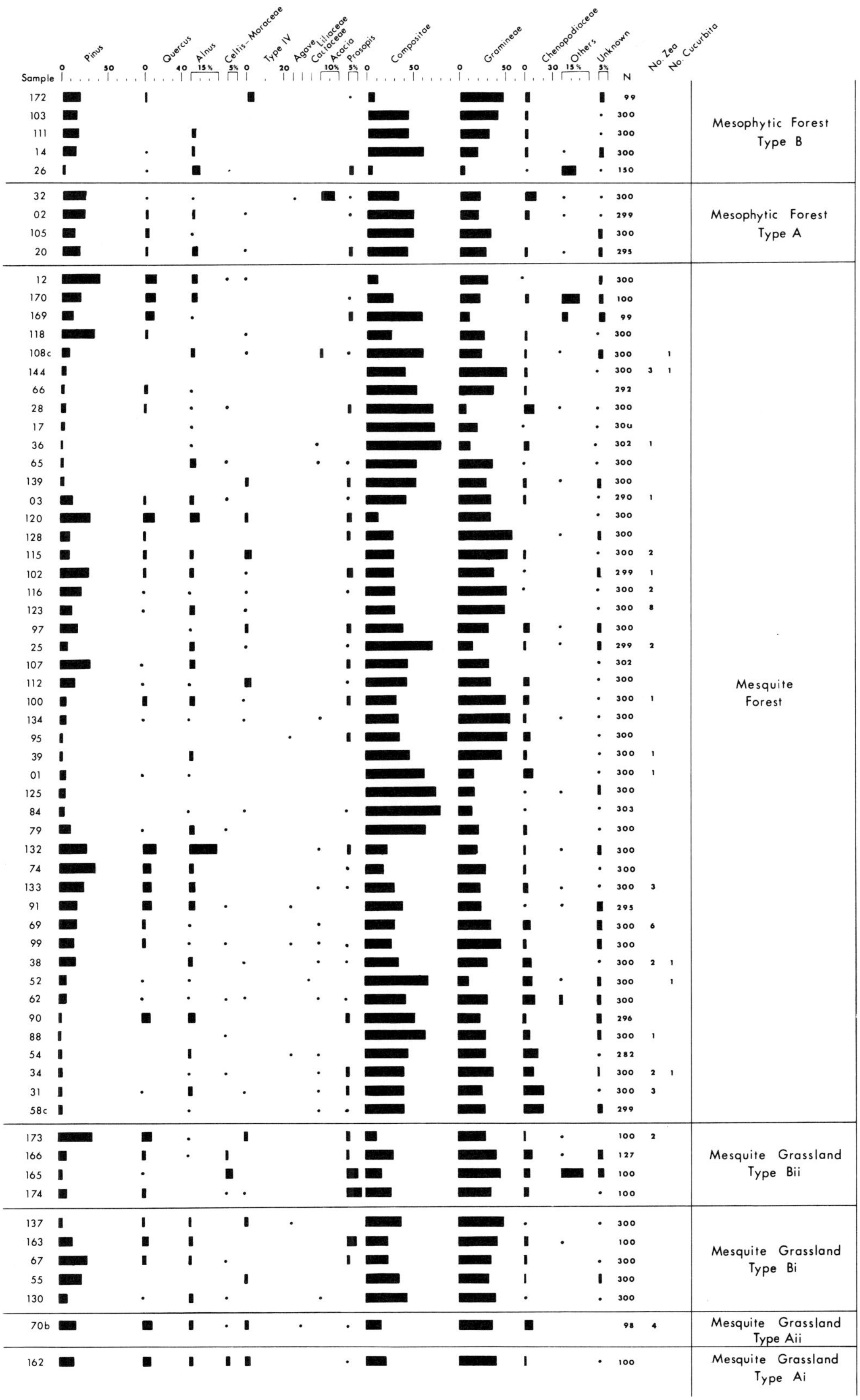

Fig. 15.3. Pollen spectra (cultigens excluded) of modern pollen rains of uncultivated valley lowland plots.

occur at cultivated, in contrast to noncultivated, locations. However, more than one third of the samples from cultivated locales do not reflect the ecological factors conditioned by cultivation in this fashion palynologically. Thus a sample that does *not* produce a statistically significant frequency of cultigen pollen cannot be assessed as either likely or not likely to derive from a cultivated plot. The ubiquity of cultigen pollen in a series of samples, however, is an excellent index of the ecological conditions occurring at cultivated locations. Data indicate that if one obtains a group of samples from the same population, the proportion of samples containing *any* cultigen pollen is quite high if that population represents cultivated locations and low if the population represents noncultivated locales. We are impressed that the latter conclusion should be so clearly evidenced in the Valley of Oaxaca since the majority of the uncultivated plot samples that demonstrate this effect derive from hedgerows and grazed land surrounded by cultivated fields. We take this to be excellent evidence that any statistically significant constancy of occurrence of cultigen pollen is a conservative basis for determining whether or not a group of samples—modern or fossil—derive from a district in which field cultivation is practiced.

The Pine Forest Zone. Fig. 15.2 illustrates the pollen frequency distinctions that occur at cultivated and uncultivated sampling locations in the Pine, Oak, and Thorn Forest plant zones. In the Pine Forest zone, where pine is dominant or codominant on uncultivated plots, the pine pollen frequency value is characteristically greater than or equal to 40% and the oak pollen frequency is less than or equal to 13%. At cultivated plots, of course, the flora is dominated by cultivated plants, and cultivated plots in the Pine Forest zone characteristically yield lower frequencies of pine pollen. This is true even in situations in which the vegetation completely surrounding the cultivated plot is dominated by pine. The occurrence of pine pollen ≥ 40% *and* oak pollen ≤ 13%, then, appears to evidence those ecological conditions that foster the maintenance of a forest dominated or codominated by pine quite clearly. However, it seems apparent that the controlling variable is not so much the local dominance of pine trees as the local occurrence of forest canopy structure. Presumably, the pine pollen frequency would also be < 40% if the sampled location were an opening in the pine dominated forest created naturally, instead of by human agency. Thus, the ecological condition reflected by the characteristic pollen frequencies of relatively undisturbed plots of the A and B(i) vegetation patterns of the Pine Forest (pine ≥ 40% *and* oak ≤ 13%) is that of canopy structure. Specifically, such pollen frequencies are not indications of temperature, precipitation, or effective moisture values in the Valley of Oaxaca today. We see no reason to presume they would reflect such parameters in a fossil pollen record either.

Oak pollen values > 13% characteristically occur at uncultivated plot locations where oak is a dominant tree. This is also true at cultivated plot locations in situations where the vegetation surrounding the field is dominated by oak. Such oak pollen frequency values occur in both the Pine Forest and the Oak Forest if oak is the dominant taxon of the district vegetation. These oak frequency values, then, cannot be reasonably interpreted as reflecting ecological factors conditioning the structure of the local canopy. Nor, since such values are recovered at both Pine Forest and Oak Forest locations, are they likely to reflect annual temperature, annual precipitation, or effective moisture values. Apparently, they reflect the occurrence of the specific ecological conditions that allow oak trees to become dominant in the district flora. This seems true whether or not oak is actually dominant at the sampling locus since oak is not dominant at the cultivated locations but these also produce oak pollen frequency values greater than 13%. The exact nature of those ecological conditions remains unknown to us. However, our investigation would not be enhanced if they were known since we do not require knowledge of them for evaluation and interpretation of the fossil pollen record.

It is of interest to note that pine pollen frequency values have a great deal more variability at sampling locations where pine occurs as a subdominant, B(ii) association, than at locations where pine occurs as a dominant or codominant taxon. Also, the pine pollen frequencies of some samples from locations where pine is subdominant are significantly lower than the pine pollen frequencies of some sample locations in the Oak Forest where pine trees are absent altogether. This reinforces our awareness that the frequency value of pine pollen in a sample is not a valid index of the frequency of pine trees at, or in the immediate area of, the location sampled for pollen rain. Rather, it is an index of ecological interactions occurring at the sampling location. This finding is not unique to Oaxaca. Martin (1963) and Mehringer (1967) documented the occurrence of higher pine pollen frequencies at greater distances from the pine pollen sources in the Chihuahuan Desert Grassland zone of Arizona and the Mohave Desert zone of Nevada. In those situations the controlling ecological variables apparently are those that affect the production and dispersal of pollen by the flora of the sampled locus. Martin found that pine pollen frequency was higher at a distance from the pine forest at a playa lake location where no local flora occurred at all. Mehringer obtained higher pine pollen frequency values at a distance from the Pine Forest in Desert Scrub locations where the dominant and subdominant floristic elements were adapted to zoögamous pollination and so did not disperse pollen in the same fashion as pine.

At least in arid and semiarid lands, then, it would seem one must be circumspect in evaluating and interpreting the significance of pine pollen frequency values in samples of pollen rain. The assumption that higher frequencies of pine pollen in a pollen record reflect greater numbers of pine trees (viz., Ohngemach 1973) is particularly suspect. The Oaxaca data indicate, however, that other statistical qualities of a pollen rain may be used to diagnose pine tree frequency if one is dealing with *a group* of samples ostensibly drawn from the same population. Apparently, if pine is a dominant or codominant member of the flora at the sampling loci, a group of surface sediment pollen rain samples will evidence

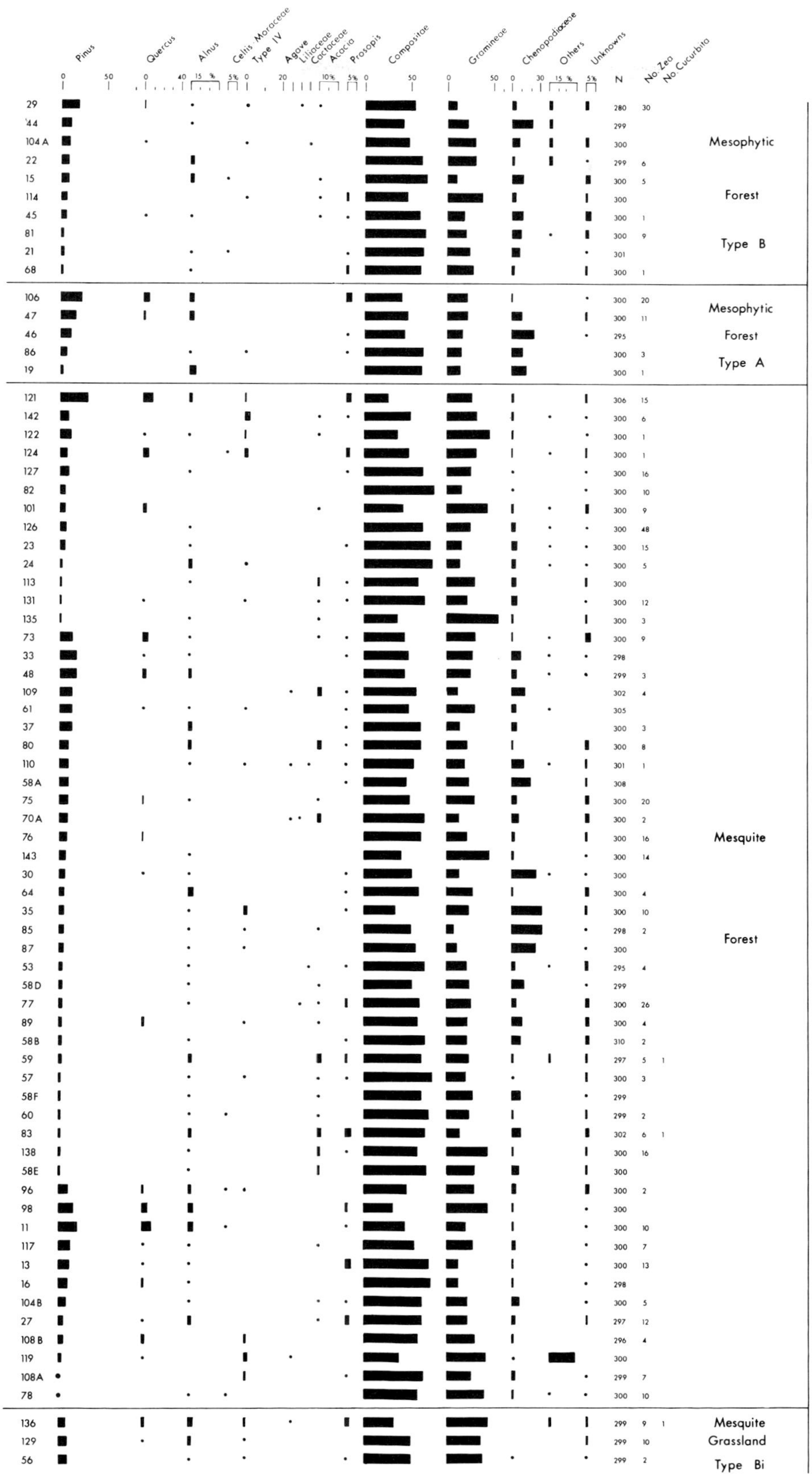

Fig. 15.4. Pollen spectra (cultigens excluded) of modern pollen rains of cultivated valley lowland plots.

little variation in pine pollen frequency values, and pine pollen will, statistically, account for at least 40% of the pollen rain (at the 0.95 level of confidence). Gish (1976) has published surficial sediment pollen records from Wisconsin that indicate this proposition may also be valid for the temperate latitudes of North America.

The Thorn Forest Zone. The floras of Thorn Forest sampling locations are quite varied in the Valley of Oaxaca, and one might thus anticipate that their pollen rains would be similarly variable. This expectation is not fulfilled by the frequency values of pollen samples collected at Thorn Forest locations. Thorn Forest pollen rain samples, taken as a group, produce a significantly greater variety of types of pollen than samples recovered from other zones. However, the variety of pollen types is not statistically greater in any given Thorn Forest sample than in any sample drawn at random from the valley as a whole. Again, then, we may identify a pollen frequency distinction that can only be applied to evaluation and interpretation of groups of samples ostensibly drawn from the same population; the occurrence of a larger or smaller frequency of pollen types does not allow diagnosis of any specific sample as reflecting the ecological characteristics of the Thorn Forest zone.

One of the pollen types occurring in the pollen rain of the Valley of Oaxaca is characterized by a tricolporate aperture morphology, subprolate shape, and tectate, finely striate, sculpturing. The pollen of 463 plants from the valley and environs has been compared with this pollen type, and also the pollen descriptions published in a number of sources, but we have been unable to identify the botanical taxon involved. The morphology of this pollen most closely approximates that of certain genera of the Anacardiaceae (cashew family). The most common member of this family in the Thorn Forest zone (*Rhus standleyi*; R. Mollis, HBK) is represented in our modern pollen reference collection but does not seem to be the taxon that produces this pollen type. We identify it simply as Type IV pollen. Schoenwetter (1972) has also observed this pollen type in surface samples from the Thorn Forest and Short Tree Forest zones in Sonora. Neither Fish (1977, 1978) nor Tsukada (Tsukada and Deevey 1967) have observed it in samples from the Tropical Forest of Belize or Guatemala.

Though we do not know the name of the plant that produces Type IV pollen, we are aware of the ecological conditions responsible for its occurrence in statistical quantity in the pollen rain of the Valley of Oaxaca. Type IV pollen regularly occurs at frequency values greater than or equal to 5% (at the 0.95 level of confidence) only in samples collected from Thorn Forest zone locations. Type IV pollen occurs irregularly at this frequency value at sample locations in the Oak Forest zone. In those cases, however, the oak pollen frequency is greater than 13% irrespective of the Type IV pollen frequency value. Since this is only true of sampling locales where oak is dominant, a pollen rain of this sort would not be misdiagnosed as deriving from the Thorn Forest zone.

In terms of our classification of ecological conditions in the Valley of Oaxaca (Table 15.2), samples containing ≥ 5% Type IV pollen reflect low-moderate effective moisture values and the presence of a layered, scattered canopy situation. It would appear that such samples (unlike those in which oak pollen frequency is > 13%) reflect ecological parameters immediate to the sampling locus rather than surrounding the sampling station. The single sample available from a cultivated location within the Thorn Forest zone is not similar to those from uncultivated loci in regard to Type IV pollen frequency.

The Low-Elevation Zones. The Mesquite Grassland, Mesquite Forest, and Mesophytic Forest zones occur at the low elevations of the valley. In sharp contrast to the plant zones of higher elevations, their distribution is not principally controlled by temperature and precipitation values. The Mesquite Grassland zone is restricted to the southern portion of the Mitla arm of the valley where, as a consequence of modern prevailing wind direction, rain shadow effects are most intensive. C. E. Smith (1978) does not reconstruct the aboriginal occurrence of Mesquite Grassland, as his position is deliberately conservative. It would be submerged into the Mesquite Forest zone in his assessment. If temperature and precipitation values were like those presently obtained, season for season, but the prevailing wind direction was changed from its present southeasterly flow during the monsoon period to a northeasterly or southern flow, the Mesquite Grassland probably would not exist as a plant zone in the Valley of Oaxaca. Instead, the plant associations characteristic of the Mesquite Grassland zone would occur in scattered locations throughout the valley within the Mesquite Forest and the Thorn Forest zones where rain shadow effects combined with local topography and sedimentology to produce the specific ecological parameters that allow the survival of these associations.

The distributions of the Mesquite Forest and the Mesophytic Forest plant zones are controlled by the distributions of the substrate required for their development and maintenance: alluvium. Though individual taxa common to these zones—or even common associations—may survive in districts where alluvium is not the normal substrate, the zonal vegetation patterns cannot exist as such except on deep soils. In the Valley of Oaxaca, only alluviated districts provide such soils because of the geology of the region and the geometry of its basins of deposition. Furthermore, the Pine Forest, Oak Forest, and Thorn Forest zones cannot exist as such on expanses of deep soils irrespective of temperature and precipitation values. If the climatic regime were to become so markedly modified—say by temperature decrease—that the Mesquite Forest and Mesophytic Forest zones became decimated, they would not be replaced by Thorn Forest, Oak Forest, or Pine Forest plant communities or associations we can identify today. Because the alluvial substrate would remain in place, they would be replaced by other zonal vegetation patterns adapted to that form of substrate conditions.

Reconstruction of paleoecological conditions in the Valley of Oaxaca must take these facts into account. Whatever paleotemperature or paleoprecipitation values might have occurred in former times, Mesquite Forest and Mesophytic Forest could never have occurred at a geographic location where an alluvial substrate did not occur. Similarly, Mesquite Grassland could not have occurred as a zonal vegetation pattern except at a location where an intensive rain shadow effect could be felt during a monsoonal growing season.

There are four archaeological sites from which samples of pollen rain attributable to the preceramic period have been collected in the Valley of Oaxaca. Three are located in the Thorn Forest plant zone today, near its upper elevational margin, in a district of broken topography. None of these sites could possibly have been locations supporting Mesophytic Forest, Mesquite Forest, or Mesquite Grassland zonal vegetation patterns at the time of occupancy. The fourth archaeological site of concern is located in the district of alluvial substrate in the Mitla arm of the valley. Presently, the site area is a cultivated field so the vegetation pattern it supports must be reconstructed from knowledge of its edaphic, topographic, and hydrological condition rather than from observation of flora. Since the site occurs at a locus of alluvial substrate where the water table is deeper than 9 m and within the district of intensive rain shadow effect, it is classified as a cultivated locus of the Mesquite Grassland plant zone—probably a member of plant association B(ii). If the water table was more elevated at this site in the past, but annual precipitation and temperature values were essentially similar to those that occur today at this elevation, the site might once have supported a Mesophytic Forest vegetation pattern of the plant community A type. If the direction of the prevailing wind during the monsoon season were different in the past, but temperature, precipitation, and water table height were essentially as they are today, the site might have supported a Mesquite Forest plant zone vegetation pattern. Under any other set of paleoecological conditions, however, the site could only have supported (1) the same sort of vegetation pattern existing today (though potentially uncultivated) or (2) a vegetation pattern that does not exist today at all in the Valley of Oaxaca.

To control the fossil pollen record of the preceramic period in the Valley of Oaxaca, then, it is necessary that we determine only a few—quite specific—palynological characteristics of the modern pollen rains of the Mesquite Grassland, Mesquite Forest, and Mesophytic Forest plant zones. It is only necessary for us to deal with those pollen rains at all because of their potential to control evaluation and interpretation of the fossil pollen rain from one archaeological site. At that site, it is only necessary that we control a few possibilities because its location obviates many possible relationships between modern and fossil pollen rain records. It is impossible, for example, for the site location to ever have supported what we can today observe as Mesophytic Forest plant community B vegetation.

There are three significant possibilities we consider it important to control: (1) the possibility that ecological conditions presently occurring in the Mesquite Grassland zone occurred at the site prehistorically, (2) the possibility that ecological conditions presently associated with Mesophytic Forest vegetation of the plant community A type occurred at the site during its period of occupancy, and (3) the possibility that ecological conditions now prevalent in the Mesquite Forest zone formerly occurred at the site. To assess these we must attempt to identify those characteristics of the pollen rain that specify a lack of canopy coverage; high effective moisture values; and low, dense canopy coverage, respectively. These are the ecological conditions distinguishing such locations today according to the classification scheme of Table 15.2.

We remind our readers that all of the modern pollen samples available from Oaxaca that might reflect such ecological conditions were collected at locations that have been highly influenced by human activities. The Mesophytic Forest, Mesquite Forest and Mesquite Grassland zones have been extensively farmed and lumbered for at least 2500 years under varying technologies. During the past 500 years, they have been the principal grazing districts for domestic livestock as well. We cannot assume such activity has had no effect—or even slight effect—on the pollen rain of our sampling locations, but we feel justified in assuming that the pollen rains are more directly affected at cultivated plots than uncultivated ones. In essence, we have made the assumption that uncultivated plot pollen rains in these zones offer prospect of variation in response to such ecological parameters but that cultivated plots, because of direct human management of their ecology, do not. We therefore do not consider the pollen rain records of cultivated plots of the Mesophytic Forest, Mesquite Forest, or Mesquite Grassland plant zones (Fig. 15.4) useful for control purposes and have ignored them in this frequency analysis.

Pollen frequency values of the low-elevation uncultivated plots (e.g., hedgerows, unfarmed ejido, mound surfaces, or relict stands) are diagramed on Fig. 15.3. It will be immediately noted that the pollen frequency values of some individual samples from these zones mimic characteristic pollen values of higher elevation plots. Three samples contain ≥ 40% pine pollen and < 13% oak pollen, and four samples contain > 13% oak pollen. This does not obviate use of this sample series as controls, however. All of the samples of Fig. 15.3 were collected from districts of unbroken topography and/or alluvial substrate. Even if a low-elevation fossil pollen sample produced this sort of pollen record, we would not evaluate or interpret that record as a reflection of ecological conditions now occurring in the Pine Forest or Oak Forest zones. Such conditions cannot occur—nor can they be expressed in these fashions palynologically—except in districts of broken topography and nonalluvial substrate. Whatever combination of ecological parameters might cause certain low-elevation pollen rains to mimic those of high-elevation plots, we assume it cannot possibly be the same combination. Since mimic pollen rains of the low-elevation plots are so rare, the responsible ecological parameters seem to be highly localized.

The occurrence of mimic pollen rains, however, and the fact that they derive from plots more likely than not to be somehow affected by human agency, requires us to take a conservative posture. We also note the degree of variability in pollen frequencies of different samples drawn from the same vegetation pattern type in this regard. It seems to us wholly injudicious even to consider seriously analysis of such samples in other than population terms.

The samples recovered from the Mesquite Grassland zone are so few that we must consider them as a single population in any case. As a group they exemplify ecological conditions in which canopy coverage is minimal, so any palynological characteristic unique to this population of samples would indicate that ecology if observed in a fossil pollen record from a low-elevation site. In terms of pollen frequency, this population of samples has one unique characteristic: the frequency of occurrence of *Celtis*–Moraceae pollen.

The pollen of *Celtis* (hackberry) may be morphologically distinguished from that of *Morus* (mulberry) and other taxa of Moraceae on the basis of sculpturing characteristics. When Moraceae and *Celtis* pollen are poorly preserved, though, details of surface sculpture are easily obscured by erosion, corrosion, or twisting and folding of the pollen grains. Because the majority of modern and fossil pollen we have observed from Oaxaca is more poorly preserved than reference specimens, we found it necessary to lump certain morphologically similar taxa together in our analyses. The category "*Celtis*–Moraceae pollen" represents such a lumping. In the Valley of Oaxaca, species of Moraceae tend to be more limited in distribution than *Celtis*. *Celtis* occurs in all the plant zones along watercourses; aboriginally, it was probably most prevalent in the Mesophytic Forest and Mesquite Grassland zones. Moraceae species are most commonly encountered today in the Thorn Forest zone and in plant community A of the Mesquite Grassland zone; aboriginally they were probably quite common in the Mesophytic Forest zone as well. The occurrence of *Celtis*–Moraceae pollen, then, is anticipated in any of the plant zones of the valley. One or another of the source plants occur within reasonable distance of any sampling location, and the source plants produce quantities of pollen that are dispersed freely into the air, rather than transported by insect or animal vectors.

As it turns out, though, *Celtis*–Moraceae pollen is not regularly recovered in samples of the modern pollen rain. It is absent in 78% of the Mesquite Forest zone samples and in 89% of the Mesophytic Forest zone samples. But it is present in 64% of the Mesquite Grassland zone samples. We do not know why these proportions occur; certainly they could not have been anticipated from our understanding of the floristic attributes of these plant zones. But the fact of their occurrence allows us to recognize that the *Celtis*–Moraceae pollen values of Mesquite Grassland samples sets them apart—as a group—from the samples of the Mesquite Forest and Mesophytic Forest. If a population of fossil pollen rain samples yielded a high frequency of occurrence of *Celtis*–Moraceae pollen, then, we may reliably interpret this as indicating the absence of canopy coverage—probably a result of rain shadow effect.

In terms of pollen frequency values the high effective moisture conditions of plant community A of the Mesophytic Forest zone are identifiable in terms of the ratio of the Compositae pollen frequency to the Gramineae pollen frequency. In the population of samples from this plant community, the frequency of Compositae pollen is 44% $\pm$ $2p_L$ (at the 0.95 level of confidence) and that of Gramineae pollen is 28% $\pm$ $2p_L$; the ratio is about 1.6:1. In the population of Mesquite Forest samples, the frequency of Compositae pollen is 45% $\pm$ $2p_L$ and that of Graminae pollen is 35% $\pm 2p_L$; the ratio is about 1:1. Though the frequency of Compositae pollen is not significantly different in the two populations, the frequency of Gramineae pollen is significantly greater in the one reflecting the low, dense canopy condition. Therefore low-elevation pollen rains that evidence a higher Compositae–Gramineae ratio value (based on pollen frequency) reflect higher effective moisture conditions.

Summary. In summary, certain pollen frequencies serve to characterize specific ecological parameters today and may be used to identify the occurrence of those parameters in fossil pollen rains of the preceramic horizon. If a population of fossil pollen rain samples is collected from a high-elevation site, the occurrence of pine pollen greater than or equal to 40% and oak pollen equal to or less than 13% indicates the presence of a high, closed canopy. The occurrence of Type IV pollen greater than or equal to 5% indicates the presence of low-moderate effective moisture values and a layered, scattered canopy. If a population of fossil pollen samples is collected from a low-elevation site, the occurrence of *Celtis*–Moraceae pollen in a high proportion of the samples indicates the absence of a tree or shrub canopy; the occurrence of Gramineae pollen at a frequency value of 35% $\pm$ $2p_L$ indicates the presence of a low, dense canopy; and the occurrence of a Compositae–Gramineae pollen ratio greater than or equal to 1.6:1 indicates high effective moisture values.

Certain pollen frequencies are also indicators of floristic conditions. If the population of pollen rain samples consistently contains pollen of cultigens, it indicates the population is influenced by ecological conditions specific to cultivated plots. If the frequency of pine pollen in the population has a high standard deviation, it indicates that pine is not a dominant in the flora even though the canopy may be high and closed. If the frequency of oak pollen is greater than 13%, it indicates the population derives from a district in which oak is the dominant member of the flora.

Multivariate Discriminant Function Analyses

The principal difficulty encountered in the prior analysis is obvious: There are ecological conditions we can identify in the Valley of Oaxaca—and we desire to reconstruct for the preceramic period—that are not evidenced as characteristic variations in pollen frequency values. No pollen frequency variations we have observed, for example, relate directly to annual temperature or annual precipitation variations occurring in the valley. It is possible to make certain assumptions that would allow a greater degree of interpretation, of course. For example, we might assume that fossil pollen records that evidence the occurrence of oak as a dominant floral element reflect ecological conditions that occur today only in the Oak Forest plant zone. This assumption is not valid if applied to modern pollen rain, as oak may today be dominant in the Pine Forest zone—plant association

B(ii). But it can be argued that oak is a dominant in the Pine Forest zone today as a result of modern lumbering practices. Since this could not be the case during the preceramic period, the assumption could be valid when applied to fossil pollen rains. Using the assumption, a population of fossil pollen rains that contained more than 13% oak pollen would indicate the presence of low-moderate annual temperatures, moderate conditions of effective moisture, and a low, moderately closed canopy condition.

In order to differentiate even the majority of ecological conditions in which we are interested, however, this approach would demand piling untestable assumptions on other untestable assumptions. Such a house of cards would be of little value, and it would violate the whole purpose of our investigation of modern pollen rains. If we were content with the acceptance of assumptions, there would be no reason to study modern pollen rains at all. The tedious and laborious analysis of the preceding section would have been a senseless exercise.

To identify the palynological distinctions between groups of pollen samples that reference different ecological conditions, we have turned to discriminant function analysis.

Discriminant function analysis is one of a number of multivariate statistical techniques pollen analysts apply to the study of the relationship of pollen frequency values to climatic, vegetational, and ecological patterns (Bernabo and Webb 1977; Birks *et al.* 1975; Burton 1973a, 1973b; Smith 1974, 1975; Webb 1974a, 1974b; Webb and Bryson 1972; Webb and McAndrews 1976; West 1978). All operate on the same body of basic assumptions about the relationship of modern or paleoenvironmental conditions to biological records expressed as frequency values and utilize the same basic mathematical formulation of that model (see Webb and Clark 1977:95–98). The different statistical techniques used—multiple regression analysis, principal components analysis, canonical correlation analysis, etc.—offer distinctive means of establishing calibration of the relationship of modern environmental conditions to biological records that ostensibly reflect the environmental patterns specified. Selection of the most appropriate technique depends on the kinds of data that characterize the environmental conditions sampled and the purpose of the investigation.

Most modern pollen records that have been subjected to multivariate study have been collected from the surficial deposits of lakes or bogs. Unlike the terrestrial surface pollen records collected in Oaxaca, they ostensibly represent the pollen rain of a basin of deposition and thus refer to a mosaic of plant associations. Though terrestrial surface pollen records incorporate pollen transported to the collection locus over long distances also, most of the pollen involved is thought to represent vegetational and ecological conditions of the sampling locus itself. The observations of vegetational, edaphic, topographic, and geomorphological conditions at the sampling locus of such terrestrial modern pollen records, then, constitutes a body of information that presumably controls evaluation of such modern pollen records quite closely. Each sample may be unequivocally assigned to a particular position in the classification scheme used to assess the ecology of the study area. Each set of samples assigned to the same position constitutes a group that is demonstrably distinct from any other group.

Discriminant function analysis is the most appropriate technique for examining the relationship between modern pollen rain frequency values and ecological conditions when data of this character are available, for the technique is expressly designed to examine the relationship of chosen dependent variables in the determination of group membership (Cooley and Lohnes 1971:13, 243ff.; Klecka 1975:434ff.; Sokol and Rohlf 1969:488ff.; Van de Geer 1971:242ff.; Veldman 1967:268–280). Here pollen frequencies are used as dependent variables, and the units of the ecological classification schemes (Table 15.2) comprise the groups to which given pollen records are assigned. In a very general sense, discriminant function analysis is the antithesis of the more familiar cluster analysis. In a cluster analysis the similarities in a body of data are isolated to allow the formation of groups. In a discriminant analysis the differences of data known to refer to distinctive groups are isolated so that each group is characterized as uniquely as possible.

The general approach of discriminant analysis is two staged. In the first stage an initial set of data is subjected to a detailed examination that identifies the set of mathematical statements that maximally separate the data values assigned to each group. These statements are the discriminant functions; they express the maximum distinctions that may be recognized between the data of each group in a form amenable to statistical evaluation. Specifically, an *F*-test reveals the probability that a given discriminant function could be generated by chance, assuming a multivariate normal distribution of data. The discriminant functions allow identification of weighting values that may be applied to each dependent variable (in this case pollen frequencies) to generate statements that maximally segregate the different groups of the classification. Thus a model is revealed that can be utilized as a device for classifying new sets of data (in this case fossil pollen spectra) in terms of the groups identified for the initial data set.

The second stage of the analysis tests the strength of the model developed in the first stage. Here the discriminant weights established for the model are applied to the initial data set, and the resulting assignments to specific ecological groups are compared against the original classification. The proportion of "correct" classifications that results from application of the model serves as an indication of model efficiency or strength.

A given body of pollen records may be used to generate a number of different models, some of which may be more efficient than others. This is accomplished by (1) combining (collapsing) two or more pollen taxon frequency values or by (2) removing certain pollen taxa from consideration in the calculation of frequency values. The decision to collapse or delete is justified by either the nature of the data set or the purposes of the analysis. For example, a series of plant taxa may be represented in the pollen record that are recognized to reference the same ecological pattern. In Oaxaca, *Alnus, Betula,* and *Salix* are restricted to riparian habitats. Since they all could be considered to reference riparian ecology, one might justify combining the frequency values of the three as a single dependent variable for purposes of an analysis that seeks maximal discrimination of riparian habitat pollen values from any other sort. Alternatively, pollen taxa may be eliminated from consideration; the effect will be to modify the frequency values of the remaining pollen taxa. In Oaxaca, it can be demonstrated that the occurrence, but not the frequency, of *Zea* and *Cucurbita* pollen reflect ecological parameters specifically correlated with agricultural behavior. If the purpose of the investigation is to produce a model of pollen rain–ecological relationships uninfluenced by agricultural activities, pollen frequency values for those taxa might be excluded from the calculations as irrelevant to its purpose.

Our problem in undertaking discriminant analysis of the modern pollen record of the Valley of Oaxaca, then, was that of identifying the pollen sum that could be used to generate an efficient mathematical model for each of the ecological classification schemes recognized on Table 15.2. The first, analytic, stage of the discriminant function procedure was used for this purpose.

Our original assumption, based on the results of the frequency analysis, was that certain pollen values would characterize and discriminate the pollen rain of cultivated plots in the valley. A discriminant function analysis confirmed this (Table 15.3). The first stage (analytic) of the analysis documented that three pollen taxa played significant roles in effecting the discrimination between the pollen records of cultivated and uncultivated plots: *Zea, Cucurbita,* and Compositae.

To create a pollen sum usable for the generation of a mathematical model for each of the classification schemes, raw pollen counts were converted into percentages after excluding the values for *Zea, Cucurbita,* and Compositae. The percentage values then served as input for the discriminant analysis of all the modern surface pollen records. The models generated on the basis of pollen counts that excluded these three taxa—and therefore presumably allow classification of ecological conditions uninfluenced by agricultural behavior—are said to be based on an *adjusted* pollen sum.

The Adjusted Sum Discriminant Function Analysis

Effective Moisture Groups. There are six effective moisture categories in the ecological classification, but these are segregated into two nonhierarchical sets: Lowland Effective Moisture (subdivided into high, moderate, low and very low categories) and Highland Effective Moisture (subdivided into high, moderate, and low categories). The sets are nonhierarchical because the conditioning variable of substrate depth uniquely separates the forms of forests adapted to lowland and highland conditions in the valley. The low-moderate effective moisture category of the highland set thus is not hierarchically ranked in relation to the low effective moisture category of the lowland set. It is necessary to generate a separate mathematical model for each set for the sake of accuracy.

TABLE 15.3
Discriminant Function Analysis of the Samples of Cultivated and Uncultivated Plots, Standard Sum[a]

		Predicted group membership	
Actual group	Number	Cultivated	Uncultivated
Cultivated samples	99.	67. 67.7%	32. 32.3%
Uncultivated samples	81.	16. 19.8%	65. 80.2%

[a]Percentage of grouped cases correctly classified is 73.33%.

Table 15.4 documents the strength of the model relevant to the valley lowlands effective moisture categories. The classification strength of this model is weaker than we consider preferable. However, we consider it to be a reasonably accurate representation of overall model strength. There is a pronounced tendency, represented by about 30% incorrect classifications, to misclassify pollen records of the high effective moisture group as members of the very low effective moisture group and vice versa. This indicates that the model has some utility as a control for accurate interpretation as regards the low effective moisture category, but it is not very reliable otherwise.

The valley highland effective moisture model is somewhat stronger (Table 15.5), but this may be an artifact of the comparatively small number of samples of each group. This model has a tendency to misclassify pollen records of the high and moderate effective moisture categories, but is markedly reliable in the classification of pollen records of the low-moderate effective moisture category. As a control for accurate interpretation, it is most useful for that purpose.

Annual Rainfall Groups. The annual rainfall groups may be hierarchically ranked, but since there are only five categories in the classification scheme the information content of this model is less than that of the two effective moisture models when taken together. The overall classification strength of the annual rainfall model (Table 15.6) is higher than that for either effective moisture model. Since the distinction between the meristic categories is usually quite clear, the annual rainfall model is more usable as a control on the interpretation of pollen records. The primary weakness of this model is in the area of distinguishing between the moderate and high-moderate categories of annual rainfall.

Annual Temperature Groups. The annual temperature model consists of only four hierarchically ranked groups and so has the lowest information content of all our ecological classifications. Indeed, since this model is essentially elevation dependent, it provides rather little control for purposes of interpretation of archaeological records. Knowledge of the elevation of the archaeological site should be about as informative an index of relative paleotemperature as this model can provide from palynological data.

The overall strength of the model is quite high (see Table 15.7d). However, there is a striking tendency for samples of the moderate group to be misclassified as members of the high group. This is probably explained by the fact that the

TABLE 15.4
Discriminant Function Analysis of Valley Lowland Samples of Effective Moisture Categories, Adjusted Sum[a]

		Predicted group membership		
Actual group	Number of cases	High	Low	Very Low
High effective moisture	24.	16. 66.7%	1. 4.2%	7. 29.2%
Low effective moisture	14.	1. 7.1%	10. 71.4%	3. 21.4%
Very low effective moisture	103.	27. 26.2%	14. 13.6%	62. 60.2%

[a]Percentage of grouped cases correctly classified is 62.41%.

TABLE 15.5
Discriminant Function Analysis of Valley Highland Samples of Effective Moisture Categories, Adjusted Sum[a]

		Predicted group membership		
Actual group	Number of cases	High	Moderate	Low Moderate
High effective moisture	22.	14. 63.6%	8. 36.4%	0. .0%
Moderate effective moisture	3.	1. 33.3%	2. 66.7%	0. .0%
Low-Moderate effective moisture	10.	0. .0%	1. 10.0%	9. 90.0%

[a]Percentage of grouped cases correctly classified is 71.43%.

TABLE 15.6
Discriminant Function Analysis of Annual Rainfall Categories, Adjusted Sum[a]

		Predicted group membership				
Actual group	Number of cases	Low	Low Moderate	Moderate	High Moderate	High
Low annual rainfall	127.	97. 76.4%	18. 14.2%	0. .0%	0. .0%	12. 9.4%
Low-Moderate annual rainfall	13.	4. 30.8%	8. 61.5%	0. .0%	0. .0%	1. 7.7%
Moderate annual rainfall	8.	0. .0%	0. .0%	5. 62.5%	2. 25.0%	1. 12.5%
High-Moderate annual rainfall	6.	0. .0%	0. .0%	2. 33.3%	3. 50.0%	1. 16.7%
High annual rainfall	22.	2. 9.1%	0. .0%	0. .0%	1. 4.5%	19. 86.4%

[a]Percentage of grouped cases correctly classified is 75.00%.

high group includes Mesquite Forest samples and the moderate group includes Mesquite Grassland samples. The occurrence of *Acacia* and *Prosopis* as characteristic taxa in both vegetation patterns probably influences pollen distribution patterns similarly and produces the overlap observed palynologically.

These results were encouraging, but not wholly satisfactory. First, adjustment of the pollen sum through exclusion of the Compositae pollen taxon concerned us as (1) frequency analyses previously used (Flannery and Schoenwetter 1970; Schoenwetter 1974) indicate that this taxon could be of significant utility in paleoenvironmental interpretation, (2) the taxon occurs in much greater frequency in some fossil than in modern pollen records, and (3) the frequency analysis of page 194 indicates that the Compositae–Gramineae ratio is relevant in interpretation of effective moisture values. Second, though the frequency values of *Zea* and *Cucurbita* are significant in the discrimination of cultivated from uncultivated plots in this model, they are much less significant than the Compositae frequency values. Third, we anticipated that the models would have significantly greater strength than this analysis indicates.

In order to improve the situation, it would be necessary to examine in greater detail the specific contribution made

TABLE 15.7
Discriminant Function Analysis of Annual Temperature Categories, Adjusted Sum[a]

Actual group	Number of cases	Predicted group membership: Low	Low Moderate	Moderate	High
Low annual temperature	22.	17. 77.3%	3. 13.6%	0. .0%	2. 9.1%
Low-Moderate annual temperature	3.	0. .0%	3. 100.0%	0. .0%	0. .0%
Moderate annual temperature	24.	1. 4.2%	0. .0%	13. 54.2%	10. 41.7%
High annual temperature	127.	12. 9.4%	0. .0%	1. .8%	114. 89.8%

[a]Percentage of grouped cases correctly classified is 83.52%.

by each pollen taxon to a model developed without the constraints of an adjusted pollen sum. We therefore undertook a second discriminant function analysis using the same groups but making no allowances for the differences that might exist between pollen records from cultivated and uncultivated loci. These models are said to be based on a *standard* pollen sum.

The Standard Pollen Sum Discriminant Function Analysis

Effective Moisture Groups. Tables 15.8 and 15.9 document the advantage of utilizing the standard pollen sum in creation of a mathematical model for discriminating the effective moisture ecological categories. The results of Table 15.8, particularly, indicate this model has extraordinary strength for interpretation of fossil pollen records from highland localities, such as are the concern of this report. The strength of this model is substantially, but not significantly, higher for interpretation of fossil pollen records from lowland locations as well (Table 15.9), but difficulties are still encountered in that about one fourth of the samples from Mesquite Forest locations are classified as if they derived from riparian plots in the Mesophytic Forest. We suggest that to

TABLE 15.8
Discriminant Function Analysis of Valley Highland Samples of Effective Moisture Categories, Standard Sum[a]

Actual group	Number of cases	Predicted group membership: High	Moderate	Low Moderate
High effective moisture	22.	20. 90.0%	2. 9.1%	0. .0%
Moderate effective moisture	3.	0. .0%	3. 100.0%	0. .0%
Low-Moderate effective moisture	10.	1. 10.0%	0. .0%	9. 90.0%

[a]Percentage of grouped cases correctly classified is 91.43%.

TABLE 15.9
Discriminant Function Analysis of Valley Lowland Samples of Effective Moisture Categories, Standard Sum[a]

Actual group	Number of cases	Predicted group membership: High	Low	Very Low
High effective moisture	24.	16. 66.7%	1. 4.2%	7. 29.2%
Low effective moisture	14.	1. 7.1%	10. 71.4%	3. 21.4%
Very low effective moisture	103.	25. 24.3%	7. 6.8%	71. 68.9%

[a]Percentage of grouped cases correctly classified is 68.79%.

some degree this may be explained as a result of the decimation of the Mesophytic Forest and its conversion to agricultural land. The result would be one of blocking and infilling minor tributary drainageways in the Mesquite Forest zone (essentially the district of upper alluvium on the valley floor) that would otherwise constitute riparian habitats. In sampling the modern surface—particularly where that surface has been converted to agriculture—a number of samples of the Mesquite Forest zone may have been collected at locations that would, if undisturbed, accommodate riparian growth. Palynologically, such samples would be expected to yield indications of locally high effective moisture values.

Annual Rainfall Groups. The classification strength of this model (Table 15.10) is higher than that of the model that uses an adjusted pollen sum; but, as a whole, this model is not of appreciably greater value as a control for interpretation of fossil pollen records. Internally, the classification strength of this model is the same as that of the adjusted pollen sum model in regard to the high, high-moderate, and moderate categories. It is stronger than the adjusted sum model in correctly classifying low annual rainfall pollen records but weaker in correctly classifying low-moderate annual rainfall records. For present purposes, however, these particular strengths and weaknesses are significant in evaluation of fossil pollen records. Using this model, the most likely misclassification that would not also occur when using the adjusted sum model occurs when a pollen record is classified to the low-moderate annual rainfall category. If that occurs, the only likely proper alternative classification is that of the low annual rainfall category. In the adjusted sum model, a record misclassified to the low-moderate annual rainfall category is most likely to be attributable to the low annual rainfall category also. But in the adjusted sum model, the low annual rainfall category has not the classification strength that exists in this model. Thus one is less likely to actually misjudge among alternative options if the standard sum model is used.

Annual Temperature Groups. The overall classification strength of this model (Table 15.11) is not appreciably higher than that of the adjusted sum model, though it is somewhat stronger. The significant advantage of this model is that it allows a greater probability of correct classification of pollen records ascribed to the moderate annual temperature category. Compared to the adjusted sum model, however, it is appreciably weaker in correctly ascribing pollen records to the low annual temperature category.

TABLE 15.10
Discriminant Function Analysis of Annual Rainfall Categories, Standard Sum[a]

		Predicted group membership				
Actual group	Number of cases	Low	Low Moderate	Moderate	High Moderate	High
Low annual rainfall	127.	106. 83.5%	13. 10.2%	0. .0%	0. .0%	8. 6.3%
Low-Moderate annual rainfall	13.	5. 38.5%	7. 53.8%	0. .0%	0. .0%	1. 7.7%
Moderate annual rainfall	8.	0. .0%	0. .0%	5. 62.5%	3. 37.5%	0. .0%
High-Moderate annual rainfall	6.	0. .0%	0. .0%	2. 33.3%	3. 50.0%	1. 16.7%
High rainfall	22.	2. 9.1%	0. .0%	0. .0%	2. 9.1%	18. 81.8%

[a]Percentage of grouped cases correctly classified is 78.98%.

TABLE 15.11
Discriminant Function Analysis of Annual Temperature Categories, Standard Sum[a]

		Predicted group membership			
Actual group	Number of cases	Low	Low Moderate	Moderate	High
Low annual temperature	22.	16. 72.7%	4. 18.2%	0. .0%	2. 9.1%
Low-Moderate annual temperature	3.	0. 0.%	3. 100.0%	0. .0%	0. .0%
Moderate annual temperature	24.	1. 4.2%	0. .0%	15. 62.5%	8. 33.3%
High annual temperature	127.	10. 7.9%	0. .0%	3. 2.4%	114. 89.8%

[a]Percentage of grouped cases correctly classified is 84.09%.

When we initiated the comparison of classification strengths of the models based on the two forms of pollen sum, we anticipated learning which sum would produce the better set of classification models. Failing that, we anticipated that evaluation of the mathematical relevance of different independent variables contributing to the two sets of models would allow us to identify the independent variables that could be selected to create a set of models based on the "best" pollen sum. The results of the comparison, however, indicate that it is not judicious to proceed with such strategies (Table 15.12). Apparently, given the distribution of sample numbers in the various categories of the models produced, it is wisest simply to recognize that any pollen sum will work to produce a set of models in which classification strength is better for some categories and worse for others. While we could select a single pollen sum that would provide a set of models having higher overall classification strength than either of the two assessed so far, we would not necessarily find that set of models more useful as controls for interpretation of the fossil pollen records than the two sets now available. Our concern here is not mathematical elegance or simplicity. To create a more elegant (i.e., stronger overall) set of models would require making assumptions about the proper integration of independent variables into groups and/or the appropriate elimination of certain variables from the analysis. This would have to be accomplished wholly on theoretical grounds. Frankly, we suspect such an approach. Creation of a single set of models based on another pollen sum would identify a simpler set of control information for use in interpretation, but we do not consider such simplicity particularly advantageous at this juncture. Simplicity would be advantageous if we were attempting to interpret fossil pollen records from a number of widely separated geographic localities, but we do not face this problem.

Our decision, then, has been to control the interpretation of the fossil pollen records representing the preceramic horizons of occupation in the Valley of Oaxaca on the basis of *both* sets of models provided through discriminant function analysis. Overall, the standard pollen sum models are stronger—particularly in regard to effective moisture classification (Table 15.12)—and so serve to provide a first approximation for interpretive purposes. Where the classification strengths of this set of models is weaker than the strengths of the set based on the adjusted pollen sum, however, the outcome of classification through the latter set of models is taken into serious consideration. Interpretation also can be controlled in some ways by frequency analysis.

TABLE 15.12
Classification Strengths of the Models, Comparison of Adjusted and Standard Sums

	Adjusted pollen sum	Standard pollen sum
Valley highland effective moisture	71.43	91.43
Valley lowland effective moisture	62.41	68.79
Annual rainfall	75.00	78.98
Annual temperature	83.52	84.09

We view our usage of the two models, and of frequency analysis where feasible, as a means of improving the accuracy of interpretation of the fossil pollen records presently available. But readers may be concerned that the level of accuracy obtained is not sufficiently adequate and therefore the conclusions drawn are not sufficiently well evidenced to be trusted. What, then, is our assessment of the validity of the interpretations presented?

The *weakest* models we have utilized to establish any interpretation correctly classify the control data about 63% of the time. In statistical terms, the probability of a Type I error of inaccurate classification is about 0.27. In terms that may be more familiar, this is about the same level of accuracy that exists if one accepts a radiocarbon date within one standard deviation. Most of our models correctly classify the control data with less probability of a Type I error. We therefore assess the *accuracy* of our interpretations as at least equal to, and usually more than equal to, interpretations archaeologists are accustomed to make on the basis of radiocarbon estimations of antiquity.

In all fairness, however, we must emphasize again that our confidence in the validity of our conclusions depends on our conviction that the control data of modern pollen rain records is appropriate to our present purposes. That is, that where similarities between modern pollen rains and fossil pollen rains are patterned and unlikely to be an effect of chance, they are palynological responses to the same environmental or cultural cause. If our conclusions prove to be inaccurate upon future analysis of additional data, it will be a result of an inappropriate application of the assumption of the principle of uniformitarianism. We recognize that this assumption is challengeable in the present work and that it is legitimate to argue that fossil pollen records recovered from these archaeological contexts may be so biased as a result of human activities that their similarities to the control samples are fortuitous. We leave it to those who would challenge the accuracy of our interpretations on such grounds to explain the patterned character of the results as best they can.

Some readers may be concerned that the interpretations are invalid as a result of low numbers of control samples in certain categories of the models (e.g., 3, 6, or 8) or low numbers of observed pollen in some of the fossil pollen records. These concerns are legitimate, but the character of the models itself alleviates apprehension in the former case, and the control data provide an index on which one may evaluate the likelihood that the latter case creates real problems. The first stage of the discriminant functions analysis, which establishes the calibration between environmental conditions and populations of pollen records, takes the number of samples of a category into account. Obviously, when a larger number of samples of a population of pollen records is available, the calibration can be estimated with greater precision. But precision and accuracy are not synonymous terms. While we would prefer a situation where we had more control pollen records of each category, so the precision of our classifications could be improved, we are not overly concerned about this deficiency. The classification strength of the models is an estimate of their accuracy *given* the number of samples in each category. Even our weakest models seem wholly adequate for purposes of interpretation despite the admittedly imprecise character of some discriminant functions.

Appropriate sample size depends in large part on the amount of variance exhibited in the population. The method we utilize for computing adequate sample size is

$$n = \frac{t^2 s^2}{E^2},$$

where n is the sample size, t is the selected confidence interval, s is the standard deviation of the population being sampled, and E is the amount of allowed error. When we set the allowed error at 5% and the t value at 1.96, we computed n for the population of modern surface samples from the Valley of Oaxaca. The result ($n = 35$) documents the low standard deviation of the population and demonstrates that the 100–400 grain pollen counts of our surface samples were between 3 and 12 times necessary size. Given this analysis, it is reasonably clear that though the fossil pollen counts are normally 100 grains or less, patterned similarities between fossil and modern pollen records are not likely to be a fortuitous result of low sample size. Since interpretation depends on such similarity, low sample size seems unlikely to adversely influence our conclusions.

There is a statistical deficiency in our analysis that we only learned how to correct after the editing of this volume was so far advanced we could not do so prior to publication. Discriminant function analysis requires the assumption that the data being analyzed have a multivariate normal distribution (Klecka 1980:10). Raw palynological data (as in Table 15.24 in Appendix 15.1) are binomially distributed, as each pollen grain can only be identified to one taxon. The percentage data of a pollen sample (as in Fig. 15.2) approach a normal distribution as the number of observations for that sample is increased. Percentage data, however, are constrained since the sum of all the frequency values must equal 100%. So percentage data cannot have a multivariate normal distribution. When we undertook this analysis we were not aware that a transformation algorithm exists that allows the constraint to be removed from percentage data (see Winer 1962 and Neter and Waserman 1974 for discussions), and we elected to assume that the percentage data sufficiently approximated a multivariate normal distribution that interpretation of the discriminant function classifications was justified. We recognized this assumption for what it was and did not make it because we preferred to do so. But we found what we understood to be the only alternative to be worse: basing interpretation of the fossil pollen record solely on the frequency variations analysis. If the assumption we have elected to accept is wrong, the result of an analysis of transformed data will be that the discriminant function models will have different (probably lower) classification strengths, and some fossil pollen records will have different classifications from those identified below.

This result would, of course, mean that some of the chronological and paleoenvironmental interpretations presented below have a lower probability of accuracy than as yet unknown alternatives. Also, it would mean that some of the chronological and paleoenvironmental interpretations we present have a lesser probability of accuracy than discriminant function analysis actually allows. We consider the latter to be the more significant problem.

As is clear from the text of the project description section of this chapter, the raw palynological data on which this study is grounded have been available since 1970. Our problem has never been *if* that data lent themselves to chronological and paleoenvironmental interpretation, and the delay in publication has not been created by concern that the pollen data were somehow inadequate. Prior to 1977, the problem was identification of an analytic method that provided objective grounds for assessing the degree to which the data of any specific site were representative of the time-space units relevant to archaeological interpretation. The discriminant function analytic method fulfills this need, in our opinion, better than any of the other univariate and multivariate analytic methods known to us. We are therefore distressed by our recent recognition that we have not applied the method in a way that allows maximal use of its analytic potential. We are further distressed that the reason for this state of affairs is our failure to fully research the availability of means for eliminating a major assumption we imposed on the analysis.

FOSSIL POLLEN RECORDS

As mentioned above, the fossil pollen records associated with evidence that allow their relegation to the preceramic horizon were not the sole focus of our palynological research. During 1966–1969 we were also concerned with fossil pollen records associated with archaeological materials of the Formative, Classic, and Postclassic horizons. Thus our strategy for identifying credible data had to be oriented toward a suite of hundreds of potential fossil pollen records, rather than the limited series reported on here, as well as the large collection of surface sample records. Further, when the strategy was formulated it was assumed that the whole mass of information would be completely analyzed and reported on—at least in preliminary form—in 1970 or 1971.

Schoenwetter's appointment to the Arizona State University faculty in 1967 and Fish's position as a full-time graduate student precluded investment of a great proportion of their time and energy in this effort during those years. So it was decided that each fossil pollen count would constitute an attempt to identify and tabulate 100 pollen grains. The combining of multiple counts of the same sample has elevated the pollen sum beyond this standard in some cases, but the widely appreciated standard of a 200-grain pollen sum (Barkley 1934) was not employed.

Our rationale for this strategy was not entirely logistic. We were also very much aware that it was preferable to treat our data in population terms and therefore recognized advantages to the recovery of larger numbers of samples from the same population. As our research time was limited, the 100-grain sum strategy allowed us to observe more samples and thus potentially record more of the variability in the population of pollen rain of any given antiquity. Also, as has been appreciated by Faegri and Iverson (1975:187–190), we recognized that the appropriate size of a given sample depends on the assumptions one makes of the nature of the true distribution of variables in the population (normal, binomial, hypergeometric, etc.), the degree of exactness of correspondence between sample statistics and true values relevant to the investigation, and the practical experience one gains indicating how many grains should be counted in order that patterned regularities in the data are not hidden by sampling errors. We are satisfied that pollen sums greater than $n = 50$ are adequate in the present instance.

The fossil pollen records presented us with no insuperable problems of identification, but we recognize that our opinions in this regard are biased by an unusual set of prior experiences. As archaeologists, we are accustomed to situations in which the data base on which interpretation rests is recognizably depauperate and its quality arguable. Archaeologists traditionally accept the requirement of making the best of a bad situation. Pollen analysts working in such natural science disciplines as geology and botany tend carefully to seek out and apply their research to situations in which the major portion of extraneous and disturbing variables can be eliminated from consideration. Basically, the research tradition of the natural sciences is grounded in the experimental method. For better or worse, the research tradition of the behavioral sciences is grounded in the comparative method.

Thus we were not particularly surprised or disturbed to discover that the pollen we observed was normally poorly preserved relative to the pollen normally observed in lacustrine and orogenic deposits such as peat. We have made botanical identifications of fragments of pollen grains, corroded and eroded pollen grains, and extremely distorted pollen grains that other palynologists might consider unidentifiable. We have accepted pollen counts deriving from samples containing very little pollen as equivalent to those deriving from samples containing a great deal of pollen, which other palynologists refuse to do (e.g., Bryant 1975; King *et al.* 1975) when dealing with archaeological contexts. We do not argue that we are right in doing so and others are wrong—though Wright (1974:14) has challenged the credibility of any palynologically based paleoecological reconstruction derived from archaeological context data, and one of us (Schoenwetter 1981) has seized the thrown gauntlet. Rather, we assert that use of the comparative method is wholly defensible in any form of archaeological study and ours *is* an archaeological study. We deal with pollen grains rather than artifacts, but we do so as a means to the end of behavioral interpretation. Our concern with chronology and paleoecological reconstruction is no different from that of the geologist or botanist, but this does not require us to abjure a defensible research method simply because it is not the one traditionally employed by the botanist or geologist. Frankly, we consider debate over what method might produce "better" forms of data to be unscientific and anti-intellectual anyway. In our judgment the test of a scientific method is not the character of data produced but the quality of insights the produced data provide regarding the sense-observable world. Certainly, scientific methods must be rational, defensible, replicable, and (presumably) improvable. But they need not be mired in the traditions of any given discipline. And, one hopes, we will ultimately see the time when truly *inter*disciplinary scientific methodology has actually evolved and disciplinary traditions are altogether obviated as a relevant variable in evaluations of methodology.

Contexts

Pollen samples that may be confidently assigned to the preceramic horizon (>4000 B.P.) by virtue of associations and stratigraphic position derived from four archaeological and two paleontological localities. In terms of palynological sampling and study, the most thoroughly investigated locus is Guilá Naquitz Cave. In terms of significance and yield of chronological and paleoecological information, however, the Guilá Naquitz records take no priority over those from other locations.

The two paleontological localities, Arroyo de los Cuajilotes and San Baltazar Guelavila, provided opportunities for the recovery of sediment samples in direct association with mastodon remains. None of the four samples, however, yielded adequate pollen for analysis and the extracts provided strong indications of chemical alteration. Virtually all organic remains (pollen, spores, plant tissue fragments, etc.) were absent, and the sediments were very highly mineralized without being indurated. Since the mastodon remains were articulated on discovery, it seems likely that the chemical alterations of the sediments were induced through changes resulting from large-scale modifications in paleoenvironment, such as climatic changes. If this interpretation is correct, it argues that these paleontological records are likely to have no less an antiquity than the horizons preceding the last glacial maximum (i.e., 25,000 years ago), and they may be very much older.

Three of the localities sampled to obtain pollen records of the preceramic horizon are cave or rockshelter sites of the ignimbrite cliffs in the Mitla district of the Valley of Oaxaca. At Martínez Rockshelter (Fig. 15.5), sediment samples were collected as a single stratigraphic column extending through the various strata of cave fill into the decomposed ignimbrite of the cave floor. The collection was made subsequent to the principal archaeological excavations of Martínez Rockshelter from a standing profile at the margin of the deposits that had been intensively studied archaeologically.

At Guilá Naquitz (Fig. 15.6), sediment samples were collected from profiles exposed during the process of archaeological excavation. These samples were not collected as stratigraphic column sequences, but as representatives of individual strata and the excavation levels exposed as work progressed. These samples are directly associated with the archaeological, faunal, and macrofossil plant remains analyzed from the site. However, there are relatively few samples and their stratigraphic interrelationships are less directly expressed than is true of samples from stratigraphic columns.

At Cueva Blanca (Fig. 15.7), a series of samples was collected as excavation proceeded in close association with the archaeological record, and a stratigraphic column series was also collected from marginal deposits. However, all but two

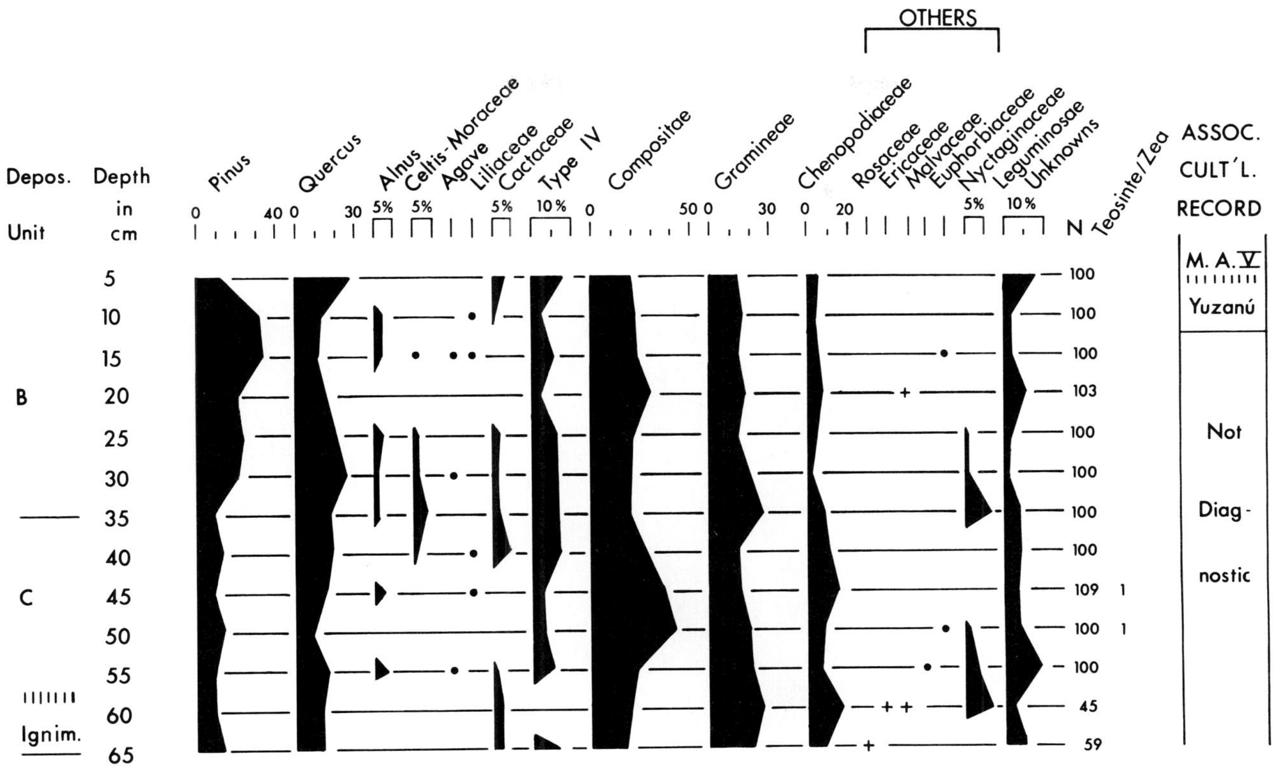

Fig. 15.5. Pollen sequence from Martínez Rockshelter.

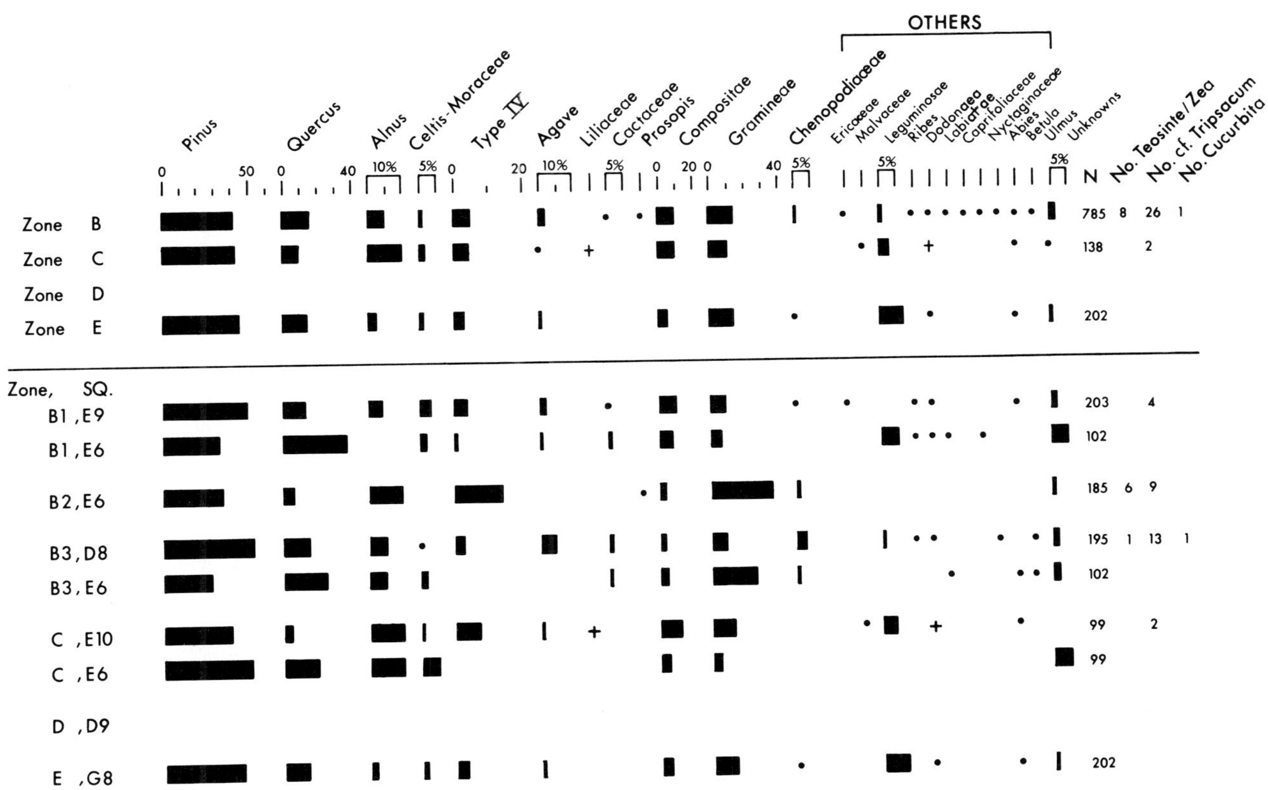

Fig. 15.6. Pollen spectra from Guilá Naquitz: upper, population spectra of the stratigraphic zones; lower, pollen spectra of individual samples.

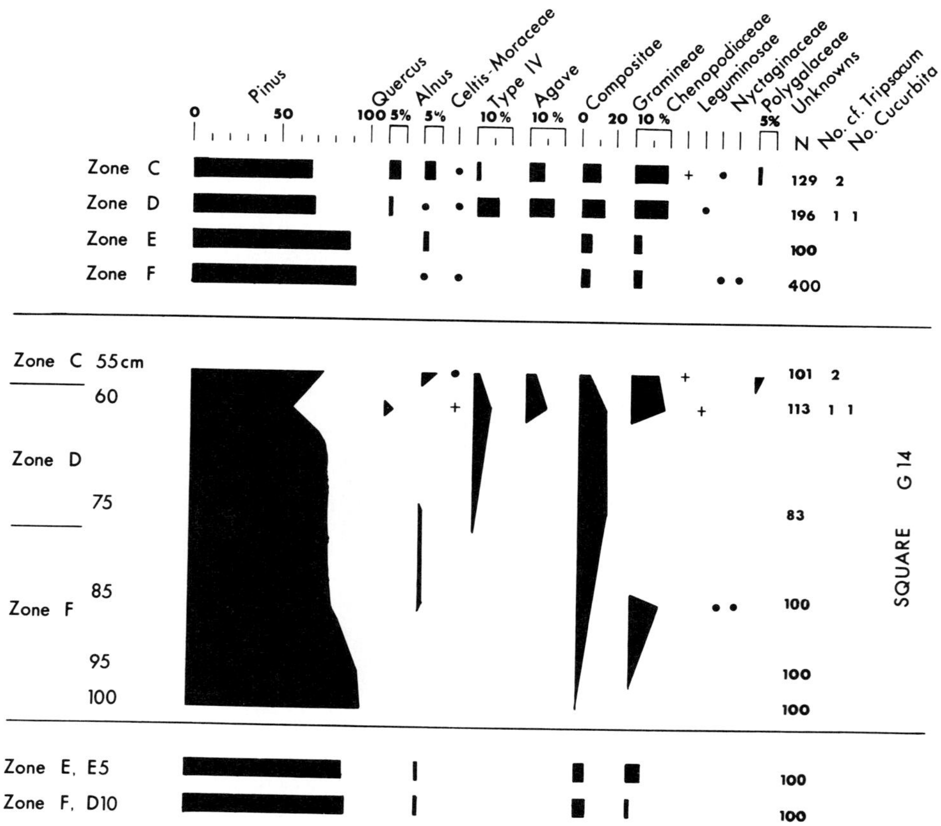

Fig. 15.7. Pollen spectra from Cueva Blanca: upper, population spectra of the stratigraphic zones; lower, pollen spectra of individual samples.

samples in the former series failed to yield sufficient pollen for analysis, and the marginal position of the latter series insures that the majority of samples derived from the culturally sterile, ancient decomposed ignimbrite deposits of the cave.

The fourth archaeological locality was the site of Gheo-Shih. This open site has not been successfully radiocarbon dated because *in situ* prehistoric charcoal or macrofossil organic materials were not preserved. Three pollen samples (Fig. 15.8) were collected by Fish in unquestionable direct association with archaeological remains of the preceramic horizon. Associated diagnostic artifacts indicate temporal placement somewhat younger than those of Guilá Naquitz.

The outcome of the sampling program, then, is as follows:

1. A stratified series of samples represent the ancient, precultural, decomposed ignimbrite deposit at Cueva Blanca (Zone F). Radiocarbon dates from a superimposed level suggest these samples date prior to 11,000 ± 400[2] B.P. Remains of Pleistocene fauna occur in this deposit, including extinct forms.

2. A single sample was collected from the earliest cultural deposit at Cueva Blanca (Zone E), which yielded modern fauna and radiocarbon assays ranging from 10,050 ± 350 to 11,000 ± 400 B.P.

3. A stratified series of samples from Martínez Rockshelter represent a sequence through basal decomposed ignimbrite, two superimposed preceramic strata, and an upper zone (A) that was basically late preceramic but whose uppermost 5 cm had later been mixed with Postclassic material. While the lower zones were clearly preceramic, they were hard to assign to a phase because the number of temporally sensitive artifacts (such as projectile points) was so low. Radiocarbon dating was not attempted as charcoal was not preserved and the organic staining of the deposits appears to be the result of decomposition of modern rootlets.

4. A suite of seven productive samples represent three of the four preceramic depositional strata at Guilá Naquitz cave (B, C, and E). Human occupation of the cave during the development of these deposits is documented by quantities of archaeological materials, but the artifactual record consistently suggests seasonal encampments by quite small groups of people through most of the period represented. Acceptable radiocarbon dates range from 8620 ± 160 to 10,700 ± 350 B.P. for the upper three preceramic strata. The oldest occupation stratum—Zone E—ostensibly, is somewhat older.

[2]All radiocarbon assays are given in Libby half-life and are in radiocarbon—not sidereal—years.

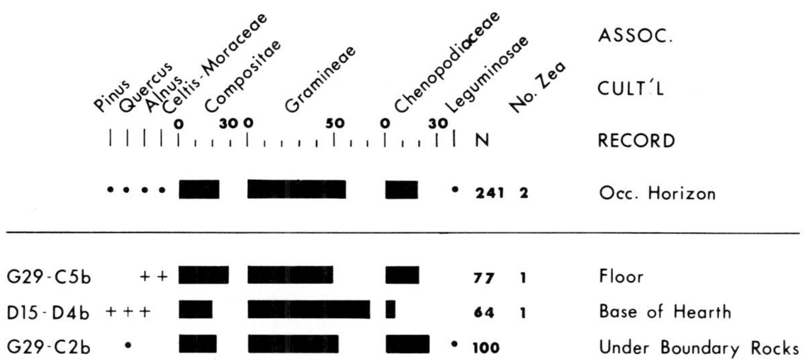

Fig. 15.8. Pollen spectra from Gheo-Shih: upper, population spectrum; lower, pollen spectra of individual samples.

5. Two productive samples from the stratigraphic column in Cueva Blanca were recovered from marginal deposits of Zones C and D. Acceptable radiocarbon dates for Zone D are 4245 ± 105 and 5245 ± 190 B.P.

6. Three samples from Gheo-Shih are apparently, but not securely, dated to 6000–7000 B.P. horizon. These samples have no stratigraphic order and are potentially noncomparable with those from the cave sites.

Chronology

The absolute ages of most of the pollen records from the cave sites are known by virtue of associated artifacts and radiocarbon dates (Table 15.13). Cueva Blanca F has been assigned to the late Pleistocene. Cueva Blanca E and Guilá Naquitz E–B have been assigned to the Naquitz phase, while Gheo-Shih is the type site of the Jícaras phase. Cueva Blanca C and D belong to the Blanca phase, while Lower Zone A of the Martínez Rockshelter has been tentatively assigned to the Martínez phase (see Chapter 3).

A pollen chronology formed by placing the pollen records in the sequential order indicated by their associated dates, however, could be more apparent than real. Sequential changes in pollen records measure changing conditions in the physical environment through time. As environmental conditions vary, they affect the production, dispersal and preservation of pollen through their more direct effects on the adaptive interrelationships of plants in the local ecosystem. But there are other factors that affect pollen rains also. The pollen producing plants may enter into different adaptive interrelationships with each other or with other organisms as the result of such things as successional patterns, pest infestations, or the introduction of new organisms or population changes in the ecosystem. Ecosystem modifications that result from changes in biotic factors have just as much effect on the adaptive interrelationships of plants as modifications of the physical environment, such as climatic changes. Further, human activities—culturally conditioned or not—affect the adaptive interrelationships of organisms in any ecosystem of which man is a part to some degree, so man's behavior may cause variation in pollen records in the same way as do the behavior of an herbivore or pathogen or a change in climate. Man also, usually unconsciously, has a potentially direct effect on pollen distribution and preservation mechanisms. This potential is exaggerated where people manipulate plants and plant parts.

The point is that though pollen rains are response functions indexing various aspects of the state of an ecosystem

TABLE 15.13
Temporal Parameters of Preceramic Horizon Pollen Records Established by Association

Martínez Rockshelter	Cueva Blanca	Guilá Naquitz	
Zone A			Monte Albán V/preceramic mixed
	Zone C Zone D		4200–5200 B.P.
		Zone B Zone C	8500–10,000 B.P.
	Zone E	Zone E	10,000–11,500 B.P.
	Zone F		Pleistocene

at any given time, the pollen record itself provides no indication of the component or components of the ecosystem that it indexes. That is to say, the pollen observed in a particular sample may differ from that in another sample for any number of reasons. One may assume that a particular component of the ecosystem—say, climate—is responsible for the distinction and interpret the difference in climatic terms. But there is nothing about the pollen records themselves that supports such an assumption. There is also nothing about a pollen record that is informative regarding the geographic area represented in the pollen spectrum, either. The ecosystem variable represented may be, for example, meteorological conditions. But the meteorological conditions occurring within a 50-m radius of the collecting station may not be characteristic of those occurring within a 1-km radius or those of the district or region from which the sample derives. Normally, the geographic extent of the conditions represented by the pollen record of a given sample is assessed by comparison. Pollen records dating to the same period from different locations are compared within and between districts and regions. Insofar as these records are similar, we normally conclude that the conditions responsible for them have broad geographic extent. Insofar as they are different, we conclude that the geographic extent is limited. The methodology is essentially similar to that accomplished by archaeologists when comparing artifact assemblages to determine the geographic extent of an archaeological culture, and it has the same sorts of difficulties. The probable factor or factors (physical, cultural, or biotic) responsible for the principal characteristics of a pollen record are usually worked out from an assessment of the depositional context of the samples.

If we construct the temporally controlled sequence of pollen records from the cave sites indicated by Table 15.13, we can identify palynological variations through most of the temporal span of the preceramic horizon. A gap of some 4000–5000 years exists, but that is not the principal problem. The principal problem is that there is no information inherent in the pollen sequence that would identify whether or not the observed variations represent highly localized conditions. We wish to assess the chronological relationships of pollen records from other sites (Martínez Rockshelter and Gheo-Shih) with those from Cueva Blanca and Guilá Naquitz. To accomplish this we must assess the potential of the pollen records from Cueva Blanca and Guilá Naquitz to provide information representing geographically extensive ecosystem conditions. If the variations in the pollen records we have were to represent, for example, changes in biotope conditions at the mouths of the caves, the information available would not be adequate for our problem.

The mathematical models we developed from analysis of the surface samples may be applied to the pollen records representing the chronological units of Table 15.13 precisely because they are regional in scope. Those that have maximal geographic breadth are the annual rainfall and the annual temperature models. The annual rainfall and annual temperature models that have the most strength are those based on the standard pollen sum. These two models are independent in the sense that they do not measure ecological conditions that are intimately related. Thus if the application of one model to the fossil data produces results consistent with those obtained from the application of the second model, redundant tests will have been performed.

If a model functions as we presume it should, any two fossil pollen records of the same antiquity (e.g., the pollen records from Zone B and Zone C at Guilá Naquitz Cave) should be classified by the model to the same category. Further, if the model is applied to a series of pollen records from deposits not assignable to phase (e.g., the lower Martínez Rockshelter records), the classification of those records could allow an assessment of comparability that can be used for biostratigraphic correlation. In Table 15.14, we present the results of application of the annual temperature model to the pollen records of Cueva Blanca, Guilá Naquitz, and Martínez Rockshelter. Before evaluating the outcome, however, some stratigraphic information about Martínez Rockshelter must be presented.

Three strata are visually evident in the section sampled for pollen analysis at Martínez Rockshelter. The uppermost of these (Zone A) shows no visible subdivisions, but excavation revealed that its lower portion was late preceramic and its upper portion contained Monte Albán V materials (ca. A.D. 1300) mixed with redeposited preceramic tools. It was sampled at depths of 5 and 10 cm. The second stratum (Zone B) is sedimentologically like that above but contains few temporally diagnostic artifacts. It was sampled at 5-cm intervals between depths of 15 and 30 cm. An erosion contact occurring at 35-cm depth, which was also sampled, separates the third stratum. The third stratum (Zone C) is lighter in color than the first and second and it contains different kinds of artifacts. Though these also include few temporally diagnostic forms, the artifact content of Zone C is more comparable to the assemblage recovered from Guilá Naquitz Cave, while that of Zone B is more like the assemblage recovered from Cueva Blanca (Flannery 1970). But the resemblances are not sufficiently detailed to allow correlation. The fourth stratum, sampled at 65-cm depth, consists of decomposed ignimbrite that has no color and contains no artifactual record. It grades gradually upward into Zone C, to the effect that the sample collected at 60-cm depth could equally well be assigned to either. The lowermost stratum is sedimentologically correlative with Zone F at Cueva Blanca.

In Table 15.14 we can observe that the expected results of application of the annual temperature model were obtained. Samples assigned the same date on the basis of archaeological similarity (e.g., Cueva Blanca Zones C and D) were classified to the same ecological category of the model. Table 15.14 also reveals a potential for biostratigraphic correlation of the records from Martínez Rockshelter with those from the other caves. The comparability of classifications suggests that the 15-cm level at Martínez Rockshelter dates to the Zone C–Zone D horizon of occupation at Cueva Blanca and the 40- and 35-cm (erosion contact) level dates to the Zone B–Zone C horizon of occupation of Guilá Naquitz.

This result was sufficiently encouraging to make a redundant test worthwhile. If the potential points of biostratigraphic correlation between Martínez Rockshelter

TABLE 15.14
Annual Temperature Classification of the Fossil Pollen Records

Years B.P.	Cueva Blanca	Guilá Naquitz	Martínez R.S.		Zone
700			5 cm	Low	ZONE A
±4000			10 cm	High	ZONE A
±4200–5200	High (Zone C) High (Zone D)		15 cm	High	ZONE B
			20 cm 25 cm 30 cm	Low Low Low	ZONE B
±8500–9500		Moderate (Zone B) Moderate (Zone C)	35 cm 40 cm	Moderate Moderate	ZONE C
10,000–11,500	High (Zone E)	Low (Zone E)	45 cm 50 cm 55 cm 60 cm	High High High High	ZONE C
>12,000	Low (Zone F)		65 cm	Low (ignimbrite)	

and the other two sites are real, one would assume they are recognizable through evaluation of more than one geographically extensive ecosystem condition. They appear to be reflected through similarity of classification in regard to annual temperature. We proceeded to test their classification in regard to annual rainfall, using the standard sum model (Table 15.15).

The apparent result of this test was one of disconfirmation of the suggestions provided by the prior test. Cueva Blanca pollen records from cultural horizons were classified similar to each other but not similar to records from any stratum at Martínez Rockshelter. The same effect occurred in respect to two of the three other potential points of biostratigraphic correlation. The intriguing thing about this test is that so many potential points of correlation yield negative results in comparison with the prior test. One of the possible explanations for this is that the classification scheme of the annual rainfall model is too site-specific to be of value as a control. The different caves are all located at about the same elevation, so their annual rainfall values are similarly influenced by adiabatic lapse rate. But they all face different directions and thus are more or less affected by rain shadows of highly localized character. The problem here could be that the pollen records reflect local annual rainfall conditions so precisely that differences controlled by geographically minor variables influence the result.

TABLE 15.15
Annual Rainfall Classification of the Fossil Pollen Records

Years B.P.	Cueva Blanca	Guilá Naquitz	Martínez R.S.		Zone
700			5 cm	Low Moderate	ZONE A
±4000			10 cm	Low Moderate	ZONE A
±4200–5200	Low (Zone B) Low (Zone C)		15 cm 20 cm 25 cm	Low Moderate Low Moderate Low Moderate	ZONE B
			30 cm	High	ZONE B
			35 cm	Moderate	
±8500–9500		Low Moderate (Zone B) Low Moderate (Zone C)	40 cm 45 cm 50 cm 55 cm	Low Moderate Low Moderate Low Moderate Low Moderate	ZONE C
10,000–11,500	High (Zone E)	High (Zone E)	60 cm	Low	
>12,000	High (Zone F)		65 cm	Low Moderate (ignimbrite)	

Pursuing this line of thought, we applied the third model to these data. We hesitate to place much confidence in this model for purposes of biostratigraphic correlation because it classifies results only in terms of the three valley highland categories of effective moisture. The information potential of the model is therefore lower than the annual rainfall model, which differentiates five categories. A second problem with the valley highland effective moisture model is that it references an ecological condition covering a smaller geographic area. Finally, it is not as precise a model as the others because it is based on a smaller number of samples. However, application of the model produced almost the same results (Table 15.16) as the first test.

We consider the fact that the outcomes of the first and third tests suggested the same points of biostratigraphic correlation to be very significant. This is especially meaningful when it is recognized that the overall classification strengths of the two models are not comparable, and the probability of correct classification of particular samples is much higher in some instances than others when the annual temperature model is applied. We believe it extremely unlikely that the comparability of outcomes in the first and third tests could be an artifact produced by mathematical manipulations of the data. We also consider it significant that application of the annual rainfall model produces results not comparable to the other two at three out of four potential points of correlation. This seems to us no chance effect. It is likely that the palynological expression of the ecological condition modeled in that case is somewhow intrinsically different from the palynological expression of the other two ecological conditions.

The biostratigraphic correlations indicated by Tables 15.14 and 15.16 suggest an integrated chronology for the three cave sites that, with no apparent gaps, spans the period from ± 12,000 to ± 4000 B.P. According to the evidence provided by comparison of artifactual records, the site of Gheo-Shih dates to the 6000–7000 B.P. horizon. Though Gheo-Shih is located on the upper alluvium of the valley floor, not in the foothills district where the caves are found, biostratigraphic comparison of the pollen record from Gheo-Shih with those pollen records from the caves attributable to the appropriate time interval should provide an independent check on the absolute data attributed to Gheo-Shih.

If Gheo-Shih has been accurately dated, it should yield a pollen record classified to the high annual temperature category. Samples from Cueva Blanca dated 4200–5200 B.P. have this classification, as does the 15-cm level Zone B record from Martínez Rockshelter. There should be a similar comparability of classification in regard to effective moisture between the roughly contemporary records from Gheo-Shih, Cueva Blanca, and Martínez Rockshelter. In the case of Gheo-Shih, it is necessary to apply the lowland effective moisture model to obtain accurate classification as it is a lowland site. But one would anticipate classification of all these presumably contemporary pollen records to the high effective moisture category.

These predictions are exactly confirmed when the Gheo-Shih pollen record is classified according to the standard sum annual temperature and effective moisture models (Appendix 15.3). On these grounds, stratigraphic correlation of the Gheo-Shih pollen data with that from Cueva Blanca and the upper portion of Zone B at Martínez Rockshelter serves independently to support the artifactually evidenced assessment of absolute chronology for this site. It should be noted, however, that the effective moisture conditions occurring at the time Gheo-Shih was occupied apparently persisted

TABLE 15.16
Valley Highland Effective Moisture Classification of the Fossil Records

Years B.P.	Cueva Blanca	Guilá Naquitz		Martínez R.S.	
700			5 cm	Low	ZONE A
± 4000			10 cm	High	
± 4200–5200	High (Zone C)		15 cm	High	ZONE B
	High (Zone D)		20 cm	High	
			25 cm	High	
			30 cm	High	
± 8500–9500		Low Moderate (Zone B) Low Moderate (Zone C)	35 cm	Low Moderate	ZONE C
10,000–11,500			40 cm	High	
			45 cm	High	
			50 cm	High	
			55 cm	High	
	High (Zone E)	High (Zone E)	60 cm	High	
>12,000	High		65 cm	High (ignimbrite)	

through the horizon of preceramic occupation at Cueva Blanca and lasted into the subsequent archaeological phase. Biostratigraphic correlation supports the dating of Gheo-Shih to the 6000–7000 B.P. horizon but does not disconfirm any alternate hypothesis that would date Gheo-Shih between 4000 and 7000 B.P.

In summary, palynological evidence of biostratigraphic correlations among the four preceramic sites offer substantive, independent support of hypotheses of chronological relationship evidenced through radiocarbon dating and temporal assessment of the artifact assemblages represented at the different sites (Table 15.17). The biostratigraphic correlations are developed through application of the mathematical models of palynological reflections of ecological conditions that are controlled by analysis of modern pollen rain variations. Thus the correlations are not made in terms of index fossils or temporally diagnostic assemblages of pollen types. They are made in terms of ecologically conditioned responses of the total pollen rain at unique intervals of time. Because this form of biostratigraphic evidence is less convincing than that based on index fossil types or assemblage types, two redundant tests were performed using separate mathematical models. One confirmed the original test and one did not. We maintain that the lack of confirmation in this case results from a lack of comparability in the precision of the three tests in identification of ecological conditions. Essentially, the mathematical model that allows identification of annual rainfall parameters seems to be too sensitive to variations imposed by rain shadow effects. The effective moisture and annual temperature parameters are less precise and therefore allow correlations within a district at a given time horizon.

The earliest temporal horizon identifiable from the palynological record available appears to be that during which the ignimbrite of the cave walls at Cueva Blanca and Martínez Rockshelter were broken down to create the Cueva Blanca F and Martínez 65-cm strata. This horizon was initiated prior to the earliest evidence of human occupation of the Valley of Oaxaca and, judging from its stratigraphic relationship to radiocarbon dates, occurred earlier than 12,000 B.P. The existence of burned and broken remains of Pleistocene fauna in this zone at Cueva Blanca, however, indicates human occupations of the district before the termination of this interval.

The following horizon is represented palynologically by records associated with the cultural remains of Zone E at Guilá Naquitz, the Zone E deposit at Cueva Blanca, and the deposits sampled at 40–60 cm at Martínez Rockshelter. The Zone D deposits at Guilá Naquitz that might also be referable to this horizon failed to yield a sufficient number of pollen grains for analysis. Radiocarbon assays for Zone E at Cueva Blanca range from 10,050 ± 350 to 11,000 ± 400 B.P. Judging by artifact similarity, these dates probably apply to an earlier interval within this horizon than is represented at Zones D and E of Guilá Naquitz. This is also attested to by the ^{14}C dates for Zone E at Guilá Naquitz, which range from 9790 ± 240 to 10,700 ± 350 B.P. We consider the pollen records of this horizon, which indicate high annual temperature and high effective moisture conditions, to date primarily to the 9500–12,000 B.P. interval.

The 40-cm level sample from Martínez Rockshelter provides evidence that annual temperature values had declined before the end of this interval, though effective moisture values remained constant within the parameters we may presently measure. However, the evidence provided by a single pollen record is not normally considered sufficient for recognition of an ecologically distinctive temporal horizon. The pollen record of this sample, therefore, has not been isolated as representative of a unique temporal unit.

The third temporal horizon of the sequence is represented palynologically in the sample collected at 35-cm depth at Martínez Rockshelter and in the pollen record of Zones C and B at Guilá Naquitz. Radiocarbon assays associated with the latter establish the antiquity of this horizon as roughly 8500–9500 B.P. Moderate annual temperature conditions and low-moderate effective moisture conditions apparently occurred at this time, though the former represent a continuation of annual temperature conditions evidenced near the end of the preceding time interval.

TABLE 15.17
Biostratigraphically Defined Chronology of the Cave Deposits

Chronological horizon	Antiquity B.P.	Site where evidenced
7	4000–6500	Martínez Rockshelter (10 and 15 cm) Cueva Blanca (Zones C and D)
6	6500–8000	Martínez Rockshelter (20–30 cm)
5 } 4 }	7500–8500	{ Martínez Rockshelter erosion Martínez Rockshelter deposition
3	8500–9500	Martínez Rockshelter (35 cm) Guilá Naquitz (Zones C and B)
2	9500–12,000	Martínez Rockshelter (40–60 cm) Guilá Naquitz (Zone E) Cueva Blanca (Zone E)
1	>12,000	Cueva Blanca (Zone F) Martínez Rockshelter (65 cm)

The 35-cm level at Martínez Rockshelter is represented sedimentologically by a strong disconformity indicative of an interval of erosion. At Guilá Naquitz Cave, deposition essentially ceased for thousands of years subsequent to the formation of Zone B. At Cueva Blanca, deposition was interrupted between the time of deposition of Zone E, ca. 10,000 B.P., and the time of deposition of Zone D, ca. 5000 B.P. Our opinion is that these geomorphological events are interrelated and indicate the occurrence of more than one paleoenvironmental condition between 5000 and 8500 B.P. The oldest of these, we believe, resulted in the deposition of sediments at Martínez Rockshelter that were subsequently removed by erosion and thus made unavailable for pollen analysis. Following the erosion interval, the 20–30-cm level deposits at Martínez Rockshelter were laid down.

The fourth temporal horizon of our chronology, then, is identified as the interval of deposition at Martínez Rockshelter that was stripped away, and the fifth temporal horizon is identified as the erosional interval itself. These units of the chronology are evidenced sedimentologically, not by palynological data, and so do not provide botanical evidence of paleoecological conditions. Dating is obscure, but it seems unlikely that the two intervals encompass a great deal of absolute time. Since the estimated antiquity of the 15-cm level in the Martínez Rockshelter sequence could be ca. 6000–7000 B.P. as a consequence of correlation with the Gheo-Shih pollen record, and since 15 cm of deposition occurred at Martínez Rockshelter subsequent to the erosional interval but prior to the estimated date of 7000 B.P., it would appear probable that the fourth and fifth intervals of the chronology together span not more than the 1000-year interval between 7500 and 8500 B.P.

The sixth interval of the chronology is represented by the pollen records of the 30- and 20-cm levels at Martínez Rockshelter. The principal evidence of a distinctive chronological unit is the consistent classification of these records to the low annual temperature category and the difference in evidenced effective moisture value between these records and the record of the 35-cm level in the same sequence. The only information available for chronometric dating of this horizon is its relative position between the date indicated for the pollen records from Gheo-Shih (6000–7000 B.P.) and that indicated for the erosional, fifth, interval (7500–8500 B.P.). The pollen records of the sixth horizon, then, probably reflect some portion or all of the 6500–8000 B.P. interval.

The seventh horizon of the chronology is represented in the 15-cm sample from Martínez Rockshelter, the pollen record of Gheo-Shih, and the pollen of Zones C and D at Cueva Blanca. Identical palynological records are recovered from the 10- and 5-cm level samples at Martínez, indicating a relatively long time span for this interval if the cross dating of the late preceramic horizon evidence at Martínez with Yuzanú on typological grounds (see Chapter 3) is satisfactory. The interval evidences conditions of high effective moisture and high annual temperature values and appears to date to 4000–6500 B.P.

Paleoenvironmental Reconstructions

The paleoenvironmental reconstruction most significant to this study is focused on the issue of changes in vegetation that occurred through the preceramic period in the geographic area of the cave occupations. This focus is intentionally specific, as it offers the best prospect for comparison with paleoenvironmental reconstructions based on other forms of data datable to the Archaic horizon provided by the Oaxaca Project. There is, however, a significant value to the question of paleoenvironmental changes that occurred more generally during this time period in the Valley of Oaxaca that reflect variations in paleoenvironment occurring synchronously over much of highland Mesoamerica. Forms of paleoenvironmental reconstruction amenable to interpretation in such terms are relevant to systematic analysis of Mesoamerican cultural adaptations.

Given the character of the palynological data and methods of analysis available as a result of these investigations, micro- and macroscale paleoenvironmental reconstructions must be addressed separately. Further, it is not possible to either generalize from the former to achieve solution of the latter, or to apply the generalized model of the latter to the specific issues of the former. As we anticipate that the desire to do so may prove irresistible to many archaeologists, it is pertinent to explain why these restrictions apply.

Since they derive from terrestrial deposits, the fossil pollen records available to us are probably reflections of ecological conditions occurring within a very limited radius of the locus of deposition. The modern surface samples were selected to approximate this situation as closely as possible in order to function effectively as controls. Thus the mathematical models of palynological expression of ecological conditions used for interpretation are models of localized conditions that cannot be generalized to encompass broad areas without extreme loss of confidence. We know, however, that a number of the paleoecological conditions expressed in the reconstructions are controlled by climatic phenomena that are unlikely to have affected the immediate environs of the cave locations without having a more general effect in the Valley of Oaxaca and much of highland Mesoamerica. Insofar as the paleoecological changes that occur in the reconstruction of the cave district record indicate phenomena of such magnitude, they may be abstracted to identify more general paleoenvironmental patterns and the horizons of change.

Such abstraction does not represent a generalization from the specifics of the localized record of paleoecological variation, as distinctive factors are required to establish a paleoenvironmental reconstruction of broader geographic scale. For example, the local-level reconstruction might identify a change from low annual temperatures accompanying high effective moisture values to high annual temperatures accompanying high effective moisture values in the caves district. This cannot be generalized directly to the statement that such a paloenvironmental change may have occurred throughout the Valley of Oaxaca and over highland Mexico. To accomplish a paleoenvironmental reconstruction of this order, it must first be expressed in the climatic–floristic terms of the control models. Then, one must use that expression of the information as the basis for deduction of the kind of local meteorological system that could operate to produce such parameters. Finally, one may extend a valley-specific meteorological reconstruction more widely only subsequent to consideration of how such a meteorological system might express the processual annual variation of major Northern Hemisphere cyclonic and anticyclonic systems.

Paleoenvironmental reconstructions of this sort can be developed from the fossil pollen record made available by this study. Indeed, an earlier draft of this report expressed both a valley-specific paleometeorological reconstruction and a paleoclimatic reconstruction of pressure cell displacements following the method outlined above. We have been advised, however, that they be deleted from the present report for a number of reasons. First and foremost, they are inappropriate here because they cannot be supported or denied by other evidence collected from Oaxaca. Thus their presentation would violate one of the principal methodological positions of the research design established for the suite of studies incorporated in the Prehistoric Human Ecology of the Valley of Oaxaca Project.

Second, acceptance of those reconstructions demands acceptance of the proposition that all horizons of time between 12,000 and 4000 B.P. are in fact represented in the fossil pollen records from these few cave sites. Though we contend that biostratigraphic evidence documents this conclusion, reviewers have informed us of their suspicion that our chronology may be more ingenious than credible. The fact is that there are neither radiocarbon assay dates nor temporally diagnostic artifacts from Martínez Rockshelter to support the argument based on biostratigraphy. Given this situation, and our recognition that the majority of readers are not likely to be familiar with the theoretical and methodological issues involved in argument from biostratigraphic evidence, we have considered it judicious not to press the point at this time.

Third, the character of the paleometeorological and paleoclimatic reconstructions presented in the draft version of this report resulted in an attack on matters that are often regarded as articles of faith in Mesoamerican archaeology. For example, the reconstructions indicate that during the entirety of the 12,000–4000 B.P. period, there was only one interval (9500–8000 B.P.) during which seasonal drought occurred. Present archaeological reconstructions of the behavior of preceramic populations of Mexico (e.g., MacNeish 1978; Niederberger 1979) not only assume that the modern monsoonal precipitation pattern of Mexico has persisted throughout the Holocene but also assume this environmental parameter played a significant conditioning role in both the early development of agriculture and in the establishment of preceramic settlement patterns. Given a lack of independent supporting evidence from Oaxaca and the prospect that readers would assess the palynological record from Oaxaca as an incomplete record of the preceramic, it is prudent to compile and present a complete and integrated pattern of supportive evidence for such reconstructions from other locations. Such effort would be beyond the scope of this report, would change its methodological emphasis, and would detract from its purposes.

In the present report, then, we have dealt only with the reconstruction of paleovegetation patterns. The pollen records of the caves are interpreted by reference to the control pollen records to identify cave district–specific ecological conditions. As we have modern floristic referents for these, we have a basis for identifying the paleovegetation of the cave district of the Valley of Oaxaca at different horizons. Our comprehension of interactive relationships existing today between such vegetation patterns as forest types, community types, and association types and hydrographic, substrate, and topographic variables has served as the basis for reconstructions of the character and distributions of vegetation patterns on other portions of the landscape of the valley. These extensions from the immediate record of cave district vegetation are similarly dependent on acceptance of a uniformitarian position.

We consider it critical to advise the reader, however, that only those paleovegetation reconstructions that we find consistent with our paleometeorological and paleoclimatic reconstructions are presented in this report. Other reconstructions of paleovegetation are possible if one uses the traditional methods of interpretation of pollen spectra. Furthermore, other paleovegetation reconstructions may be made using the method adopted here if different assessments of the relationship of paleoecological conditions to floristic composition are adopted. Thus the paleovegetation reconstructions offered in this report are not actually divorced from those we have made of paleometeorology and paleoclimate, though the latter are not presented here.

We believe the three to be integrable to this degree because all are evidenced through their effect on the expressions of ecology recorded as populations of pollen rain. Using the method of traditional pollen analysis, investigators adopt the thesis that climatically induced and meteorologically induced changes in ecological relationships are causes of vegetation change. Thus changes in vegetation are responsibly interpreted in paleoclimatic and/or paleometeorological terms. The method of this analysis provided a direct reconstruction of changing ecological conditions and requires no assessment of the causes of such variation. The changes may be interpreted as reconstructions of vegetation *or* reconstructions of seasonal variations in weather *or* as reconstructions of climatic systems *or* in other terms appropriate to problem orientations or in any combinations thereof quite independently. We are not trying to say that meteorological and climatic changes are not causal variables resulting in vegetational change. Indeed, we are most assuredly convinced that this is the case. Rather, we are saying that the method we use for interpretation of these pollen records in meteorological and climatic terms does not depend on assessment of that causal relationship. We have therefore not considered it necessary to ignore the implications of our paleometeorological and paleoclimatic reconstructions when advancing a particular vegetational reconstruction as the one we find most credible.

Paleoenvironment: Before 12,000 B.P. This horizon is represented in the cave pollen records of Zone F at Cueva Blanca and the 65-cm level of Martínez Rockshelter. As shown in Table 15.18, both models provide the same reconstruction of effective moisture conditions for this interval. The two models reconstruct annual temperature conditions differentially, however. Review of Tables 15.7 and 15.11 indicates that though both models classify low-moderate annual temperature records accurately 100% of the time, the standard sum model misclassifies low annual temperature records more often than the adjusted sum model. The probability that the reconstruction provided by the standard sum model is erroneous is therefore somewhat greater than the probability that the reconstruction provided by the adjusted sum model is erroneous. Low-moderate annual temperature values, then, are the preferred reconstruction.

The two models also provide different reconstructions of the annual rainfall value at Cueva Blanca and Martínez Rockshelter. The strength of the adjusted sum model in correctly classifying low-moderate annual rainfall pollen records is greater than that of the standard sum model in correctly classifying high-moderate annual rainfall pollen records. The more reliable reconstruction for Martínez, therefore, is one of low-moderate annual rainfall conditions during this interval. But this reconstruction is only about 60% reliable, while the reconstruction of high annual rainfall conditions for ostensibly contemporary samples from Cueva Blanca would be over 80% reliable. We thus prefer this reconstruction.

Our paleoecological reconstruction for the cave district on this horizon, then, is one of high annual rainfall values, high effective moisture values, and low-moderate annual temperature values. The rainfall and effective moisture levels would have been about those presently conditioning the expression of Pine Forest forms of vegetation, while annual temperature values would have been on the order of those presently influencing the expression of Oak Forest zone vegetation. Pollen frequency analysis allows inference of the local occurrence of a high, closed forest canopy dominated by pine in the caves district.

The evidence of low-moderate annual temperature at this elevation documents relatively warmer conditions than now occur in Pine Forest locales. This, in turn, suggests that winter temperatures on the valley floor, while markedly cooler than today's, were probably sufficiently high to allow the occurrence of Mesophytic Forest vegetation in protected habitats, though the general vegetation pattern of the lower and upper alluvium would probably have been a mix of the more-frost-tolerant Thorn Forest and Mesquite Forest taxa and Oak Forest. Oak Forest would have mantled the lower piedmont and lower mountain slopes with Pine Forest associations occurring in canyon habitats. Cloud Forest associations very likely occurred at high elevations in the mountains.

Paleoenvironment: 12,000–9500 B.P.. Paleoecological conditions indicated for this period by the two models (Table 15.19) are somewhat distinctive. The adjusted sum model reconstructs conditions as more variable, and varying in different degrees for some parameters, then the standard sum model. A problem is also introduced by the fact that the sequence of samples from Martínez Rockshelter covers a span of time, while the single samples of this horizon from Guilá Naquitz and Cueva Blanca ostensibly refer to one interval.

If one takes the lower overall classification strength of the adjusted sum model into account, it is fairly clear that high effective moisture values probably obtained throughout this period. High annual temperatures probably prevailed in the earlier part of the period, but were reduced to a low-moderate or moderate level later, during the Zone E occupations at Cueva Blanca and Guilá Naquitz. It seems reasonable to assume a similar paleoecological pattern for the Zone D occupation at Guilá Naquitz. High annual rainfall values seem most probable for the early part of the period, dropping to low-moderate annual rainfall values at the time temperature level fluctuated.

When the caves were first occupied, local vegetation was of a Pine Forest type but probably strongly influenced by the high annual temperature average in the direction of piñon and oak dominance. Habitat variations in the caves district would have allowed the occurrence of frost-tolerant Thorn Forest associations mixed with Oak Forest associations. Oak Forest would have mantled the slopes below the caves, but the degree of admixture with Thorn Forest associations probably presented a visual impression for the Oak Forest quite different from anything seen today. Mesophytic Forest probably dominated the upper as well as lower alluvial districts, and the Mesquite Forest and Thorn Forest very likely were mixed in a complex mosaic of associations on the lower piedmont.

Paleoenvironment: 9500–8000 B.P. By the time of Zone C occupation of Guilá Naquitz, effective moisture values had

TABLE 15.18
Paleoecological Reconstructions, >12,000 B.P.

Sample	Model	Valley highland effective moisture	Annual temperature	Annual rainfall
Cueva Blanca (Zone F)	Adjusted sum	High	Low Moderate	High
	Standard sum	High	Low	High
Martínez Rockshelter (65 cm)	Adjusted sum	High	Low Moderate	High Moderate
	Standard sum	High	Low	Low Moderate

TABLE 15.19
Paleoecological Reconstructions, 12,000–9500 B.P.

Sample	Model	Effective moisture	Annual temperature	Annual rainfall
Cueva Blanca (Zone E)	Adjusted sum	High	Low Moderate	High
	Standard sum	High	High	High
Guilá Naquitz (Zone E)	Adjusted sum	High	Low Moderate	High Moderate
	Standard sum	High	Low	High Moderate
Martínez Rockshelter (40 cm)	Adjusted sum	Moderate	Low Moderate	Moderate
	Standard sum	High	Moderate	Low Moderate
Martínez Rockshelter (45 cm)	Adjusted sum	Moderate	Low	High
	Standard sum	High	High	Low Moderate
Martínez Rockshelter (50 cm)	Adjusted sum	Moderate	Low	Low Moderate
	Standard sum	High	High	Low Moderate
Martínez Rockshelter (55 cm)	Adjusted sum	High	Low Moderate	High Moderate
	Standard sum	High	High	Low Moderate
Martínez Rockshelter (60 cm)	Adjusted sum	High	Low	High
	Standard sum	High	High	Low

dropped from high to low-moderate levels, annual temperatures were unchanged from moderate levels, and annual rainfall values were reduced from high to moderate levels (Table 15.20). Judging by the frequency of Type IV pollen, Thorn Forest became strongly established in the cave district. Mixed Oak Forest and Pine Forest plant associations would have mantled the slopes shortly above the caves and occurred in the cave area in canyon habitats with cool air drainage. Vegetation pattern change would have been most marked on the upper alluvium and lower piedmont slopes as the preexisting Mesophytic Forest relinquished territory to Mesquite Forest and Mesquite Grassland patterns.

Paleoenvironment: 8000–6500 B.P. The most probable assessment of Table 15.21, considering the classification strengths and weaknesses of the relevant portions of the two discriminant models, is that effective moisture values and annual rainfall values were significantly elevated in the area of the caves during this interval relative to the 9500–8000 B.P. horizon, and annual temperatures significantly reduced. The annual rainfall and effective moisture values were about those enjoyed in the Pine Forest zone today, with annual temperatures about those now experienced in the Oak Forest zone. Frequency analysis of the pollen spectra indicates that the tree canopy was neither high and closed nor layered and that oak was the dominant floristic element in the local flora of the caves district.

The replacement of Thorn Forest by Oak Forest vegetation in the caves district would see corresponding changes on the lower piedmont slopes and upper alluvium as earlier Mesquite Grassland and Mesquite Forest vegetation patterns were widely replaced by Thorn Forest and Mesophytic Forest patterns. The slopes above the caves would have lost a good deal of their earlier mixed Pine Forest–Oak Forest character and developed as Pine Forest zone plant associations. Frequency analysis indicates the modern sort of Pine Forest zone B association is represented in the oldest pollen records of the Cueva Blanca D stratum. A stand of this type probably grew immediately in front of the cave entrance on the slopes of this canyon.

TABLE 15.20
Paleoecological Reconstructions, 9500–8000 B.P.

Sample	Model	Effective moisture	Annual temperature	Annual rainfall
Martínez Rockshelter (35 cm)	Adjusted sum	Low Moderate	Moderate	Moderate
	Standard sum	Low Moderate	Moderate	Moderate
Guilá Naquitz (Zone B)	Adjusted sum	Low Moderate	Moderate	Low
	Standard sum	Low Moderate	Moderate	High Moderate
Guilá Naquitz (Zone C)	Adjusted sum	Low Moderate	Moderate	Moderate
	Standard sum	Low Moderate	Moderate	High Moderate

TABLE 15.21
Paleoecological Reconstructions, 8000–6500 B.P.

Sample	Model	Effective moisture	Annual temperature	Annual rainfall
Martínez Rockshelter (20 cm)	Adjusted sum	High	Low Moderate	High
	Standard sum	High	Low	Low Moderate
Martínez Rockshelter (25 cm)	Adjusted sum	High	Low Moderate	High Moderate
	Standard sum	High	Low	Low Moderate
Martínez Rockshelter (30 cm)	Adjusted sum	High	Low Moderate	High Moderate
	Standard sum	High	Low	High
Cueva Blanca Square G14 (85 cm)	Adjusted sum	High	Low Moderate	High
	Standard sum	High	High	High
Cueva Blanca Square G14 (95 cm)	Adjusted sum	High	Low Moderate	High
	Standard sum	High	Low	High
Cueva Blanca Square G14 (100 cm)	Adjusted sum	High	Low Moderate	High
	Standard sum	High	Low	High

There is no direct archaeological evidence that Guilá Naquitz or Cueva Blanca were occupied at this time. Evidence of occupation occurs at Martínez Rockshelter but is not diagnostic to cultural phase for this interval. There is, however, a discriminant functions similarity between the pollen records of stratigraphic Zone B1 at Guilá Naquitz and those of this horizon at Martínez Rockshelter. There is also a similarity between the pollen records obtained at the 75-cm level in Square G14 at Cueva Blanca, which is the oldest record of the D stratum at that site, and the Martínez records attributed to the 6500–8000 B.P. horizon. It is interesting to note that though this Zone D pollen record from Cueva Blanca and the Zone B1 pollen records from Guilá Naquitz are like each other, the former is unlike the other Zone D and Zone C pollen records from Cueva Blanca, and the latter are unlike the pollen records from Zone B2 at Guilá Naquitz.

One of the implications of these data is that some portion of the youngest preceramic occupation of Guilá Naquitz, evidenced at least in the B1 horizon at Squares E6 and E9, occurred in the context of different paleoenvironmental conditions from the other occupations of Zone B and Zone C. More than one pattern of environmental exploitation might, then, be expressed in the archaeological record of Zone B at this site. A second implication is that portions of the B2 and generalized B zone deposits may date to the interval of time represented by the erosional unconformity at Martínez Rockshelter and the sediments removed by that geomorphological process. This, in turn, implies that sedimentation in the caves was more or less rapid at different intervals in the preceramic (and the deposits were subject to erosion if local conditions fostered that process coincident with a slower deposition rate), but deposition never was completely halted for an interval as long as a half millennium.

Paleoenvironment: 6500–4000 B.P. Though this chronological horizon is defined by a single set of biostratigraphic indicators, it is divisible into an earlier paleoenvironmental pattern, reflected at Gheo-Shih and the C and D zone occupations at Cueva Blanca, and a later pattern, reflected in the 10- and 15-cm levels at Martínez Rockshelter (Table 15.22). On the earlier horizon, annual rainfall values were low and annual temperature values high, but effective moisture values were also high. Major vegetation patterns and vegetation zone distributions in the valley would have remained much as they had been in the preceding millennium until ca. 5000 B.P., except Oak Forest and Pine Forest B vegetation patterns may have expanded their upper elevational limits at the territorial expense of Pine Forest A associations. Limited amounts of Mesquite Forest may have developed on the more xeric habitat positions of the upper alluvium. The low annual rainfall, however, probably had a marked effect on the distributions, depths, and frequencies of pond and swamp habitats on the valley floor and the character of riparian associations on the upper alluvium and lower piedmont slopes. Far less open and surface flow water would have been the rule than occurred in the 8000–6500 B.P. interval. Thus, while annual rainfall values were high enough during the growing season to be more effective for vegetation than is the case today, general reduction in rainfall quantities would have affected the hydrographic system.

Pollen frequency values of the samples from Gheo-Shih indicate the local vegetation pattern was not Mesquite Grassland (*Celtis*–Moraceae pollen only occurs in one sample), but the average Compositae–Gramineae ratio of 1:2.5 argues for the occurrence of a form of Mesquite Forest with a quite open canopy. Our feeling is that this site, located between two watercourses on the upper margin of the upper alluvium, supported atypical vegetation when occupied. As a result of minimal substrate depth, low annual precipitation had a more marked effect at this location and created a relatively xeric habitat to which Mesquite Forest was better adapted than the Mesophytic Forest that then dominated the upper alluvium zone. This was mitigated to a degree by summer runoff waters from the piedmont (then mantled by

TABLE 15.22
Paleoecological Reconstructions, 6500–4000 B.P.

Sample	Model	Effective moisture	Annual temperature	Annual rainfall
Martínez Rockshelter (10 cm)	Adjusted sum	High	Low Moderate	High
	Standard sum	High	High	Low Moderate
Martínez Rockshelter (15 cm)	Adjusted sum	High	Low Moderate	High Moderate
	Standard sum	High	High	Low Moderate
Cueva Blanca (Zone C)	Adjusted sum	High	Low Moderate	High
	Standard sum	High	High	Low
Cueva Blanca (Zone D)	Adjusted sum	High	Low Moderate	High
	Standard sum	High	High	Low
Gheo-Shih	Adjusted sum	Low	High	Low
	Standard sum	High	High	Low

Thorn Forest) that kept the drainages bounding the site wet during the summer. Those streamlets may even have overflowed their banks in that season. At the end of the summer season, Gheo-Shih would have been a gathering station for mesquite pods and grass and weed seeds. As one of the few low-elevation areas of open canopy conditions available, it also might have been an attractive late summer-winter camping location in terms of relative relief from insects.

Pollen frequency values from samples associated with the C and D occupations at Cueva Blanca indicate the local existence of Pine Forest vegetation in the canyon and on the talus in front of the site (also reported in Lindsay 1968). It seems not unlikely that this site was also attractive as a result of unusual habitat conditions, as cool air drainage seems to have maintained a Pine Forest plant association at an unusually low elevation at this locality. This could have provided an attractive fall hunting camp location relatively close to a later summer macroband settlement such as Gheo-Shih.

At the elevation of the caves, a mix of Thorn Forest and Oak Forest plant associations apparently developed between 5000 and 4000 B.P., and Thorn Forest mantled the lower piedmont and invaded xeric niches on the upper alluvium. Mesquite Grassland and Mesquite Forest probably ceased to exist as vegetation categories, though the species of their respective plant associations occurred in the Mesophytic Forest and the Thorn Forest in situations to which they were adapted. Valley-floor lacustrine and swamp niches would have been revived and even more extensive than was the case in the 8000–6500 B.P. interval. Earlier, relatively lower annual temperatures would have slowed runoff of winter precipitation from the highest elevations (where it was greatest). During the 5000–4000 B.P. interval of higher temperatures, winter runoff from high elevations would have been immediate and swelled the lowland wetlands appreciably. Above the elevation of the caves, Oak Forest vegetation would have extended almost to the crests of the highest margins of the valley. The significant reduction in quantity and extent of Pine Forest resources in the valley may have been dramatic and extraordinary from the perspective of human populations adapted to hunting–gathering strategies operant for the previous 4000–5000 years.

Summary. The sequence of vegetation changes that apparently occurred during preceramic occupation of the valley in the caves district is as follows:

1. Before 12,000 B.P., the caves district was Pine Forest with a high, closed canopy incorporating or not distant from Cloud Forest taxa. Zone F occupation is evidenced at Cueva Blanca at this time.
2. From about 12,000 to 10,000 B.P., a mixed high-elevation mountain forest with a lower, more open canopy incorporating oak, piñon, and Oak Forest and Thorn Forest associations prevailed. This was the time of the Zone C occupation at Martínez Rockshelter.
3. From 10,000 to 9500 B.P., the vegetation was as previously but with piñon and high-elevation species of oak dominant and less Thorn Forest elements in the local vegetation. This vegetation pattern occurred at the time of the Stratum E deposition at Cueva Blanca and probably in the interval of the Zone E and Zone D occupation of Guilá Naquitz.
4. From 9500 to 9000 B.P., there was a return to the conditions described for the 12,000–10,000 B.P. horizon. The uppermost Zone C occupation of Martínez Rockshelter and, possibly, part of the Zone D occupation of Guilá Naquitz occurred at this time.
5. From 9000 to 8000 B.P., the Thorn Forest vegetation pattern became strongly established, with Oak Forest and Pine Forest associations in canyon habitats controlled by cool-air drainage. Mesquite Grassland occurred in the most xeric niches. Cultigens were locally grown (Schoenwetter 1974). This vegetation pattern is associated with Strata C and B occupation of Guilá Naquitz.
6. From 8000 to 6000 B.P., the Oak Forest vegetation pattern became strongly established, with Pine Forest B occurring in cool canyon habitats. Guilá Naquitz may have had

some occupancy (Stratum B1) during the early part of this interval, but the early B stratum occupation of Martínez Rockshelter is the only securely known occupation of the area. The basal D zone deposition of Cueva Blanca is also referable to this horizon.

7. From 6500 to 5000 B.P., the vegetation was as previously but with the Pine Forest associations of canyon floors controlled more by humidity levels than cool-air drainage. Cueva Blanca Zone D and Zone C occupations that yielded pollen data are of this horizon. There is no evidence of deposition at Guilá Naquitz or Martínez Rockshelter. The vegetation pattern at the lower piedmont–upper alluvium transition site of Gheo-Shih was open canopy Mesquite Forest. Cultigens were locally grown.

8. From 5000 to 4000 B.P., a mixed low-elevation mountain forest of Thorn Forest and Oak Forest association prevailed. Upper elevation Oak Forest associations were established in canyon floor niches, with Thorn Forest associations in most xeric habitats. The upper levels of Martínez Rockshelter were occupied by preceramic populations. Occupations of some lower alluvium locations during this interval by agricultural villagers is possible. There are no available pollen (or archaeological) records of such villages, but it is arguable that they were established near the margins of swamps or ponds in Mesophytic Forest associations having a strong component of tropical taxa.

Plant Resources

The pollen records of Guilá Naquitz Cave and Gheo-Shih document probable utilization of cultivated food resources during the 9500–8000 and the 6500–5000 B.P. intervals. Other palynological evidence of cultivated food resources or their wild relatives occurs in the form of cf. *Zea mexicana* in the 45-cm level deposits at Martínez Rockshelter (estimated to date 10,000–9500 B.P.; see Table 15.25, in Appendix 15.2) and the cf. *Tripsacum* and the *Cucurbita* pollen observed in the C and D zone records of Cueva Blanca (6500–5000 B.P.; see Table 15.27 in Appendix 15.2). The reliability of our Tribe Maydeae (i.e., corn, *Tripsacum*, or teosinte) identifications (discussed in Schoenwetter 1974) is constrained; there may be different species contributing this pollen to the record, or all such pollen may derive from a single species. The plants producing this pollen may have been wild, cultivated but not genetically affected by human selection, domesticated, or some combination thereof. The *Cucurbita* pollen may derive from one or more wild, cultivated, or domesticated species.

There are two lines of evidence that argue that whether or not the plants involved were genetic domesticates they were cultivars. First, there is the evidence that leads to recognition that ecological conditions were inappropriate for the growth of such plants as competitors with local flora at the times involved. Schoenwetter (1974) has explored this argument in regard to the Maydeae pollen from Guilá Naquitz, and in that case supports this position with the evidence provided by associated macrofossils of cultivated plants. In the Martínez Rockshelter and Cueva Blanca records, the Maydeae and *Cucurbita* pollen are evidenced as part of a local flora even more obviously inappropriate. Considering the cultural associations of these records, it seems much more likely to us that they represent the results of human behavioral patterns of cultivation, storage and/or consumption than the results of natural dispersal and preservation of pollen.

Second, there is the evidence provided through comparison of these fossil pollen records and the modern surface controls. In the modern surface pollen record the principal clue to the existence of cultivation behavior is constancy of occurrence of pollen of cultigens in the pollen rain of cultivated plots. This is the pattern expressed by Maydeae and *Cucurbita* pollen in the fossil record. Guilá Naquitz, Gheo-Shih, and Cueva Blanca are occupation site locations; Martínez Rockshelter is a specialized activity site. Maydeae pollen is observed in 10 of the 16 pollen records of the 9500–5000 B.P. horizon in the former three sites. Maydeae pollen is thus more consistently associated with evidence of human residency on the preceramic horizon than it is associated with uncultivated plot locations in the modern agricultural, grazed, disturbed Valley of Oaxaca.

Overall, we find the two independent, but convergent, lines of palynological evidence convincing. Therefore, analysis of the pollen record leads us to conclude that one or more members of the Maydeae tribe of grasses has been a cultivar in the Valley of Oaxaca at least since 10,000 B.P., and a member of the genus *Cucurbita* has also been a cultivar at least since 8500 B.P. Recovery of these pollen types at Cueva Blanca, which was probably a fall or winter season camping site in the Pine Forest zone, indicates the use of effective storage and transportation technology for these cultivars by 5200 B.P.

Utilization of wild plant food resources is indicated palynologically by the recovery of *Agave*, Ericaceae, Liliaceae, and Cactaceae pollen in records of the 12,000–9500 B.P. horizon; *Agave,* Ericaceae, Cactaceae, and *Ribes* pollen on the 9500–8000 B.P. horizon; *Agave* and Cactaceae pollen on the 8000–6500 horizon; *Agave* pollen on the 6500–5000 horizon; and *Agave* and Liliaceae pollen on the 5000–4000 B.P. horizon. *Agave* is represented at all sites as well as at all periods of the record, which may be an index of its resource significance. The Cactaceae pollen, which is prominent in the records of both Martínez Rockshelter and Guilá Naquitz but drops out of the sequence after 6500 B.P., is not identifiable to genus. But it is *not* pollen of the prickly pear (nopal) genus; it is probably pollen of a pitahayo or *órgano* cactus. Ericaeae and *Ribes* pollen, probably manzanita and currant, are cooccurrents early in the occupational sequence; their disappearance from the sequence after 8000 B.P. may reflect wild food harvesting strategy variation responsive to climatic change. Liliaceae pollen, possibly representing the harvest of yucca fruit, has a bimodal distribution: its occurrence only very early and very late in the sequence—and only at Martínez Rockshelter—is unique, and we attribute this to human manipulation.

The contrast in variety and patterning of records of ostensible cultivar pollen and possible wild food resource pollen seems to us of particular relevance. There are a number of wild food resource pollen types, but only two cultivar pollen types. Both cultivar pollen types occur consistently throughout the preceramic at occupation sites. Only one wild food resource pollen type (*Agave*) occurs consistently, but it has a higher constancy of occurrence than the cultivar types. The other wild food resource pollen types consistently occur on some horizons but not on others during the preceramic horizon.

It should also be realized that the wild food resource pollen record identified here is necessarily minimal. Pollen referable to other potential wild food resources—mesquite, oak, piñon, mint, nopal, *guaje,* hackberry, nanche, and grasses—have not been considered because such pollen is not more prevalent in the fossil record than in the surface control samples. But it is probable that all of those wild food resources were utilized when and where available and that this behavior is reflected palynologically to some degree. We simply are unable to identify that degree and are therefore unable to isolate it.

The overall impression given by the palynological record is that behavioral patterns responsible for the distribution of Maydeae, *Cucurbita*, and *Agave* pollen are different from those behavior patterns responsible for the distributions of the pollen of other plant food resources in the preceramic sequence. Our suspicion is that this is related to storage and transportation technology more than any other factor and that this technology remained highly constant for those three plants during the whole of the preceramic horizon. Alternatively, behavioral change seems indicated as regards collection, processing, and storage strategies directed toward other plant food resources. Rather less of this variation seems related to environmental change than might be expected. Possibly, the observed variations indirectly reflect organizational changes in the nature of the work parties and camp groups represented in the archaeological record associated with the pollen spectra.

Palynological evidence on the availability and distribution of plant resources is of two forms. On the one hand, there is the evidence that allows reconstruction of the character and distributions of vegetation zones at any given period in the Archaic. If one assumes a catchment area of a specific radius and knows the location and date of a given occupation, as well as the probable season of occupation, these reconstructions allow one to retrodict the quantity and quality of wild plant resources available to a human population of a given size. Palynological evidence that allows reconstruction of hydrographic patterns is also useful. During those intervals of high water tables (e.g., 8000–6500 and 5000–4000 B.P.); swamp, pond, and streamside plant resources (e.g., sedge and cattail) and animal resources (e.g., river crabs, turtles, frogs, and migratory waterfowl) would have been plentiful.

On the other hand, palynological evidence providing reconstructions of meteorological conditions and the distributions of vegetation patterns allows conclusions to be drawn regarding the agricultural potential of given sites or areas at different horizons in the Archaic. Agricultural potential is maximized in alluviated areas where an open canopy of natural vegetation reduces the demand for labor investment in land clearance. Where land clearance is necessary, the occurrence of a dry season aids the process substantially, as it allows burning of slash that otherwise would require haulage. Climatic conditions favoring longer growing seasons increase agricultural potential, unless annual precipitation values are low. Partly, this is true because such conditions allow more than one crop to be planted and matured at the lowest elevations in a calendar year. But it is also true because, at higher elevations, it allows a crop to be planted earlier or later than the period of vigorous growth of weedy annual competitor plants.

At the elevation of the caves, arable lands are mostly confined to canyon-bottom alluviums. Ecological conditions favored closed canopy forest associations in this niche except during the 10,000–8000 B.P. interval. There is no evidence of cultivation in the cave district earlier, and the pollen record of more recent cultivars in this district during the preceramic is best assessed as introduced as stored and transported foods.

In the district of upper alluvium the period of highest agricultural potential was 8000–6500 B.P. During this interval forest canopy in the district would have been open, and average annual rainfall values were higher than occur today. During the 6500–5000 B.P. interval the agricultural potential of this district decreased as a result of canopy closure and a modification of rainfall receipts. But xeric locations, such as Gheo-Shih, would remain fairly open. Because average temperatures were milder and rainfall was greater, it may even have been possible to plant crops at Gheo-Shih in late August for a late November–mid-December harvest. This would have minimized competition from weedy annuals also. The yield would be larger if pot irrigation was utilized, but there is no archaeological evidence for this technique prior to 3000 B.P.

Today, partly as a result of land clearance and partly because of the occurrence of a monsoonal rainfall pattern, the lower alluvium district has the highest agricultural potential. This was not the case during the preceramic. Between 12,000 and 4000 B.P., the district would have been heavily forested and/or would have had expanses of surface water. In the 9500–8000 B.P. interval, conditions ameliorated and patchy areas of Mesquite Forest would have developed where summer season dry farming on a slash-and-burn basis might have been profitable. The agricultural potential of this district was never as high during the preceramic, however, as was its potential for providing subtropical and subaquatic resources in great quantity.

The history of preceramic horizon occupation of the Valley of Oaxaca ostensibly was terminated between 4500 and 3300 B.P. Presumably, agricultural populations either invaded the area or developed out of indigenous hunter–gatherer populations and settled in permanent or semipermanent villages in the lower alluvium district. At least, this is where the early villages of the Oaxacan Formative are found, which are thought to date to 3900–3300 B.P.

What sort of vegetation conditions existed at the time and in the district of these terminal preceramic–earliest Formative settlements?

There is no direct palynological evidence, as no sediment samples referable to this horizon have been analyzed. But a hypothesis may be put forth based on recognition that the change to modern climatic conditions probably did not occur earlier than 3500 B.P. At least, the palynological evidence from other areas of the Northern Hemisphere uniformly dates that transition between 3500 and 2000 B.P., not before (Bryson and Wendland 1970).

The implication is that terminal preceramic–earliest Formative settlements may have been established in the Valley of Oaxaca *before* the modern monsoonal climate was established. Villages in the lower alluvium district, if any existed between 5000 and 3500 B.P., would have been located in a district of dense sub- and semitropical forest broken by areas of swamp and pond. Agricultural productivity in this environment would only have been high if populations were large enough and sufficiently organized to command the labor for land clearance, except for locales on the fluctuating margins of the swamps. Niederberger (1976, 1979) describes occupations of the Zohapilco Phase in the Valley of Mexico that are argued to reflect a sedentary adaptation of the latter sort. Depending on the type of architecture utilized, the occurrence of high water tables would limit the placement of village locations. Hills and ridges elevated above the valley floor would probably have been preferred for houses, and it is in such localities that Oaxaca's earliest (Espiridión phase) village remains have been found.

ACKNOWLEDGMENTS

Suzanne K. Fish, as mentioned in the preceding, played a prominent role in the collection and processing of ecological and palynological data. Without her effort this report could not have been produced. Robert J. Burton also deserves special acknowledgment, as it was his patient and persistent discussion that influenced our first halting steps into the maze of multivariate analysis. Kent Lightfoot, Carol Legard, and Richard Romero deserve our thanks for their excellent draftsmanship of the map and figures of this report. R. M. Lombardi and M. Raring-Hart typed the two principal drafts of this report. In the course of that effort, they learned to spell "palynological," "paleoecological," and "multivariate" much better than we did.

Appendix 15.1: Surface Sample Observations

TABLE 15.23
Surface Samples: Ecological Observations

Sample number	Valley arm	Vegetation	Topographic position	Elevation (m)	Substrate	Cultivated	Irrigated	Crop type
001	Etla	Mesquite Forest	Piedmont	1577	Sandy			
002	Etla	Mesophytic Forest A	Upper Alluvium	1562	Sandy			
003	Etla	Mesquite Forest	Lower Alluvium	1559	Sandy			
004	Etla	Pine Forest A	Mountain	2239	Sandy	x		Maize mixed
005	Etla	Pine Forest A	Mountain	2239	Sandy			
006	Etla	Pine Forest B(I)	Mountain	2057	Sandy			
007	Etla	Pine Forest B(I)	Mountain	2057	Sandy	x		Maize
008	Etla	Pine Forest B(II)	Mountain	1894	Sandy			
009	Etla	Pine Forest B(II)	Mountain	1894	Sandy	x		Maize mixed
010	Etla	Pine Forest B(II)	Mountain	1894	Sandy	x		Beans mixed
011	Etla	Mesquite Forest	Piedmont	1711	Sandy	x		Maize mixed
012	Etla	Mesquite Forest	Piedmont	1711	Sandy			
013	Etla	Mesquite Forest	Upper Alluvium	1701	Sandy	x		Maize mixed
014	Etla	Mesophytic Forest B	Lower Alluvium	1693	Sandy			
015	Etla	Mesophytic Forest B	Upper Alluvium	1701	Sandy	x		Maize mixed
016	Etla	Mesquite Forest	Piedmont	1708	Clayey	x		Maize mixed
017	Etla	Mesquite Forest	Piedmont	1708	Clayey			
019	Etla	Mesophytic Forest A	Piedmont	1767	Clayey	x	Canal	Maize mixed
020	Etla	Mesophytic Forest A	Piedmont	1739	Clayey			
021	Etla	Mesophytic Forest B	Upper Alluvium	1601	Clayey	x		Maize mixed
022	Etla	Mesophytic Forest B	Lower Alluvium	1602	Clayey	x		Maize
023	Etla	Mesquite Forest	Upper Alluvium	1605	Clayey	x		Maize mixed
024	Etla	Mesquite Forest	Piedmont	1609	Clayey	x		Maize
025	Etla	Mesquite Forest	Piedmont	1609	Clayey			
026	Etla	Mesophytic Forest B	Upper Alluvium	1674	Sandy			
027	Etla	Mesquite Forest	Piedmont	1591	Sandy	x		Maize
028	Etla	Mesquite Forest	Piedmont	1591	Sandy			
029	Etla	Mesophytic Forest B	Upper Alluvium	1620	Sandy	x		Maize
030	Etla	Mesquite Forest	Upper Alluvium	1630	Sandy	x		Maize
031	Central	Mesquite Forest	Piedmont	1529	Sandy			
032	Central	Mesophytic Forest A	Upper Alluvium	1559	Sandy			
033	Central	Mesquite Forest	Upper Alluvium	1580	Sandy	x		Maize
034	Central	Mesquite Forest	Upper Alluvium	1527	Clayey			
035	Central	Mesquite Forest	Upper Alluvium	1525	Clayey	x		Maize
036	Central	Mesquite Forest	Upper Alluvium	1540	Clayey			
037	Central	Mesquite Forest	Piedmont	1703	Sandy	x	Canal	Maize
038	Central	Mesquite Forest	Piedmont	1661	Sandy			
039	Central	Mesquite Forest	Piedmont	1567	Sandy			
040	Central	Pine Forest A	Mountain	2632	Sandy			
041	Central	Pine Forest A	Mountain	2167	Sandy	x		Maize
042	Central	Pine Forest B(I)	Mountain	2167	Sandy			
043	Central	Pine Forest B(II)	Piedmont	1724	Sandy			
044	Zaachila	Mesophytic Forest B	Upper Alluvium	1530	Silty	x		Other
045	Zaachila	Mesophytic Forest B	Upper Alluvium	1530	Silty	x		Maize mixed
046	Zaachila	Mesophytic Forest A	Upper Alluvium	1529	Clayey	x	Pot	Maize mixed
047	Zaachila	Mesophytic Forest A	Piedmont	1580	Clayey	x		Other
048	Zaachila	Mesquite Forest	Piedmont	1650	Sandy	x		Maize
049	Zaachila	Mesquite Forest	Piedmont	1650	Sandy			
050	Zaachila	Pine Forest B(II)	Mountain	1800	Sandy			
051	Zaachila	Pine Forest B(II)	Mountain	1800	Sandy			
052	Zaachila	Mesquite Forest	Upper Alluvium	1520	Clayey			
053	Zaachila	Mesquite Forest	Upper Alluvium	1520	Clayey	x		Maize mixed
054	Zaachila	Mesquite Forest	Upper Alluvium	1505	Silty			
055	Zaachila	Mesquite Grassland B(I)	Mountain	1600	Clayey			
056	Zaachila	Mesquite Grassland B(I)	Mountain	1600	Clayey	x		Maize mixed
057	Zaachila	Mesquite Forest	Piedmont	1528	Clayey	x		Maize mixed
058A	Zaachila	Mesquite Forest	Upper Alluvium	1485	Silty	x	Pot	Other
058B	Zaachila	Mesquite Forest	Upper Alluvium	1485	Silty	x	Pot	Maize
058C	Zaachila	Mesquite Forest	Upper Alluvium	1485	Silty		Pot	
058D	Zaachila	Mesquite Forest	Upper Alluvium	1485	Silty	x	Pot	Beans
058E	Zaachila	Mesquite Forest	Upper Alluvium	1485	Silty	x	Pot	Other
058F	Zaachila	Mesquite Forest	Upper Alluvium	1485	Silty	x	Pot	Other
059	Zaachila	Mesquite Forest	Upper Alluvium	1470	Sandy	x		Maize mixed

TABLE 15.23 (*continued*)
Surface Samples: Ecological Observations

Sample number	Valley arm	Vegetation	Topographic position	Elevation (m)	Substrate	Cultivated	Irrigated	Crop type
060	Zaachila	Mesquite Forest	Upper Alluvium	1470	Sandy	x		Maize
061	Zaachila	Mesquite Forest	Upper Alluvium	1495	Sandy	x		Maize mixed
062	Zaachila	Mesquite Forest	Piedmont	1499	Sandy			
063	Zaachila	Mesquite Forest	Upper Alluvium	1492	Silty			
064	Zaachila	Mesquite Forest	Upper Alluvium	1596	Sandy	x		Maize mixed
065	Zaachila	Mesquite Forest	Piedmont	1500	Sandy			
066	Central	Mesquite Forest	Mountain	1900	Clayey			
067	Central	Mesquite Grassland B(I)	Mountain	1775	Sandy			
068	Zaachila	Mesophytic Forest B	Upper Alluvium	1503	Sandy	x		Maize mixed
069	Zaachila	Mesquite Forest	Piedmont	1520	Clayey			
070A	Zaachila	Mesquite Forest	Piedmont	1481	Sandy	x		Maize mixed
070B	Zaachila	Mesquite Grassland A(II)	Piedmont	1480	Unknown			
071	Zaachila	Pine Forest A	Mountain	1800	Clayey	x		Maize
072	Zaachila	Pine Forest A	Mountain	1800	Clayey			
073	Zaachila	Mesquite Forest	Piedmont	1590	Sandy	x		Maize
074	Zaachila	Mesquite Forest	Piedmont	1590	Sandy			
075	Zaachila	Mesquite Forest	Upper Alluvium	1420	Clayey	x		Maize
076	Zaachila	Mesquite Forest	Upper Alluvium	1460	Clayey	x		Maize
077	Zaachila	Mesquite Forest	Lower Alluvium	1495	Clayey	x		Maize
078	Central	Mesquite Forest	Piedmont	1590	Clayey	x		Maize
079	Central	Mesquite Forest	Piedmont	1590	Clayey			
080	Central	Mesquite Forest	Piedmont	1590	Sandy	x		Maize
081	Central	Mesophytic Forest B	Upper Alluvium	1548	Clayey	x		Maize mixed
082	Zaachila	Mesquite Forest	Piedmont	1501	Clayey	x		Maize mixed
083	Zaachila	Mesquite Forest	Upper Alluvium	1550	Clayey	x		Maize mixed
084	Zaachila	Mesquite Forest	Piedmont	1525	Clayey			
085	Zaachila	Mesquite Forest	Upper Alluvium	1500	Clayey	x	Canal	Maize mixed
086	Zaachila	Mesophytic Forest A	Upper Alluvium	1487	Clayey	x		Maize mixed
087	Zaachila	Mesquite Forest	Piedmont	1501	Clayey	x		Maize mixed
088	Central	Mesquite Forest	Lower Alluvium	1562	Clayey			
089	Mitla	Mesquite Forest	Mountain	1800	Sandy	x		Maize mixed
090	Mitla	Mesquite Forest	Mountain	1800	Sandy			
091	Mitla	Mesquite Forest	Mountain	1913	Sandy			
092A	Mitla	Pine Forest B(I)	Mountain	2096	Clayey	x		Maize mixed
092B	Mitla	Pine Forest B(I)	Mountain	2100	Unknown			
093	Mitla	Pine Forest B(I)	Mountain	2096	Clayey			
094	Mitla	Pine Forest B(II)	Mountain	2000	Sandy			
095	Mitla	Mesquite Forest	Piedmont	1709	Sandy			
096	Mitla	Mesquite Forest	Piedmont	1709	Sandy	x		Maize
097	Mitla	Mesquite Forest	Upper Alluvium	1660	Clayey			
098	Mitla	Mesquite Forest	Piedmont	1555	Clayey	x		Agave
099	Mitla	Mesquite Forest	Piedmont	1555	Clayey			
100	Mitla	Mesquite Forest	Piedmont	1800	Clayey			
101	Mitla	Mesquite Forest	Piedmont	1800	Clayey	x		Maize mixed
102	Mitla	Mesquite Forest	Piedmont	1710	Clayey			
103	Mitla	Mesophytic Forest B	Lower Alluvium	1646	Clayey			
104A	Mitla	Mesophytic Forest B	Lower Alluvium	1646	Clayey	x		Maize mixed
104B	Mitla	Mesquite Forest	Piedmont	1680	Clayey	x		Maize mixed
105	Mitla	Mesophytic Forest A	Piedmont	1720	Clayey			
106	Mitla	Mesophytic Forest A	Piedmont	1720	Clayey	x		Maize
107	Zaachila	Mesquite Forest	Piedmont	1515	Clayey			
108A	Hierve	Mesquite Forest	Mountain	1700	Clayey	x		Maize mixed
108B	Hierve	Mesquite Forest	Mountain	1700	Clayey	x		Maize mixed
108C	Hierve	Mesquite Forest	Mountain	1700	Clayey			
108D	Hierve	Thorn Forest A	Mountain	1700	Unknown			
108E	Hierve	Thorn Forest A	Mountain	1700	Unknown			
108F	Hierve	Thorn Forest A	Mountain	1700	Unknown			
108G	Hierve	Thorn Forest B	Mountain	1700	Unknown			
109	Zaachila	Mesquite Forest	Upper Alluvium	1507	Sandy	x		Maize
110	Zaachila	Mesquite Forest	Upper Alluvium	1510	Sandy	x		Maize
111	Zaachila	Mesophytic Forest B	Lower Alluvium	1500	Sandy			
112	Zaachila	Mesquite Forest	Upper Alluvium	1527	Clayey			
113	Zaachila	Mesquite Forest	Upper Alluvium	1527	Clayey	x		Maize

TABLE 15.23 (*continued*)
Surface Samples: Ecological Observations

Sample number	Valley arm	Vegetation	Topographic position	Elevation (m)	Substrate	Cultivated	Irrigated	Crop type
114	Mitla	Mesophytic Forest B	Upper Alluvium	1577	Silty	x		Maize
115	Mitla	Mesquite Forest	Piedmont	1630	Sandy			
116	Mitla	Mesquite Forest	Piedmont	1604	Sandy			
117	Mitla	Mesquite Forest	Upper Alluvium	1645	Clayey	x	Canal	Maize mixed
118	Mitla	Mesquite Forest	Piedmont	1700	Clayey			
119	Mitla	Mesquite Forest	Piedmont	1700	Clayey	x		Maize
120	Mitla	Mesquite Forest	Piedmont	1720	Clayey			
121	Mitla	Mesquite Forest	Piedmont	1720	Clayey	x		Agave
122	Mitla	Mesquite Forest	Piedmont	1620	Clayey	x		Agave
123	Mitla	Mesquite Forest	Piedmont	1620	Clayey			Maize
124	Zaachila	Mesquite Forest	Piedmont	1527	Sandy	x		Maize
125	Zaachila	Mesquite Forest	Piedmont	1527	Sandy			
126	Zaachila	Mesquite Forest	Piedmont	1545	Sandy	x		Maize mixed
127	Zaachila	Mesquite Forest	Piedmont	1575	Sandy	x		Maize mixed
128	Zaachila	Mesquite Forest	Piedmont	1575	Sandy			
129	Zaachila	Mesquite Grassland B(I)	Piedmont	1550	Clayey	x		Maize
130	Zaachila	Mesquite Grassland B(I)	Piedmont	1550	Clayey			
131	Zaachila	Mesquite Forest	Piedmont	1560	Sandy	x		Maize
132	Mitla	Mesquite Forest	Piedmont	1773	Silty			
133	Mitla	Mesquite Forest	Piedmont	1773	Silty			
134	Mitla	Mesquite Forest	Piedmont	1595	Clayey			
135	Mitla	Mesquite Forest	Piedmont	1595	Clayey	x		
136	Mitla	Mesquite Grassland B(I)	Piedmont	1610	Clayey	x		Maize mixed
137	Mitla	Mesquite Grassland B(I)	Piedmont	1610	Clayey			
138	Mitla	Mesquite Forest	Upper Alluvium	1632	Clayey	x		Maize
139	Mitla	Mesquite Forest	Piedmont	1624	Clayey			
141	Mitla	Domestic House Floor	Upper Alluvium	1650	Clayey			
142	Mitla	Mesquite Forest	Piedmont	1600	Sandy	x		Maize
143	Mitla	Mesquite Forest	Upper Alluvium	1600	Clayey	x		Maize
144	Mitla	Mesquite Forest	Lower Alluvium	1600	Clayey			
145	Mitla	Domestic House Floor	Piedmont	1700	Clayey			
146	Mitla	Domestic House Floor	Piedmont	1700	Clayey			
147	Mitla	Domestic House Floor	Piedmont	1700	Clayey			
148	Mitla	Pine Forest A	Mountain	3000	Unknown			
149	Mitla	Pine Forest A	Mountain	2750	Unknown			
151	Mitla	Pine Forest B(II)	Mountain	2400	Unknown			
152	Mitla	Oak Forest	Mountain	2000	Unknown			
153	Mitla	Oak Forest	Mountain	1900	Unknown			
157	Hierve	Thorn Forest B	Mountain	1700	Unknown			
158	Mitla	Thorn Forest B	Mountain	1750	Unknown			
159	Mitla	Thorn Forest B	Mountain	1775	Unknown			
160	Mitla	Thorn Forest B	Mountain	1800	Unknown			
162	Mitla	Mesquite Grassland A(I)	Piedmont	1680	Unknown			
163	Mitla	Mesquite Grassland B(I)	Mountain	1795	Unknown			
165	Mitla	Mesquite Grassland B(II)	Piedmont	1450	Unknown			
166	Mitla	Mesquite Grassland B(II)	Upper Alluvium	1650	Unknown			
167	Mitla	Thorn Forest B	Mountain	1500	Unknown			
168	Mitla	Oak Forest	Mountain	2200	Unknown			
169	Etla	Mesquite Forest	Upper Alluvium	1600	Unknown			
170	Zaachila	Mesquite Forest	Upper Alluvium	1500	Unknown			
171	Mitla	Thorn Forest B	Upper Alluvium	1660	Unknown	x		Agave
172	Mitla	Mesophytic Forest B	Lower Alluvium	1640	Unknown			
173	Mitla	Mesquite Grassland B(II)	Upper Alluvium	1650	Unknown			
174	Etla	Mesquite Grassland B(II)	Lower Alluvium	1675	Unknown			

TABLE 15.24
Surface Samples: Palynological Observations

	Picea	*Pinus*	*Quercus*	*Alnus*	*Betula*	*Celtis*–Moraceae	**Leguminosae**	*Prosopis*	*Acacia*	Rosaceae	Type IV	Type VIII	*Dodonaea*	*Agave*	Cactacaea	Liliaceae	Compositae	*Artemisia*	Gramineae	Chenopodiaceae	*Polygonum*	*Salvia*	Zygophyllaceae	Nyctaginaceae	*Zea*	*Cucurbita*	Others	Unknowns	Total
001		16	3	1		0	0	0	0	0	0	0	0	0	0	0	190		59	29	0	0	0	0	2	0	0	1	301
002		64	6	5		0	3	0	0	1	1	0	0	0	0	0	148		57	13	0	0	0	0	1	0	0	1	300
003		39	5	6		1	2	0	0	0	0	0	0	0	0	0	125		103	7	0	0	0	0	1	0	0	2	291
004		77	1	4		0	0	0	0	0	0	0	0	0	0	0	132		62	13	0	0	0	0	58	0	0	1	348
005		152	17	0		0	3	0	0	0	0	0	0	0	0	0	86		42	0	0	0	0	0	0	0	0	0	300
006		147	37	5		3	3	0	0	0	3	0	0	1	0	0	42		50	3	0	0	0	0	0	0	0	6	300
007		80	47	4		0	0	0	0	0	0	0	0	0	0	0	81		82	3	0	0	0	0	9	0	1	2	309
008		148	66	3		1	1	0	0	0	1	0	0	0	0	0	27		53	0	0	0	0	0	2	0	0	2	304
009		75	56	1		0	1	0	0	0	0	0	0	0	0	0	82		83	0	0	0	0	0	19	0	0	2	319
010		67	48	4		0	10	0	0	0	0	0	0	0	0	0	40		124	2	0	0	0	0	0	0	0	5	300
011		55	32	8		1	2	0	0	0	0	0	0	0	0	0	134		60	6	0	0	0	0	10	0	0	2	310
012		123	37	10		1	0	0	0	0	2	0	0	0	0	0	33		86	3	0	0	0	0	2	0	0	5	302
013		34	2	3		0	8	0	0	0	0	0	0	0	0	0	207		38	7	0	0	0	0	13	0	0	1	313
014		43	1	5		0	0	0	0	2	0	0	0	0	0	0	180		57	6	0	0	0	0	3	0	0	6	303
015		23	0	6		1	0	0	1	0	0	0	0	0	0	0	200		28	33	0	0	0	0	5	0	0	8	305
016		26	6	3		0	0	0	0	0	0	0	0	0	0	0	216		39	6	0	0	0	0	0	0	0	2	298
017		11	0	3		0	0	0	0	0	0	0	0	0	0	0	223		59	3	0	0	0	0	0	0	0	1	300
019		8	0	11		0	1	0	0	0	0	0	0	0	0	0	190		44	47	0	0	0	0	0	0	0	0	301
020		55	4	9		0	3	2	0	0	1	0	0	0	0	0	128		83	5	1	0	0	0	0	0	0	4	295
021		8	0	1		1	1	0	0	0	0	0	0	0	0	0	190		71	27	0	0	0	0	0	0	0	2	301
022		24	0	10		0	2	0	0	0	0	0	0	0	0	0	187		61	5	0	7	0	0	6	0	0	3	305
023		14	0	1		0	2	0	0	0	0	0	0	0	0	0	213		48	19	0	0	0	0	15	0	1	2	315
024		8	0	6		0	0	0	0	0	1	0	0	0	0	0	221		44	16	0	0	0	0	5	0	1	3	305
025		20	0	6		1	1	2	0	0	0	0	0	0	0	0	213		45	4	0	0	0	0	2	0	3	4	301
026		11	2	78		0	2	5	0	0	0	0	0	0	0	0	14		12	0	0	0	0	0	0	0	25	1	150
027		16	2	4		0	6	0	1	0	0	0	0	0	0	0	187		66	11	0	0	0	0	12	0	0	4	309
028		15	5	1		1	2	2	0	0	0	0	0	0	0	0	212		25	33	0	0	0	0	0	0	1	3	300
029		57	4	1		0	3	1	1	0	1	0	0	0	0	1	162		31	13	0	0	0	0	30	0	0	5	310
030		20	1	1		0	1	0	0	0	0	1	0	0	0	0	155		39	80	0	0	0	0	0	0	0	2	300
031		14	1	7		1	1	0	1	0	0	0	0	1	0	0	127		78	66	0	0	0	0	3	0	0	3	303
032		75	1	3		0	2	0	21	0	0	0	0	1	0	0	99		62	32	0	0	0	0	3	0	1	3	303
033		55	2	2		0	1	0	0	1	0	0	0	0	0	0	145		57	32	0	0	0	0	0	0	0	3	298
034		16	0	3		0	4	0	1	0	0	0	0	1	0	0	127		109	33	0	0	0	0	2	1	0	6	303
035		15	0	2		0	1	0	0	0	4	0	0	0	0	0	103		72	99	0	0	0	0	10	0	0	4	310
036		6	0	2		0	0	0	1	0	0	0	0	0	0	0	240		36	14	0	0	0	0	1	0	0	3	303
037		40	0	6		0	2	0	0	0	0	0	0	0	0	0	186		44	22	0	0	0	0	3	0	0	0	303
038		53	0	7		0	2	0	1	0	1	0	0	0	0	0	118		91	24	0	0	0	0	2	1	0	3	303
039		5	0	4		0	0	0	0	0	0	0	0	0	0	0	141		139	9	0	0	0	0	1	0	0	2	301
040		151	19	24		1	11	0	0	0	0	0	0	0	0	0	59		26	4	0	0	0	0	0	0	0	4	299
041		122	28	10		1	0	0	0	0	0	0	0	1	0	0	40		72	14	0	0	0	0	2	0	0	11	301
042		206	13	18		4	4	0	0	0	0	0	0	0	0	0	35		17	1	0	0	0	0	0	0	0	2	300
043		55	41	10		0	0	0	1	0	0	0	0	0	0	0	121		70	1	0	0	0	0	0	0	0	1	300
044		31	0	1		0	1	0	0	0	0	0	0	0	0	0	129		63	68	0	0	0	0	0	0	0	6	299
045		17	1	1		0	2	0	1	0	0	0	0	1	0	0	180		55	33	0	0	0	0	1	0	0	9	301

TABLE 15.24 (*continued*)
Surface Samples: Palynological Observations

	Picea	*Pinus*	*Quercus*	*Alnus*	*Betula*	*Celtis*–Moraceae	Leguminosae	*Prosopis*	*Acacia*	Rosaceae	Type IV	Type VIII	*Dodonaea*	*Agave*	Cactacaea	Liliaceae	Compositae	*Artemisia*	Gramineae	Chenopodiaceae	*Polygonum*	*Salvia*	Zygophyllaceae	Nyctaginaceae	*Zea*	*Cucurbita*	Others	Unknowns	Total
046		35	0	0		0	4	0	0	0	0	0	0	0	0	0	131		51	72	0	0	0	0	0	0	0	2	295
047		47	5	7		0	0	0	0	0	0	0	0	0	0	0	141		64	32	0	0	0	0	11	0	0	4	311
048		54	8	5		0	0	0	0	0	0	0	0	0	0	0	131		78	19	0	0	0	1	3	0	0	3	302
049		154	29	11		0	2	0	0	0	0	0	0	0	0	0	22		71	7	0	0	0	0	0	0	0	4	300
050		183	33	5		0	1	0	0	0	0	0	0	1	0	0	29		34	7	0	0	0	0	0	0	0	7	300
051		93	69	11		1	2	0	0	0	0	0	0	0	0	0	41		74	6	0	0	0	0	0	0	0	3	300
052		20	1	2		0	1	0	0	0	0	1	0	0	3	0	203		36	29	0	0	0	0	0	1	0	4	301
053		12	0	2		0	1	0	0	0	0	0	0	0	1	0	199		63	11	0	0	0	1	4	0	0	5	299
054		12	0	4		0	0	0	2	0	0	0	0	2	0	0	130		84	47	0	0	0	0	0	0	0	1	282
055		7	0	2		0	1	0	2	0	3	0	0	0	0	0	109		101	5	0	0	0	0	0	0	0	5	301
056		29	0	1		0	3	0	0	0	3	0	0	0	0	0	147		114	1	0	0	0	0	2	0	0	1	301
057		75	0	0		0	0	0	0	0	5	0	0	0	0	0	220		56	3	0	0	0	0	3	0	0	4	303
058A		26	0	0		0	1	0	0	0	0	0	0	0	0	0	142		72	63	0	0	0	0	0	0	0	4	308
058B		9	0	1		0	2	0	0	0	0	0	0	0	0	0	199		66	26	0	0	0	0	2	0	0	7	312
058C		13	0	2		0	1	0	1	0	0	0	0	0	0	0	125		86	64	0	0	0	0	0	0	0	7	299
058D		10	0	2		0	0	0	1	0	0	0	0	0	0	0	155		86	43	0	0	0	0	0	0	0	2	299
058E		5	0	1		0	0	0	4	0	0	0	0	0	0	0	170		88	28	0	0	0	0	0	0	0	4	300
058F		7	0	1		0	0	0	1	0	0	0	0	0	0	0	183		82	23	0	0	0	0	0	0	0	2	299
059		9	0	4		0	4	0	5	0	0	0	4	0	0	0	187		75	5	0	0	0	0	5	1	0	4	303
060		6	0	3		1	0	0	1	0	0	0	0	0	0	0	208		72	4	0	0	0	0	2	0	0	4	301
061		36	1	2		0	0	3	0	0	1	0	1	0	0	0	146		94	16	0	0	0	0	0	0	0	5	305
062		20	2	1		2	0	2	1	0	1	0	1	0	0	0	130		93	40	0	0	0	0	0	0	3	4	300
063		1	0	2		2	0	0	0	1	2	0	0	1	0	0	123		148	12	1	0	0	0	5	0	1	5	304
064		16	0	7		0	2	0	0	0	0	0	0	0	0	0	177		85	6	0	0	0	0	4	0	0	7	304
065		8	0	10		1	0	3	2	0	0	0	0	0	0	0	164		107	2	0	0	0	0	0	0	0	3	300
066		10	10	1		0	0	0	0	0	0	0	0	1	0	0	157		107	6	0	0	0	0	0	0	0	0	292
067		93	12	4		3	4	0	0	0	0	0	0	0	0	0	74		105	4	0	0	0	0	0	1	0	1	301
068		6	0	3		0	4	0	0	0	0	0	0	0	0	1	184		86	12	0	0	0	0	1	0	0	4	301
069		54	6	3		0	0	0	4	0	0	0	0	0	0	0	96		106	24	0	0	0	0	6	0	0	7	306
070A		23	0	0		0	0	0	4	0	0	0	0	1	0	3	199		39	24	0	0	0	0	2	0	0	7	302
070B		18	10	2		1	0	1	0	0	2	0	0	0	0	1	17		37	9	0	0	0	0	4	0	0	0	102
071		46	4	6		0	0	0	1	1	0	0	0	0	0	1	173		48	10	0	0	0	0	22	0	0	10	322
072		154	10	1		0	1	0	0	1	0	0	0	0	0	0	84		34	6	0	0	0	0	0	0	0	9	300
073		39	15	2		0	1	1	2	0	0	0	0	0	0	0	126		94	7	1	0	1	0	9	0	0	11	309
074		118	22	6		0	1	0	0	0	0	0	0	0	0	0	56		88	6	0	0	0	0	0	0	0	3	300
075		26	4	3		0	0	0	3	0	0	0	0	0	0	0	148		92	18	0	0	0	0	20	0	0	6	320
076		22	4	0		0	0	0	0	0	0	0	0	0	0	0	187		70	12	0	0	0	0	16	0	0	5	316
077		10	0	2		0	4	0	2	0	0	0	0	0	0	1	181		78	16	0	0	0	0	26	0	0	6	326
078		2	0	2		1	0	0	0	0	0	0	0	0	0	0	171		116	4	1	0	0	0	10	0	0	3	310
079		32	2	6		2	0	0	0	0	0	0	0	0	0	0	189		62	4	0	0	0	0	0	0	0	3	300
080		28	0	4		0	0	1	4	0	0	0	0	0	0	0	184		66	6	0	0	0	0	8	0	0	7	308
081		9	0	0		0	0	0	0	0	0	0	0	0	0	0	194		59	30	0	0	0	1	9	0	0	7	309
082		22	0	0		0	0	0	0	0	0	0	0	0	0	0	226		48	2	0	0	0	0	10	0	0	2	310
083		6	0	4		0	8	2	6	0	0	0	0	0	0	0	198		42	30	0	0	0	0	6	1	0	6	309

TABLE 15.24 (*continued*)
Surface Samples: Palynological Observations

	Picea	*Pinus*	*Quercus*	*Alnus*	*Betula*	*Celtis*–Moraceae	Leguminosae	*Prosopis*	*Acacia*	Rosaceae	Type IV	Type VIII	*Dodonaea*	*Agave*	Cactacaea	Liliaceae	Compositae	*Artemisia*	Gramineae	Chenopodiaceae	*Polygonum*	*Salvia*	Zygophyllaceae	Nyctaginaceae	*Zea*	*Cucurbita*	Others	Unknowns	Total
084		12	0	2		0	1	0	0	0	2	0	0	0	0	0	238		43	2	0	0	0	0	3	0	0	0	303
085		16	0	2		0	0	0	1	0	1	0	0	0	0	0	153		20	103	0	0	0	0	2	0	0	2	300
086		21	0	3		0	2	0	0	0	1	0	0	0	0	0	195		45	33	0	0	0	0	3	0	0	0	303
087		14	0	2		0	0	0	0	0	1	0	0	0	0	0	168		32	82	0	0	0	0	0	0	0	1	300
088		10	0	0		1	0	0	0	0	0	0	0	0	0	0	174		90	20	0	0	0	0	1	0	0	5	301
089		10	8	0		0	0	0	1	0	2	0	0	0	0	0	170		68	34	0	0	0	0	4	0	0	7	304
090		10	24	10		0	4	0	0	0	0	0	0	0	0	0	162		72	8	0	0	0	0	0	0	0	6	296
091		54	30	1	1	1	0	0	0	0	0	0	0	2	0	0	118		70	2	0	0	0	0	0	0	1	7	295
092A		94	30	12	1	3	4	0	0	0	2	0	0	0	0	0	118		28	8	0	0	0	0	0	0	0	0	300
092B	1	137	31	2	1	3	0	0	0	0	2	0	0	0	0	0	10		12	1	0	1	0	0	0	0	0	0	200
093		120	42	9		0	0	0	0	0	0	0	0	0	0	0	81		45	0	0	0	0	0	0	0	0	3	300
094		46	60	4		0	2	0	1	0	0	0	0	0	0	0	86		84	8	0	0	0	0	0	0	0	9	300
095		7	3	2		0	3	1	0	0	0	0	0	2	0	0	107		154	20	0	0	0	0	0	0	0	1	300
096		34	8	5		1	0	0	0	0	1	0	0	0	0	0	140		88	18	0	0	0	0	2	0	0	5	302
097		55	0	2		0	0	6	0	0	8	0	0	0	0	0	120		95	8	0	0	1	0	0	0	0	5	300
098		44	16	12		0	0	8	0	0	0	0	0	0	0	0	64		144	8	0	0	0	0	0	0	0	4	300
099		47	10	2		2	2	0	1	0	2	0	0	1	0	0	83		138	6	0	0	0	0	0	0	0	6	300
100		18	12	7		0	0	4	0	0	1	0	0	0	0	0	99		150	8	0	0	0	0	1	0	0	1	301
101		21	8	0		0	0	0	1	0	0	0	0	0	0	0	123		135	6	0	0	0	1	9	0	0	5	309
102		62	8	7		0	0	10	0	0	2	0	0	0	0	0	90		114	2	0	0	0	0	0	1	0	4	300
103		45	0	0		0	0	0	0	0	0	0	0	0	0	0	126		120	6	0	0	0	0	0	0	0	3	300
104A		28	2	0		0	2	2	0	0	1	0	0	0	1	0	144		90	22	2	0	0	0	0	0	0	6	300
104B		20	0	1		0	0	2	1	1	0	0	0	0	0	0	183		66	24	0	0	0	0	5	0	0	2	305
105		40	10	2		0	0	0	0	0	0	0	0	0	0	0	146		98	0	0	0	0	0	0	0	0	4	300
106		70	18	8		0	6	4	0	0	0	0	0	0	0	0	122		66	4	0	0	0	0	20	0	0	2	320
107		64	2	0		0	4	0	0	0	0	0	0	0	0	0	132		98	0	0	0	0	0	0	0	0	2	302
108A		3	0	0		0	0	3	0	0	6	0	0	0	0	0	191		81	12	0	0	0	0	7	0	0	3	306
108B		14	4	0		0	0	0	0	0	6	0	0	0	0	0	176		92	4	0	0	0	0	0	0	0	4	300
108C		23	0	5		0	2	0	4	1	3	0	0	0	0	0	181		70	4	0	0	0	1	0	1	0	6	301
108D		31	7	1		2	1	2	0	0	10	0	0	0	0	0	28	2	7	2	0	0	0	0	0	0	0	7	100
108E	1	69	33	8	1	3	2	0	0	0	14	3	0	0	0	0	115		11	2	0	2	0	0	1	0	9	12	285
108F		41	22	9		6	0	0	0	0	8	4	0	0	0	0	125		59	6	0	2	0	0	0	0	0	20	302
108G	1	46	29	5		16	7	0	0	4	53	3	0	1	0	0	180		23	13	1	4	0	0	1	0	5	11	402
109		40	0	0		0	2	0	6	0	0	0	0	1	0	0	171		44	38	0	0	0	0	4	0	0	0	306
110		28	0	2		0	1	0	0	0	1	1	0	2	2	0	158		58	44	0	0	0	0	1	0	0	4	302
111		50	0	6		0	0	0	0	0	0	0	0	0	0	0	132		94	12	4	0	0	0	2	0	0	2	302
112		45	1	2		0	1	0	0	0	0	0	0	0	0	0	133		102	15	0	0	0	0	0	0	0	1	300
113		6	0	2		0	2	0	4	0	0	0	0	0	0	0	178		90	14	0	0	0	0	0	0	0	4	300
114		21	0	0		0	5	0	3	0	2	0	0	0	0	0	135		114	15	0	0	0	0	0	0	0	5	300
115		26	8	4		0	0	0	0	0	10	0	0	0	0	0	86		160	4	0	0	0	0	1	0	0	2	301
116		70	3	3		0	0	2	0	0	1	0	0	0	0	0	62		156	1	0	0	0	0	2	0	0	2	302
117		37	2	1		1	0	0	2	0	0	0	0	0	0	0	158		83	14	0	0	0	0	7	0	0	2	307
118		104	8	0		0	0	0	0	0	2	0	0	0	0	0	89		88	6	0	0	0	0	0	0	0	3	300
119		8	2	0		0	0	0	0	1	6	0	0	1	0	0	112		128	2	0	0	0	1	0	0	39	0	300

TABLE 15.24 (*continued*)
Surface Samples: Palynological Observations

	Picea	*Pinus*	*Quercus*	*Alnus*	*Betula*	*Celtis*–Moraceae	Leguminosae	*Prosopis*	*Acacia*	Rosaceae	Type IV	Type VIII	*Dodonaea*	*Agave*	Cactacaea	Liliaceae	Compositae	*Artemisia*	Gramineae	Chenopodiaceae	*Polygonum*	*Salvia*	Zygophyllaceae	Nyctaginaceae	*Zea*	*Cucurbita*	Others	Unknowns	Total
120		98	34	14	1	0	6	0	0	0	4	0	0	0	0	0	40		102	0	0	0	0	0	0	0	0	1	300
121		94	26	4		0	8	0	0	0	4	0	0	0	0	0	76		82	8	0	0	0	0	15	0	0	4	321
122		36	2	2		0	0	0	2	0	4	0	0	0	0	0	104		142	6	0	0	0	0	1	0	0	2	301
123		38	2	8		0	0	0	0	0	2	0	0	0	0	0	95		154	0	0	0	0	0	8	0	0	1	308
124		24	14	0		2	4	2	0	0	6	0	0	0	0	0	146		92	4	2	0	0	0	1	0	0	4	301
125		18	0	0		0	0	0	0	0	0	0	0	0	0	0	222		52	2	1	0	0	1	0	0	0	4	300
126		20	0	1		0	0	0	0	0	0	0	0	0	0	0	188		76	12	0	0	0	1	48	0	0	2	348
127		28	0	1		0	1	0	0	0	0	0	0	0	0	0	188		78	2	0	0	0	0	16	0	0	2	316
128		30	4	0		0	4	0	0	0	0	0	0	0	0	0	86		170	0	0	0	0	2	0	0	0	4	300
129		26	2	4		0	0	0	0	0	1	0	0	0	0	0	151		111	0	0	0	0	0	10	0	0	4	309
130		26	2	6		2	0	0	2	0	0	0	0	0	0	0	136		122	2	0	0	0	0	0	0	0	2	300
131		6	2	0		0	2	0	2	0	2	0	0	0	0	0	198		64	22	0	0	0	0	12	0	0	2	312
132		86	42	56		0	4	0	2	0	0	0	0	0	0	0	41		60	4	0	0	0	1	0	0	0	4	300
133		82	26	10		0	1	0	1	1	0	0	0	0	0	0	94		70	12	0	0	0	1	3	0	0	2	303
134		18	0	0		0	0	0	2	1	2	0	0	0	0	0	105		164	6	0	0	0	0	0	0	0	2	300
135		6	0	2		0	0	0	2	0	0	0	0	0	0	0	108		171	6	0	0	0	0	3	0	0	5	303
136		24	12	8		0	0	5	0	0	6	0	0	4	0	0	96		136	0	3	0	0	1	9	1	0	4	309
137		11	10	4		0	0	0	0	0	6	0	0	2	0	0	115		148	2	0	0	0	0	0	0	0	2	300
138		6	0	2		0	1	0	4	0	0	0	0	0	0	0	145		132	6	0	0	0	0	16	0	0	4	316
139		12	0	0		0	0	4	0	0	4	0	0	0	0	0	173		94	8	1	0	0	0	0	0	0	4	300
141		45	4	5	1	8	20	28	2	3	0	0	0	0	0	0	124		24	32	1	0	2	0	6	1	1	10	317
142		26	0	0		0	2	0	1	0	10	0	0	0	0	0	150		97	10	0	0	0	2	6	0	0	2	306
143		22	0	2		0	0	0	0	0	0	0	0	0	0	0	122		142	10	0	0	0	0	14	0	0	2	314
144		14	0	0		0	0	0	0	0	0	0	0	0	0	0	123		153	8	0	0	0	0	3	1	0	2	304
145		15	5	6		0	3	7	2	0	1	0	0	0	0	0	136		66	53	0	0	0	0	27	0	0	6	327
146		21	7	6		4	6	10	2	0	7	0	0	0	0	1	138		58	40	0	0	0	0	30	0	0	0	330
147		15	6	8		2	9	6	0	0	2	0	0	0	0	0	135		65	44	0	0	1	0	18	0	0	7	318
148		121	51	6		0	0	0	0	0	1	0	0	0	0	1	11		3	0	0	3	0	0	0	0	0	3	200
149		160	22	6		1	0	0	0	0	1	0	0	0	0	0	5		3	1	0	0	0	0	0	0	1	0	200
151		73	64	27		4	5	0	0	0	0	0	0	0	0	0	15		12	0	0	4	0	0	0	0	0	3	207
152		114	36	6	1	4	2	0	0	0	4	0	0	0	0	0	15		12	0	0	4	0	0	0	0	0	3	201
153	1	108	81	16	2	5	10	0	0	0	6	0	0	0	0	0	48	1	14	1	0	6	0	0	0	0	1	4	303
157		39	8	2		1	0	0	0	0	7	0	0	0	0	0	30		5	3	0	3	0	0	0	0	0	2	100
158	1	76	23	6	1	3	2	2	1	1	40	0	0	1	0	0	64		44	12	2	8	0	1	0	0	7	8	302
159	1	84	23	12	1	0	4	1	0	0	59	5	0	0	0	1	62		31	8	0	2	0	1	0	0	1	5	300
160	1	66	25	18	3	6	4	0	0	0	41	5	0	0	1	3	75		26	11	1	5	0	0	0	0	3	9	302
162		16	9	1	1	2	1	0	0	0	3	0	0	0	0	0	22		41	3	0	0	0	0	0	0	0	1	100
163		15	7	1	1	0	5	0	0	0	0	0	0	0	0	0	24		42	4	0	1	0	0	0	0	0	0	100
165		4	1	0		0	6	0	0	0	4	0	0	0	0	0	18		45	7	0	2	0	0	0	0	10	3	100
166		11	5	1		2	2	0	0	0	0	0	0	0	0	0	38		52	12	0	1	0	0	0	0	0	3	127
167		26	10	8	2	1	1	0	0	4	0	1	0	0	0	0	18		25	4	0	0	0	0	0	0	0	1	101
168		35	39	5		0	0	0	0	0	1	0	0	0	0	0	7		8	2	1	0	0	0	0	0	0	1	99
169		12	9	1		0	2	0	0	1	0	0	0	0	0	0	59		10	0	0	0	0	0	0	0	2	3	99
170		21	11	3		0	1	0	0	5	0	0	0	0	0	0	28		22	4	0	4	0	0	3	0	0	2	104
171		9	2	1		3	5	0	0	0	2	0	0	1	0	0	57		19	0	0	0	0	0	2	0	0	1	102
172		18	2	0		0	1	0	0	0	3	0	0	0	0	0	21		46	4	0	3	0	0	0	0	1	1	100
173		38	11	1		0	0	2	0	0	2	0	0	0	0	0	13		30	2	0	0	1	0	2	0	0	0	102
174		9	4	0		1	0	8	0	0	9	0	0	0	0	0	28		35	5	0	0	0	0	0	0	0	1	100

Appendix 15.2: Fossil Sample Observations

TABLE 15.25
Pollen Observed, Martínez Rockshelter Samples[a]

Pollen type	Mixed Zone A	Zone A		Zone B			Zone B–C	Zone C				Weathered ignimbrite	
	5 cm	10 cm	15 cm	20 cm	25 cm	30 cm	35 cm	40 cm	45 cm	50 cm	55 cm	60 cm	65 cm
Ulmus	—	—	—	—	—	—	—	—	—	—	—	—	—
Picea	—	—	—	—	—	—	—	—	—	—	—	—	—
Abies	—	—	—	—	—	—	—	—	—	—	—	—	—
Pinus	12	32	33	21	24	20	7	13	9	13	9	4	8
Quercus	29	13	11	15	20	25	17	18	16	9	16	6	14
Alnus	—	2	2	—	2	1	1	—	3	—	3	—	—
Betula	—	—	—	—	—	—	—	—	—	—	—	—	—
Celtis–Moraceae	—	—	1	—	1	1	3	1	—	—	—	—	—
Leguminosae	—	—	—	—	1	1	6	—	—	1	3	3	—
Prosopis	—	—	—	—	—	—	—	—	—	—	—	—	—
Rosaceae	—	—	—	—	—	—	—	—	—	—	—	—	1
Type IV	8	2	5	2	6	6	6	7	3	3	5	—	4
Type VIII	—	—	—	—	—	—	—	—	—	—	—	—	—
Dodonaea	—	—	—	—	—	—	—	—	—	—	—	—	—
Agave	—	—	1	—	—	1	—	—	—	—	1	—	—
Tricolpate Cactaceae	3	—	—	—	2	1	1	3	—	—	1	1	1
Platyopuntia	1	1	—	—	—	—	—	1	—	—	—	—	—
Liliaceae	1	1	1	—	—	—	—	1	1	—	1	—	—
Ericaceae	—	—	—	—	—	—	—	—	—	—	—	1	—
Ribes	—	—	—	—	—	—	—	—	—	—	—	—	—
Compositae	20	22	23	31	22	19	19	27	40	41	23	9	10
Gramineae	14	17	15	18	15	20	27	14	16	20	22	12	13
Chenopodiaceae	6	4	5	8	5	3	9	11	17	8	7	8	5
cf. *Salvia*	—	—	—	—	—	—	—	—	—	—	—	—	—
Nyctaginaceae	—	2	1	—	—	—	—	—	—	1	—	—	—
Onagraceae	—	—	—	—	—	—	—	—	—	—	—	—	—
cf. *Jatropha*	—	—	—	—	—	—	—	—	—	—	1	—	—
Polygalaceae	—	—	—	—	—	—	—	—	—	—	—	—	—
cf. *Sphaeralcea*	—	—	—	1	—	—	—	—	—	—	—	—	
cf. *Sidalcea*	—	—	—	1	—	—	—	—	—	—	—	—	—
cf. *Zea mays*	—	—	—	—	—	—	—	—	—	1	—	—	—
cf. *Zea mexicana*	—	—	—	—	—	—	—	—	1	—	—	—	—
cf. *Tripsacum*	—	—	—	—	—	—	—	—	—	—	—	—	—
Cucurbita	—	—	—	—	—	—	—	—	—	—	—	—	—
Unknowns	7	4	2	6	2	2	4	4	3	4	8	1	3
Total	101	100	100	103	100	100	100	100	109	101	100	46	59
Verrucated spore									P				

[a]P, present.

TABLE 15.26
Pollen Observed, Guilá Naquitz Samples[a]

Pollen type	Zone B1			Zone B2			Zone B3		Zone C		Zone E	
	Square E9 proc. 1967	Square E9 proc. 1977	Square E6 proc. 1967	Square E6 1967 count	Square E6 1977 new count	Square D8 proc. 1967	Square D8 proc. 1977	(?) proc. 1977	Square E10 proc. 1967	Square E6 proc. 1977	Square G8 proc. 1967	Square G8 proc. 1977
Ulmus	—	—	—	P	—	1	—	1	P	—	—	—
Picea	—	—	—	—	—	—	—	—	—	—	—	—
Abies	—	—	—	—	—	—	1	—	P	—	—	—
Pinus	46	54	33	31	34	49	54	29	39	20	48	44
Quercus	10	16	37	4	7	8	20	26	5	8	6	23
Alnus	2	6	—	9	9	4	6	5	10	4	8	2
Betula	1	—	—	—	—	—	—	1	1	—	1	1
Celtis–Moraceae	5	1	2	—	—	—	1	2	2	2	—	3
Leguminosae	—	—	5	2	—	2	—	—	4	—	14	—
Prosopis	—	—	—	—	1	—	—	—	—	—	—	—
Rosaceae	—	—	—	—	—	—	—	—	—	—	—	—
Type IV	5	3	1	26	—	2	2	—	7	—	5	1
Type VIII	—	—	—	—	—	—	—	—	3	—	—	—
Dodonaea	—	1	1	—	—	—	1	—	—	—	—	1
Agave	1	3	1	—	—	1	2	2	1	—	1	1
Tricolpate Cactaceae	—	1	1	—	—	3	—	3	P	—	—	—
Platyopuntia	—	—	—	—	—	—	—		—	—	—	—
Liliaceae	—	—	—	—	—	—	—	—	2	—	—	—
Ericaceae	1	—	—	—	—	—	—	—	—	—	—	—
Ribes	1	—	1	—	—	—	1	—	P	—	—	—
Compositae	14	6	8	5	2	4	2	4	12	2	5	7
Gramineae	11	8	5	20	30	9	8	27	13	2	11	17
Chenopodiaceae	—	1	—	—	2	3	3	1	—	—	—	1
cf. *Salvia*	—	—	1	—	—	—	—	1	—	—	—	—
Nyctaginaceae	—	—	1	—	—	—	—	—	—		—	—
Caprifoliaceae	—	1	—	—	—	—	—	—	—	—	—	—
Onagraceae	—	—	—	—	—	—	—	—	—	—	—	—
cf. *Jatropha*	—	—	—	—	—	—	—	—	—	—		—
Polygalaceae	—	—	—	—	—	—	—	—	—	—	—	—
cf. *Sphaeralcea*	—	—	—		—	—	—	—	1	—	—	
cf. *Sidalcea*	—	—	—	—	—	—	—	—	—	—	—	—
cf. *Zea mays*	1	—	—	—	1	—	—	—	—	—	—	—
cf. *Zea mexicana*	—	—	—	2	4	1	—	—	P	—	—	—
cf. *Tripsacum*	3	1	—	P	9	13	—	—	2	—	—	—
Cucurbita	—	—	—	—	—	1	—	—	—	—	—	—
Unknowns	2	2	5	1	2	1	2	2	—	1	2	—
Total	103	104	102	100	101	102	103	104	102	39	101	101
Verrucated spore		P	P		P							P

[a]P, present; proc., sample processed.

TABLE 15.27
Pollen Observed, Cueva Blanca Samples[a]

Pollen type	Zone C		Zone D			Zone E	Zone F		
	Square G14 45 cm proc. 1977	Square G14 55 cm proc. 1967	Square G14 60 cm proc. 1967	Square G14 75 cm proc. 1967	Square G14 85 cm 1968	Square E5 100–110 cm 1968	Square G14 95 cm 1968	Square G14 100 cm 1968	Square D10 110 cm 1968
Ulmus	—	—	—	—	—	—	—	—	—
Picea	—	—	—	—	—	—	—	P	—
Abies	—	—	—	—	—	—	—	P	—
Pinus	9	76	58	65	81	89	97	99	91
Quercus	4	—	2	—	—	P	—	P	—
Alnus	—	4	P	1	1	1	—	P	1
Betula	—	—	—	—	—	—	—	—	—
Celtis–Moraceae	—	1	2	—	—	—	—	—	—
Leguminosae	—	P	2	—	—	—	—	—	—
Prosopis	—	—	—	—	—	P	—	—	—
Rosaceae	—	—	—	—	—	—	—	—	—
Type IV	—	1	10	1	—	P	—	—	—
Type VIII	—	—	—	—	—	—	—	—	—
Dodonaea	—	—	—	—	—	—	—	—	—
Agave	2	—	1	—	—	P	—	—	—
Tricolpate Cactaceae	—	P	—	—	—	—	—	—	—
Platyopuntia	—	P	—	—	—	—	—		—
Liliaceae	—	—	—	—	—	—	—	—	—
Ericaceae	—	—	—	—	—	—	—	—	—
Ribes	—	—	—	—	—	—	—	—	—
Compositae	7	6	16	9	8	6	2	1	7
Gramineae	3	6	8	7	8	4	1	—	1
Chenopodiaceae	—	2	—	—	—	—	—	—	—
cf. *Salvia*	—	—	—	—	—	—	—	—	—
Nyctaginaceae	1	—	—	—	1	—	—	—	—
Caprifoliaceae	—	—	—	—	—	—	—	—	—
Onagraceae	—	—	—	—	—	—	P	—	—
cf. *Jatropha*	—	—	—	—	—	—	—	—	—
Polygalaceae	—	—	—	—	1	—	—	—	—
cf. *Sphaeralcea*	—	—	—	—	—	—	—	—	—
cf. *Sidalcea*	—	—	—	—	—	—	—	—	—
cf. *Zea mays*	—	—	—	—	—	—	—	—	—
cf. Zea mexicana	—	—	—	—	—	—	—	—	—
cf. *Tripsacum*	—	2	1	—	—	—	—	—	—
Cucurbita	—	—	—	—	—	—	—	—	—
Unknowns	—	2	—	—	—	—	—	—	—
Total	26	100	100	83	100	100	100	100	100

[a]P, present; proc., sample processed.

TABLE 15.28
Pollen Observed, Gheo-Shih Samples, 1967

Pollen type	Square G29 c5b floor	Square D15 d4b hearth floor	Square G29 c2b under boundary rocks
Ulmus	—	—	—
Picea	—	—	—
Abies	—	—	—
Pinus	—	1	—
Quercus	—	1	1
Alnus	1	1	—
Betula	—	—	—
Celtis–Moraceae	—	—	—
Leguminosae	1	—	1
Prosopis	—	—	—
Rosaceae	—	—	—
Type IV	—	—	—
Type VIII	—	—	—
Dodonaea	—	—	—
Agave	—	—	—
Tricolpate Cactaceae	—	—	—
Platyopuntia	—	—	—
Liliaceae	—	—	—
Ericaceae	—	—	—
Ribes	—	—	—
Compositae	22	12	21
Gramineae	38	45	52
Chenopodiaceae	15	3	25
cf. *Salvia*	—	—	—
Nyctaginaceae	—	—	—
Onagraceae	—	—	—
cf. *Jatropha*	—	—	—
Polygalaceae	—	—	—
cf. *Sphaeralcea*	—	—	—
cf. *Sidalcea*	—	—	—
cf. *Zea mays*	1	1	—
cf. *Zea mexicana*	—	—	—
cf. *Tripsacum*	—	—	—
Cucurbita	—	—	—
Unknowns	—	—	—
Total	78	64	100

Appendix 15.3: Discriminant Functions Classifications of Fossil Pollen Spectra

TABLE 15.29
Standard Pollen Sum, Total Populations

Sample	Lowland effective moisture	Highland effective moisture	Annual temperature	Annual rainfall
Guilá Naquitz				
Zone B		Low Moderate	Moderate	High Moderate
Zone C		Low Moderate	Moderate	High Moderate
Zone E		High	Low	High
Martínez Rockshelter				
Floor + 65 cm		High	Low	Low Moderate
Floor + 60 cm		High	High	Low Moderate
Floor + 55 cm		High	High	Low Moderate
Floor + 50 cm		High	Low	Low Moderate
Floor + 45 cm		High	Low	Low Moderate
Floor + 40 cm		High	Low	High
Floor + 35 cm		Low Moderate	Moderate	Moderate
Floor + 30 cm		High	Moderate	Low Moderate
Floor + 25 cm		High	High	Low Moderate
Floor + 20 cm		High	High	Low Moderate
Floor + 15 cm		High	High	Low Moderate
Floor + 10 cm		High	High	Low
Floor + 05 cm		High	Low	Low Moderate
Cueva Blanca				
55 cm		High	High	Low
65 cm		High	High	Low
75 cm		High	High	High
85 cm		High	High	High
95 cm		High	Low	High
100 cm		High	Low	High
Zone E 100–110 cm		High	High	High
Zone F 100–110 cm		High	Low	High
Gheo-Shih				
Zone B	High		High	Low

TABLE 15.30
Adjusted Pollen Sum, Total Populations

Sample	Lowland effective moisture	Highland effective moisture	Annual temperature	Annual rainfall
Guilá Naquitz				
Zone B		Low Moderate	Moderate	Low
Zone C		Low Moderate	Moderate	Moderate
Zone E		High	Low Moderate	High Moderate
Martínez Rockshelter				
Floor + 65 cm		High	Low Moderate	High Moderate
Floor + 60 cm		High	Low Moderate	High
Floor + 55 cm		High	Low Moderate	High Moderate
Floor + 50 cm		High	Low Moderate	High
Floor + 45 cm		High	Low Moderate	High Moderate
Floor + 40 cm		High	Low Moderate	High Moderate
Floor + 35 cm		Moderate	Moderate	Moderate
Floor + 30 cm		Moderate	Low Moderate	Moderate
Floor + 25 cm		Moderate	Low	High
Floor + 20 cm		Moderate	Low	Low Moderate
Floor + 15 cm		High	Low Moderate	High Moderate
Floor + 10 cm		High	Low	High
Floor + 05 cm		High	Low Moderate	High Moderate
Cueva Blanca				
55 cm		High	Low Moderate	High
65 cm		High	Low Moderate	High
75 cm		High	Low Moderate	High
85 cm		High	Low Moderate	High
95 cm		High	Low Moderate	High
100 cm		High	Low Moderate	High
Zone E 100–110 cm		High	Low Moderate	High
Zone F 100–110 cm		High	Low Moderate	High
Gheo-Shih				
Zone B	Low		High	Low

TABLE 15.31
Standard Pollen Sum, Cueva Blanca Individual Samples

OC-30 sample	Lowland effective moisture	Highland effective moisture	Annual temperature	Annual rainfall
Zone C				
Square G14, 55 cm		High	High	Low
Zone D				
Square G14, 60 cm		High	High	Low
Square G14, 75 cm		High	High	High
Square G14, 85 cm		High	High	High
Zone E				
Square E5, 100–110 cm		High	High	High
Zone F				
Square G14, 95 cm		High	Low	High
Square G14, 100 cm		High	Low	High
Square G14, 100–110 cm		High	Low	High

TABLE 15.32
Adjusted Pollen Sum, Cueva Blanca Individual Samples

OC-30 sample	Lowland effective moisture	Highland effective moisture	Annual temperature	Annual rainfall
Zone C				
Square G14, 55 cm		High	Low Moderate	High
Zone D				
Square G14, 60 cm		High	Low Moderate	High Moderate
Square G14, 75 cm		High	Low Moderate	High
Square G14, 85 cm		High	Low Moderate	High
Zone E				
Square E5, 100–110 cm		High	Low Moderate	High
Zone F				
Square G14, 95 cm		High	Low Moderate	High
Square G14, 100 cm		High	Low Moderate	High
Square G14, 100–110 cm		High	Low Moderate	High

TABLE 15.33
Standard Pollen Sum, Guilá Naquitz Individual Samples

OC-30 sample	Lowland effective moisture	Highland effective moisture	Annual temperature	Annual rainfall
Zone B				
Square E9		High	Low	High
Square E9 reprocessed sample		High	Low	High
Square E6		High	Low	Low
Zone B2				
Square E6		Low Moderate	Moderate	Moderate
Square E6 recount		High	High	Low
Zone B3				
Square D8		High	High	Low
Square D8 reprocessed sample		High	Low	High
Zone B3		High	Low	High
Zone C		Low Moderate	Moderate	High Moderate
Zone E				
Square G8		High	High	Low Moderate
Square G8 reprocessed sample		High	Low	High

TABLE 15.34
Adjusted Pollen Sum, Guilá Naquitz Individual Samples

OC-43 sample	Lowland effective moisture	Highland effective moisture	Annual temperature	Annual rainfall
Zone B1				
Square E9		High	Low	Low
Square E9 reprocessed sample		High	Low Moderate	High Moderate
Square E6		High	Low Moderate	High Moderate
Zone B2				
Square E6		High	Moderate	Moderate
Square E6 recount		High	Low	High
Zone B3				
Square D8		High	Low Moderate	High
Square D8 reprocessed sample		High	Low Moderate	High Moderate
Zone B3		High	Low Moderate	High Moderate
Zone C		Low Moderate	Moderate	Moderate
Zone E				
Square G8		High	Low Moderate	High Moderate
Square G8 reprocessed sample		High	Low Moderate	High Moderate

TABLE 15.35
Standard Pollen Sum, Gheo-Shih Individual Samples

OS-70 sample	Lowland effective moisture	Highland effective moisture	Annual temperature	Annual rainfall
Square D15, d4b	High		High	Low
Square G29, c2b	High		High	Low
Square G29, c5b	High		High	Low Moderate

TABLE 15.36
Adjusted Pollen Sum, Gheo-Shih Individual Samples

OS-70 sample	Lowland effective moisture	Highland effective moisture	Annual temperature	Annual rainfall
Square D15, d4b	Very Low		High	Low
Square G29, c2b	Very Low		High	Low
Square G29, c5b	Low		Moderate	Low Moderate

16

Comparing the Preceramic and Modern Microfauna

Kent V. Flannery
Jane C. Wheeler

INTRODUCTION

Another way of establishing the relevance of the modern environment for interpreting environments of the past is by comparing ancient and modern microfauna. For example, small rodent species are generally regarded as having habitat preferences that make them more sensitive environmental indicators than wide-ranging species such as the white-tailed deer or the cottontail rabbit. In addition, some species of small birds, reptiles, and land snails are considered useful for paleoenvironmental reconstruction.

Small animal remains have proved to be valuable indicators of ancient habitats even when brought into the site by man. However, they may be still more useful when it can be shown that a sample of microfauna directly reflects the species composition of the environment, without human intervention. Such is the case, for example, with small rodents derived from owl pellets. The barn owl (*Tyto alba*), a species virtually worldwide in its distribution, frequently selects caves as its nesting place. At dusk, the owl flies out over an area generally no greater than 2–3km in diameter, where it hunts small rodents from the air. Swallowing them whole, the owl returns to its roost in the cave where the mice are digested and a capsule of hair forms around the complete but disarticulated skeleton. This pellet of bones and fur is then regurgitated, and large samples of the local rodent population, unbiased by human selection, accumulate on the floor of the cave.

It was the presence of just such rodent remains from disintegrated owl pellets in archaeological cave sites of the Levant that allowed Tchernov (1968) to produce his classic study, "Succession of Rodent Faunas during the Upper Pleistocene of Israel." Closer to Oaxaca, a sequence of rodents from MacNeish's Tehuacán caves allowed one of us to suggest that the late Pleistocene environment of the Early Ajuereado phase had been significantly different from that of Holocene times (Flannery 1967). In turn, the environments of most later periods could be shown to approximate today's except for some recent changes wrought by man. This was done by means of a difference of proportions test that measured the discrepancies in species composition between Pleistocene and Holocene samples.

In the case of Guilá Naquitz, only four categories of microfauna with which to evaluate the preceramic environment were available: small rodents, songbirds, small lizards, and land snails. All are discussed in this chapter, with the small rodents providing the largest sample.

MODERN RODENT SAMPLES

Our first step was to obtain a large sample of the modern wild rodent population to use as a control. We were fortunate enough to find a large cave, in the same cliff as Guilá Naquitz and only a few hundred meters to the northeast, that was occupied by barn owls in 1966. The floor of this cave, numbered OC-45 in our survey and nicknamed "Cueva de los Afligidos" by our Mitla workmen (Fig. 16.1), was literally covered with modern owl pellets. In an hour of collecting, we recovered nearly 500 pellets and spent the rest of the day cleaning and extracting the bones. For purposes of identification to species, the Cueva de los Afligidos sample included 466 complete rodent crania and more than 300 isolated mandibles.

FIGURE 16.1. Site OC-45, Cueva de los Afligidos, which produced nearly 500 owl pellets.

Next came the task of identifying the remains, and again we were fortunate because our colleague, Emmet T. Hooper of the University of Michigan Museum of Zoology, had spent several seasons trapping small rodents in Oaxaca and had collected excellent comparative material. In all, 10 species of rodents were identified with the aid of specialists in the Museum of Zoology (Table 16.1). One of these was the common black rat (*Rattus rattus*), an Old World species introduced at the time of the Spanish Conquest and represented by only 3 crania and 3 mandibles. Since this commensal species was not present in Precolumbian times, we ignored it for statistical purposes and concentrated on the 9 species of indigenous wild rodents.

One of the most common wild species was the Mexican spiny pocket mouse, *Liomys irroratus irroratus*, which amounted to 31.1% of the sample. This lone representative of the family Heteromyidae frequents arid and semiarid regions of Mexico, where it feeds on seeds and tender herbaceous plants. Nocturnal in its habits, it carries seeds, leaves, grass, and fiber back to its nest in its large cheek pouches. This is one of the rodents that occasionally made its nest in Guilá Naquitz; an entire mummified spiny pocket mouse was found curled up in its burrow in Square F6 of Zone D (Fig. 16.2).

The remaining eight species belonged to the family Cricetidae. Coues' rice rat, *Oryzomys couesi aztecus* (sometimes lumped with *Oryzomys palustris*), made up 6.3% of our sample. This gregarious, very active rodent feeds on the succulent parts of grasses and sedges, seeds, fruit, and invertebrates; it may become a crop pest in wet areas. Its low frequency in our sample may reflect the fact that there are very few wet areas in the eastern Valley of Oaxaca.

Two species of harvest mouse were present but rare. The western harvest mouse (*Reithrodontomys megalotis alticolus*) made up 1.1% of our sample, while the fulvous harvest mouse (*Reithrodontomys fulvescens helvolus*) made up only 0.4%. Both are rodents occupying stands of short grass, where they eat seeds and green shoots; they do not thrive on cultivated land and have probably retreated in the face of land clearance in Oaxaca.

TABLE 16.1
Rodents from Recent Owl Pellets in OC-45, Cueva de los Afligidos[a]

Family	Scientific name	Crania	Mandibles (left)	Mandibles (right)	Percentage of indigenous wild rodents
Heteromyidae	*Liomys irroratus irroratus*	144	21	37	31.1
Cricetidae	*Oryzomys couesi aztecus*	29	15	14	6.3
	Reithrodontomys megalotis alticolis	5	2	2	1.1
	Reithrodontomys fulvescens helvolus	2	—	—	0.4
	Peromyscus maniculatus fulvus	47	11	10	10.2
	Peromyscus melanophrys melanophrys	5	—	—	1.1
	Baiomys musculus musculus	9	—	2	1.9
	Sigmodon hispidus mascotensis	207	78	91	44.7
	Neotoma mexicana isthmica	15	7	9	3.2
Muridae	*Rattus rattus*	3	2	1	
		466	136	166	

[a]Collection was carried out in 1966.

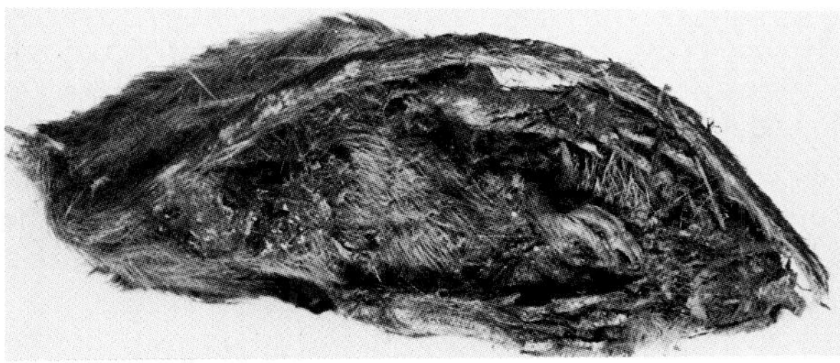

FIGURE 16.2. Mummified spiny pocket mouse from Zone D, Square F6 (total length of specimen, 8.5 cm).

Two species of white-footed mouse were present in our sample, one more common than the other. The deer mouse, *Peromyscus maniculatus fulvus*, reached 10.2%. The plateau mouse, *Peromyscus melanophrys melanophrys*—a large, long-tailed species that prefers rocky habitats in the more arid regions of southern Mexico—made up only 1.1% of our sample. Both species are virtually omnivorous, eating seeds, nuts, berries, fruit, insects, and carrion. Deer mice are known to gnaw on deer carcasses in North America, and it is probable that most of the gnawing of deer bone in Guilá Naquitz was done by *Peromyscus* after the abandonment of each living floor.

The southern pygmy mouse, *Baiomys musculus musculus*, made up 1.9% of the Cueva de los Afligidos sample. This tiny mouse, which lives in dense grass and eats mostly green vegetation, was also one of the species making occasional nests in Guilá Naquitz. A complete, mummified pygmy mouse was found in its burrow in Square D9 of Zone B2.

The hispid cotton rat, *Sigmodon hispidus mascotensis*, was the most common indigenous wild rodent, reaching 44.7%. This is not surprising, since "wherever cotton rats occur they usually are the most numerous mammals" (Hall and Kelson 1959:671). These omnivorous and enormously fecund rats—whose habitat preferences include tall weeds or grass—eat vegetation, insects, unwary songbirds and small mammals, quail eggs, and a wide variety of other foods. *Sigmodon* does well in cultivated fields and on fallow land, frequently increasing in response to agricultural land clearance.

Finally we come to the Mexican wood rat, *Neotoma mexicana isthmica*, which represented 3.2% of our sample. This species loves rocky canyons and areas of organ cactus and prickly pear, getting some of its water from the aforementioned spiny succulents. Hooper trapped *Neotoma* on the volcanic tuff mesas near Yagul and Caballito Blanco, only a few kilometers to the west of Guilá Naquitz, in what would seem to be ideal wood rat habitat: dense thorn–cactus brushland where the rats could eat roots, stems, leaves, seeds, juicy cactus fruits, and invertebrates.

A word should be said about the habits of the genus *Neotoma*. In western North America, some resident species, known as "pack rats," live in caves and construct huge nests that sometimes disturb archaeological deposits (Heizer 1960:136). So far, we have seen no caves so disturbed by the Mexican wood rat, which belongs to a different set of species, the *mexicana* group. While there are occasional isolated rodent burrows in our caves and MacNeish's, neither Oaxaca nor Tehuacán produced pack rat nests of the magnitude seen in the caves of California and Nevada. Whether this is due to a different lifestyle, lower population density, or both is unclear. Certainly the wood rat is not a dominant species in the southern Mexican highlands today,[1] and we found no large *Neotoma* nests in Guilá Naquitz. We did, however, find some evidence that wood rats were occasionally eaten in prehistoric times, just as they are eaten today in rural areas around Tlacolula and San Juan Guelavía in the eastern Valley of Oaxaca. As wild rats go, *Neotoma* is somewhat larger than the other species and is considered "cleaner" in its habits and therefore more edible by the Zapotec.

[1]This was not true during the late Pleistocene, when *Neotoma* was one of the most common rodents present in Oaxaca and Tehuacán (Flannery 1967 and unpublished data).

PRECERAMIC RODENT SAMPLES

The next step in our analysis was to identify the small rodents from Zones B–E in Guilá Naquitz, a task that proved relatively easy given the large amounts of comparative material we had accumulated. However, because of the nature of the criteria used for species identification, we relied heavily on crania and mandibles and made no attempt to go to subspecies level as we had with the modern material. The distribution of identified remains throughout Zones B–E is given in Table 16.2.

Most small rodent remains appeared in contexts that suggested that they had been deposited in the form of owl pellets, presumably at times when there were no people in the cave. Such contexts were ones in which nearly complete (but totally disarticulated) small rodent skeletons were found, sometimes with bits of caked fur of the type characteristic of owl pellets. However, we also found at least two mummified mice that appear to have died in their burrows (Table 16.2). In addition, five mandibles of small rodents had been burned: three mandibles of *Neotoma* (Zone D, Square B9; Zone C, Square C10; Zone B2+3, Square C8), one mandible of *Sigmodon* (Zone B3, Square D9), and one mandible of *Liomys* (Zone E, Square D11). These mandibles may have been burned accidentally after the owl pellet disintegrated, or they may be from rodents actually cooked and eaten. We note that while *Neotoma* was only 16.3% of the preceramic rodent sample, it accounts for 60% of our small sample of burned mandibles. This suggests that at least some of the wood rats were eaten, a not unreasonable suggestion in light of the fact that *Neotoma* is still eaten today by some valley Zapotecs.

With the possible exception of Zone D, none of the cave living floors produced samples that we felt were adequate to stand alone. We have therefore added up all the small rodent remains from Zones B–E, producing a single Naquitz phase sample of 49 minimum individuals (Table 16.3). In this sample, cotton rats are the dominant species (53.1%), followed by wood rats (16.3%), spiny pocket mice (12.2%), and plateau mice (8.2%). Lower in frequency were deer mice (4%) and rice rats, western harvest mice, and pygmy mice (each 2%). No fulvous harvest mice were recovered. Significantly, the Naquitz phase levels did not produce a single species that cannot be found in the Mitla area today; moreover, 8 of the 9 indigenous wild rodents known from modern owl pellets also occurred in preceramic levels.

COMPARING ANCIENT AND MODERN SAMPLES

In order to determine how different the rodent samples from preceramic Guilá Naquitz and modern Cueva de los Afligidos were, we subjected our data to a two-sample test (Dixon and Massey 1969:249) for difference in proportions among the various species. This test was run by Robert G. Reynolds, who has provided the following description. The test produces a *Z* score by means of the formula

$$Z = \frac{\overline{X}_1 - \overline{X}_2}{\sqrt{\overline{p}\,(1 - \overline{p})\,(1/N_1 + 1/N_2)}}$$

where

$$\overline{p} = \frac{(N_1\overline{X}_1 + N_2\overline{X}_2)}{(N_1 + N_2)}$$

N_1 is the number of owl pellets in the OC-45 sample, N_2 is the number of owl pellets in the Guilá Naquitz Zones B–E sample, $\overline{X}_1$ is the species proportion for OC-45, and $\overline{X}_2$ is the species proportion for Guilá Naquitz.

The assumption made is that both samples were randomly chosen from the same target population. If this is the case, the *Z* has a normal distribution. We return to this assumption below.

The results of the two-sample test are given in Table 16.4, where it may be seen that for six of the species involved, the difference between the modern and preceramic samples is insignificant at the .01 level. For three of the species, however, the difference in proportions is statistically significant. Spiny pocket mice, which made up only 12.2% of the preceramic sample, constitute 31.1% of the modern sample. The plateau mouse, amounting to 8.2% of the preceramic rodents, makes up only 1.1% of the modern sample. Finally, the wood rat—the second most common preceramic rodent, at 16.3%—contributes only 3.2% to the modern sample.

Do these differences mean that during the Naquitz phase, local habitats were significantly different from today's? For various reasons, we think not. First of all, the preceramic sample is only one-tenth the size of the modern sample, and some small rodent species are known to exhibit periodic fluctuations in population density that could affect the makeup of ancient owl pellet samples. Second, we know the modern environment of the eastern Valley of Oaxaca has been significantly modified by human activity. Thousands of years of land clearance for agriculture, combined with 400 years of browsing by goats, have undoubtedly changed the suitability of the environment for some rodent species; for example, the pocket gopher (*Orthogeomys grandis*), a very common item in Formative village debris, has apparently disappeared from the Valley of Oaxaca. Thus it may well be that spiny pocket mice have increased and plateau mice decreased as a function of human modification of the landscape.

Finally, we return to the assumption that the preceramic sample is a random one. While most preceramic rodents may indeed have come from owl pellets, at least two (including one spiny pocket mouse) were found dead in their burrows. Moreover, as stated above, as many as three of the wood rats were burned and may therefore have been trapped and eaten by man. If the occupants of the cave were eating wood rats, it invalidates our assumption of randomness and helps to explain the higher frequency of *Neotoma* in preceramic times.

TABLE 16.2
Rodent Remains from Guilá Naquitz Cave

Stratigraphic zone	*L. irroratus*[a]			*O. couesi*			*R. megalotis*			*R. fulvescens*			*P. maniculatus*			*P. melanophrys*			*B. musculus*[b]			*S. hispidus*			*N. mexicana*		
	crania	mandibles L	mandibles R	crania	mandibles L	mandibles R	crania	mandibles L	mandibles R	crania	mandibles L	mandibles R	crania	mandibles L	mandibles R	crania	mandibles L	mandibles R	crania	mandibles L	mandibles R	crania	mandibles L	mandibles R	crania	mandibles L	mandibles R
B1	—	—	—	—	—	—	—	—	—	—	—	—	—	—	—	—	—	1	—	—	—	2	1	3	—	1	—
B2	—	—	—	—	—	—	—	—	—	—	—	—	—	—	—	—	—	—	1	1	1	—	1	2	—	—	—
B2 + 3	—	—	—	—	—	—	—	—	—	—	—	—	—	—	1	—	—	—	—	—	—	1	—	—	—	1	—
B3	—	—	—	—	—	—	—	1	—	—	—	—	—	—	—	—	—	—	—	—	—	—	1	1	—	1	1
C	—	2	1	—	1	—	—	—	—	—	—	—	—	—	—	—	—	1	—	—	—	1	5	1	—	3	1
D	1	2	1	—	—	—	—	—	—	—	—	—	1	—	—	—	—	2	—	—	—	—	9	4	1	—	1
E	—	1	2	—	—	—	—	—	—	—	—	—	—	—	—	—	—	—	—	—	—	3	4	2	1	1	1
	1	5	4	0	1	0	0	1	0	0	0	0	1	0	1	0	0	4	1	1	1	7	21	13	2	7	4

[a]Includes a complete, mummified spiny pocket mouse in Zone D, Square F6.
[b]Includes a complete, mummified pygmy mouse in Zone B2, Square D9.

TABLE 16.3
Rodents from Zones B–E at Guilá Naquitz Cave

Family	Scientific name	Minimum no. of individuals	Percentage of all rodents
Heteromyidae	*Liomys irroratus*	6	12.2
Cricetidae	*Oryzomys couesi*	1	2.0
	Reithrodontomys megalotis	1	2.0
	Reithrodontomys fulvescens	0	0.0
	Peromyscus maniculatus	2	4.0
	Peromyscus melanophrys	4	8.2
	Baiomys musculus	1	2.0
	Sigmodon hispidus	26	53.1
	Neotoma mexicana	8	16.3
		49	

TABLE 16.4
Results of a Two-Sample Test for Significance of Difference in Proportions between Rodent Samples from Cueva de los Afligidos (Modern) and Guilá Naquitz Zones B–E (Preceramic)

Scientific name	Percentage of all wild rodents: preceramic	Percentage of all wild rodents: modern	Z score	Significance at .01 level
Liomys irroratus	12.2	31.1	2.76	Significant
Oryzomys couesi	2.0	6.3	1.22	Not significant
Reithrodontomys megalotis	2.0	1.1	−0.55	Not significant
Reithrodontomys fulvescens	0.0	0.4	0.44	Not significant
Peromyscus maniculatus	4.0	10.2	1.40	Not significant
Peromyscus melanophrys	8.2	1.1	−3.57	Significant
Baiomys musculus	2.0	1.9	−0.05	Not significant
Sigmodon hispidus	53.1	44.7	−1.12	Not significant
Neotoma mexicana	16.3	3.2	−4.23	Significant

In summary, we see sufficient possibility for human intervention and sampling error so that we cannot ascribe the differences in our two rodent samples to natural modification of the habitats now present near the cave. In our opinion, the small rodents from Guilá Naquitz reinforce the conclusion we draw from Schoenwetter and Smith's pollen study (Chapter 15): Except for those changes brought about by man, the environment of the Naquitz phase was *generally* similar to today's.

SONGBIRDS

In addition to small rodent skeletons, the sample of owl pellets from Cueva de los Afligidos contained the remains of a few small birds, evidently songbirds that had been active during the hours when the barn owls were hunting. While songbirds are notoriously hard to identify from postcranial remains, there were 16 crania in the sample that could be tentatively identified with the help of Richard Zusi of the Smithsonian Institution. These are given in Table 16.5.

Ten of 16 crania belonged to a grosbeak, apparently the blue grosbeak (*Guiraca caerulea*). This species includes a local race (*G. caerulea eurhyncha*) that is native to Oaxaca (Blake 1953:560–561). It is not clear why grosbeaks were such favored prey for owls, but it may have something to do with their hours of activity.

The remaining crania are divided among four families. Two are orioles of the genus *Icterus*, several species of which are native to Oaxaca. Two more are unidentified finches. One is a flycatcher, possibly of the genus *Empidonax*, which is represented in Oaxaca by several species. One is a tanager, possibly *Tangara* sp. or a closely related genus.

In the preceramic levels at Guilá Naquitz, only three songbirds could be linked to ancient owl pellets by their depositional context. One of these, appropriately enough, was a grosbeak (Zone C, Square E7); the other two were too fragmentary to identify (Zone D, Square C9; Zone B2, Square C9). Five other fragments of unidentified songbirds were founds in Zones B–E, but their contexts suggested they might have been brought into the cave by humans rather than owls. All are listed with the other bird remains in Chapter 22.

In summary, the songbirds from owl pellets, while very few in number, do nothing to upset the notion of general environmental similarity between the Naquitz phase and modern times.

TABLE 16.5
Small Birds from Recent Owl Pellets in OC-45, Cueva de los Afligidos

Family	Scientific name	Common name	No. of crania
Tyrannidae	cf. *Empidonax* sp.	Flycatcher	1
Icteridae	*Icterus* spp.	Oriole (2 species)	2
Thraupidae	cf. *Tangara* sp.	Tanager	1
Emberizidae	*Guiraca* cf. *caerulea*	Blue(?) grosbeak	10
Fringillidae	Unidentified fringillids	Finches	2

SMALL LIZARDS

While there are many types of lizards native to the Valley of Oaxaca, two species are particularly common in the Guilá Naquitz area: a whiptail or racerunner of the family Teidae (*Cnemidophorus mexicanus*) and the Texas spiny lizard (*Sceloporus olivaceus*) of the family Iguanidae. We are indebted to Richard Etheridge of San Diego State College for his help in identifying these lizards and to Steven and Robert Hole for hunting down an adequate sample of modern specimens in 1967.

It is clear that these lizards occasionally got into Guilá Naquitz by themselves; for example, we found a complete, mummified *Cnemidophorus* in Zone A, Square I5 (Fig. 16.3). However, the occupants of the Tehuacán caves occasionally ate racerunners of a somewhat larger species (Flannery 1967), so we cannot rule out the possibility that some of ours were brought to the cave by man. Two mandibles of *Cnemidophorus* appeared in Zone B1, Square C10, while isolated mandibles of *Sceloporus* were found in Zone D, Square E8, and in Zone B1, Square C10 (Fig. 16.4). It is not unlikely that the three lizards that clustered in Square C10 of Zone B1 were eaten and discarded as a group. All are recorded in Table 16.6.

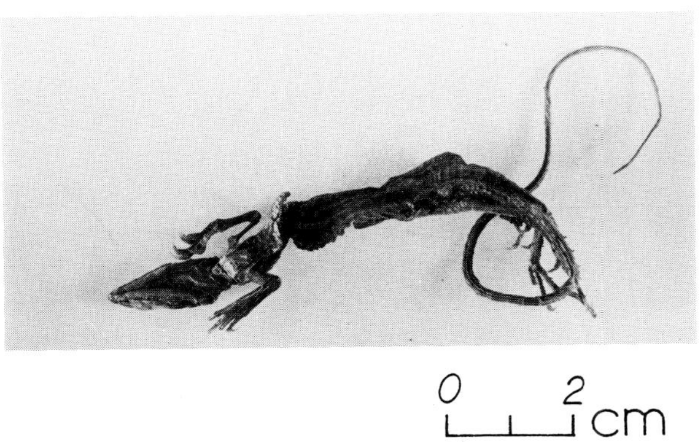

FIGURE 16.3. Mummified whiptail, or racerunner, lizard from Zone A, Square I5.

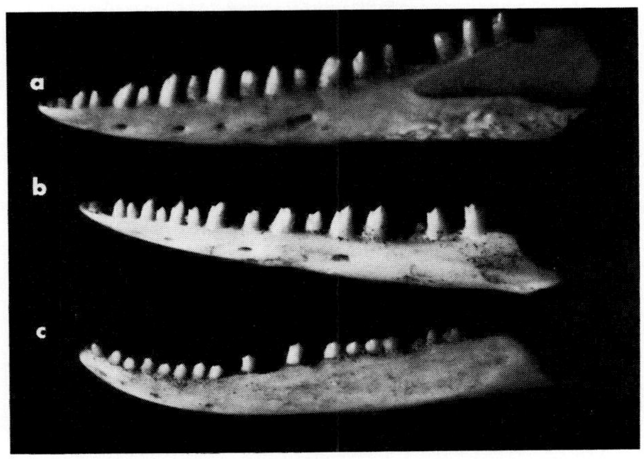

FIGURE 16.4. Lizard mandibles from Guilá Naquitz: a, *Cnemidophorus* (length, 12 mm); b, *Cnemidophorus*; c, *Sceloporus* (length, 10.5 mm).

MOLLUSKS

Mollusks have frequently proven useful in paleoenvironmental reconstruction, but the sample from Guilá Naquitz is so small as to require only a few paragraphs of comment. There are two large land snails native to the region,

TABLE 16.6
Reptile Remains from Guilá Naquitz Cave[a]

Item	Stratigraphic zone				
	E	D	C	B2 + 3	B1
Cnemidophorus cf. *mexicanus* mandibles	—	—	—	—	2 (C10)
Sceloporus cf. *olivaceus* mandibles	—	1 (E8)	—	—	1 (C10)
Unidentified reptile maxilla fragments	—	1 (E8)	—	—	—
Reptile vertebra	2 (D11, F11)	3 (E6, E7, G6)	4 (C10, D11, E5, F10)	1 (D10)	—

[a]1 × 1-m squares in which items were found are given in parentheses.

Euglandina and *Oxystyla*. Neither resembles the common garden snail of the United States, and *Euglandina* in particular is so large and exotic looking that it could be mistaken for a marine shell by an unwary excavator (Fig. 16.5). The likelihood for confusion is increased by the fact that *Euglandina* likes carrion and will burrow into a burial, where it may eventually die and remain to give the impression of a shell offering. In our excavations we found no evidence that either land snail was eaten, and most *Euglandina* we recovered appear to have entered the site in search of discarded meat. Whole specimens occurred in Guilá Naquitz Zone B3 (Square C9) and Zone B2+3 (Square C8).

FIGURE 16.5. Land snail, *Euglandina* sp., from the Mitla region.

In addition, two small unidentified freshwater snails showed up in Zone C, Square C9 (Table 16.7). They resemble the snails that nowadays frequent small pools of water in the rocky arroyos below Guilá Naquitz, as well as the Río Mitla into which the arroyos empty. Since there is no evidence these snails were ever eaten or used as ornaments, it is possible that they were introduced into the cave in containers of drinking water that were filled at the Río Mitla or one of its tributary arroyos or springs. Such drinking water could have been transported in bottle gourds or deerskin bags. The presence of two snails in Square C9 suggests that a container may have been emptied in that area of Zone C.

TABLE 16.7
Mollusk Remains from Guilá Naquitz Cave[a]

Item	Stratigraphic zone		
	C	B3	B2 + 3
Euglandina sp.	—	1 (C9)	1 (C8)
Unidentified freshwater snail	2 (C9)	—	—

[a]1 × 1-m squares in which items were found are given in parentheses.

SUMMARY AND CONCLUSIONS

Examination of small rodents, songbirds, lizards, and mollusks from Guilá Naquitz Zones B–E does not suggest that the Naquitz phase cave occupants (human or animal) witnessed or exploited habitats unlike those observed in the area today. Rather, it supports the notion that plant associations resembling Thorn Forest A and B and Mesquite Grassland A and B occurred near the cave and that, except for the obvious deterioration brought on by man, the environment of the Naquitz phase was not unlike today's. This in turn reinforces our conclusion that the present landscape is relevant to any interpretation of the preceramic environment.

V

ANALYSIS OF SUBSISTENCE

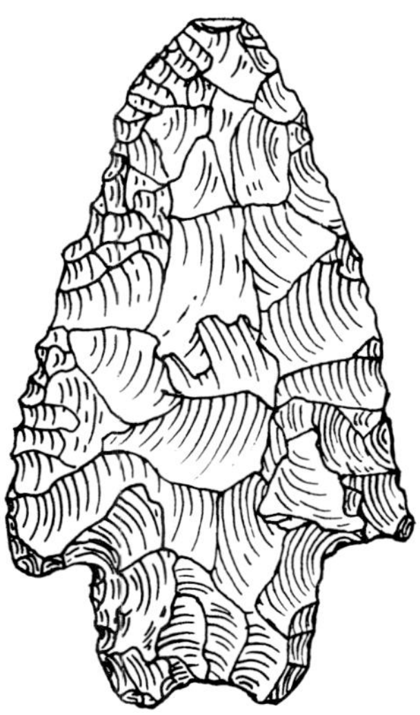

17

The Quantification of Subsistence Data: An Introduction to Part V

Kent V. Flannery

INTRODUCTION

All the chapters in this section deal with some aspect of subsistence at Guilá Naquitz. Our starting point is the information provided by plant and animal debris recovered from the cave. This debris was collected by fine-screening each 1 × 1-m square within each living floor, with the counts for each square constituting our raw data. We also conducted a seven-year series of plant censuses designed to tell us the frequency with which the Guilá Naquitz plants occurred in the wild so that we could evaluate the subsistence strategy of the cave occupants.

In order to be further manipulated, our data on plants and animals have been quantified in various ways. That quantification depends on a set of assumptions that are the subject of this introduction. We feel the assumptions are reasonable, but they are ones that archaeologists often fail to discuss in print.

Because these issues are usually left implicit, in recent years an increasing number of young archaeologists have come to believe that they belong to the first generation to consider them. They are mistaken. More than 20 years ago, Heizer (1960) published an article entitled, "Physical Analysis of Habitation Residues," in which he summarized half a century of efforts to interpret plant and animal remains. Heizer's paper makes it clear that archaeologists have long been concerned with the processes by which sites form, the ways in which plants and animals enter the deposits, and the validity of the assumptions that lead to quantification of food debris. Animal disturbance, variable rates of natural decay, and problems of refuse sampling all come under his scrutiny. It is yet another example of a set of questions discussed long ago, incompletely resolved, allowed to become implicit, and then "rediscovered" by a later generation of archaeologists.

THE NATURE OF THE GUILA NAQUITZ FOOD RESIDUES

One of the first questions to be resolved is, Were all the plants and animals brought to the cave by man? At Guilá Naquitz, as in all cave sites of which I am aware, there was significant evidence of rodent activity. Many piñon nuts showed signs of gnawing (Fig. 17.1), many seed coats of *susí*

FIGURE 17.1. Piñon nut hull from Guilá Naquitz, showing rodent gnawing marks (actual length of specimen, 16 mm).

had been chewed open, and many deer bone splinters had been rodent gnawed as well. Obviously, we needed to determine whether certain plants (1) had actually been brought into the cave by rodents or (2) had been gnawed by rodents who moved into the cave after the human occupants had moved out.

Because of the clarity of the Guilá Naquitz living floors, rodent burrows were extremely easy to recognize (see Fig. 5.17). All such burrows were isolated, left on a pedestal, and cross-sectioned, and then their contents were examined. Interestingly enough, none contained acorns, *susí* nuts, or any of the plants we frequently found gnawed. More often than not they contained caches of small grass seeds such as *Heteropogon* or occasional desiccated kernels of Postclassic maize types, evidently introduced from the upper levels of the cave at the time the burrows were dug. In other words, the burrows we excavated provided no evidence that rodents were bringing into the cave any of the plants regularly used by preceramic humans. They did, however, suggest that in the process of digging their burrows the rodents could move a small object, such as a corn kernel or a bean, from an upper level to a lower level.

Two burrows contained the mummified remains of actual rodents—a spiny mouse (*Liomys*) in Square F6 and a pygmy mouse (*Baiomys*) in Square D9 (see Chapter 16 for more details). Since spiny mice are known to carry seeds back to the nest in their cheek pouches, they may be responsible for many of the caches of grass seeds. We found no nests of the larger pack rat (*Neotoma*), the nemesis of North American cave excavators.

A second line of evidence was the distribution of various plant species on the living floors themselves (see Chapter 26). Rather than being distributed across the living floor in a series of small pockets, as would be the case with rodent caches, the plants showed a pattern of large discard areas such as would result from human food processing activity.

A third line of evidence was the rodent gnawing on deer bone splinters. White-footed mice (*Peromyscus* spp.), reasonably common in the Guilá Naquitz area, are known to gnaw freshly discarded carcasses and bones for their fat and salt content. Clearly, however, they did not kill the deer and bring them into the cave! In this case, our best explanation is that local rodents gnawed on the bone scraps left behind when the human occupants of the cave were absent.

In addition, some plants showed evidence of human food preparation that ruled out introduction by animals. Maguey, which must be baked underground for days, was a case in point. And given the small home range of rodents, it seems very unlikely they would have brought hackberries and mesquite pods to the cave from groves more than 3 km away.

In summary, on the basis of all the evidence available to us from rodent burrows, plant and animal distributions on living floors, signs of food preparation, and other considerations, we see no evidence that local wild rodents introduced anything larger than grass seeds into Guilá Naquitz. However, those rodents did enter the cave when humans were absent, gnawing on discarded acorns, *susí* nuts, deer bone splinters, and other items. While this may have affected the location of the individual items they gnawed, we doubt that it significantly affected the overall distributions reflected in our debris contour maps (Chapter 26). Plant and animal remains were simply too numerous, and activity areas too clear, to be seriously affected by the occasional rodent activity for which we see evidence.

Some rodents did make burrows in the cave, and their burrowing occasionally moved small plant remains from upper levels to lower levels. We doubt that the number of plants so moved was large. However, in light of such disturbance, we feel we should be cautious in interpreting small plant remains found in the preceramic levels. For example, an isolated bean of a domestic race, if found in a Naquitz phase level, should not be uncritically accepted as evidence for early cultivation, especially if that race is well represented in the Postclassic levels above (see Chapter 21). Only when the evidence is relatively abundant, comes from squares unaffected by rodent burrows, and/or includes evidence for human processing should it be accepted. Such is the case with early domestic cucurbits, whose rinds are abundant and whose seeds had frequently been roasted (see Chapter 20).

Having established that the bulk of the food remains in the cave were introduced by humans, we must next face the question, How representative are these remains? Was the species most common in the cave also the most common in the diet? Did a species that represents 15% of the plant remains also represent 15% of the plant diet?

Such questions would be difficult to answer even if the residues of all plant remains were comparable. Unfortunately, they are not wholly comparable: Some species are represented by uneaten specimens, others by the debris of eaten specimens, and still others by their inedible parts. Heizer (1960:111) anticipated this problem when he said, "mere identification of species of plants and animals which anciently served for food, no matter how great in bulk, can serve only to tell us *what* was being eaten. . . . [W]e must be able to deal with this information on a quantitative basis, and we should try to understand how foods were secured."

Figure 17.2 gives examples of food procurement and processing routines for the Guilá Naquitz plants. Piñon nuts and *susí* nuts have similar routines. It is likely that at the time they were harvested, some were hulled and eaten on the spot; these specimens would be unrepresented in the cave, and no means of quantifying them exists. Still others would have been brought back to the cave, perhaps stored for some time, then hulled and eaten. It is the discarded, inedible seed coats of piñon and *susí* that were recovered by excavation.

Wild onions were pulled up by the roots, and some might have been eaten on the spot. Others were brought back to the cave, perhaps to be used as flavoring for meals that included relatively bland foods such as prickly pear stems (nopales). The leaves and bulb coat would be trimmed off and the bulb eaten, leaving no trace. Obviously, any bulbs recovered archaeologically would be leftovers; it was much more common to find bulb coats from which the bulbs had been removed.

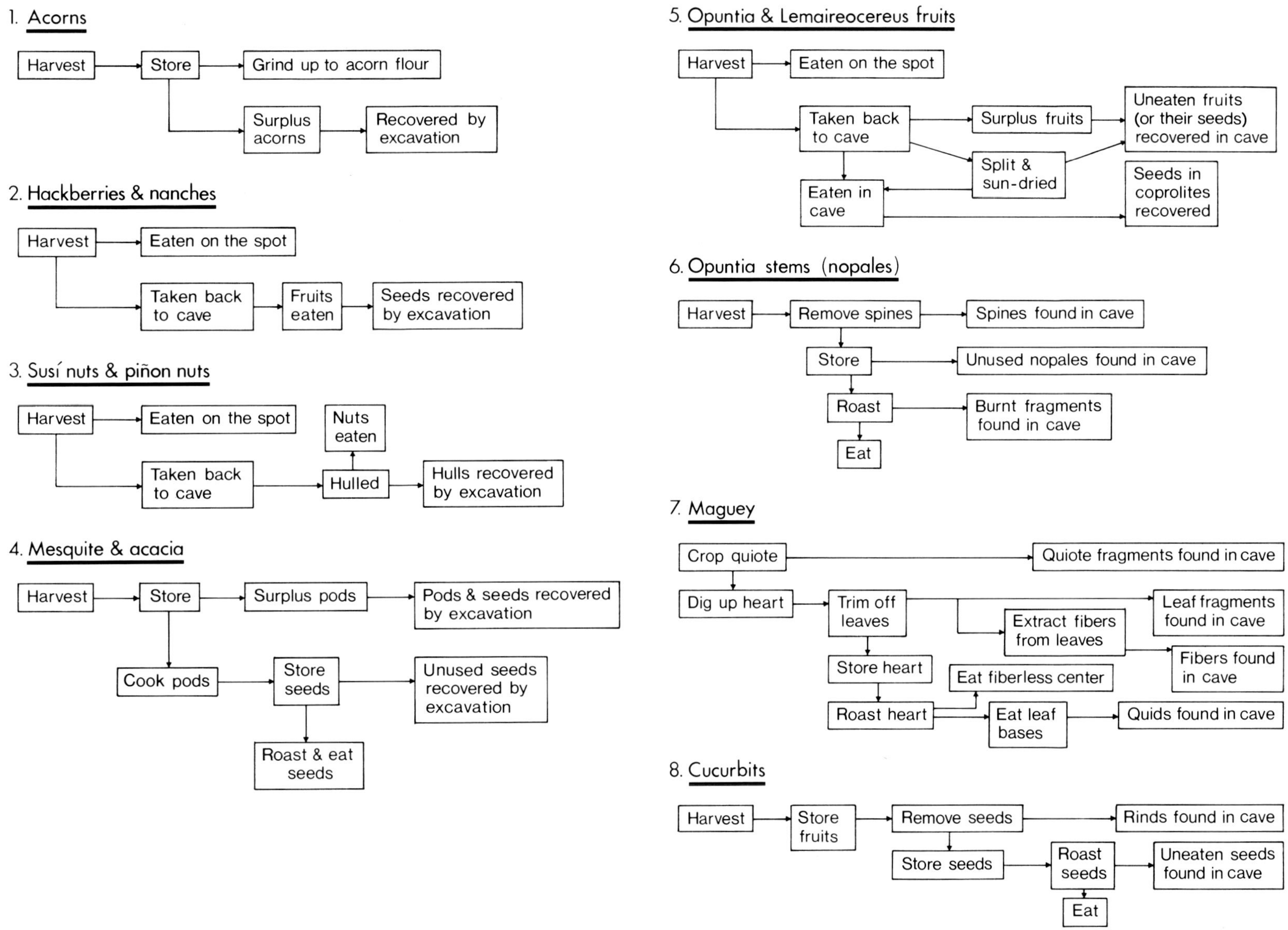

FIGURE 17.2. Procurement and processing routines for major Guilá Naquitz plants.

Maguey (*Agave* spp.) is a plant whose harvest and processing produce several different kinds of archaeological remains. We have already seen that agaves were used when the *quiote*, or inflorescence, had been sent up, and that the *quiote* was frequently cropped. *Quiotes* were sometimes brought back to the cave, where inflorescence fragments and seed capsules occur in the refuse. The heart of the plant was dug up and the leaves trimmed off, sometimes to be used for fiber. Both leaf fragments (including spines) and actual maguey fiber cordage occurred in the cave. Because it is inedible when raw, maguey could not be eaten on the spot like some vegetal foods. It was brought back to the cave, where a roasting pit was prepared and the heart cooked for 24–72 hours. The fiberless core of the heart could then be eaten, leaving no archaeological trace (see Chapter 24); next eaten were the leaf bases, each of which would produce a quid. If these quids were expectorated in the cave, we found them in the refuse (Fig. 17.2); when "chews" of maguey were taken along on trips away from the cave, the quids would never be found.

Acorns present a somewhat different problem. Because they cannot be eaten raw, they had to be brought back to the cave after harvesting, perhaps to be stored in baskets or in pits like those found in Zone D. We presume that many would have been made into acorn flour by grinding in a mortar or with a one-hand mano and slab metate; this flour would be unrecoverable by our archaeological methods. Only the surplus acorns, those left behind unused, would survive in the refuse. One could argue that one reason acorns were so abundant at the cave is that they all had to be brought back for processing. However, it is also likely that if so many were left over, many must have been harvested.

Oak galls represent another case where only uneaten, leftover specimens would be archaeologically recovered.

Hackberries and nanches are both fruits that can be eaten raw, and many were probably consumed on the spot during the harvest; one could therefore argue that they may be chronically underrepresented in the cave. Those that were brought back and eaten in the cave left behind either an incredibly resistant white seed (in the case of the hackberry) or a fragile but readily identifiable brown seed (in the case of the nanche).

Since *guaje* seeds can be eaten raw, they probably were frequently eaten on the spot and may therefore be underrepresented in the cave. The refuse found consists mainly of pods that have been opened and the seeds removed, so obviously some *guajes* were brought back to the cave.

Wild beans cannot be eaten raw. It appears that their pods were collected and brought back to the cave, where the seeds could either have been pounded into flour or softened by some method unknown to us (perhaps soaked in water or boiled). For the most part, beans are represented by the valves of pods that have been opened and discarded, but there are also occasional unused seeds in the cave refuse. Some local wild beans also have an edible root, but only those tubers left behind uneaten would survive archaeologically.

There are several procurement and preparation routines by which mesquite could get into the cave. Although very fresh green mesquite pods can be chewed (producing a quid), it is doubtful that many were eaten on the spot this way. Most were probably brought back to the cave and stored; a few surplus pods may have been left uneaten in the refuse. Most pods, however, were probably cooked up to produce the edible syrup (or *miel*) favored by some Zapotec villages today, and such pods would not survive archaeologically. However, even in the case of those pods, the seeds may have been saved and stored to be roasted at a later date. We recovered many uneaten seeds that could have come from such pods.

Acacia pods may have been processed much as mesquite, but the plant is much less palatable and was rare in the cave refuse; our remains consist mainly of pods that were left behind uneaten.

Fruits of the organ cactus (*Lemaireocereus*) and prickly pear (*Opuntia*) probably had similar routines. Many were probably eaten on the spot by slitting open the leathery skin and peeling it away to reveal the juicy fruit below. Those brought back to the cave appear to have had their spines removed, either plucked off or singed off over the campfire. Occasionally, a desiccated uneaten fruit appeared in the cave debris, but most uneaten fruits evidently rotted, leaving only the seeds. Prickly pear supplies can be made to last several weeks by splitting the fruit open and sun-drying it (Pennington 1963:117–118), but eventually even those fruits would spoil. It is the tough, indigestible seeds that survive archaeologically, and some found in Guilá Naquitz could even have resulted from the disintegration of human coprolites. Most, however, probably came from uneaten or partially eaten fruits.

Prickly pear stems, like maguey, could reach the cave from many points in the processing routine. Since they are unlikely to have been eaten on the spot, much of this processing probably took place in or near the cave. The tender young nopales had their spines removed (some of these were left in the refuse) and must have been stored at least briefly (some that were unused dried out and were left behind). Most evidently were cooked, one common method being to roast them over the fire on a wooden skewer. Burnt fragments of both nopal and skewer were found in the cave (see Chapter 10).

Finally, cucurbits went through yet another routine. We have no reason to believe that either the wild form or the early cultivar had edible fruit, so it is likely that all those harvested were brought back to the cave. There the edible seeds could be removed, with the discarded rinds left broken in the refuse. The seeds could be stored, either before or after roasting; obviously, it is only the uneaten leftovers that survived archaeologically. Many of the cucurbit seeds from Guilá Naquitz show evidence of roasting (Chapter 20).

The various procurement and processing routines considered above make it clear that we would be overconfident to assume that the Guilá Naquitz plant remains give us the exact proportion of the diet that each species made up. What can be done about this? How can the residues of plants that were eaten (such as *susí* seed coats) be compared to the uneaten surpluses of other plants (such as acorns)? And what of those species that may often have been eaten raw in the field? How can they be compared to species such as the maguey, which had to be processed for days back at the cave? Can we assume that half of all hackberries were eaten on the spot, and multiply the hackberry seeds in the cave by 2 to correct for underrepresentation? Can we assume that 10% of all acorns harvested went uneaten, and multiply by 10 to get the true number harvested?

I believe that the answer to all these questions is, "No." In my opinion, we have no solid basis for "adjusting" any of the figures on plant remains. The procurement and processing routines are too different, and involve too many additional assumptions, to permit fiddling around with the data. Almost certainly, the errors we would introduce would be as great as any already inherent in the data. The raw figures from the cave, however imperfect, at least reflect the activities of the prehistoric occupants, rather than the fantasies of the archaeologist.

In the chapters that follow, therefore, we use the actual numbers of plant remains as recovered from the cave. These numbers are not sacred, but they are the only ones we have. Their use does not imply that the percentage of acorns or nanches in the cave refuse reflects the exact percentage of acorns or nanches in the diet. It reflects a much simpler assumption: that the cave remains give us some idea of human preference and use. Regardless of how the plants were procured and processed, if there are many acorns in the cave, it is because many were harvested. If there are few acacia pods, it is because few were harvested.

In addition to problems of plant species comparability, there is a problem of comparability between plant and animal debris. No matter how carefully remains are recovered and counted, we have no reliable way of assessing the proportion of meat versus plants in the diet. Coprolites come the closest to providing this information, for they sometimes allow the determination that a given meal was, let us say, 40% meat and 60% plant. Even in this case, however, we would need a lot of coprolites, from many different individuals and many different meals, before we could see an overall pattern for even one limited season.

A faunal analyst can find the humerus of a deer and conclude that at least one individual deer was killed. A botanist, however, cannot find six acorns and conclude that one entire individual tree was harvested. To be sure, we could estimate the amount of meat on a humerus (rather than the whole deer) and restrict ourselves to that; but we would still have no guarantee that the occupants of the cave ate six acorns for every deer humerus. About all we can do is attempt to model the relationship, and that is what we do later in Part V.

THE FORMAT FOR PART V

Part V proceeds with two bodies of raw data: the plants and animals recovered by excavation and the results of a seven-year plant census conducted near Guilá Naquitz.

The plant remains were organized by C. E. Smith (Chapter 19), who sent the beans to Lawrence Kaplan (Chapter 21) and the cucurbits to Thomas W. Whitaker and Hugh C. Cutler (Chapter 20). Animal bones were studied by K. Flannery and Jane C. Wheeler (Chapter 22). This gave us the quantitative data from which further studies could proceed.

The plant censuses (Chapter 18) gave us data on the density of various species in the Guilá Naquitz environment, as well as their seasonality and annual variation. They also provided the relationship between archaeological plant and animal debris and 100-g edible portions (Chapter 24).

Nutritionists J. R. K. Robson and J. N. Elias (Chapter 23) analyzed the Guilá Naquitz foods from the standpoint of calories, protein, carbohydrates, fats, and other elements and proposed hypothetical diets for several living floors. Their data could then be combined with the plant census data to estimate the productivity of the Guilá Naquitz environment (Chapters 18 and 24).

Finally, all of the studies above could be combined to reconstruct the average diet of the cave occupants, estimate the number of hectares necessary to support the persons seen there, and discuss the environmental limitations of their preceramic lifeway (Chapter 24).

There is one additional set of studies we would like to have but do not: an input–output analysis comparing the calories and protein expended to procure the Guilá Naquitz foods with the calories and protein obtained from them. This would help us to construct a model for the hunting–gathering strategy of the cave occupants. Unfortunately, we felt we had no reliable way of calculating the occupants' energy expenditure during such activities as Precolumbian deer hunting, acorn harvesting, and maguey roasting. The alternative approach we have decided on is to make the assumption that a relationship exists between area harvested and energy expended; this allows us to calculate "effort" in terms of square meters harvested (see Part VII).

18

Wild Food Resources of the Mitla Caves: Productivity, Seasonality, and Annual Variation

Kent V. Flannery

In any study of hunting–gathering cultures, some of the most basic questions concern the relationship between the human population and its wild food resources. Botanical studies by C. E. Smith (1978) have given us some idea of the original vegetation of the eastern Valley of Oaxaca. Kirkby, Whyte, and Flannery (Chapter 4) have divided the present-day vegetation of the Mitla cave area into four idealized facies. In Chapter 15, Schoenwetter and Smith have established the relevance of today's vegetation for interpreting the past. The plant and animal remains from Guilá Naquitz indicate which species the occupants of the cave chose to use.

These data, however, leave a number of important questions unanswered. What is the productivity of the native vegetation of the Mitla region? How many people could it support, and for how long? How close would the four to five persons occupying Guilá Naquitz have come to exhausting the locally available wild foods? What effect might annual variation in rainfall have had on those wild foods? What effect might seasonal variation in availability have had? In this chapter, we make a start toward answering those questions.

PLANT CENSUSES, 1966–1976

During 7 of the first 10 years of the Oaxaca project, we conducted censuses of the wild plants available in the neighborhood of the Mitla caves. Our purpose was to provide estimates for the numbers of edible plant parts per hectare for all four of the vegetational facies defined by Kirkby, Whyte, and Flannery.

Smith (1978) has described the extent to which centuries of agriculture, grazing, and charcoal burning have modified the original vegetation of the Valley of Oaxaca. Fortunately for our study, although there is probably no totally undisturbed area of the valley, the region of the Mitla caves is one of the least modified. As Smith (1978:22) puts it, "only in the Mitla arm of the valley, toward Guilá Naquitz Cave, are there some remnants of the original vegetational cover. Even here, it is obvious that this thorn–scrub is disturbed and, in many instances, has only recently returned to the area." Undoubtedly, one of the factors allowing the original vegetational cover to return to the area was the conversion of the Rancho El Fuerte to largely unused ejido land.

We decided that our censuses should be done in the stands of least-disturbed native vegetation, relying on Smith's advice in selecting these. Once such areas had been identified, Kirkby and Whyte established a series of sampling transects in the four facies of vegetation defined in Chapter 4. Each transect was a strip 3 m wide and at least 300 m long, chosen at random, but with its long axis recorded with reference to one of the many local landmarks (such as a prominent rock outcrop) so that it could be relocated year after year. In this way, we could compare the productivity of a given transect in rainy years and dry years. Transects were also sampled at several different seasons of the year, depending on the duration of the field season and the availability of personnel.

The sampling procedure involved walking each transect slowly, with one person counting the number of plants of each species present and a second person recording the number of fruits or edible parts per plant. The latter figure was obtained by having a group of Zapotec workmen harvest each plant. This task varied considerably in difficulty; harvesting a small *susí* bush by hand might take only a few minutes, while harvesting all the acorns from an oak tree might take hours.

Although the censuses were aimed primarily at vegetation, we recovered data on the density of cottontail rabbits as well. Cottontails have relatively small home ranges (frequently no greater than 100 m in diameter) in which they leave small piles of dung pellets as territorial markers. By counting the small piles of pellets we could frequently determine the number of home ranges within a transect.

While Smith, Kirkby, Whyte, and Schoenwetter inaugurated the plant censuses, virtually every member of the Oaxaca project participated over the years. I am grateful to our Mitla workmen for performing what was often a hot, thorny task, and I am particularly indebted to Susan Lees for taking on the unglamorous assignment of counting rabbit pellets. In all, the area covered by our transects was as follows:

Thorn Forest A	940 m^2
Thorn Forest B	1000 m^2
Mesquite Grassland A	1200 m^2
Mesquite Grassland B	1000 m^2
Total area sampled	5340 m^2

While it is difficult to make such a calculation, I estimate that our transects sampled about 5% of the least-disturbed native vegetational areas between Guilá Naquitz and the Río Grande de Mitla.

Table 18.1 compares four facies of vegetation in the Mitla cave area on the basis of density of plant and animal taxa; note that the figures in this first table reflect numbers of plants (e.g., oak trees), rather than edible parts (e.g., acorns), per 100 m^2. The ranges reflect differences among the various transects within each facies. Also clearly indicated in the table are the differences in resources among the four vegetational facies. If you want acorns, *guajes,* or *tepeguajes,* you had better look to Thorn Forest A; if you want mesquite pods and hackberries, you will be better off in Mesquite Grassland B. A few plants, such as the prickly pear, are present in all facies but in varying densities.

In Tables 18.2–18.5 we have taken the average number of plants per 100 m^2 for each vegetational facies (Column 2) and multiplied it by the average number of edible parts per plant per year (Column 3) to produce an average number of edible parts per 100 m^2 (Column 4); in Column 5 this figure is converted to edible parts per hectare. All averages represent numerical means for all transects in a facies for all years in which sampling was done.

Tables 18.2–18.5 not only spell out in more detail the differences in resources among the four vegetational facies, they also provide the basis from which many of our later calculations are done. For example, the figures in Column 5 can be converted to calories and protein to help us estimate how many people could be supported by each vegetational facies (Chapter 24).

In addition to indicating the large quantities of wild plant foods per hectare that are available in an average year, Tables

TABLE 18.1
Four Facies of Vegetation in the Eastern Valley of Oaxaca Compared with Regard to Density of Plant and Animal Taxa, 1966–1976

	Density per 100 m^2			
Taxon	Thorn Forest A	Thorn Forest B	Mesquite Grassland A	Mesquite Grassland B
Oak (*Quercus*)	0.11– 0.43	—	—	—
Tepeguaje (*Cassia*)	0.43– 2.45	—	—	—
Guaje (*Leucaena*)	0.00– 3.94	—	—	—
Nanche (*Malpighia*)	0.16– 0.21	0.00– 0.10	—	—
Wild bean (*Phaseolus*)	0.05– 0.15	0.00– 0.10	—	—
Susí (*Jatropha*)	7.23– 7.55	0.60– 9.50	0.00– 0.42	—
Agave potatorum	4.57– 8.72	0.00– 0.40	0.00– 0.08	—
Agave marmorata	0.32– 2.66	0.00– 0.30	0.00– 0.08	—
Catclaw (*Acacia*)	2.87– 2.98	0.00– 2.00	0.25– 0.42	—
Myrtillocactus	0.21– 2.77	0.20– 5.90	0.33– 4.08	—
Lemaireocereus	0.40–27.0	0.15– 0.80	0.08– 3.50	—
Biznagre (*Ferrocactus*)	0.00– 1.28	0.40–19.9	0.08– 2.33	—
Prickly pear (*Opuntia*)	0.32–40.0	0.20– 2.60	0.75– 6.00	0.90– 2.50
Huizache (*Acacia*)	0.00– 0.06	0.00– 0.40	0.00– 0.60	0.40– 1.50
Cholla (*Opuntia*)	2.13–52.6	3.30–18.8	2.92– 7.83	0.50– 4.10
Chilillo (*Mammillaria*)	0.11– 2.13	8.40–21.8	10.7 –53.4	0.00– 0.10
Wild onion (*Allium*)	—	0.00– 0.002	0.00– 0.002	—
Wild *Cucurbita*	—	0.00– 0.01	0.00– 0.02	—
Mesquite (*Prosopis*)	—	0.00– 0.10	0.75– 4.42	3.70–10.5
Hackberry (*Celtis*)	—	—	0.00– 0.02	0.00– 0.04
Cottontail (*Sylvilagus*)	0.032	0.015	0.01	0.01

TABLE 18.2
Thorn Forest A: Estimated Number of Edible Plant Parts per Hectare per Year, Based on 1966–1976 Plant Censuses

Name of plant	No. plants per 100 m²	No. edible parts per plant per yr		No. edible parts per 100 m²	No. edible parts per ha
Oak (*Quercus*)	0.22	Acorns:	1,401.2	308.2	30,820
		Galls:	1.3	0.29	29
Tepeguaje (*Cassia*)	1.7	Pods:	1,730.00	2,941.0	294,100
		Seeds:	13,840	23,528	2,352,800
Guaje (*Leucaena*)	1.4	Pods:	142.2	199.1	19,910
		Seeds:	1,422.0	1,991.0	199,100
Catclaw (*Acacia*)	2.9	Pods:	20.0	58.0	5,800
Huizache (*Acacia*)	0.03	Pods:	35.0	1.05	105
Susí (*Jatropha*)	7.4	Nuts:	51.0	377.4	37,740
Agave potatorum	7.2	Mature:	17%	1.22	122
Agave marmorata	1.2	Mature:	24%	0.29	29
Nanche (*Malpighia*)	0.2	Fruits:	18.0	3.6	360
Prickly pear (*Opuntia*)	11.8	Nopales:	19.0	224.2	22,420
		Fruits:	16.5	194.7	19,470
Cholla (*Opuntia*)	14.9	Fruits:	6.7	99.8	9,980
Myrtillocactus	0.8	Fruits:	934.4	747.5	74,750
Lemaireocereus	9.4	Fruits:	15.8	148.5	14,850
Ferrocactus	0.7	Fruits:	1.0	0.7	70
Mammillaria	1.1	Fruits:	3.0	3.3	330
Wild bean (*Phaseolus*)	0.1	Pods:	7.0	0.7	70
		Seeds:	35.0	3.5	350
		Tuber:	1.0	0.1	10

18.2–18.5 suggest that the occupants of Guilá Naquitz were not collecting the various plant species in the same frequency in which they occur in the wild. For example, *tepeguaje* seeds are the most common edible item in Thorn Forest A (more than 2 million/ha), but were relatively infrequent in the cave refuse. Far more important were the 30,820 acorns/ha and the 37,740 *susí* nuts. Certain plants, such as *Agave potatorum* (122/ha), were low in frequency but important because each provided so much food. Certain others, such as *Myrtillocactus* fruits (74,750/ha), may be absent from the cave because they ripen at the wrong time of the year (see below).

SEASONAL VARIATION

One of the characteristics of the Oaxaca environment is great seasonal variation in the availability of foodstuffs. Certain plants become available at the start of the rainy season, others toward the end of the rainy season, and still others during the heart of the dry season.

Figure 18.1 follows the eastern Valley of Oaxaca through a complete year, beginning with the month of April. At this point, late in the dry season, there are few plant foods available except for maguey hearts and prickly pear stems (nopales). Assuming the rains begin on time in May, the month of June can see a considerable improvement in resources. In Mesquite Grassland B, hackberries and mesquite pods can appear; elsewhere, the spring rains can lead to the flowering of wild bean, cucurbit, and onion plants. While the latter three genera are not edible at this stage, the appearance of the flowers is a critical event because it indicates to foragers the places where the plants will be available later on.

June is usually one of the two rainiest months, and it is during June, July, and August that the resources of the Mesquite Grassland are likely to be at their peak. In the Thorn Forest, nanche and organ cactus fruits (*Lemaireocereus*) and *guaje* and acacia pods appear, and mature agaves begin to send up their inflorescences.

September is usually the other month of peak rains, bringing the rainy season plants to the end of their cycle and providing the autumn plants with the moisture necessary for maturation. During September, October, and November the resources of Thorn Forest A begin to outstrip those of the Mesquite Grassland. Now plant collectors can return to the places where they earlier saw the flowers of wild onion, bean, and cucurbit and harvest the edible plants. Now they can dig up agave hearts or crop the inflorescence and leave the plant for a later season. Prickly pear fruit, nopales, *Randia,* and *Aeschynomene* are ripe. By November (and perhaps earlier in certain years) the great nut resources of the upper piedmont—acorns, piñon nuts, and *susí*—are ready for harvest and some of the oak trees have edible galls.

This autumn season, when the rains have ended and the resources of Thorn Forest A are at peak abundance, appears to have been the main period of occupation at Guilá Naquitz. The plants recovered in the cave suggest that the occupants arrived while hackberries and mesquite pods were still available, August–September(?), and stayed until the end of the acorn and piñon nut harvest, December–January(?). Obviously, the length and timing of that period would vary from year to year depending on variation in rainfall (see below).

TABLE 18.3
Thorn Forest B: Estimated Number of Edible Plant Parts per Hectare per Year, Based on 1966–1976 Plant Censuses

Name of plant	No. plants per 100 m²	No. edible parts per plant per yr		No. edible parts per 100 m²	No. edible parts per ha
Catclaw (*Acacia*)	1.0	Pods:	20.0	20.0	2,000
Huizache (*Acacia*)	0.2	Pods:	35.0	7.0	700
Susí (*Jatropha*)	5.1	Nuts:	45.0	229.5	22,950
Agave potatorum	0.2	Mature:	20%	0.04	4
Agave marmorata	0.15	Mature:	20%	0.03	3
Nanche (*Malpighia*)	0.05	Fruits:	10.0	0.5	50
Prickly pear (*Opuntia*)	1.4	Nopales:	19.0	26.6	2,660
		Fruits:	11.0	15.4	1,540
Cholla (*Opuntia*)	10.3	Fruits:	4.4	45.3	4,530
Myrtillocactus	2.2	Fruits:	59.9	131.7	13,170
Lemaireocereus	0.4	Fruits:	8.5	3.4	340
Ferrocactus	7.3	Fruits:	4.4	32.1	3,210
Mammillaria	15.1	Fruits:	0.7	10.6	1,060
Wild bean (*Phaseolus*)	0.05	Pods:	7.0	0.35	35
		Seeds:	35.0	1.75	175
		Tuber:	1.0	0.05	5
Wild onions (*Allium*)	0.02	Bulbs:	1.0	0.02	2
Mesquite (*Prosopis*)	0.05	Pods:	32.0	1.6	160
Wild *Cucurbita*	0.005	Seeds:	75.0	0.38	38

TABLE 18.4
Mesquite Grassland A: Estimated Number of Edible Plant Parts per Hectare per Year, Based on 1966–1976 Plant Censuses

Name of plant	No. plants per 100 m²	No. edible parts per plant per yr		No. edible parts per 100 m²	No. edible parts per ha
Catclaw (*Acacia*)	0.3	Pods:	20.0	6.0	600
Huizache (*Acacia*)	0.3	Pods:	35.0	10.5	1,050
Susí (*Jatropha*)	0.02	Nuts:	40.0	0.8	80
Agave potatorum	0.04	Mature:	20%	0.008	0.8
Agave marmorata	0.04	Mature:	20%	0.008	0.8
Prickly pear (*Opuntia*)	3.6	Nopales:	19.0	68.4	6,840
		Fruits:	24.6	88.6	8,860
Cholla (*Opuntia*)	5.6	Fruits:	5.9	33.0	3,300
Myrtillocactus	1.5	Fruits:	504.8	757.2	75,720
Lemaireocereus	1.2	Fruits:	16.7	20.0	2,000
Ferrocactus	1.0	Fruits:	5.4	5.4	540
Mammillaria	31.6	Fruits:	0.34	10.7	1,070
Wild onions (*Allium*)	0.05	Bulbs:	1.0	0.05	5
Mesquite (*Prosopis*)	2.2	Pods:	80.0	176.0	17,600
Wild *Cucurbita*	0.04	Seeds:	75.0	3.0	300
Hackberry (*Celtis*)	0.01	Fruits:	2522	25.52	2,552

TABLE 18.5
Mesquite Grassland B: Estimated Number of Edible Plant Parts per Hectare per Year, Based on 1966–1976 Plant Censuses

Name of plant	No. plants per 100 m^2	No. edible parts per plant per yr		No. edible parts per 100 m^2	No. edible parts per ha
Huizache (*Acacia*)	1.0	Pods:	35.0	35.0	3,500
Prickly pear (*Opuntia*)	1.6	Nopales:	21.0	33.6	3,360
		Fruits:	42.0	67.2	6,720
Cholla (*Opuntia*)	2.3	Fruits:	5.9	13.6	1,360
Mammillaria	0.05	Fruits:	1.4	0.07	7
Mesquite (*Prosopis*)	8.33	Pods:	132.0	1099.6	109,960
Hackberry (*Celtis*)	0.02	Fruits:	3004	60.08	6,008

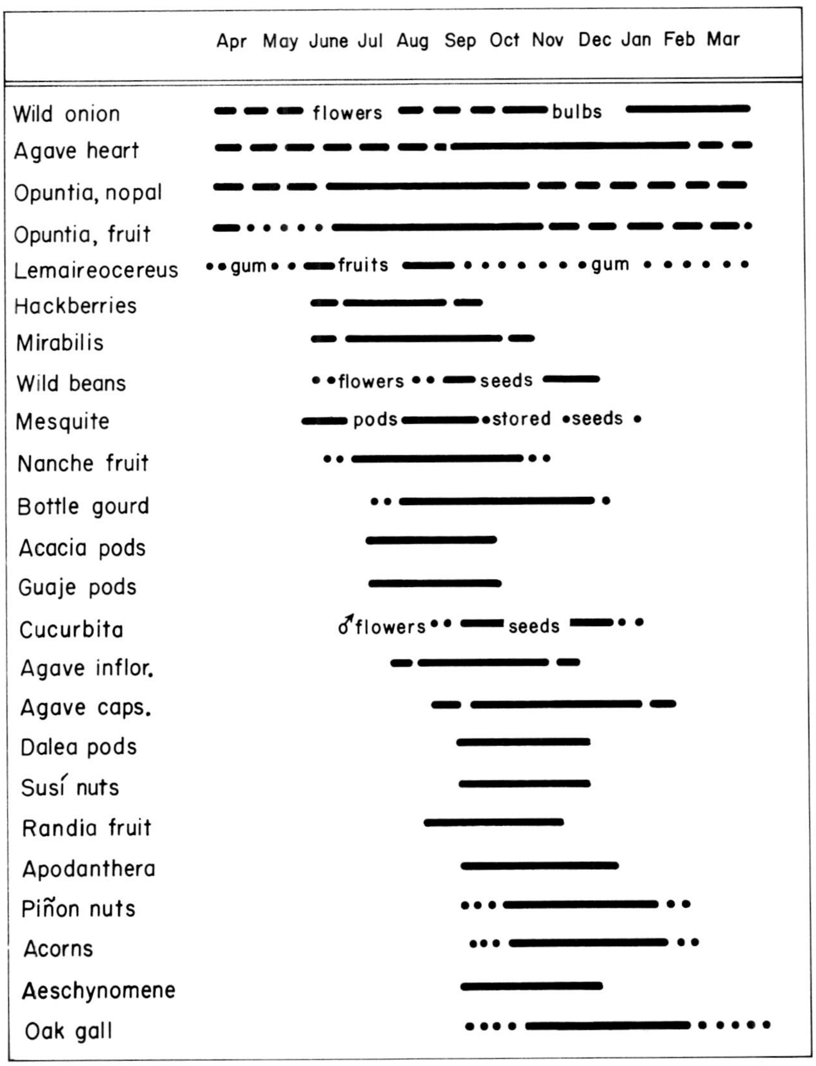

FIGURE 18.1. Seasonal availability of various plants in the Guilá Naquitz area. Solid lines indicate periods of peak availability, dashed or dotted lines indicate periods of occasional availability.

TABLE 18.6
Seasonal Differences in Productivity of Thorn Forest A[a]

Name of plant	Autumn (November)	Early spring (March)
Oak (*Quercus*)	1401.2	—
Tepeguaje (*Cassia*)	1730.0	—
Guaje (*Leucaena*)	142.2	—
Susí (*Jatropha*)	51.0	—
Catclaw (*Acacia*)	20.0	—
Huizache (*Acacia*)	35.0	—
Prickly pear (*Opuntia*)	10.5	22.5
Myrtillocactus	—	934.4

[a]Figures given are the average numbers of nuts, pods, or fruits per plant, based on 1966–1976 plant censuses.

Seasonal differences in productivity in the Guilá Naquitz area are striking, as Tables 18.6 and 18.7 reveal. Table 18.6 shows the differences between November and March harvests in Thorn Forest A. Acorns, *guaje* seeds, and *susí* nuts, usually abundant in the fall, have virtually all been eaten by spring; only the tuna, or prickly pear fruit, is present in both seasons. One of the few foods available in March is the small red fruit of the organ cactus the Zapotec call *bids-žob* (*Myrtillocactus*), and it is present at the rate of more than 900 per plant. Not a single fruit of this genus was found at Guilá Naquitz, which strongly suggests that the cave was not occupied in February or March. To be sure, the *bids-žob* (or *garambullo*, as it is called in Spanish) is one of those fruits that might have been eaten on the spot rather than brought back to the cave; but the same could be said for the hackberry, which was abundant in the cave even though it comes from much farther away.

Hackberries came largely from Mesquite Grassland B, represented in Table 18.7. Here it is clear that August is a month of abundant mesquite and hackberry, while February looks bleak. Once again, prickly pear fruit is available in both seasons. And the pods of the milkweed-like *binya'a* may survive until February, although most are pretty dry by then.

To summarize, the June–August period of the rainy season would be harvest time for the Mesquite Grassland, and autumn (September–December) harvest time for the Thorn Forest. There would be little plant food to eat in the Guilá Naquitz area between January and April.

ANNUAL VARIATION

Seasonal differences are only one source of variation in food resources for the eastern Valley of Oaxaca. There are also differences in productivity that result from annual variations in rainfall. In the Tlacolula region, only about two years out of every four have rainfall approaching the "average" figure of 500–650 mm; one year in four is likely to be drier than average (under 400 mm at times) and one in four wetter than average (over 800 mm at times; see Kirkby 1973; Smith 1978:9).

TABLE 18.7
Seasonal Differences in Productivity of Mesquite Grassland B[a]

Name of plant	Summer (August)	Winter (February)
Mesquite (*Prosopis*)	132	—
Hackberry (*Celtis*)	3004	—
Prickly pear (*Opuntia*)	34	50
Binya'a (Asclepiadaceae)	—	10

[a]Figures given are the average numbers of pods or fruits per plant, based on 1966–1976 plant censuses.

We were fortunate in that the period of our plant censuses included not only average years but also one markedly dry year (1967) and one markedly wet year (1968). This gave us a chance to observe the differences in productivity caused by annual variation within the same survey transects, and these are given in Table 18.8.

In December 1967—the dry season following a dry year—there were only 1279 acorns per oak tree and 8 tunas (fruits) per prickly pear within our Thorn Forest A transects. In February 1969—the dry season following a wet year—there were twice as many acorns (2688) per oak tree and more than three times as many tunas (27) per prickly pear. In September 1967—toward the end of the rainy season in a dry year—there were only 1386 pods per *tepeguaje* and 11 nuts per *susí* bush. In August 1968—during the rainy season of a wet year—there were roughly 50% more *tepeguaje* pods (2074) and nine times as many *susí* nuts (91) per plant. In the case of wild nanche fruits, the differences included the timing of maturation; trees that had not even ripened yet in September 1967 already had 30 nanches per tree in August 1968. Such annual variations in maturation time warn us not to try to specify too rigidly the months when Guilá Naquitz was occupied (see above).

Figures like those in Table 18.8 do more than illustrate the importance of annual variation in rainfall in a semiarid environment; they also provide us with figures that can be used in later analyses, such as Reynolds' simulation of preceramic adaptation in the eastern Valley of Oaxaca (Chapter 31). They also underscore the point that the occupants of Guilá Naquitz were not adapting to a static, predictable set of resources but to an environment whose productivity varied along two axes: *predictably* from season to season, and *unpredictably* from year to year.

ESTIMATING THE PRODUCTIVITY OF FOOD ANIMALS

As indicated in Table 18.1, we were able to estimate cottontail rabbit densities in the Guilá Naquitz area on the basis of dung pellet piles marking home ranges. Our estimates range from 1.0 to 3.2 rabbits per hectare, depending on vegetational facies.

TABLE 18.8
Observed Differences in Productivity between Drier-than-Average and Wetter-than-Average Years in Thorn Forest A near Mitla, Oaxaca[a]

	Rainy season, dry year		Dry season following dry year	Rainy season, wet year	Dry season following wet year
Name of plant	July 1967	Sept. 1967	Dec. 1967	Aug. 1968	Feb. 1969
Oak (*Quercus*)	Not yet ripe	Not yet ripe	1278.7	Not yet ripe	2687.5
Tepeguaje (*Cassia*)	<50	1386.2	<100	2073.9	<100
Guaje (*Leucaena*)	<10	113.1	<10	171.2	<10
Nanche (*Malpighia*)	In flower[b]	Not yet ripe	<10	30.0	None left
Susí (*Jatropha*)	Not yet ripe	11.0	<2	91.0	<5
Opuntia (fruits)	In flower	6.1	8.1	20.0	27.0

[a]Where figures are given, they are in average number of nuts, fruits, or pods per plant.
[b]Cultivated nanches on the valley floor near San Pablo Huitzo were already ripe by July 1967.

Unfortunately, our censuses told us nothing about the most important animal hunted by the occupants of the cave—the white-tailed deer. Deer can still be found in the mountains surrounding the Valley of Oaxaca, but their numbers have been severely reduced by centuries of environmental degradation and unrestricted hunting with firearms. Therefore, we are forced to produce an estimate of deer density based on wildlife studies conducted in similar environments elsewhere in Mexico.

Leopold (1959:508) has estimated that populations of white-tailed deer may have reached 30–40 individuals per square mile in favorably wooded areas of highland Mexico before overhunting, land clearance, overgrazing, and logging reduced them.[1] For the basis of our studies, we have converted the lower of these two figures (30 deer per mi^2) to its metric equivalent (12 deer per km^2) and used it as our estimate for deer populations in the undisturbed thorn forest of preceramic Oaxaca. Such an estimate, it should be noted, is conservative by comparison with some parts of North America.

Leopold (1959:508) states that "it is the thickets of scrub oak, manzanita, mesquite, acacia, and the second-growth *monte* of the tropics that form the deer's castle—not the forest proper." This would serve as a reasonably good description of Thorn Forest A, which was probably the best deer habitat in the eastern Valley of Oaxaca. While the deer presumably ate green grass during the rainy season, they usually rely heavily during all periods of the year on "browse," the twig tips, berries, and growing shoots of trees and shrubs. Among the favored types of browse listed by Leopold (1959:508–509) are many native to the Guilá Naquitz area, such as the morning glory tree, madroño, manzanita berry, and acorn. For these reasons, we consider our estimate of preceramic deer populations reasonable. This is fortunate, since the present population is too badly decimated to provide any meaningful figures.

We also relied on Leopold to help us in our estimates for the original density of bobwhite quail in the Guilá Naquitz area. Figures are not easy to pin down, because quail coveys aggregate or disperse depending on season and quality of range. Winter coveys "consist of 8 to 20 birds averaging about 12. Each covey has its own home area, which is usually about a quarter mile or less in radius" (Leopold 1959:247). That amounts to 1 covey per 0.20 mi^2, or an average of 60 quail per mi^2. Converting this to the metric system, and taking into account our own quail sightings in Thorn Forest A, we have settled on a figure of 1 quail per 4 ha.

As for the other animals, we unfortunately have only rough estimates. Based on our encounters with doves and pigeons in Thorn Forest A, we would estimate that there is at least one individual of each species per hectare. Mud turtles are common in mudholes and spring-fed pools in the Guilá Naquitz area, but since no such pools were crossed by our census transects, we have no basis for calculating the number of turtles per hectare.

CONVERTING RAW PRODUCTIVITY TO CALORIES AND PROTEIN

Earlier in this chapter we considered the facies of vegetation surrounding Guilá Naquitz in terms of food productivity per hectare under various conditions of annual and seasonal variation. In this section we use the nutritional data provided by Robson and Elias in Chapter 23 to convert those productivity figures into calories and protein.

Table 18.9 gives the productivity of a hectare of Thorn Forest A in kilograms of edible portion, estimated total kilocalories, and estimated grams of protein. With three exceptions, the figures in Column 2 are based on our 1966–1976 plant censuses. Figures on deer and quail densities are based on studies by Leopold (1959). No figures on piñon density are available, so we have made the assumption that the ratio of piñon nuts to acorns in Thorn Forest A was similar to the

[1]"In the Gavilán basin west of Pacheco, Chihuahua, I estimated a density of perhaps 30 to 40 whitetails per square mile in 1938 before the area was invaded by loggers. Populations of like density were found in the virgin mountains of southern Durango in 1946, on the protected pine–oak mesas of the Sierra de Tamaulipas in 1945, and in isolated sections of the Sierra del Carmen of Coahuila in 1953" (Leopold 1959:508).

ratio of piñon nuts to acorns in Guilá Naquitz Cave. Since piñon nuts were a relatively minor part of the refuse, any error introduced by this assumption should be minor.

The table makes clear that prickly pear fruit, nopales, and agave hearts contain most of the kilograms of food in the thorn forest. The cave refuse does not suggest, however, that they were the plants most heavily used during the season when Guilá Naquitz was occupied (see Chapter 19). There are several possible explanations for this. One is that they simply were not preferred foods, and were therefore not as heavily used when other plants were available. The thought of trying to eat 589 kg of prickly pear stems, even flavored with wild onions, is not particularly exciting. Perhaps the best that can be said for nopales and agave hearts is that they are dependable foods that would be available even in seasons when there is precious little else to eat.

Another possibility is that some items, such as prickly pear fruit, are underrepresented in our refuse because they were often eaten on the spot rather than brought back to the cave.[2] This possibility could be raised with regard to all the fresh fruits (tunas, nanches, hackberries) but probably not with regard to plants such as nopales and agave hearts that require considerable processing.

When one turns to the kilocalorie column in Table 18.9, two other plants, acorns and *susí* nuts, emerge as more imporant than one would have guessed from their sheer weight per hectare alone. These same plants, along with *guajes,* stand out as important in the protein column as well. While acorns require processing—a fact that probably helps to account for why so many were brought back to the cave—*guaje* seeds and *susí* nuts can be eaten raw. This means that, like the fruits already discussed above, they may be underrepresented in the cave because they were often eaten on the spot.

As for the animals of Thorn Forest A, it is interesting that cottontails, because of their density, do not lag far behind deer in terms of biomass, calories, and protein per hectare. Our suspicion is that while Naquitz phase hunters were probably willing to travel greater distances for deer, and were delighted to kill one, in the long run they probably ate rabbit more frequently.

Table 18.10 reveals the resources of Thorn Forest B to be similar in composition to those of Thorn Forest A but less abundant. Mesquite Grassland A (Table 18.11) is a real zone of transition, where a few of the resources of the Thorn Forest (such as *susí* and agave) are still available and some of the resources of Mesquite Grassland A are reasonably good, and it has the highest wild cucurbit density of any of the vegetational facies.

Finally we come to Mesquite Grassland B (Table 18.12)—a zone rich in resources but, as previously indicated, coming ripe at an earlier season than the Thorn Forest. Prickly pear fruits still provide the greatest weight of edible portion per

[2]While tunas are tasty, and probably were often eaten on the spot, there is a limit to how many you can eat: They are famed in Mexican song and legend for causing constipation. Since acorn flour is notorious for causing diarrhea, perhaps the combination of the two in the diet kept everything stable.

TABLE 18.9
Estimated Productivity of One Hectare of Thorn Forest A[a]

Item	No. found in 1 ha	Kilograms of edible portion represented	Estimated total kilocalories	Estimated total protein (g)
Piñon nuts*	1,340	0.2	1,270	26
Acorns	30,820	98.6	283,968	2,396
Guaje seeds	199,100	24.9	17,928	2,017
Susí nuts	37,740	18.9	112,077	3,326
Agave hearts				
potatorum	122	341.0	429,660	1,057
marmorata	29	377.0	475,020	1,169
Nanche fruits	360	1.3	897	22
Prickly pear				
nopales	22,420	588.50	70,620	11,358
fruits	19,470	973.50	350,460	10,806
Wild beans	350	0.04	52	3
Deer*	.12	2.40	3,024	504
Cottontails	3.2	1.90	2,565	399
Quail*	0.25	0.02	35	5.2

[a]Estimates are based on 1966–1976 plant censuses and published data from other sources, nutritional data are from Chapters 23 and 24, and estimates for asterisked items are based on sources other than 1966–1976 plant censuses in the eastern Valley of Oaxaca.

TABLE 18.10
Estimated Productivity of One Hectare of Thorn Forest B[a]

Item	No. found in 1 ha	Kilograms of edible portion represented	Estimated total kilocalories	Estimated total protein (g)
Susí nuts	22,950	11.5	68,195	2,024
Agave hearts				
potatorum	4	11.2	14,112	35
marmorata	3	39.0	49,140	121
Nanche fruits	50	0.18	124	3
Prickly pear				
nopales	2,660	69.8	8,376	1,347
fruits	1,540	77.0	27,720	855
Wild beans	175	0.02	24	2
Wild onions	2	0.02	8	<1
Mesquite pods	160	0.27	802	15
Wild cucurbit seeds	38	0.006	26	2
Deer*	0.12	2.4	3,024	504
Cottontails	1.5	0.9	1,200	187
Quail*	0.25	0.02	35	5.2

[a]Estimates are based on 1966–1976 plant censuses and published data from other sources, nutritional data are from Chapters 23 and 24, and estimates for asterisked items are based on sources other than 1966–1976 plant censuses in the eastern Valley of Oaxaca.

TABLE 18.11
Estimated Productivity of One Hectare of Mesquite Grassland A[a]

Item	No. found in 1 ha	Kilograms of edible portion represented	Estimated total kilocalories	Estimated total protein (g)
Susí nuts	80	0.04	237	7
Agave hearts				
potatorum	0.8	2.2	2,772	7
marmorata	0.8	10.4	13,104	32
Prickly pear				
nopales	6,840	179.6	21,552	3,466
fruits	8,860	443.0	159,480	4,917
Wild onions	5	0.06	23	1
Mesquite pods	17,600	29.4	87,318	1,676
Wild cucurbit seeds	300	0.05	214	17
Hackberries	2,552	0.46	152	2
Deer*	0.12	2.4	3,024	504
Cottontails	1.0	0.6	810	126
Quail*	0.25	0.02	35	5.2

[a]Estimates are based on 1966–1976 plant censuses and published data from other sources, nutritional data are from Chapters 23 and 24, and estimates for asterisked items are based on sources other than 1966–1976 plant censuses in the eastern Valley of Oaxaca.

TABLE 18.12
Estimated Productivity of One Hectare of Mesquite Grassland B[a]

Item	No. found in 1 ha	Kilograms of edible portion represented	Estimated total kilocalories	Estimated total protein (g)
Prickly pear				
nopales	3,360	88.2	10,584	1,702
fruits	6,720	336.0	120,960	3,730
Mesquite pods	109,960	183.6	545,292	10,465
Hackberries	6,008	1.1	363	4
Deer*	0.12	2.4	3,024	504
Cottontails	1.0	0.6	810	126
Quail*	0.25	0.02	35	5.2

[a]Estimates are based on 1966–1976 plant censuses and published data from other sources, nutritional data are from Chapters 23 and 24, and estimates for asterisked items are based on sources other than 1966–1976 plant censuses in the eastern Valley of Oaxaca.

hectare, as they have in every zone; but when it comes to estimated calories, nothing can compare to mesquite. More than 100,000 pods per hectare, yielding half a million kilocalories and more than 10,000 g of protein, make this the key species of its vegetational facies. Hackberries are abundant, and it is likely that many more were eaten on the spot than ever showed up in the cave; but their contribution in vitamins was probably greater than that for protein and calories.

19

Preceramic Plant Remains from Guilá Naquitz

C. Earle Smith, Jr.

The preceramic levels at Guilá Naquitz produced some 21,705 identifiable plant remains, not counting grass and wood (Table 19.1). Most numerous of all these plant remains were acorns (8424), followed by mesquite seeds (4603), hackberry seeds (3112), and prickly pear seeds (1819). Literally dozens of other plants were represented, though in lower frequency.

Plant remains from all living floors have been sorted and identified in comparison with modern plant specimens from the area. For the purpose of this chapter, material from all preceramic levels is reported under identifications grouped by an arrangement known as the Engler and Prantl system of classification. As with all other series with which I have worked, classification by usage is hazardous, because we do not know what use many of the remains served. Furthermore, some plants can be readily classified in several use categories. Thus, the botanical scheme is the most practical way of arranging the identifications. However, in some cases I have also indicated the use to which a particular plant may have been put.

CATALOGUE OF THE PLANT REMAINS

Pinaceae (Pine Family)

Pinus sp. (Fig. 19.1): The seeds of short-needled pines with stubby cones, known as piñons, are a rich source of oil and protein when they are available. Piñons were present in all preceramic levels except Zone B2+3. They were most plentiful in the oldest level and gradually diminished in number upward: Zone E (155), Zone D (94), Zone C (80), Zone B3 (4), Zone B2 (2), and Zone B1 (31).

Typhaceae (Cattail Family)

Typha sp.: Two fragments of cattail inflorescence were found in Zone B1. Almost certainly, these were brought into the cave along with a supply of leaves of this plant to be used in the preparation of mats. Also, the pollen is sometimes gathered for food (it is produced very copiously) and the rhizomes are also eaten. At the time that these fragments were brought in, water may have been impounded someplace nearby, but it is more likely that these were brought up from the river margin in the lower valley.

Graminae (Grass Family)

Numerous masses of grass, apparently brought into the cave for bedding, furnish little identifiable material. In order to identify grass specimens, nearly mature inflorescences are needed. The only preceramic specimens identifiable to genus are some examples of *Heteropogon* sp. from Zones B2 and B3.

Inasmuch as no significant quantity of caryopses was recovered from the fill in spite of a search for them, I can only conclude that grasses did not furnish any significant amount of food for the people inhabiting Guilá Naquitz.

Bromeliaceae (Pineapple Family)

Plants of the pineapple family were used in the Oaxaca Valley as a source of leaf fibers.

Hechtia sp.: Leaf fragments were present in Zones B1 (6) and B2 (5).

Tillandsia usneoides L. (Fig. 19.2): Fragments of Spanish moss may have had some ceremonial significance, as it is used at the present time for decoration for special fiestas. It must

TABLE 19.1
Plant Remains (Exclusive of Grass and Wood) Recovered at Guilá Naquitz

Plants	Stratigraphic zones							
	E	D	C	B3	B2 + 3	B2	B1	Totals
Piñon nuts	155	94	80	4	—	2	31	366
Cattails	—	—	—	—	—	2	—	2
Hechtia	—	—	—	—	—	5	6	11
Tillandsia	3	1	2	—	—	4	8	18
Onion bulb coats	—	2	—	—	—	—	1	3
Agave quids	6	49	196	18	8	45	46	368
Agave leaves	2	4	20	1	—	3	10	40
Agave capsules	1	8	—	—	—	1	—	10
Agave inflorescence	—	—	1	1	—	—	—	2
Agave, other	—	1	—	—	—	—	1	2
Acorns	1846	3182	1570	421	95	418	892	8424
Oak galls	1	—	—	—	—	—	—	1
Hackberry seeds	1166	416	969	65	149	107	240	3112
Mirabilis	—	1	—	—	—	—	—	1
Avocado seeds	—	—	—	—	—	2(?)	—	2(?)
Acacia pods	1	1	—	1	—	—	—	3
Aeschynomene	17	4	11	—	—	—	—	32
Guaje pods	—	11	247	5	6	43	105	417
Bean pod valves	161	17	43	26	9	104	111	471
Bean seeds	—	—	—	—	2	—	15	17
Mesquite pods	—	3	7	2	3	—	16	31
Mesquite seeds	32	59	554	236	276	1613	1833	4603
Dalea pods	812	21	—	—	—	—	—	833
Nanche seeds	6	49	196	18	8	45	46	368
Susí **seed coats**	98	237	68	15	3	12	22	455
Cardiospermum	—	10	—	—	—	3	—	13
Cotton seeds	—	—	—	—	—	1(?)	—	1(?)
Lemaireocereus	—	—	—	—	—	—	1	1
Opuntia nopales	3	28	49	—	2	31	45	158
Opuntia fruits	2	—	2	—	1	1	4	10
Opuntia seeds	135	348	751	—	360	82	143	1819
Thevetia seeds	—	—	—	—	—	—	1	1
Milkweed stems	—	—	1	—	—	—	—	1
Chile stems	—	—	—	—	—	—	2(?)	2(?)
Randia fruits	—	1	3	—	1	—	—	5
Cucurbit rinds	—	4	35	—	3	5	8	55
Cucurbit seeds	—	1	5	1	—	2	8	17
Cucurbit, other	—	1	10	—	—	4	5	20
Apodanthera	—	1	1	—	—	—	2	4
Bottle gourd	—	—	3	—	1	1	2	7
	4447	4554	4824	814	927	2535	3604	21,705

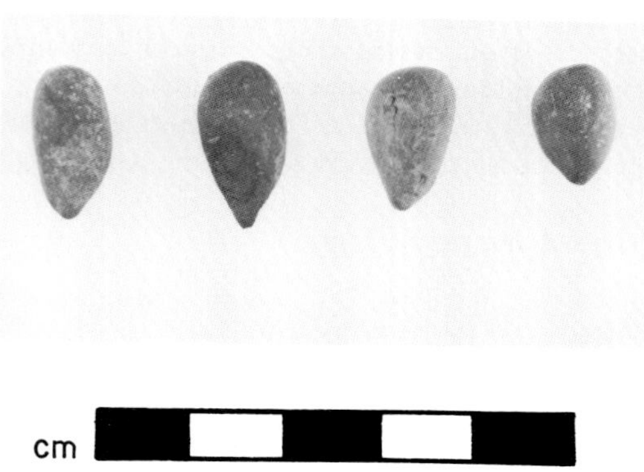

Fig. 19.1. Piñon nuts (*Pinus* sp.).

have been gathered higher on the mountain slope, either on an expedition for this purpose or when some other product was being sought, because it does not grow at the elevation of the cave. *Tillandsia usneoides* requires higher rainfall and a more humid atmosphere than that of the cave area today. Fragments were present in Zones B1 (1), B2 (1), C (1), and D (1).

Tillandsia sp.: Broad-leaved tillandsias were probably used for leaf fibers. Leaf fragments were present in Zones B1 (7), C (1), and E (3); seed pods were present in Zone B2 (3).

Liliaceae (**Lily Family**)

Allium sp. (Fig. 19.3): Outer bulb coats of a wild onion (*Allium*) were present but infrequent in preceramic levels: Zone B1 (1) and Zone D (2).

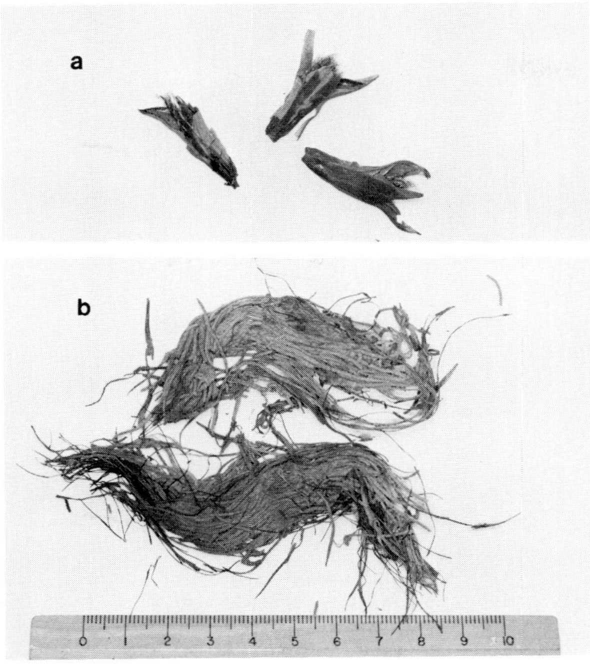

Fig. 19.2. Spanish moss (*Tillandsia usneoides*): a, seed pods that probably fell from a plant brought into the cave for fiber extraction; b, mats of Spanish moss found in Zone B1 that may have been used for bedding, much as oak leaves were (scale in cm).

Amaryllidaceae (Amaryllis Family)

Agave spp. (Fig. 19.4): Magueys have always played an important role in the domestic economy of the Mexican people. In preceramic times, they formed a portion of the food supply and must have been used for their stout leaf fibers. *Agave* remains recovered are as follows: Zone B1, 46 quids, 10 leaf fragments, and 1 bundle of fiber; Zone B2, 45 quids, 3 leaf fragments, and 1 capsule; Zone B2+3, 8 quids; Zone B3, 18 quids, 1 leaf fragment, and 1 inflorescence fragment; Zone C, 196 quids, 1 spine, 20 leaf fragments, and 8 capsules or capsule valves; and Zone E, 6 quids, 2 leaf fragments, and 1 capsule fragment.

Fagaceae (Oak Family)

Quercus sp. (Fig. 19.5): It is obvious from the amounts of material recovered that acorns must have constituted a major food resource for the people inhabiting Guilá Naquitz Cave. Although these were available only during the fall, they are easily storable and may have furnished food during the dry portion of the year. *Quercus* remains recovered are as follows: Zone B1, 892 acorns, and 10 leaves; Zone B2, 418 acorns; Zone B2+3, 95 acorns; Zone B3, 421 acorns; Zone C, 1570 acorns and 10 leaves; Zone D, 3182 acorns; and Zone E, 1846 acorns.

Oak gall: Certain wasps of the family Cynipidae puncture the bark or leaf of the oak tree and lay their eggs in the wound. The irritation stimulates a hypertrophy or enlargement on the bark or leaf, and the wasp larva feeds on this.

Fig. 19.3. Wild onion bulb coats (*Allium* sp.) from which the bulbs were removed.

A single gall appeared in Zone E of Guilá Naquitz, presumably having been collected at the same time acorns were brought back to the cave. Whether the occupants used such galls is uncertain (the tannic acid content is quite high).

Ulmaceae (Elm Family)

Celtis sp. (Fig 19.6): Fruit of the local species of hackberry (*Celtis*) was apparently used consistently, being the third most common plant at Guilá Naquitz: Zone B1 (240), Zone B2 (107), Zone B2+3 (149), Zone B3 (65), Zone C (969), Zone D (416), and Zone E (1166).

Nyctaginaceae (Four-O'clock Family)

Mirabilis sp.: A single seed of Mirabilis was found in Zone D. Because plants of this group grow wild in the area, this is not too surprising. However, it is not clear why the seed was present, as they are not known to be used at all nowadays.

Lauraceae (Laurel Family)

Persea americana Mill.: Only 2 cotyledons of avocado were found in the lower levels of the cave (Zone B2; 2 cotyledons). Although these were different in shape from the majority of cotyledons in upper strata, this isolated find should be regarded with caution, as we have no assurance that these specimens are not intrusive from a Postclassic level (Smith 1969).

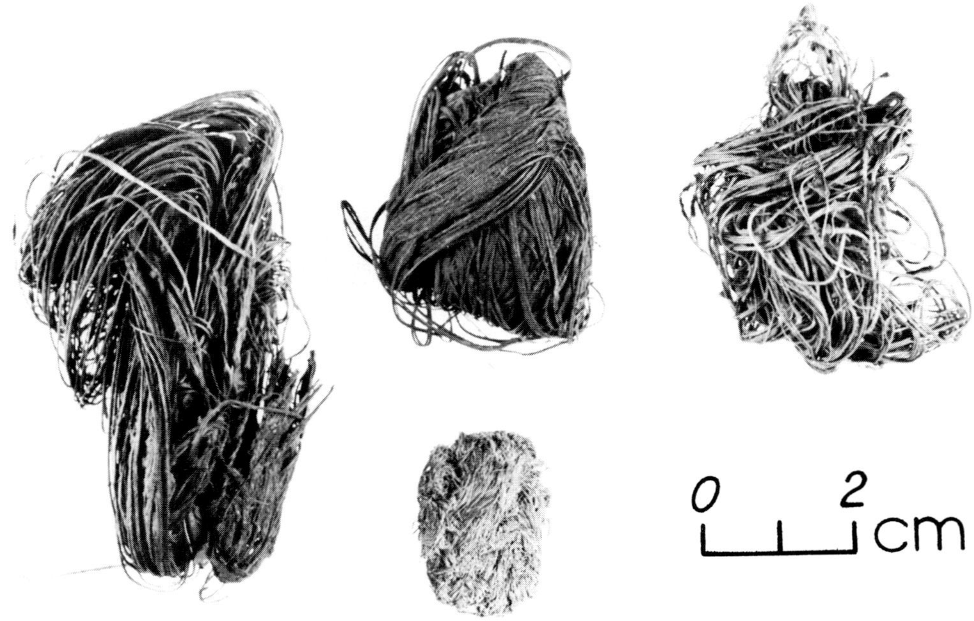

Fig. 19.4. Quids of chewed maguey (*Agave* spp.) common in all levels.

Fig. 19.5. Acorns (*Quercus* spp.). At least two species are represented by these specimens.

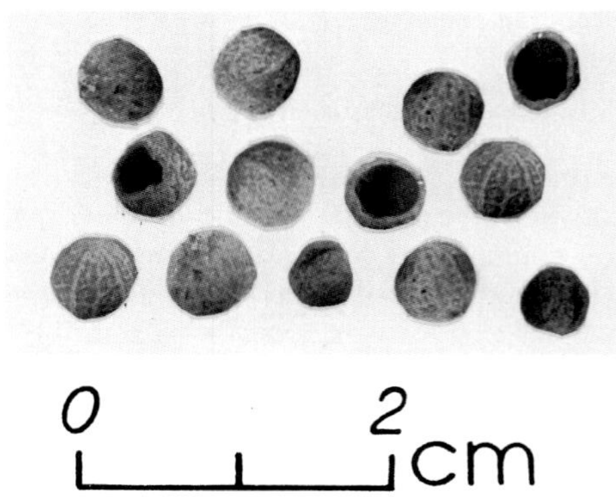

Fig. 19.6. Seeds from hackberry fruits (*Celtis* sp.).

Leguminosae (Pea Family)

Acacia sp.: The early inhabitants of Guilá Naquitz Cave did not find the pods of these plants as attractive as did later people. Only one pod fragment was found in each of three strata: Zones B3, D, and E.

Aeschynomene sp. (Fig. 19.7): The following loments of this shrub were found: Zone C (11), Zone D (4), and Zone E (17). It is possible that these were used for food, but we have no current record of a use for this plant.

Leucaena sp. (Fig. 19.8): From the number of pod fragments recovered, we can assume that *guaje* was one of the regularly used food resources in the area. Pod fragments recovered are as follows: Zone B1 (105), Zone B2 (43), Zone B2+3 (6), Zone B3 (5), Zone C (247), and Zone D (11).

Phaseolus sp.: The bean remains for the preceramic levels of Guilá Naquitz Cave have been studied by Lawrence Kaplan (Chapter 21). He has found these levels to contain pod valves and seeds of a wild, vining, hypogeal germinating *Phaseolus* bean that cannot be identified to species and has therefore been designated Guilá Naquitz Type 1 (Fig. 19.9). Of this bean, Kaplan (1981:248) has recently written, "the Mitla wild bean is marked as a *Phaseolus* species by its general morphology and by hooked trichomes on the pod epidermis but its embryo structure excludes it from *P. vulgaris*." Zones B1–E

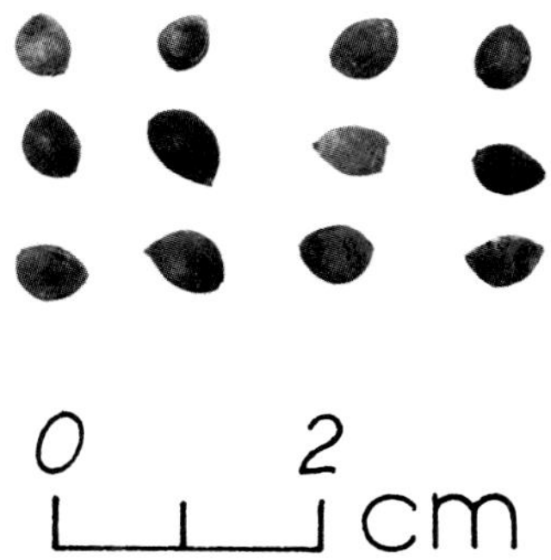

Fig. 19.7. Seeds of *Aeschynomene* sp.

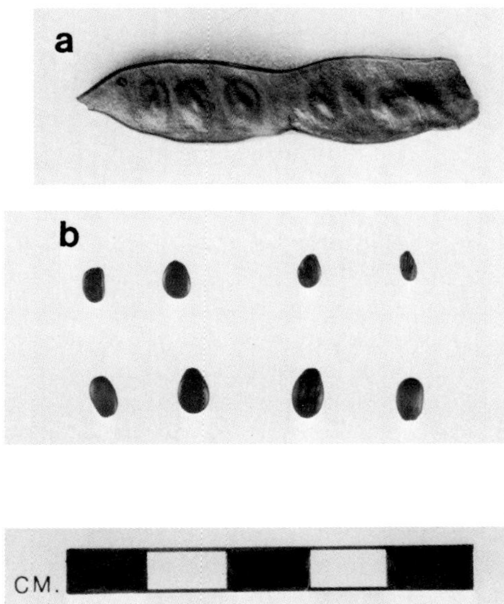

Fig. 19.8. Remains of *guaje* (*Leucaena* sp.): a, fragment of pod; b, seeds.

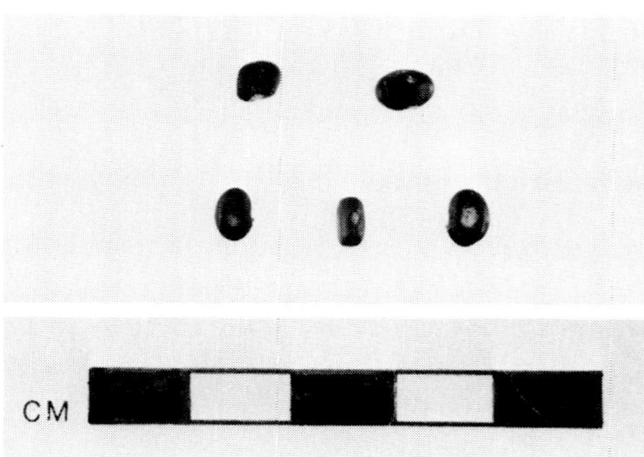

Fig. 19.9. Wild black runner beans (see Kaplan, Chapter 21).

produced 471 pod valves of which more than 168 are identifiable as Guilá Naquitz Type 1: Zone B1, 15 seeds and Zone B2+3, 2 seeds. Four specimens of the common bean, *P. vulgaris*, were also found but should be regarded with caution because they occurred in contexts where the possibility of their being intrusive from Postclassic levels cannot be ruled out.

Prosopis juliflora (Sw.) DC. (Fig. 19.10): Gathering the pods of mesquite has apparently been a traditional early rainy season activity from early times. These certainly furnished a major food resource during later occupations. The following were found: Zone B1, 1833 seeds and 16 pods; Zone B2, 1613 seeds; Zone B2+3, 276 seeds and 3 pods; Zone B3, 236 seeds and 2 pods; Zone C, 554 seeds and 7 pods; Zone D, 59 seeds and 3 pods; and Zone E, 32 seeds.

Dalea sp. (Fig 19.11): Pods of this plant are small, but the punctate dots and the shape make the identification certain. So far as I know, there is no current use for these fruits. They may well have had a medicinal use for the people who gathered them. Pods were found in Zones D (21) and E (812).

Malpighiaceae (Malpighia Family)

Malpighia sp. (Fig. 19.12): The fruits of shrubs and trees of this group are known as nanches. Even today they are a

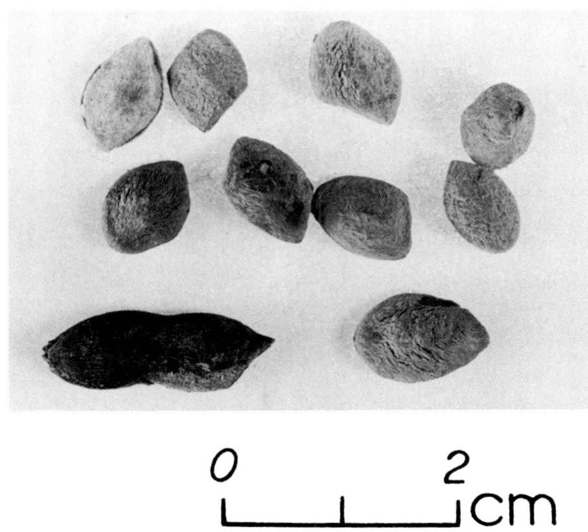

Fig. 19.10. Seeds (top, center, and lower right) and pod fragment (lower left) of mesquite (*Prosopis juliflora*).

popular fruit when they are available. Seeds were found in the following zones: Zone B1 (46), Zone B2 (45), Zone B2+3 (8), Zone B3 (18), Zone C (196), Zone D (49), and Zone E (6).

Euphorbiaceae (Spurge Family)

Jatropha neodioica (Fig. 19.13): The seeds of *susí* are prominent in the remains found in the cave. These seeds are widely eaten by Zapotec shepherd boys today, and nutritional

Fig. 19.11. Numerous seed pods of *Dalea*, a leguminous plant, found in Zone E. They may have been eaten or perhaps were considered to be medicinal (scale in cm).

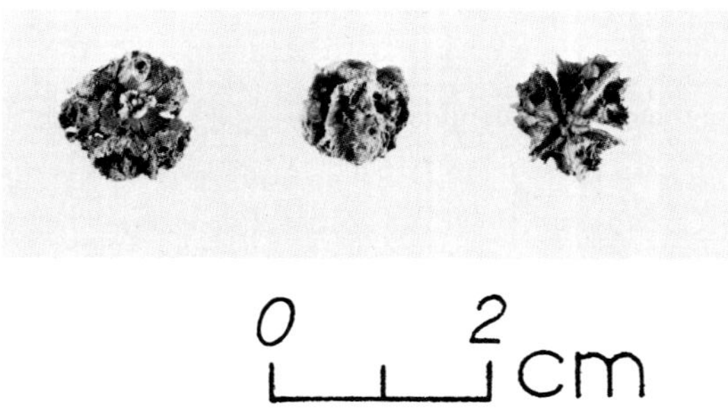

Fig. 19.12. Nanche seeds (*Malpighia* sp.).

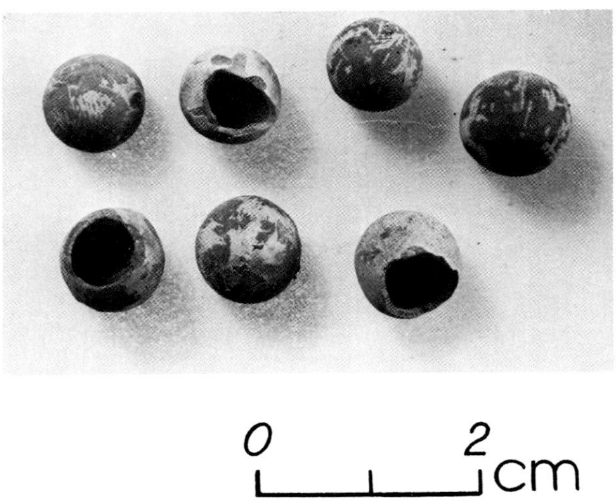

Fig. 19.13. Seed coats from which the nuts of the *susí* (*Jatropha neodioica*) have been removed.

analyses (Chapter 23) show that they were one of the richest plant protein sources available to the preceramic cave dwellers. Seeds were recovered in the following zones: Zone B1 (22), Zone B2 (12), Zone B2+3 (3), Zone B3 (15), Zone C (68), Zone D (237), and Zone E (98).

Sapindaceae (Soapberry Family)

Cardiospermum sp.: Balloon vine seeds were found in Zones B2 (3) and D (10). Because this plant grows in the area, the seeds may have been casual introductions or they may have had some use of which I am unaware. I know of no current use for them.

Malvaceae (Mallow Family)

Gossypium sp.: A single cotton seed was found in Zone B2. Lacking any other evidence for cotton at this time, I must assume that the seed was intrusive from a Postclassic level.

Cactaceae (Cactus Family)

Lemaireocereus sp.: A ball of plant gum, probably obtained from a plant of this genus, was found in Zone B1. Burning of fragments produced no odor of the resins normally associated with copal or conifer gums and resins. It may have been used as an adhesive for hafting projectile points or may have been saved for some medicinal or ceremonial use, but the material furnished no direct evidence for any of these uses.

Opuntia sp. (Fig. 19.14): Stems (nopales) and fruits (tunas) of these plants have always been popular with the inhabitants of Mexico, if we may judge from the archaeological remains that have been recovered to date. I have no doubt that they formed an important part of the diet of the people of Guilá Naquitz Cave. The following fragments were recovered: Zone B1, 45 nopales, 143 seeds, and 4 tunas; Zone B2, 31 nopales, 82 seeds, and 1 tuna; Zone B2+3, 2 nopales, 360 seeds, and 1 tuna; Zone C, 49 nopales, 751 seeds, and 2 tunas; Zone D, 28 nopales and 348 seeds; and Zone E, 3 nopales, 135 seeds, and 2 tunas.

Apocynaceae (Dogbane Family)

Thevetia sp.: A single *Thevetia* seed was found in Zone B1. It probably has no significance, but the plants are known to be poisonous.

Asclepiadaceae (Milkweed Family)

A single stem fragment from Zone C may represent this family. Because of the bast fibers, asclepiadaceous stems are used for tying or for the extraction of the fiber for spinning.

Solanaceae (Potato Family)

Capsicum annuum L.: Only two stem fragments of chile were found in Zone B1 and these could conceivably be intrusive from the Postclassic level immediately above. They should be regarded with caution and not taken as evidence for domestication at this date.

Rubiaceae (Madder Family)

Randia sp. (Fig. 19.15): Fruit of the local *Randia* is known as "mora" and is eaten out-of-hand. Apparently use of the

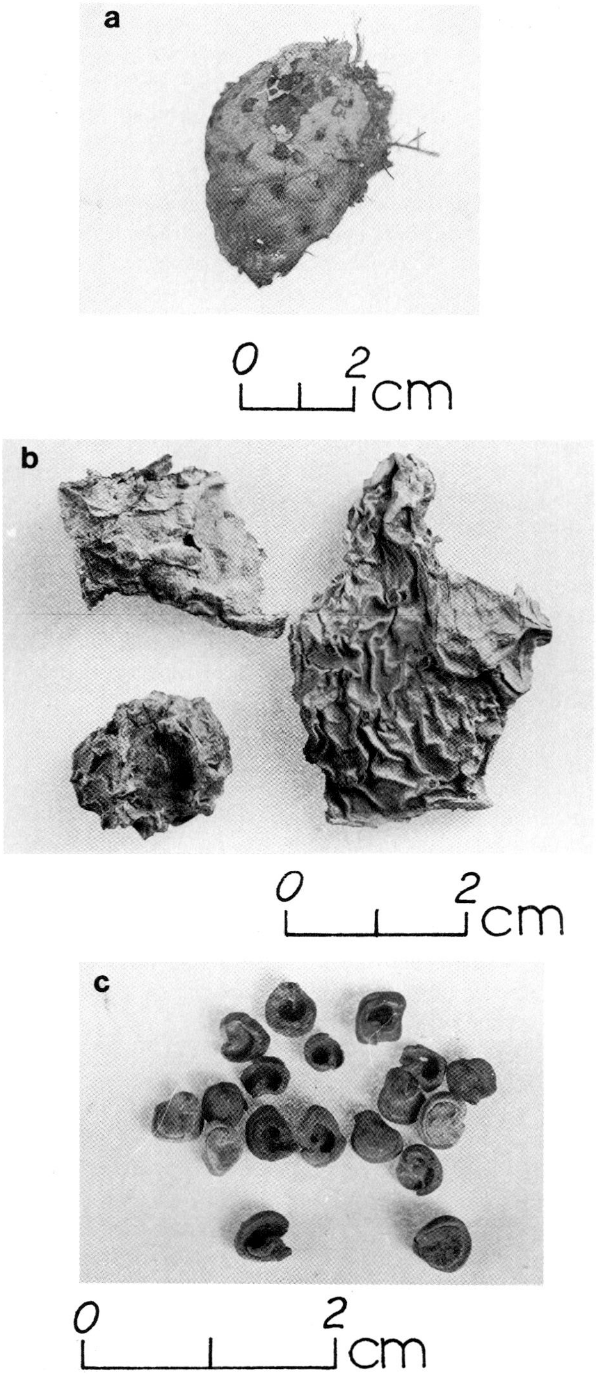

Fig. 19.14. Remains of prickly pear (*Opuntia* sp.): a, small nopal (stem section) from Zone B1; b, dried skins of tunas (fruits); and c, loose seeds from fruits. (Note different scales.)

fruit has a long history as the fragments found in the following zones show: Zone B2+3 (1), Zone C (3), and Zone D (1).

Cucurbitaceae (Squash Family)

Cucurbita pepo L.: Remains of the cultivated squash or pumpkin are rare in the preceramic levels of the cave, but

Fig. 19.15. The dried fruit of the mora (*Randia* sp.).

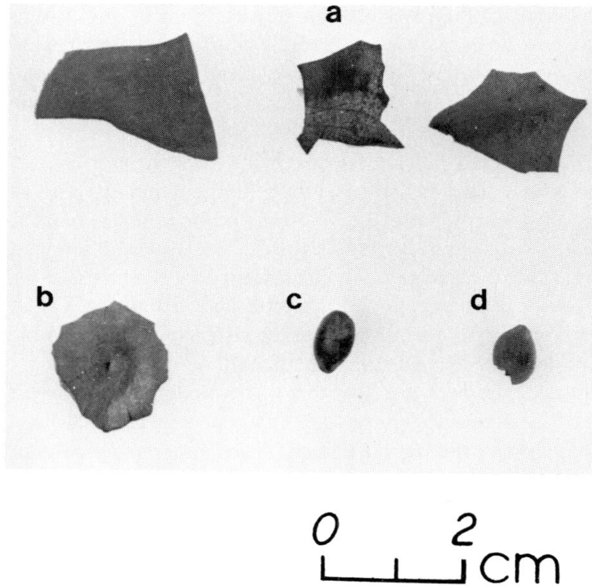

Fig. 19.16. Cucurbit remains from Zone D: a, rind fragments; b, peduncle; c, *Apodanthera* seed; and d, *Cucurbita pepo* seed.

fragments identified by Whitaker and Cutler as belonging to this species were found (Chapter 20). It had been thought that *C. pepo* originated far to the northward in northeastern Mexico where its closest wild relative, *C. texana*, can still be found (Cutler and Whitaker 1961). However, the presence of nine *C. pepo* seeds (Fig. 19.16d) and stems in Zone B, four *C. pepo* seeds and stems in Zone C, and single seed in Zone D definitely establish the cultivation of this plant in the preceramic period of Oaxaca at a surprisingly early date. This in turn has caused Whitaker and Cutler (Chapter 20) to call for some rethinking of the northern-origin hypothesis.

Apodanthera sp. (Fig 19.16c): Seeds of this wild plant were found in the following zones: Zone B1 (2), Zone C (1), and Zone D (1). It is possible that these were harvested as a food, for many kinds of wild cucurbits seem to have been used in the preceramic period. However, it is no longer eaten by the Mitla Zapotec.

Other cucurbits: Numerous fragments of *Cucurbita* rind (19.16a) were found in Zones B and C, but they cannot be identified to species. They may have come from wild fruit locally collected for edible seeds or for the meager fruit flesh, which is sometimes used as a soap.

Lagenaria siceraria (Mol.) Stadl.: The bottle gourd was widely used as a container by preceramic hunters and gatherers in Mesoamerica and may have been one of the oldest domesticated plants there. Seven fragments showed up in preceramic Guilá Naquitz: Zone B1 (2), Zone B2 (1), Zone B2+3 (1), and Zone C (3).

INTERPRETATION OF THE PLANT REMAINS

In order to understand more properly the relationship of the people of the caves to their subsistence economy and to the environment, we must carefully consider the time period with which we are dealing. If we round off the interpretations possible from the ^{14}C dates, we can say that we are dealing with the period from 7000 to 8000 B.C. For this time period in North America, relatively little evidence for the way plants were used has come to light. In the temperate area of the continent, Danger Cave, Utah, perhaps approaches it most closely (Jennings 1957). In Mexico, the lower levels of Coxcatlán Cave in the Tehuacán Valley represent this time period. Otherwise, our information concerning the way of life of these early people must be reconstructed from the more durable artifacts that they left behind.

One of the first observations to be made on the current lot of remains is the immediate contrast in number of items used between the period discussed here and the much later Postclassic levels represented in several caves near Guilá Naquitz. From a total of 51 recognized species in use in these later levels, the people of the preceramic levels were using only 23 species (of which one or two are doubtful, because they occur as only a single small fragment). In part, this is because a few species were not locally available to them. These species became available to later people through introduction from another area. In part, though, we find that these people were not completely utilizing the resources of their immediate environment. A number of the cacti, for instance, have apparently always been available nearby. However, relatively little use was made of the arboreal cacti, unless the material was consumed *in situ* during harvest operations. Certainly, little of it found its way into the cave deposits. Similarly, some of the legume pods that were later found in some quantity, were missing conspicuously from the earlier cave levels. Even the grasses, which are normally expected as a starch component in a diet, were lacking in the remains recovered in the early levels. Because of the milling or other preparation that most grasses require, these almost certainly would have been brought into the shelter of the cave if they were used in abundance.

The diet of these people, as represented by the cave remains, was a relatively simple one. Virtually no starch component is seen in the plant remains. That which is present is represented by the remains of acorns, which are actually more plentiful toward the lower levels of the cave deposit. These fruits, locally gathered from the oaks that are at and above the cave level today (approximately 1900 m) might have been held in storage to furnish a supply during the large portion of the year when they would not have been available.

Carbohydrate intake was not deficient, though, because the sugars present in the mesquite pods, available primarily in May, and the maguey hearts, usually harvested during the late summer and fall, provide a good share of carbohydrates. There is also sugar present in the wild fruits, such as the *Celtis* berries, nanches, and tunas that were utilized.

Plant protein materials were represented in the diets of the Guilá Naquitz people by several items. *Guaje* pods are found in all of the levels, although they are never abundant. *Susí* nuts were a good protein resource and they were obviously heavily used in early levels. Mesquite, though, was obviously a most important food plant, and the seeds are rich in protein. Furthermore, these people were harvesting beans in sufficient quantity to drop moderate amounts of valves in the debris in all levels.

Minor nutritional elements were well taken care of by the green *guaje* pods, the nopales, and, probably, by other leafy plants. It is probable that these people made casual use of many other local plants that they gathered and ate while on other business outside of the caves. These plants would not show in the remains within the caves and would not be a prime food source for the inhabitants.

Vegetable fats are largely lacking in their food. Piñon nuts were apparently used sparingly, and they are more abundant in earlier levels. However, they were not a major component in the gathered food items. The two avocado cotyledons in Zone B1 may be intrusive due to animal activity. Even if they were not, they are such meager evidence of use as to be practically unimportant.

AGRICULTURE IN THE PRECERAMIC

The major evidence in support of the use of cultivation is the unmistakable presence of *Cucurbita pepo* remains in Zones B, C, and D and of bottle gourds in Zones B and C. It reinforces the discovery of early remains of these same species in the Tehuacán Valley, where the evidence was not nearly so convincing (Cutler and Whitaker 1967). At that time, I suggested that several plants were in primitive cultivation, although little proof could be assembled to support this other than the long time needed for noticeable changes to occur in the tree fruits. We now have evidence of sufficient weight to state unequivocably that cultivation was being practiced in the Oaxaca Valley at the time that Zone D was being deposited, or around 8000 B.C. This fits well into the long time scale needed for the gradual realization of the value of agriculture to a people living on a gathering base. The full development of a social order in which people aggregated into villages with an economy fully geared to cultivated crops could not come about in the short period that many have granted for this development.

Even though the circumstantial evidence for agriculture this old in the New World was presented in the Tehuacán Valley report, none of the plant remains of that area definitely confirmed it (Smith 1967). The biological facts indicated by the selection of avocados, *cosahuicos*, and *chupandillas* for greater size ensured that a knowledge of reaction of plants under cultivation was already old by 5000 B.C. In South America, the recovery of cultivated *Phaseolus vulgaris* (common beans) and *Capsicum chinense* (pepper) from early levels in Guitarrero Cave support the Mexican evidence (Smith, in Lynch, 1980). In the valleys of southern Mexico, this knowledge of the reaction of plants under cultivation must have been coupled with a thorough understanding of the need for water over the long (November–May) dry period. No perennial plant could have been grown in this area away from an available water table or without the help of artificial water supply. Knowledge of the effects of selection is now abundantly proven by the cultivated plant remains from Guilá Naquitz Cave dating to about 8000 B.C.

The most important point is not the age of the cultivars in terms of years. Biologically, the effects of natural selection operate at all times under all conditions. Humans, in the course of their daily affairs, discovered that they could influence the course of natural selection through judicious selection of seed for planting. Even under the best of circumstances (i.e., a potential crop plant of infinite genetic variability), the complex interrelationships of gene action frequently mean the shifting of whole blocks of characteristics with an attempt to change only a single characteristic, such as flavor. For artificial selection to sort out effectively these complexities, a great deal of time and many generations of plants are needed. Thus, the most important fact presented by the presence of one cultivar in Guilá Naquitz Cave at about 8000 B.C. and two by 7000 B.C. is that a very long period of time before this has been needed for the human selection for those characteristics that make these recognizable as cultivated varieties.

Since Guilá Naquitz was apparently used by a very small group of people (perhaps no more than 4–5), and perhaps for no more than one season of the year, we do not know if their relatively conservative use of plants was characteristic of the region as a whole. During the preceramic levels, only domestic cucurbits were added to the local wild plants utilized. We find that the percentage of usage changes slightly, because acorns, *Celtis* berries and *susí* nuts were obviously more important in the earliest levels. The use of nopal and maguey remained relatively stable during this time period, while mesquite gradually increased in use, as did beans. It is not possible to attribute these shifts to changing vegetational patterns because we have no evidence that such changes in vegetation occurred.

VEGETATIONAL HISTORY OF THE GUILA NAQUITZ AREA

The plant remains from the caves near Mitla, Oaxaca, have once again furnished us with a history of the vegetation over a relatively long period of time. This, in turn, provides a check on possible major changes in the climatic patterns and also some measure of the use of land.

Schoenwetter and Smith (Chapter 15) feel that the pollen data from Guilá Naquitz indicate significant fluctuations in preceramic climate during the period of occupation. However, apparently none of these fluctuations was of sufficient magnitude to move the vegetational facies seen in the area outside the walking distance used by the occupants of the cave. As a result, the plant remains in the cave have remained the same in species content from the earliest level to the uppermost level. This, of course, excludes the addition of those species that were introduced into the area as cultivated crop plants. Native plant species have always formed an important part of the subsistence of the cave dwelling people. The same native plant species seem to have been valued throughout the history of the cave deposit.

Tree species are less immediately sensitive to climatic changes than shorter-lived species. However, many of the tree species are demanding in their requirements, so that a *major* climatic shift would have caused a change in the native flora over the time period encompassed by the cave deposits. Many of the species of this part of Mexico are particularly sensitive to shifts in mean annual temperatures. A prolonged period of cooler climate would have brought the frost line further down the slopes of the mountains and caused a resultant change in the species composition of the flora. Similarly, any marked shift in the total annual precipitation would have had an effect on the species composition of the native vegetation.

We can be sure that such major variations in climatic pattern did not occur, because the species of native vegetation appearing in the cave deposit did not change. This is further supported by the pattern of native vegetation found in the Tehuacán Valley to the north and east (Smith 1967). In the Tehuacán Valley, the vegetational history covered a similar period from about 8000 B.C. to about A.D. 1500, with no major change found in the species composition.

SUMMARY

The plant remains from the preceramic levels of Guilá Naquitz Cave indicate that the occupants were living on a relatively selected set of species in spite of the greater variety of edible plant materials available to them in the native vegetation. Apparently, food grain was not used because we have no evidence for the gathering of sizable amounts of grass seed or the seeds of other plants normally used in this manner. The starch component of the diet seems to have been furnished largely by acorns, which may have been stored after the autumn gathering for use throughout the year.

Vegetable oils and fats were also largely lacking in the diets of these people. Piñon nuts, found only in small quantities, must have been the major source.

Vegetable proteins were available from a number of leguminous plants, such as mesquite, *guaje*, *susí*, and wild runner beans, that were extensively used. Minor elements are accounted for by the many fragments of plants such as nopal or prickly pear.

The macroscopic plant remains suggest that this area has undergone no *major* climatic change over the period since about 8000 B.C., although the pollen data are reported to show significant periodic fluctuations. All of the species represented in the early cave levels (except for piñon) still grow in the area, although changes in the proportional abundance of some species must have been made through the continuing disturbances of clearing for cultivation. This pattern is supported by similar evidence that covered a similar time period in the Tehuacán Valley.

Evidence provided by the seed of *Cucurbita pepo* in Zone D of Guilá Naquitz Cave, supported by more abundant remains in Zones C and B, conclusively shows that a form of agriculture was being practiced by the inhabitants sometime between 7800 and 8700 B.C. While the early date for agriculture in the New World is interesting, the significance lies in the biological fact that much time and much human selection were required to produce this cultivated variety of squash.

20

Cucurbits from Preceramic Levels at Guilá Naquitz

Thomas W. Whitaker
Hugh C. Cutler

Some 103 specimens of cucurbits were recovered from preceramic levels at Guilá Naquitz Cave. More than half these specimens were fragments of rind, stem, blossom, or peduncle that could be identified only as *Cucurbita* sp. (Figs.20.1a–d, 20.2, and 20.3b). Much of this material appears to be that of wild species of *Cucurbita* growing in the area and probably collected for the edible seeds (Fig. 20.4). However, there are 9 seeds and 6 peduncles identifiable as *Cucurbita pepo*, the summer squash, a cultivar (Figs. 20.1e, 20.3a, 20.5, and 20.6a). Seven specimens were identifiable as *Lagenaria siceraria*, the bottle gourd, also a cultivar. Four seeds of the wild coyote melon, *Apodanthera*, were also recovered (Figs. 20.1f and 20.6b). Much of the cucurbit material was charred (Whitaker and Cutler 1971). In addition to the macroscopic material we examined, pollen analyst James Schoenwetter (personal communication, 1980) reports one pollen grain of *Cucurbita* cf. *moschata* from Zone B3 of the cave.

The cucurbit specimens recovered from Guilá Naquitz were not as abundant or diversified as those found in other Mexican archaeological sites, for example, Coxcatlán Cave in Puebla (Cutler and Whitaker 1967) and the Ocampo Caves in Tamaulipas (Whitaker *et al.* 1957). There are good reasons for the lack of numbers and diversification of the cucurbits at Guilá Naquitz. According to Flannery (personal communication, 1980) it is a small cave, probably used only on a seasonal basis by very small groups of people. Also, the area in the vicinity of the cave is one of poor soils and is plagued by chronic drought conditions. If we assume that one or more of these conditions occurred, the impoverishment of the cucurbit flora in the stratigraphy of the cave could be anticipated.

In spite of their depauperate nature, the cucurbit collections from Guilá Naquitz Cave offer several important features that support previous studies of the ecology of these species and of the changes brought about in a species after domestication. Also, the relatively rare occurrence of *Lagenaria* in the strata of the cave is worthy of comment.

Cucurbita pepo (Table 2.2) is the dominant cultivated species of *Cucurbita* found in the cave. There are barely traces of the other cultivated species, such as the aforementioned pollen grain of *Cucurbita* cf. *moschata* found in Zone B3 and a seed of *C. ficifolia* found by us in Zone A, a Postclassic level of the cave. It is not surprising to find evidence of *C. pepo* at this latitude, and in preceramic strata, because *C. pepo* is the dominant squash at higher elevations in Central Mexico and much of Guatemala. It is surprising, however, to find a seed of this species in strata dated at 8750–7840 B.C., making it one of the oldest archaeological specimens of cultivated *Cucurbita* on record. Prior to the identification of the cucurbit specimens from Guilá Naquitz it was thought that *C. pepo* originated and was probably brought under domestication in Northern Mexico and the southern United States. The abundant and early remains of *C. pepo* at the Ocampo Caves and copious evidence that it was the earliest cucurbit in the Southwest are the main archaeological pillars of this idea. The presence of a wild *Cucurbita, C. texana*, along the watercourses of a few streams in South Central Texas bolstered the idea of a northern origin and domestication of *C. pepo* because *C. texana* is obviously closely related to the gourd-like forms of *C. pepo*. Some have speculated that it may be a primitive *C. pepo* or a progenitor of this species. The identification and antiquity of the *C. pepo* from the archaeological remains in Guilá Naquitz suggest that

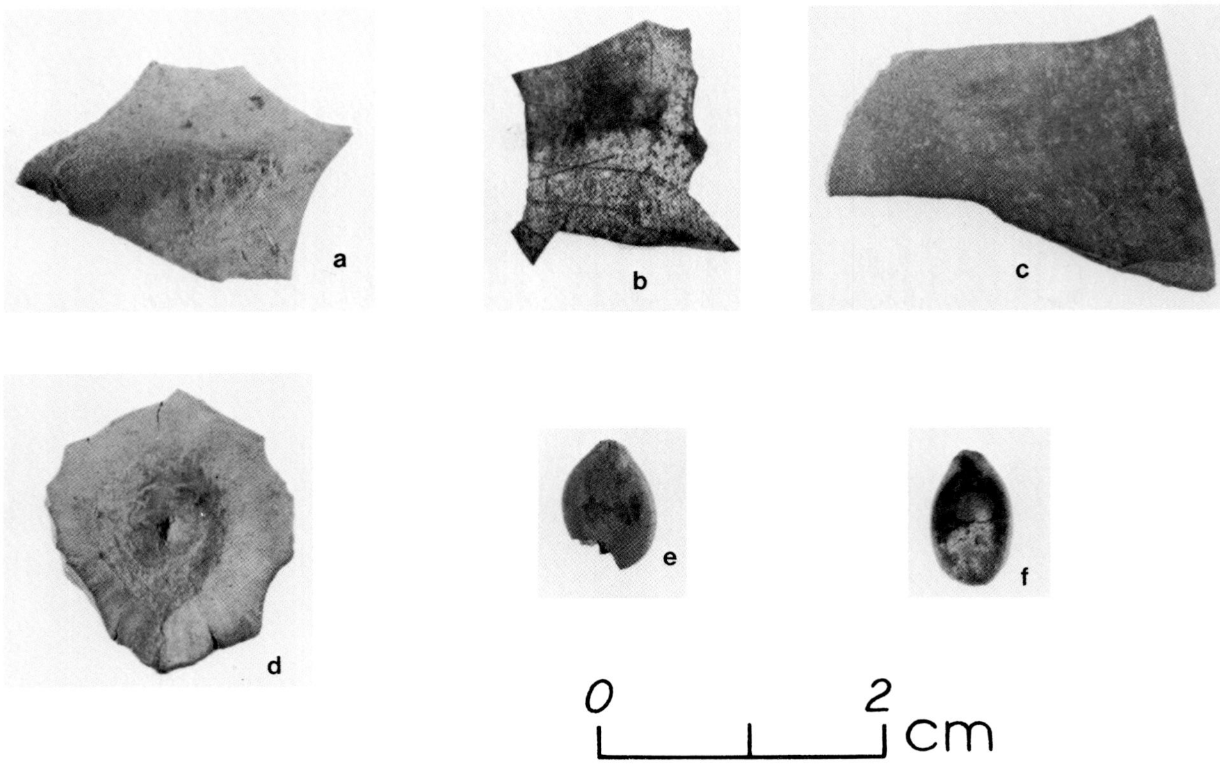

Fig. 20.1. Cucurbit remains from Zone D: a–c, rind fragments; d, peduncle; e, *Cucurbita pepo* seed dating to 7840 B.C.; and f, *Apodanthera* seed.

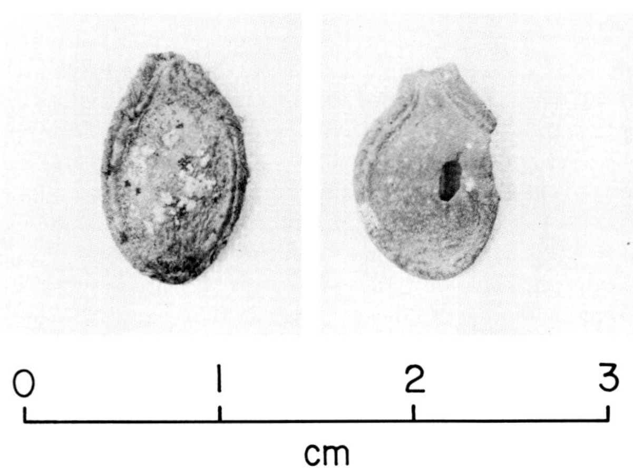

Fig. 20.2. *Cucurbita* sp. seeds from Zone B1, Square C9.

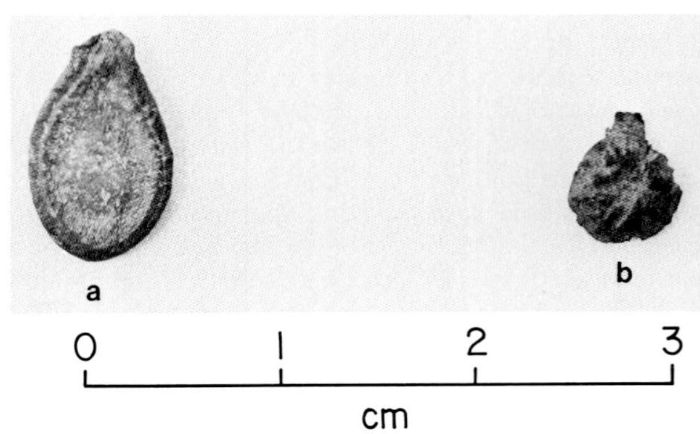

Fig. 20.3. Cucurbit remains from Zone B1, Square E5: a, *C. pepo* seed; b, rind fragment.

a northern origin of *C. pepo* may have to be modified or qualified.

The dominance of *C. pepo* at Guilá Naquitz is most likely explained by the distinct ecological preferences of the cultivated species of *Cucurbita*. One would normally expect to find *C. mixta* and *C. moschata* cultivated at the latitude of the cave (16°57′N). The cave, however, is located at an altitude of about 1926 m (6319 feet) in an arid region (500–700 mm annual rainfall). The days are predominantly warm, with cool nights and minimum moisture for plant growth. Such conditions are well suited to cultivars of *C. pepo*. On the other hand, most cultivars of *C. moschata* and *C. mixta*, usually grown at this latitude, thrive under high temperatures and moist, humid conditions, but they do not tolerate cool temperatures.

Our measurements of seed size of *C. pepo* (Table 20.1) suggest there is a trend toward selection, perhaps largely unconscious, by the early cultivators for larger size seed and

Fig. 20.4. Peduncle fragments of *Cucurbita* sp., wild type, from Zone C. The specimen on the left is from Square E8.

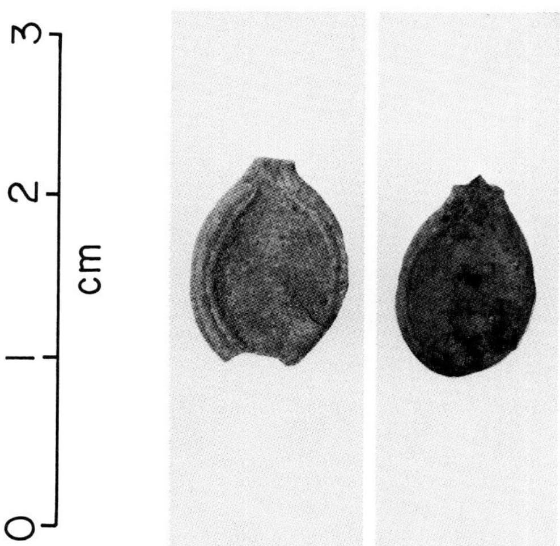

Fig. 20.5. *Cucurbita pepo* seeds from Zone C.

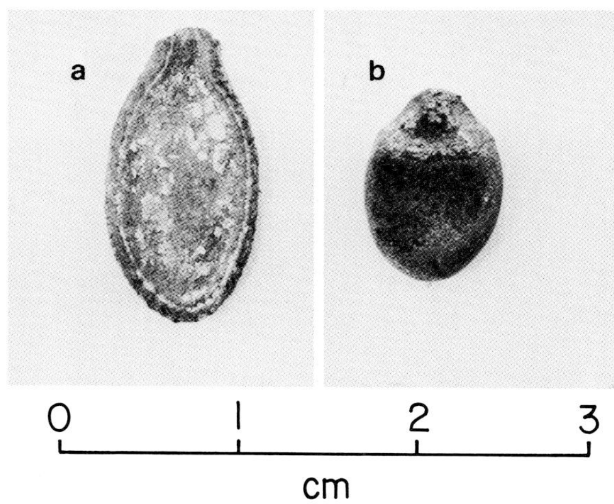

Fig. 20.6. Cucurbit seeds from Zone B1, Square E7: a, *C. pepo*; b, *Apodanthera* sp.

presumably larger fruit. Improvement in the size, and perhaps the quality of their cultivars with increasing sophistication of agriculture is characteristic of primitive cultivation.

It is highly probable that the poor showing of *Lagenaria* (Table 20.2) in the plant remains of the cave can also be attributed to ecological preferences. *Lagenaria* is better adapted

TABLE 20.1
Dimensions of *Cucurbita pepo* Seeds from Guilá Naquitz

Zone	Square	Dimensions (mm)
B1	C11	12 × 7, 10 × 7, 11 × 7[a]
B1	E5	11 × 6
B1	E7	17 × 8.5
C	F9	12 × 8, 11 × 8, 11 × 8
D	F5	10 × 7

[a]All three seeds in this square are probably immature.

to moist, humid conditions than to an arid, cool environment. Alternative explanations are (1) that *Lagenaria* was chiefly in the immature state and hence no remains would be anticipated or (2) that as the cave was occupied mostly on a seasonal basis, perhaps the people did not use the cave as living quarters long enough to break their gourd vessels. It is clear, however, that gourds did not enter into the economy of the people using the cave in a major way, either for culinary, household, or storage purposes.

The El Riego Cave in the Tehuacán Valley (Cutler and Whitaker 1967) is similar to Guilá Naquitz in that there was relatively little bottle gourd (*Lagenaria*). Also, at El Riego there was more wild *Cucurbita* species and more *C. pepo* than at other sites in the Tehuacán Valley.

One other point deserves comment: There is a large body of material consisting of rinds, seeds, and peduncles, mostly fragments. This material is undoubtedly that of species of

TABLE 20.2
Distribution of Cucurbit Remains in Guilá Naquitz

Item	Stratigraphic zone						Totals
	D	C	B3	B2 + 3	B2	B1	
C. pepo seeds	1	3	—	—	—	5	9
C. pepo peduncles	—	2	—	—	—	4	6
Cucurbita sp. rind fragments	4	35	—	3	5	8	55
Cucurbita sp. seeds	—	2	1	—	2	3	8
Cucurbita sp. peduncles	1	4	—	—	—	—	5
Cucurbita sp. stem ends and blossom ends	—	4	—	—	4	1	9
Lagenaria rind fragments	—	3	—	1	1	2	7
Apodanthera seeds	1	1	—	—	—	2	4
	7	54	1	4	12	25	103

Cucurbita and is so indicated in the tables. This material appears to be that of wild species of *Cucurbita* growing in the area and probably collected mainly for their edible seeds. The rind fragments of cultivated and noncultivated species of cucurbits are indistinguishable; thus rinds as listed in the tables may contain material of both cultivated and noncultivated species.

Only 4 seeds of *Apodanthera* were found (Table 20.2). It is uncertain whether this small sample represents material brought to the cave by man to use as food or whether it might have been introduced by rodents. However, Flannery (personal communication, 1980) reports that no cucurbit seeds were found in any of the rodent burrows identified in the cave. It is also significant that seeds of *Apodanthera* are found in most archaeological caves in Mexico wherever a species of this genus is growing in the immediate vicinity (Whitaker and Cutler 1971:124).

The total sample of preceramic cucurbits is listed by stratigraphic zone in Table 20.2. In addition, Tables 20.3–20.8 give the exact location of every cucurbit specimen by 1 × 1-m square on each living floor of the cave. It was thought that this more detailed information should be provided because Guilá Naquitz has yielded some of the oldest cultivated cucurbit remains from any archaeological site.

TABLE 20.3
Distribution of Cucurbit Remains in Zone D, Guilá Naquitz

Item	Distribution through 1 × 1-m squares				Totals
	E7	F5	F8	G8	
C. pepo seeds	—	1[a]	—	—	1
Cucurbita sp. rind fragments	1	—	1	2	4
Cucurbita sp. peduncles	—	—	—	1	1
Apodanthera seeds	—	1	—	—	1

[a]This *C. pepo* seed came from the immediate vicinity of radiocarbon sample GX0783 (7840 ± 240 B.C.).

TABLE 20.4
Distribution of Cucurbit Remains in Zone C, Guilá Naquitz

Item	Distribution through 1 × 1-m squares									Totals
	E7	E8	E9	E10	F5	F6	F9	F10	G9	
C. pepo seeds	—	—	—	—	—	—	3	—	—	3
C. pepo peduncles	—	—	2	—	—	—	—	—	—	2
Cucurbita sp. rind fragments	1	1	2	1	1	1	25	2	1	35
Cucurbita sp. seeds	1	—	—	—	—	—	1	—	—	2
Cucurbita sp. peduncles	—	1	—	1	—	—	2	—	—	4
Cucurbita sp. stem ends and blossom ends	—	—	—	—	—	—	4	—	—	4
Lagenaria rind fragments	—	—	—	—	—	—	2	1	—	3
Apodanthera seeds	—	—	—	—	—	—	—	1	—	1

TABLE 20.5
Distribution of Cucurbit Remains in Zone B3, Guilá Naquitz

Item	Distribution through 1 × 1-m squares	
	E7	Total
Cucurbita sp. seeds	1	1

TABLE 20.6
Distribution of Cucurbit Remains in Zone B2+3, Guilá Naquitz

Item	Distribution through 1 × 1-m squares			
	C10	E6	F5	Total
Cucurbita sp. rind fragments	1	1	1	3
Lagenaria rinds	1	—	—	1

TABLE 20.7
Distribution of Cucurbit Remains in Zone B2, Guilá Naquitz

Item	Distribution through 1 × 1-m squares						
	C6	C11	D8	D9	E9	E10	Totals
Cucurbita sp. rind fragments	1	1	—	1	1	1	5
Cucurbita sp. seeds	—	1	1	—	—	—	2
Cucurbita sp. stem ends and blossom ends	—	1	—	—	3	—	4
Lagenaria rind fragments	—	—	—	—	1	—	1

TABLE 20.8
Distribution of Cucurbit Remains in Zone B1, Guilá Naquitz

Item	Distribution through 1 × 1-m squares								
	B9	C8	C9	C11	D5	E5	E7	E11	Totals
C. pepo seeds	—	—	—	3	—	1	1	—	5
C. pepo peduncles	2	—	—	2	—	—	—	—	4
Cucurbita sp. rind fragments	1	1	1	2	1	1	—	1	8
Cucurbita sp. seeds	—	—	2	—	—	—	—	1	3
Cucurbita sp. stem ends and blossom ends	1	—	—	—	—	—	—	—	1
Lagenaria rind fragments	—	—	—	—	1	—	—	1	2
Apodanthera seeds	—	—	—	1	—	—	1	—	2

21

Preceramic *Phaseolus* from Guilá Naquitz

Lawrence Kaplan

INTRODUCTION

Bean remains were abundant in preceramic levels at Guilá Naquitz. Zones B1 through E yielded a total of 471 pod valves and more than 17 seeds of the genus *Phaseolus,* leaving no doubt as to the use of beans by the preceramic occupants of the cave. Of the 471 pod valves, more than 168 could be identified as belonging to a wild, hypogeal germinating *Phaseolus* that cannot be identified to species and has been designated Guilá Naquitz Type 1 (Kaplan 1981). Some 17 seeds from preceramic levels could also be assigned to this type. Guilá Naquitz Type 1, whose pattern of germination would be more like the runner bean than the common bean, was the dominant (and possibly the only) bean used by the Naquitz phase residents of the cave.

Four small specimens of the common bean, *Phaseolus vulgaris,* also were recovered by fine-screening preceramic levels of the cave. However, all were from squares close enough to areas of later human or animal disturbance so that the possibility of their being intrusive cannot be ruled out. In view of this (and in view of their rarity relative to Guilá Naquitz Type 1), the most reasonable position to take would seem to be to regard these *P. vulgaris* specimens with caution. Certainly it would be unwise to claim such an early date for the domestication of the common bean on such scanty and possibly intrusive material.

All *Phaseolus* bean remains at Guilá Naquitz consist of dried, uncharred pods and seeds (Table 21.1).

DESCRIPTION OF GUILA NAQUITZ TYPE 1

Guilá Naquitz Type 1 is a wild *Phaseolus* sp. that cannot at present be identified to species; however, the morphology of the meristematic embryo shows that this type is not an ancestor or close relative of *P. vulgaris.*

The dormant meristematic embryo, that part of the seed that comprises the embryonic root, stem, epicotyl, and foliar leaves, reveals the germination pattern in its overall structure (Fig. 21.1). Germination in *Phaseolus* results either in raising the cotyledons above the soil surface or not doing so. *Phaseolus vulgaris,* its known wild relatives, and many other species germinate *epigeally;* that is, the cotyledons are raised. *Phaseolus coccineus,* its known relatives, and many other species are called *hypogeal* since their cotyledons remain below the soil surface.

Since there are about 100 American *Phaseolus* species and only 2 germination patterns, it is evident that 2 species that share the same germination pattern are not necessarily closely related. The embryo structure of Guilá Naquitz Type 1 is that of a hypogeal germinator. All of the commonly cultivated *Phaseolus* species are epigeal except for *P. coccineus* (e.g., cultivar scarlet runner and the Mexican *ayecote* and *botil*).

Seed color is dark reddish brown or medium reddish brown in a ratio of 1.6:1 dark to medium. There are no intermediates. This color intensity variation is most likely due to a single gene difference and is likely to be found within a single cultivar or, as is more likely in this case, within a local population of a wild species. No variegation or color patterning in the seed coat can be discerned. *Phaseolus obvallatus* Schlecht., a wild hypogeal species that is allied to *P. coccineus,* has variegated seeds.

However, the nonvariegated coat of Guilá Naquitz Type 1 is worthy of note. Seed coats of domesticates of *P. vulgaris* and *P. coccineus* may be variegated or self-colored. However, the seed coats of most wild relatives of these two species (and

TABLE 21.1
Distribution of Bean Remains throughout Lower Stratigraphic Zones at Guilá Naquitz Cave

Stratigraphic zones	Total pod valves	Valves clearly identifiable as Guilá Naquitz Type 1	Seeds, Guilá Naquitz Type 1	Possible fragments, other types[a]
B1	111		15	X
B2	104	58	—	
B2 + 3	9		2	
B3	26		—	
C	43	Several	—	
D	17	11	—	X
E	161	99	—	X
	471	168 +	17	

[a]All of these occurred in contexts where the possibility of intrusion cannot be ruled out.

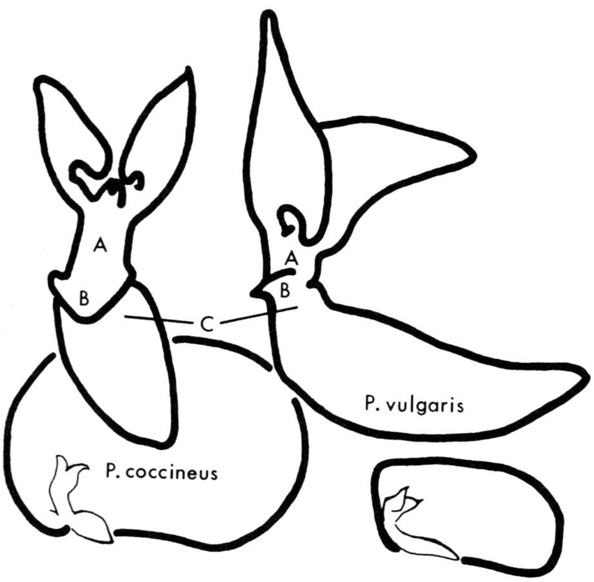

Fig. 21.1. Embryo structure in *Phaseolus:* A, epicotyl; B, point of cotyledonary attachment; and C, hypocotyl. During germination of *P. vulgaris*, the hypocotyl elongates; that of *P. coccineus* does not.

of *P. lunatus,* the lima and sieva beans) are almost always intensely fleck patterned or otherwise marked. Such cryptically colored or patterned seed coats may be advantageous in protecting seeds from bird predation. The wild pods too are commonly splotched with purple and difficult to detect visually. Once humans had begun to harvest, plant, and cover the planted seed with soil, cryptic coloration would cease to be advantageous. At this point in the process of domestication there would be a distinct advantage favoring seeds that were more easily detected visually by humans, since the wild type seed coat is extremely difficult to see against the soil as a background and might often be left uncovered during planting.

The pods show a fairly high degree of twisting, which is the mechanism for seed dispersal. A randomly chosen sample of 20 pod valves showed an average of 3.8 turns per valve for an average extended valve length of 4.58 cm (Table 21.2). Pods of cultivars either twist not at all or do so loosely. The degree of twisting in Guilá Naquitz Type 1 (Fig. 21.2) is typical of wild species and would preclude allowing the pods to hang on the plants until all were mature. The small size of the seeds (Table 21.2, Fig. 21.3) would make stripping of individual pods less economical and just more troublesome than collecting the entire pods for threshing or storage at the cave. Gathering of pods in this manner is more consistent with the vining growth habit than with the bush type in which the entire plant would be pulled.

Assuming that Guilá Naquitz Type 1 was used for food, I would conclude that this unidentified *Phaseolus* species was not a cultivar but that it was vining and was gathered by stripping the pods from the plant. The record of this wild bean at Guilá Naquitz is unique in two respects. The only other reported occurrence of wild beans in archaeological remains is in the Ocampo caves of Tamaulipas (Kaplan and MacNeish 1960). In those caves, wild runner bean pods were present in early levels but disappeared when domesticated beans appeared. The wild beans of Guilá Naquitz continued to be collected long after domesticated beans were available; many were present in Zone A.

NATIVE WILD BEANS OF THE GUILA NAQUITZ AREA

In 1968 I carried out field collections at Guilá Naquitz, Hierve el Agua, and other localities above the Valley of Oaxaca. Two *Phaseolus* species were recovered: *P. anisotrichos* Schlect. and *P. heterophyllus* Willd.[1] Neither of these is represented in the archaeological materials. *Phaseolus anisotrichos,* a vining species, is widespread in Mexico and

[1]*Macroptilium heterophyllum* (Willd.) Marechal & Baudet.

TABLE 21.2
Measurements of Guilá Naquitz
***Phaseolus* Type 1**

Pod	
Avg. length in cm	4.58
Avg. width in cm	.89
No. valves[a]	660
Seed	
Avg. length in cm	.52
Avg. width in cm	.43
Avg. thickness, cm	.27
No. seeds[a]	68

[a]Measurements were taken not only on the 168 valves and 17 seeds in preceramic levels, but also on the 492 valves and 51 seeds of GN 1 in Postclassic levels.

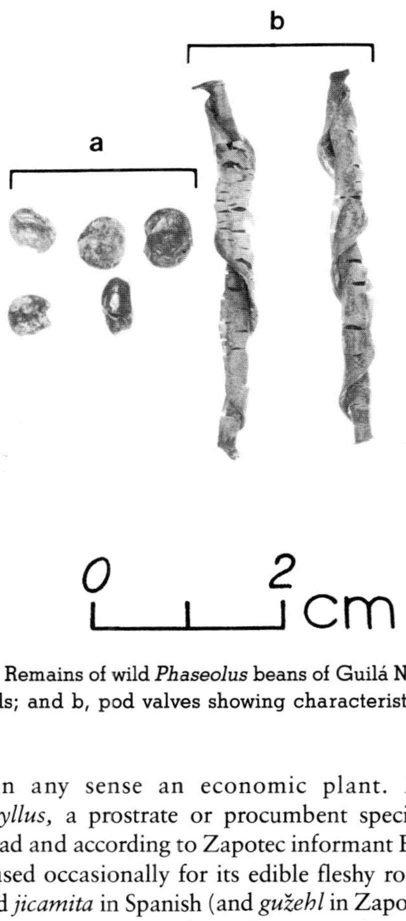

Fig. 21.2. Remains of wild *Phaseolus* beans of Guilá Naquitz Type 1: a, seeds; and b, pod valves showing characteristic twisting.

is not in any sense an economic plant. *Phaseolus heterophyllus,* a prostrate or procumbent species, is also widespread and according to Zapotec informant Eligio Martínez is used occasionally for its edible fleshy root. Locally it is called *jicamita* in Spanish (and *gužehl* in Zapotec according to Messer 1978:58). This information from Mr. Martínez is most gratifying since it long has appeared to me that this plant must be a potential, if not actual, food source. The roots are fleshy, about the size of a garden radish, and not unpleasant to the taste.

The plants are locally abundant along road sides and in disturbed open areas. In such places significant quantities may be gathered quickly with only a sharpened stick or other simple digging tool. Very likely the plants have always grown on talus slopes and other disturbed areas and were used as a food resource. Unfortunately no direct archaeological evidence of early use is available.

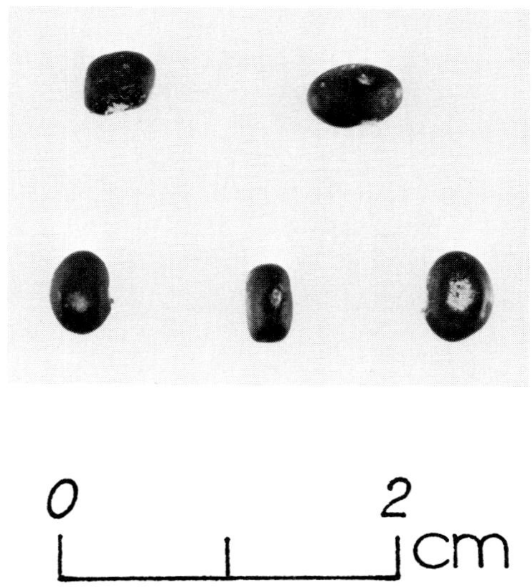

Fig. 21.3. Seeds of Guilá Naquitz Type 1.

It should be noted that the species of wild beans that can be found near Guilá Naquitz seem to be quite variable from year to year. In 1968 the only species I was able to find in the vicinity of the cave was *P. anisotrichos* Schlect. Between 1971 and 1973, Messer (1978) was able to find *P. heterophyllus* and *P. atropurpureus* DC.[2] near Guilá Naquitz (identifications by Rogers McVaugh, University of Michigan Herbarium).

POSSIBLE GENETIC INFLUENCE OF GUILA NAQUITZ TYPE 1 ON DOMESTICATES

Because Guilá Naquitz Type 1 is hypogeal it cannot be a direct ancestor of *P. vulgaris* cultivars. All *P. vulgaris* cultivars and all of the putative wild *P. vulgaris* collections of Howard Scott Gentry, formerly of the U.S. Department of Agriculture, are epigeal. Guilá Naquitz Type 1 conforms to the germination pattern of *P. coccineus,* but there is insufficient evidence to evaluate it as a possible ancestor of cultivars of this species. Collections of possible wild *P. coccineus* and such relatives as *P. obvallatus* and *P. formosus* H.B.K. (Piper 1926) are

[2]*Macroptilium atropurpureum* (DC.) Urb.

mostly recorded from areas cooler and more humid than Guilá Naquitz.

Lamprecht (1968) has maintained that a *P. vulgaris* × *P. coccineus* cross is successful whereas the reciprocal is not because the pollen parent (*P. coccineus*) has pollen large enough that sufficient cytoplasm is carried over with the nucleus to enable functioning of the heterozygous hypogeal/epigeal genotype. The relevance of the observations of Lamprecht and others (Yarnell 1965) on the interspecific cross is that *P. vulgaris* may be the seed parent where the pollen parent is hypogeal. Since Guilá Naquitz Type 1 appears in a cultural context over a long period of time in an area in which wild *P. vulgaris* has been collected, any wild extant species that is hypogeal, whose seeds and pods bear resemblance to those of Guilá Naquitz Type 1, and that has pollen at least as large as that of *P. coccineus* (50 microns or greater) might be investigated as a possible contributor to variation in an epigeal species such as *P. vulgaris*.

22

Animal Food Remains from Preceramic Guilá Naquitz

Kent V. Flannery
Jane C. Wheeler

INTRODUCTION

More than 360 identifiable bone fragments from Zones B–E of Guilá Naquitz belonged to animals that had been hunted or trapped for food. The exact number cannot be specified because, as explained in Chapter 16, in the case of some small rodents and lizards it is frequently difficult to decide whether a given specimen got into the cave through human intervention or by another means.

Our analysis of these animal remains was methodologically similar to the study of the fauna from the Tehuacán caves (Flannery 1967:157). All identifiable fragments were checked against one or more of the following collections: (1) skeletons of animals collected by Flannery in the Tehuacán Valley or by Flannery and Wheeler in the Valley of Oaxaca; (2) skeletal material in the collections of the Instituto de Biología, Universidad Nacional Autónoma de México, Mexico City; (3) skeletons of Mexican animals in the collections of the Museum of Zoology, University of Michigan; and (4) skeletons of Mexican animals in the U.S. National Museum, Smithsonian Institution. We are grateful to several specialists who aided us in our species identification of both modern and ancient fauna. These include Emmet T. Hooper of the University of Michigan (rodents and lagomorphs), Richard Zusi of the Smithsonian Institution (birds), and Richard Etheridge of San Diego State College (lizards).

Archaeological faunal analysts are currently in disagreement as to what constitutes the best way to report animal remains: by raw count of bones, raw weight of bones, minimum individual animals, minimum individuals adjusted by number of bones per species, or some combination of these. In order that our Oaxaca data be comparable to the Tehuacán data, we present them both as raw counts and as minimum individuals. Our minimum number of individuals was calculated using the method of White (1953), with the added modification of checking to see if all left units matched all right units (see Flannery 1967:157 for details).

Basically, the two methods of presentation serve different purposes. The raw counts, which are given by square, tell us which parts of the body were represented and where they occurred on each living floor. The minimum number of individuals is a calculation that must be made before we can begin estimating the amount of meat consumed per living floor or the area of hunting territory needed to account for the remains on each living floor (Chapter 24). There are more complex and sophisticated ways of presenting the data, but frankly, we are not yet convinced that they are significantly better than the raw counts. In our opinion, faunal analysis would be better served by more rigorous ways of determining how each bone got into the site than by more complex ways of calculating the percentage each species contributed.

ANIMALS HUNTED OR TRAPPED

The animals hunted or trapped at Guilá Naquitz were all species that are common in the area today, or would have been common until they were reduced by overhunting with firearms. No extinct forms or rare species were found in any of the preceramic levels. The list of animals that presumably served as food is as follows:

Large mammals
  White-tailed deer (*Odocoileus virginianus*)
  Collared peccary (*Dicotyles tajacu*)
Small mammals
  Mexican cottontail rabbit (*Sylvilagus cunicularius*)
  Eastern cottontail rabbit (*Sylvilagus floridanus connectens*)
  Raccoon (*Procyon lotor*)
  and possibly some wild rodents such as the wood rat (*Neotoma mexicana*; see Chapter 16)
Reptiles
  Mud turtle (*Kinosternon integrum*)
  and possibly some lizards such as *Cnemidophorus* and *Sceloporus* (see Chapter 16)
Birds
  Bobwhite quail (*Colinus virginianus*)
  Band-tailed pigeon (*Columba fasciata*)
  Mourning dove (*Zenaidura macroura*)
  Ground dove (*Columbigallina* cf. *passerina*)
  Red-tailed (?) hawk (*Buteo* cf. *jamaicensis*)
  Barn owl (*Tyto alba*)
  and possibly an occasional songbird (see section below on Birds)

Obviously, there are a number of edible species conspicuous by their absence from this list. The opossum (*Didelphis marsupialis*), the gray fox (*Urocyon cinereoargenteus*), several genera of skunks (*Mephitis, Conepatus,* and *Spilogale*), the jackrabbit (*Lepus mexicanus*), the pocket gopher (*Orthogeomys grandis*), the white-winged dove (*Zenaida asiatica*), the Montezuma quail (*Cyrtonyx montezumae*), and the chachalaca (*Ortalis poliocephala*) are all common to the area and have been found in other caves in Oaxaca and Tehuacán. We do not for one minute assume that these food sources were ignored by the occupants of Guilá Naquitz. However, their remains did not appear in the modest sample we recovered from what is, after all, a rather small cave.

Of all these species, the four that seem to have been most important to the residents of Guilá Naquitz were the white-tailed deer, the two species of cottontail, and the mud turtle. These four animals have therefore been given their own set of tables showing their distribution throughout the various living floors of the cave (Tables 22.1–22.7).

MAMMALS AND REPTILES

The white-tailed deer, whose bones occurred on every living floor at Guilá Naquitz, was perhaps the major source of meat during the Naquitz phase (Fig. 22.1a, b). Elsewhere (Chapter 18) we have estimated its aboriginal density in Thorn Forest A as roughly 12 animals per km^2, and it appears that 2 or more deer were killed almost every time the cave was occupied. Presumably the hunting was done with spears or atlatl darts tipped with Lerma, Almagre, or Pedernales points. As for the season of hunting, we have relied on the same criteria provided by Leopold (1959) and Villa (1954) and utilized by Flannery (1967:157) in his study of the Tehuacán fauna.

Villa's study of deer from various regions of Mexico showed that "the first bucks to lose their antlers are those from the south—Oaxaca and Guerrero—where the breeding season is earlier" (Leopold 1959:512). Bucks from Oaxaca lose their antlers in February or March (occasionally in early April) and begin to grow new ones in April or May. During the rainy season (May to September) their antlers are "in velvet" and would appear spongy and immature in archaeological deposits. Not until the end of October do they reach their *forma decidua definitiva*, fully grown and ossified right to the tips of the antler tines (Villa 1954:459). There are five occurrences of such fully hardened antlers in the preceramic levels at Guilá Naquitz—in Zones D, C, B3, B2+3, and B1. No fragments of antlers in velvet, or of frontal bones with the pedicels of recently shed antlers, appeared in the deposits. Thus, most bucks at Guilá Naquitz must have been killed between October and February.

Leopold (1959:510–511) provides evidence that fawning may span the period from June to August (rarely as late as September) in order for nursing does to take advantage of the rainy season vegetation. Late-term deer foetuses (which have shown up in the caves in both Oaxaca and Tehuacán) might therefore be assigned to the period May–August. Only Zone B1 produced such a foetus at Guilá Naquitz, suggesting an occupation that might have begun as early as July or August. Other than this, the deer remains do nothing to alter our conclusion (based on plant remains) that the preceramic living floors fall mainly in the period from late August to December.

Cottontail rabbits of two species were also an important source of meat for the occupants of the cave (Fig. 22.1c–e). Indeed, while cottontails probably did not provide nearly as many grams of meat as deer did, they may have been a more frequent item on the menu simply because of their greater numbers in the environment around the cave (Fig. 22.2). Elsewhere (Chapter 18) we have estimated that as many as 320 cottontails per km^2 may be present even today in Thorn Forest A. The data from Guilá Naquitz suggest that there were numerous occasions on which as many as 2–4 rabbits were butchered and discarded together. This may mean that there were significant periodic efforts to hunt or trap cottontails—each effort resulting in the capture of several animals—rather than a continuous low-level effort, such as the daily setting of trap lines. For example, 3 left humeri of cottontail appeared in Squares C11–C12 of Zone B1, 4 mandibles appeared in Square D9 of Zone B2, and 3 right mandibles appeared in Squares B9–C10 of Zone C. There are perhaps 7 other cases where duplicate bones from at least 2 different rabbits appear in the same or adjacent squares, reinforcing our impression that several rabbits may have been caught and processed at the same time.

The mud turtle (*Kinosternon integrum*), also utilized in preceramic Tehuacán, was the fourth most common species

TABLE 22.1
Total Counts of Deer, Cottontail, and Mud Turtle Bones from Zone E, Guilá Naquitz Cave

Animal	Distribution through 1 × 1-m squares														
	C8	C10	C11	D9	D10	D11	E10	E11	E12	F9	F10	F11	G4	G5	H10
White-tail deer															
minimum no. individuals, 2[a]															
teeth	—	—	—	—	—	—	—	—	—	—	1	—	—	—	—
ramus, mandible	—	—	—	—	—	—	1	—	—	—	—	—	—	—	—
humerus	—	—	—	—	—	—	—	—	1	—	—	—	—	—	—
radius	—	—	1	—	—	—	—	—	—	—	—	—	—	—	—
ulna	—	—	—	—	—	—	—	—	—	—	—	—	—	1	—
femur	—	—	—	—	—	—	—	1	—	—	—	—	—	—	—
metapodial	—	—	1	—	1	—	—	—	—	—	1	—	—	—	—
phalanx	—	—	—	—	—	—	—	—	—	—	1	—	—	—	—
vertebra	1	—	—	—	—	1	1	—	—	—	—	—	—	—	—
rib	—	1	—	—	—	—	—	—	—	1	—	—	—	—	—
other	—	—	1	—	1	—	—	—	—	—	—	—	—	—	—
Eastern cottontail															
minimum no. individuals, 2															
maxilla	—	—	—	—	—	1	—	—	—	—	—	—	—	—	—
ramus, mandible	—	2	—	—	—	1	—	—	—	—	—	—	—	—	—
humerus	—	—	—	—	—	—	—	1	—	—	—	1	—	—	—
scapula	—	—	—	—	—	1	—	—	—	—	—	—	—	—	—
innominate	—	1	1	—	1	—	—	—	—	—	—	—	—	—	—
femur	—	2	—	—	—	1	—	—	—	1	—	—	—	—	—
tibia	—	—	—	—	—	—	—	—	1	—	—	—	—	—	—
calcaneum	—	—	—	—	1	1	—	—	—	—	1	—	—	—	—
metapodial	—	—	—	—	—	1	—	—	—	—	—	—	—	—	—
Mexican cottontail															
minimum no. individuals, 1															
femur	—	—	—	—	1	—	—	—	—	—	—	—	—	—	—
Sylvilagus spp., 2 juveniles															
various bones	—	1	—	—	1	5	—	2	—	—	—	—	2	—	—
Mud turtle															
minimum no. individuals, 1															
scutes	—	—	3	—	—	—	—	1	—	—	1	—	—	—	—
limb bones	—	—	—	—	1	—	1	—	—	—	—	—	1	—	—

[a]Worn molars and fused bones indicate an adult; unfused bones and unerupted teeth indicate a juvenile. Lack of antlers may indicate a female and her fawn were killed.

eaten at Guilá Naquitz (Fig. 22.1f). There is not much meat on this small turtle, so its popularity may be due to the fact that it can be caught easily, even by a child. The vast majority of the shell fragments in the cave (perhaps 80%) are carbonized, and in two squares (D7 and E6) of Zone B1 there were clear signs that the turtle's carapace had been deliberately burned. This strongly suggests that mud turtles were turned upside down and roasted in the shell over hot coals. This custom is still carried out today with river turtles in Chiapas, where "there exists the cruel practice of throwing them alive on the fire so that they are cooked whole, and then picking out the edible parts" (Álvarez del Toro 1952:89).

Table 22.8 lists the other mammal remains (exclusive of rodents) from preceramic Guilá Naquitz. The collared peccary is represented by three bone fragments from Zone D and two from Zone C; most are bones from the extremities of the body and would not have had much meat on them, so the bulk of the carcass must have been discarded outside the cave. Peccaries may once have been common in Oaxaca, but today they are hardly ever found close to human settlements. A single femur of raccoon from Zone C shows that that animal, still common along the Río Mitla, was occasionally taken by Naquitz phase hunters.

As we have stated in Chapter 16, three mandibles of wood rat from Zones D, C, and B2+3 were burned, suggesting that *Neotoma* was sometimes eaten. However, it is impossible for us to specify how many wood rats were brought to the cave by man and cooked and how many were brought to the cave by owls and later burned accidentally.

A single human bone (third finger, left hand) appeared in Zone B1. Since no burials were found in Guilá Naquitz, we can only surmise that the occupants of that living floor had disturbed a burial out on the talus slope, or even further away.

In Chapter 16 we also discuss the problems of interpreting lizard remains from Guilá Naquitz. None of the fragments of *Cnemidophorus* or *Sceloporus* were burned, but the possibility that these small lizards were eaten cannot be ruled out, especially in the case of Zone B1.

BIRDS

The variety of birds eaten by the preceramic occupants of the cave is more striking than the absolute number of bird bones; both are given in Table 22.9. The impression given is that the occupants did not systematically pursue any one

TABLE 22.2
Total Counts of Deer, Cottontail, and Mud Turtle Bones from Zone D, Guilá Naquitz Cave

Animal	Distribution through 1 × 1-m squares																
	B9	C7	C9	C10	D5	D7	D9	D10	E6	E7	E8	F4	F5	F6	F8	G5	G6
White-tail deer																	
minimum no. individuals, 2[a]																	
antler fragments	—	—	(?)	1	1	—	—	3	—	1	—	—	—	—	—	—	—
teeth	—	—	—	—	—	—	—	—	—	—	—	—	1	—	—	2	—
ramus, mandible	—	—	—	—	—	—	—	—	—	—	—	—	1	—	—		
femur	—	—	—	1	—	—	—	—	—	—	—	—	—	—	—	—	—
calcaneum	—	—	—	—	—	—	—	—	—	1	—	—	—	—	—	—	—
metapodial	—	—	—	—	—	1	1	—	—	—	—	1	—	—	—	—	—
vertebra	—	—	1	—	—	—	—	—	—	—	1	—	1	—	—	—	—
rib	—	—	1	—	—	—	—	—	—	—	1	—	1	1	—	—	—
other	—	—	—	—	—	—	3	—	—	—	—	—	3	—	—	—	1
Eastern cottontail																	
minimum no. individuals, 3																	
maxilla	—	—	—	—	—	—	—	—	1	—	—	—	—	—	—	—	—
ramus, mandible	—	—	—	—	—	1	—	1	—	—	—	—	—	—	—	—	—
humerus	—	—	—	—	—	—	—	—	2	—	—	—	—	—	—	—	—
scapula	—	—	—	—	—	—	—	1	—	—	—	—	—	—	—	—	—
ulna	—	—	1	—	—	—	—	—	—	—	—	—	—	—	—	—	—
innominate	—	—	—	—	—	1	—	—	—	—	—	—	1	—	—	—	1
femur	—	—	—	—	—	—	1	—	—	—	—	—	—	—	—	—	—
calcaneum	—	—	—	—	—	—	—	1	1	—	2	—	—	—	—	—	—
Mexican cottontail																	
minimum no. individuals, 1																	
ramus, mandible	—	—	—	—	—	—	—	—	—	—	1	—	—	—	—	—	—
calcaneum	—	—	—	—	—	—	—	—	—	—	—	—	—	—	—	—	1
Sylvilagus spp., 1 juvenile																	
various bones	—	—	—	—	—	—	2	—	—	1	1	—	—	—	1	—	—
Mud turtle																	
minimum no. individuals, 1																	
scutes	—	—	—	—	—	3	—	—	—	—	—	—	—	—	—	—	—

[a]Mature antler indicates adult male; several unfused bones and unerupted teeth indicate a juvenile.

species or set of species but took whatever happened to be available near the cave. On three occasions, it appears that they may have eaten a barn owl that was living in the cave when they arrived. At least, the broken-up remains do not appear to be those of an owl that died in the cave of natural causes.

As at Tehuacán, quail, pigeons, and doves were favored. The band-tailed pigeon, found in Zones E and B2+3, frequently spends the dry season in flocks of 10–100 foraging for acorns and manzanita berries (Leopold 1959:293), so it should have been abundant in Thorn Forest A during the autumn. The mourning dove, or *huilota*, resident in small numbers all year, is considerably more abundant in Oaxaca from September to March when huge numbers migrate to Mexico from the north. In Chapter 18, we have estimated an aboriginal density of 1 quail per 4 ha for Thorn Forest A. The occupants of Guilá Naquitz also killed an occasional hawk.

Fragments of unidentified songbirds turned up in three zones (E, D, and B2+3). We assume that these birds were eaten, but we cannot rule out the possibility that songbird bones from owl pellets were occasionally kicked around the cave and wound up elsewhere in secondary context. Three birds definitely found in owl pellets were among the remains from Zones D, C, and B2 and are listed in Table 22.9.

SUMMARIES OF INDIVIDUAL LIVING FLOORS

Zone E

During the deposition of Zone E, the occupants of the cave killed and butchered at least 2 individual deer (represented by 17 identifiable fragments). The presence of worn molar teeth and fully fused limb bones indicates that at least 1 deer was an adult; some unfused bones and unerupted teeth indicate that at least 1 was a juvenile. The lack of antler fragments on this living floor suggests that the 2 minimum individuals may have been a female and her fawn.

The Zone E occupants also killed a minimum of 5 cottontail rabbits: at least 2 eastern cottontails, 1 Mexican cottontail, and 2 juveniles not identifiable to species. In all, a total of 31 bones could be assigned to *Sylvilagus*. Two left femora were found in Square C10, suggesting that 2 eastern cottontails may have been butchered at the same time. The bulk of the eastern cottontail bone (17 fragments) appeared in adjoining squares, C10 and D11.

TABLE 22.3
Total Counts of Deer, Cottontail, and Mud Turtle Bones from Zone C, Guilá Naquitz Cave

Animal	Distribution through 1 × 1-m squares																
	B9	B10	C8	C9	C10	D8	D10	D11	E5	E6	E7	E9	E10	F9 (upper)	F9 (lower)	F10	G9
White-tail deer																	
minimum no. individuals, 2[a]																	
antler fragments	—	—	—	—	1	—	1	—	—	—	—	1	—	—	1	1	—
teeth	—	—	—	—	1	—	—	—	—	—	—	—	—	—	—	—	—
scapula	—	—	—	1	—	—	—	—	—	—	—	—	—	—	—	—	—
humerus	1	—	—	—	—	—	—	—	—	—	—	—	—	—	—	—	—
femur	—	—	—	—	1	—	—	—	—	—	—	—	—	—	—	—	—
naviculocuboid	—	—	—	—	—	—	—	—	—	—	—	1	—	—	—	—	—
metapodial	1	—	—	—	—	—	1	—	—	1	1	—	—	—	—	—	—
phalanx	—	—	—	—	—	1	—	—	1	—	—	—	—	—	—	—	—
vertebra	—	—	—	—	1	—	—	—	—	—	—	—	—	—	—	—	—
rib	—	1	—	—	1	—	—	—	—	—	—	1	—	—	1	—	—
other	1	1	—	—	2	—	3	1	1	2	—	—	1	1	—	1	1
Eastern cottontail																	
minimum no. individuals, 3																	
maxilla	1	—	—	—	—	—	—	—	—	—	—	—	—	—	—	—	—
ramus, mandible	2	—	—	—	—	—	—	1	—	—	—	—	—	—	—	—	—
humerus	—	1	1	—	—	—	—	1	—	—	—	1	—	—	—	—	—
scapula	—	—	1	—	—	—	—	—	—	1	—	—	—	—	—	—	—
innominate	—	—	—	—	—	—	—	—	2	—	—	—	1	—	—	—	—
tibia	1	—	—	1	—	—	—	—	—	—	—	—	—	—	—	—	—
calcaneum	—	1	—	—	—	—	—	—	—	—	1	—	1	—	—	—	—
metapodial	1	—	—	—	—	—	—	3	—	—	—	—	—	—	—	—	—
Mexican cottontail																	
minimum no. individuals, 2																	
maxilla	—	—	—	—	—	—	1	—	—	—	—	—	2	—	—	—	—
ramus, mandible	—	—	—	—	1	—	—	—	—	—	—	—	—	—	—	—	—
Sylvilagus spp., 1 juvenile																	
various bones	—	—	—	—	—	—	1	—	—	—	—	—	—	—	—	—	1
Mud turtle																	
minimum no. individuals, 1																	
scutes	—	1	—	—	—	—	—	2	—	—	—	—	—	—	—	—	—
limb bones	—	—	1	—	—	—	—	—	—	—	—	—	1	—	—	—	—

[a]Mature antlers and a fused metapodial indicate an adult; unfused metapodial and phalanx indicate a juvenile.

TABLE 22.4
Total Counts of Deer, Cottontail, and Mud Turtle Bones from Zone B3, Guilá Naquitz Cave

Animal	Distribution through 1 × 1-m squares				
	C6	C7	C9	D9	E5
White-tail deer					
minimum no. individuals, 1					
antler	1	—	—	—	—
carpal	2	—	—	—	—
metapodial	—	—	—	—	1
rib	1	—	—	—	—
other	1	—	—	2	—
Eastern cottontail					
minimum no. individuals, 3					
scapula	—	—	—	—	1
humerus	—	—	—	2	1
tibia	—	—	—	1	—
Sylvilagus spp.					
various bones	—	—	2	—	—
Mud turtle					
minimum no. individuals, 1					
scutes	—	—	1	—	—

At least one mud turtle, represented by eight fragments (three of them in C11) was roasted and eaten in this zone. The occupants also ate at least one band-tailed pigeon, one mourning dove, two songbirds, and a barn owl (perhaps a former resident of the cave).

One mandible of spiny pocket mouse from Zone E was burned, but we are uncertain whether it represents food or an accidentally burned rodent from an owl pellet. The same living floor produced two vertebrae of an unidentified small reptile.

Zone D

During the deposition of Zone D, the occupants of the cave killed and butchered a minimum of 2 deer, represented by 29–30 identifiable fragments. At least 1 adult male is reflected in the numerous fragments of mature antler scattered throughout the zone, and he must have been killed during the October-to-February season when antlers are fully hardened. Several unfused bones and unerupted teeth indicate that at least 1 juvenile deer was also present, and presumably, this younger animal could have been killed as early in the year as August or September. One shaft fragment from what is probably a deer long bone shows a series of cut marks, perhaps from butchering (Fig. 22.3b).

The occupants of Zone D also seem to have eaten at least 5 rabbits: 3 eastern cottontails (15 bones), 1 Mexican cottontail (2 bones), and 1 juvenile cottontail not identifiable to species (5 bones). The presence of 2 right femora in Square D9 and 2 right calcanea in nearby E8 suggests that several rabbits may have been butchered at the same time by the same person.

Included in the remains from this occupation are three fragments of collared peccary, representing at least one individual. The Zone D residents also roasted at least one mud turtle, three of whose scutes appeared in Square D7. Two stray songbird bones were found in Zone D, but since this occupation also produced an owl pellet with fragments of songbird, we hesitate to make too much of them.

A burned wood rat mandible appeared in Zone D, either another animal eaten or an accidentally burned fragment from an owl pellet. The zone also produced a Texas spiny lizard and four fragments of unidentified small reptiles (see Table 16.6, Chapter 16).

TABLE 22.5
Total Counts of Deer, Cottontail, and Mud Turtle Bones from Zone B2 + 3, Guilá Naquitz Cave

Animal	Distribution through 1 × 1-m squares					
	C8	C10	D10	E6	F4	F5
White-tail deer						
minimum no. individuals, 1						
antler	—	—	1	—	—	—
ramus, mandible	1	—	—	—	—	—
radius	—	—	—	—	—	1
phalanx	1	—	—	—	—	—
other	1	—	—	—	—	1
Eastern cottontail						
minimum no. individuals, 2						
maxilla	—	—	1	—	—	—
ramus, mandible	1	—	—	—	1	—
innominate	—	2	—	—	—	—
Mexican cottontail						
minimum no. individuals, 1						
maxilla	1	—	—	—	—	—
Sylvilagus spp.						
various bones	—	1	—	1	—	—
Mud turtle						
minimum no. individuals, 1						
scutes	1	1	1	—	—	—

TABLE 22.6
Total Counts of Deer, Cottontail, and Mud Turtle Bones from Zone B2, Guilá Naquitz Cave

Animal	Distribution through 1 × 1-m squares						
	C6	C9	C11	D5	D8	D9 (upper)	D9 (lower)
White-tail deer							
minimum no. individuals, 1							
humerus	—	1	—	—	—	—	—
tibia	—	1	—	—	—	1	—
naviculocuboid	—	1	—	—	—	—	—
vertebra	—	1	—	—	—	—	—
rib	1	—	—	—	—	—	2
other	1	—	1	—	—	5	—
Eastern cottontail							
minimum no. individuals, 1							
ramus, mandible	—	—	—	—	—	—	1
Mexican cottontail							
minimum no. individuals, 2							
ramus, mandible	—	—	—	—	—	2	—
scapula	—	—	—	—	—	2	—
Sylvilagus spp.							
various bones	—	—	—	1	—	—	—
Mud turtle							
minimum no. individuals, 1							
scutes	—	4	7	—	3	—	3

Zone C

During the deposition of Zone C, the occupants of the cave killed and butchered at least 2 deer, represented by 36 identifiable fragments. This is one case where the minimum number of individuals is probably conservative, and the actual number of deer killed may well be higher. At least 1 of the deer was a juvenile, as indicated by an unfused distal metapodial and proximal phalanx. The presence of a fused metapodial indicates that at least 1 other deer was mature, and a series of antler fragments suggest that 1 individual was a buck killed between October and February. A first phalanx from Square D8 displays cut marks on the lateral side of the proximal end at right angles to the long axis of the bone, perhaps reflecting an effort to separate the inedible hoof from the metapodial, or cannon bone.

The occupants also hunted or trapped a minimum of 6 cottontail rabbits, represented by 28 identifiable bones. There were at least 3 eastern cottontails, 2 Mexican cottontails, and 1 juvenile rabbit not identifiable to species. Two left innominates in Square E5 suggest that 2 eastern cottontails may have been butchered at one time. Equally suggestive is the evidence from Squares B9 and C10, which adjoin. We found 2 right mandibles of eastern cottontail in B9 and a right mandible of Mexican cottontail in C10; if all resulted from the same butchering episode, then 3 rabbits of 2 different species were processed at the same time.

Also during the occupation of Zone C, a collared peccary was killed and butchered. All that remained of it were two metapodials found in two adjoining squares, indicating how close an archaeologist can come to recovering *no* evidence of an animal that contributed perhaps 14 kg of meat. The Zone C residents also ate a raccoon, as indicated by a single femur in Square F9. At least one mud turtle, represented by five fragments, was roasted as well. One phalanx of barn owl may represent a former occupant of the cave, eaten on the arrival of the Zone C microband.

We also found a burned wood rat mandible in this zone, either another animal eaten or an accidentally burned fragment from an owl pellet. Finally, four reptile vertebrae scattered through Zone C defied our best efforts to identify them.

Lower Zone B

As explained in Chapter 5, Zones B3 and B2 appear to have been genuine living floors, while Zone B2+3 was a small unit where the two floors had been so compacted in postdepositional times that they could no longer be separated. This means that while we can present raw counts of animal bones for all these zones, we cannot realistically estimate minimum numbers of individuals for Zone B2+3; presumably, the bones in that zone came either from B3, B2, or both.

At least one deer was killed during the Zone B3 occupation; it was represented by eight bones, most of them concentrated in Square C6. An antler tine indicates that this might have been a buck killed during the October-to-February dry season. At least three eastern cottontails were butchered in Zone B3 (five to seven bone fragments, including three humeri in Square D9). Mud turtle was represented by only one scute, which (unless it is intrusive from Zone B2) would suggest only one individual.

No bird or reptile bones were found in Zone B3, but there was one burned mandible of cotton rat that might be attributable to food remains.

TABLE 22.7
Total Counts of Deer, Cottontail, and Mud Turtle Bones from Zone B1, Guilá Naquitz Cave

Animal	Distribution through 1 × 1-m squares																				
	B9	B10	C8	C9	C10	C11	C12	D5	D6	D7	D8	D10	E6	E7	E10	E11	F4	F5	F6	F7	G4
White-tail deer																					
minimum no. individuals, 2[a]																					
antler fragments	—	—	—	—	—	—	—	—	—	—	—	—	—	1	—	1	—	1	—	—	—
scapula	—	—	—	—	—	—	—	—	—	—	—	—	—	—	—	—	—	1	—	—	—
radius	—	—	—	—	—	1	—	—	—	—	—	—	—	—	—	—	—	1	—	—	—
femur	—	—	—	—	—	1	1	—	—	—	—	—	—	—	—	—	1	—	—	—	—
tibia	—	—	—	—	—	1	—	—	—	—	—	—	—	—	—	—	—	1	—	—	—
astragalus	—	—	—	—	—	—	1	—	—	—	—	—	—	—	—	—	—	—	1	—	—
naviculocuboid	—	—	—	—	—	1	—	—	—	—	—	—	—	—	—	—	—	—	—	—	—
metapodial	—	—	—	1	—	1	1	—	1	—	—	—	—	—	—	—	—	—	—	—	—
vertebra	—	—	—	—	1	—	—	—	—	—	—	—	—	—	—	—	—	—	—	—	—
rib	—	—	—	—	—	—	—	—	—	—	—	1	—	—	—	2	—	1	—	—	—
other	—	—	—	—	—	1	—	4	—	2	—	2	—	—	—	2	1	—	1	—	—
foetus fragments	—	—	—	—	—	1	1	—	—	—	—	—	—	—	—	—	—	—	—	—	—
Eastern cottontail																					
minimum no. individuals, 2																					
ramus, mandible	—	—	—	—	—	—	1	—	—	—	—	—	—	—	—	1	—	—	—	—	—
humerus	—	—	—	—	—	—	1	—	—	—	—	—	—	1	—	1	—	1	—	—	—
innominate	1	1	—	—	—	—	1	—	—	—	—	—	—	—	—	1	—	—	—	—	—
femur	—	—	—	—	—	1	—	—	—	—	—	—	—	—	—	—	—	—	—	—	—
tibia	—	—	—	—	—	—	1	—	—	—	—	—	—	1	—	—	—	—	—	—	—
metapodial	—	—	—	—	—	—	2	—	—	—	—	—	—	—	—	—	—	—	—	—	—
Mexican cottontail																					
minimum no. individuals, 3																					
ramus, mandible	—	—	—	—	—	—	1	—	—	—	—	—	—	—	—	—	—	—	—	—	—
humerus	—	—	—	—	—	—	1	—	—	1	—	—	—	—	—	—	—	—	—	1	—
Sylvilagus sp., 1 juvenile																					
various bones	—	—	1	—	—	—	1	—	—	—	—	1	—	—	—	—	1	—	—	—	2
Mud turtle																					
minimum no. individuals, 2																					
scutes[b]	—	—	2	2	1	—	—	—	1	14*	1	—	10*	—	1	—	1	—	—	—	—
limb bones	—	—	—	—	—	—	—	—	—	—	—	1	—	—	—	—	—	—	—	—	—

[a]A fragment of antlerless frontal bone and portions of foetus indicate a pregnant female; an antler indicates an adult male.

[b]One nearly complete turtle was roasted in each of the two asterisked squares.

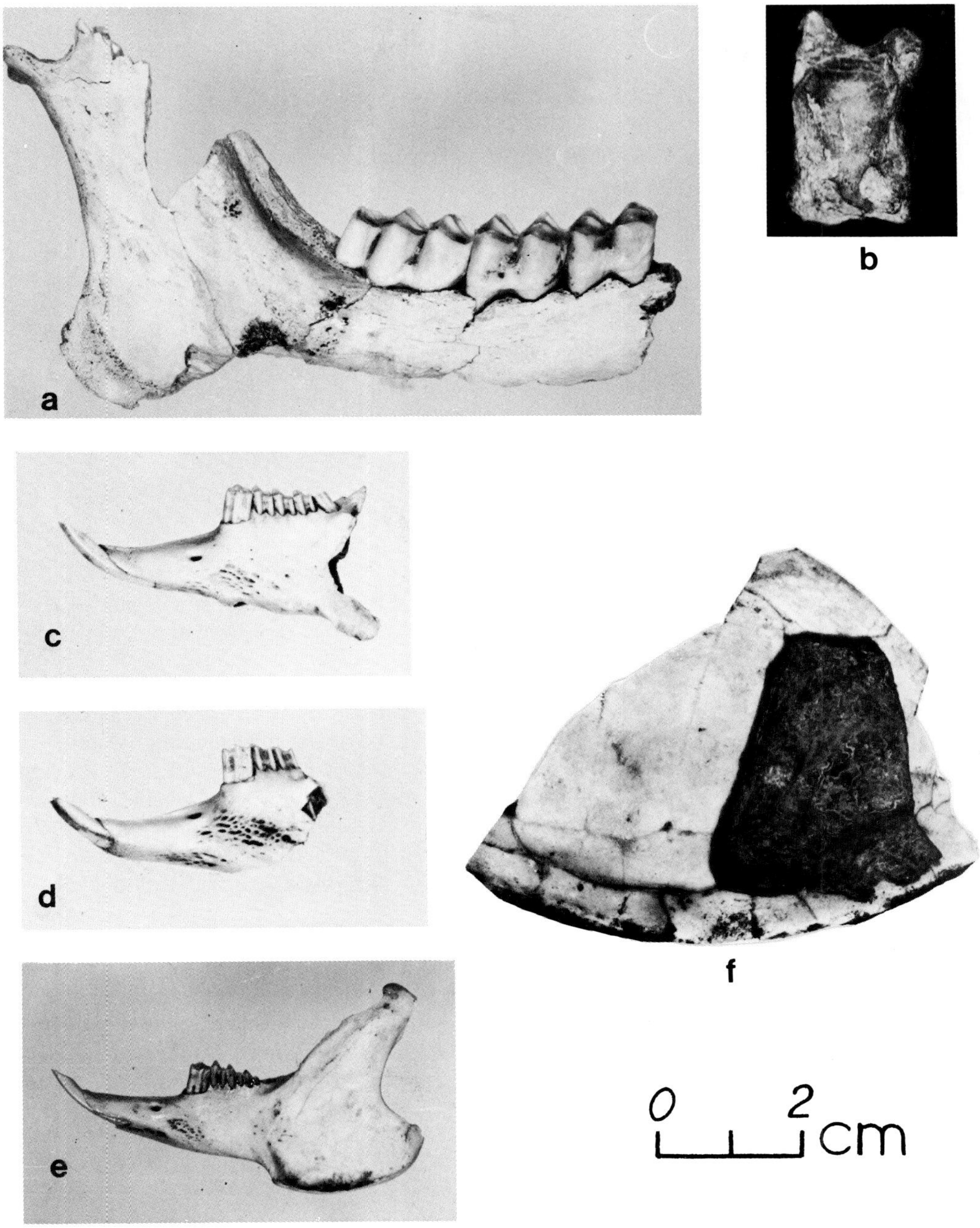

FIGURE 22.1. Faunal remains from preceramic Guilá Naquitz: a, mandible fragment of white-tailed deer (*Odocoileus virginianus*), Zone D, Square F5; b, astragalus of white-tailed deer, Zone B1, Square C12; c, mandible fragment of Mexican cottontail (*Sylvilagus cunicularius*), Zone B1, Square F7; d, mandible fragment of Mexican cottontail, Zone D, Square E8; e, mandible of eastern cottontail (*Sylvilagus floridanus connectens*), Zone C, Square D11; and f, carapace fragment of mud turtle (*Kinosternon integrum*), Zone B1, Square E10.

During the occupation of Zone B2, at least 1 deer (represented by 15 bones) was killed. Most of the fragments were concentrated in Squares C9–D9. In fact, deposits in Square D9 were so thick they could be divided into upper and lower sections; in terms of deer bone, the lower section contained only 2 ribs, while the upper contained a tibia fragment. Square C9 contained fragments from both forelimb and hindlimb. Whether this pattern reflects the order in which various parts of the deer carcass were processed and eaten (first the rib cage, then the limbs), or whether it simply reflects the order in which they were discarded, is not clear. No antler tines or other seasonal data on deer hunting were recovered in this zone.

Also in Zone B2, at least two Mexican cottontails (represented by four bones in Square D9) had been discarded, perhaps following one hunting–butchering episode. There

FIGURE 22.2. Eastern cottontail (*Sylvilagus floridanus connectens*) killed near Guilá Naquitz in 1966.

was also one bone of eastern cottontail in lower Square D9, but in view of the presence of eastern cottontails in this same square in Zone B3, we cannot totally rule out the possibility of redeposition. While the eastern cottontails of Zone B3 may have come from Thorn Forest A or B, the Mexican cottontails of Zone B2 (which prefer somewhat more open country) may have come from Mesquite Grassland A. At any rate, the distribution of cottontails in Zones B3 and B2 reinforces our notion that these animals were taken in bunches during periodic hunting or trapping episodes.

Seventeen scutes of roasted mud turtle, representing the remains of at least one individual, were found in Zone B2. Ten of these occurred in Squares C9, lower D9, and D8, all of which adjoin. We suggest that the animal may have been cooked over hot coals in or near this area, perhaps early in the occupation judging by its presence in lower D9.

Zone B2 also produced the remains of a bobwhite quail.

As for Zone B2+3, it produced only six fragments of deer, perhaps from one individual. However, it is possible we are dealing with parts of the same deer found in Zone B3 and/or Zone B2. For example, an antler fragment found in Square D10 of Zone B2+3 could be from the same buck eaten in Zone B3.

There were also eight rabbit bones in Zone B2+3, representing at least two eastern cottontails and one Mexican cottontail. However, since these bones do not duplicate any elements from Zones B3 or B2, they may simply be parts of the same animals found in those zones. The same could be said of the three mud turtle scutes in Zone B2+3, which (on the basis of circumstantial evidence) may be part of the same turtle in Zone B2.

Remains of a band-tailed pigeon and an unidentified songbird were found together in Square C8 of Zone B2+3, increasing the likelihood that this particular songbird was brought to the cave by hunters rather than owls. A burned wood rat mandible, also possibly part of the food remains, was found in this same zone. A single small reptile vertebra, not identifiable at this writing, rounds out the faunal evidence from Zone B2+3.

Zone B1

During the deposition of Zone B1, the occupants of the cave killed and butchered at least 2 individual deer (represented by 37 identifiable fragments). The antler fragments in this zone indicate that at least 1 deer was a buck killed during the period when the antlers are fully hardened (October to February). However, we also found portions of a foetal skeleton and 1 fragment of antlerless frontal bone from an adult female. This indicates that at least 1 pregnant doe was killed—presumably during the June-to-August period, when late term foetuses would be present.

Because of the seasonal differences involved, it is unlikely that both of these animals were killed at the same time. Based on the seasonality of the plants from Zone B1 (Chapter 18), we assume the pregnant doe would have been killed first, during the July–August season of mesquite harvests, while the buck would have been killed later, during the October–December season of *susí* and acorn harvests. In fact, the butchered remains of the two deer are somewhat spatially discrete, with the foetus in Squares C11–C12 and the antler fragments in E7, E11, and F5. It is worth mentioning that this evidence for July–August deer hunting fits well with the large numbers of mesquite seeds in Zone B1 (Chapter 19), since July–August is usually peak mesquite season as well.

The Zone B1 occupants also killed a minimum of 6 rabbits, represented by 25 identifiable bones. There were at least 2 eastern cottontails, 3 Mexican cottontails, and 1 juvenile

TABLE 22.8
Distribution of Additional Mammal Remains at Guilá Naquitz[a]

Animal	Stratigraphic zone D	Stratigraphic zone C	Stratigraphic zone B1
Collared peccary (*Dicotyles tajacu*)	1 right mandible (F8) 1 carpal/tarsal (E8) 1 right calcaneum (E7)	2 metapodials (C9, D10)	—
Raccoon (*Procyon lotor*)	—	1 right femur (F9)	—
Human (*Homo sapiens*)	—	—	left third metacarpal (E10)

[a]Squares in which specimens were found are indicated in parentheses.

TABLE 22.9
Distribution of Bird Remains at Guilá Naquitz[a]

Name	Stratigraphic zone					
	E	D	C	B2 + 3	B2	B1
Bobwhite quail (*Colinus virginianus*)	—	—	—	—	1 humerus (D9)	—
Band-tailed pigeon (*Columba fasciata*)	1 tibia (D10)	—	—	1 humerus (C8)	—	—
Mourning dove (*Zenaidura macroura*)	1 humerus (D10)	—	—	—	—	—
Ground dove (*Columbigallina* cf. *passerina*)	—	—	—	—	—	1 humerus (C12)
Red-tailed(?) hawk (*Buteo* cf. *jamaicensis*)	—	—	—	—	—	1 femur (B9)
Barn owl (*Tyto alba*)	1 fragment(?) (C8)	—	1 phalanx (D10)	—	—	2 bones (F5)
Songbirds in owl pellets	—	1 (C9)	Grosbeak (E7)	—	1 (C9)	—
Other songbirds	2 (E10, F9)	2 (E6, F6)	—	1 (C8)	—	—

[a]Squares in which specimens were found are given in parentheses.

FIGURE 22.3 Deer bone fragments with possible butchering cuts on them, preceramic Guilá Naquitz: a, from Zone B3, Square C6 (actual length, 6.8 cm) and b, from Zone D, Square F8.

not identifiable to species. It is not clear how many separate hunting incidents were represented, but there are hints that more than 1 animal was butchered at a time (2 mandibles in C12; 2 humeri in C12; 2 more humeri in E7–F7). If these finds imply the taking of several rabbits at one time, then the Zone B1 hunts (unlike those of some earlier levels) were ones in which both local species of cottontail were captured.

The evidence for use of mud turtle is clearer in Zone B1 than for any previous level. Some 34 fragments (mostly carbonized carapace scutes) were present, representing at least 2 individual turtles. Fourteen of these were concentrated in Square D7, where 1 turtle was evidently roasted. Ten more were concentrated in Square E6, where a second turtle was evidently roasted. The unusually concentrated nature of these remains may mean that the roasting of the 2 turtles took place only a short time before abandonment of the living floor. This would have lessened the scattering and destruction of carbonized scutes by the occupants as they walked back and forth.

At least three birds were killed during the deposition of Zone B1. Two bones were those of a barn owl, possibly a former resident of the cave killed early in the occupation. The other two bones belong to a hawk and a small ground dove. We assume that all these birds could have served as food, but we suggest that both hawks and owls may have been killed for their feathers as well.

Zone B1 is one of the few levels in which there is fairly conclusive evidence that the cave occupants ate small lizards. Clustered together in Square C10 were two mandibles of Mexican whiptail (*Cnemidophorus*) and one mandible of Texas spiny lizard (*Sceloporus*) in a context that suggested discarded food refuse. Realistically, however, their actual contribution in terms of grams of meat would have been very small (see Table 16.6, Chapter 16).

23

Nutritional Significance of the Guilá Naquitz Food Remains

J. R. K. Robson
J. N. Elias

INTRODUCTION

As part of the program of investigations of Guilá Naquitz and other caves of the Mitla region, plant censuses were carried out over a 10-year period (See Chapter 18). During some of these censuses, edible parts of wild plants from the cave area were frozen and flown back to the Indigenous Foods Laboratory at the University of Michigan. Included were acorns, baked *Agave* heart, hackberries, *guaje* seeds, mesquite pods, nanche fruits, *Opuntia* fruits (tunas), *Opuntia* stems (nopales), cucurbit seeds, and *susí* nuts. Nutritional contents of all these foods were determined by Robson and J. E. Konlande at the Indigenous Foods Laboratory.

Unfortunately, certain of the foods eaten by the prehistoric occupants of the cave could not be collected, either because they no longer grow in the area (as in the case of piñon) or because they were not ripe during the period when someone was available to bring the refrigerated plants to the United States (as in the case of wild onions). In the case of these additional food plants, we have used the best available published data from other New World areas (Table 23.1).

In this chapter, we attempt to assess the nutritional significance of the food remains from preceramic Guilá Naquitz.

DISCUSSION

In living populations, the assessment of nutritional status depends on the collection of data from a variety of sources. This epidemiological approach (World Health Organization [WHO] 1963) takes into account *inter alia,* vital statistics, anthropometry, clinical signs of malnutrition, measurement of nutrient levels in the tissues or tissue fluids, as well as appraisement of food availability and dietary intake. This approach is needed because of inherent weaknesses in individual data items collected. These weaknesses include the influence of recent meals on nutrient levels in the blood, inaccuracies in reporting dietary intakes, problems in weighing food consumed, as well as lack of knowledge of the nutrient composition of the foods consumed. In the epidemiological approach, data from various sources of information are collected. Each data item helps to support (or refute) other information. For example, an observation from a diet survey that foods containing vitamin C were not consumed (or were not available) means little by itself; but the presence of sore gums, while not itself diagnostic of scurvy, would be a significant finding in the presence of supporting biochemical or dietary data. Similarly, evidence of large-scale parasitic infection may require reappraisal of a population's nutritional requirements (Berlin and Markell 1977).

In the science of archaeology, however, the opportunities for the epidemiological approach are obviously reduced, and the interpretation of data resulting from the recovery of desiccated, unprocessed food materials from various strata of Guilá Naquitz poses some particularly formidable difficulties. At first sight, it would seem impossible to make a scientifically accurate appraisal of the nutritional value of these foods, but their recovery is sufficiently important, however, to justify attempts to judge their nutritional significance.

Two major problems have to be surmounted: The first concerns the availability of the nutrients in the foods found in the cave, and the second relates to the nutritional requirements of the population living at Guilá Naquitz at that time. Such requirements are influenced by the physique, activity, and health of the individuals comprising that particular community. The best model that can be used in these circumstances is that of contemporary aboriginal populations whose food intake is also largely dependent on hunting and gathering.

As Robson has summarized elsewhere (Robson and Yen 1976), studies of Papuans (Bailey 1963; Hipsley and Kirk

1965; Oomen 1961), and aborigines in Australia (McArthur 1960) and the Philippines (Robson and Yen 1976; Schlegel and Guthrie 1973) suggest that nutritional adequacy can be achieved on intakes significantly lower than the estimates for requirements made by the WHO (WHO 1973). Whether this reflects physiological adaptation, overestimation of nutrient and energy requirements, or underreporting of the food intake is not clear at this time.

In the light of these reports, however, as well as evidence of the relatively small stature of prehistoric and historic burial populations in the Valley of Oaxaca[1], we are assuming that the average nutrient and energy intake of the occupants of Guilá Naquitz was generally in keeping with those of contemporary aboriginal populations. The smaller body size implies lower demands for basal metabolism as well as for physical activity. This suggests that the requirements of the prehistoric peoples of Guilá Naquitz would be lower than those reported for the current U.S. population.

Definitive evaluation of the prehistoric diet in this location is difficult because of the lack of knowledge of the dietary practices of the populations inhabiting the caves. One example of our deficiencies in knowledge is that it is not known whether the food materials recovered represent the full range of food resources exploited during the various occupations. Worldwide ethnographic evidence of subsistence practices in hunter–gatherer groups suggest that it is probable that some foods would have been consumed away from the cave. Furthermore, some of the foods brought to Guilá Naquitz may not have been preserved for archaeological recovery. We are not able to provide answers to either of these questions.

Even if it is assumed that preservation of food at the cave was not only complete, but also that the foods recovered truly represented the total spectrum of foods consumed at that site, further information is needed before the nutritional value of the foods can be properly evaluated. It is well known, for example, that both the duration of storage and the preservation techniques used can significantly affect the nutritional value of a food *as consumed* (Caldwell 1972; Konlande and Robson 1972; Somogyi and Schiebe 1966; Walker *et al.* 1975). It would be necessary, therefore, to know how long the foods were in storage before they were eaten and also what techniques were employed to prepare the foods. The effect of the composition of the diet also has to be considered, because the interaction of nutrients present in the individual foods can significantly increase or decrease the nutritional value of the total meal. Phytates and oxalates in particular influence the availability of calcium, while the former also affect the uptake of zinc in the body (Reinhold 1972; Reinhold *et al.* 1973; Walker 1951). In addition, the assimilation of the iron in the food by the body during digestion and absorption can be greatly influenced by the other dietary components (Jacobs 1969).

The methods used in preparing the foods also need to be taken into consideration, as these too can affect nutrient value (Konlande and Robson 1972); other techniques of preparation may be equally important. For example, the addition of alkali to water used for soaking maize prior to grinding (Katz *et al.* 1974), or the addition of ashes to a food (Calloway *et al.* 1974; Kuhnlein and Calloway 1979; Townsend *et al.* 1973), may increase its mineral content. In the case of maize, biological enoblement results from the alkalinization because niacin and various essential amino acids are rendered more available for absorption and utilization by the body (Bressani and Scrimshaw 1958). Among the Aguaruna, fermentation of masticated, boiled manioc root to make a beer converts starch to simple sugars and alcohol, thereby effecting more rapid absorption by the body (Berlin and Markell 1977).

No satisfactory adjustment factors are known that could be applied to these dietary data to make conclusions more scientifically sound. The best that can be done under the circumstances is to assume that an inevitable, but unknown nutrient loss occurs that reduces the availability of the nutrients to the body.

The next step in assessing the nutritional value of a diet necessitates matching nutrient availability to nutritional requirements. Reference has been made to the higher requirements of the U.S. population and the increases in dietary allowances that have been proposed for the United States. The 1974 estimate for the United States includes a safety factor to cover unforeseen needs. It is based on current knowledge of the average requirements of the U.S. population plus two standard deviations of the mean, and it is adequate for 97.5% of the population (National Academy of Sciences [NAS] 1974). For the purposes of this discussion we are assuming that the inflationary safety factor in the dietary allowances and the deflationary losses of nutrients occurring in the preparation and storage of food by the inhabitants of Guilá Naquitz cancel one another out and that the raw values for nutrient composition and the dietary allowances of the U.S. can be equated.

The botanical evidence suggests that Guilá Naquitz was seasonally occupied as part of a regional strategy to exploit certain local wild food resources. The phenology of the plants recovered from the excavation indicated an autumn occupation between the months of August and December (see Fig. 18.1). Such a prehistoric scheme for resource exploitation is unlikely to have placed acquisition of balanced nutrition foremost in the minds of the indigenous population. Nevertheless, a successful exploitative strategy would have had reasonable nutrition as its latent function, inasmuch as there can be no long-term health advantage in ignoring nutrient-rich foods in favor of nutritionally inferior varieties. Thus, assuming prehistoric man to have been rational in his search for food, it is reasonable for us to focus attention, as well, on those species that could have provided both the abundance and nutrient density to support the occupation of Guilá Naquitz.

Proximate analyses reveal that within the Guilá Naquitz inventory there are a number of nutritionally important foods that may have been exploited as primary dietary staples (Table 23.1 and Fig. 23.1). Both piñon nuts (*Pinus* sp.) and cucurbit seeds (*Cucurbita* sp.) are potentially important contributors of protein, fat, carbohydrates, and calories. *Susí* nuts (*Jatropha* sp.) lack carbohydrates but are rich in fat and protein, which accounts for their high caloric density. The

[1]Unpublished studies by Richard G. Wilkinson suggest a population generally between 4′10″ and 5′4″ during the Formative period.

TABLE 23.1
Nutritional Content of Various Wild Plants Used at Guilá Naquitz, Based on a 100-g Portion

Name of plant	Moisture (%)	Fat (g)	Protein (g)	Carbohydrate (g)	Fiber (g)	Ash (g)	Kilocalories
Piñon nuts[a]	3.1	60.50	13.00	20.5	1.10	2.90	635
Onions[a]	89.1	0.10	1.50	8.7	0.60	0.60	38
Agave heart	64.0	0.04	0.31	31.5	2.98	1.20	126
Acorns	37.4	9.33	2.43	43.3	6.32	1.21	288
Hackberries	89.3	0.46	0.33	7.5	0.33	2.10	33
Guaje seeds	81.6	1.00	8.10	8.3	0.50	0.50	72
Beans[a]	69.0	0.60	7.80	21.2	1.50	1.40	118
Mesquite pods	13.4	1.00	5.70	45.7	31.60	2.64	297
Nanche fruits	73.0	0.06	1.69	17.6	4.13	3.48	69
Susí nuts	4.0	52.50	17.60	3.3	19.30	3.29	593
Opuntia, nopal	92.2	0.34	1.93	0.5	0.59	4.50	12
Opuntia, fruits	90.2	0.53	1.11	6.3	1.42	0.43	36
Cucurbit seeds	6.31	18.60	33.50	37.1	0.90	3.63	427

[a]Values taken from Watt and Merrill (1963). All other values were determined by the Indigenous Foods Laboratory, University of Michigan, using specimens collected within 5 km of Guilá Naquitz.

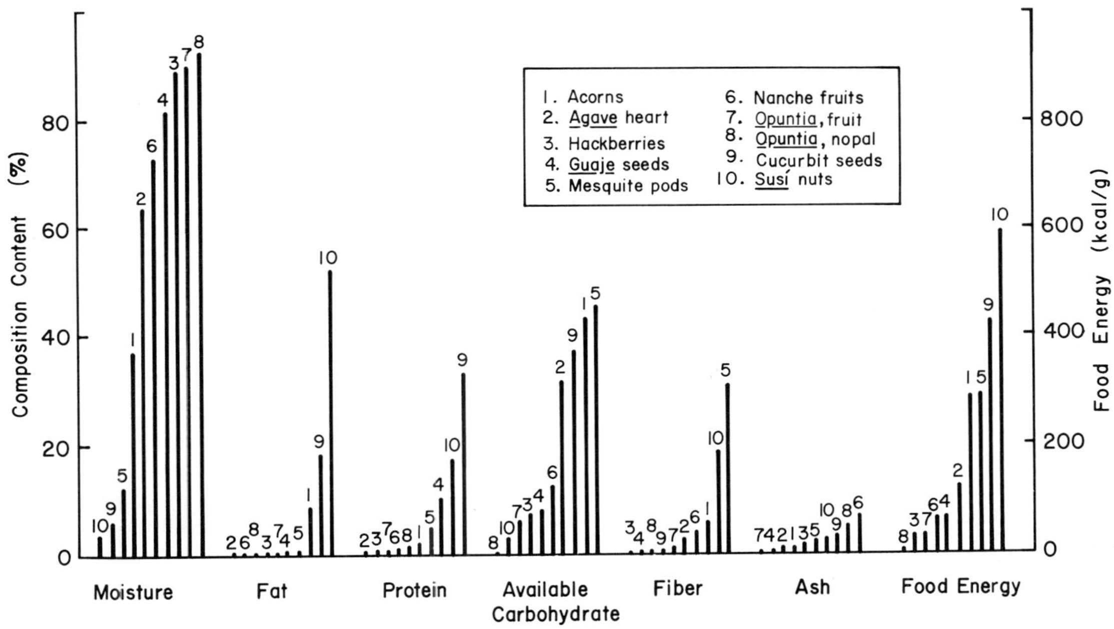

Fig. 23.1. Proximate composition of selected plant foods from Guilá Naquitz.

various leguminous seeds—such as beans (*Phaseolus* sp.), mesquite (*Prosopis* sp.), *Dalea*, and *guaje* (*Leucaena* sp.)—with their relatively high protein and carbohydrate content, could help to balance the lack of carbohydrates in *susí*. Similarly, both acorns (*Quercus* sp.) and agave (*Agave* spp.), which contain large amounts of carbohydrates, would raise the caloric content of the diet significantly.

In addition to the above-mentioned foods, the animal foods identified at Guilá Naquitz (Table 23.2) may be considered important contributors to the staples. Availability of individual foods within this primary group of plant and animal species may have varied from year to year, but there would appear to be sufficient nutritional overlap to permit substitution of one food for another without seriously affecting overall nutrient availability.

One mineral, calcium, seems especially noteworthy in view of the relatively limited consumption of dairy products characteristic of most Mesoamerican peoples. It has been pointed out that the practice of soaking corn in limewater prior to milling provides an important source of dietary calcium (Bressani and Scrimshaw 1958). However, there are other vegetable foods that could be important sources of this mineral. Nopales have been found to contain calcium within the range of 46–105 mg per 100 g edible portion. Cholla, also of the *Opuntia* species, contains as much as 281 mg of calcium (Calloway *et al.* 1974; Cravioto *et al.* 1945; Food and Agriculture Association [FAA] 1972.) The presence of relatively large amounts of nopales at Guilá Naquitz suggests that this food may have been an important source of dietary calcium to the occupants of the cave.

TABLE 23.2
Nutritional Content of Various Wild Animals Eaten at Guilá Naquitz Cave, Based on a 100-Gram Portion[a]

Animal	Moisture (%)	Fat (g)	Protein (g)	Carbohydrate (g)	Fiber (g)	Ash (g)	Kilocalories
Deer	74.0	4.0	21.0	0	0	1.0	126
Raccoon	54.8	14.5	29.2	0	0	1.5	255
Cottontail	73.0	5.0	21.0	0	0	1.0	135
Pigeon	58.0	22.1	18.6	0	0	1.5	279
Quail	65.9	6.8	25.0	0	0	1.6	168
Turtle	80.0	1.0	16.0	0	0	1.2	79

[a]Data are from Platt (1962) and Watt and Merrill (1963).

In conclusion, the archaeological evidence from Guilá Naquitz suggests that the seasonal occupation of this cave was part of a successful regional strategy to maximize nutrient intake through exploitation of local wild food resources. In the various occupational levels, major sources of plant refuse always include a high proportion of those species one would expect to serve as primary dietary staples on the basis of proximate nutritional analysis. One may venture to speculate that contemporaneous occupations in other ecological zones within the valley will provide evidence for a similar strategy in those zones yielding species whose phenology is complementary to that of the species exploited at Guilá Naquitz.

HYPOTHETICAL DAILY INTAKES FOR GUILA NAQUITZ

Tables 23.3–23.5 are based on the plant remains identified in three levels at Guilá Naquitz. They present three hypothetical daily intakes that would easily meet the 1974 Recommended Dietary Allowance (RDA) for protein and calories. Each diet includes 150 g of meat and 1350 g total daily food intake. Ahrens and Boucher (1978) found that a simulated American diet based on a U.S. Department of Agriculture (USDA) household consumption survey provided 1583.4 g of food/person/day, including 107.5 g of protein, 136.9 g of fat, 298.9 g of carbohydrates, and a total of 2858 kcal (Table 23.6). The proportion of calories provided by protein, fat, and carbohydrates, respectively, was 15%, 43%, and 42%. The hypothetical diets shown in Tables 23.3–23.5 compare favorably with the simulated American diet, as they provide similar proportions of nutrients and easily exceed the U.S. caloric levels on less total food. This latter property would enable the occupants of Guilá Naquitz to achieve nutritional adequacy on less bulk. That indigenous foods are so easily able to provide significantly higher quantities of nutrients than contemporary diets suggests that the occupants probably did not experience difficulty in meeting their nutritional needs during the season when they were at Guilá Naquitz.

In addition to the requirements for protein and energy, there are, of course, requirements for other nutrients, particularly vitamins and minerals. Mixtures of several fruits and vegetables could easily have provided these, and the variety

TABLE 23.3
Hypothetical Daily Intake for an Occupant of Zone E, Guilá Naquitz, Based on Plant and Animal Remains Recovered from that Level[a]

Zone E	Fat (g)	Protein (g)	Carbohydrate (g)
150 g deer or cottontail	6.8	31.5	—
150 g piñon nuts	90.8	19.5	30.8
550 g acorns	51.3	13.4	238.2
300 g beans	1.8	23.4	63.6
50 g *susí* nuts	26.3	8.8	1.7
100 g *Opuntia* fruit	0.5	1.1	6.3
50 g hackberries	0.2	0.2	3.8
1350 g	177.7	97.9	344.4
Kilocalories[b]	1599.3	391.6	1377.6
Percentage of kilocalories	47.5	11.6	40.9

[a]The diet assumes consumption of 1350 g of food per day.
[b]Total kilocalories equal 3368.5.

TABLE 23.4
Hypothetical Daily Intake for an Occupant of Zone C, Guilá Naquitz, Based on Plant and Animal Remains Recovered from that Level[a]

Zone C	Fat (g)	Protein (g)	Carbohydrate (g)
150 g deer or cottontail	6.8	31.5	—
200 g piñon nuts	121.0	26.0	41.0
150 g *Agave* heart	0.1	0.5	47.3
400 g acorns	37.3	9.7	173.2
300 g mesquite pods	3	17.1	137.1
50 g *guaje* seeds	0.5	4.1	4.2
40 g nanche fruits	—	0.7	7.0
50 g *Opuntia* fruit	0.3	0.6	3.2
10 g hackberries	0.1	—	0.8
1350 g	169.1	90.2	413.8
Kilocalories[b]	1521.9	360.8	1655.2
Percentage of kilocalories	43.0	10.2	46.8

[a]The diet assumes consumption of 1350 g of food per day.
[b]Total kilocalories equal 3537.9.

TABLE 23.5
Hypothetical Daily Intake for an Occupant of Zone B1, Guilá Naquitz, Based on Plant and Animal Remains Uncovered from that Level[a]

Zone B1	Fat (g)	Protein (g)	Carbohydrate (g)
150 g raccoon	21.8	43.8	—
300 g mesquite pods	3.0	17.1	137.1
350 g acorns	32.7	8.5	151.6
100 g *guaje* seeds	1.0	8.1	8.3
150 g beans	0.9	11.7	31.8
50 g *Agave* hearts	—	0.2	15.8
50 g nanche fruits	—	0.8	8.8
200 g *susí* nuts	105.0	35.2	6.6
1350 g	164.4	125.4	360.0
Kilocalories[b]	1479.6	501.6	1440.0
Percentage of Kilocalories	43.2	14.7	42.1

[a]The diet assumes consumption of 1350 g of food per day.
[b]Total kilocalories equal 3421.2.

TABLE 23.6
Simulated American Diet Based on a USDA Household Consumption Survey, Provided for Comparison with Tables 23.3–23.5[a]

Food	Fat (g)	Protein (g)	Carbohydrate (g)
592.0 g dairy	44.3	27.4	41.7
39.4 g fats and oils	32.6	0.2	0.9
90.9 g cereals	1.6	6.6	64.5
150.3 g bakery	11.9	12.1	79.8
250.8 g meat, poultry, and fish	34.0	47.3	0.8
48.6 g eggs	5.6	6.3	0.5
411.4 g other	7.7	7.7	110.9
1583.4 g	136.9	107.5	298.9
Kilocalories[b]	1232.1	430.0	1195.6
Percentage of kilocalories	43.1	15.0	41.8

[a]Data are from Ahrens and Boucher (1978).
[b]Total kilocalories equal 2857.7.

reflected in the Guilá Naquitz material seems more than adequate (see Tables 23.3–23.5). It has been noted elsewhere (Robson 1978) that the advent of farming often resulted in a more localized way of life with fewer opportunities to exploit abundant, albeit scattered, wild food resources. There is no reason to suppose that the occupants of Guilá Naquitz felt any such dietary constraints in view of the diversity of foods recovered.

24

Food Procurement Area and Preceramic Diet at Guilá Naquitz

Kent V. Flannery

Having estimated the productivity of the Guilá Naquitz environment (Chapter 18) and determined the nutritional makeup of the most common foods eaten at the cave (Chapter 23), we can now move to questions about the procurement area needed to sustain the preceramic population. In this chapter we convert the food resources of the Guilá Naquitz area to calories and protein, estimate the harvest area necessary to support the cave occupants during their stay, and reconstruct their diet from archaeological remains.

THE RELATION BETWEEN FOOD GENERA AND 100-G PORTIONS

MacNeish (1967) was the first to attempt to convert the remains from Mexican dry caves into standard measures of food in the metric system. Using "a liter can (courtesy of Pemex Oil Company) and a measuring cup marked in liquid ounces (33.8 ounces per liter)," he tried to estimate liters of vegetal foods in his Tehuacán caves (1967:296). This rough-and-ready procedure yielded some of MacNeish's best lines—"The flesh of avocados, guavas, and prickly pear and other cactus fruits averaged about 30 examples per smelly, sticky liter" (1967:297)—but by his own admission it was "based upon calculations that are fraught with difficulties" (1967:296).

Not the least of these difficulties was the fact that archaeological remains vary greatly in the degree to which they correspond to the original foodstuffs. Some plant remains left behind on a cave floor, such as an acorn or a wild onion bulb, represent the part that was edible. Others, such as a hackberry seed or a maguey quid, represent the inedible part that was thrown away. To calculate liters of acorns and onions might be meaningful; to calculate liters of quids and inedible seeds would not be. And even in the case of the acorns and onions, we are faced with a problem: We are measuring not the specimens they ate but the ones left over—the ones they did not eat (Figure 17.2).

"Ideally," said MacNeish (1967:296), "we would have liked to have this data transferred . . . into food energy units such as proteins, carbohydrates, fats, and minerals. Unfortunately, since nutritional tools and nutritionists were unavailable for this task, we were not able to proceed beyond the first step." As of 1981, however, MacNeish (personal communication, 1981) was laying the ground work for just such a nutritional study in the Tehuacán Valley.

In 1969, I attempted to convert Near Eastern archaeological food remains into proteins, carbohydrates, fats, and minerals (Flannery 1969: Table 3). The first thing I discovered was that nutritionists do not want the data given to them in liters; their standard unit is the 100-g edible portion (or its multiple, the kilogram). What archaeologists need to establish, therefore, is the relationship between recoverable foodstuffs and 100-g edible portions, and that is what we attempt in this section.

PLANT FOODS

Table 24.1 gives the relationship between archaeologically recoverable plant remains and 100-g edible portions for the major species recovered from Guilá Naquitz. Data on these species, listed in phylogenetic order, result from a series of studies carried out between 1966 and 1976.

Our sample of piñon nuts (obtained from elsewhere in Mexico since they no longer occur near Guilá Naquitz) showed that 600 had to be shelled to provide a 100-g portion. Thus, an average nut weighed 0.167 g. A 100-g portion of wild onion bulbs averages 9 onions, suggesting a mean weight of 11 g for our sample of bulbs.

TABLE 24.1
Relationship between Archaeologically Recoverable Plant Remains and 100-gram Edible Portions for Various Plants Used at Guilá Naquitz

Plant	Edible part of plant	Weight of one edible part (g)	Amt. yielding 100-g edible portion	Archaeological evidence for 100-g portion
Piñon (*Pinus*)	Nut	0.167	600 nuts	600 nuts
Onion (*Allium*)	Bulb	11	9 bulbs	9 dried bulbs
Agave potatorum	Heart	2,795	100 g	6.15 quids
Agave marmorata	Heart	13,000	100 g	6.15 quids
Oak (*Quercus*)	Acorn	3.2	31 acorns	31 acorns
Oak (*Quercus*)	Gall	5	20 galls	20 dried galls
Hackberry (*Celtis*)	Fruit	0.18	550 fruits	550 seeds
Acacia farnesiana	Pod	1.67	60 pods	60 dried pods
Guaje (Leucaena)	Seed	0.125	800 seeds	80 dried pods
Wild bean (*Phaseolus*)	Seed	0.125	800 seeds	800 seeds or 320 pod valves
Mesquite (*Prosopis*)	Pod	1.67	59.9 pods	59.9 dried pods or 809 seeds
Nanche (*Malpighia*)	Fruit	3.5	28.6 fruits	28.6 seeds
Susí (Jatropha)	Nut	0.5	200 nuts	200 hulls
Lemaireocereus	Fruit	50	2 fruits	2 dried fruits
Prickly pear (*Opuntia*)	Nopal	20 (roasted) 32.5 (boiled)	5 nopales 3 nopales	4 dried nopales
Prickly pear (*Opuntia*)	Fruit	50	2 fruits	2 dried fruits or 100 seeds
Wild *Cucurbita*	Seed	0.167	600 seeds	600 dried seeds or 24 fruits

Calculations for maguey hearts were more complicated, since the hearts had to be cooked to render them edible. A local mezcal factory kindly agreed to do the baking for us in one of their ovens, and the cooked heart was then divided into 1-kg portions for eating. Baked maguey is extremely sweet—a little like fudge soaked in hard cider—and while the first mouthful is pleasant, it can cloy on you after a while. Roughly 5–6 volunteers divided up each kg and chewed it, each saving the quids of indigestible fibers left in his mouth afterwards, and depositing them in a jar. This particular experiment showed that the prehistoric cave occupants surely had no lack of roughage in their diet.

One interesting discovery was that the center of each maguey heart is without fiber: it can be eaten completely, without leaving a single archaeologically detectable trace. Outside this center begin the leaf bases, and it is these that produce the quids of fiber. Had the prehistoric occupants of the cave eaten only the fiberless core of the heart, we would have had little evidence for maguey.

Agave potatorum, the smaller species favored by the prehistoric cave dwellers, has a heart averaging 5590 g; cooking drives off a lot of moisture and reduces this to 2795 g. Some 30% of this (838 g) is fiberless and would leave no trace. The remaining 70% (1957g) has fiber and produces an average of 88 quids per kg, for a total of 172 quids. If we take the fiberless 30% of the heart into account (as we must, since it was probably eaten before the leaf bases), this means that each quid represents about 16.25 g of edible portion.

Agave marmorata, the larger maguey that is preferred for mezcal production nowadays, has a heart averaging 26,000 g; cooking reduces this to 13,000 g. Thirty percent of this (3900 g) is fiberless, while the remaining 70% (9100 g) produces quids at the rate of 88 per kg for a total of 800. Taking the fiberless 30% into account, we once again have 16.25 g of edible portion reflected in each quid.

There is so much food available on one of these magueys that we could account for all the quids in Guilá Naquitz without assuming the baking of more than 1 heart per living floor. Let us suppose that the occupants located a mature agave and cropped the inflorescence the day they arrived. The second day they might have dug the roasting pit, uprooted the maguey, and started it cooking. The third day they could have gone about other hunting–gathering tasks while the heart cooked on. By the fourth day of occupancy, they would have had 3–13 kg of edible heart available, depending on the species. Indeed, a large *Agave marmorata* would produce so much more "fudge" than a family of 4–5 persons could conveniently eat that, apart from any difference in taste, it probably was more practical to stick mainly with the smaller *A. potatorum.*

The most common plant food recovered from Guilá Naquitz was the acorn. Based on numerous random samples of local acorns, a 100-g portion averaged 31 specimens; thus an average acorn weighed 3.2 g.

The inedible, highly durable seed of the hackberry represents about 40% of the weight of each fruit; the remaining flesh averages 0.18 g. This means that some 550 fruits are necessary to produce 100 g of edible portion.

Infrequent in the cave remains were the pods and seeds of two wild legume trees, the *guaje* and the acacia. It took 60 acacia pods to produce 100 g of edible portion (the jellylike material in the fresh pod), for an average of 1.67 g/pod. In the case of *guajes,* it is the seeds that are edible. There are an average of 10 seeds/pod in Thorn Forest A; and since an average seed weighs 0.125 g, it would take 80 pods (800 seeds) to produce a 100-g portion.

Wild beans represented a problem, since some varieties are collected for their root (or *jicamita*) and others for their seeds. Since we had no archaeological evidence for roots, we decided to restrict ourselves to the small black seeds. A 100-g portion consisted of 800 seeds (0.167/seed), and since there was an average of 5 seeds/pod, 160 pods would yield a 100-g portion. Pods averaged 7 per plant in our sample.

Mesquite was perhaps the most important legume at Guilá Naquitz. It was most likely used for the jellylike material in the fresh pods, although it is also possible that the seeds were saved to be roasted and eaten at a later time. A 100-g portion of the jelly would require 59.9 pods (1.67 g of edible portion per pod). However, much of the evidence in Guilá Naquitz consisted of loose seeds. In our sample there was an average of 13.5 seeds/pod; thus, 809 seeds would constitute evidence for 59.9 pods, or a 100-g portion.

The edible flesh of the wild nanche fruit averages 3.5 g, so 28.6 fruits would provide a 100-g portion. (It should be pointed out that this is somewhat less than the weight of the cultivated nanche, which has been selected for size.)

Susí was represented in the cave primarily by the spherical hull that once surrounded the small, edible seed. This seed weighs only 0.5 g and tastes like a Brazil nut; it is rich in protein (Chapter 23). Some 200 nuts make up a 100-g portion.

Lemaireocereus (organ cactus) fruit was rare at Guilá Naquitz. Each fruit weighs about 50 g (a good bit of which is water), so two fruits would yield a 100-g portion.

The prickly pear was represented by both fruits (tunas) and tender young stem sections (nopales). Each fruit weighs 50 g, with two fruits contributing a 100-g portion. In our sample, fruits had an average of 50 seeds, so 100 seeds could also be used as evidence for a 100-g portion. As for the nopales or stem sections, a lot depends on whether they were prepared by roasting or boiling. The average uncooked nopal is 50 g; roasting reduces it to 40% (20 g), while boiling reduces it to 65% (32.5 g). Thus, archaeological evidence for a 100-g portion would consist of 5 nopales if roasted, 3 nopales if boiled. In Table 24.1 we have averaged these two figures to produce an estimate of 4 nopales. However, it should be noted that we have direct archaeological evidence that nopales were roasted on sticks at Guilá Naquitz (Chapter 10), while no evidence for boiling exists.

Finally, we weighed a sample of wild cucurbit seeds. They average 0.167 g, and a 100-g portion consists of roughly 600 seeds. Since there was an average of 25 seeds/fruit, it would take about 24 fruits to produce 100 g of cucurbit seeds.

In Table 24.2, we have totaled up the major food plants recovered in Zones B–E of Guilá Naquitz and converted them to grams of edible portion. In Table 24.3 these same food plants are ranked in order of the weight of edible portion that each contributed to the diet. Here the simplifying assumption made is that the relative abundance of a particular plant in the cave is a reasonably accurate reflection of its relative abundance in the diet. We have already discussed the pitfalls of this assumption in Chapter 17, as well as the fact that we have no basis for assuming anything else.

Both tables point up the overwhelming abundance of acorns, a product of Thorn Forest A. Acorns amount to nearly 27 kg, or 62.9% of the edible portion represented by the archaeological plant remains; nothing else comes close. Baked agave heart, at roughly 6 kg, is far back in second place. Two products of the prickly pear, nopales and *Opuntia* fruits, are in third and fourth place. Nanche fruits, at roughly 1.3 kg,

TABLE 24.2
Major Food Plants Recovered in Zones B–E of Guilá Naquitz, Converted to Grams of Edible Portion

Plant	Total recovered in Zones B–E		Weight of edible portion (g)
Piñon nuts	366		61.1
Wild onions	3		33.0
Agave quids	368		5983.7
Acorns	8424		26,956.8
Hackberries	3112		560.2
Acacia pods	3		5.0
Guaje pods	417 (= 4170 seeds)		521.3
Bean pod valves	471	1194.5 seeds	149.3
Bean seeds	17		
Mesquite pods	31	372 pods	621.2
Mesquite seeds	4603		
Nanche fruits	368		1288.0
Susí nuts	455		227.5
Nopales	158		4147.5
Opuntia fruits	10	46 fruits	2300.0
Opuntia seeds	1819		
Cucurbit rinds	55	± 150 seeds	± 25.0
Cucurbit seeds	17		

TABLE 24.3
Food Plants Recovered in Zones B–E of Guilá Naquitz, Ranked in Order of Weight of Edible Portion Represented

Plant	Grams of edible portion represented	Percentage of total	Rank
Acorns	26,956.8	62.9	1
Agave	5,983.7	14.0	2
Nopales	4,147.5	9.7	3
Opuntia fruits	2,300.0	5.4	4
Nanches	1,288.0	3.0	5
Mesquite pods	621.2	1.4	6
Hackberries	560.2	1.3	7
Guaje seeds	521.3	1.2	8
Susí nuts	227.5	0.5	9
Beans	149.3	0.3	10
Piñon nuts	61.1	0.1	11
Wild onions	33.0	0.08	12
Cucurbit seeds	25.0	0.05	13
Acacia pods	5.0	0.01	14
	42,879.6		

are in fifth. These five plant products account for 95% of the weight of edible portion represented by the plant remains.

Sixth and seventh places on the list are occupied by two products of the Mesquite Grassland, mesquite pods (621 g) and hackberries (560 g). *Guajes* (521 g), *susí* nuts (228 g), and wild beans (149 g), all products of the Thorn Forest, are in eighth through tenth places. None of the other plants reach 100 g.

A point clearly made by Table 24.3 is that the food remains left behind in Guilá Naquitz cannot represent more than a small fraction of the plants collected by the occupants. They total less than 43 kg of edible portion, a quantity that would not have supported a family of 4–5 persons for more than a week or so.

While the ranking of these plants in order of sheer weight is interesting, it is not the whole story. Additional factors emerge as we convert that weight to calories and protein, using the analyses presented by Robson and Elias in Chapter 23.

In Table 24.4 we have taken the 13 most common plants recovered from Guilá Naquitz and ranked them in order of caloric contribution. Here the 42,880 g of edible portion yield 91,828 kcal, and the contribution of acorns looms even larger than it did when weight alone was considered. At 77,636 kcal, acorns account for 84.5% of the total. Whatever problems there may be with our assumptions about the extent to which the cave remains reflect reality, it would seem safe to conclude that the occupants of Guilá Naquitz derived a lot of their caloric energy from acorns.

Baked agave hearts, at 7539 kcal, are a distant second (8.2%) but probably were an energy source that fluctuated less from season to season. And then comes a series of interesting changes in rank: mesquite pods and *susí* nuts, only sixth and ninth in weight of edible portion, are third and fourth in caloric contribution. Another plant ranking higher in caloric contribution than in sheer weight is the piñon nut. Still other plants, such as nopales, tunas, and hackberries, rank lower in caloric contribution than in sheer weight.

In Table 24.5 we have taken those same 13 plants and ranked them in order of their contribution in grams of protein. Acorns are still number one on the list (contributing 655 g out of a total of 948.6) owing to their great abundance, though their percentage contribution (69%) is not as overwhelming as was the case with calories. A number of plants move up in the ranking when plant protein is considered: Nopales (8.4%) are in second place, followed by *guajes, susí* nuts, and mesquite pods. Baked agave heart, although high in calories, is only eighth on the list in protein.

A word should perhaps be said here about two plants, *susí* nuts and cucurbit seeds, that Robson and Elias' studies (Chapter 23) show to be higher in protein than any other local plant. *Susí* nuts, which are abundant in Thorn Forest A and were widely used, rank fourth in both protein and calorie contribution in Tables 24.4 and 24.5. However, they grow on perennial shrubs that could not be quickly and easily domesticated. Wild cucurbits, which are not naturally abundant in the area and are less frequent in the cave remains, rank tenth in protein contribution. However, because they are weedy annuals that do well in disturbed soils, they lend themselves to commensalism and domestication. There was probably no plant available to the preceramic cave occupants whose domestication would have been less complicated, and at the same time, whose increased use would have done more to raise the level of plant protein in the diet.

TABLE 24.4
Food Plants Recovered in Zones B–E of Guilá Naquitz, Ranked in Order of Their Contribution in Kilocalories

Plant	Kilocalories contributed	Percentage of total	Rank
Acorns	77,636	84.5	1
Agave	7,539	8.2	2
Mesquite pods	1,845	2.0	3
Susí nuts	1,349	1.5	4
Nanches	889	0.9	5
Opuntia fruits	828	0.9	6
Nopales	498	0.5	7
Piñon nuts	388	0.4	8
Guaje seeds	375	0.4	9
Hackberries	185	0.2	10
Beans	176	0.2	11
Cucurbit seeds	107	0.1	12
Wild onions	13	0.01	13
	91,828		

TABLE 24.5
Food Plants Recovered in Zones B–E of Guilá Naquitz, Ranked in Order of Their Contribution in Grams of Protein

Plant	Grams of protein contributed	Percentage of total	Rank
Acorns	655.0	69.0	1
Nopales	80.0	8.4	2
Guaje pods	42.2	4.4	3
Susí nuts	40.0	4.2	4
Mesquite pods	35.4	3.7	5
Opuntia fruits	25.5	2.7	6
Nanches	21.8	2.3	7
Agave	18.5	2.0	8
Beans	11.6	1.2	9
Cucurbit seeds	8.4	0.9	10
Piñon nuts	7.9	0.8	11
Hackberries	1.8	0.2	12
Wild onions	0.5	0.05	13
	948.6		

ANIMAL FOODS

The animal foods require a different set of calculations, which we have summarized in Table 24.6. What we have done is to give the average weight of an adult specimen of each of the major genera found at Guilá Naquitz, based on our own Oaxaca faunal collections and on data given by Leopold

TABLE 24.6
Weights of Animals Eaten at Guilá Naquitz

Animal	Average weight of 1 adult (kg)	Percentage of weight that is usable meat[a]	Usable meat on 1 adult (kg)	Amt. needed for 100-g portion
Deer (*Odocoileus*)	40.0	50	20.00	.005 deer
Peccary (*Dicotyles*)	20.0	70	14.00	.007 peccary
Raccoon (*Procyon*)	5.0	70	3.50	.03 raccoon
Cottontail (*Sylvilagus*)	1.2	50	0.60	.17 rabbit
Hawk (*Buteo*)	1.0	70	0.70	.14 hawk
Owl (*Tyto*)	1.0	70	0.70	.14 owl
Pigeon (*Columba*)	0.3	70	0.20	.5 pigeon
Dove (*Zenaidura*)	0.15	70	0.10	1.0 dove
Quail (*Colinus*)	0.15	70	0.10	1.0 quail
Turtle (*Kinosternon*)	0.50	40	0.20	0.5 turtle
Lizard (*Sceloporus*)	0.03	50	0.015	6.6 lizards

[a]Percentages are from White (1953).

(1959). In the third column of Table 24.6, we give the percentage of the animal's weight that is usable meat, based on studies by White (1953). The figures in Columns 2 and 3 can then be used to produce Column 4, the usable meat on one adult of each genus. Column 5 gives the amount needed for a 100-g portion, which can vary from as many as six to seven animals (in the case of small lizards) to as little as five to seven one-thousandths of an animal (in the case of deer and peccary).

Table 24.6 makes it clear how important the white-tailed deer was to the occupants of Guilá Naquitz, with each adult providing enough meat for two hundred 100-g portions. If, for the sake of argument, we allowed each individual living at Guilá Naquitz 230 g (a quarter of a pound) of meat per day, a family of 4–5 persons could, by drying and preserving the meat, have made an adult deer last 2–3 weeks. And that means that if, for example, they were only at Guilá Naquitz for 4 months in the autumn, they would probably only need to kill 6–8 deer during that period to meet their venison needs.

HARVEST AREAS REPRESENTED BY THE CAVE REMAINS

The plant and animal remains at Guilá Naquitz, like those from any cave, raise a number of questions. What proportion of the total food collected by the cave occupants actually survived to be found by archaeologists? How big a harvest area would be needed in order to account for the plants and animals left behind on each living floor? How much harvesting effort is represented by the "leftover" plants, that is, those that were collected and then left behind in the cave uneaten? In this section we consider some of those questions.

Zone E

The most common plants and animals recovered in Zone E are given in Tables 24.7 and 24.8, along with the areas that would need to have been harvested in order to produce them. It will be immediately apparent that there were great differences in harvest area represented.

Many of the plants present, such as nopales and mesquite seeds, could have been procured by harvesting less than 10 m^2 in the vegetational facies of their greatest abundance. Others, such as wild runner beans, represent the yield of more than a hectare (10,000 m^2) according to our 1966–1976 censuses. Still others, such as acorns and nanches, occupy an intermediate position: they represent the harvest of 200–600 m^2. There are several possible explanations for these discrepancies. One possibility, already discussed, is that the plant remains in the cave do not reflect the exact frequency in which plants were collected. Another likely possibility is that the cave occupants were willing to search over a larger area to get certain plants than they were for others. Another possibility is annual variation; Kaplan (Chapter 21) concludes that wild beans seem to vary greatly in abundance from year to year, and they might have been unusually plentiful during the Zone E occupation. Still another possibility is that early efforts at domestication could have produced densities of beans higher than one would find in the wild. (It should be remembered, however, that the species of runner bean common in preceramic Guilá Naquitz has left no cultivated descendants; so if anyone did experiment with its domestication, that experiment ultimately failed.)

Whatever the case, it is interesting that one of the staples left behind in Guilá Naquitz E—acorns—represents the yield of 600 m^2 in today's Thorn Forest A. In other words, when the occupants moved out, they left behind, uneaten, the harvest of perhaps one entire oak tree. This suggests that they felt no shortage of acorns.

As for the animals from Zone E, they also vary greatly in the estimated area necessary to procure them. Three genera (cottontail, pigeon, and dove) could be obtained in an area of 1.0–1.6 ha. Two white-tailed deer, on the other hand, would require 17.2 ha of Thorn Forest range on the average. This is the largest area required to account for any of the genera found at Guilá Naquitz; it is nearly 287 times the area that would have to be harvested to account for the acorns found in Zone E.

TABLE 24.7
The Area that Would Have to Have Been Harvested in Order To Produce the Most Common Plants Recovered in Zone E

Plant	No. found on cave floor	Harvest area represented (ha) Thorn Forest A	Thorn Forest B	Mesquite Grassland A	Mesquite Grassland B
Acorns	1846	0.06	—	—	—
Piñon nuts	155	0.12	—	—	—
Nanches	6	0.02	0.12	—	—
Bean pod valves	161	1.15	2.3	—	—
Susí nuts	98	0.003	0.004	1.23	—
Agave quids	6	0.0002	0.006	0.03	—
Acacia pods	1	0.0002	0.0003	0.0006	0.0003
Opuntia					
nopales	3	0.0001	0.001	0.0004	0.0009
fruits	2 }	0.0003	0.003	0.0005	0.0007
seeds	135 }				
Mesquite seeds	32	—	0.01	0.0001	<0.0001
Hackberries	1166	—	—	0.46	0.19

TABLE 24.8
Animals Hunted by the Occupants of Zone E, Guilá Naquitz

Animal	Minimum no. of individuals	Edible portion (kg)	Kilocalories provided	Protein provided (g)	Average area needed to procure (ha)
Deer	2	40.0	50,400	8,400	17.2
Cottontail	5	3.0	4,050	630	1.6
Mud turtle	1	0.2	158	32	(?)
Pigeon	1	0.2	558	37	1.0
Dove	1	0.1	279	19	1.0
Owl(?)	1	0.7	1,176	130	(?)
Songbirds	2	0.1	168	19	<1.0
	13	44.3	56,789	9,267	

Zone D

Tables 24.9 and 24.10 present the plants and animals recovered in Zone D of Guilá Naquitz. For the most part, the plants reflect small collecting areas. Acorns, bean pod valves, and nanches represent harvest areas of 1000–1400 m^2 in Thorn Forest A; most of the other plants represent the produce of less than 100 m^2. Perhaps the largest harvest area required is that for cucurbits—2500 m^2 if collecting was done in Mesquite Grassland A, 2 ha if done in Thorn Forest B. However, such figures would apply only to wild cucurbits. Since we have some phenotypic evidence for domestication in Zone D, these figures may simply reflect the fact that cucurbits were being grown near the cave in densities higher than those seen in the wild.

The animal remains from Zone D show a pattern similar to E. The two white-tail deer represent a 17.2-ha range, the five cottontails about 1.6 ha.

Zone C

Plants and animals from Zone C are presented in Tables 24.11 and 24.12. Nanches and bean pods reflect the largest harvest area of any of the plants, 3100–5400 m^2 in Thorn Forest A, while hackberries reflect an area of 1600 m^2 in Mesquite Grassland B. Most other plants could be accounted for by harvesting an area of 600 m^2 or less in the vegetational facies of their greatest abundance. However, as in Zone D, cucurbits show the largest harvest area: up to 1 ha in Mesquite Grassland A, up to 4 ha in Thorn Forest B. And once again, this is perhaps best explained by their being cultivated plants, grown near the cave in densities greater than those characteristic of wild cucurbits.

Animal remains continue the pattern seen in Zones E and D: six cottontails reflect 1.9 ha of home range; 2 deer reflect 17.2 ha.

Lower Zone B

Tables 24.13–24.18 present the common plants and animals recovered in Zones B3, B2+3, and B2. Obviously, the figures on Zone B2+3 should be taken with a grain of salt because of its uncertain relationship to Zones B2 and B3. At any rate, the pattern seen in B2+3 differs in no significant way from that seen in the rest of Lower Zone B.

Most plants are present in quantities that reflect a harvest area of 200 m^2 or less. An exception to this would be bean

TABLE 24.9
The Area that Would Have to Have Been Harvested in Order To Produce the Most Common Plants Recovered in Zone D

		Harvest area represented (ha)			
Plant	No. found on cave floor	Thorn Forest A	Thorn Forest B	Mesquite Grassland A	Mesquite Grassland B
Acorns	3182	0.10	—	—	—
Piñon nuts	94	0.07	—	—	—
Guaje pods	11	0.0006	—	—	—
Nanches	49	0.14	0.98	—	—
Bean pod valves	17	0.12	0.24	—	—
Susí nuts	237	0.006	0.01	2.96	—
Agave quids	49	0.002	0.05	0.25	—
Acacia pods	1	0.0002	0.0004	0.0006	0.0003
Opuntia					
nopales	28	0.001	0.01	0.004	0.008
seeds	348	0.0004	0.005	0.0008	0.001
Wild onion	2	—	1.0	0.4	—
Cucurbit rinds	4	—	2.0	0.25	—
Mesquite					
pods	3	—	0.05	0.0004	<0.0001
seeds	59				
Hackberries	416	—	—	0.16	0.07

TABLE 24.10
Animals Hunted by the Occupants of Zone D, Guilá Naquitz

Animal	Minimum no. of individuals	Edible portion (kg)	Kilocalories provided	Protein provided (g)	Average area needed to procure (ha)
Deer	2	40.0	50,400	8,400	17.2
Collared peccary	1	14.0	17,640	2,940	(?)
Cottontail	5	3.0	4,050	630	1.6
Mud turtle	1	0.2	158	32	(?)
Songbirds	2	0.1	168	19	<1.0
	11	57.3	72,416	12,021	

TABLE 24.11
The Area that Would Have to Have Been Harvested in Order To Produce the Most Common Plants Recovered in Zone C

		Harvest area represented (ha)			
Plant	No. found on cave floor	Thorn Forest A	Thorn Forest B	Mesquite Grassland A	Mesquite Grassland B
Acorns	1,570	0.05	—	—	—
Piñon nuts	80	0.06	—	—	—
Guaje pods	247	0.01	—	—	—
Nanches	196	0.54	3.92	—	—
Bean pod valves	43	0.31	0.61	—	—
Susí nuts	68	0.002	0.003	0.85	—
Agave quids	196	0.007	0.20	1.01	—
Opuntia					
nopales	49	0.002	0.02	0.007	0.01
fruits	2	0.0009	0.01	0.002	0.003
seeds	751				
Cucurbit rinds	35	—	2–4	0.5–1.0	—
Mesquite					
pods	7	—	0.30	0.003	0.0004
seeds	554				
Hackberries	969	—	—	0.38	0.16

TABLE 24.12
Animals Hunted by the Occupants of Zone C, Guilá Naquitz

Animal	Minimum no. of individuals	Edible portion (kg)	Kilocalories provided	Protein provided (g)	Average area needed to procure (ha)
Deer	2	40.0	50,400	8,400	17.2
Peccary	1	14.0	17,640	2,940	(?)
Raccoon	1	3.5	8,925	1,022	(?)
Cottontail	6	3.6	4,860	756	1.9
Mud turtle	1	0.2	158	32	(?)
Barn owl	1	0.7	1,176	130	(?)
	12	62.0	83,159	13,280	

TABLE 24.13
The Area that Would Have to Have Been Harvested in Order To Produce the Most Common Plants Recovered in Zone B3

		Harvest area represented (ha)			
Plant	No. found on cave floor	Thorn Forest A	Thorn Forest B	Mesquite Grassland A	Mesquite Grassland B
Acorns	421	0.01	—	—	—
Piñon nuts	4	0.003	—	—	—
Guaje pods	5	0.0003	—	—	—
Nanches	18	0.05	0.36	—	—
Bean pod valves	26	0.19	0.37	—	—
Susí nuts	15	0.0004	0.0007	0.19	—
Agave quids	18	0.0006	0.02	0.09	—
Acacia pods	1	0.0002	0.0004	0.0006	0.0003
Mesquite pods / seeds	2 / 236	—	0.12	0.001	0.0002
Hackberries	65	—	—	0.03	0.01

TABLE 24.14
Principal Animals Hunted by the Occupants of Zone B3, Guilá Naquitz

Animal	Minimum no. of individuals	Edible portion (kg)	Kilocalories provided	Protein provided (g)	Average area needed to procure (ha)
Deer	1	20.0	25,200	4,200	8.6
Cottontail	3	1.8	2,430	378	0.9
Mud turtle	1	0.2	158	32	(?)
	5	22.0	27,788	4,610	

pod valves, which reflect a harvest area of up to 7400 m^2 in Zone B2 and 1900 m^2 in Zone B3. This is the same phenomenon we saw in Zone E, and although it could result from sampling error, it could also mean that an attempt was being made to grow these beans near the cave at densities higher than those normally seen in the wild.

The Lower B living floors provide evidence for no more than one individual deer per zone (8.6 ha of range) and about three cottontails per zone (0.9 ha). Bird remains tend to reflect a procurement area of 1.0–4.0 ha. One could perhaps characterize the living floors of Lower B by saying that they contain many of the same animals as Zones E, D, C, and B1 but in somewhat smaller numbers.

Upper Zone B

Tables 24.19 and 24.20 present the data on plants and animals from Zone B1. Acorns and piñon nuts reflect a harvest area of 200–300 m^2 in Thorn Forest A, while hackberries reflect a harvest area of 400 m^2 in Mesquite Grassland B. Most plants can be accounted for with much smaller areas,

TABLE 24.15
The Area that Would Have to Have Been Harvested in Order To Produce the Most Common Plants Recovered in Zone B2 + 3

Plant	No. found on cave floor	Harvest area represented (ha)			
		Thorn Forest A	Thorn Forest B	Mesquite Grassland A	Mesquite Grassland B
Acorns	95	0.003	—	—	—
Guaje pods	6	0.0003	—	—	—
Nanches	8	0.02	0.16	—	—
Bean pod valves	9	0.06	0.13	—	—
Susí nuts	3	0.0001	0.0001	0.04	—
Agave quids	8	0.0003	0.008	0.04	—
Opuntia					
nopales	2	0.0001	0.0008	0.0003	0.0006
fruits	1 }	0.0004	0.005	0.0009	0.001
seeds	360 }				
Mesquite					
pods	3 }	—	0.15	0.001	0.0002
seeds	276 }				
Hackberries	149	—	—	0.06	0.02

TABLE 24.16
Principal Animals Hunted by the Occupants of Zone B2 + 3, Guilá Naquitz

Animal	Minimum no. of individuals	Edible portion (kg)	Kilocalories provided	Protein provided (g)	Average area needed to procure (ha)
Deer	1	20.0	25,200	4,200	8.6
Cottontail	3	1.8	2,430	378	0.9
Mud turtle	1	0.2	158	32	(?)
Pigeon	1	0.2	558	37	1.0
Songbird	1	0.05	84	9	<1.0
	7	22.25	28,430	4,656	

TABLE 24.17
The Area that Would Have to Have Been Harvested in Order To Produce the Most Common Plants Recovered in Zone B2

Plant	No. found on cave floor	Harvest area represented (ha)			
		Thorn Forest A	Thorn Forest B	Mesquite Grassland A	Mesquite Grassland B
Acorns	418	0.01	—	—	—
Piñon nuts	2	0.001	—	—	—
Guaje pods	43	0.002	—	—	—
Nanches	45	0.13	0.90	—	—
Bean pod valves	104	0.74	1.49	—	—
Susí nuts	12	0.0003	0.0005	0.15	—
Agave quids	45	0.002	0.05	0.23	—
Opuntia					
nopales	31	0.001	0.01	0.005	0.009
fruits	1 }	0.0001	0.002	0.0003	0.0004
seeds	82 }				
Mesquite seeds	1613	—	0.75	0.007	0.001
Hackberries	107	—	—	0.04	0.02

TABLE 24.18
Principal Animals Hunted by the Occupants of Zone B2, Guilá Naquitz

Animal	Minimum no. of individuals	Edible portion (kg)	Kilocalories provided	Protein provided (g)	Average area needed to procure (ha)
Deer	1	20.0	25,200	4,200	8.6
Cottontail	3	1.8	2,430	378	0.9
Mud turtle	1	0.2	158	32	(?)
Quail	1	0.1	168	25	4.0
	6	22.1	27,956	4,635	

TABLE 24.19
The Area that Would Have to Have Been Harvested in Order To Produce the Most Common Plants Recovered in Zone B1

Plant	No. found on cave floor	Harvest area represented (ha)			
		Thorn Forest A	Thorn Forest B	Mesquite Grassland A	Mesquite Grassland B
Acorns	892	0.03	—	—	—
Piñon nuts	31	0.02	—	—	—
Guaje pods	105	0.005	—	—	—
Nanches	46	0.13	0.9	—	—
Bean pod valves	111	0.8	1.59	—	—
Susí nuts	22	0.0006	0.001	0.28	—
Agave quids	46	0.002	0.05	0.23	—
Opuntia					
nopales	45	0.002	0.02	0.007	0.01
fruits } seeds	4 } 143	0.0004	0.005	0.0008	0.001
Wild onions	1	—	0.5	0.2	—
Cucurbit rinds	8	—	2.0	0.5	—
Mesquite					
pods } seeds	16 } 1,833	—	0.95	0.009	0.001
Hackberries	240	—	—	0.09	0.04

TABLE 24.20
Principal Animals Hunted by the Occupants of Zone B1, Guilá Naquitz

Animal	Minimum no. of individuals	Edible portion (kg)	Kilocalories provided	Protein provided (g)	Average area needed to procure (ha)
Deer	2	40.0	50,400	8,400	17.2
Cottontail	6	3.6	4,860	756	1.9
Mud turtle	2	0.4	316	64	(?)
Ground dove	1	0.1	279	19	1.0
Hawk	1	0.7	1,176	130	(?)
Barn owl	1	0.7	1,176	130	(?)
	13	45.5	58,207	9,499	

although nanches reflect an area of some 1300 m^2. Once again, some of the largest areas are associated with cucurbits and runner beans. Since there is genetic evidence for domestic cucurbits in Zone B, their high frequency might be explained that way. No such evidence exists for beans, but since this is the third instance in which their frequency is higher than expected, we are understandably suspicious.

The animals of Zone B1 continue the pattern seen in earlier levels. Six cottontails reflect an area of perhaps 1.9 ha; two white-tailed deer would have required about 17.2 ha of range.

A SUMMARY OF THE LIVING FLOORS

While no two living floors at Guilá Naquitz are identical, an examination of Tables 24.7–24.20 reveals a number of similarities.

1. The plant remains left behind rarely reflect the harvest of an area greater than a tenth of a hectare (1000 m^2); more often than not, they reflect the harvest of an area ranging from a few square meters to a few hundred square meters.

2. This is true whether we are dealing with the residue of plants that were eaten or plants that were left uneaten. Acorns, present in the form of uneaten specimens, can generally be accounted for by the harvest of 100–1000 m^2. *Susí* nuts, present in the form of discarded hulls from eaten specimens, can generally be accounted for by the harvest of 60 m^2 or less. Agave hearts, represented by masticated and expectorated quids, can generally be accounted for by the harvest of 70 m^2 or less. Hackberries, represented by the expectorated seeds from eaten fruits, can generally be accounted for by the harvest of 100–1900 m^2.

3. Another way to view the quantities of plants left behind is to consider them in terms of individual plants harvested, by reference to Tables 18.2–18.5. For example, the number of uneaten acorns left behind in Zone E has already been described in the section titled, "Harvest Areas Represented by the Cave Remains," as equivalent to the yield of 1 oak tree. The acorns left behind in Zone D represent the yield of more than 2 average oaks. The hulls from *susí* nuts eaten during the Zone D occupation amount to the yield of nearly 5 bushes. On the other hand, even the 1166 hackberry seeds expectorated in Zone E amount to no more than half the yield of an average bush. And we have already pointed out that no floor—not even Zone C with its 196 quids—requires us to postulate the cooking of more than one maguey heart. These great differences in productivity per plant among the various genera suggest that it may be more reasonable to compare them in terms of square meters of landscape harvested.

4. It would appear from Tables 24.7–24.19 that it would be most practical to harvest piedmont genera in Thorn Forest A and floodplain genera in Mesquite Grassland B. It is simply the case that more edible parts can be obtained per square meter in those vegetational facies than in Thorn Forest B or Mesquite Grassland A.

5. It seems unlikely the occupants of Guilá Naquitz would have been willing to forage for wild cucurbits over an area 10 times greater than that over which they foraged for other plants. Therefore, when cucurbits occur in numbers suggesting a harvest area of 1–2 ha, it seems more likely we are dealing with cultivated populations growing more densely than their wild relatives. This is supported by genetic evidence for domestic *Cucurbita pepo* as early as Zone D.

6. Runner beans also occasionally occur in numbers suggesting a harvest area of 1–2 ha. This raises the possibility that they, too, might have been cultivated populations growing more densely than in the wild. However, no genetic evidence supports this—the *Phaseolus* in preceramic Guilá Naquitz is a species that has never been successfully domesticated. This does not mean that no attempts were made to grow it near the cave; it simply means that if such attempts were made, they were eventually given up in favor of other species of beans.

7. Procurement areas for animals are remarkably similar from floor to floor. This may, in part, result because animals were treated in terms of minimum individuals, which smooths out some of the variations in bone debris among living floors.

THE SUSTAINING AREA FOR THE GUILA NAQUITZ POPULATION

Armed with the figures given in this and previous chapters, we can now consider some of the more interesting questions of procurement. How large an area did the occupants of Guilá Naquitz exploit? How many people could that area support and for how long? How close would the four to five persons occupying Guilá Naquitz have come to exhausting the locally available wild foods?

There is no straightforward answer to the first question, but it can be approached by considering some of the possible parameters. We know from the chipped-stone tools at Guilá Naquitz that its occupants frequently used chert from the Matadamas quarries, some 50 km away. It is very unlikely, however, that they traveled that far on a regular basis for plant foods. There are simply too many alternative caves and camp sites they could have moved to when food resources were that far from Guilá Naquitz.

The willingness of hunters and gatherers to travel long distances for game is well established, but ethnographic data from mid-latitude environments do not suggest that women regularly travel long distances for plants. In the Kalahari, for example, Lee (1968) reports that women are unlikely to forage farther than 10 km from their base camp, even though small groups of men may travel farther on hunting trips. At Guilá Naquitz, where we may be dealing with a family of no more than four to five persons, it is even less likely that plant collecting trips were long. Indeed, almost all the plants seen in the cave could today be collected between Guilá Naquitz and the Río Mitla, a distance of only 3–4 km.

There are several ways one can approach the modeling of a sustaining area. One is by the site-catchment method pioneered by Vita-Finzi and Higgs (1970), who drew circles of 5-km radius around each site and then tried to determine how the resources of that 7900-ha circle could be used to support the population seen at the site. Such a method takes the size of the sustaining area as given and reconstructs the resource use. An alternative approach takes the resource use as given, based on plant and animal remains, and reconstructs the size of the sustaining area (Flannery 1976). The latter approach is taken here.

DAILY PLANT CONSUMPTION AT GUILA NAQUITZ

I begin by reconstructing the "average" daily plant consumption of a resident of Guilá Naquitz. Since Robson and

Elias (Chapter 23) have already attempted something similar, I should make explicit how my effort differs from theirs.

Robson and Elias have presented hypothetical daily intakes for occupants of Zones E, C, and B1 at Guilá Naquitz (Tables 23.3–23.5), assuming a daily intake of 1350 g, including 150 g of meat. They compare their hypothetical intakes to a simulated American diet based upon a USDA household consumption survey (Table 23.6). Americans are among the world's best-fed people, but Robson and Elias make the point that the hypothetical Guilá Naquitz diets "compare favorably with the simulated American diet, as they provide similar proportions of nutrients and they easily exceed the U.S. caloric levels on less total food."

Robson and Elias' goal was to see whether the mix of foods eaten at Guilá Naquitz could be compared favorably with that of a well-nourished contemporary population, and they conclude that it probably could be. My goal is different: I want to see how many people a hectare of the Guilá Naquitz environment could support and for how long. And since I am dealing with a population in which children have to be averaged in with adults, I assume an average daily intake of less than 1350 g. While nutritionists are not universally agreed on "minimum daily requirements" for Third World populations, a frequent rule of thumb is that every individual ought to get at least 2000 cal and 40 g of protein every day. This is somewhat less than the 2858 cal and 108 g of protein in the simulated American diet, but as Robson and Elias point out in their chapter, many Third World populations achieve nutritional adequacy on intakes significantly lower than World Health Organization (WHO) recommendations.

Another difference between my scheme and that of Robson and Elias is that while they constructed hypothetical plant-and-meat intakes to fit specific living floors at Guilá Naquitz, I simply average Zones E–B and present a single model for plant intake during the Naquitz phase as a whole. In that model, presented in Table 24.21, I have made two assumptions: (1) Minimum daily intake was 1 kg/person, and (2) the various plants were eaten in the proportions in which they were recovered by excavation.

As might be expected, the diet is dominated by acorns (629 g, representing the flour from 196.5 seeds). Whether anyone could really eat more than a pound of acorn flour a day is a good question. Other plants, such as prickly pear fruit, seem underrepresented (1.1 fruits/day), even when one remembers that these are averages over the span of a season of occupation. Thus, my diet may exaggerate the importance of acorns relative to other plants, but I prefer to leave it as is because I have no reliable basis for "adjusting" any of the figures.

One other feature of Table 24.21 deserves comment: Our 1-kg daily plant consumption produces the minimum 2000 cal but *not* the minimum 40 g of protein. Indeed, to reach 40 g of protein, one would literally have to double the plant diet, bringing it to 2 kg. This emphasizes the crucial role of meat in the Guilá Naquitz diet. The addition of a mere 100 g of venison or rabbit to the plant diet shown in Table 24.21 raises the total protein to 43 g and the total cal to 2265 or more. Although the occupants of Guilá Naquitz, like so many mid-latitude hunter–gatherers, relied heavily on plant food, it would be a mistake to underestimate the importance of even small quantities of meat eaten on a regular basis.

CALCULATING THE SUSTAINING AREA

Even more to the point of this section is Column 6 of Table 24.21, which gives the area that would need to be harvested to produce the amounts of plants listed in Column 5. For most plants, including the apparent staple foods, that area would seem to be less than 100 m^2. We have already discussed (above) the reasons why the apparently larger cucurbit and bean areas might be deceptive. Wild onions are also deceptive because they tend to occur in clusters near springs and pools of water, although their overall density is low; anyone who knew where they grew could get a good supply without having to search an area of 2000 m^2.

If we assume that Guilá Naquitz was occupied by five persons, we can multiply Column 6 by 5 to produce the harvest area needed per day: 300 m^2 for acorns, 200 m^2 for nopales and piñon nuts, and perhaps 1200 m^2 for nanches. Multiply these figures by 7 and we have the harvest area needed per week: still only 2100 m^2 for acorns, 1400 m^2 for nopales and piñon nuts, and 8400 m^2 for nanches. In other words, a week's plant food for a family of 5 might involve harvest areas of less than 1 ha.

Multiply the figures in Column 6 by 150 and we have the harvest area needed to support our family of 5 for a month: 9000 m^2 for acorns, 6000 m ^{2} for nopales and piñon nuts, 1500 m^2 for *susí* nuts, and perhaps 3.6 ha for nanches. In other words, except for a handful of plants such as nanches, we are still talking about harvest areas of less than 1 ha.

Now let us suppose that our family of 5 were to occupy Guilá Naquitz for 4 months (September–December), or 120 days. Multiply the figures in Column 6 by 600 and we have the harvest area required for those 4 months: 3.6 ha for acorns, 2.4 ha for nopales and piñon nuts, 4.2 ha for maguey hearts, 6000 m^2 for *susí* nuts, and perhaps 14.4 ha for nanches. Again, except for a handful of plants such as nanches, we are talking about a harvest area of about 2.5 ha in Thorn Forest A. As for the products of Mesquite Grassland B, the corresponding figures would be 6000 m^2 for mesquite pods and 7.2 ha for hackberries.

To put these figures in perspective, consider that the density we have settled on for white-tailed deer is 1 adult per 8.6 ha, so the 2 minimum deer in Zone D represent a procurement area of 17.2 ha. Within an area that size one could find virtually all the plant foods a family of 5 would need during 4 months.

To put the figures further in perspective, however, consider what we would find if we drew a circle with a 5-km radius around Guilá Naquitz, as recommended by Vita-Finzi and Higgs (1970) in their site-catchment analyses. Such a circle would consist of 7900 ha, of which approximately 25% (1975 ha) would be Thorn Forest A. Such an expanse of Thorn

TABLE 24.21
Reconstruction of the "Average" Daily Plant Consumption of a Resident of Guilá Naquitz Zones B–E[a]

Plant	No. of grams consumed	No. of kilocalories represented	No. of grams of protein	Average amount needed	Area harvested (ha)[b]
Acorns	629	1,812	15.3	196.5 acorns	0.006
Agave	140	176	0.4	8.6 quids	0.007
Nopales	97	12	1.9	3.7 nopales	0.004
Opuntia fruits	54	19	0.6	1.1 fruits	<0.001
Nanches	30	21	0.5	8.6 fruits	0.024
Mesquite pods	14	42	0.8	8.4 pods	<0.001
Hackberries	13	4	0.04	72.2 berries	0.012
Guaje seeds	12	9	0.9	96.0 seeds	<0.001
Susí nuts	5	30	0.9	10.0 nuts	<0.001
Beans	3	4	0.2	24.0 beans	0.069
Piñon nuts	1	6	0.1	6.0 nuts	0.004
Wild onions	1	0	0.02	<1.0 bulb	<0.200
Cucurbit seeds	1	4	0.3	6.0 seeds	0.020
	1,000	2,139	22.00		

[a]The values are based on the assumptions that (1) the daily intake was 1 kg of the various plant species and (2) that these were eaten in the proportions in which they were recovered by excavation. Figures may not add exactly to totals given due to a slight rounding error.

[b]Areas are given for each plant's zone of maximum productivity (see Tables 24.7–24.19).

Forest would contain an estimated 195 metric tons of acorns, 673 metric tons of *Agave potatorum* hearts, 1923 metric tons of prickly pear fruits, and 4.7 metric tons of venison—more food, in other words, than a family of 5 could eat in their lifetimes.

Regardless of how many errors may be inherent in our model, it is safe to say that the occupants of Guilá Naquitz could not have exhausted the resources of a 5-km-radius catchment circle. First of all, the resources are not available all year; there are only about 4 months during which the major plants of Thorn Forest A come ripe, and during those months a family of 5 would not even exhaust a circle with a 1-km radius. In the case of some of their staple foods, such as acorns and agave hearts, our figures raise two interesting possibilities:

1. The most important factor limiting their harvest was not acreage, but time. The season of availability would end long before the occupants of Guilá Naquitz could have harvested all the acorns, *susí* nuts, nanches, mesquite pods, or hackberries within walking distance of the cave.
2. It is likely that the occupants of the cave would have run out of deer long before they ran out of plant foods. By killing 2 deer they would, on the average, have exhausted the venison from a 17.2-ha area that still had plenty of plants left.

We may now respond to the questions we raised at the start of this section. How large an area did the occupants of Guilá Naquitz exploit? For plant foods, it was perhaps no more than 5–15 ha. For deer, it was at least 17.2 ha, and probably a good bit more. For raw materials like chert, a circle at least 50 km in radius was needed. How many people could the area support and for how long? It could support many times the number we see at Guilá Naquitz and it could do so until the season of availability ended and the remaining wild plants rotted, dried out, or were eaten by animals. How close would the occupants have come to exhausting the locally available wild foods? They would not have come very close—not unless they were sharing the area with a lot of neighboring bands, for which we have no archaeological evidence.

CHANGES IN DIET THROUGH TIME

It will be clear from the preceding pages that there was a certain stability to the use of food resources at Guilá Naquitz. In general, the 15 most common plant and animal genera used in Zone B1 were the same 15 used in Zone E. However, there were some apparent changes through time that may be of significance, and we focus on those changes in this section.

One of the most obvious changes, of course, is the increasing evidence for plant domestication through time. Cucurbits were not even represented in Zone E; by Zone B1, phenotypically domestic *Cucurbita pepo* was present. Evidence for *Zea* pollen in Zone C and intriguing numbers of *Phaseolus* beans (albeit phenotypically wild) in at least four zones contribute to the overall picture. However, none of the aforementioned genera ever accounted for more than 4% of the plant remains by raw count.

Just as interesting are changes in the frequencies of five local wild plants, several of which were among the most common genera in the cave. Included are three products of the Thorn Forest (acorns, piñon, and *susí*) and two products of the Mesquite Grassland (mesquite and hackberry). Seemingly, the use of mesquite pods increased through time, while the use of acorns, piñon nuts, *susí*, and hackberry declined.

Raw data on the zone-by-zone fluctuations of the various plants are given in Tables 24.7–24.19. There are at least two ways the changes can be summarized. (1) In Table 24.22 we consider the 14 most common plant remains in the cave and follow each item's increase or decrease in terms of the percentage it contributes to the total remains from that group of 14 plants. (2) In Table 24.23 we consider the same 14 items in terms of the area that would have to be harvested in order to account for the amount found on a given living floor.

Acorns amount to 51% of the remains (of the group of 14 items) in Zone E and remain abundant through Zones D (71%) and B3 (52%). After this, they drop to 17–25% in Zones B2–B1. In terms of area harvested, this involves a presumed reduction from 0.06 ha in Zone E to 0.03 ha in Zone B1. In other words, the occupants of Zone B1 can be interpreted as harvesting half as many acorns (relative to other plants) as the occupants of Zone E. Piñon nuts, presumably from the same environment as the acorns, show a decline from 4% in Zone E to 1% in Zone B1. *Susí* nuts, also abundant in Thorn Forest A, decline from 3–5% in Zones E–D to 1% in Zone B1.

The two products of Mesquite Grassland B have different histories. Mesquite, limited to 1% of the remains in Zones E and D, starts increasing in Zone C and reaches 52–64% in Zones B2–B1. In terms of area harvested this represents a tenfold increase, from less than 0.0001 ha in Zone E to 0.001 ha in Zone B1. Hackberries, on the other hand, decline from 32% in Zone E to 7% in Zone B1. This represents a presumed reduction from 0.19 ha harvested in Zone E to 0.04 ha harvested in Zone B1.

While there are fluctuations in the other plant frequencies at Guilá Naquitz, none seem to show as clear a trend through time as the five mentioned above, and several of the minor plants provide such low percentages that it would be hard to rule out sampling error.

There are several ways these changes in frequency could be interpreted. Superficially, it could be argued that they represent a decrease in the use of Thorn Forest vegetation, coupled with an increase in the use of Mesquite Grassland. Such a shift could have had several causes: (1) the environment might gradually have changed, decreasing the area of Thorn Forest available; (2) the group occupying the cave might have begun to spend a greater part of the year in the Mesquite Grassland; or (3) the group might have begun to arrive at the cave earlier in the year (when more Mesquite Grassland products were available) and leave earlier (before the products of the Thorn Forest were fully mature).

Various objections can be leveled at this superficial interpretation, however. For one thing, the fact that use of hackberry *decreases* while use of mesquite *increases* suggests that we are not dealing with a simple increase in the use of Mesquite Grassland. Second, we see no reason why the group would move to Guilá Naquitz while they were still using the Mesquite Grassland intensively, since it would increase their daily travel distance. Third, the pollen and microfaunal records do not document a systematic increase of Mesquite Grassland at the expense of Thorn Forest. Fourth, the changes in frequency we see are not synchronous: mesquite and *susí* percentages change as early as Zone C, while hackberry percentages do not become permanently lower until Zone B3, and acorn percentages do not fall dramatically until Zone B2. This being the case, it is possible that the various percentage changes are at least partially independent of one another.

The interpretation we prefer is that the percentage fluctuations we see have to do with changes in harvesting strategy over time. Given the annual variability of the Oaxaca environment discussed in Chapter 18, it would be unrealistic to suppose that the strategy of these hunter–gatherers remained static throughout the Naquitz phase. Moreover, we suspect that as agriculture developed during Zone D and subsequent levels, the occupants of Guilá Naquitz may have made scheduling changes in their use of wild plant genera and modified the quantities in which they used them (Flannery 1968).

We therefore see the sequence of plant remains from Guilá Naquitz as reflecting three interrelated processes: (1) a degree of long-term stability and conservatism in the genera used, not unexpected among hunter–gatherers; (2) a long-term series of gradual adjustments in the frequencies of the plants used, leading either to greater efficiency or to greater resilience in the face of annual variations; and (3) a later series of changes in the "mix" of plants used, which followed a gradually increasing commitment to agriculture. In Chapter 31, Reynolds designs a computer simulation of these processes and compares it to the actual sequence of plant remains from Guilá Naquitz.

TABLE 24.22
Changes in the Frequency of the 14 Most Common Plant Items found at Guilá Naquitz[a]

Plant	Stratigraphic zone					
	E	D	C	B3	B2	B1
Acorns	51	71	33	52	17	25
Piñon nuts	4	2	2	+	+	1
Guaje pods	+	+	5	+	2	3
Nanches	+	1	4	2	2	1
Bean pod valves	4	+	1	3	4	3
Susí nuts	3	5	1	2	+	1
Agave quids	+	1	4	2	2	1
Acacia pods	+	+	+	+	+	+
Opuntia						
nopales	+	1	1	+	1	1
seeds	4	8	16	+	3	4
Wild onions	+	+	+	+	+	+
Cucurbit rinds	+	+	1	+	+	+
Mesquite seeds	1	1	12	29	64	52
Hackberries	32	9	20	8	4	7

[a]The numbers given for each zone are the percentages that each item contributes to the total debris from the group of 14 items. Percentages may not add up to 100 due to rounding off. A plus sign indicates a percentage below 1.0.

TABLE 24.23
Six Stratigraphic Zones at Guilá Naquitz Compared in Terms of the Area that Would Have To Be Harvested To Account for the Quantities of 14 Major Plant Items Found in Each Zone[a]

Plant	Stratigraphic zone						Vegetational Facies
	E	D	C	B3	B2	B1	
Acorns	0.06	0.10	0.05	0.01	0.01	0.03	Thorn Forest A
Piñon nuts	0.12	0.07	0.06	0.003	0.001	0.02	
Guaje pods	—	0.0006	0.01	0.0003	0.002	0.005	
Nanches	0.02	0.14	0.54	0.05	0.13	0.13	
Bean pods	1.15	0.12	0.31	0.19	0.74	0.8	
Susí nuts	0.003	0.006	0.002	0.0004	0.0003	0.0006	
Agave quids	0.0002	0.002	0.007	0.0006	0.002	0.002	
Acacia pods	0.0002	0.0002	—	0.0002	—	—	
Opuntia							
nopales	0.0001	0.001	0.002	—	0.001	0.002	
fruits	0.0003	0.0004	0.0009	—	0.0001	0.0004	
Wild onions	—	0.4	—	—	—	0.2	Mesquite Grassland A
Cucurbits	—	0.25	0.5–1.0	—	—	0.5	
Mesquite	<0.0001	<0.0001	0.0004	0.0002	0.001	0.001	Mesquite Grassland B
Hackberries	0.19	0.07	0.16	0.01	0.02	0.04	

[a]Figures given are in hectares, based on the vegetational zone of the plant's major productivity (see Tables 24.7–24.19).

VI

SPATIAL ANALYSIS OF LIVING FLOORS

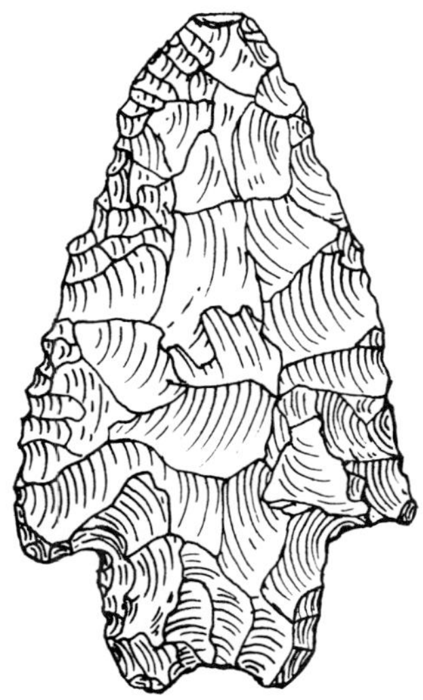

25

Spatial Analysis of Guilá Naquitz Living Floors: An Introduction to Part VI

Kent V. Flannery

In Chapter 5 we discussed the locations of all features and artifacts on the preceramic living floors at Guilá Naquitz. In this chapter we consider the distributions of the much more numerous plant remains, animal bones, and flint debitage across those same living floors. Our ultimate goal is to see whether the cave floor can be divided into activity areas, giving us better insight into the organization of labor and space in a Naquitz phase microband camp.

There are several ways in which this problem has been approached at southern Mexican preceramic sites. At Yuzanú in the Yanhuitlán Valley of Oaxaca, Lorenzo (1958:Fig. 14) plotted the exact location of every piece of flint by Cartesian coordinates. At Cueva del Texcal in the Valsequillo Basin of Puebla, García Moll (1977:Láminas 34–57) indicated the density of tools and animal bones per 1 × 1-m square by differentially shading each square. At Coxcatlán Cave in the Tehuacán Valley, Fowler and MacNeish (1972: Figs. 112, 115, 116, and others) made use of a program written by Robert Vierra, whereby a Cal. Comp. line plotter printed out symbols for artifacts, plants, and animal bones on a grid plan of each living floor. Over this printout, Fowler and MacNeish went on to place patches of zip-a-tone over clusters of items that they felt represented activity areas.

On the one hand, the approach we have taken in Chapters 26–28 shares some attributes of the studies by Fowler and MacNeish, Lorenzo, and García Moll. On the other hand, for reasons we discuss below, our approach adds several steps that we felt were necessary to achieve our goals.

GRAPHIC DISPLAY TECHNIQUES

The method chosen by Lorenzo—piece-plotting the exact location of every tool as a three-dimensional point on the living floor—is widely used by Old World Paleolithic specialists. For a population of 58 flints, such as occurred at Yuzanú (Lorenzo 1958:28), this method may well be ideal. In Figs. 5.32–5.38, we have presented our artifacts in much the same way; note, however, that none of our living floors had more than 50 or 60 artifacts.

One logistic problem with piece-plotting arises when the numbers of items begin to creep into the hundreds or even thousands. Should we have piece-plotted the 466 pieces of flint debitage in Zone B1? Some Old World prehistorians would (and they perhaps deserve a medal for doing so), but their resultant map is often a mass of unreadable, overlapping dots. And how about piece-plotting 3182 acorns from Zone D? We shudder to think what the map would look like. It is for these reasons that we decided to record our acorns, mesquite seeds, and unretouched flakes by square as Fowler and MacNeish and García Moll did.

A related logistic problem arises when one attempts to analyze piece-plotted data for randomness, clustering, or regular spacing (see below). When data are collected as a series of plotted points, a method such as nearest-neighbor analysis must be used (Whallon 1974a). Performing a nearest-neighbor analysis on Lorenzo's 58 flints would be reasonable, but analyzing 3182 acorns by that method could take the rest of your life. Such data are much more convenient to handle by grid unit.

There are several methods of presenting grid unit data graphically, each with its strengths and weaknesses. After viewing the Cal. Comp. printout of their living floors, MacNeish and Nelken-Terner (1972:12) concluded that "the computer had a sort of personality of its own, for, if it was fed the exact location of an artifact or ecofact, then it printed it in that position, but if it was given only the square location, then it printed diagonally across the square." Reference to Fig. 106 of Fowler and MacNeish (1972) will provide the reader with a classic case of such diagonal printing; thanks to the Cal. Comp. plotter, the artifacts appear to be flying east in V-formation like a flock of Canadian geese. Within the diagonal lines we find symbols for *Acacia* seeds, *Agave* leaves, and other plants, which we assume must indicate the squares where those items were most abundant. Unfortunately, in this system there is no way to indicate how widely *Acacia* seeds were scattered across the floor, or how concentrated their distribution was in the square where the symbol occurs.

García Moll's differential shading of grid squares according to the density of artifacts or animal bones gives the reader a sense of this distribution; for example, his Lámina 38 shows us where most of the blades, knives, and use-retouched flakes of the Texcal II phase were found (García Moll 1977:71).[1] Even this technique has some drawbacks, however, since our 1 × 1-m squares may unintentionally divide a single concentration into as many as four arbitrary units. And it is hard to compare densities, for in some Cueva del Texcal diagrams the darkest shading represents 8 items per square, in others 16 per square, and so on.

We decided that what we wanted to use were *contour lines* on which the actual density intervals for each item could be recorded. What we needed, therefore, was a program for converting the counts in our arbitrary 1 × 1-m squares into density contours that would model the actual distribution of items on each living floor. Charles S. Spencer discovered that the geography department of the University of Michigan had just such a computer program, and in Chapter 26 we apply it to the 17 most abundant items in the preceramic levels of Guilá Naquitz. While this program is not without its problems either, we feel it does the best job of graphically presenting the distribution of a given item on a living floor of any program we had access to.

[1]The reader should note that each of these diagrams in the Cueva del Texcal report represents a whole phase, even when that phase spans several levels. For example, Lámina 38, showing the Texcal II phase, lumps "Levels III, IV and V," even though the latter were distinct cultural strata (García Moll 1977:19). Thus the distributions would only be meaningful if the occupants used every living floor the same way during a given phase.

DEFINING ACTIVITY AREAS

García Moll's square shading, Robert Vierra's Cal. Comp. plots of Fowler and MacNeish's data, and Spencer's contour maps of our data all share one significant characteristic. All are *descriptive*; none are *analytic.* All show us where tools, plants, or bones were found: none demonstrates statistically significant associations among those items.

For us to define plausible activity areas, we still have to demonstrate such associations. Because Coxcatlán Cave, Cueva del Texcal, and Guilá Naquitz were excavated by arbitrary 1 × 1-m units, we cannot assume that two items were associated simply because they were found in the same square. Our arbitrary square boundaries may have separated two items that were actually associated, or joined two that were unrelated.

It is a great temptation to define activity areas by inspection, a temptation few of us can resist. Archaeologists also like to assume that the activity for which a tool was designed was carried out on the very spot where the tool was found. From time to time we all fall prey to what Binford (1981) has called "the Pompeii premise," a tendency to assume that prehistoric people left everything just where they used it. Commenting on his Lámina 38, which covers Levels III–V of the Texcal II phase, García Moll (1977:87) suggests that "the area where knives and blades are most abundant may represent the place where game caught during the day was cut up, while the area with scrapers of two types may represent the area of hide preparation."

Fowler and MacNeish have gone farther than this. Wherever they saw, by visual inspection, what seemed to be a significant cluster of items, they have covered it with a specific type of zip-a-tone and designated it an "activity area." Figure 116 of Fowler and MacNeish (1972) can serve as a convenient example; it shows five activity areas for Zone XIII of Coxcatlán Cave, labeled A through E, which include such diverse activities as "hunting and trapping," "woodworking," "plant collecting, growing and processing," "food and skin preparation," "flint knapping," and "refuse."

For several reasons, we have resisted the temptation to do this with either our Chapter 26 contour maps or our Chapter 5 artifact plots. To begin with, as suggested above, the significance of the association of items within each cluster has yet to be demonstrated by appropriate statistical testing. Second, we cannot realistically believe that activities such as hunting, trapping, or plant collecting and growing took place *inside* Coxcatlán Cave. These may well be the functions for which tools such as projectile points, nets, and digging sticks were designed, but when these artifacts wound up on the living floor they were all refuse.

Even deer and rabbit bones do not necessarily indicate a butchering area; so many of the bones from Guilá Naquitz and Coxcatlán Cave were burned that the activity that deposited them on the floor might best be labeled "eating." To be sure, the occupants of Zone XIII of Coxcatlán Cave stalked deer, trapped rabbits, grew squash, and dug up maguey hearts, but they did most of those things outside.

What they did inside the cave was cook, eat, sleep, make love, and discard the kinds of small refuse that did not get in their way. In other words, we feel that Fowler and MacNeish have correctly inferred the range of activities engaged in by the occupants of Zone XIII, but we suggest that only a small percentage of those activities actually took place on the living floor.

A third point is that, given the amount of walking around that must have taken place on living floors such as Zone XIII, we find it hard to believe that activity areas would have remained as discrete and nonoverlapping as the patches of zip-a-tone in Fowler and MacNeish's drawing indicate; many suggest gaps of up to a meter or two between areas. Fourth and last, some of the items in the Zone XIII activity areas seem to us to be strange bedfellows. A subheading marked "Textile Activities" lists a square knot, a coiled basket, and *Agave* quids. Now, we know what Fowler and MacNeish had in mind here: maguey fiber, knots, and basketry can all be considered components of weaving activity. But on a cave living floor, maguey quids are the expectorated refuse of a meal; a knot in plant fiber may represent the remains of a torn rabbit snare; and a discarded basket may be the worn-out container in which *Setaria* seeds were winnowed. This reinforces our conviction that activity areas on a living floor must be defined not on the basis of the functions for which the items were originally manufactured, but on the basis of the subsequent activities that got them left behind on the floor.

What we would love to see done with the Coxcatlán Cave data is the sort of thing that Whallon (Chapter 27) and Reynolds (Chapter 28) have done with our Guilá Naquitz data in the pages that follow. Whallon produces a Pearson's *r* matrix for Zones E, D, C, and B1, showing that certain variables do tend to be strongly associated over blocks of four to eight contiguous squares. Reynolds produces a multidimensional scaling for both the variables and the 1 × 1-m squares on those same four living floors, segregating apparent work areas, refuse discard areas, and activity pathways. While many of our basic assumptions about hunting–gathering activities are the same as Fowler's, MacNeish's, and García Moll's, an important conclusion to be drawn from Chapters 27 and 28 is this: you probably cannot detect most cave activity areas and pathways with the naked eye. They often overlap, their borders are usually blurred, and defining them takes a statistical program that will detect them even after they have been walked, worked, slept, and excreted on during a whole season of foraging.

THE DATA SET

The studies in Chapters 27 and 28 define activity areas on the basis of association—by showing, in other words, that an increase in Variable A is a good predictor of an increase in Variable B or a decrease in Variable C. Obviously, to be statistically significant such associations have to involve relatively large numbers of occurrences for each variable. For this reason, Whallon's and Reynolds' studies are restricted to the four most extensive living floors (Zones E, D, C, and B1). They are also restricted to the most numerous variables present on those living floors, a number that varies between 15 and 17 depending on the floor. Included are *Agave* (maguey) quids, *Agave* leaves, hackberry seeds, cucurbit remains, *susí* nut seed coats, *guaje* pods, nanche seeds, *Opuntia* (prickly pear) seeds and stems, bean pod valves, piñon nut hulls, mesquite seeds, acorns, turtle bones, deer bones, cottontail bones, and unretouched flint flakes (debitage). Once activity areas have been defined on the basis of these statistically adequate variables, we can take a look at how the less numerous variables (such as flint cores, one-hand manos, cordage, or fire drills) are distributed with regard to the defined areas (Chapter 29).

In Tables 25.1–25.7 we give the square-by-square raw counts for the 17 most numerous variables in all 7 preceramic stratigraphic zones at Guilá Naquitz. Spencer used 6 of these zones, Whallon and Reynolds 4, while Zone B2+3 was omitted for reasons suggested in Chapter 5. One reason for presenting our raw data here, rather than burying it somewhere on a microfiche, is the realization that our studies are not the only ones possible with these data. Spencer's, Whallon's, and Reynolds' studies are mutually reinforcing, since they arrive at many of the same conclusions while manipulating the same data in different ways. However, other ways of manipulating such data will surely be proposed by our fellow archaeologists, and when this happens we want the data to be readily accessible.

SOURCES OF SPATIAL CO-OCCURRENCE AND SEPARATION

All archaeologists who interpret the spatial patterning of data on living floors do so on the basis of assumptions about the use of space by ancient peoples. In this case, our assumptions are based in part on ethnographic descriptions of Native American hunter–gatherers and in part on the nature of the eastern Valley of Oaxaca environment and its indigenous plants and animals (see Chapters 4 and 17). Let us briefly consider some of the possible reasons why the 15–17 variables used in our analyses might tend to co-occur or be spatially separate on a living floor at Guilá Naquitz. We restrict ourselves to 10 observations, all of which we believe to be supported either by ethnographic analogy or by the nature of the plant and animal species involved.

1. Sexual division of labor is typical of hunters and gatherers; some foods are collected and/or processed by men, others by women (see for example Lee and DeVore 1968). If space at Guilá Naquitz was conceptually divided into men's and women's work areas, the refuse from such areas might be archaeologically separate. To give one example, women might have played the major role in processing acorns, while men might have played the major role in butchering deer. (However, because of the high percentage of burned bone

TABLE 25.1
Distribution of 17 Major Categories of Debris throughout 1 × 1-m Squares in Zone E of Guilá Naquitz

Item	Distribution through 1 × 1-m squares																									Totals
	C8	C10	C11	D5	D9	D10	D11	E6	E7	E8	E9	E10	E11	E12	F7	F8	F9	F10	F11	G4	G5	G8	G10	H6	H9	
Agave quids	—	—	—	—	—	—	—	—	—	—	—	1	—	—	2	1	—	—	1	1	—	—	—	—	—	6
Agave leaves	—	—	—	—	—	—	—	—	—	—	—	—	—	—	—	—	—	1	—	—	—	—	—	—	1	2
Hackberry seeds	2	218	1	4	2	755	128	2	9	—	—	25	1	—	1	—	—	1	1	4	11	—	1	—	—	1166
Cucurbit remains	—	—	—	—	—	—	—	—	—	—	—	—	—	—	—	—	—	—	—	—	—	—	—	—	—	0
Susí seed coats	—	—	—	—	—	1	1	—	3	2	—	8	2	—	28	—	7	2	1	35	2	5	1	—	—	98
Guaje pods	—	—	—	—	—	—	—	—	—	—	—	—	—	—	—	—	—	—	—	—	—	—	—	—	—	0
Nanche seeds	—	—	—	—	—	1	—	—	—	—	—	—	—	—	1	—	—	—	—	4	—	—	—	—	—	6
Opuntia seeds	—	—	—	—	—	1	6	18	—	—	—	14	43	—	—	—	1	—	—	—	52	—	—	—	—	135
Opuntia nopales	—	—	—	—	—	—	—	—	—	—	—	—	—	—	—	—	—	—	—	1	2	—	—	—	—	3
Bean pod valves	—	—	—	—	8	5	1	—	—	—	—	20	2	—	—	—	6	98	2	—	—	—	19	—	—	161
Piñon nut hulls	—	—	—	—	1	5	21	—	—	—	—	17	79	—	—	—	4	1	3	22	2	—	—	—	—	155
Mesquite seeds	—	—	—	—	1	8	6	4	—	—	—	3	4	—	2	—	1	—	1	1	1	—	—	—	—	32
Acorns	1	12	2	—	—	48	54	—	92	—	98	191	58	—	163	10	270	59	40	210	44	12	5	463	14	1846
Turtle bones	—	—	3	—	—	1	—	—	—	—	—	1	1	—	—	—	—	1	—	1	—	—	—	—	—	8
Deer bones	1	1	3	—	—	2	1	—	—	—	—	2	1	1	—	—	1	3	—	—	1	—	—	—	—	17
Cottontail bones	—	6	1	—	—	4	11	—	—	—	—	—	3	1	—	—	1	1	1	2	—	—	—	—	—	31
Flint debitage	—	3	23	2	4	1	6	—	—	—	—	3	4	—	—	—	1	1	3	9	—	—	—	—	—	60

TABLE 25.2
Distribution of 17 Major Categories of Debris throughout 1 × 1-m Squares in Zone D of Guilá Naquitz

Item	Distribution through 1 × 1-m squares																							Totals
	B9	C7	C8	C9	C10	D5	D7	D8	D9	D10	E5	E6	E7	E8	F4	F5	F6	F7	F8	G5	G6	G7	G8	
Agave quids	—	—	—	—	—	—	—	—	—	—	1	1	1	4	—	4	12	—	11	2	6	1	6	49
Agave leaves	—	—	—	—	—	—	—	—	—	—	—	—	—	—	—	3	—	—	—	—	—	1	—	4
Hackberry seeds	1	—	11	6	31	—	4	—	1	77	6	111	18	22	4	12	47	—	—	27	37	—	1	416
Cucurbit remains	—	—	—	—	—	—	—	—	—	—	—	—	1	—	—	1	—	—	1	—	—	—	3	6
Susí seed coats	—	—	—	—	—	—	2	2	4	1	1	—	32	13	2	50	32	—	51	2	22	—	23	237
Guaje pods	—	—	—	—	—	—	—	—	—	—	—	—	—	8	—	—	—	—	1	—	—	—	2	11
Nanche seeds	—	—	—	—	—	—	—	—	—	—	—	1	—	33	—	—	—	—	12	—	2	—	1	49
Opuntia seeds	—	—	—	—	—	—	—	—	—	36	—	100	—	11	—	12	90	—	—	90	9	—	—	348
Opuntia nopales	—	—	—	—	—	—	—	—	—	—	—	16	—	—	—	3	—	—	3	—	3	3	—	28
Bean pod valves	—	—	—	—	—	—	1	1	1	—	—	1	—	5	—	—	3	—	—	—	4	—	1	17
Piñon nut hulls	—	—	—	—	—	—	6	—	2	11	2	10	8	1	—	22	23	5	—	2	—	2	—	94
Mesquite seeds	—	—	—	—	—	—	1	9	23	—	—	10	—	6	—	1	3	—	1	1	—	—	4	59
Acorns	3	4	1	1	1	2	288	183	68	17	6	372	547	261	7	302	647	—	111	180	133	—	48	3182
Turtle bones	—	—	—	—	—	—	3	—	—	—	—	—	—	—	—	—	—	—	—	—	—	—	—	3
Deer bones	—	—	—	3	2	1	1	—	4	3	—	—	2	2	1	7	1	—	—	2	1	—	—	30
Cottontail bones	—	—	—	1	—	—	2	—	3	3	4	—	1	4	—	1	—	—	1	—	2	—	—	22
Flint debitage	3	18	4	5	1	9	67	17	28	9	1	2	10	19	5	10	18	—	9	3	20	17	5	280

TABLE 25.3
Distribution of 17 Major Categories of Debris throughout 1 × 1-m Squares in Zone C of Guilá Naquitz

Item	Distribution through 1 × 1-m squares																					Totals
	B9	B10	C8	C9	C10	C11	D8	D10	D11	E5	E6	E7	E8	E9	E10	F4	F5	F6	F9	F10	G9	
Agave quids	—	—	—	—	—	—	2	—	—	—	2	6	14	2	3	—	3	6	132	5	21	196
Agave leaves	—	—	—	—	—	—	—	—	—	—	—	—	—	1	1	—	—	—	18	—	—	20
Hackberry seeds	18	27	36	18	29	—	—	7	31	8	10	6	—	662	32	4	5	19	17	40	—	969
Cucurbit remains	—	—	—	—	—	—	—	—	—	—	—	2	2	4	2	—	1	1	35	2	1	50
Susí seed coats	—	—	—	—	—	—	—	—	—	1	4	8	5	22	5	2	—	5	8	—	8	68
Guaje pods	—	—	—	—	—	—	—	—	—	—	3	9	4	5	13	4	—	—	204	4	1	247
Nanche seeds	—	—	—	—	—	—	—	2	—	—	16	2	5	47	17	26	3	27	16	35	—	196
Opuntia seeds	—	—	—	—	5	—	—	3	19	—	53	—	—	26	71	—	324	250	—	—	—	751
Opuntia nopales	—	—	—	—	—	—	—	—	—	—	1	—	2	—	1	—	—	—	31	3	11	49
Bean pod valves	—	—	—	1	—	—	2	1	1	—	1	—	—	25	6	—	—	6	—	—	—	43
Piñon nut hulls	—	—	1	—	—	—	—	1	9	19	1	—	—	23	16	2	—	—	5	3	—	80
Mesquite seeds	—	—	—	—	4	—	14	—	47	1	2	25	12	112	36	—	5	10	205	57	24	554
Acorns	1	—	16	3	7	1	5	8	8	86	126	94	41	641	281	12	—	163	59	9	9	1570
Turtle bones	—	1	1	—	—	—	—	—	2	—	—	—	—	—	1	—	—	—	—	—	—	5
Deer bones	3	2	—	1	7	—	1	5	1	2	3	1	—	3	1	—	—	—	3	2	1	36
Cottontail bones	5	2	2	1	1	—	—	2	5	2	1	1	—	1	4	—	—	—	—	—	1	28
Flint debitage	11	7	5	24	16	—	28	15	38	10	19	37	8	41	23	—	2	16	43	27	5	375

TABLE 25.4
Distribution of 17 Major Categories of Debris throughout 1 × 1-m Squares in Zone B3 of Guilá Naquitz

Item	Distribution through 1 × 1-m squares												Totals
	C6	C7	C9	C11	D5	D6	D8	D9	E5	E7	E8	F7	
Agave quids	—	—	—	—	—	—	—	2	10	6	—	—	18
Agave leaves	—	—	—	—	—	—	—	1	—	—	—	—	1
Hackberry seeds	—	—	22	5	—	—	—	17	16	5	—	—	65
Cucurbit remains	—	—	—	—	—	—	—	—	—	1	—	—	1
Susí seed coats	—	—	1	—	—	—	—	12	—	—	2	—	15
Guaje pods	—	—	—	—	—	—	—	1	—	4	—	—	5
Nanche seeds	—	—	—	—	—	—	—	18	—	—	—	—	18
Opuntia seeds	—	—	—	—	—	—	—	—	—	—	—	—	0
Opuntia nopales	—	—	—	—	—	—	—	—	—	—	—	—	0
Bean pod valves	—	—	—	—	—	—	—	4	—	22	—	—	26
Piñon nut hulls	—	—	—	—	—	—	—	3	1	—	—	—	4
Mesquite seeds	—	—	1	1	—	—	—	134	1	99	—	—	236
Acorns	—	—	56	—	—	—	69	248	24	—	7	17	421
Turtle bones	—	—	1	—	—	—	—	—	—	—	—	—	1
Deer bones	5	—	—	—	—	—	—	2	1	—	—	—	8
Cottontail bones	—	—	2	—	—	—	—	3	2	—	—	—	7
Flint debitage	8	2	15	—	8	8	6	5	7	8	2	—	69

TABLE 25.5
Distribution of 17 Major Categories of Debris throughout 1 × 1-m Squares in Zone B2 + 3 of Guilá Naquitz

Item	Distribution through 1 × 1-m squares						Totals
	C8	C10	D10	E6	F4	F5	
Agave quids	—	1	—	4	—	3	8
Agave leaves	—	—	—	—	—	—	0
Hackberry seeds	74	23	5	34	—	13	149
Cucurbit remains	—	1	—	1	—	1	3
Susí seed coats	—	2	—	—	1	—	3
Guaje pods	—	—	—	1	—	5	6
Nanche seeds	4	3	—	1	—	—	8
Opuntia seeds	—	32	—	169	—	159	360
Opuntia nopales	—	—	—	—	—	2	2
Bean pod valves	2	1	—	6	—	—	9
Piñon nut hulls	—	—	—	—	—	—	0
Mesquite seeds	4	33	42	—	3	194	276
Acorns	21	41	16	3	9	5	95
Turtle bones	1	1	1	—	—	—	3
Deer bones	3	—	1	—	—	2	6
Cottontail bones	2	3	1	1	1	—	8
Flint debitage	16	11	30	36	9	11	113

mentioned above, most bone concentrations at Guilá Naquitz are referred to by the more ambiguous category "animal processing and/or eating.")

2. Some plants, such as acorns and maguey, need a lot of processing to render them edible; others, such as nanches and hackberries, can be eaten raw. Those genera needing more processing might leave spatially restricted refuse and have features associated with them. Those eaten raw might simply be scattered around the cave wherever people happened to sit.

3. A corollary of Point 2 above is that the processing of some plants and animals (such as maguey and deer) leaves such conspicuous debris that large amounts of it would probably have been thrown out of the cave. The processing of other plants and animals (such as hackberries and cottontail rabbits) might leave such inconspicuous refuse that much of it might be overlooked or swept into a corner.

4. Plants come ripe at different seasons. Those genera collected in August and September might have been processed in one area, those collected in October and November in

TABLE 25.6
Distribution of 17 Major Categories of Debris throughout 1 × 1-m Squares in Zone B2 of Guilá Naquitz

Item	Distribution through 1 × 1-m squares													Totals
	C5	C6	C7	C9	C11	D5	D8	D9	D11	E5	E9	E10	E11	
Agave quids	1	—	—	2	—	—	7	9	4	1	10	11	—	45
Agave leaves	—	—	—	—	—	—	—	1	—	—	1	1	—	3
Hackberry seeds	—	1	2	18	9	—	6	13	4	19	4	31	—	107
Cucurbit remains	—	1	—	—	3	—	1	1	—	—	4	1	—	11
Susí seed coats	—	—	—	—	—	—	7	5	—	—	—	—	—	12
Guaje pods	—	—	—	—	—	—	1	4	6	—	29	3	—	43
Nanche seeds	—	—	—	4	2	—	—	29	2	4	4	—	—	45
Opuntia seeds	—	—	—	—	32	—	—	—	48	—	—	2	—	82
Opuntia nopales	—	—	—	—	—	—	—	1	10	—	4	16	—	31
Bean pod valves	—	—	2	4	—	—	—	31	64	—	3	—	—	104
Piñon nut hulls	—	—	—	1	—	—	1	—	—	—	—	—	—	2
Mesquite seeds	—	2	3	9	63	—	266	216	815	9	89	141	—	1613
Acorns	—	—	—	154	6	1	48	105	14	4	59	27	—	418
Turtle bones	—	—	—	4	7	—	3	3	—	—	—	—	—	17
Deer bones	—	2	—	4	1	—	—	8	—	—	—	—	—	15
Cottontail bones	—	—	—	—	—	1	—	5	—	—	—	—	—	6
Flint debitage	—	33	2	19	32	12	31	25	12	5	21	7	2	201

another area. However, such seasonal differences could be obscured if work areas based on sexual division of labor were well established (that is, women might process plants in the same area of the cave regardless of season).

5. Some plants, such as hackberries and *susí* nuts, are quite seasonally restricted in their availability. Others, such as *Agave* hearts and prickly pear stems, are available over a much longer period. Such plants might be collected and processed at different times or in different work areas.

6. Plants come from different vegetational zones. Those brought back from a trip to Thorn Forest A (such as acorns, piñon nuts, or wild beans) might have been processed in one area, while those brought back from a trip to Mesquite Grassland B (such as hackberries or mesquite) might have been processed elsewhere. However, such vegetational zone differences might also be obscured by sexual division of labor (see Point 4).

7. A corollary of Points 4 and 6 is that plants collected on the same day and in the same vegetational zone might be processed together, unless that processing required significantly different techniques. Such activity could result in a pattern whereby plants available in August in the Mesquite Grassland would be separate from those available in November in the Thorn Forest.

8. Unretouched flint flakes (debitage) could result from the making of tools for many different tasks, including both men's and women's work, and the processing of both plants and animals. In other words, we seriously question the assumption that the making of flint tools was an exclusively male activity (as, for example, in García Moll 1977:87). Unless one individual made all the tools for all tasks (which we doubt), we might therefore expect debitage to occur in a variety of contexts rather than being associated only with one activity.

9. As we are reminded by Thomas (1971, 1972, 1973), some tasks are performed jointly by men and women. Men and women also may have eaten at the same hearth in the cave, discarding their food debris in similar ways. Such communal activity could reduce the evidence for spatial partitioning and produce areas where sexual division of labor could not be inferred from the debris.

10. Even if the cave was conceptually divided into work areas, people could not be expected to confine themselves to those areas throughout the occupation. Merely by walking back and forth in the cave, both men and women presumably blurred the boundaries of their activity areas by kicking debris around—particularly the finer fraction, which had not been thrown out on the talus. The question then becomes, to what extent could work areas still be recognized even after such blurring? We feel that Spencer's computer graphics show some differential use of the cave floor that is evident even by inspection, while Whallon's and Reynolds' analyses recover still other activity areas that had been considerably blurred.

TABLE 25.7
Distribution of 17 Major Categories of Debris throughout 1 × 1-m Squares in Zone B1 of Guilá Naquitz

Item	Distribution through 1 × 1-m squares																											
	B9	B10	B11	C7	C8	C9	C10	C11	C12	D5	D6	D7	D8	D10	D11	E5	E6	E7	E9	E10	E11	F4	F5	F6	F7	G4	G5	Totals
Agave quids	—	—	—	—	1	1	2	5	—	—	—	3	—	1	1	—	1	—	8	—	10	2	—	9	2	—	—	46
Agave leaves	—	—	—	—	—	—	—	—	—	—	—	—	—	—	—	—	—	10	—	—	—	—	—	—	—	—	—	10
Hackberry seeds	18	15	—	—	35	32	41	13	2	—	—	—	4	22	2	8	3	—	10	—	17	7	3	2	—	—	6	240
Cucurbit remains	4	—	—	—	1	3	—	7	—	1	—	—	—	—	—	2	—	1	—	—	2	—	—	—	—	—	—	21
Susí seed coats	—	1	—	—	2	—	4	—	—	—	1	2	—	2	—	—	—	—	—	—	7	—	—	—	3	—	—	22
Guaje pods	—	—	—	—	2	—	—	2	—	—	1	—	—	7	—	—	—	—	61	—	4	—	—	18	9	—	1	105
Nanche seeds	—	—	—	—	—	—	—	1	—	—	3	—	10	—	—	10	—	—	5	—	6	1	—	—	—	10	—	46
Opuntia seeds	—	—	—	—	2	1	18	32	—	—	—	—	—	—	1	—	—	—	—	—	—	—	—	64	—	20	5	143
Opuntia nopales	—	—	—	—	—	—	—	—	—	—	—	—	—	—	—	—	—	—	32	—	4	1	—	6	—	2	—	45
Bean pod valves	—	—	—	—	2	3	—	88	—	—	—	—	—	12	—	—	—	—	2	—	4	—	—	—	—	—	—	111
Piñon nut hulls	—	—	—	—	2	—	—	—	—	—	—	—	—	—	—	—	—	—	—	—	26	—	—	—	—	3	—	31
Mesquite seeds	1	1	—	—	—	1	14	116	—	—	—	—	—	—	13	35	12	182	90	140	1066	51	2	76	—	25	8	1833
Acorns	5	17	—	—	68	225	89	3	—	22	—	8	6	42	2	19	—	—	131	—	115	4	—	9	98	15	14	892
Turtle bones	—	—	—	—	2	2	1	—	—	—	1	14	1	1	—	—	10	—	—	1	—	1	—	—	—	—	—	34
Deer bones	—	—	—	—	—	1	1	7	4	4	1	2	—	3	—	—	—	1	—	—	5	2	4	2	—	—	—	37
Cottontail bones	1	1	—	—	1	—	—	1	9	—	—	1	—	1	—	—	—	2	—	—	3	1	1	—	1	2	—	25
Flint debitage	37	3	12	2	8	—	3	—	30	11	17	28	37	10	5	17	30	35	19	10	26	10	60	13	11	26	6	466

26

Spatial Variation of Debris at Guilá Naquitz: A Descriptive Approach

Charles S. Spencer
Kent V. Flannery

INTRODUCTION

Excavation of six living floors (Zones E–B1) at Guilá Naquitz provided us with a plethora of plant remains, animal bones, and flint debitage (see Tables 25.1–25.7). At the time of the excavation it was already intuitively clear that classes of items were not uniformly distributed over the living floors, but the quantity and complexity of debris made it impossible to visualize those distributions clearly or to guess which items would turn out to be statistically associated. In this chapter, we subject the Guilá Naquitz data to a program explicitly designed to sort through and simplify this distributional complexity so as to make it more amenable to the generation of hypotheses about the use of the cave living floors. In Chapters 27 and 28, Whallon and Reynolds subject the same data to measures of association, confirming some of our conclusions and discovering other patterns that were not detectable by visual inspection of our results.

The first problem facing any archaeologist with a large data set is how to screen the data in a preliminary fashion to check for patterns of variation that might bear on the general research problem. This initial data screening should involve limited manipulations of the original data, and it ought to precede more elaborate statistical procedures. In many cases, simple descriptive measures and visual inspection can give archaeologists a better understanding of their data and thus guide them in subsequent analyses.

Choosing the right kind of data-screening procedure requires a consideration of both the kinds of data involved and the nature of the research problem. In the case of Guilá Naquitz, the data consisted of frequencies of various items per grid unit (1 × 1-m square) of occupation space. Our research problem focused on the organization of activities that occurred in that space. We wished to know whether the flint flakes, plant remains, and animal bones were scattered randomly throughout the cave or whether they tended to occur in distributional clusters. If such clusters did occur, we then wanted to know what sorts of activities they represented and how these activities were related to the social and economic organization of the inhabitants of Guilá Naquitz. A full answer to this question, of course, would demand an extended quantitative analysis that is beyond the scope of this chapter; but we knew that a program of data screening could provide us with an initial understanding of the kinds of spatial variation that could be explored in further analyses, and that is what we seek to present here.

DESCRIPTIVE MEASURES

Before we began our preliminary analysis of the Guilá Naquitz data, we transferred the distributions of our 17 major items of debris (Tables 25.1–25.7) to computer files. The data from each occupation floor were assigned to a separate file, in which the 17 categories of debris were regarded as variables, and the 23 grid squares occupied in the preceramic as cases. It was then possible to use the Michigan Interactive Data Analysis System (MIDAS) to generate some simple descriptive measures of the variables' distributions across the grid squares of each living floor (Fox and Guire 1976). The first step in our data-screening procedure entailed an examination of these statistics to determine if there were any indications of clustering in the spatial distributions of the variables. In the interests of brevity, we do not go through all the measures for each zone here but use as an example a representative occupation floor, Zone D. Table 26.1 contains a series of descriptive measures for this zone.

If we look at the minimum column in Table 26.1, we can see that for each variable there is at least one square on the living floor with none of the items in question. A perusal of the maximum column, on the other hand, reveals that many of the variables have rather high frequencies in at least one square. This immediately indicates that our items are not evenly distributed on the living floor but tend to occur more often in some squares than in others.

If the items are, in fact, concentrated in a few squares, we would expect to find considerable divergence about the mean values for each item per square. The mean values in Table 26.1 were computed by dividing the total quantities for each category by the total number of squares occupied in the preceramic. Note that although Zone D material was not found in all 23 excavation squares, we nonetheless divided the totals by 23. This is because the entire Naquitz phase residential area can presumably be considered the overall unit of habitation space in Zone D times (and in other time periods as well), even though Zone D debris may have been only discarded in certain areas within the cave.

The standard deviation about the mean was computed for each item in the usual manner:

$$S = \sqrt{\frac{\Sigma_{i=1}^{n} (X_i - \overline{X})^2}{n - 1}}$$

The standard deviation column in Table 26.1 contains high values for all the variables; in each case, the standard deviation is greater than the mean. This implies that the items in each category tend to exhibit an uneven distribution in the cave.

The coefficient of variation is another measure of dispersion, one that is calculated by dividing the standard deviation by the mean for each item. Though we will not do so here, the resulting ratio can be converted to a percentage by multiplying by 100 (Thomas 1976:83). The coefficient of variation expresses the dispersion of a particular category of debris in terms relative to the mean value of the category. As such, the coefficient of variation is a "pure measure" (Thomas 1976:84); results are expressed in proportions rather than in the particular units (e.g., counts or weights) of the variables in question. The absolute magnitude of a standard deviation, of course, depends in part on the magnitude of the frequencies used in its calculation; consequently, it is often difficult to assess the relative dispersion of variables that occur with widely differing frequencies in a data set. On the other hand, by computing the coefficient of variation, we generate a relative measure of dispersion that can be used to compare all the variables in a data set such as the one from Zone D.

As Table 26.1 shows, the coefficients of variation are all above 1.0; some even range from 2.0 to 4.0. These are rather high values for this coefficient. Thomas (1976:84) has suggested that when the coefficient of variation is greater than 0.1, it is likely that the underlying distribution is multimodal in shape. In our case, with coefficients of variation that are all much higher than 0.1, it appears that the items in Zone D were distributed in such a way that some squares had few or no items while other squares had many.

Because the descriptive statistics suggested that the distributions of the variables across the grid squares may have been multimodal in shape, we decided to generate a series of histograms for all the variables in each zone of the cave. Visual inspection of these histograms revealed that most of the distributions were, in fact, bimodal or trimodal in shape. As an example, let us consider the histogram for the distribution of acorns in Zone D.

This histogram (Table 26.2) shows at least a bimodal distribution of acorns among the grid squares of the cave, clearly making a distinction between squares that had many acorns and squares that had few or none. To be less conservative, we could even suggest that a trimodal distribution is represented. The first mode comprises squares with low acorn densities; these would include the 15 squares falling in the 2 lowest intervals, those with midpoints of 0.0 and 92.429 acorns. The second mode comprises squares with moderate acorn densities; included here are 6 squares covering the 3 intervals that have midpoints of 184.86, 277.29, and 369.71 acorns. The third mode comprises 2 squares with high acorn densities, one in each of the 2 intervals that have midpoints of 554.57 and 647.00 acorns. This multimodal pattern is *not* what we would expect to see if acorns had been randomly scattered across the cave floor. Instead, the histogram suggests that the acorn distribution in Zone D was characterized by clustering.

Yet another relative measure of clustering is provided by the *variance–mean ratio* (Table 26.1). To compute this ratio, one simply divides the variance (the square of the standard deviation, the formula for which was given earlier) by the mean. The variance–mean ratio can be used with grid-count data (such as we have here) to assess tendencies toward nonrandom spacing; the approach is based on the Poisson distribution, which has a variance equal to its mean. Thus, a random distribution will have an expected variance–mean ratio close to 1.0, a clustered distribution will produce a variance–mean ratio larger than 1.0, and an evenly spaced distribution will yield a variance–mean ratio smaller than 1.0 (Carr 1984:140; Dacey 1973:321). Table 26.1 reveals that all 17 variables in Zone D have variance–mean ratios larger than 1.0, which suggests that these spatial distributions all exhibit nonrandom clustering.

TABLE 26.1
Descriptive Measures for Zone D[a]

Variable	n	Minimum	Maximum	Mean	Standard deviation	Coefficient of variation	Variance–mean ratio
Agave quids	23	0.0	12.0	2.13	3.52	1.65	5.82
Agave leaves	23	0.0	3.0	0.17	0.65	3.74	2.49
Hackberry seeds	23	0.0	102.0	17.70	26.61	1.50	40.01
Cucurbits	23	0.0	3.0	0.26	0.69	2.64	1.83
Susí nut hulls	23	0.0	51.0	10.30	16.41	1.59	26.14
Guaje pods	23	0.0	8.0	0.48	1.70	3.56	6.02
Nanche seeds	23	0.0	33.0	2.13	7.18	3.37	24.20
Opuntia seeds	23	0.0	100.0	15.13	32.04	2.12	67.85
Nopales	23	0.0	16.0	1.22	3.42	2.81	9.59
Bean pod valves	23	0.0	5.0	0.74	1.39	1.88	2.61
Piñon nuts	23	0.0	23.0	4.09	6.71	1.64	11.01
Mesquite seeds	23	0.0	23.0	2.57	5.32	2.08	11.01
Acorns	23	0.0	647.0	138.35	186.05	1.34	250.20
Turtle bones	23	0.0	3.0	0.13	0.63	4.80	3.05
Deer bones	23	0.0	7.0	1.30	1.72	1.32	2.28
Rabbit bones	23	0.0	4.0	0.96	1.36	1.43	1.93
Flint debitage	23	0.0	67.0	12.13	14.19	1.17	16.60

[a]Explanation of column headings is as follows: n, number of squares; minimum, smallest number of items in each category that were found in any square; maximum, greatest number of items in each category that were found in any square; mean, average number of items in each category per square; standard deviation, measure of dispersion calculated in usual manner (see text); coefficient of variation, standard deviation divided by the mean to provide a relative measure of dispersion; variance–mean ratio, the variance divided by the mean to yield a relative measure of clustering.

CONTOUR MAPPING

The next step in the data-screening procedure for Guilá Naquitz involved the GEOG:CONTUR program of the Michigan Terminal System (MTS; Department of Geography, University of Michigan 1974). This program generates the best-fitting contour map from a lattice of input densities. The output can be routed to a plotter, yielding high-quality maps that facilitate visual inspection of the distributional patterns and thus permit first-level generalizations regarding the nature of these patterns.

One's first encounter with GEOG:CONTUR can be a frustrating experience, as the available documentation is far from illuminating in either style or content. The purpose of this section is to present the various steps one must follow in order to implement the program successfully. Most of the specific commands we discuss here are relevant only to the GEOG:CONTUR program of MTS, but we feel it is useful to go through the procedure in some detail to give the reader a feeling for the kinds of analytical decisions and simplifying assumptions we were required to make. Those who are not concerned about the details of the contour mapping program can skip ahead to the next section.

TABLE 26.2
Histogram of Acorn Distribution in Zone D[a]

Midpoint	Count for acorns
0.	11 + XXXXXXXXXXX
92.429	4 + XXXX
184.86	2 + XX
277.29	3 + XXX
369.71	1 + X
462.14	0 +
554.57	1 + X
647.00	1 + X
	23

[a]Interval width is 92.429 acorns. The midpoint of each interval is presented, along with the actual count of grid squares containing acorns that fall in that interval. The histogram is depicted just as the computer drew it, using a series of Xs each of which represents one grid square.

Since we have already used the acorn distribution in Zone D to illustrate how descriptive measures were established, we use that same distribution to illustrate the procedure for generating a contour map. The first thing we did was convert the field designations for the excavation squares into a format acceptable to the program. Figure 26.1 shows how each square was given a set of coordinates corresponding to the midpoint of that square. (Actually, only the first three digits of each coordinate are given in Fig. 26.1; two trailing zeros have been omitted for both the x and y coordinates.) Each square could then be referenced by these coordinates for the purposes of coding and keypunching.

The values for the acorn distribution in Zone D were recorded as follows. On each punch card, the first 10 digits referred to the x and y coordinates for the midpoint of the square in question; the last 3 digits represented the frequency of a particular variable (in this case, acorns) in that square, taken from Table 25.2. These data were then copied into an MTS line file called "ACORNS."

Because Zone D material did not occur in all grid squares of the cave, it was necessary to create a file of zero densities for the squares in Guilá Naquitz in which no Zone D material was found. This was done as described above, except that 000 was entered after the 10-digit coordinate designation. This information was copied into a line file called "ZNDUNEX."

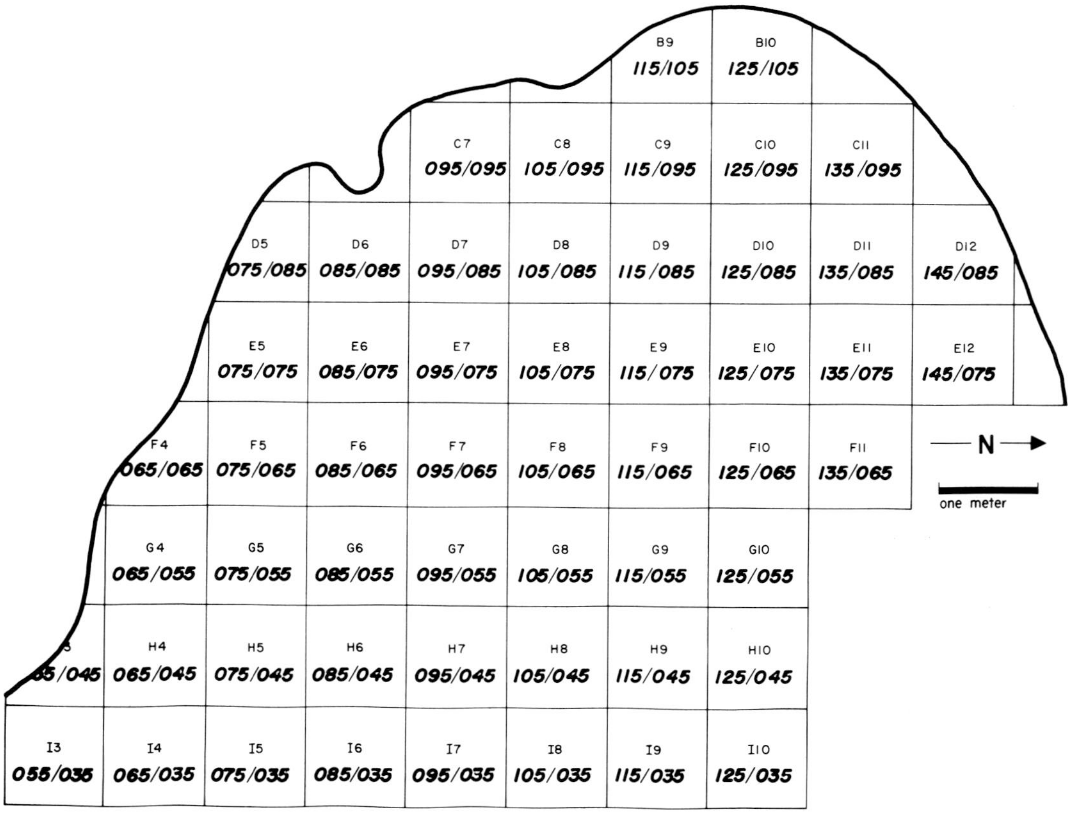

Fig. 26.1. Guilá Naquitz: relationship of original 1 × 1-m squares to grid coordinates used for GEOG:CONTUR program.

It was also necessary to indicate the location of the cave wall. This was done by creating a line file called "ZERODENS," which consisted of x and y coordinates for approximately 100 points around the actual edge of the cave, followed by a density entry of 000.

Prior to running GEOG:CONTUR, it was necessary to concatenate the above 3 files. The concatenated files were then recopied into the file ACORNS. Still more preparatory work was necessary before running GEOG:CONTUR however. One routine in the program generates an interpolated lattice based on the input data (which are obviously not on a perfectly symmetrical lattice). The final contour map is drawn based on this interpolated lattice. Therefore, there had to be a line file already created at the time GEOG:CONTUR was to be run to accept the interpolated lattice and the calculation of error between the input lattice and the interpolated lattice. This file was called "CAVECON." It was also necessary to create a line file for the plot description generated by the GEOG:CONTUR program. In our acorn example, this plot file was called "ACORNPLTD."

Finally, in order for the GEOG:CONTUR program to draw the contour map of the acorn distribution on a base map of the Guilá Naquitz cave floor, it was necessary to create a line file containing the coordinates of that floor. This, however, had to be done in a different manner from that described above for the acorn densities. The outer edge of the cave floor was broken up into 12 open-ended polygons, each containing 7 vertices. This information was copied into a file called "CAVEFLR."

At this point, GEOG:CONTUR could be run. The program is designed to be interactive and is best run on a terminal with graphics capability. In our example of the acorn distribution from Zone D, the GEOG:CONTUR program was started by issuing the following MTS command:

```
$RUN GEOG:CONTUR 2=*SOURCE* 3=*SINK*
5=ACORNS 6=CAVECON 9=ACORNPLTD
```

The program first requested the title of the map. "Guilá Naquitz Zone D Acorns" was entered as the response. The next request was whether or not the data were already on a lattice, that is, in an equally spaced rectangular array. The proper answer to this query, of course, was no, entered simply as "N." The format of the data was then requested. The specification was a normal Fortran-type format: (2F5.3, F3.0).

The interpolation subroutine then began. The configuration of original data points was transformed into a rectangular, regularly spaced array. The program gave the size of the grid and the number of rows and columns in the transformed data matrix, then requested a series of confirmations. The interpolated lattice and the computation of error between this lattice and the original data matrix were written into the CAVECON file.

The program then asked if we had a base map onto which we wanted the contour map to be drawn. On receiving an affirmative answer, the program requested the number of polygons in the map and the average number of points in each polygon. As mentioned above, in the Guilá Naquitz case these answers were 12 and 7, respectively. The name of the file containing the polygons was then requested; the answer was "CAVEFLR." The program then went on to ask for the format of the polygons. This required a typical Fortran-type format specification: (I2, 14F5.3). The I2 in the format refers to a two-digit integer in the first two columns of each record in CAVEFLR that refers to the number of coordinates contained in that record.

Next, the program presented the minimum and maximum values for the lattice and asked if the lattice should be *smoothed*. This is a subroutine of the program that literally smooths out local anomalies in the lattice using a procedure akin to trend-surface analysis. The program also asked if the lattice should be splined. *Splining* is essentially the reverse of smoothing: it amplifies variability within the lattice. For the sake of consistency, the smoothing and splining options were *never* chosen when running GEOG:CONTUR with the Guilá Naquitz data.

At this stage, the interpolated lattice was in final form and the generation of the contour map itself could begin. The program suggested values for the various parameters of the contour map and asked if these values were acceptable. For aesthetic as well as analytical reasons, it is often necessary to request a parameter value different from that suggested by the program. This does *not* result in an alteration of the accuracy or validity of the contour map, because the interpolated lattice (on which the contour map is based) remains exactly the same whether the contour interval is 1.0 or 20.0. The contour map generated is always the best-fitting representation of the interpolated lattice, regardless of the exact parameter values chosen.

Often the program will suggest a minimum contour value of zero. We recommend running the program with a listing of the input density file (e.g., ACORNS) handy and changing the value of the minimum contour level to the lowest (or close to the lowest) value in the file. We do not recommend changing the suggested maximum contour value.

The program will also suggest a contour interval. It has been our experience that the interval chosen by the program will produce a contour map with as many as 11 or 12 levels. We feel this number of levels is far too many for most purposes; the resulting map is extremely cluttered, making interpretation rather difficult. We have found that a contour interval that yields between 4 and 8 contour levels is usually adequate. As an example of this, we can turn to the contour map eventually produced for our Zone D acorns (Fig. 26.2). This map, which used 7 contour lines at intervals of 100 acorns, is uncluttered and easy to read.

Having generated the contour map according to our specifications, the program then asked to display the map on the screen. In general, it is a good idea to have the contour map displayed, as this allows an assessment of the results before the plot description is copied into the plot file. Next,

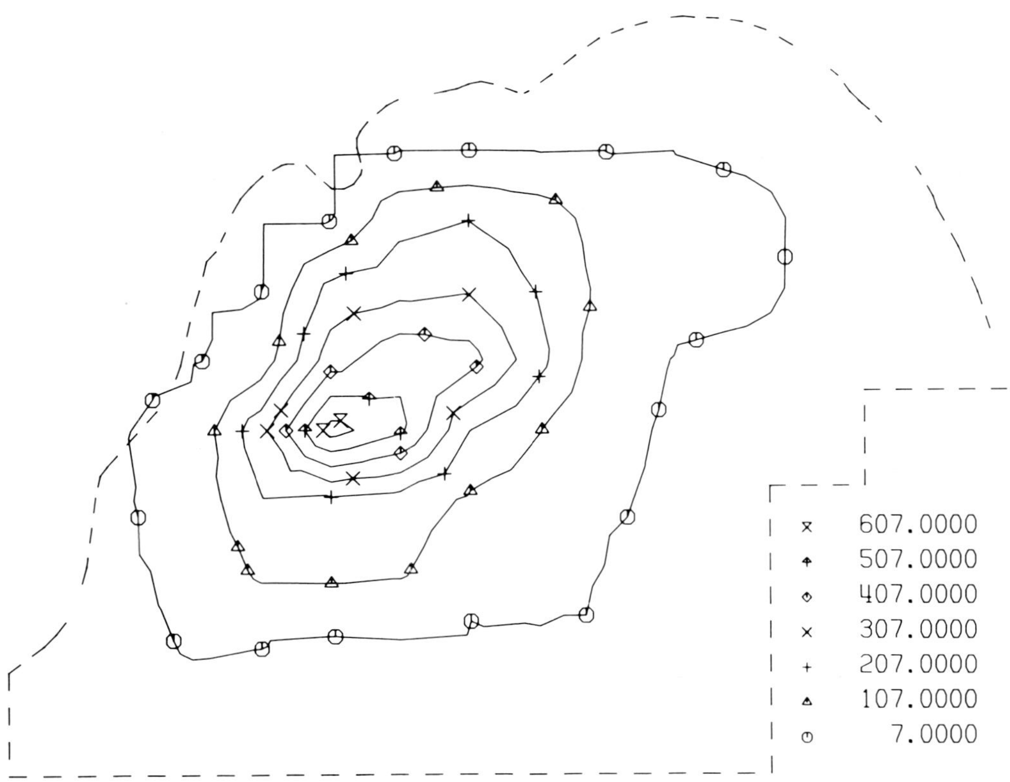

Fig. 26.2. An example of the density contour maps printed out by the GEOG:CONTUR program: the distribution of acorns on the Zone D living floor (compare with the artist's drafted version, Fig. 26.12). Numbers represent counts of acorns per 1 × 1-m square.

the program asked if the contour map should be copied into the plot file (e.g., ACORNPLTD). As our response was affirmative, the program went on to ask if a contour map including the outline of the base map was desired and whether we wanted the plot labeled (this includes the title of the map as well as symbols for each contour level and their associated values).

The execution of the program was then terminated, and the plot description for the generated contour map resided in the plot file (ACORNPLTD). We could access this file at any time for visual display on a graphics terminal, and we could route the plot file to a plotter to get a hard copy (Fig. 26.2).

INTERPRETING THE CONTOUR MAPS

In the sections that follow, we examine in detail the contour maps for Zones E–B1 at Guilá Naquitz that were produced by the procedures described above. These contour maps clearly show that no variable was homogeneously distributed across a living floor; all variables show spatial clustering, and many show multiple clusters that indicate where individual piles of items such as flint flakes, hackberry seeds, or maguey quids were discarded. These clusters, in turn, can be interpreted in terms of the use of the cave by the men and women of the Naquitz phase.

An additional step we have taken is to have one of our artists, Lisa Klofkorn, re-draft the contour maps (Figs. 26.3–26.47). Our reasons for doing this are apparent if the reader compares Fig. 26.2, the computer-drawn map of Zone D acorns, with Fig. 26.12, the artist's rendering of the same map. First, the artist can insert the actual contour values so that reference to a key is unnecessary. Second, features such as hearths and storage pits can be added to the drawing, which aids in our interpretation of the variable distributions. Third, by using solid lines and dashed lines we can display two variables on one map, reducing the number of illustrations necessary as well as pointing out variables with similar distributions.

We should note that in the zone-by-zone interpretations that follow we occasionally refer to a concentration of debris in terms of the original grid designations (for example, "the concentration of acorns in Square C9"). We have included Fig. 26.1 so that readers will have ready access to the map of designated grid units at Guilá Naquitz as they examine the results of the contour mapping program presented in Figs. 26.2–26.47.

One further comment about the contour maps should be made: The products of the GEOG:CONTUR program should not be expected to look like the maps one would obtain if one were to draw the contour lines freehand based on Tables 25.1–25.7. Peaks may be divided between two adjacent high-density squares, rather than centered in the square with the single highest frequency. If five (items) is chosen as the lowest contour value, isolated occurrences of only one or two items will not be shown on the map. However, in each case the map will be the best-fitting representation of the interpolated lattice as determined by the computer.

ZONE E

Flint debitage occurred primarily in two scatters (Fig. 26.3). The larger of these scatters (marked by the 13-flake contour) covered the northwest quadrant of the cave, while the smaller scatter (marked by the 4-flake contour) lay against the south wall. Notice that in Fig. 5.32 each of these two scatters included a hammerstone, possibly for percussion flaking. This raises the possibility that at least two different individuals in two different activity areas did some flint knapping during the course of the occupation.

Acorns were strewn throughout the cave, but their distribution shows four peaks (Fig. 26.4). In the center of the living floor we see two scatters defined by the 70-acorn contour, with one scatter reaching a peak of more than 200 acorns per square. In the southeast quadrant of the living floor we see two more scatters defined by the 70-acorn contour, with one scatter exceeding 265 acorns per square. Tentatively, these scatters suggest places where individual basketloads of acorns were unloaded on the living floor. However, it is not clear whether 4 separate basketloads were involved, or only 2. In the latter case, it might only be the 200-acorn contours that represent basketloads, with the nearby 70-acorn contours representing processing areas beside them.

Susí nut hulls had a distribution so similar to that of acorns that they have been included on the same map (Fig. 26.4). While there were fewer *susí* nuts, the 1-*susí* nut contour fits comfortably within the 5-acorn contour, and the two densest *susí* nut scatters (19–25-nut contours) partially overlap two of our 70-acorn contours. Since *susí* nuts are common in the oak facies of Thorn Forest A, it makes sense that they might have been collected and shelled at the same general time as acorns.

Pinõn nut hulls had a distribution not unlike that of flint debitage (Fig. 26.5): One large scatter (marked by the 49-nut contour) occurred in the northwest quadrant of the cave, while a smaller scatter (marked by the 13-nut contour) lay against the south wall. Since piñon nuts can be eaten raw after removing the hull, what we probably have here are the piles of hulls left by two different persons who ate piñon nuts as they sat working in their respective activity areas.

Mesquite seeds were not numerous. Their major distribution was in the northwest quadrant of the cave, with a minor extension toward the south (Fig. 26.5). Since mesquite requires the kind of processing that might (on the basis of ethnographic analogy) have been women's work, the distribution of these seeds is our first strong clue that the northwest quadrant of the cave may have included women's work space.

Opuntia seeds were strewn over much of the cave floor, but their distribution shows two peaks not unlike those for piñon nuts (Fig. 26.6). The densest scatter (marked by the

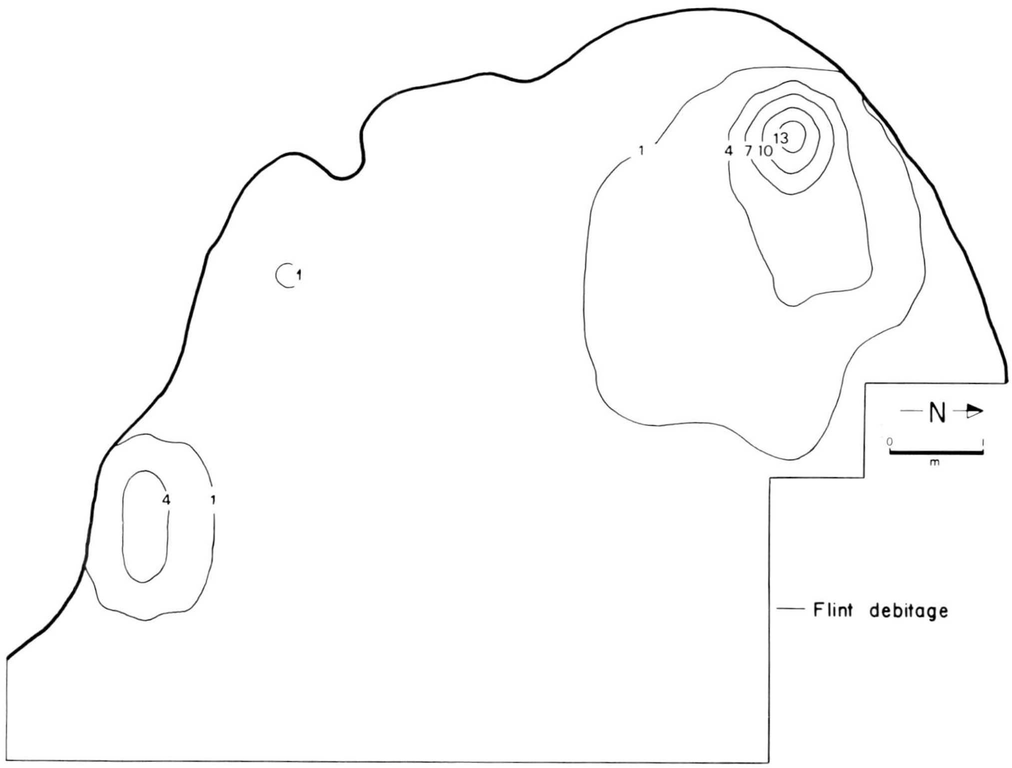

Fig. 26.3. Density contours of flint debitage in Zone E.

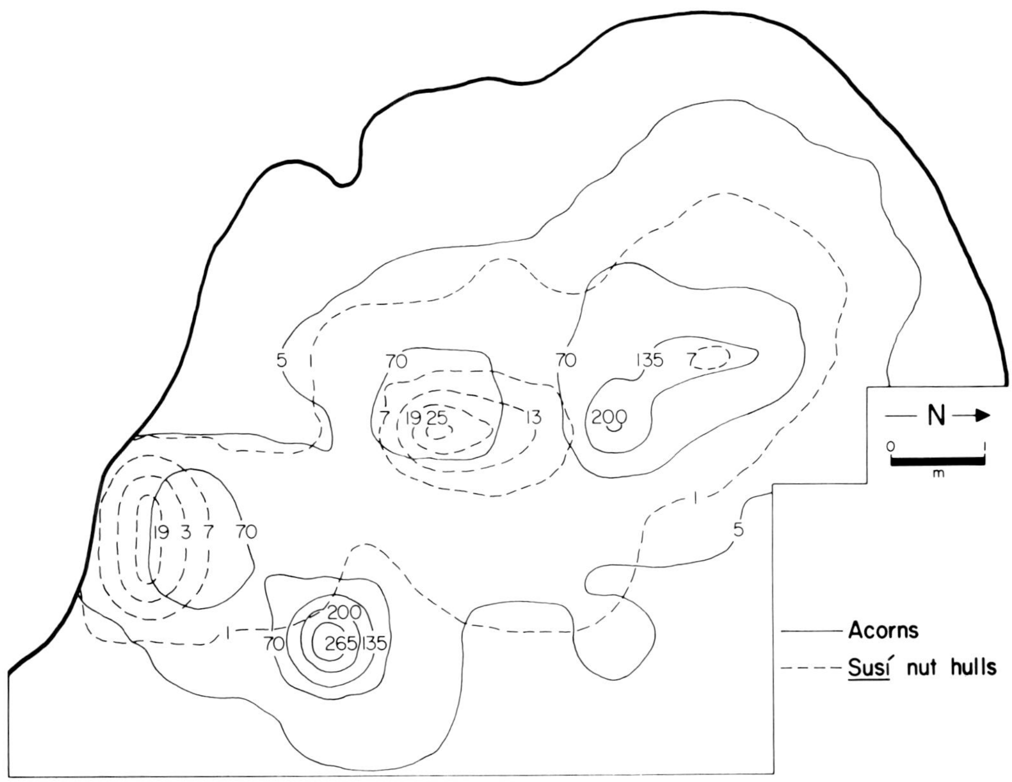

Fig. 26.4. Density contours of acorns and *susí* nut hulls in Zone E.

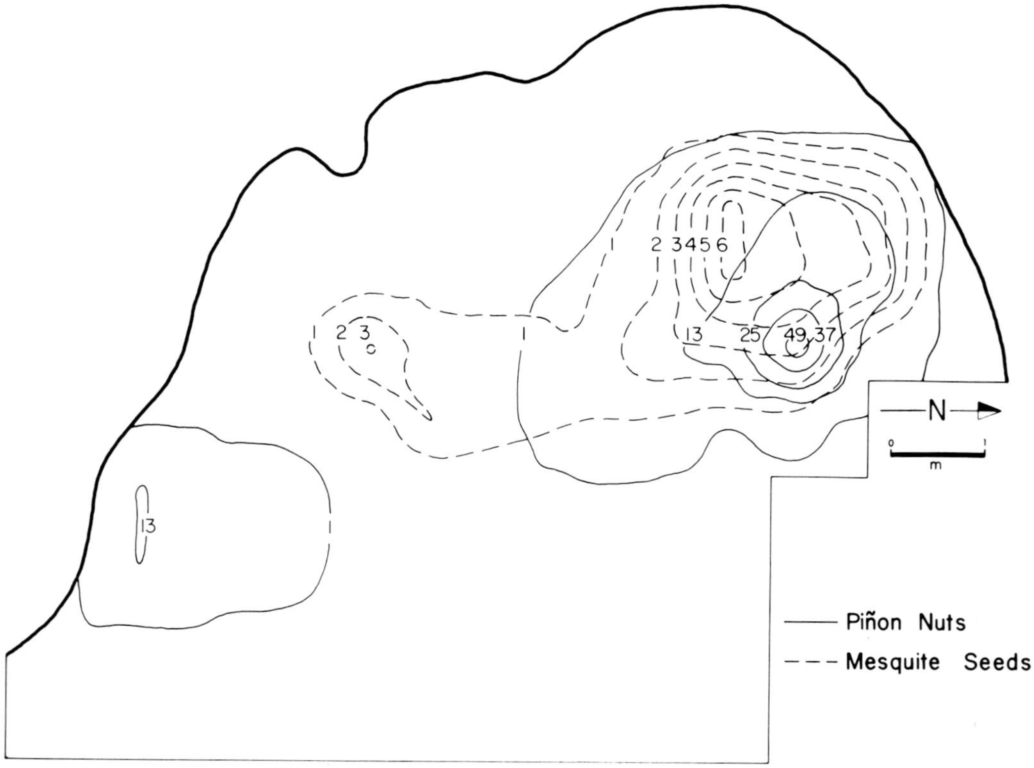

Fig. 26.5. Density contours of piñon nuts and mesquite seeds in Zone E.

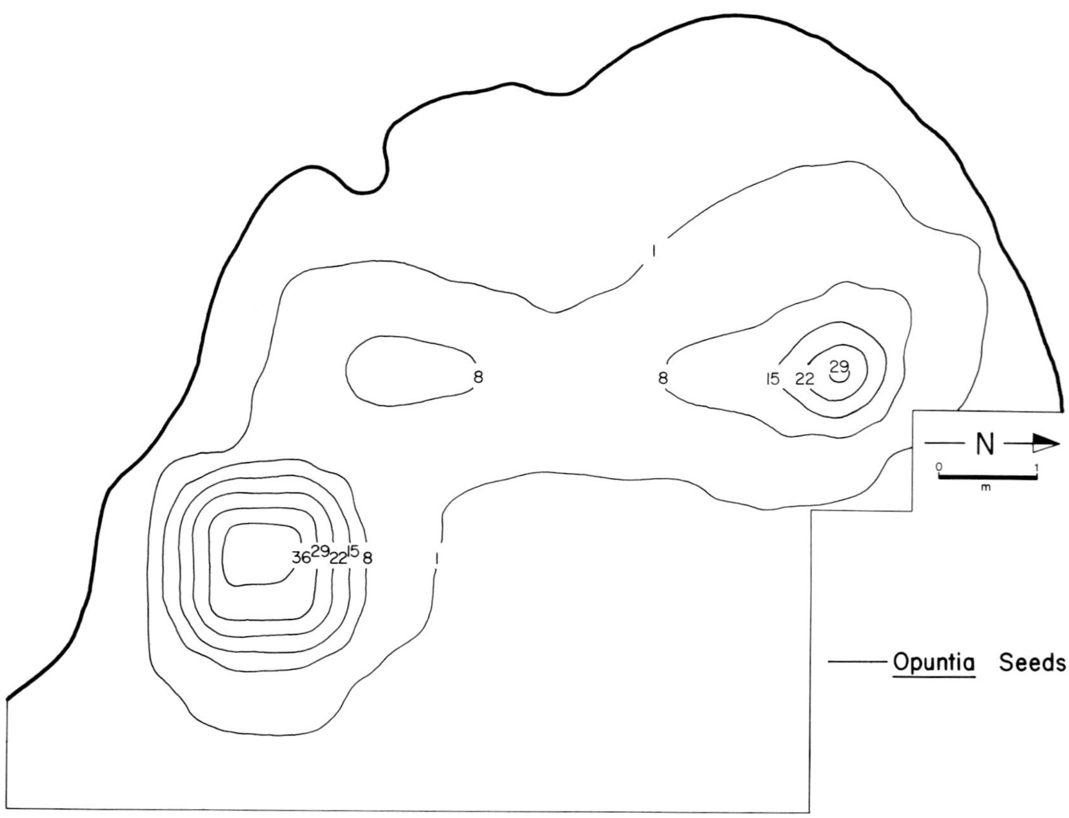

Fig. 26.6. Density contours of prickly pear seeds in Zone E.

36-seed contour) lay in the southeast quadrant of the cave, while the second densest (marked by the 29-seed contour) occurred in the northwest quadrant. Our 1-seed contour may simply reflect the inadvertent spreading around of seeds from these two clusters as people moved about the cave.

Prickly pear fruits are usually eaten raw. Could we have here another case where two persons ate fruits as they sat working in their respective activity areas? Possibly, but the situation is not like that of piñon hulls, which are discarded as the nuts are eaten. No one picks the seeds out of tunas; they are swallowed right along with the fruit (see Chapter 17). What our scatters of prickly pear seeds presumably represent, therefore, are places where clusters of uneaten fruits disintegrated. Perhaps the most overripe tunas, the ones getting soft and rotten on the bottom of the pile, simply went uneaten until they were trodden into the living floor.

Hackberry seeds were concentrated mainly in the northwest quadrant of the cave (Fig. 26.7), and it is useful to compare their distribution with that of mesquite seeds (Fig. 26.5). Both are products of the Mesquite Grassland, and their major distributions on the floor virtually overlapped. This could mean that both mesquite and hackberry were brought back by the same person and/or during the same trip downslope from the cave. A significant difference, however, is that the hackberry seeds may have resulted from eating raw fruits and spitting out the seeds, while the mesquite seeds may be a by-product of processing the jelly-like content of the pods. It is also clear that if this is the source of the hackberry seeds, the person(s) working in the northwest quadrant did most of the eating; relatively few seeds were discarded near the south wall of the cave.

Bean pod valves occurred predominantly in one scatter that peaked at more than 59 valves/square (Fig. 26.8). This suggests that regardless of how many trips were made to collect runner beans, all were threshed and discarded in one pile near the northern limits of the excavation. No other product has the same distribution, which suggests that the threshing of beans might have been incompatible with some of the other plant processing activities.

Deer and turtle bones were not particularly numerous in Zone E, with no square having more than three fragments of either animal (Fig. 26.9). The largest scatter (such as it was) occurred in the northwest quadrant of the living floor, with peaks in Squares C11 and F10. The apparent overlap between this scatter and the major concentration of flint debitage (Fig. 26.3) suggests an area where animal foods were processed. It cannot simply be a man's work area, however, because it also has evidence for the processing of a wide variety of plant genera. Perhaps, therefore, it would be best to regard the northwest quadrant of the cave as an area for general food preparation.

Two more bones appeared in Squares G4 and G5 near the south wall of the cave (Table 25.1). Because of the way the GEOG:CONTUR program works, these isolated bones did not receive contour lines. However, we have indicated their locations with a black star (deer) and a white star (turtle) in Fig. 26.9 to make a point about one difference between the GEOG:CONTUR program and the Q-mode analysis carried out by Reynolds (Chapter 28). In that Q-mode analysis, Squares G4 and G5 emerge as so similar in content that they can be regarded as belonging to a small activity area; the two bones, although not picked up by our contour program, contribute to the similarity of the squares. We mention this because it emphasizes the point that there may be statistically detectable associations among materials on a living floor that will not necessarily be revealed by density contours and which cannot be inferred by simple inspection of the distribution of materials.

Although no statistical association can be demonstrated, notched flakes were found near the 3 deer bones in Square F10 and the deer bone in G5 (Fig. 5.32).

Cottontail bones had a distribution similar to that of flint debitage (Fig. 26.10). The largest scatter (marked by the seven-bone contour) occurred in the northwest quadrant of the cave, our possible food preparation area. A smaller scatter (one-bone contour) lay near the south wall of the cave in an area with minor concentrations of plants that can be eaten raw (*susí* nuts, piñon nuts, hackberry, and prickly pear fruit). The fact that the latter area had flint and animal bone but virtually no evidence of plants that need extensive processing suggests that it might have been a man's work space, but the evidence is far from definite.

ZONE D

Flint debitage was concentrated in the center of the Zone D living floor, but with two separate peaks (Fig. 26.11). The largest of these peaks (marked by the 44-flake contour) was focused on the area of Features 2 and 3. In this area, the 14-flake contour line extended north to Feature 10, suggesting that all three features were in an area of significant flint chipping. To the east lay a second scatter of flint flakes (marked by the 14-flake contour) somewhat less impressive than the first scatter. While most of the retouched flint tools in Zone D were enclosed within the 4-flake contour line, they were no more numerous in the Feature 2–3 area than anywhere else (Fig. 5.33). Presumably, then, many of the tools flaked in the Feature 2–3 area were used elsewhere.

Acorns occurred over most of the living floor, but a glance at Fig. 26.12 shows an impressive peak centered on Square F6. This distribution is about what one would expect if a single large pile of acorns (500–600 per square) gradually became spread out over a wider area of the cave floor during the course of the occupation. Since the acorns we recovered were uneaten, they may represent the unused remainder of a pile that was once much larger.

Zone D had three pits (Features 2, 3, and 10) that appeared to have been used for acorn storage. Interestingly enough, none coincided with the peak of acorn distribution. It is possible, therefore, that the acorn pile centered on Square F6 resulted from a collecting trip that was separate from those used to fill Features 2, 3, and 10. Indeed, the fact that

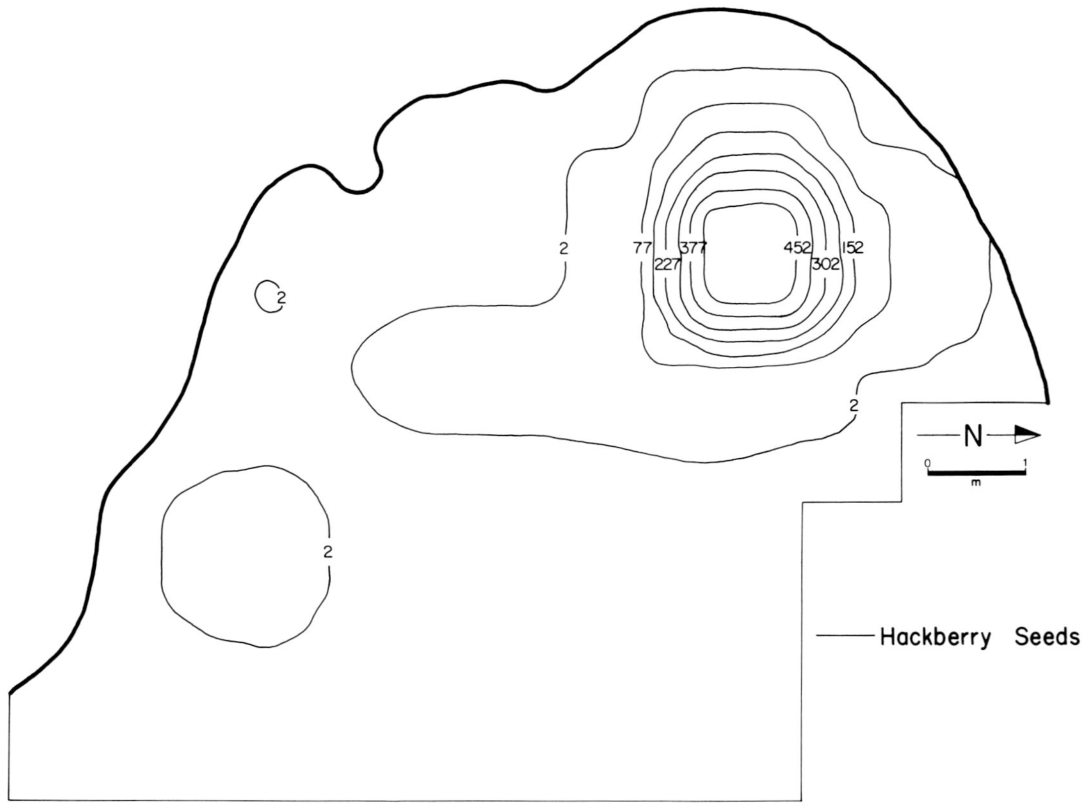

Fig. 26.7. Density contours of hackberry seeds in Zone E.

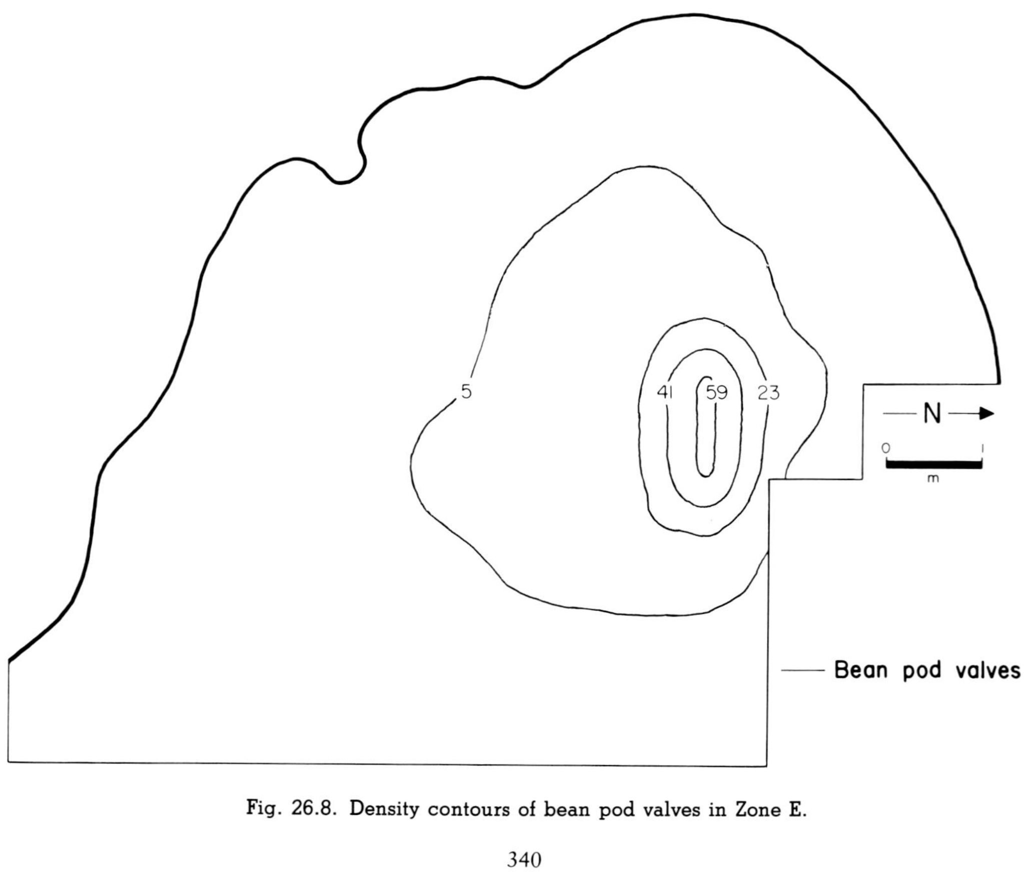

Fig. 26.8. Density contours of bean pod valves in Zone E.

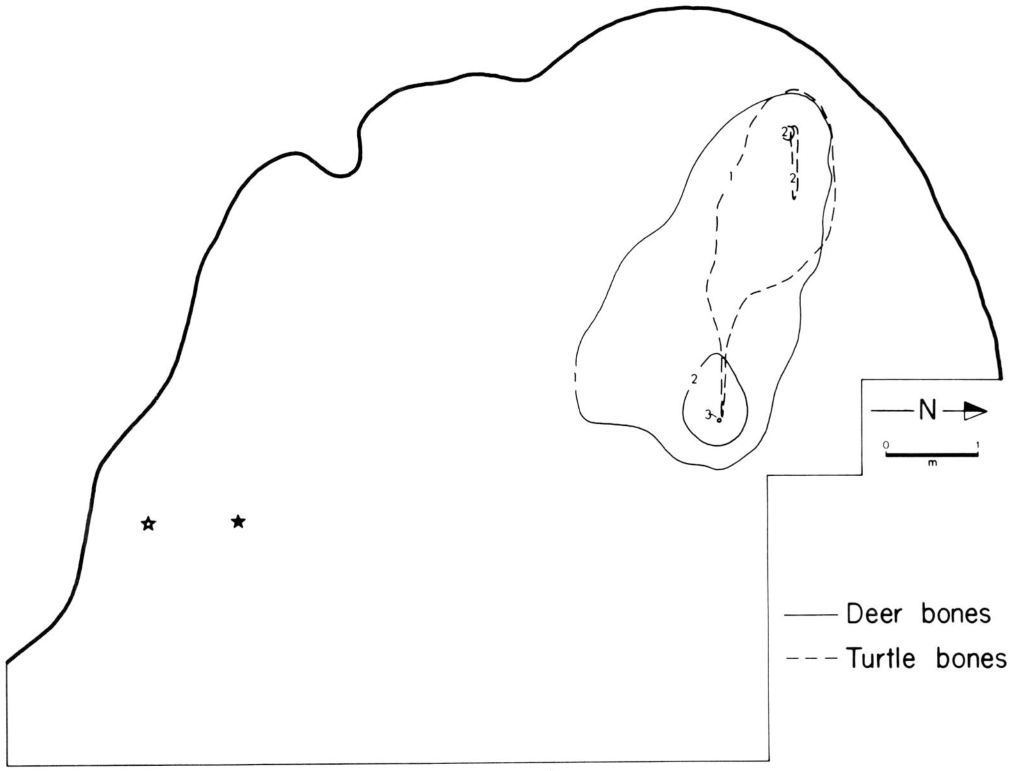

Fig. 26.9. Density contours of deer and turtle bones in Zone E.

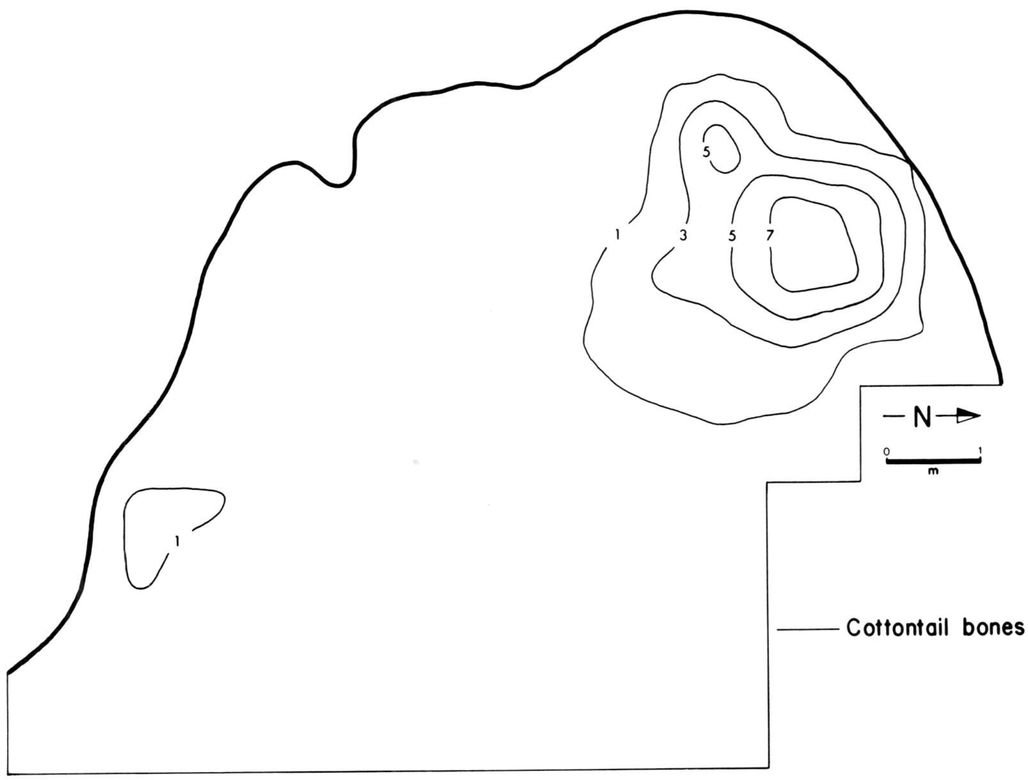

Fig. 26.10. Density contours of cottontail bones in Zone E.

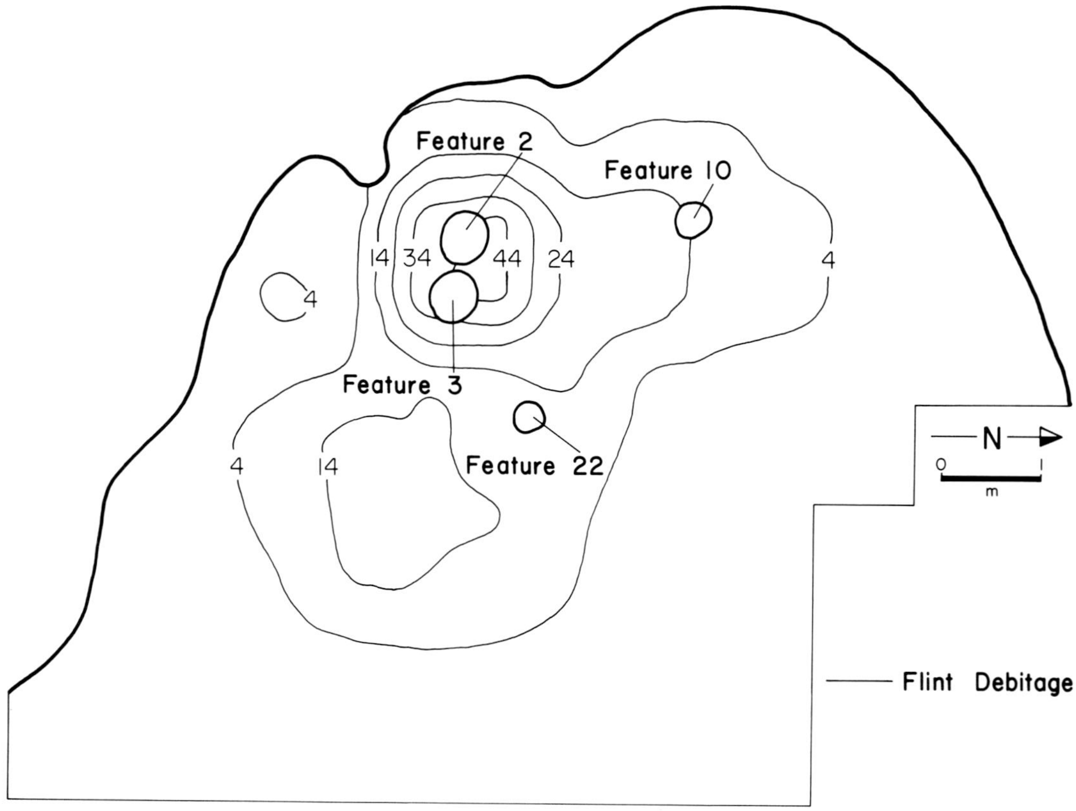

Fig. 26.11. Density contours of flint debitage in Zone D.

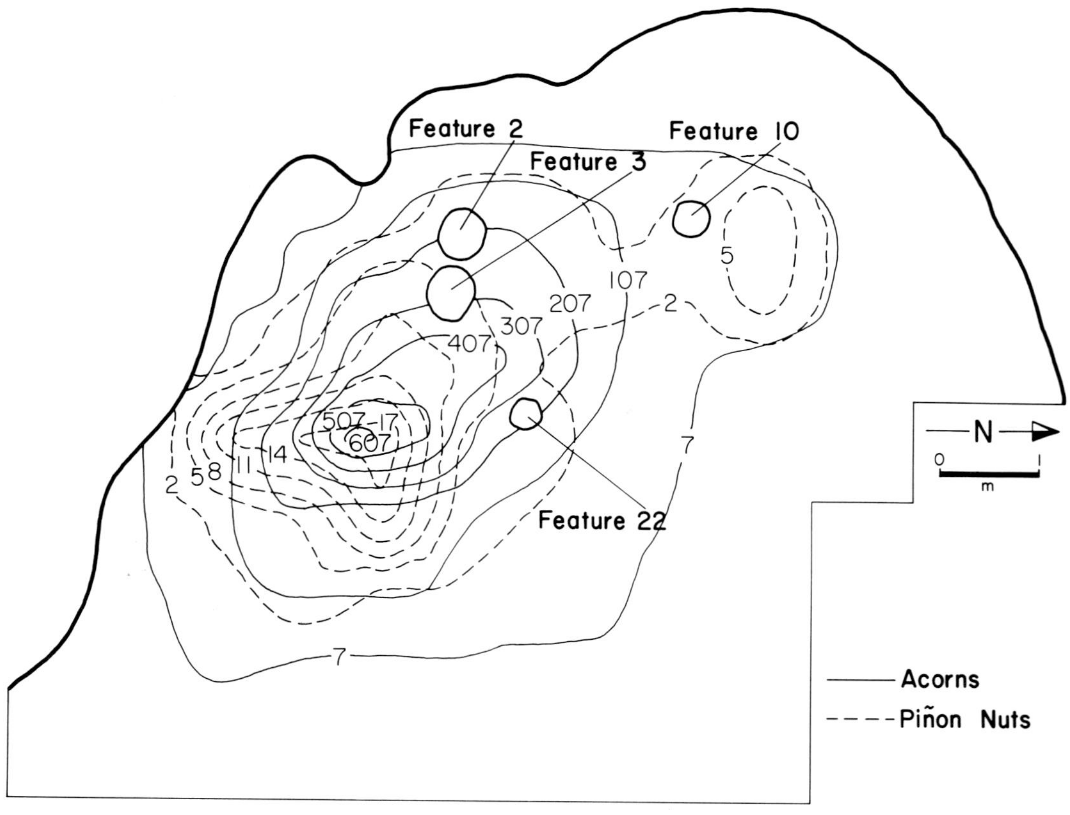

Fig. 26.12. Density contours of acorns and piñon nuts in Zone D.

thousands of these acorns never got eaten may mean that they resulted from one of the last collecting trips made before Zone D was abandoned.

Piñon nut hulls had a distribution so similar to that of acorns that they have been included in Fig. 26.12. Their peak density (17-hull contour), while much lower than that of acorns, was also centered on Square F6. In this case, however, the hulls represent food eaten, rather than food left behind uneaten.

Prickly pear stems (nopales) were found mainly in the vicinity of Features 2, 3, and 22 (Fig. 26.13). Nopales need to be cooked (some were roasted on wooden skewers), and their distribution suggests that the area just south of the Feature 22 hearth may have been used for their spine removal and peeling.

Agave quids had an interesting distribution, with two peaks of approximately equal density (Fig. 26.14). One peak (seven-quid contour) fell adjacent to Feature 22, a hearth. The second peak (nine-quid contour) occurred 2 m to the south. The lower-frequency contour lines can be seen as resulting from a gradual spreading out of quids from those two peaks.

In later levels, such as Zone B1, it will be seen that this multiple-peak distribution was typical of *Agave* quids. We suggest that the occupants of the cave chewed maguey as they worked, leaving small piles of expectorated quids that indicate the locations where various individuals sat. In the case of Zone D, such piles may reflect two individuals sitting about 2 m apart, not far from the Feature 22 hearth.

Susí nut hulls have been included in Fig. 26.14 because their distribution, while not including two separate peaks, overlapped strongly with that of *Agave* quids. The 30-hull contour covers both *Agave* peaks, and the *susí* hull peak (more than 37 per square) fits within the cluster of quids near Feature 22. Perhaps, therefore, the same two individuals mentioned above ate *susí* nuts and discarded the hulls along with their quids.

Mesquite seeds had a distribution that centered on Feature 10, a storage pit in the northwest quadrant of the cave (Fig. 26.15). Declining from a peak of more than 16 seeds per square around that feature, the higher mesquite contour lines run south to the vicinity of another storage pit, Feature 3; from there, the seeds spread out in every direction at lower frequencies.

During the course of excavation we found acorns in Feature 10, and in Chapter 5 it was interpreted as an acorn storage pit. While this was a reasonable conclusion, Fig. 26.15 suggests that the pit might have been used at an earlier time to store mesquite. It will be remembered that the mesquite harvest season is from July to September, while the acorn harvest is from November to January. Thus Feature 10 could have been used to store mesquite pods in September, emptied when the jelly was extracted and the mesquite seeds discarded nearby, then used for acorn storage in November. Clearly, we should consider the possibility that some storage pits at Guilá Naquitz had multiple, sequential functions.

Hackberry seeds had a two-peak distribution (Fig. 26.16) as interesting as that of *Agave* quids, especially in light of our data on mesquite. Like mesquite, hackberries are products of Mesquite Grassland B, and one of their distributional peaks (marked by the 49-seed contour) lay adjacent to Feature 10. The other peak (marked by the 76-seed contour) occurred just to the southeast of Feature 3; the remaining hackberry scatter can be seen as the gradual spreading out of seeds from each of the two prominent peaks.

Since hackberries are eaten raw, their distribution probably reflects the spitting out of seeds by two different individuals as they sat working in the cave—a situation similar to the one we have suggested for maguey quids. Note, however, that while the hackberry peaks seem related to Features 10 and 3, the *Agave* peaks seem related to Feature 22 and an area 2 m to its south (Fig. 26.14). We suggest, therefore, that two different harvest seasons may be involved. Perhaps early in the occupation, following harvests in Mesquite Grassland B, someone seated near Feature 10 processed mesquite while eating hackberries; a second person sat near Feature 3. Later in the occupation, following harvests in Thorn Forest A, two individuals seated near a hearth (Feature 22) chewed maguey. Of course, we could be dealing with the same two individuals at two different times of the year.

Cucurbit fragments had a distribution not isomorphic with that of any other plant—east of Feature 22 and near the mouth of the cave. One *Cucurbit pepo* seed from Square F5 did not appear in the GEOG:CONTUR printout because it was an isolated specimen; we have indicated its location in Fig. 26.16 with a black star. Since we have evidence for cucurbit domestication in Zone D (see Chapter 20), it may be that the unique distribution of that plant resulted from a separate harvest in a cultivated area near the cave, unrelated to the major wild plant harvests.

Bean pod valves were very lightly distributed over the central part of the living floor (Fig. 26.17). Their contour lines overlap those of cucurbits, but the shapes of the two areas are quite different, suggesting some differences in discard (and hence, perhaps in harvesting and/or processing as well).

Deer bones had a two-peak distribution (Fig. 26.17). The densest scatter lay near the south wall of the cave, not far from one of the *Agave* quid piles; this may have been an animal processing (i.e., butchering and/or eating) area. A more diffuse scatter surrounded Feature 10, a possible food preparation area where mesquite appears to have been processed.

It will be remembered that notched flakes were found near deer bones in Zone E. For what it is worth, three of the four notched flakes found in Zone D also occurred near scatters of deer bone, though no association can be statistically demonstrated.

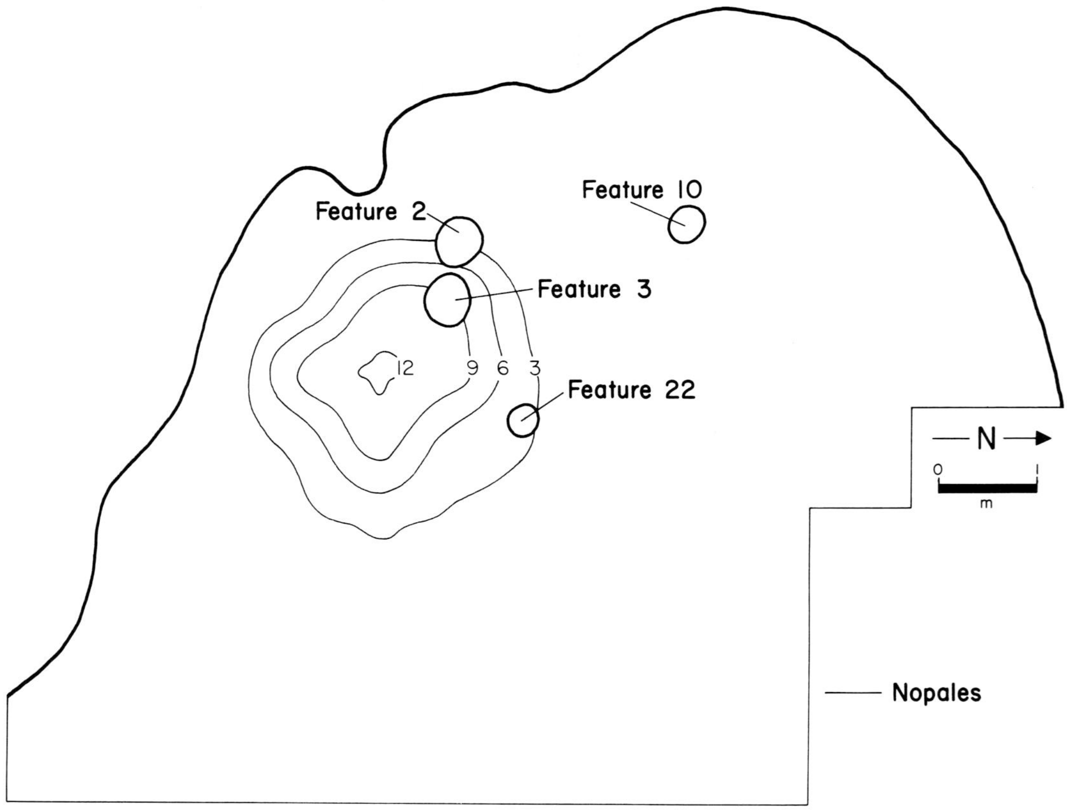

Fig. 26.13. Density contours of prickly pear stem sections (nopales) in Zone D.

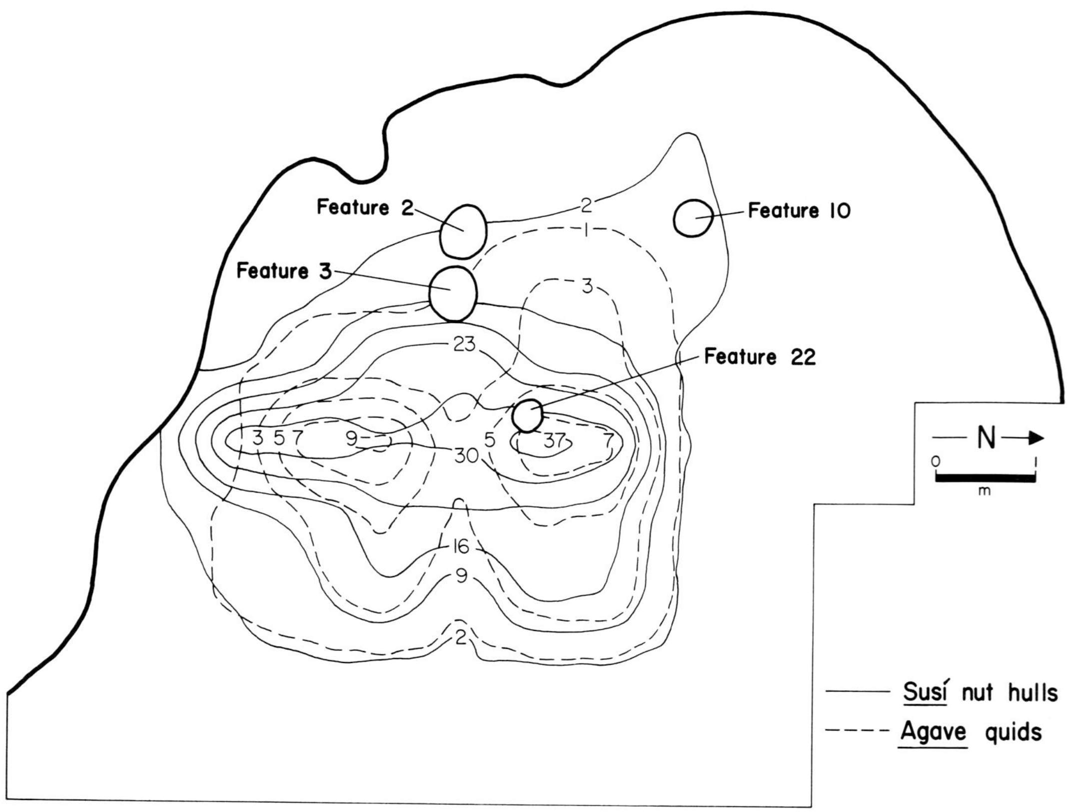

Fig. 26.14. Density contours of *susí* nut hulls and maguey quids in Zone D.

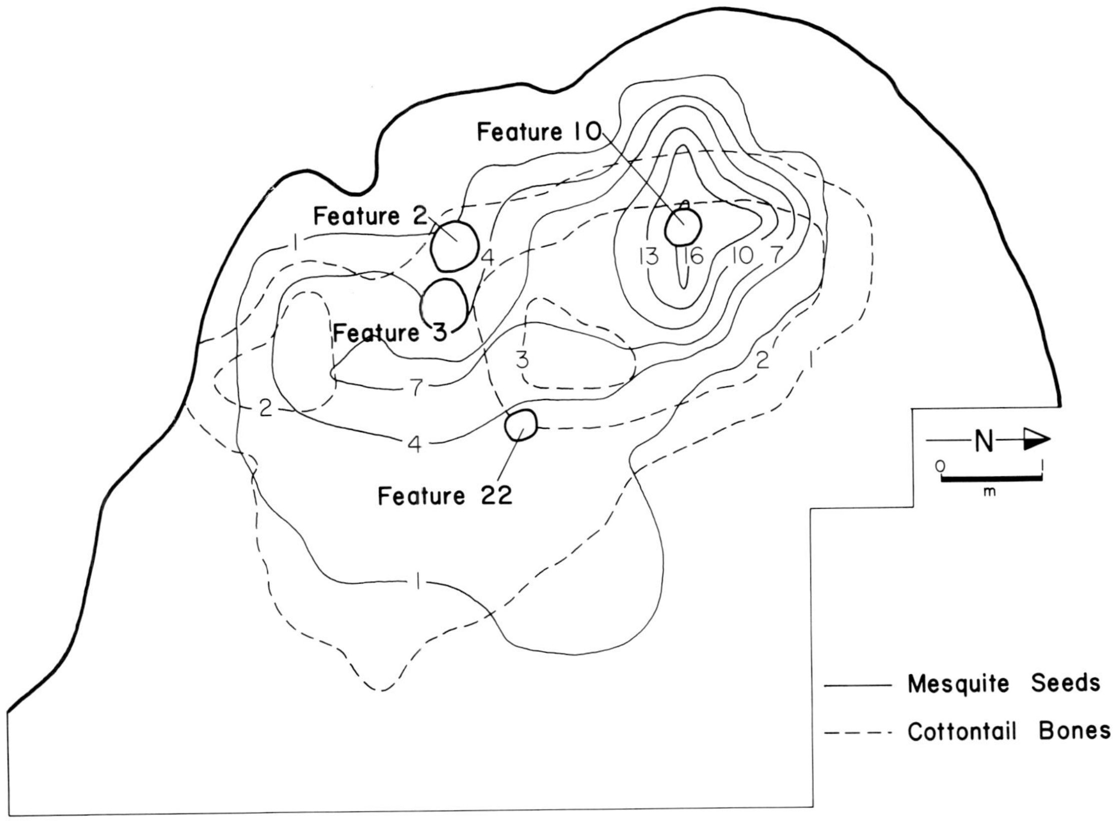

Fig. 26.15. Density contours of mesquite seeds and cottontail bones in Zone D.

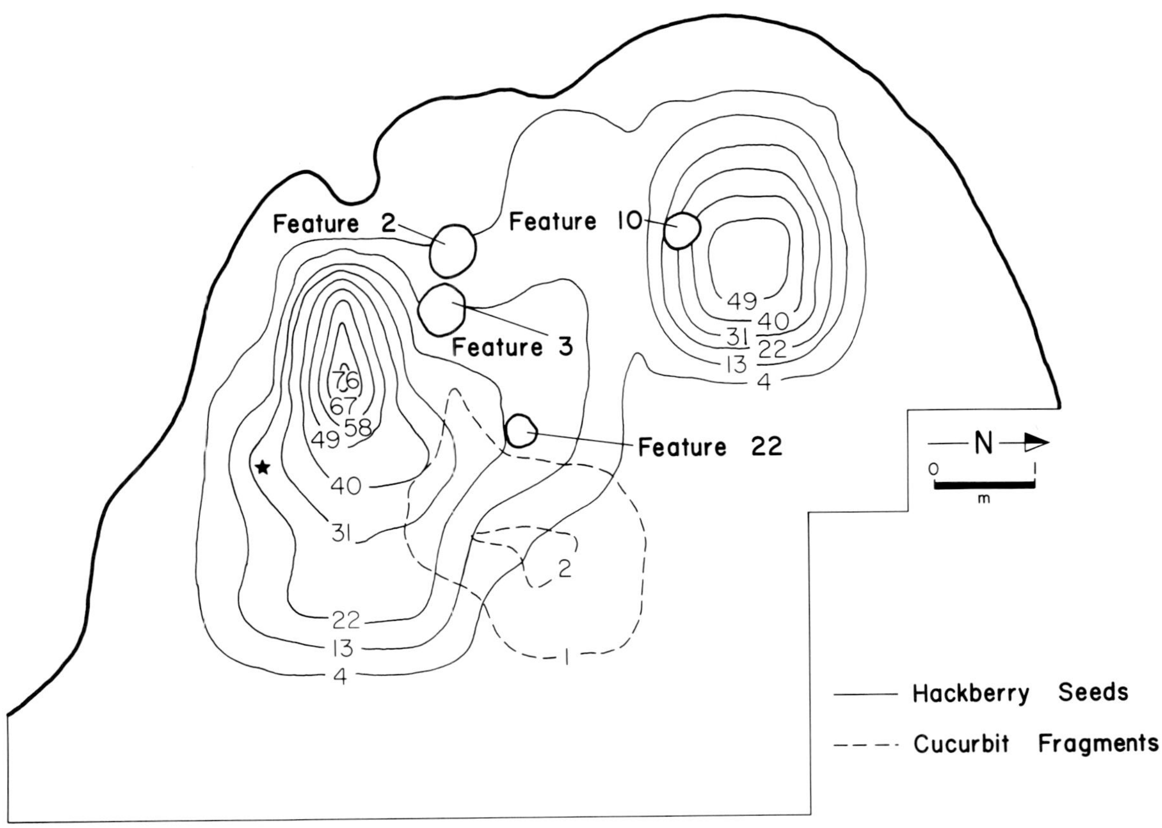

Fig. 26.16. Density contours of hackberry seeds and cucurbit fragments in Zone D.

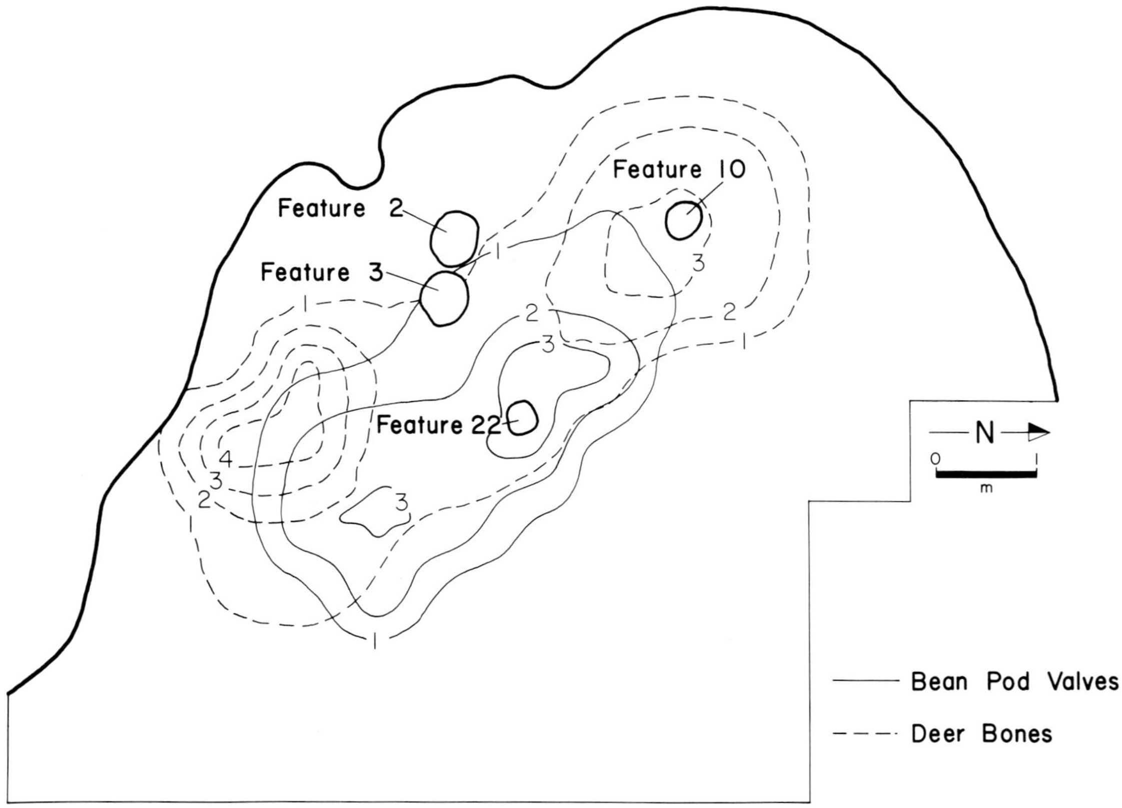

Fig. 26.17. Density contours of bean pod valves and deer bones in Zone D.

Cottontail bones showed a distribution (Fig. 26.15) not unlike that of deer. One scatter linked Features 10 and 22, with its peak density occurring between those features. A smaller scatter lay near the south wall of the cave, overlapping one of our scatters of deer bone. While no clear association can be demonstrated, a Variety C biface was also found in this area (see Fig. 6.38).

Our suspicion is that this area along the south wall of the cave may have been used primarily as a man's work space, with animals processed, quids chewed, and meat, raw fruits, and nuts eaten, but without evidence for plants needing extensive processing. That would leave the area circumscribed by Features 2, 3, 10, and 22 as a possible plant storage and plant-and-animal food preparation area, utilized mainly by women.

ZONE C

Flint debitage displayed a complex pattern in Zone C, with multiple peaks and valleys (Fig. 26.18). Perhaps the most interesting lacuna came in Square D9, where the 15-flake contour made a wide detour around an area of much lower density. It will be remembered that during the excavation (Chapter 5) we suggested that Square D9 had been the locus of a surface campfire, which could explain the low frequency of tools there.

Three peaks of debitage (more than 25 flakes/square) occurred in Squares D11, E7, and F9. In contrast to some zones, these squares also had fair numbers of retouched tools: (1) 2 steep denticulate scrapers, a notched flake, and a flake with sheen in D11, (2) a crude blade, a notched flake, and a utilized flake in E7, and (3) a sidescraper/knife, a utilized flake, and a flake core in F9 (Fig. 5.34). Square D11 also had a hammerstone, perhaps for percussion flaking.

While the Zone C pattern is rather complex for a one-sentence explanation, it would appear that there were at least three areas of flint chipping and/or discard arranged in a semicircle around our presumed campfire, equidistantly spaced at intervals of 2 m from each other and from the fire.

Acorns were widespread in Zone C, but their distribution had two clear peaks that probably represent piles of acorns left on the living floor (Fig. 26.19). The larger pile (marked by the 405-acorn contour) occurred in the north-central part of the cave near our presumed campfire, while the smaller pile (marked by the 105-acorn contour) lay near the south wall and surrounded a mortar fragment (Fig. 5.34). The remaining acorns could easily have resulted from the gradual spreading out of these two piles during the course of the occupation.

Piñon nut hulls had a distribution so close to that of acorns that we have included them in Fig. 26.19. They occurred mainly in two scatters, one of which is virtually isomorphic

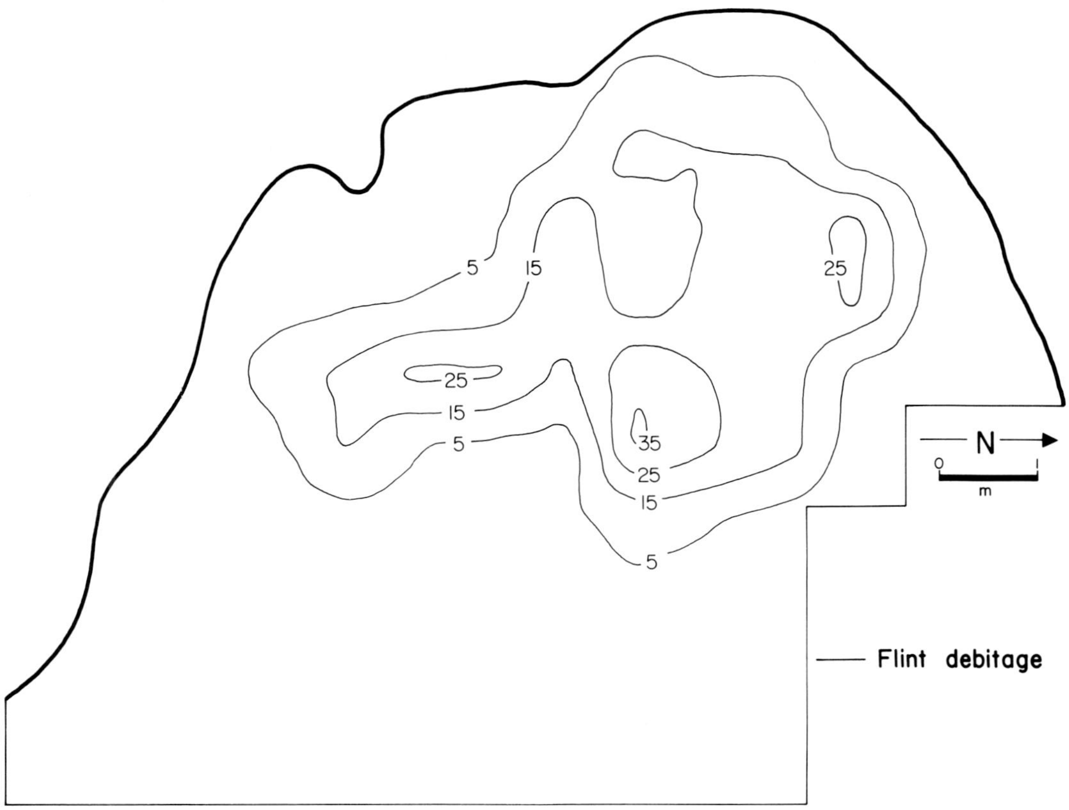

Fig. 26.18. Density contours of flint debitage in Zone C.

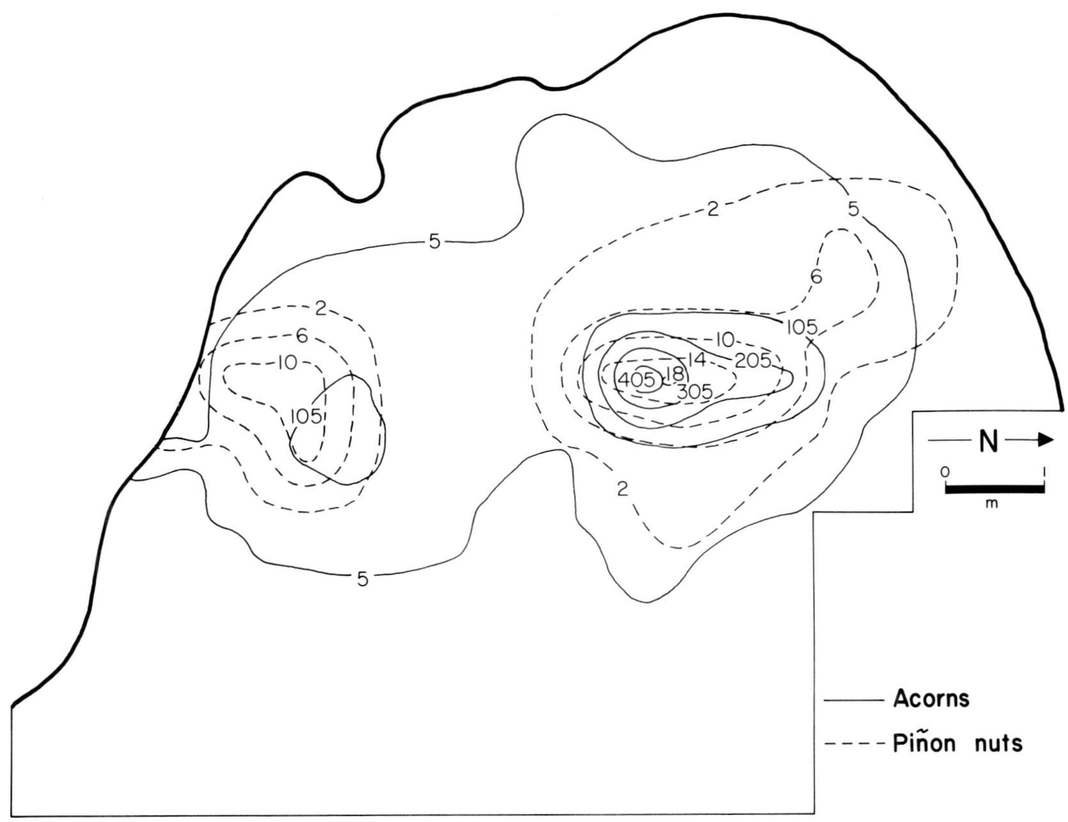

Fig. 26.19. Density contours of acorns and piñon nuts in Zone C.

with the larger acorn pile and the other of which overlaps with the smaller acorn pile. Their close association might result from their having been harvested at the same season in the same general environment.

Agave quids displayed one prominent peak that was centered in Square F9 (Fig. 26.20); outside this square, the dropoff in density was precipitous. Evidently someone seated in or near F9 had expectorated more than 100 quids—possibly the same person who had left the heap of flint debitage in that area.

Opuntia seeds have been included in Fig. 26.20 because they also displayed one prominent peak. In this case, however, the pile of disintegrating prickly pear fruits evidently lay near the south wall of the cave. There were other small scatters elsewhere in the cave, but none compared with the pile already mentioned, which reached a peak of more than 200 seeds per square.

Hackberry seeds offered us our third example of a plant whose distribution was characterized by a single impressive peak (Fig. 26.21). More than 600 seeds were found in Square E9 (Table 25.3) next to our presumed campfire and immediately adjacent to the flint debitage and *Agave* quid concentrations in Square F9. Perhaps the same individual (or individuals) responsible for the F9 concentrations had eaten more than 100 g of hackberries as well.

Nanches had a distribution not unlike that of acorns (Fig. 26.21). A modest concentration (marked by the 32-seed contour) appeared to the northeast of our presumed campfire, adding one more fruit to the piles of hackberries and *Agave* quids already found in that area. A smaller concentration (marked by the 22-seed contour) occurred along the south wall of the cave in an area where prickly pear and piñon nuts had also peaked. Thus, at least two individuals were probably responsible for most of the nanche eating in Zone C.

Mesquite seeds and guaje pods had distributions so similar that we have included them on the same map (Fig. 26.22). Each had a single peak centered in the Square E9–F9 area, surrounded by contours of lower density that extend northwest, east, and south in much the same way as the flint debitage scatter in Fig. 26.18.

In addition to increasing the already dense concentrations of debris to the east of our presumed surface campfire, the mesquite and *guaje* distributions contribute some valuable clues concerning the use of the cave during Zone C times. Both plants are tree legumes, but they grow mainly in different vegetational zones—*guajes* in Thorn Forest A, mesquite in Mesquite Grassland B. They also differ in the amount of processing required, for *guaje* seeds can be removed from the pod and eaten raw, while mesquite in Oaxaca is usually cooked up into an edible syrup. However, their seasons of harvest can overlap in August, give or take a month.

What Fig. 26.22 suggests is that the pods of two different genera of tree legumes, growing perhaps 3–4 km apart and probably collected during different harvesting trips, were both processed at virtually the same spot on the living floor. This strongly suggests that specific areas of the cave were considered appropriate for processing certain categories of plants, regardless of where the plants came from. Further, the presence nearby of a major concentration of acorns suggests that this area was considered appropriate for plant processing regardless of the season involved. On the basis of ethnographic analogy, we might further suggest that this space to the east of our presumed campfire was a food preparation area used primarily by women. As if to support this possibility, two manos were found in Square F9 (Fig. 5.34).

Bean pod valves and susí nut hulls were also concentrated in the plant processing area mentioned above, with minor extensions to the south and east (Fig. 26.23). Both *susí* bushes and runner beans can be found in the same Thorn Forest A environment with oak trees, so we would not be surprised to learn that they were collected during the same trips made to procure acorns.

Cucurbits and nopales occurred over a small area of the cave, both reaching their peak in Square F9 along with mesquite, *guajes*, and maguey quids (Fig. 26.24). Nopales require spine removal, peeling, and cooking, so it is no surprise to see them concentrated in this plant-processing area; it will be remembered that a nopal skewered on a stick, as well as three more possible burned nopal skewers, were also found here (Fig. 5.34). The opening of the cucurbits and the toasting of the cucurbit seeds were evidently done by the same people (presumably women) who processed the prickly pear stems.

By now it is becoming clear that the area along the south wall of the cave—the only other locality where debris had consistently seemed to cluster—was never used for major plant processing and hence might have been a men's work area.

Deer and cottontail bones had their own distinctive distribution, which was discontinuous (Fig. 26.25). The larger scatter virtually surrounded our presumed campfire, with the peak for deer bone actually occurring between the fire and the west wall of the cave. Lower peaks for rabbit bone appeared (1) near the west wall and (2) between the fire and the north wall. Evidently the area around the presumed Square D9 campfire had been used for the preparation of both plant and animal foods; and while most plant debris had been left between the campfire and the mouth of the cave, many small animal bones had been discarded toward the back wall of the cave.

Our second, and smaller, concentration of animal bone appeared near the south wall of the cave in the locality we have already suggested might be a men's work area. There were enough flint tools nearby to raise the possibility that a few animals were processed here before being carried to the food preparation area (Fig. 5.34), although some of the bones might be the remains of meals. Whoever worked in this space had also eaten and discarded significant quantities of raw fruit, nuts, and cooked maguey, but apparently without participating in the processing of bulky plant materials.

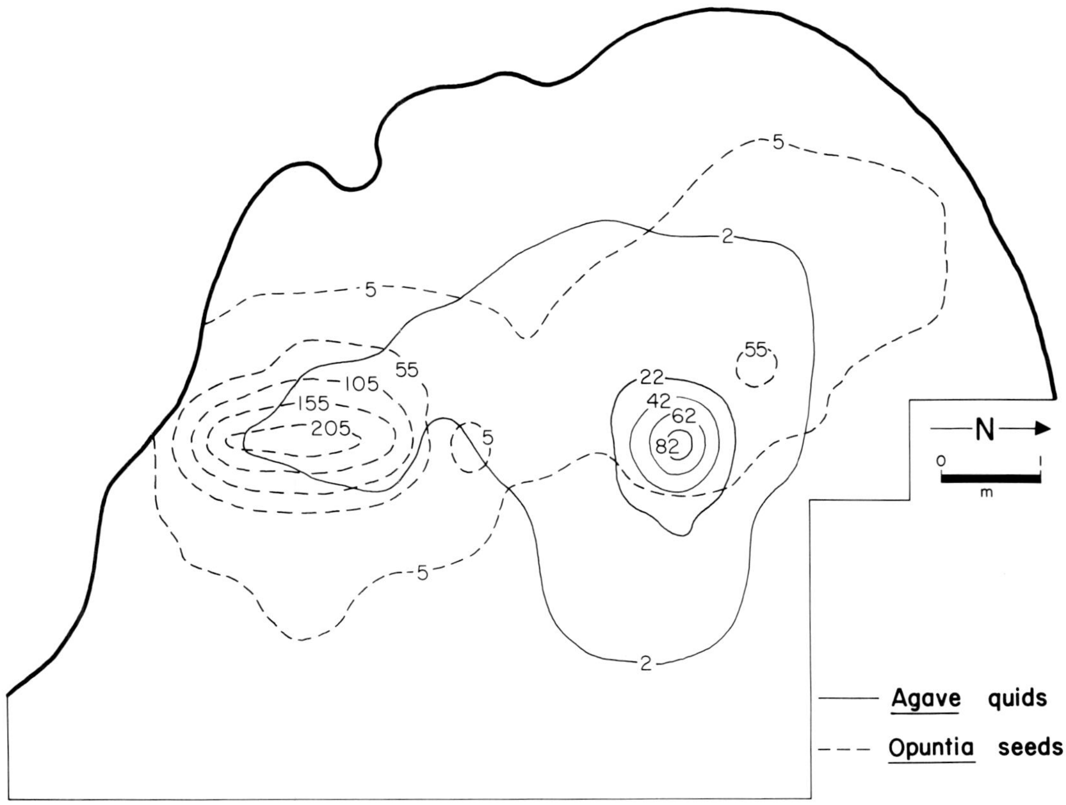

Fig. 26.20. Density contours of maguey quids and prickly pear seeds in Zone C.

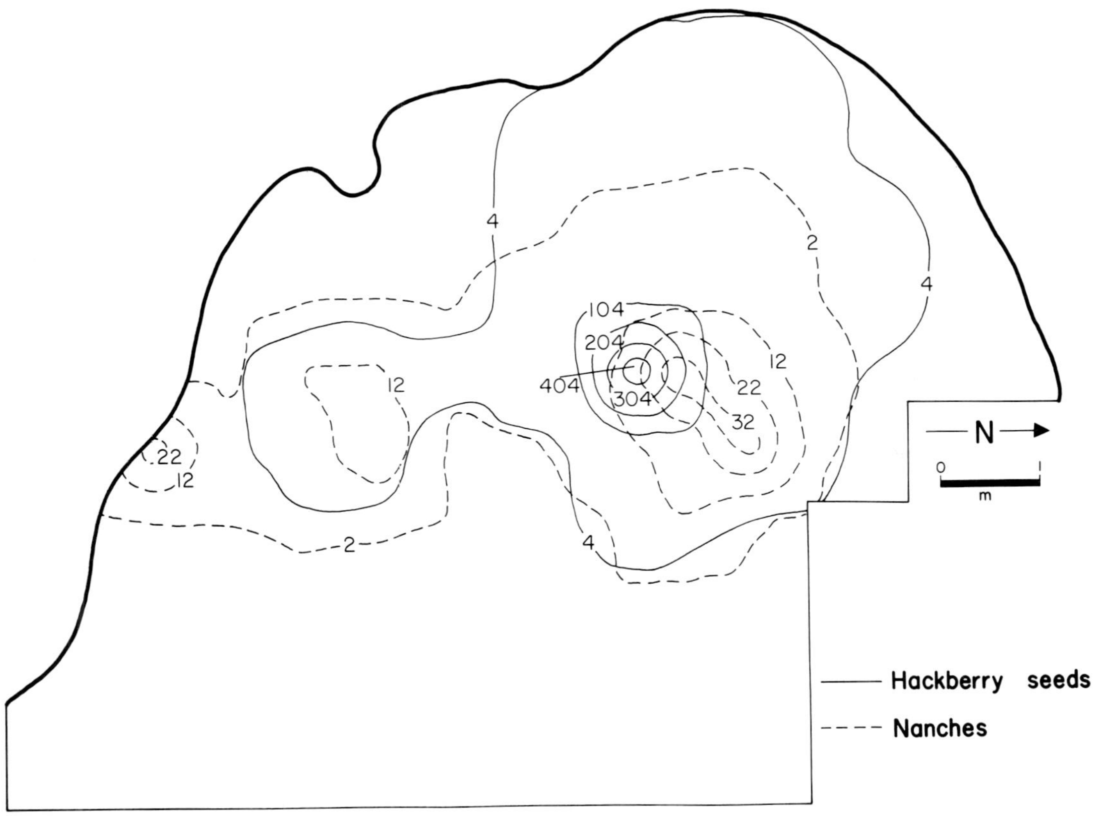

Fig. 26.21. Density contours of hackberry and nanche seeds in Zone C.

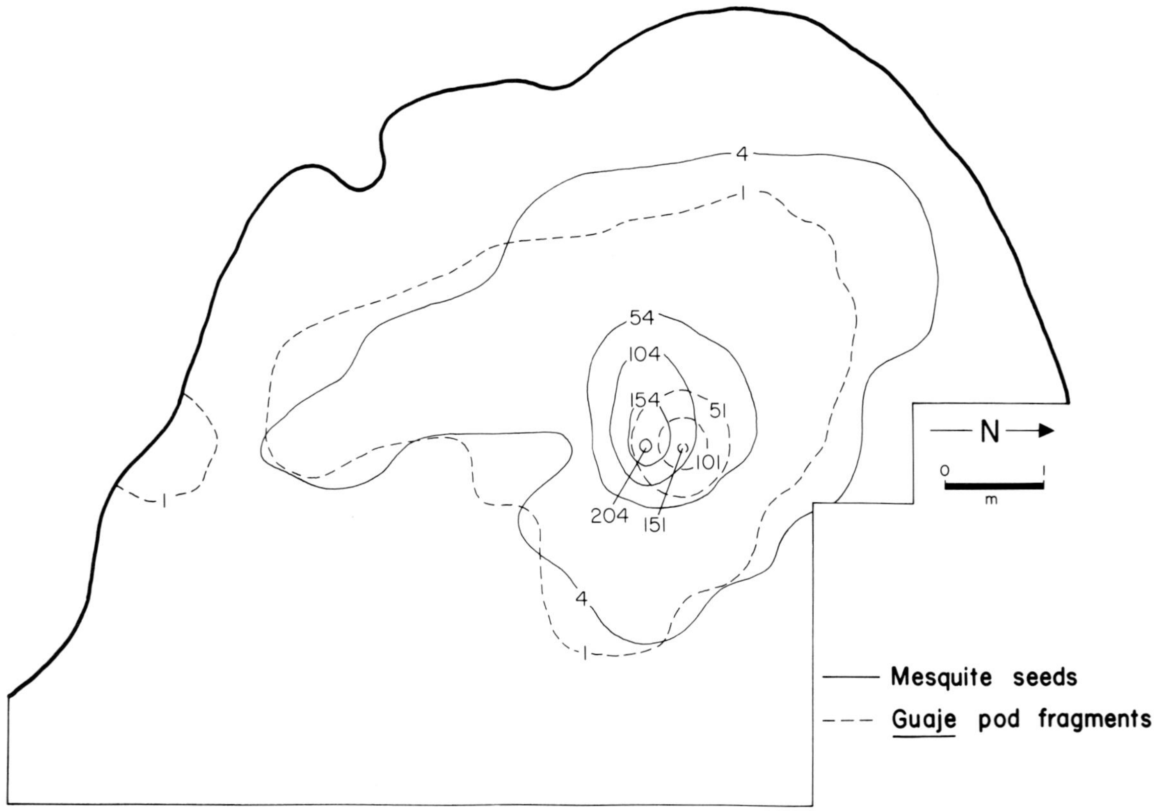

Fig. 26.22. Density contours of mesquite seeds and *guaje* pods in Zone C.

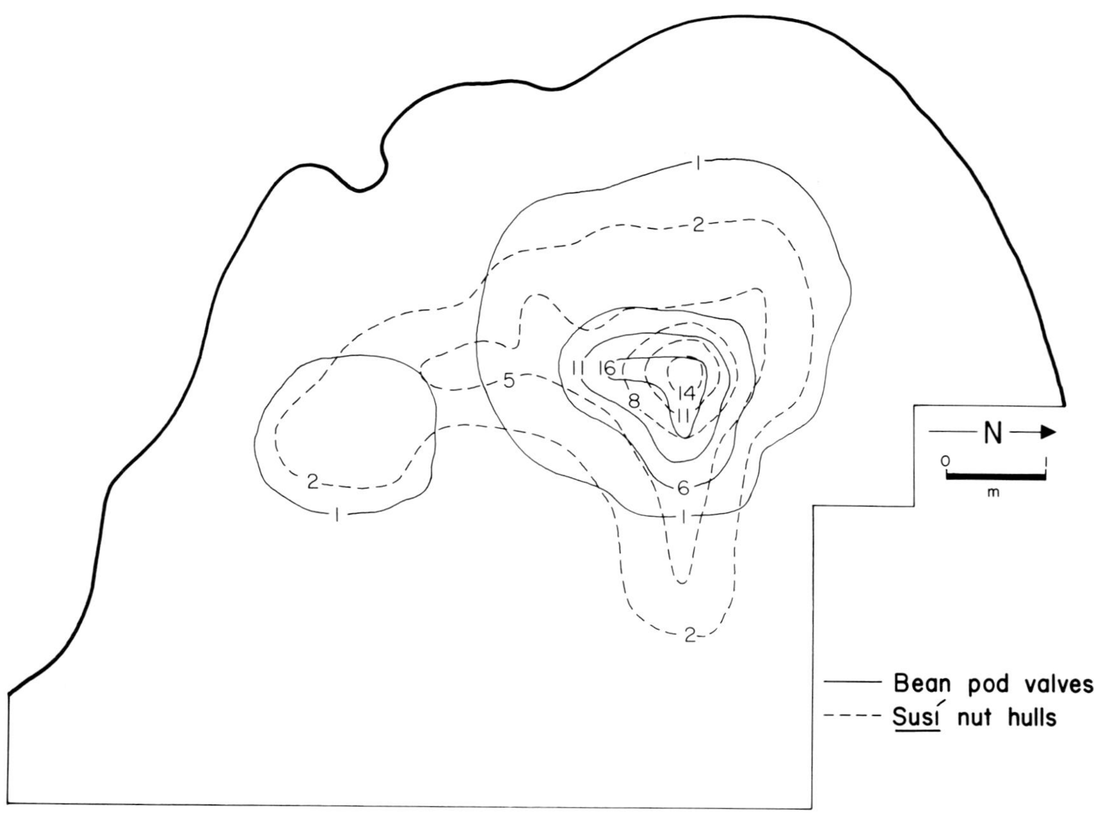

Fig. 26.23. Density contours of bean pod valves and *susí* nut hulls in Zone C.

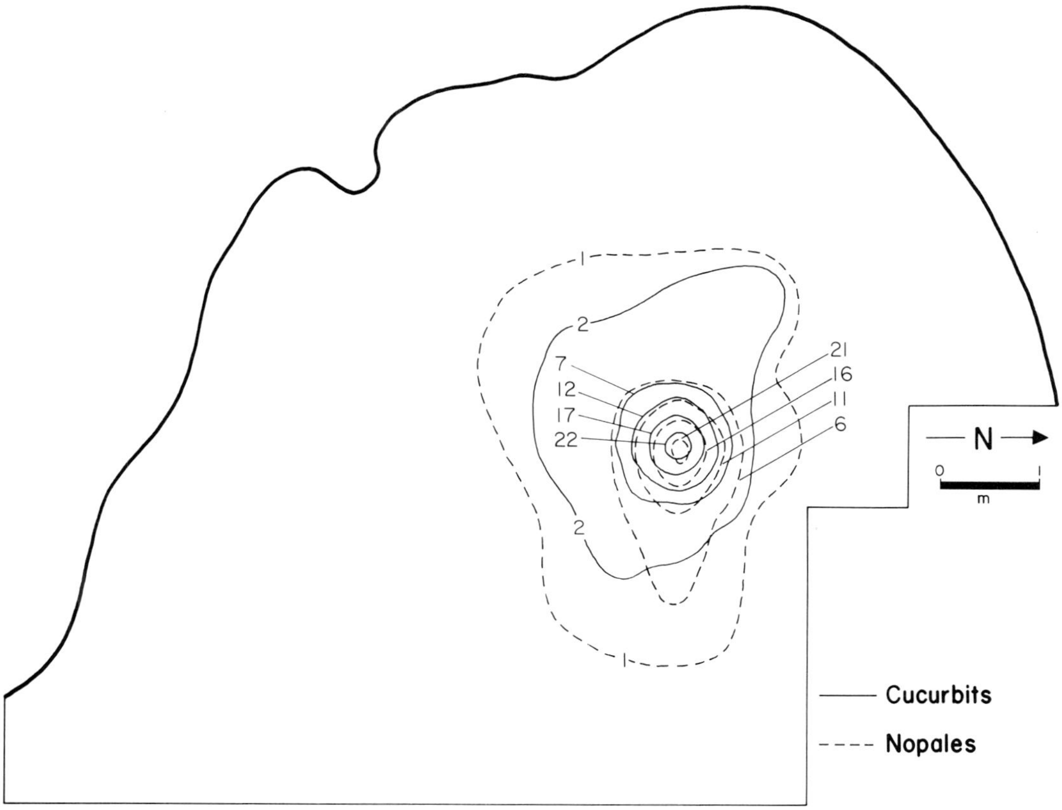

Fig. 26.24. Density contours of cucurbit fragments and prickly pear stem sections (nopales) in Zone C.

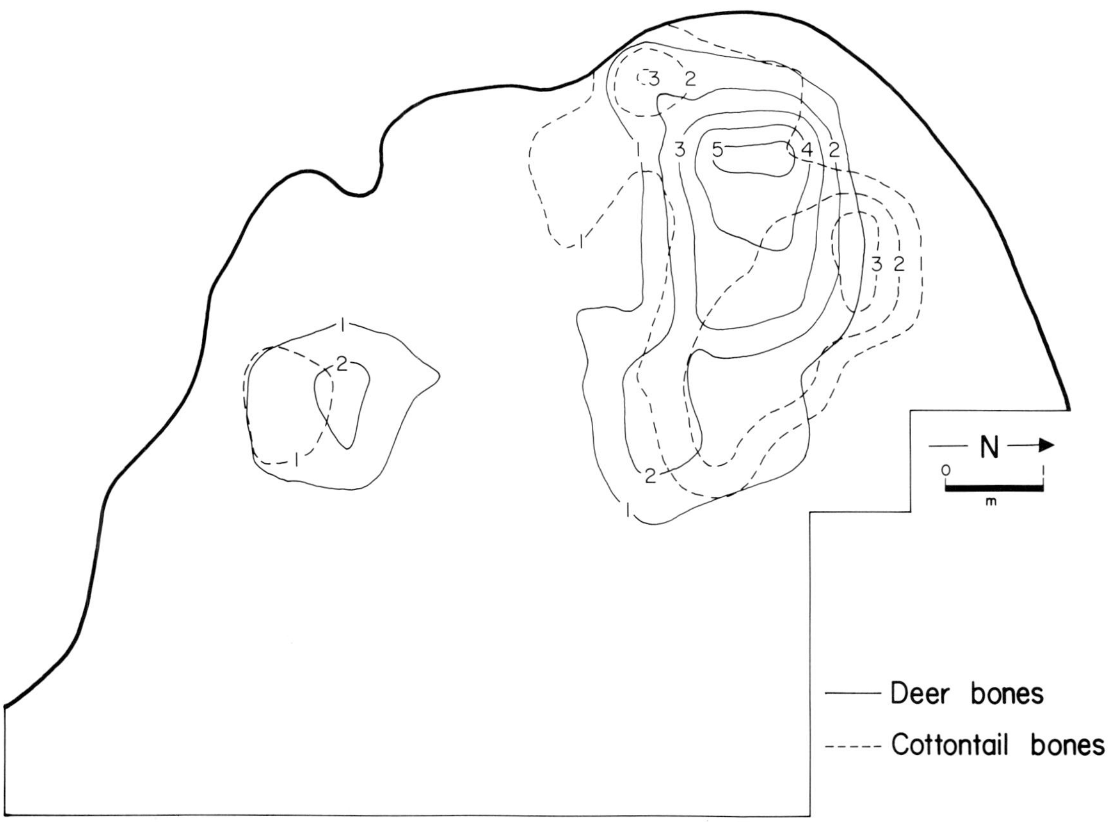

Fig. 26.25. Density contours of deer and cottontail bones in Zone C.

ZONE B3

Flint debitage had a discontinuous distribution in Zone B3, with perhaps as many as four small concentrations (Fig. 26.26). Most of these occurred in a semicircular area near the rear wall of the cave and were accompanied by core fragments and retouched tools (Fig. 5.36).

Acorns and susí nut hulls both peaked in Square D9 and have been included on the same contour map (Fig. 26.27). In the case of the acorns, we are probably dealing with a pile of several hundred left unprocessed on the cave floor, gradually expanding outward at lower densities as they were disturbed by the activities going on around them. A separate scatter of acorns near the south wall of the cave overlapped with one of the small flint clusters seen in Fig. 26.26. As for the *susí* hulls, they occurred in much smaller numbers and may have lain intermixed with the acorns mainly because they were collected in the same vegetational facies.

Hackberry seeds occurred in two separate concentrations, perhaps indicating places where various individuals ate the raw fruit and expectorated the seeds (Fig. 26.28). One scatter lay near the south wall of the cave, the other in and around Square C9; both areas had flint and acorn concentrations as well.

Agave quids also occurred in two scatters (Fig. 26.28). The densest of these was centered on Square E5 near the south wall of the cave, overlapping one of our hackberry concentrations. The other scatter was lighter and more diffuse, generally following the semicircle of flint debitage seen in Fig. 26.26. The denser concentration in E5 suggests that the same person eating hackberries near the south wall was also chewing maguey.

Mesquite seeds displayed peak concentrations in Squares D9 and E7, which are somewhat smoothed over by the contours in Fig. 26.29. These concentrations may represent two areas where mesquite was processed, and it is instructive to compare them with the distributions of *guajes* and beans (see below).

Bean pod valves and guaje pods had distributions that were literally isomorphic (Fig. 26.30). Their joint peak occurred in Square E7, where they overlapped with the second largest mesquite concentration. Thus Zone B3 illustrates one of the same phenomena noted for Zone C: The pods of three different leguminous genera were all processed and/or discarded at the same place on the living floor, regardless of the vegetational facies in which they had been harvested.

A fragment of knotted net, probably part of a collecting bag, was found in the south profile of Square E7 (Fig. 5.36). While no statistical association can be demonstrated, this fragment comes from the type of bag we assume was used to transport the pods of mesquite, *guaje*, and bean from the harvest areas to the cave, so its presence in E7 is at least interesting.

Deer and cottontail bones were not numerous in Zone B3, and their distribution was discontinuous (Fig. 26.31). The largest concentration of deer bone was in Square C6, where it overlapped with the most southwesterly debitage concentration shown in Fig. 26.26. Since nothing else was found in this square, C6 might contain the debris from an animal processing (butchering and/or eating) area.

Scatters of deer and cottontail bone also overlap with debitage in the northwest quadrant of the cave, an area already characterized by concentrations of mesquite, hackberry, acorns, and *susí* hulls. By analogy with other levels, this may have been an area of food preparation and/or the discard of processing refuse.

Parenthetically, we might add that two pieces from broken projectile points (as well as five possible segments of dart mainshaft) were also found in the northwest quadrant of the cave (Fig. 5.36). While most of these fragments of broken hunting equipment overlapped with scatters of bone or flint, the overall aspect of the area is one of general discard, rather than of a work space devoted specifically to the results of hunting.

ZONE B2

Flint debitage appeared mainly in the northwest quadrant of the cave (Fig. 26.32). Within the oval area defined by the five-flake contour, however, there were three clear peaks near Squares C6, D8, and C11. During the course of excavation, we had noted that C6 appeared to be an area where numerous chipped-stone and ground-stone tools were discarded against the back wall of the cave (see Fig. 5.37). The peaks near D8 and C11, however, could be areas where some tool preparation was done. Further reference to Fig. 5.37 shows that numerous cores and retouched tools also occurred in and around both the D8 and C11 peaks. In the paragraphs that follow, we present evidence that two different individuals (or one individual at two different times) may have been responsible for these two scatters of debitage.

A much smaller scatter of debitage lay near the south wall of the cave. As we shall see later, this scatter may have been left by still another individual.

Acorns and susí nut hulls, two products of Thorn Forest A, both were concentrated in the northwest quadrant of the cave (Fig. 26.33). Each seemed to consist of a single pile, with acorns peaking in Square C9 and *susí* in Square D8. The scatter of empty *susí* hulls, in other words, covers the D8 debitage concentration and may have resulted from nuts eaten by that flint knapper.

Nanches and Agave quids had distributions so similar that they have both been included in Fig. 26.34. Both exhibited major concentrations in the northwest quadrant of the cave and minor scatters near the south wall. Nanche seeds reached their peak (the 18-seed contour) in D9, adjacent to the debitage and *susí* nut scatter described above. *Agave* quids covered the whole area from the D8 flint scatter to the margins of the C11 flint scatter.

We have suggested that two individuals may have worked in the northwest quadrant of the cave. These scatters suggest that both may have chewed maguey as they worked, while the individual working in the D8–D9 area may have

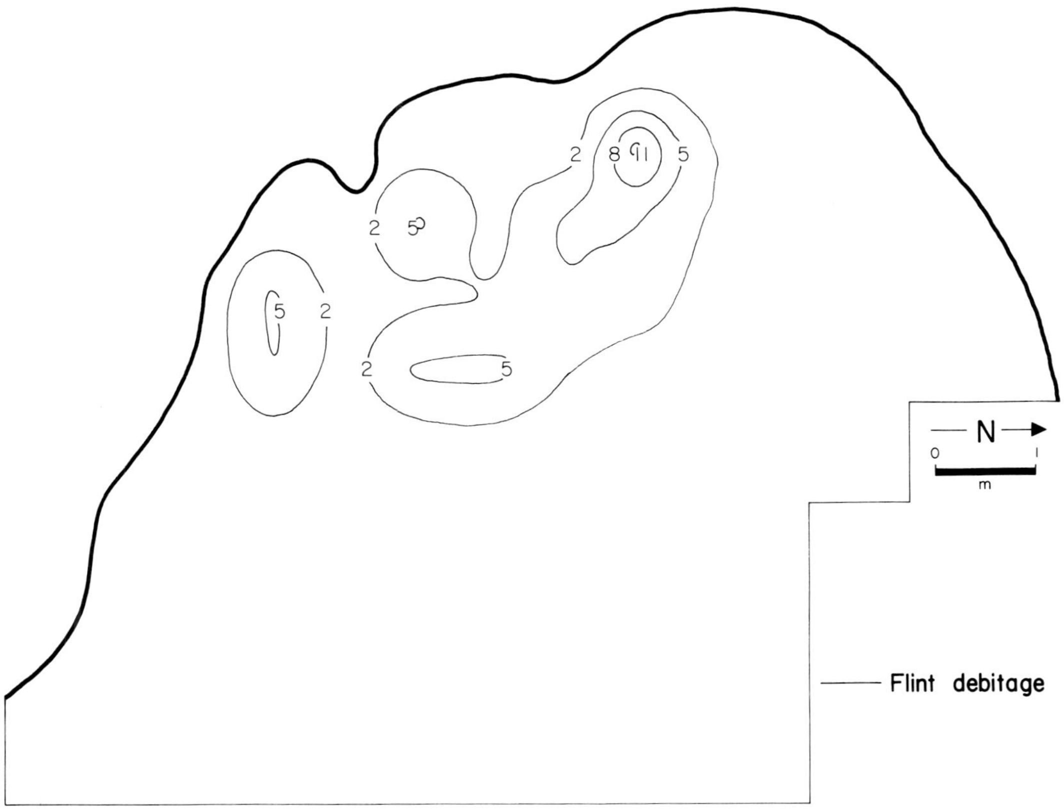

Fig. 26.26. Density contours of flint debitage in Zone B3.

Fig. 26.27. Density contours of acorns and *susí* nut hulls in Zone B3.

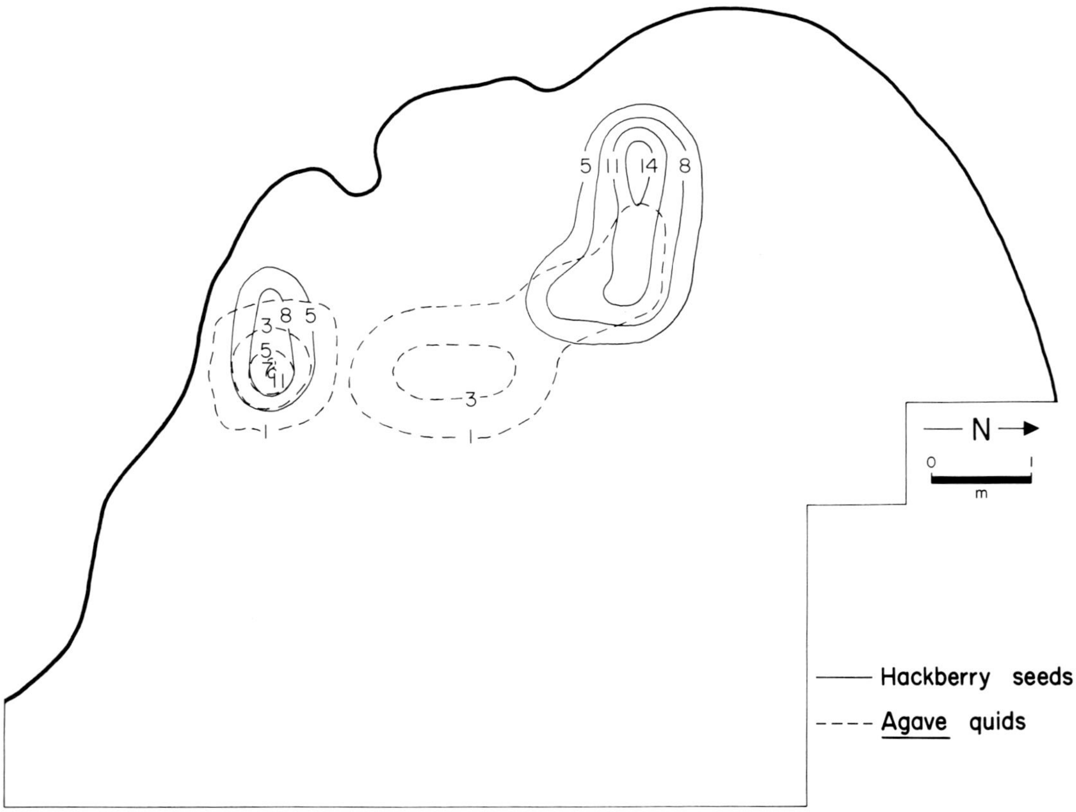

Fig. 26.28. Density contours of hackberry seeds and maguey quids in Zone B3.

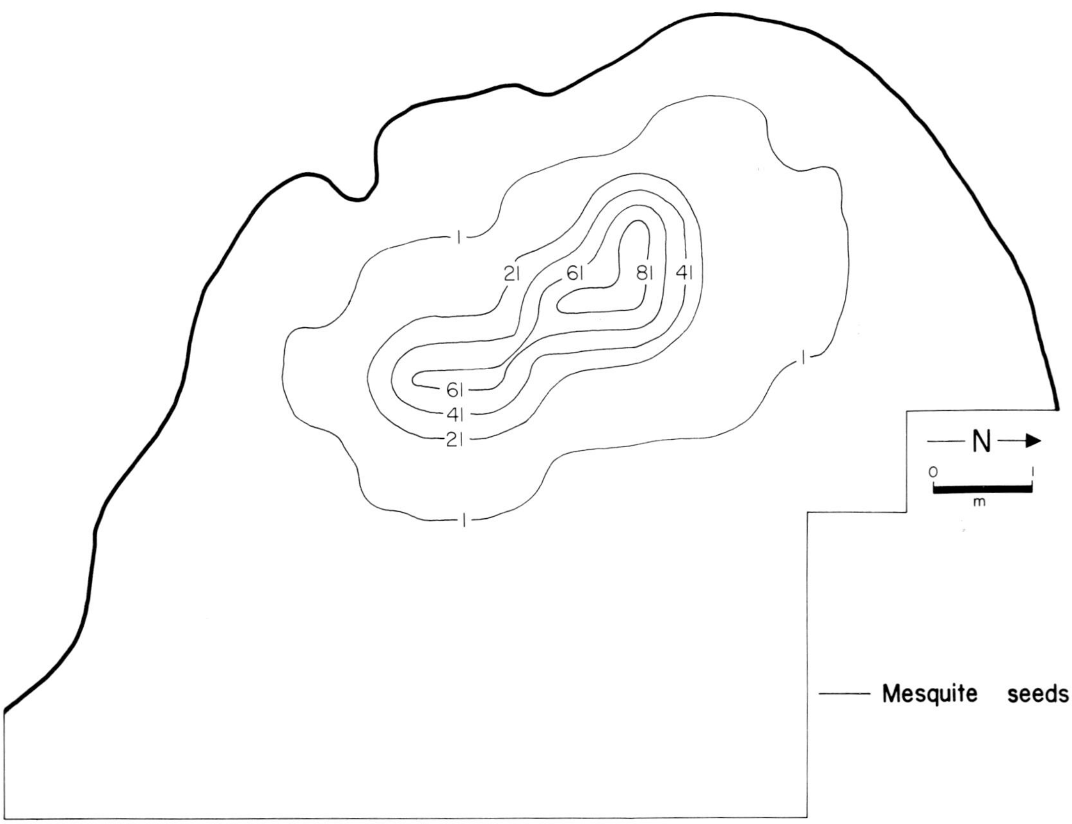

Fig. 26.29. Density contours of mesquite seeds in Zone B3.

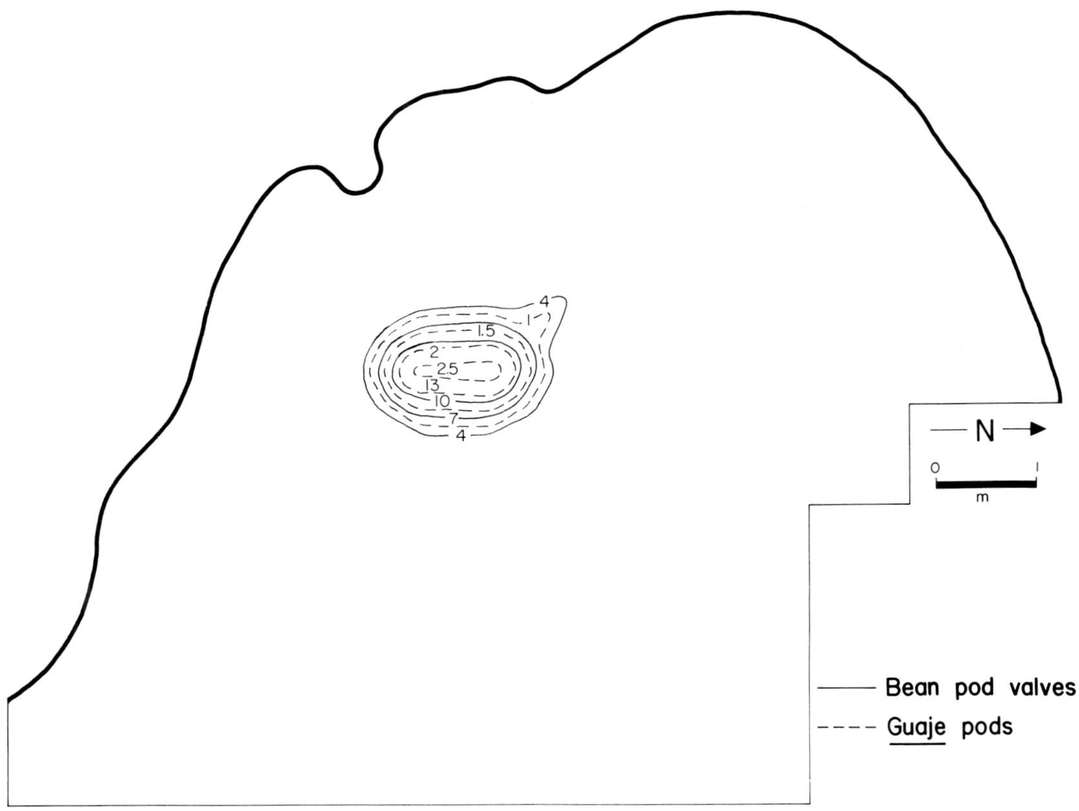

Fig. 26.30. Density contours of bean and *guaje* pods in Zone B3.

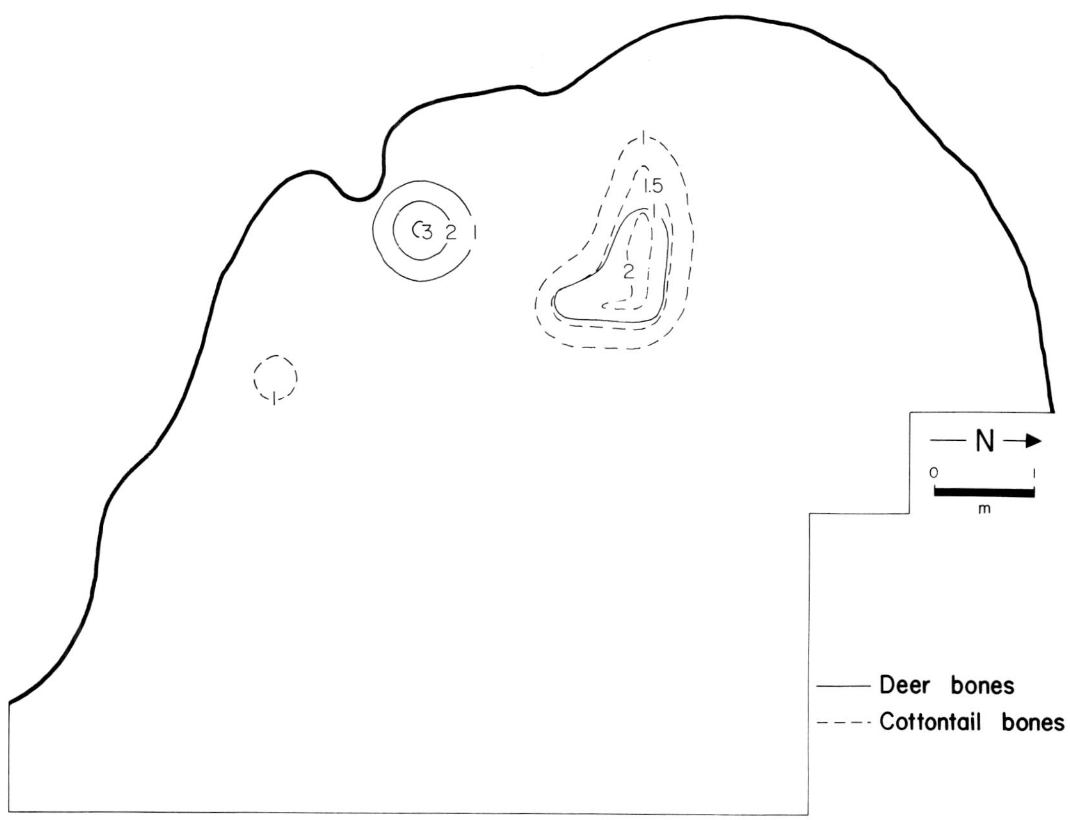

Fig. 26.31. Density contours of deer and cottontail bones in Zone B3.

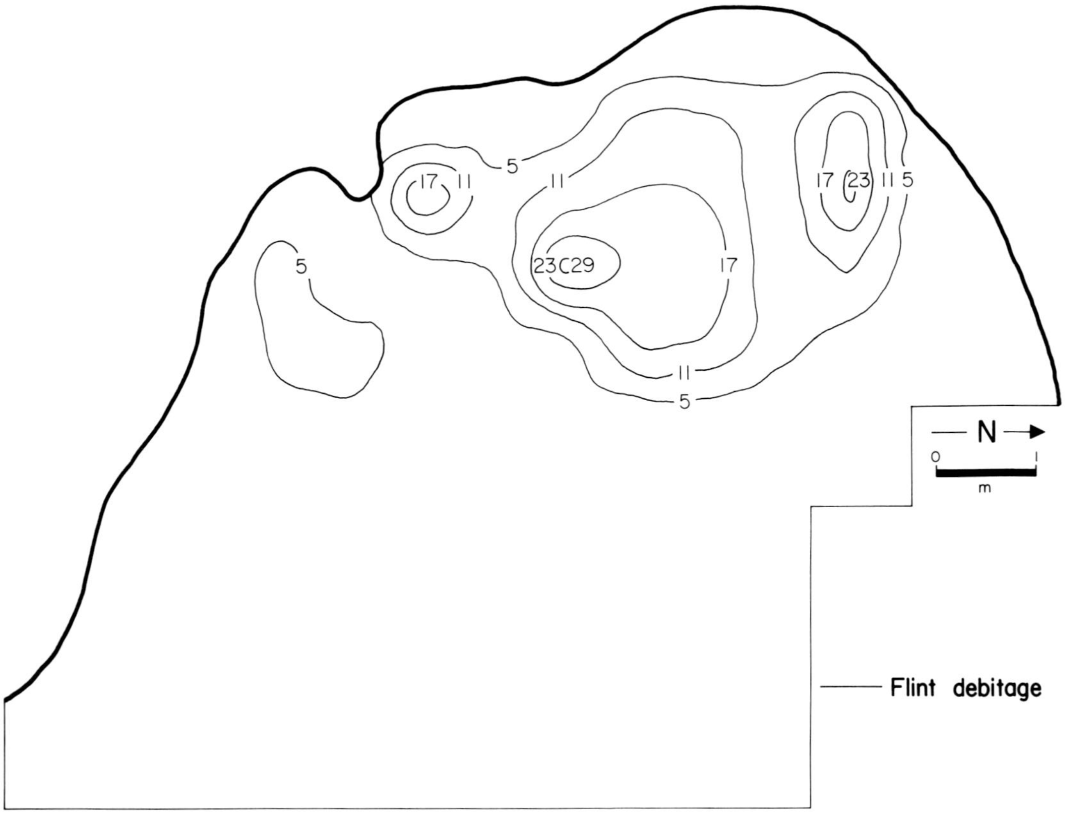

Fig. 26.32. Density contours of flint debitage in Zone B2.

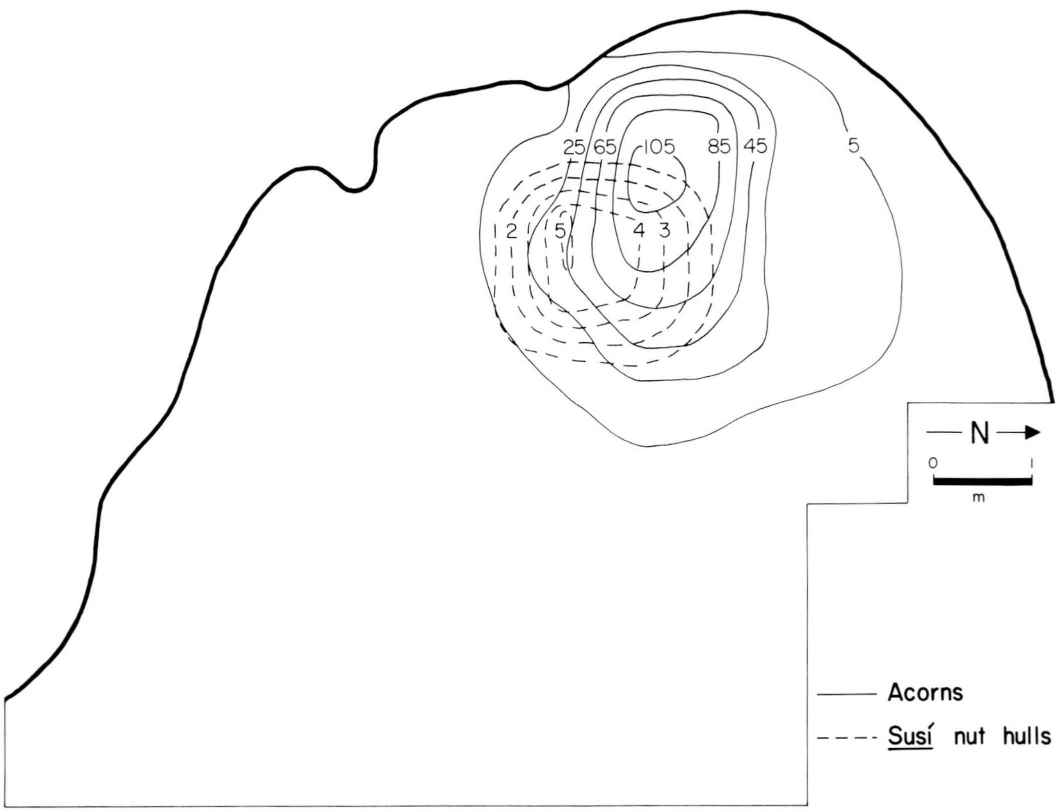

Fig. 26.33. Density contours of acorns and *susí* nut hulls in Zone B2.

Fig. 26.34. Density contours of nanche seeds and maguey quids in Zone B2.

eaten nanche fruits as well. A third individual, seated near the south wall of the cave, may have been responsible for the small nanche and maguey quid scatters there as well as the small debitage cluster in that area.

Mesquite seeds clustered strongly in the Square D11 area near the northern limits of the living floor (Fig. 26.35). If we ignore the four-seed contour drawn by the GEOG:CONTUR program (a figure that represents the yield of less than a single mesquite pod), it would appear that a major heap of pods was processed by our hypothetical individual in Square D11, while a smaller but still impressive pile was processed by the individual in the Square D8 area. Since mesquite is one of the plants needing both processing and cooking, we might now suggest that our two hypothetical individuals were women. The relatively insignificant quantities of mesquite seeds near the south wall, all of which could have come from a single stray pod, suggest the individual working there did not engage in major plant processing. Might he have been a man? We have insufficient evidence to decide.

Bean pod valves and cucurbits were concentrated in the northwest quadrant of the cave, and their distribution also suggested the presence of two individuals seated roughly 2 m apart (Fig. 26.36). One cucurbit scatter centered on Square E9 is partially overlain by a bean pod scatter centered on D9. A second cucurbit scatter centered on C11 is partially overlain by a bean pod scatter centered on D11. This second scatter occurs in an area of peak mesquite distribution that we have suggested may be a woman's plant processing area.

Zone B2 was the first living floor in our sequence where cucurbits and beans were so closely spatially patterned. However, the pattern may be fortuitous, since bean pod valves also seem to peak in two of the same squares, D9 and D11, where mesquite pods were extremely abundant. Thus these plant genera may be united primarily by the fact that they were all processed by the same two women(?) we have suggested were working in the northwest quadrant.

Deer and cottontail bones occurred mainly in a single modest scatter that peaked in Square D9 (Fig. 26.37). The individual responsible for this scatter also appears to have done a little flint chipping, processed mesquite, chewed maguey, and eaten raw fruits and nuts. Since we have already suggested this individual may have been a woman, we can now propose that the Square D9 area was devoted to the preparation of plant and animal foods. As further support for this, a metate, a mano, and an unretouched flake resting against a deer bone were found nearby in Square C9 (see Fig. 5.37). We infer from the shape of the contour lines for deer and cottontail that some bone fragments were discarded toward the cave wall to the west and northwest.

Turtle bones had a discontinuous distribution that provided a contrast to deer and rabbit (Fig. 26.38). One modest scatter overlapped with the area where most deer and cottontail had been processed (Squares C9–D9). A second scatter,

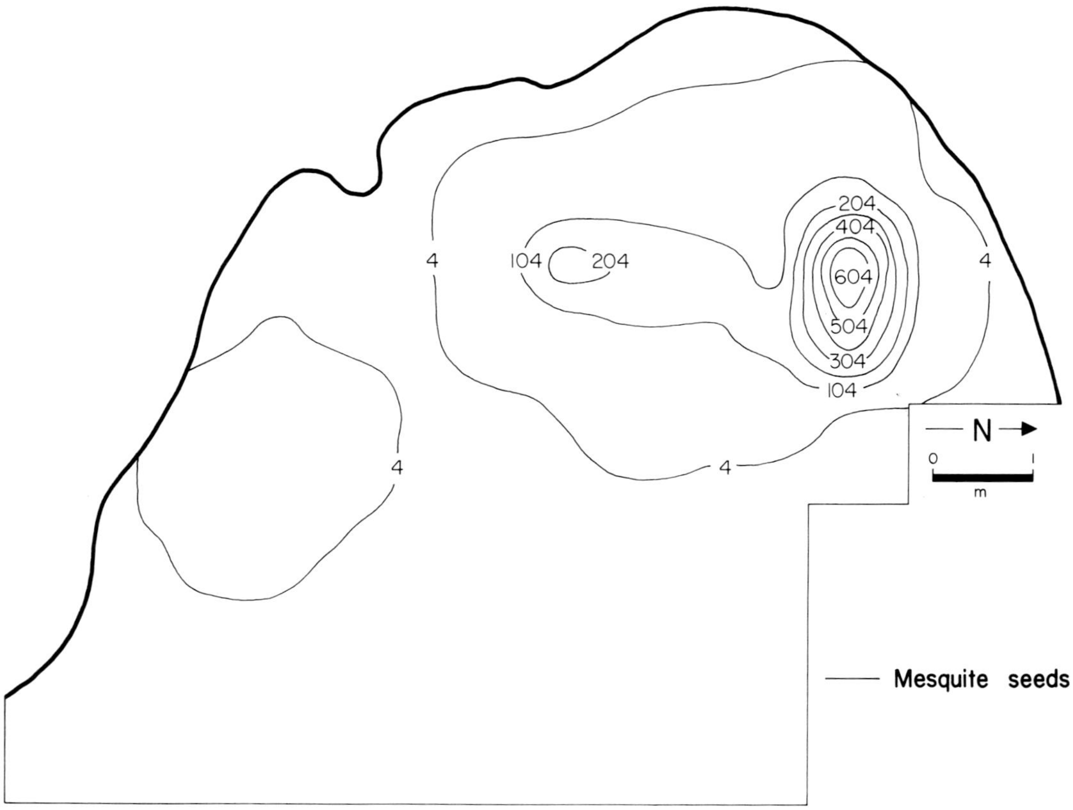

Fig. 26.35. Density contours of mesquite seeds in Zone B2.

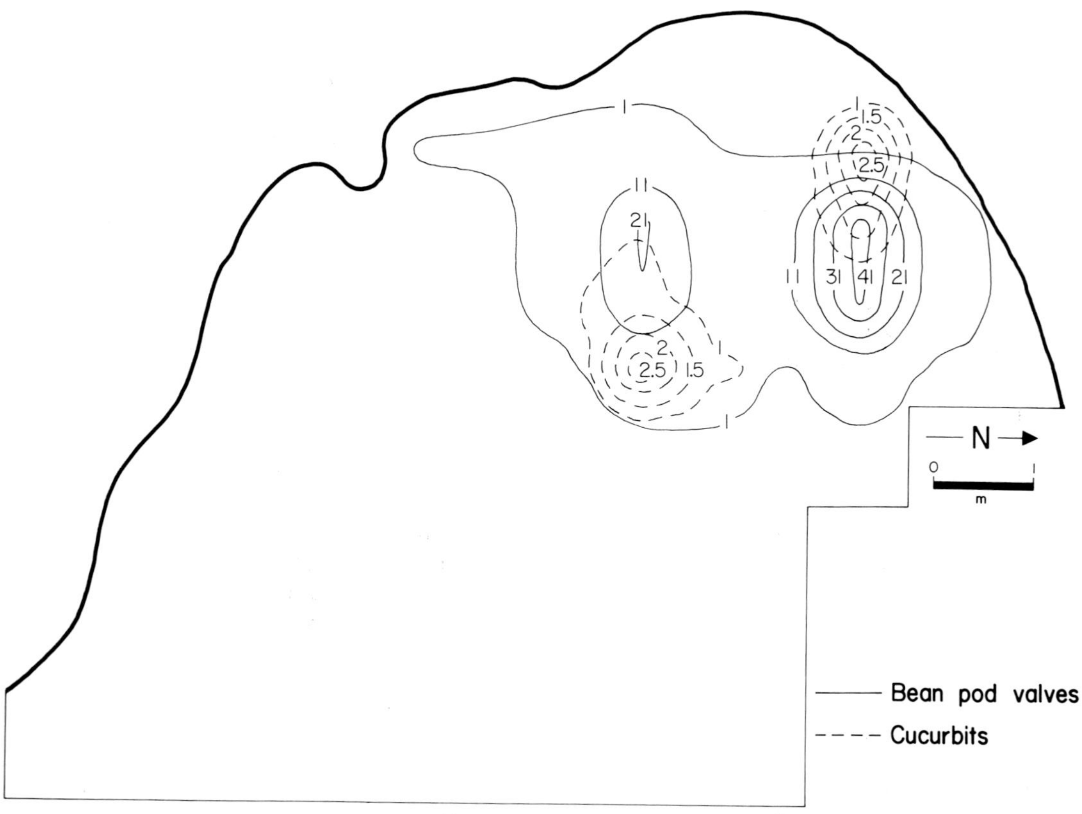

Fig. 26.36. Density contours of bean pod valves and cucurbit fragments in Zone B2.

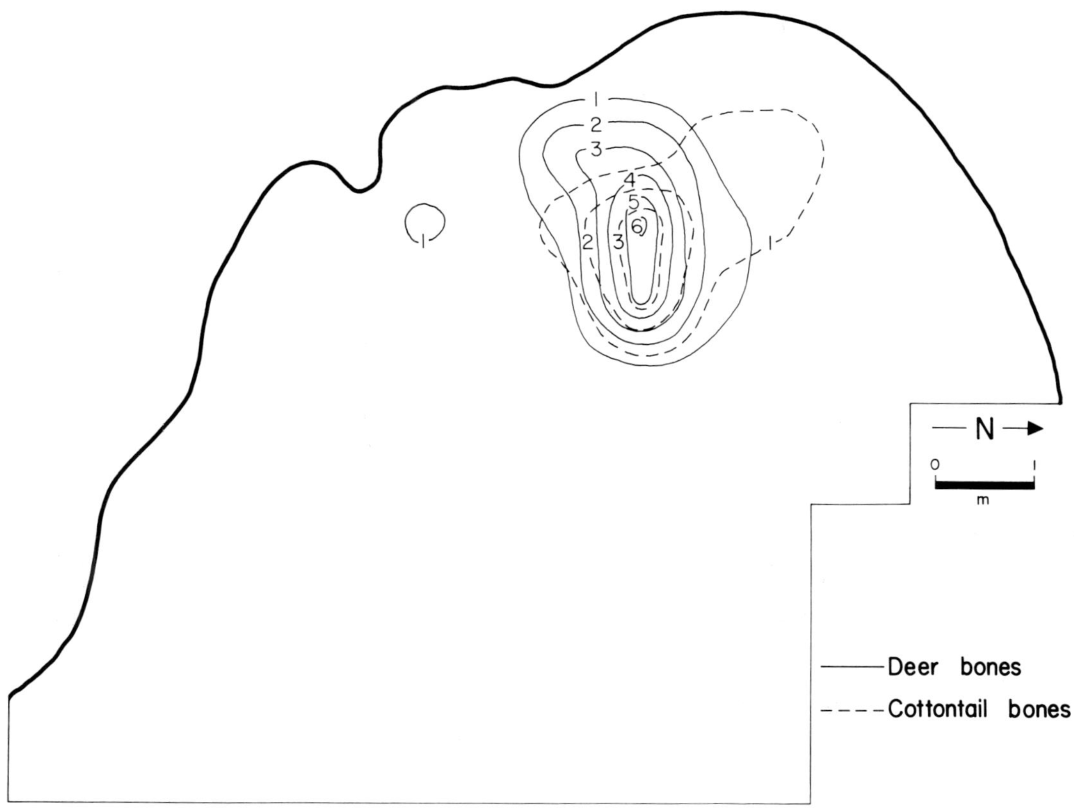

Fig. 26.37. Density contours of deer and cottontail bones in Zone B2.

Fig. 26.38. Density contours of turtle bones in Zone B2.

however, peaked in C11 and may have been associated with the second of our hypothetical individuals in the northwest quadrant. Thus, this individual may have been responsible for processing a turtle (perhaps disarticulating it after it had been roasted over coals) but not deer or cottontail.

What tool did the occupants of Guilá Naquitz use to open a mud turtle shell? Squares C11–D11 produced a Variety C biface and a chopper/knife that are possible candidates (see Fig. 5.37).

ZONE B1

Of all the living floors at Guilá Naquitz, Zone B1 appears to have had the most complicated pattern of plant, animal, and artifactual remains.

Flint debitage was abundant and covered a large area of the cave, exhibiting multiple peaks (Fig. 26.39). The densest concentration occurred in Square F5 near the south wall, but that was merely the highest peak in a long, narrow strip defined by the 25-flake contour line. Numerous core fragments, a hammerstone, and retouched tools such as steep denticulate scrapers, notched flakes, a projectile point fragment, and a Variety B biface were found either in this strip or within a meter to either side (see Fig. 5.38).

In the northwest quadrant of the cave were 2 hearths, Features 11 and 14. The 5-flake contour line swung around these hearths, defining an oval apron of flint flakes whose symmetry was broken in three places by minor concentrations of debitage (>15 flakes). These concentrations were spaced 2–3 m apart and included retouched tools such as notched flakes (Fig. 5.38). The most northern concentration, in Square C12, may represent debitage swept against the wall of the cave; nearby, in Square C11, were found 2 cores and 2 core fragments. The most western concentration, in B9, may also represent debitage swept against the cave wall. To the northeast, the third concentration (Square E11) included 4 retouched tools.

What this flint flake distribution suggests is that one or more individuals in the extreme southern part of the living floor engaged in considerable flint knapping and tool manufacture. In the northwest quadrant of the cave, perhaps as many as two individuals sitting near the hearths engaged in a modest amount of tool preparation, discarding their refuse toward the cave wall.

Acorns were widespread on the living floor, but their distribution seemed to be linked to at least four concentrations (Fig. 26.40). The three most prominent of these were in the vicinity of the hearths in the northwest quadrant. There seems to have been one major pile of unused acorns centered on Square C9, just south of Feature 14; a mano, possibly of the type used to grind acorns into flour, was recovered in C9 (see Fig. 5.38). Two smaller clusters of acorns occurred in Squares E9 and E11, to the north of Feature 11. Thus the area around Features 11 and 14 may have been used for storage and processing of acorns (among other things).

Farther out toward the center of the cave was a minor concentration of acorns centered on Square F7. This may represent the last remnant of a pile of acorns that was brought to the cave and never fully used. All other acorns in Zone B1 (e.g., those delimited by the four-acorn contour) could have resulted from the gradual spreading out of the already described concentrations during the course of the occupation.

Piñon nut hulls occurred mainly in a single concentration that overlapped the acorn pile in Square E11 (Fig. 26.40). These may represent the hulls from nuts eaten by an individual seated to the north of the Feature 11 hearth.

Agave quids in Zone B1 had one of the most interesting distributions of any plant: They occurred in five small concentrations, two in the south and three in the northwest part of the living floor (Fig. 26.41). Nothing could be clearer than that we are dealing with the piles of expectorated quids chewed by two to four different individuals working in different places in the cave. Indeed, Fig. 26.41 is almost like a map of where our Naquitz phase microband members chose to sit.

At least one person sat in or near Square F6 in the south part of the cave, accounting for one prominent pile of quids and perhaps some minor extensions to the south and west. Another individual may have sat near Feature 14, leaving a smaller pile. To the north of Feature 11 are twin concentrations in E9 and E11, resulting either from two individuals working side by side or from one individual on two different occasions. One of those individuals (in E11) also seems to have eaten piñon nuts (see above).

Opuntia seeds had a discontinuous distribution (Fig. 26.42). The largest concentration was centered near Square F6 in the southern part of the cave floor; a smaller scatter surrounded Feature 14 in the northwest quadrant. Each concentration probably represents a place where uneaten prickly pear fruit disintegrated.

Nanches had a distribution like that of agave quids: A multiplicity of small seed concentrations showed where as many as four individuals might have sat, eating the raw fruit, during the course of the occupation (Fig. 26.43). Twin concentrations near the south wall (in Squares E5 and G4) may have been left either by two individuals, or by one individual on two occasions. A third scatter lay to the southwest of Feature 11. A fourth was centered on Square E11, along with the piñon nut hulls and maguey quids already discussed.

Mesquite seeds were the most numerous plant product discovered in Zone B1; their distribution had one very clear peak in the vicinity of Square E11 (Fig. 26.44). This may represent a single pile of refuse from the processing of a large number of mesquite pods. However, it appears that once discarded, this pile was disturbed by the ongoing activities in the cave to the point where stray seeds became spread over most of the living floor at low densities (note the five-seed contour).

Since mesquite and acorns are both plants requiring a great deal of processing, we can suggest that the northwest quadrant of the cave, especially the area around Features 11

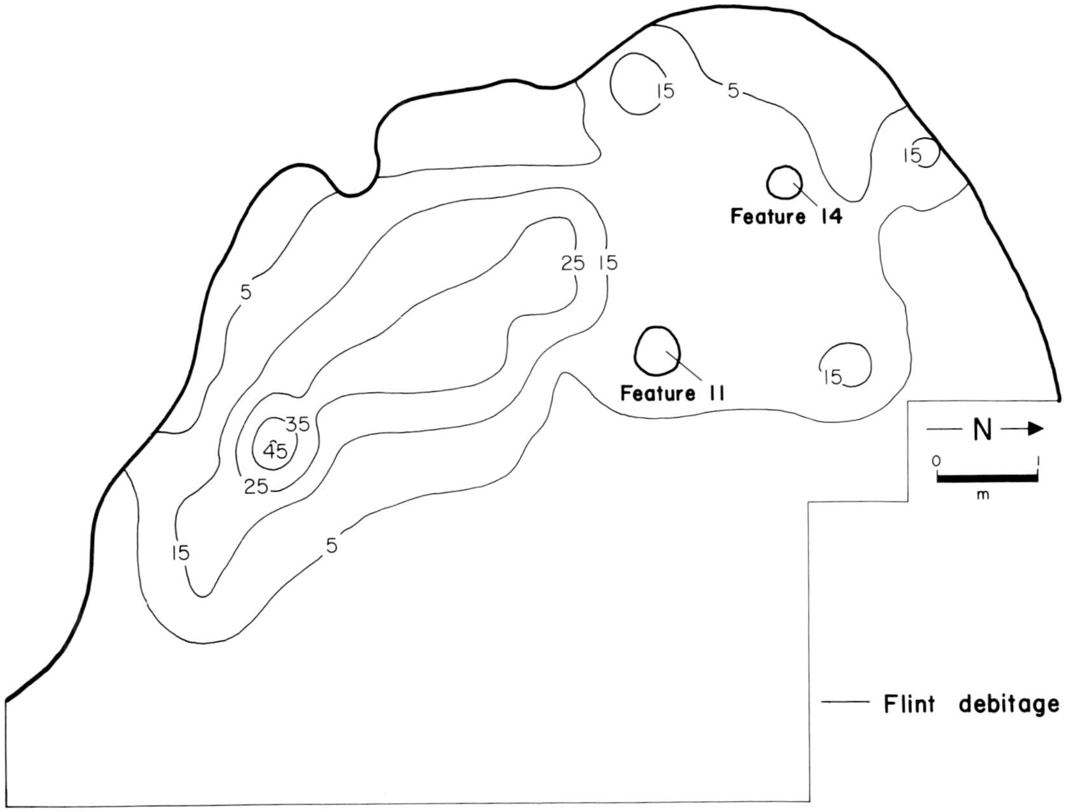

Fig. 26.39. Density contours of flint debitage in Zone B1.

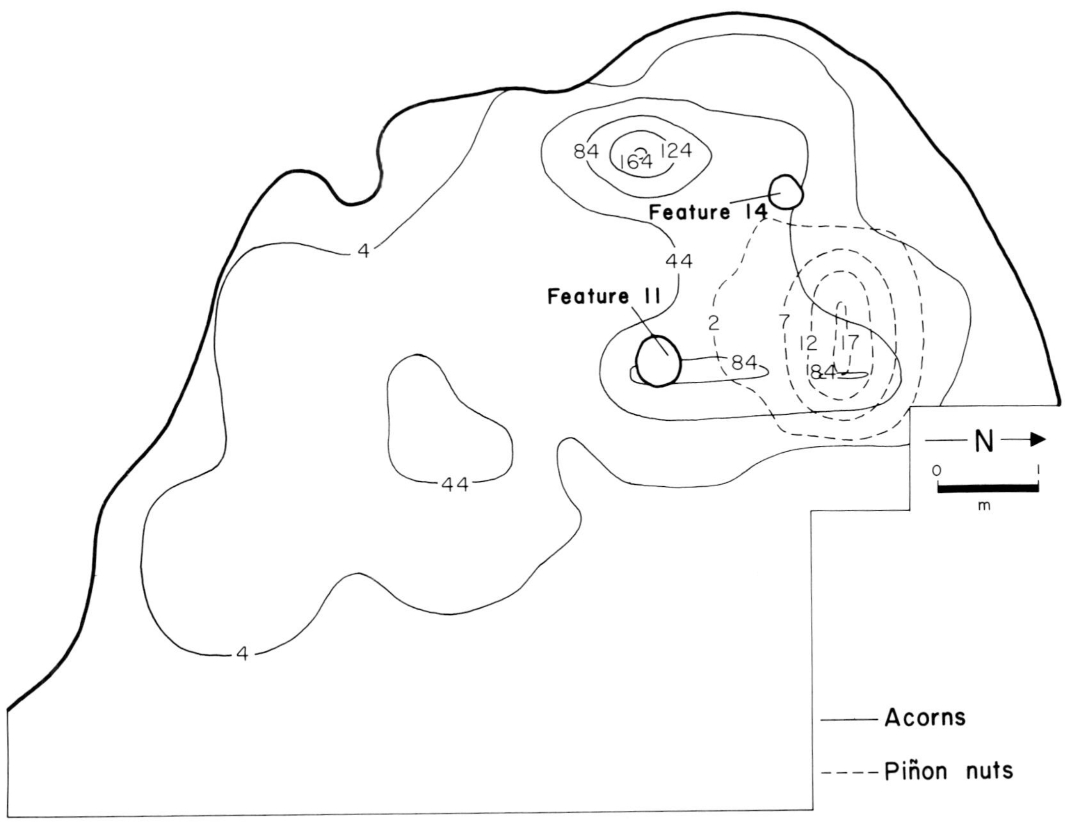

Fig. 26.40. Density contours of acorns and piñon nuts in Zone B1.

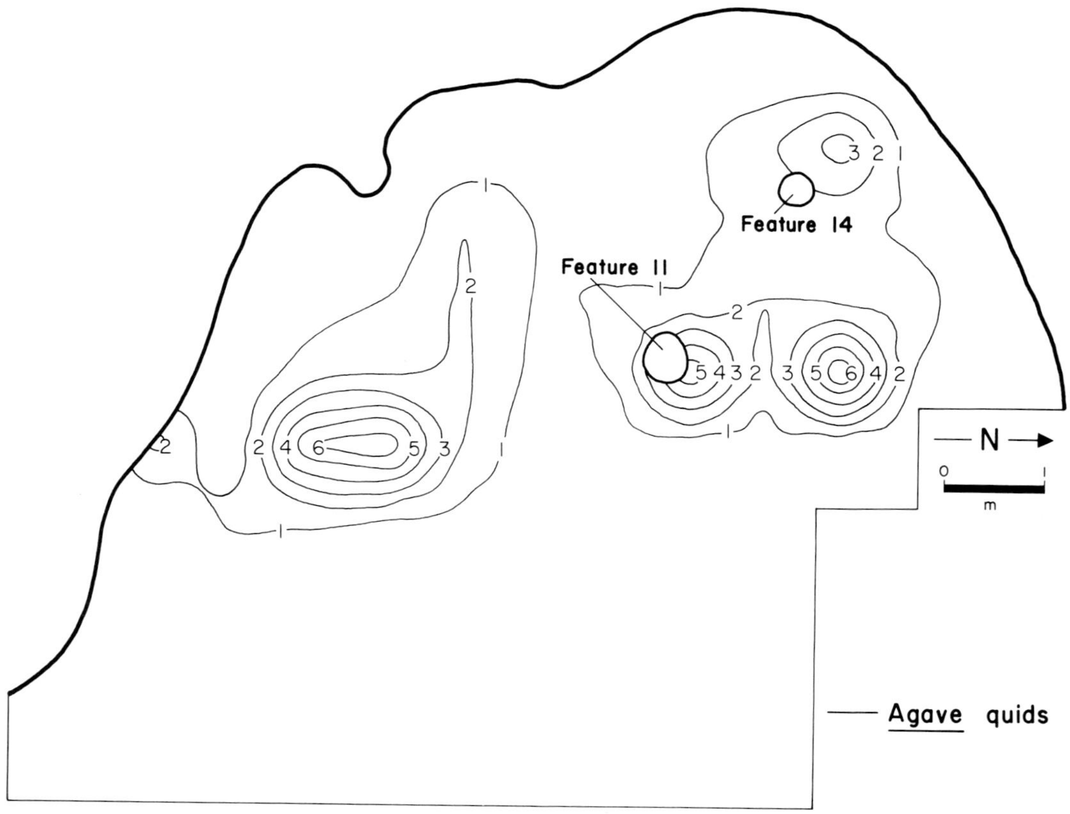

Fig. 26.41. Density contours of maguey quids in Zone B1.

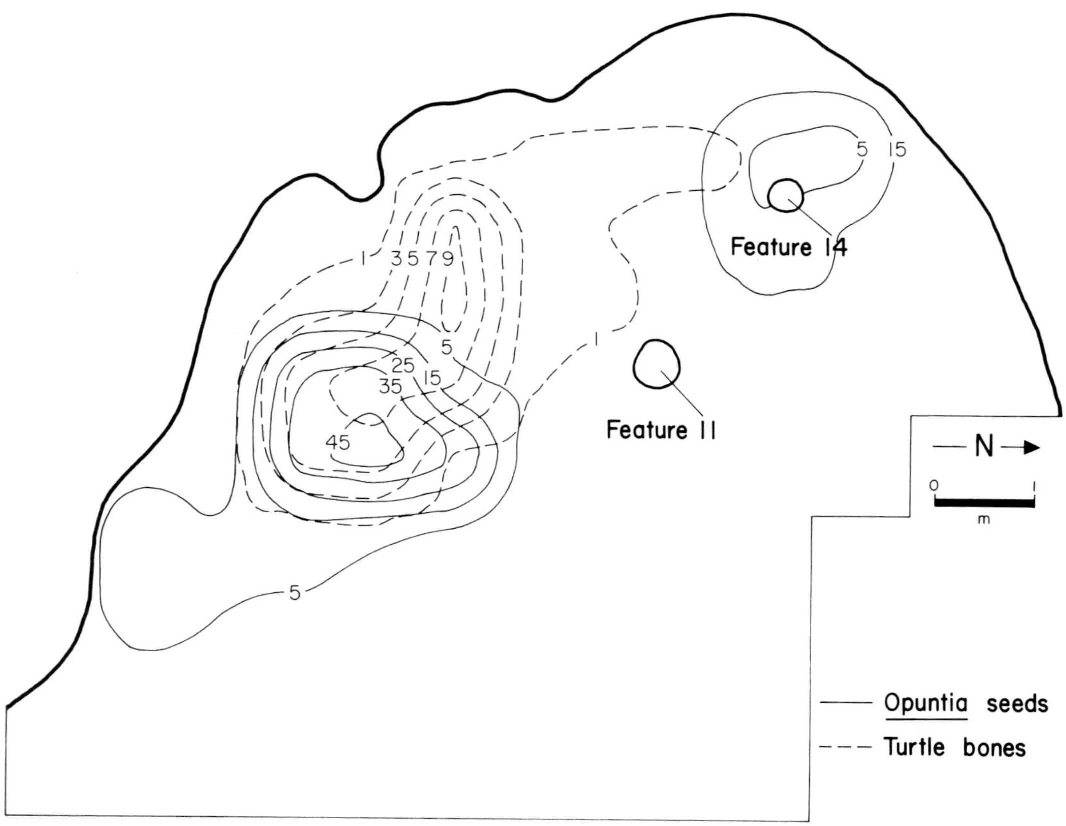

Fig. 26.42. Density contours of prickly pear seeds and turtle bones in Zone B1.

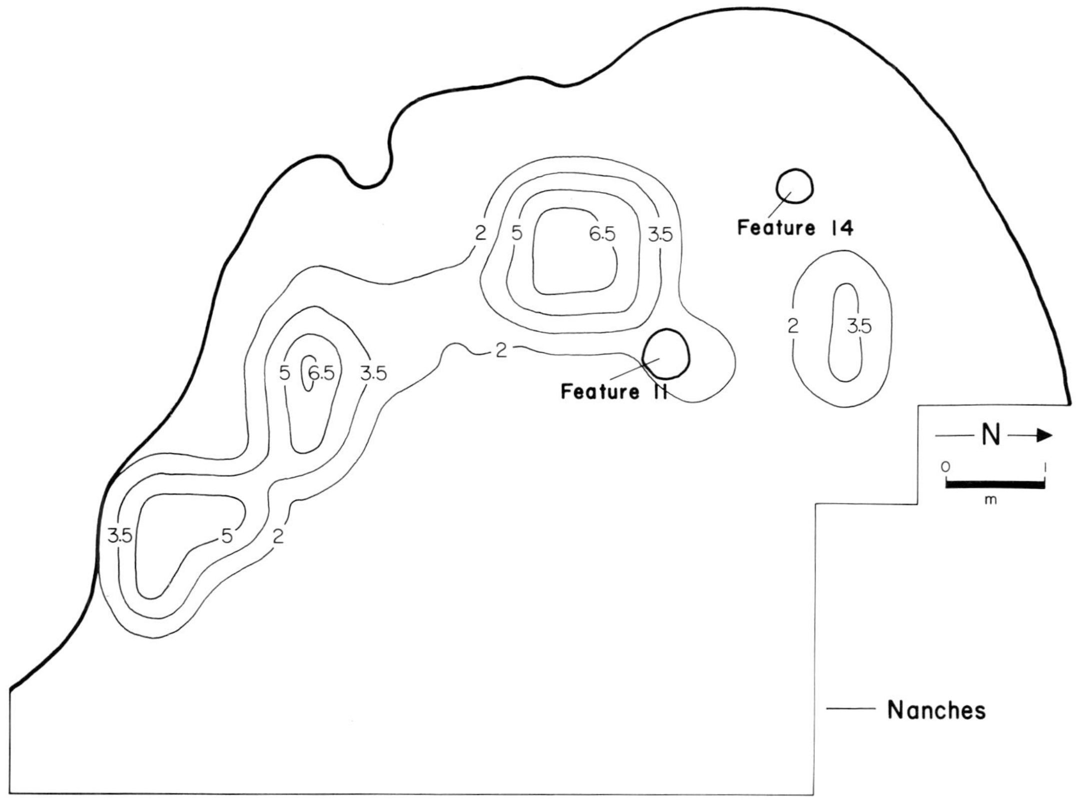

Fig. 26.43. Density contours of nanche seeds in Zone B1.

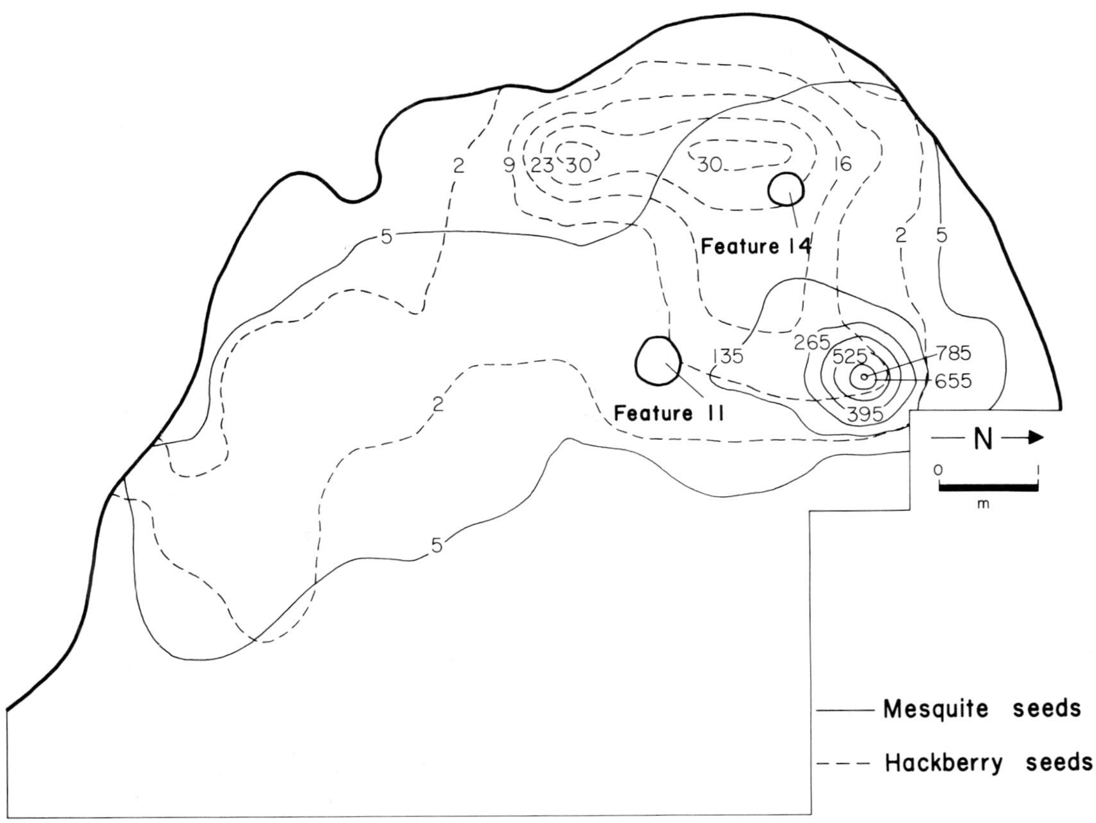

Fig. 26.44. Density contours of mesquite and hackberry seeds in Zone B1.

and 14, was used for plant processing and/or food preparation. We also note that although these two plants come from different vegetational facies and ripen at different seasons of the year, they were both processed at the same place on the living floor. We consider this further evidence that sexual division of labor (with women having their own work space set aside for bulky plant processing) played more of a role in determining the way the living floor was conceptually divided than did the season of the year, or the area from which the harvested product had come.

Hackberry seeds showed the same pattern we have seen in most other fruits that were eaten raw (Fig. 26.44). Twin peaks occurred between Squares C8 and C10, immediately to the south of Feature 14. These were, however, only the most prominent peaks in a larger scatter defined by the 9-seed contour, which also reached into the C11 mesquite concentration. Thus we might argue that anywhere from 1–3 persons, working in the northwest quadrant of the cave, ate more than 100 hackberries. The remaining distribution (an irregular strip defined by the 2-seed contour, and running to the south wall of the cave) can be accounted for through seeds having been spread around by persons walking back and forth across the living floor.

Guaje pod fragments had their peak distribution around Feature 11 (Fig. 26.45). This lay only a meter south of the peak of mesquite pod distribution (see above) and reinforced a pattern seen in earlier levels: Most legume pods, of whatever genus and vegetational facies, tended to be processed in the same area of the living floor. A smaller scatter in the Square F6 area might represent a second processing area; or, like the scatter of acorns in nearby F7, it could simply be the remnant of a once-larger pile brought back to the cave and left temporarily in F6 prior to processing.

Susí nut hulls provided us with yet another example of a food eaten raw by two to three individuals seated in different parts of the cave (Fig. 26.45). One small scatter lay immediately to the west of Feature 14, another in E11 to the north of Feature 11. Still a third scatter, even smaller, occurred in Square F7 in the southern part of the living floor.

Bean pod valves and cucurbit fragments had an extremely interesting distribution in the far northwest corner of the cave (Fig. 26.46). While the two scatters are not totally isomorphic, it is clear that the processed remains of both plants were discarded mainly in and around Square C11, to the northwest of Feature 14.

What is important about this distribution is that it strongly reinforces a trend we have seen in the sequence of Naquitz phase levels. On earlier living floors, such as Zones E, C, and B3, it appears that beans were usually threshed and discarded with other legume pods such as mesquite and *guajes*; cucurbits were processed and discarded elsewhere. Beginning in Zone B2, and growing stronger in B1, is an apparent tendency for beans to be processed and discarded with cucurbits. In other words, instead of being processed along with other wild legumes, runner beans were now being processed along with a plant known to be domestic.

This is not the first line of circumstantial evidence to suggest that these small black beans might conceivably have been cultivated. In Chapter 24 we discussed the fact that on several living floors at Guilá Naquitz, beans were present in such high numbers as to suggest that the occupants either (1) foraged over a larger area for beans than for almost any other plant or (2) attempted to increase artificially the availability of beans by cultivating them near the cave. However, it would be wise to remember that any evidence for "incipient bean cultivation" is no more than circumstantial; not only do the beans themselves show no genetic change, they also belong to a species that has left no domestic descendants. Thus—to repeat what we stated in Chapter 24—if the Naquitz phase did witness an experiment in runner bean cultivation, it was an experiment ultimately given up in favor of more promising species of beans.

Deer and cottontail bones showed two distinct areas of distribution (Fig. 26.47). The principal concentration lay in the northwest quadrant of the cave, where bones were distributed over a large, oval area enclosing the Feature 14 hearth. Given the density of plant remains in this area, we now suspect this space was at least partly devoted to the preparation of plant and animal foods. The cottontail bone peak in Square C12 suggests that some bone refuse was discarded against the wall of the cave.

A second scatter of bone, somewhat lighter in density, covered a large area near the south wall of the cave. This area had abundant evidence for the preparation of flint tools and the eating of raw fruits and nuts, but almost no evidence for the processing of bulky plants. The animal bone found here could result from butchering, eating, or a combination of both.

Given the evidence presented so far, we suspect that the northwest quadrant of the cave was primarily women's work space, while the southern area was primarily used by men. In this case, the piles of nanche seeds and maguey quids already discussed would suggest that there were at least two males and two to three females present in the cave during the deposition of Zone B1.

Turtle bones were concentrated primarily in the southwestern quadrant of the cave (Fig. 26.42). In Chapter 22 we argued that there were at least 2 individual turtles roasted over hot coals in Zone B1, one concentrated mainly in Square D7 and the other in E6. Two fragments of slab metate found in Square D6 might have been used as hammers to fracture the carapace (see Fig. 5.38).

SUMMARY AND CONCLUSIONS

In our brief examination of six living floors at Guilá Naquitz we have detected a number of repetitive patterns that may reflect regularities in the way tasks were organized in the cave. There are two ways these patterns can be used: They

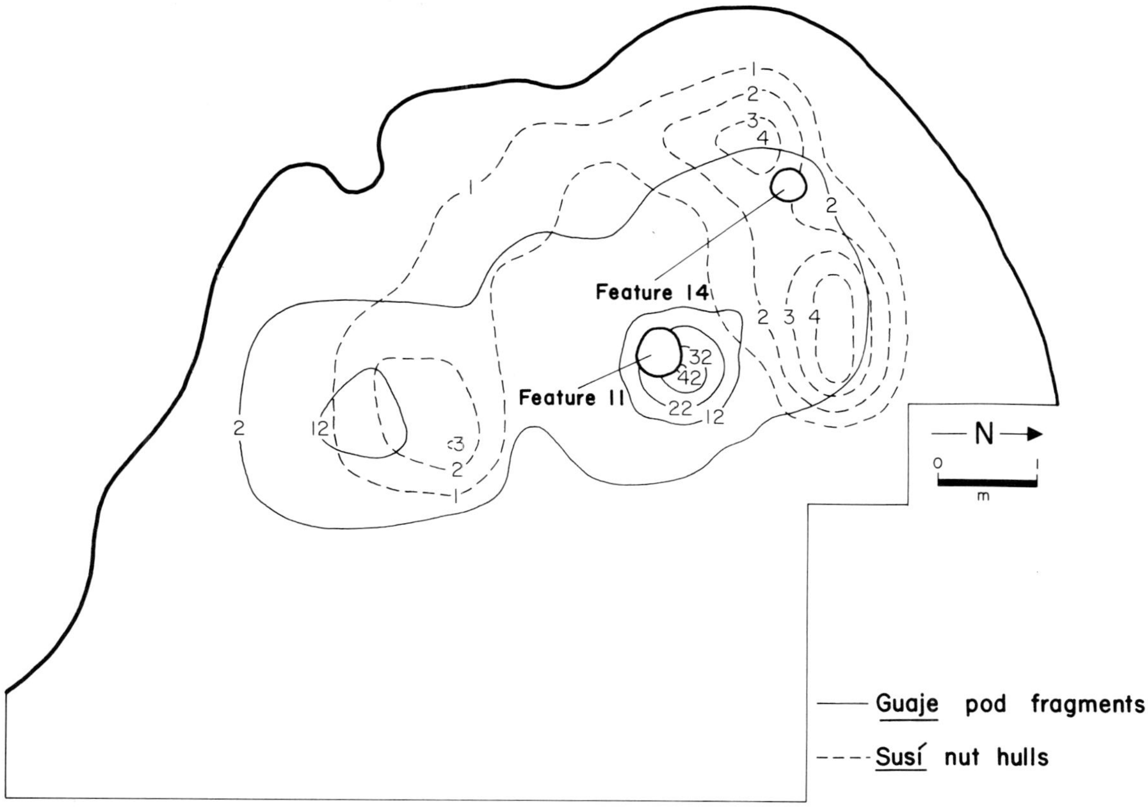

Fig. 26.45. Density contours of *guaje* pods and *susí* nut hulls in Zone B1.

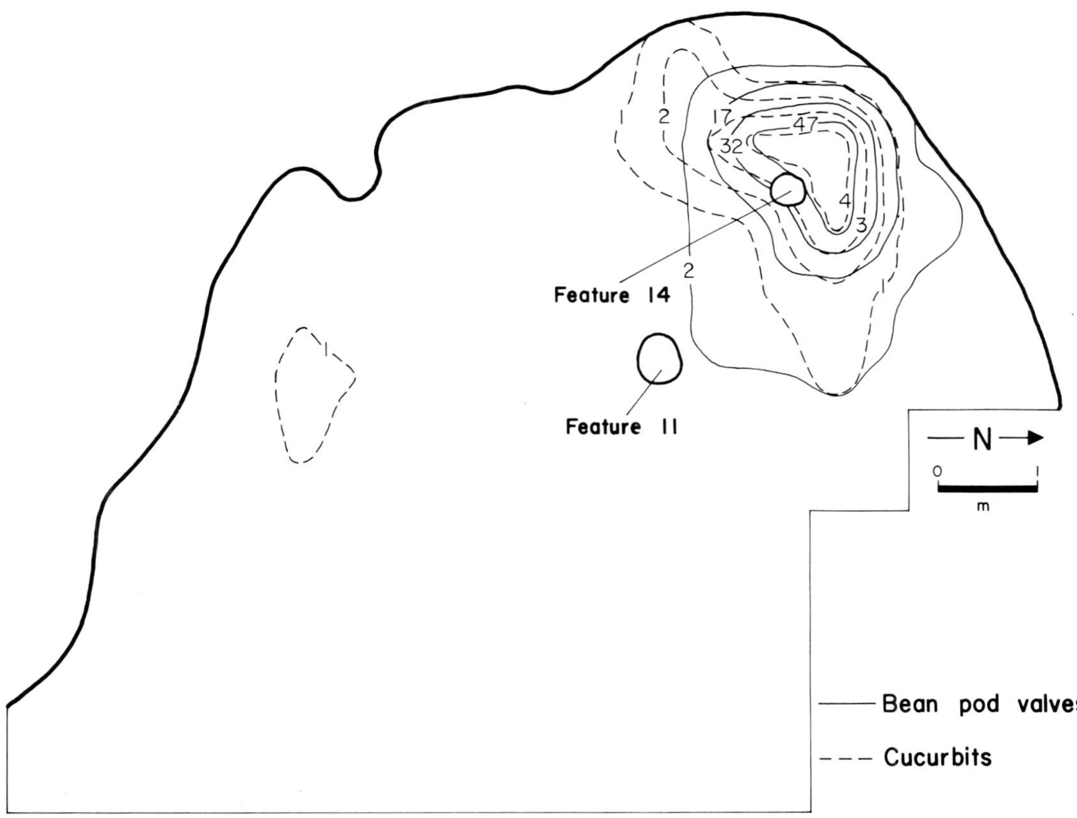

Fig. 26.46. Density contours of bean pod valves and cucurbit fragments in Zone B1.

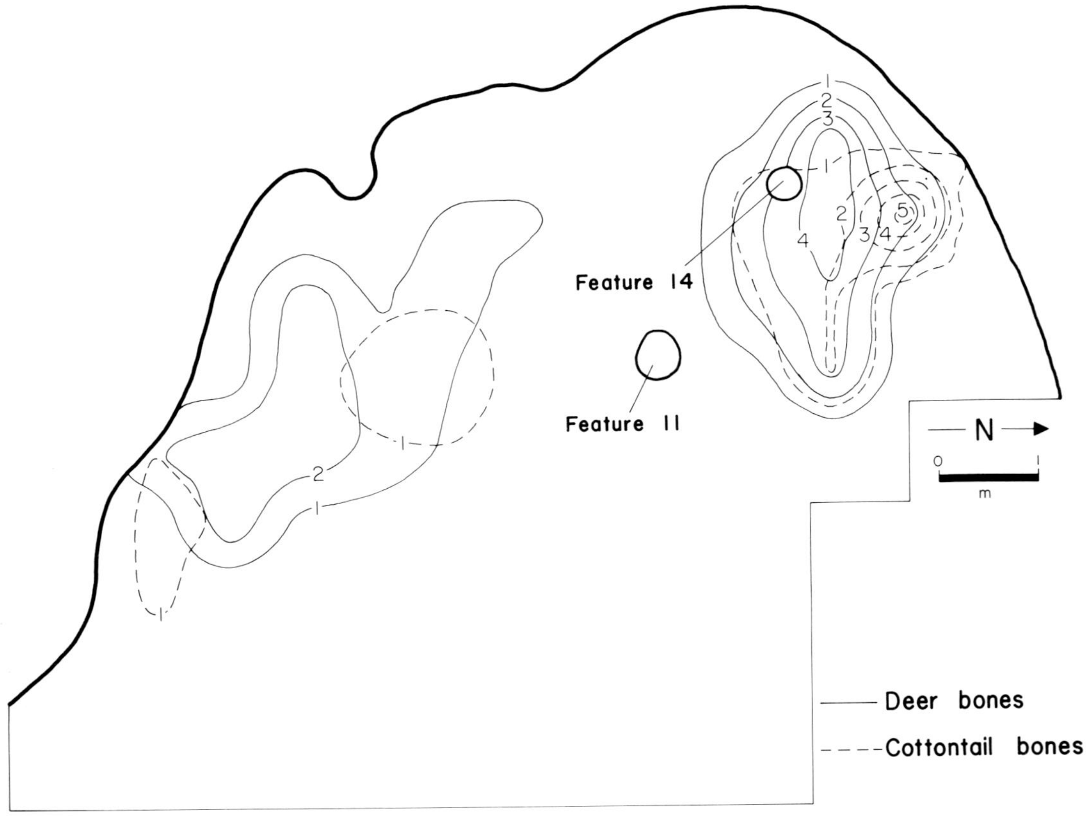

Fig. 26.47. Density contours of deer and cottontail bones in Zone B1.

can be presented as empirical generalizations in this chapter, and they can be used as background information to aid in the design of Whallon's and Reynolds' studies in Chapters 27 and 28.

1. We have suggested that sexual division of labor played a more critical role in determining how the cave floor was partitioned than did the harvest season or vegetational facies from which the products came. Once an area was set aside for plants whose processing required a substantial labor input, it was so used regardless of whether the plants were August-harvested mesquite pods from Mesquite Grassland B or December-harvested acorns from Thorn Forest A.
2. Assuming that women did most of the substantial plant processing, females seem to have used the northwest quadrant of the cave more than any other area. This was also the area most likely to have hearths, and it had at least half the storage pits. Assuming that areas combining flint knapping and butchering but *without* evidence for substantial plant processing were men's work spaces, males seem to have used the extreme south of the cave more than any other area.
3. A simple division into men's and women's work space, however, does not do justice to the complexity of activity areas. They included areas for light butchering, raw plant eating, tool manufacture, meal preparation and cooking (which included both meat and vegetables), and refuse discard. Even refuse could be subdivided, with the coarser fraction often found nearer to the cave mouth and the finer fraction discarded toward the cave wall.
4. Any a priori notion that men did all or most of the flint knapping also proved oversimplified at Guilá Naquitz. While men may have done most of the knapping on some living floors (and perhaps assumed the major responsibility for some tool categories, such as projectile points and bifaces), women definitely seem to have chipped their share of tools. Their work spaces included cores, heavy scrapers, and a variety of retouched flakes for light cutting tasks, as one might expect when food preparation is involved.
5. Acorns seem to have been left in one to two large heaps on the living floor (perhaps the result of dumping out basketloads) or occasionally kept in storage pits. Manos or mano fragments were sometimes found nearby, and may have been the tools used to produce acorn flour. Use of acorns was profligate, with hundreds frequently left behind uneaten.
6. The refuse from those plants that produced a great deal of processing debris, such as mesquite, *guajes*, and nopales, often was left in one big pile in or near a

women's work area. Such products may have been brought back in knotted net bags.

7. Judged by what is missing from the deer skeletons at Guilá Naquitz (see Chapter 22), a lot of the rough preliminary butchering of deer must have taken place outside the cave, with pieces of convenient size brought inside for further work. Much the same could be said for bulky plants such as maguey hearts and nopales: The fragments found in the cave imply many kilograms of refuse that must have been discarded outside.
8. While men may have done the heavy preliminary butchering of deer, it is evident from the bone fragments in food preparation areas that women trimmed and disarticulated smaller portions of deer and rabbit. Turtles, on the other hand, seem to have been roasted in the shell over hot coals, then broken up and disarticulated after cooking. Based on numbers of burned bones, we suspect that many of the fragments found in the cave represent discard after cooking and eating, not the remains from butchering areas.
9. Fruits eaten raw (hackberry, nanche), nuts eaten raw (piñon, *susí*), and maguey quids were not restricted to any one area of the cave floor. Their pattern suggests that all the individuals in the cave, both men and women, engaged in casual eating of these products, with each person producing his or her own small midden of fruit seeds, nut hulls, and expectorated quids. On some living floors, these small individual middens indicated the presence of at least four to five persons in the cave, but no more.
10. Prickly pear fruits evidently were left in small piles at different places on the living floor, perhaps associated with specific individuals as the small middens mentioned above were. As unused fruits "passed their prime" and deteriorated, they produced scatters of *Opuntia* seeds on the floor.
11. Cucurbits were the only plant in the cave showing genetic evidence for domestication. Their remains on the cave floor usually covered only a small area, and their contour lines were rarely isomorphic with those of other plants. If we assume that these early squashes were being grown in a small patch somewhere near the cave, then (1) their restricted scatter may reflect a short harvest period that produced only one pile of squash to process, and (2) their lack of isomorphy with other plants may indicate that harvest trips to the cucurbit garden were not accompanied by wild plant collecting.
12. On the earlier living floors, runner bean pods tended to be discarded with other legume pods such as mesquite and *guajes*. On the later living floors, bean pods tended to occur with cucurbit fragments. This could be construed as circumstantial evidence that squashes and beans were now being cultivated together, but it is not confirmed by any genetic change in the beans; they belong to a species that has left no domesticated descendants.

27

A Spatial Analysis of Four Occupation Floors at Guilá Naquitz

Robert Whallon

Four of the occupation floors at Guilá Naquitz—Zones E, D, C, and B1—were large enough and produced adequate quantities of plant, animal, and artifactual remains for a statistical analysis of spatial patterning of these materials. Unfortunately, Zones B3, B2+3, and B2 were too small for spatial analysis of this sort.

STATISTICAL METHODS

The analyses of spatial patterning on the four occupation floors consisted of three basic steps. First, the tendencies for spatial concentration of each kind of plant, animal, and artifactual material were determined for each occupation floor. The most common, or dominant, scale of such spatial concentration was established in this same step of the analysis. To assess the degrees of coincidence in spatial concentration, or clustering, among these different materials, the original grid-square data then were grouped in quadrats the size of the dominant scale of material clusters on each floor, and these grouped data were used to calculate the correlations (r) among all classes of material. Finally, the resultant matrix of correlation coefficients was ordered to see if there were sets of mutually associated materials that could be defined from these correlations. The sets of materials thus defined formed the basis for interpretations and inferences about the possible activities that might have generated the spatial patterns revealed by this analysis of the Guilá Naquitz occupation floors.

The spatial analyses were begun with a technique I have called *dimensional analysis of variance.* This statistical technique was introduced to archaeological analysis in a preliminary study of Guilá Naquitz Zone C, which was published in *American Antiquity* in 1973 (Whallon 1973). Because Guilá Naquitz was excavated in 1 × 1-m grid squares, it was impossible to consider using a technique such as nearest-neighbor analysis, which relies on exact provenience data for each item. At the time of my preliminary study, other available techniques for analyzing the distributions of materials among grid squares were based on the fitting of Poisson distributions (or other, more complex, statistical distributions) to the observed frequencies of item counts per square. With these techniques, it was difficult, however, either to assess the relative strength of patterning at different spatial scales or to define the patterns of spatial location of any unusually high or low concentrations of items. Dimensional analysis of variance, borrowed from quantitative plant ecology, offered at least a partial solution to these restrictions and provided more detailed, useful information on spatial patterning in archaeological applications. Since the technique has been presented and discussed extensively in my 1973 article and elsewhere (Riley 1974; Schiffer 1974; Whallon 1974b; Hodder and Orton 1976:33–39; Orton 1980:146–149), only the essential points are recapitulated briefly here.

AN INTRODUCTION TO DIMENSIONAL ANALYSIS OF VARIANCE

Dimensional analysis of variance was developed by quantitative plant ecologists (Greig-Smith 1952, 1961, 1964:86–93; Kershaw 1957, 1964:104–113; Thompson 1958; Greig-Smith *et al.* 1963; Pielou 1969:104–106), although in this field it is called *the analysis of pattern using contiguous quadrats.* It is a technique for detecting and defining spatial concentrations in data expressed as counts per grid unit.

The frequency distributions of counts within the quadrats of a grid may be tested for randomness or nonrandomness by comparing them to a Poisson distribution. This may be done by testing the observed variance/mean ratio for a significant departure from unity or by testing the fit of the observed distribution to a Poisson distribution with χ^2 (Dacey 1973; Greig-Smith 1964:61–64; Kershaw 1964:101–103). However, this approach to spatial analysis ignores the existence of larger scales of pattern, which may be expected to exist in many cases. Dimensional analysis of variance, on the other hand, searches for nonrandom spatial distribution at different scales, and even at multiple scales, by testing for the tendency of items to concentrate in areas of different sizes. It does this through the calculation of corrected variances and variance/mean ratios, to detect significant departures from random distribution within a series of quadrats of increasing size, as well as within the original grid units.

Dimensional analysis of variance requires a grid whose sides contain a number of units equal to a power of 2 and which is either square or rectangular, with one side exactly twice the length of the other. Statistically, corrected mean squares (variances) of item counts are calculated at different "block sizes," in which each block above the size of the original grid units is created by grouping together two blocks of the next smaller size. Successively larger blocks thus cover twice the area of the next smaller blocks and consist of 2, 4, 8, 16, 32, and so on original grid squares. The last and largest block comprises the entire area of the grid (Fig. 27.1).

These requirements create two problems that frequently have been seen to be weaknesses of the technique. First, only those specific block sizes created by the progression of quadrats doubling in size at each step actually are tested for randomness or nonrandomness of item counts, or degree of spatial concentration. Obviously, the range of area covered between successive block sizes increases geometrically. Thus, there is a progressive loss of fineness, or precision, in the detection of spatial patterning and the ability to specify the scale of such patterning. This means, for example, that the detection of significant spatial concentration of items within Block Size 32 (composed of 32 original grid units grouped together) does not mean necessarily that this is the actual size of the concentration(s) but only that the actual size lies somewhere near the middle of the range between Block Size 16 and Block Size 64 (see also Pielou 1969:105). An area of Block Size 32 is simply the best approximation of the actual size of the concentration(s) that can be made by doubling and redoubling the quadrats formed from the original units of that particular grid.

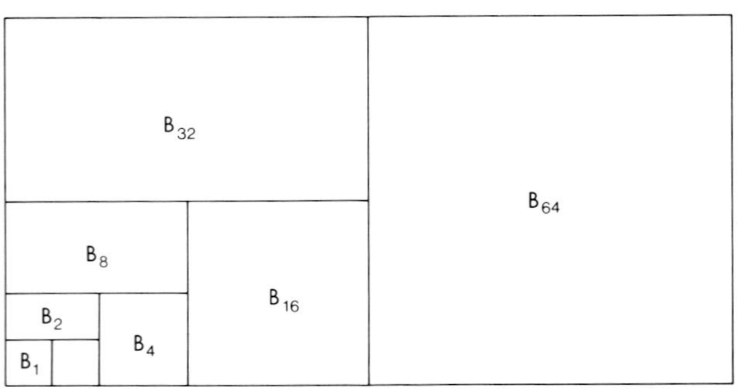

Fig. 27.1. The division of a grid into blocks of the sizes and shapes used in dimensional analysis of variance.

Second, the shapes of such best approximations are either square or rectangular, due to the grouping and regrouping of the original grid squares as described. Obviously, these shapes may be rather poor approximations of the shapes of many concentrations on an occupation floor. A great deal depends on the exact placement and orientation of the grid, as well as on the size of its basic units, or squares, and these are factors that often are established before excavation or collection are done and cannot be modified after the fact in the light of subsequent analytical results. The rectangular blocks, in particular, may provide poor representation of the shapes of spatial concentrations, especially if these concentrations themselves are elongated but oriented differently from the rectangular blocks. Thus, in square grids, better approximations sometimes may be made by doing two analyses, one grouping square blocks into horizontal rectangles and one grouping them into vertical rectangles. If either horizontal or vertical grouping provides a much better approximation to the size and orientation of elongated spatial concentrations of materials on the area under analysis, this will be evident in the much stronger and more significant patterning revealed by the analysis using that particular mode of grouping. The analysis of rectangular grids requires, of course, the use of rectangular blocks that are oriented in the same way as the whole grid, but these grids always can be treated as two separate square grids if it is suspected that rectangular grouping in the other orientation would provide better results.

Given these constraints on the size and shape of the area within which spatial concentrations can be detected and defined by dimensional analysis of variance, as well as other, probably generally less important constraints inherent in this approach (Hodder and Orton 1976:36–38; see also discussions in Riley 1974; Schiffer 1974; Whallon 1974b), it obviously must be used with some caution, prudence, and even finesse. This involves, among other things, regular checking back and forth between the results of analysis and the original data. Used thus, with care and intelligence, however, dimensional analysis of variance can be a useful technique to the archaeologist engaged in the search for meaningful spatial patterning of the materials recovered from clear occupation floors.

THE STATISTICAL TECHNIQUE

A separate dimensional analysis of variance must be made for each kind of material (item) in turn. It is a technique for defining the spatial patterning only of a single class of items, not for analyzing the spatial relationships among materials.

The evaluation of spatial relationships among different kinds of items is made at later steps in the sequence of analytical procedures, which, however, are applied to data organized in terms of the results of the preceding dimensional analyses of variance.

A dimensional analysis of variance begins with the calculation of the sums of squares, S_1, S_2, S_4, S_8, S_{16} . . . , S_T, for each block size. These sums of squares are calculated as

$$S_j = (1/j) \sum_{i=1}^{T/j} N^2_{j(i)}$$

where $N_{j(i)}$ is the item count in the ith block of j grid squares and T is the total number of squares in the original grid (Thompson 1958). In running through the entire sequence of block sizes, j = 1, 2, 4, 8, 16 . . . , T, and i = 1, 2, 3, 4, 5, . . . , T/j. At the largest block size, that comprising the total area covered by the grid, N_T is simply the total number of items found within the entire grid. In other words, to calculate the sum of squares at each block size, the count of items in each block is squared and these squared values are added up and divided by the number of original grid units that were grouped together to form a single block of that particular size.

Tendencies for spatial concentration and the scale(s) at which such concentration occurs then are found by calculating the variance in counts per block at each block size. As block size approaches the size of any actual spatial concentrations, the tendency will grow for these to be contained entirely within one or more blocks, to the exclusion of the other blocks (i.e., the between-block variance in counts will increase). As block size passes beyond this point, the precise "fit" of such concentrations within blocks will weaken and the between-block variance consequently will decrease.

This variance, or mean square, is calculated for each block size as

$$M_j = (S_j - S_{2j})/D_j$$

where D_j is the number of degrees of freedom appropriate to block size j and is defined as $D_j = T/2j$.

The results of a dimensional analysis of variance consist, therefore, of a table of block sizes, degrees of freedom, mean squares, and, occasionally, F-tests of these mean squares, corrected variance/mean ratios, and the Z-scores of these ratios. Commonly, also, and more useful for visual presentation and interpretation, the results of analysis are presented as a graph of mean square versus block size (e.g., Figs. 27.2 and 27.3). Such graphs show tendencies for concentration as distinct peaks in the line of the graph at certain block sizes. They thus provide a visual impression of the existence and degree of spatial concentration(s), as well as a clear demonstration of the scale, or size of area, at which these concentrations are detected most strongly by the analysis.

Statistically testing the significance of observed peaks in the graph of mean square versus block size is relatively difficult and has been the subject of some discussion (Thompson 1958; Greig-Smith 1961, 1964:86–87). The first suggestion, to test the significance of variances (mean squares) at larger block sizes against the variance at Block Size 1 (the original grid units) by means of an F (or variance ratio) test, has been shown to be statistically unjustifiable. The necessary assumption of normal and independent distribution of the counts among the blocks within the grid is shown to be violated as soon as a "significant" tendency for concentration (i.e., for nonrandomness of distribution) is demonstrated by the test (Thompson 1958:326; Greig-Smith 1964:86). The variances at the different block sizes cannot, in fact, be independent since they are calculated from counts obtained by continuous recombining of the counts in the original grid units rather than from independent sampling of spatial distribution at different scales (Pielou 1969:105).

The generally accepted solution in ecological analysis simply has been to use consistency of results from several analyzed samples as the criterion for significance. This is not

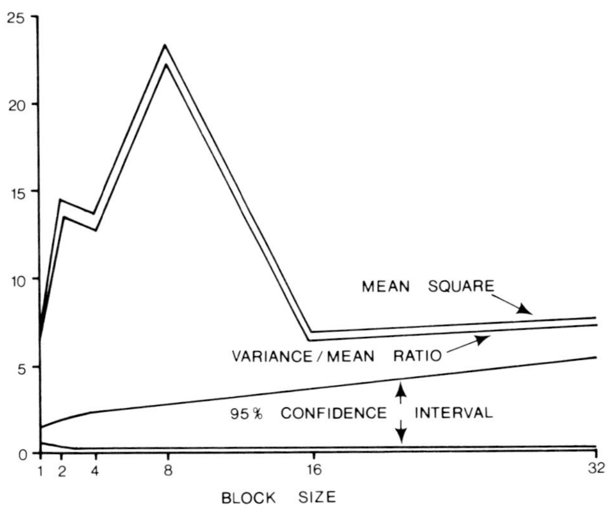

Fig. 27.2. Dimensional analysis of variance of *susí* nuts in Guilá Naquitz Zone C—graphs of mean square and of corrected variance/mean ratio against block size, with the 95% confidence interval for the variance/mean ratio. (Redrawn after Whallon 1973: Fig. 3.)

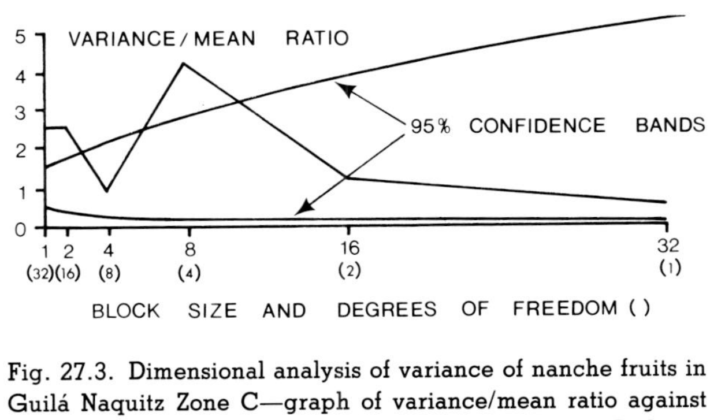

Fig. 27.3. Dimensional analysis of variance of nanche fruits in Guilá Naquitz Zone C—graph of variance/mean ratio against block size, with the 95% confidence bands for this ratio. (Redrawn after Whallon 1973: Fig. 5.)

so satisfactory for archaeology since most "samples" of occupation floors are unique. Exactly comparable situations for which relevant data are available seldom, if ever, exist.

If results are expressed, not in terms of mean squares, but as variance/mean ratios, however, confidence intervals readily can be calculated. Variance/mean ratios have a variance of 1, and if a distribution is random with unit variance, sums of squares with D_j degrees of freedom are distributed as χ^2 with D_j degrees of freedom, while variance (or mean square) is distributed as χ^2/D_j (Thompson 1958; Greig-Smith 1961, 1964:87). For example, significance bands for a 95% confidence interval are calculated with the upper limit for sums of squares (U) equal to the value of χ^2 for $p = .025$ with D_j degrees of freedom. The corresponding values for the upper and lower (L) bounds for variance (in this case, variance/mean ratio) are given by U/D_j and L/D_j. Tables for these 95% confidence intervals are given in Greig-Smith (1961, 1964:230). The variance/mean ratio at each block size is calculated by dividing the mean square at that block size by the mean number of items per original grid square (Thompson 1958:327; Greig-Smith 1961:699). This is done at all block sizes, including Block Size 1, whereby a "corrected" value for the variance/mean ratio for the original grid units is obtained.

One disadvantage of these confidence intervals is that they rapidly become very large as the number of degrees of freedom decreases (i.e., as block sizes become larger; see, for example, Figs. 27.2 and 27.3). This allows an increasingly wide range of random fluctuation in the variance/mean ratio at larger block sizes. It seems that regularly recurring peaks of variance at these larger block sizes can be dismissed as random by the rigorous application of confidence limits to the variance/mean ratio. In ecological work, therefore, it is often the case that consistency of results is preferable to confidence intervals as a means to determine "significance." This probably often will be the case in archaeological analysis also. In short, although a rigorous, statistical test of significance may be applied, the use of informed judgment frequently may be necessary or desirable in the interpretation of the results of dimensional analyses of variance.

With due caution and attention to potential problems, therefore, dimensional analysis of variance can be used successfully as the first step in the spatial analysis of archaeological materials that have been collected or excavated in grid squares over an occupation area. This technique will, with some reliability and accuracy, detect and define spatial concentrations and the scales at which they occur in such grid-count data. Further, the way in which it does this readily allows the regrouping of the original data in a form that not only is appropriate to the analysis of spatial relationships—correlations or associations—among classes of items but also represents the spatial patterning—concentrations or aggregations—of the individual item classes in this subsequent analysis. Such regrouping is done simply by reorganizing the data into counts per block for the block size at which the majority of classes show their peak variance (greatest mean square) or highest variance/mean ratio.

These regrouped data then may be used for further analysis. Here, they are used in simple correlation analysis, and the matrices of correlation coefficients are ordered to display the mutually correlated classes as clusters. These clusters, and any other systematic patterns of relationships revealed by the ordered matrices, form the bases for description and interpretation of the spatial organization of archaeological materials on the occupation floor(s) under analysis.

THE GUILA NAQUITZ OCCUPATION FLOORS

Guilá Naquitz, as described in earlier chapters, was excavated in 1 × 1-m squares. Any analysis dependent on exact provenience data, therefore, was impossible. Although contour maps of density per 1 × 1-m square showed clear tendencies for most, if not all, classes of items to be spatially aggregated, these concentrations overlapped in a manner too confusing to allow any clear definition by eye of clusters of spatially associated classes of items. Statistical analysis, therefore, was necessary, and our procedure began with dimensional analysis of variance, preliminary to correlation analysis and the ordering of the correlation matrices.

The real grid of 1 × 1-m excavation units was neither square nor rectangular. It did not fit the requirements outlined above for the application of dimensional analysis of variance. The central portion of the grid was not far from square, however, and through the addition of a few imaginary or "dummy" squares (see Whallon 1973:268, 1974b), a square grid 8 m on a side was constructed which covered the major portion of the occupation area and was of appropriate dimensions for analysis (Fig. 27.4).

Dimensional analysis of variance may be applied in cases where items are very infrequent and counts are extremely low. However, it makes sense to use it only when items are distributed within more than one or two grid units. Accordingly, on three of the four occupation floors analyzed, a few items were eliminated from the spatial analysis because they were concentrated exclusively, or virtually exclusively, within a single grid square. In addition, a few classes of items did not occur on one of the floors. Depending on the occupation floor, there were 13–16 classes of items that occurred with an adequate distribution to allow meaningful analysis. These included unretouched flint flakes, 2–3 genera of animals, and 9–12 genera of plants.

The results of the dimensional analyses of variance were quite clear, and tendencies to significant spatial concentration were seen for virtually every class of item on every occupation floor. Mean squares generally were high, and the variance/mean ratios typically also were high, all far above the upper confidence limit for random values of this ratio at the various block sizes. The graphs of mean squares and of variance/mean ratios typically showed a striking major peak, as in the case of *Jatropha* or *susí* nuts in Zone C (Fig. 27.2). In other cases, such as that of *Malpighia* (the West

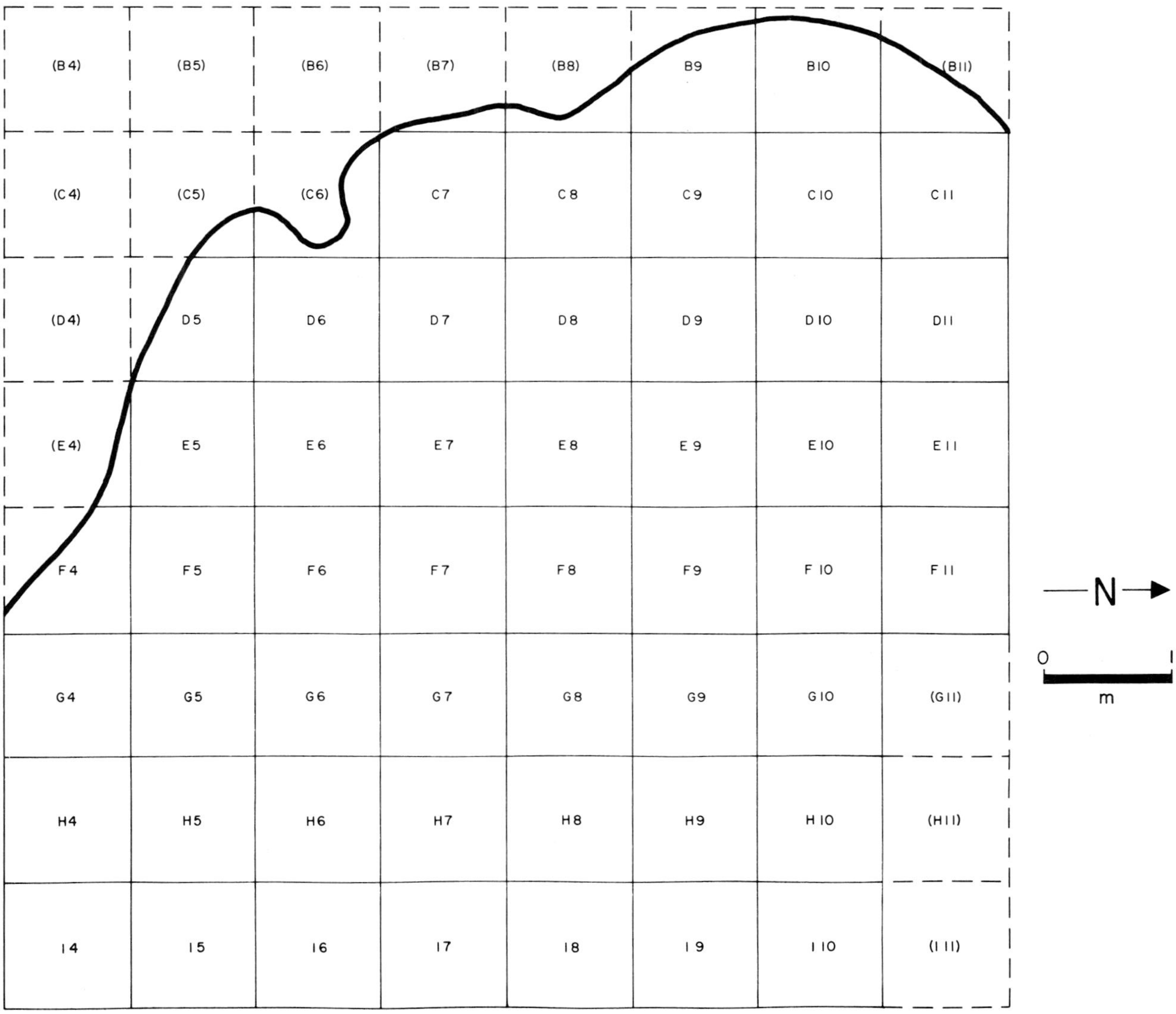

Fig. 27.4. Plan of Guilá Naquitz showing the grid used for spatial analysis using dimensional analysis of variance. "Dummy" squares added to create a square grid are shown with dashed lines. (Redrawn after Whallon 1973: Fig. 2.)

Indian cherry or nanche fruit) in Zone C (Fig. 27.3), the mean squares were small and the variance/mean ratios low, due to the very low density of this item on the occupation floor. However, the application of 95% confidence intervals generally indicated that the spatial patterns revealed by the peaks in these graphs were significant.

Once it had been determined at which block size the strongest spatial patterning occurred on each occupation floor, the data were regrouped in units of that size and correlations were calculated among all classes of items using the standard Pearson's r. The resultant correlation matrices then were ordered with Craytor's program SERIATE (Craytor and Johnson 1968).

RESULTS FOR INDIVIDUAL OCCUPATION FLOORS

Zone E

It was possible to analyze the spatial distribution of 13 materials on the occupation floor of Zone E (Table 27.1). Dimensional analysis of variance of each of these materials showed strong and regular patterning at the scale of Block Size 16, representing differential concentration of materials within blocks composed of 16 original 1 × 1-m squares. These blocks comprise a division of the grid used for analysis into fourths, or quarters, covering the north and south sides and the front and back of the cave.

TABLE 27.1
Materials Used in the Spatial Analysis of Zone E at Guilá Naquitz

Item no.	Material
1	*Agave* quids
2	Hackberry seeds
3	*Susí* hulls
4	Nanche fruits
5	*Opuntia* seeds
6	Bean pod valves
7	Piñon nuts
8	Mesquite seeds
9	Acorns
10	Turtle bones
11	Deer bones
12	Rabbit bones
13	Flint flakes

Correlation coefficients based on the Zone E data, regrouped at Block Size 16, were calculated as indices of spatial association among these 13 materials, and the correlation matrix was ordered so as to bring mutually interrelated groups of materials together. In this ordered matrix for Zone E materials (Fig. 27.5), two distinct groups of spatially associated materials clearly are evident.

The first of these groups is made up of *susí* nuts, *Agave* quids, nanche fruits, and acorns. All these materials are likely to have been brought back to the cave from plant collecting trips to Thorn Forest A. Such trips, it may be hypothesized, most likely would have been made by women. The density contours of all four of these items, combined, exhibit a series of strong spatial concentrations running north–south across the middle to the front of the cave floor (Fig. 27.6).

The second major spatial association includes piñon nuts, flint flakes, turtle, deer, and rabbit bones, mesquite seeds, and hackberry seeds (Fig. 27.5). Perhaps the most informative aspect of this grouping is the inclusion of all three genera of animals as well as flint flakes. Ultimately, these materials are likely to be the remains of men's hunting and butchering activities, but given their association with numerous plants, we may be dealing with an area of debris from food preparation. The plant remains in this association include those native to Mesquite Grassland B (mesquite, hackberry), as well as piñon nuts, which presumably came from higher elevations (perhaps Thorn Forest A). Thus, the complex of spatially associated materials on this occupation floor may represent the remains of both men's and women's activities, as well as trips to more than one vegetational zone. Spatially, this group of materials, as a whole, shows an exceptionally tightly localized concentration on the north side and at the back of the cave (Fig. 27.7).

Wild beans and *Opuntia* seeds both exhibit independent patterns of spatial distribution and, while individually concentrated in spots on the occupation floor (Figs. 26.8 and 26.6), are not associated with any other materials in these locations.

Although the composition of the two major groupings of spatially associated materials on this floor allow at least partial interpretation, and thus, "make sense" to some degree, this living floor in Guilá Naquitz does not show a clear division into men's and women's work areas as do some of the other zones. This conclusion parallels the results of Reynolds' analyses (Chapter 28).

Zone D

Fifteen different kinds of materials could be analyzed for spatial distributional pattern over Zone D at Guilá Naquitz (Table 27.2). Dimensional analyses of variance indicated that spatial patterning was strongest for most of these materials at Block Size 8, grouped horizontally. This indicated differential spatial distribution among areas composed of eight original 1 × 1-m grid squares, or among eighths of the grid used for analysis, representing the north and south sides of the site and finer subdivisions of the front–back dimension of the floor than were found for Zone E. The ordered matrix of correlation coefficients among these materials, calculated

MATERIAL	3	1	4	9	5	7	13	10	8	12	2	11	6
3. SUSI HULLS	1.00	.87	.97	.94	.49	-.08	-.10	-.13	-.22	-.21	-.24	-.26	-.29
1. AGAVE QUIDS	.87	1.00	.75	.92	.22	-.09	-.12	-.09	-.27	-.17	-.25	-.10	.19
4. NANCHE FRUITS	.97	.75	1.00	.90	.64	.02	.00	-.05	-.08	-.11	-.13	-.22	-.48
9. ACORNS	.94	.92	.90	1.00	.58	.17	.15	.14	.01	.06	.00	.04	-.11
5. OPUNTIA SEEDS	.49	.22	.64	.58	1.00	.73	.72	.65	.68	.62	.64	.45	-.59
7. PINON NUTS	-.08	-.09	.02	.17	.73	1.00	.99	.99	.98	.98	.98	.92	-.08
13. FLINT FLAKES	-.10	-.12	.00	.15	.72	.99	1.00	.99	.98	.99	.98	.92	-.08
10. TURTLE BONES	-.13	-.09	-.05	.14	.65	.99	.99	1.00	.97	.99	.98	.96	.03
8. MESQUITE SEEDS	-.22	-.27	-.08	.01	.68	.98	.98	.97	1.00	.98	.99	.90	-.14
12. RABBIT BONES	-.21	-.17	-.11	.06	.62	.98	.99	.99	.98	1.00	.99	.96	.03
2. HACKBERRY SEEDS	-.24	-.25	-.13	.00	.64	.98	.98	.98	.99	.99	1.00	.93	-.05
11. DEER BONES	-.26	-.10	-.22	.04	.45	.92	.92	.96	.90	.96	.93	1.00	.28
6. BEAN POD VALVES	-.29	.19	-.48	-.11	-.59	-.08	-.08	.03	-.14	.03	-.05	.28	1.00

Fig. 27.5. Ordered matrix of correlation coefficients among the 13 classes of material analyzed from the occupation floor of Zone E, Guilá Naquitz. Coefficients of .90 or greater have been underlined to make clusters of mutually associated materials stand out more clearly in this matrix. (The range of coefficients underlined is arbitrary and does not indicate significant, as opposed to nonsignificant, values.) The associations and clusters evident in this ordered matrix are described and discussed in the text.

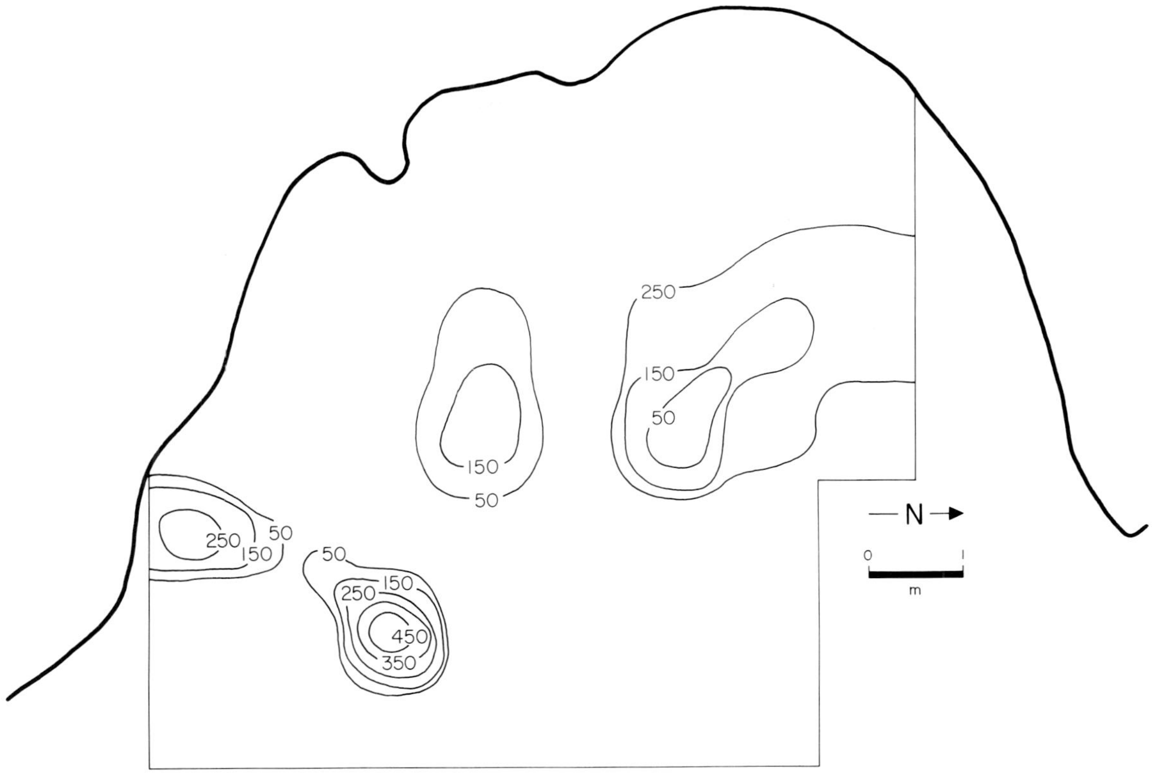

Fig. 27.6. Density contours of *susí* hulls, *Agave* quids, nanche fruits, and acorns, taken together as a group over the area subjected to spatial analysis at Guilá Naquitz, Zone E.

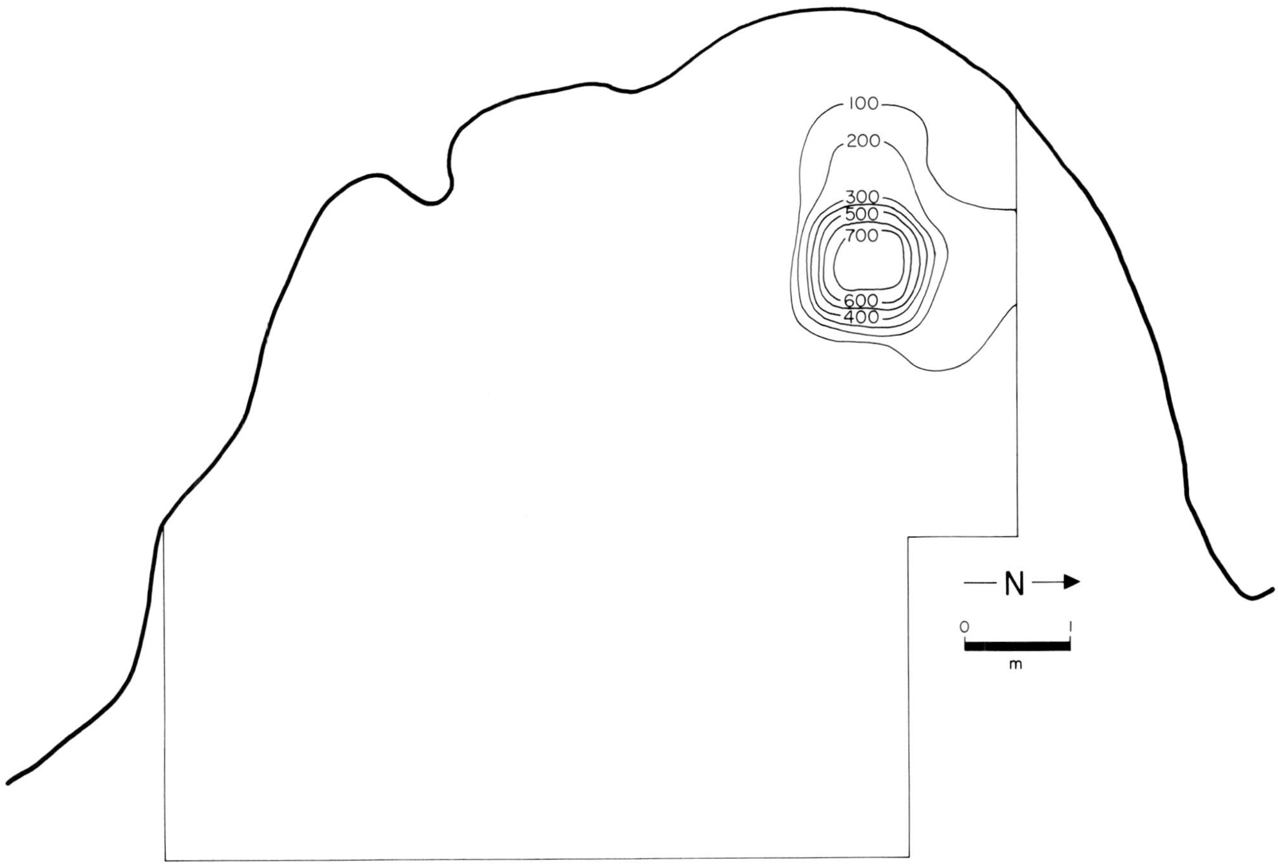

Fig. 27.7. Density contours of piñon nuts, flint flakes, turtle bones, mesquite seeds, rabbit bones, hackberry seeds, and deer bones, taken together as a group over the area subjected to spatial analysis at Guilá Naquitz, Zone E.

TABLE 27.2
Materials Used
in the Spatial Analysis
of Zone D at Guilá Naquitz

Item no.	Material
1	*Agave* quids
2	Hackberry seeds
3	Cucurbit remains
4	*Susí* hulls
5	*Guaje* pods
6	Nanche fruits
7	*Opuntia* seeds
8	*Opuntia* nopales
9	Bean pod valves
10	Piñon nuts
11	Mesquite seeds
12	Acorns
13	Deer bones
14	Rabbit bones
15	Flint flakes

on the data reorganized at Block Size 8, reveals one major group of mutually interrelated items and three minor groups of two items each (Fig. 27.8).

The major group of spatially associated materials includes wild beans, deer bones, flint flakes, hackberries, acorns, piñon nuts, and *Opuntia* seeds. This grouping includes plant materials characteristic of both the Thorn Forest (wild beans, acorns, piñon nuts) and Mesquite Grassland (hackberries), as well as items that are found in both vegetational zones. If such a group of items, found clustered together on an occupation floor, can be interpreted in behavioral terms, it might be supposed to represent the remains of both men's and women's hunting and gathering in both Thorn Forest and Mesquite Grassland areas. The combined counts of the materials included in this group show a relatively quite dense and large spatial aggregation on the occupation floor, heavily concentrated on the south side of the site, but extending also to the north, and ranging from the front to the middle of the cave (Fig. 27.9).

There are three pairs of items that, conceivably, could be meaningful groupings in this matrix. The pair based on the highest counts, and the greatest range of counts, is made up of *susí* nuts and *Agave* quids. The combined counts of these two plant materials clearly are concentrated in the front and to the south side of the site (Fig. 27.10). These might be the remains from women's collecting trips to the Thorn Forest.

The other pairs are based on much lower counts, with more restricted ranges. Mesquite seeds and rabbit bones tend to be found together on this occupation floor and might be considered to represent the results of gathering and hunting (perhaps trapping) in the Mesquite Grassland. Their combined distribution shows some tendency to concentrate slightly toward the back of the cave, but a wide scatter of low counts over much of the occupation floor makes any conclusions drawn from this association very tentative (Fig. 27.11).

The final grouping of *guaje* pods and nanche fruits is based on even lower counts but occurs on the floor as virtually a single, tight concentration in the center of the site (Fig. 27.12). Both could result from collecting in the Thorn Forest (presumably a women's activity), and their low frequency combined with their sharp spatial localization may reflect a single instance of such activity during this occupation.

Nopales show no clear pattern of spatial association on this floor. Early domestic cucurbits, likewise, are not associated with any other items. Nopales are, by and large, concentrated in one small area (Fig. 26.13). Cucurbits occur in extremely low frequency (Fig. 26.16). In both cases, the data are not adequate to draw any significant statistical conclusions about spatial distributions or associations.

MATERIAL	5	6	11	14	9	13	15	2	12	10	7	8	4	1	3
5. GUAJE PODS	1.00	.99	.87	.62	.33	.09	.19	-.11	-.11	-.56	-.59	-.84	-.60	-.43	-.22
6. NANCHE FRUITS	.99	1.00	.87	.63	.36	.11	.23	-.09	-.07	-.54	-.57	-.85	-.58	-.40	-.23
11. MESQUITE SEEDS	.87	.87	1.00	.90	.51	.32	.56	.31	.22	-.27	-.30	-.52	-.76	-.65	-.64
14. RABBIT BONES	.62	.63	.90	1.00	.58	.36	.81	.62	.52	.19	.18	.17	-.28	-.36	-.47
9. BEAN POD VALVES	.33	.36	.51	.58	1.00	.85	.72	.62	.63	.58	.55	-.30	.08	.17	-.75
13. DEER BONES	.09	.11	.32	.36	.85	1.00	.64	.72	.63	.67	.65	.03	.24	.27	-.65
15. FLINT FLAKES	.19	.23	.55	.81	.72	.64	1.00	.92	.90	.76	.75	.70	.32	.18	-.22
2. HACKBERRY SEEDS	-.11	-.09	.31	.62	.62	.72	.92	1.00	.89	.79	.79	.62	.10	.00	-.62
12. ACORNS	-.11	-.07	.22	.52	.63	.63	.90	.89	1.00	.91	.92	.86	.60	.46	-.01
10. PINON NUTS	-.56	-.54	.27	.19	.58	.67	.76	.79	.91	1.00	.99	.50	.53	.45	-.51
7. OPUNTIA SEEDS	-.59	-.57	-.30	.18	.55	.65	.75	.79	.92	.99	1.00	.53	.53	.45	-.49
8. NOPALES	-.84	-.85	-.52	.17	-.30	.03	.70	.62	.86	.50	.53	1.00	.11	-.08	-.84
4. SUSI HULLS	-.60	-.53	-.76	-.28	.08	.24	.32	.10	.60	.53	.53	.11	1.00	.98	.40
1. AGAVE QUIDS	-.43	-.40	-.65	-.36	.17	.27	.18	.00	.46	.45	.45	-.08	.98	1.00	.40
3. CUCURBITS	-.22	-.23	-.64	-.47	-.75	-.65	-.22	-.62	-.01	-.51	-.49	-.84	.40	.40	1.00

Fig. 27.8. Ordered matrix of correlation coefficients among the 15 classes of material analyzed from the occupation floor of Zone D, Guilá Naquitz. Coefficients of .90 or greater have been underlined to make clusters of mutually associated materials stand out more clearly in this matrix. (The range of coefficients underlined is arbitrary and does not indicate significant, as opposed to nonsignificant, values.) The associations and clusters evident in this ordered matrix are described and discussed in the text.

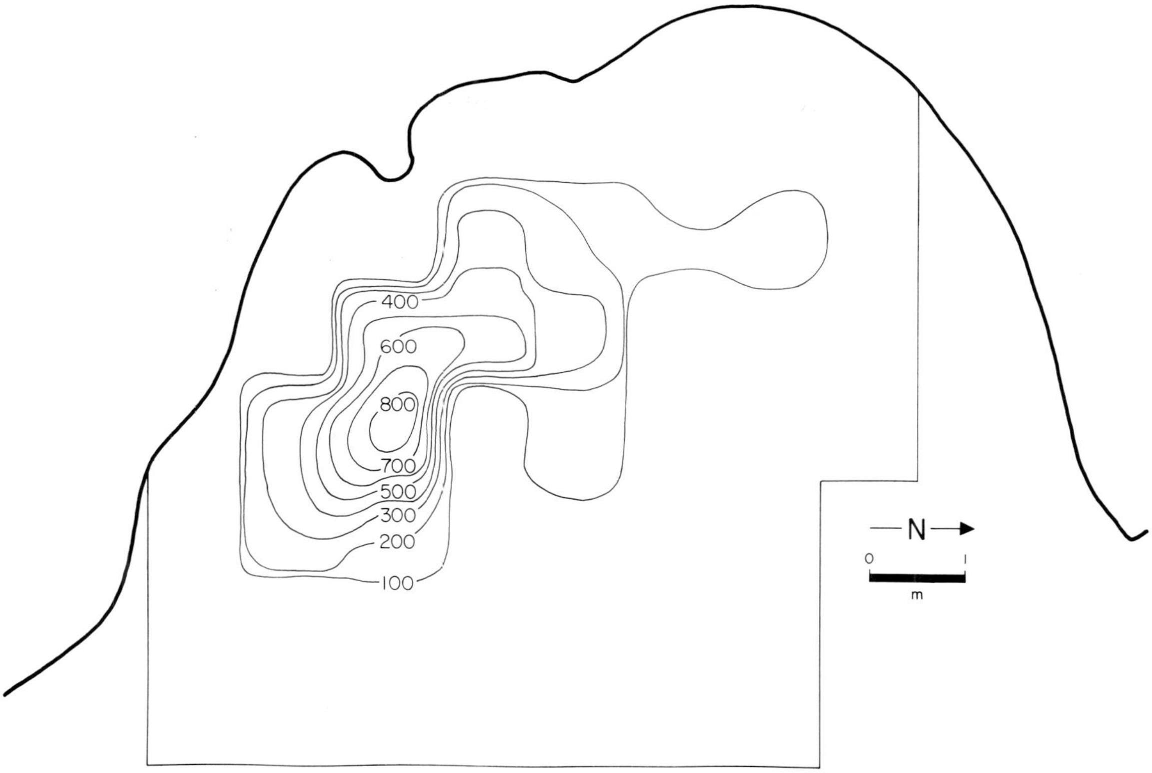

Fig. 27.9. Density contours of bean pod valves, deer bones, flint flakes, hackberry seeds, acorns, piñon nuts, and *Opuntia* seeds, taken together as a group over the area subjected to spatial analysis at Guilá Naquitz, Zone D.

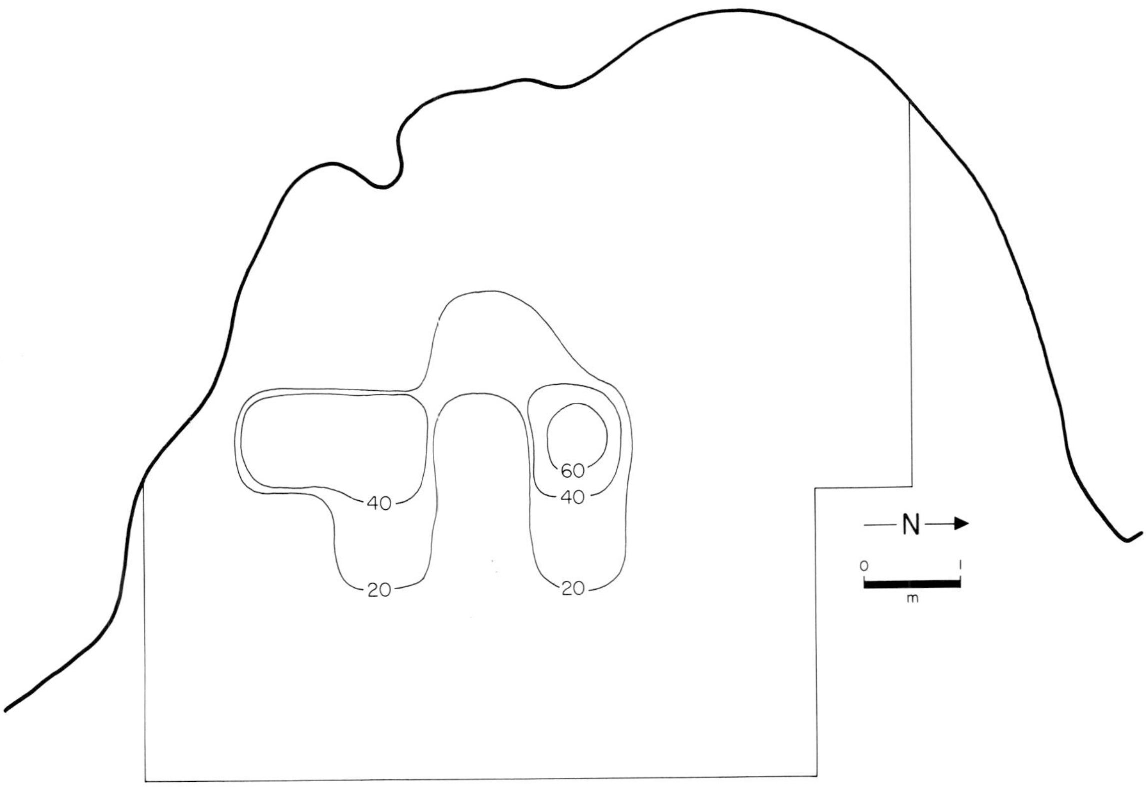

Fig. 27.10. Density contours of *susí* hulls and *Agave* quids, taken together as a group over the area subjected to spatial analysis at Guilá Naquitz, Zone D.

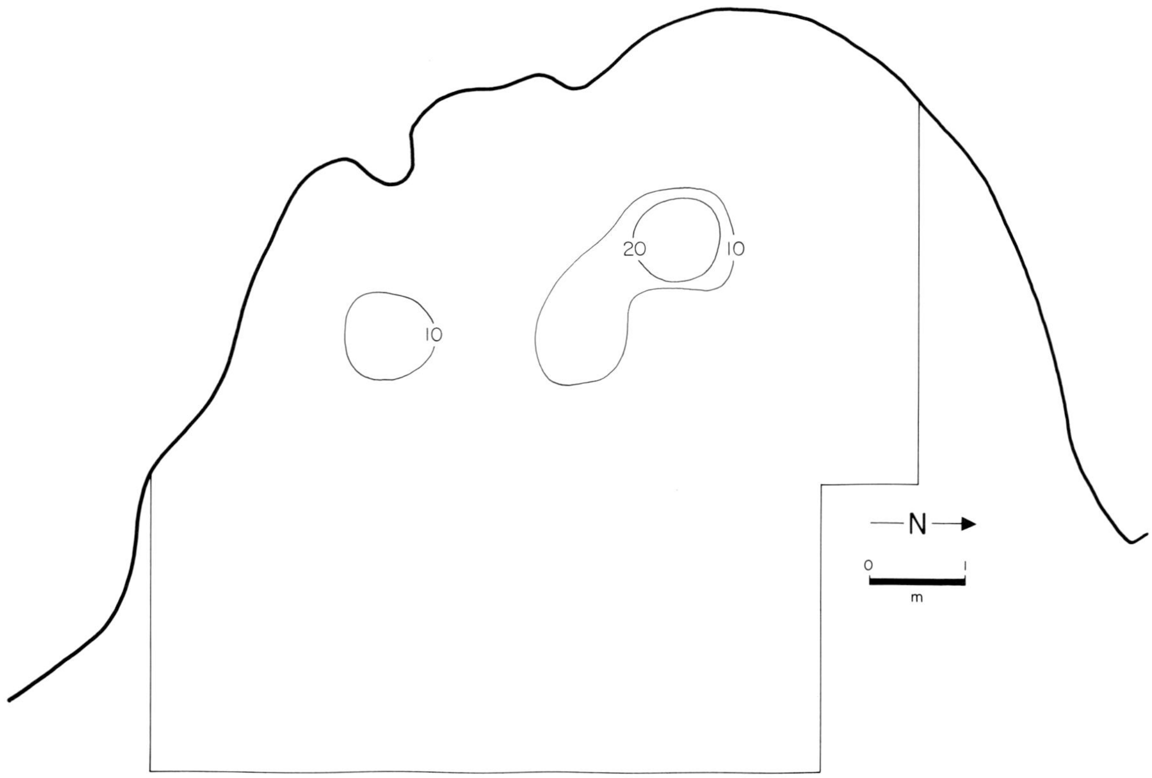

Fig. 27.11. Density contours of mesquite seeds and rabbit bones, taken together as a group over the area subjected to spatial analysis at Guilá Naquitz, Zone D.

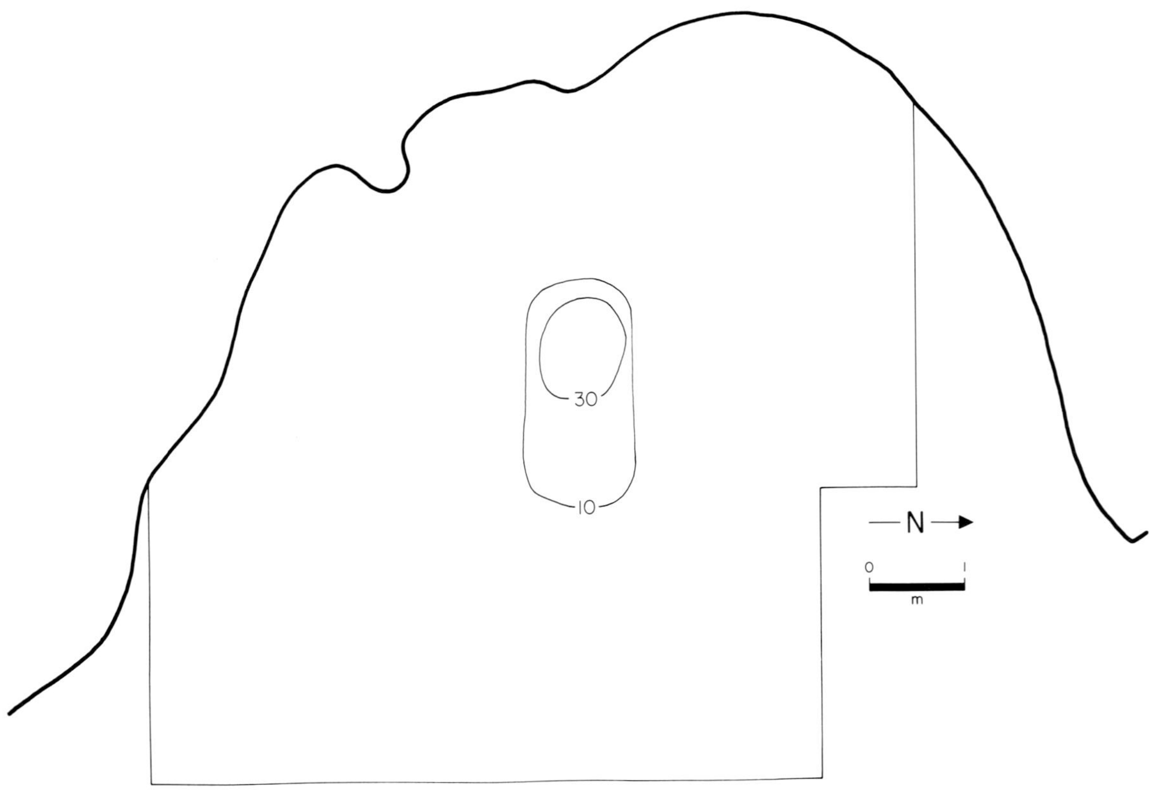

Fig. 27.12. Density contours of *guaje* pods and nanche fruits, taken together as a group over the area subjected to spatial analysis at Guilá Naquitz, Zone D.

Zone C

On the occupation floor of Zone C at Guilá Naquitz, 17 kinds of material occurred in sufficient quantities for spatial analysis (Table 27.3). The great majority showed the highest degree of spatial patterning when blocks of 8 squares were used, as in the analysis of Zone D. The ordered matrix of correlation coefficients, calculated on the data grouped at Block Size 8, exhibits 3 major groups of spatially interrelated items (Fig. 27.13).

The smallest of these groups is composed of deer, turtle, and rabbit bones. Flint flakes are also highly correlated with these animal bones. However, flint flakes also are highly correlated with one of the large groupings of plant remains and are not discretely associated with the faunal remains. The pattern of relationships, nonetheless, strongly suggests a spatial association on this occupation floor of all the materials related to hunting and butchering of animals, assumed to be men's activities. The distributional pattern of all the animal bones, taken together, shows a generally low density of occurrence but a clear pattern of two areal concentrations—primarily on the north side of the site, with a minor concentration on the south, both near the rear of the cave (Fig. 27.14).

One of the large groupings of plant remains is characterized by *susí* hulls, piñon nuts, acorns, hackberry seeds, and wild beans. Nanche fruits definitely are associated with this group, but they are not so tightly integrated into it as are the other, very highly interrelated items. Flint flakes also appear clearly associated with this group of plant remains. With the exception of hackberry, all these plants are products of Thorn Forest A, and this grouping perhaps may be considered generally to reflect gathering trips, presumably made by women, into this vegetational facies. Distributionally, there is a small, relatively low-density concentration at the back of the cave on the north side (Fig. 27.15). However, the major concentration occurs spread out north–south across the middle of the occupation floor, with two centers of peak aggregation—one on the north side and one on the south side of the site.

TABLE 27.3
Materials Used in the Spatial Analysis of Zone C at Guilá Naquitz

Item no.	Material
1	*Agave* quids
2	*Agave* leaves
3	Hackberry seeds
4	Cucurbit remains
5	*Susí* hulls
6	*Guaje* pods
7	Nanche fruits
8	*Opuntia* seeds
9	*Opuntia* nopales
10	Bean pod valves
11	Piñon nuts
12	Mesquite seeds
13	Acorns
14	Turtle bones
15	Deer bones
16	Rabbit bones
17	Flint flakes

The distribution of unretouched flint flakes (Fig. 26.18) obviously overlaps the spatial concentrations of this plant group, but it also clearly overlaps with the spatial concentrations of animal bones. Impressionistically, the overall configuration of the flint distribution appears to conform more closely to that of the bones on this floor, and it is very likely that the generally very low counts of bones from the grid squares of this level have attenuated significantly the correlations among flint flakes and each of the genera of animals represented.

The other major grouping of plant remains on this occupation floor consists of nopales, *guaje* pods, *Agave* leaves and quids, cucurbits, and, less strongly associated, mesquite seeds.

MATERIAL	9	6	2	1	4	12	7	5	11	13	3	10	17	15	14	16	8
9. NOPALES	1.00	.99	.99	.99	.98	.78	.40	.32	.01	-.07	-.04	-.11	.27	.15	-.16	-.15	-.19
6. GUAJE PODS	.99	1.00	.99	.99	.99	.80	.43	.36	.06	-.03	-.01	-.08	.30	.16	-.14	-.13	-.17
2. AGAVE LEAVES	.99	.99	1.00	.99	.99	.81	.43	.35	.05	-.03	.00	-.07	.30	.16	-.12	-.12	-.18
1. AGAVE QUIDS	.99	.99	.99	1.00	.99	.81	.47	.38	.07	-.01	.01	-.05	.32	.16	-.13	-.12	-.13
4. CUCURBITS	.98	.99	.99	.99	1.00	.86	.51	.45	.15	.06	.08	.02	.38	.20	-.06	-.07	-.13
12. MESQUITE SEEDS	.78	.80	.81	.81	.86	1.00	.74	.80	.60	.53	.57	.51	.73	.43	.35	.29	-.09
7. NANCHE FRUITS	.40	.43	.43	.47	.51	.74	1.00	.85	.66	.71	.62	.70	.68	.27	.36	.26	.55
5. SUSI HULLS	.32	.36	.35	.38	.45	.80	.85	1.00	.93	.91	.82	.83	.90	.51	.56	.52	.13
11. PINON NUTS	.01	.06	.05	.07	.15	.60	.66	.93	1.00	.97	.90	.90	.89	.55	.69	.65	.02
13. ACORNS	-.07	-.03	-.03	-.01	.06	.53	.71	.91	.97	1.00	.93	.95	.85	.51	.72	.65	.18
3. HACKBERRY SEEDS	-.04	-.01	.00	.01	.08	.57	.62	.82	.90	.93	1.00	.97	.86	.62	.88	.77	.04
10. BEAN POD VALVES	-.11	-.08	-.07	-.05	.02	.51	.70	.83	.90	.95	.97	1.00	.78	.46	.79	.65	.21
17. FLINT FLAKES	.27	.30	.30	.32	.33	.73	.68	.90	.89	.85	.86	.78	1.00	.82	.78	.80	-.03
15. DEER BONES	.15	.16	.16	.16	.20	.43	.27	.51	.55	.51	.62	.46	.82	1.00	.82	.93	-.22
14. TURTLE BONES	-.16	-.14	-.12	-.13	-.06	.35	.36	.56	.69	.72	.88	.79	.78	.82	1.00	.95	-.06
16. RABBIT BONES	-.15	-.13	-.12	-.12	-.07	.29	.26	.52	.65	.65	.77	.65	.80	.93	.95	1.00	-.13
8. OPUNTIA SEEDS	-.19	-.17	-.18	-.13	-.13	-.09	.55	.13	.02	.18	.04	.21	-.03	-.22	-.06	-.13	1.00

Fig. 27.13. Ordered matrix of correlation coefficients among the 17 classes of material analyzed from the occupation floor of Zone C, Guilá Naquitz. Coefficients of .90 or greater have been underlined to make clusters of mutually associated materials stand out more clearly in this matrix. (The range of coefficients underlined is arbitrary and does not indicate significant, as opposed to nonsignificant, values.) The associations and clusters evident in this ordered matrix are described and discussed in the text.

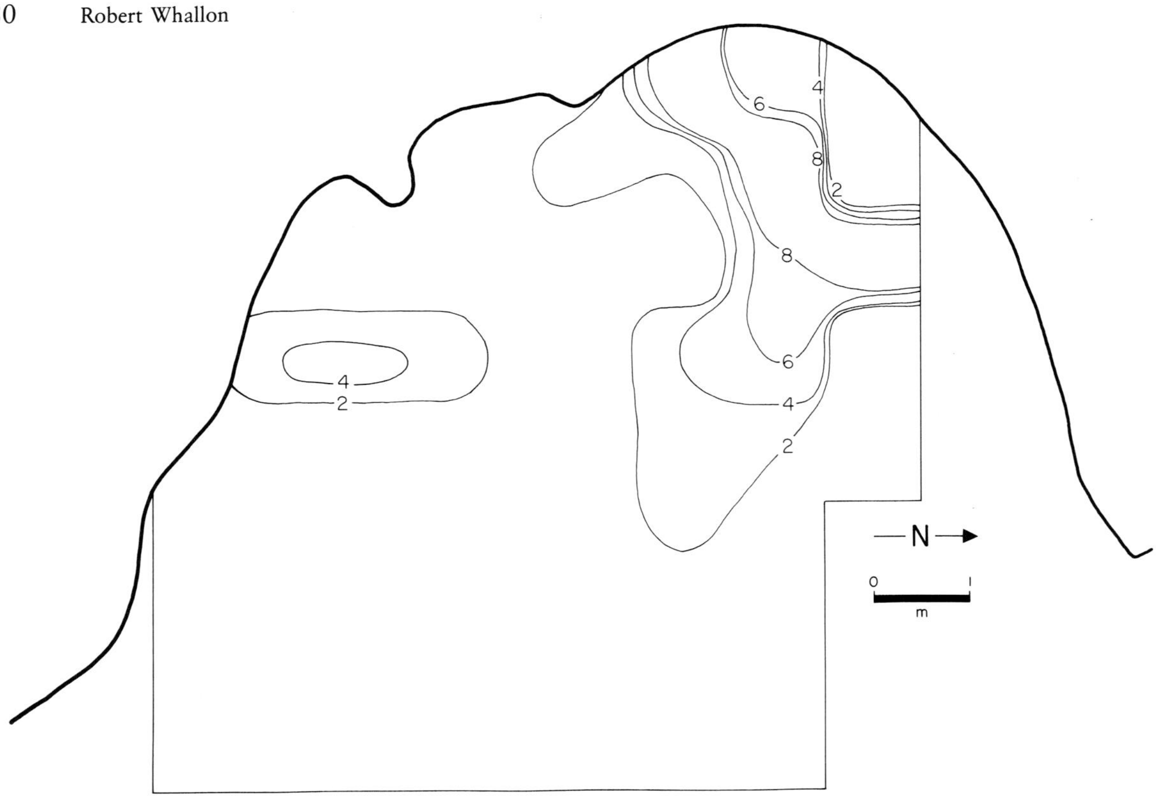

Fig. 27.14. Density contours of deer bones, turtle bones, and rabbit bones, taken together as a group over the area subjected to spatial analysis at Guilá Naquitz, Zone C.

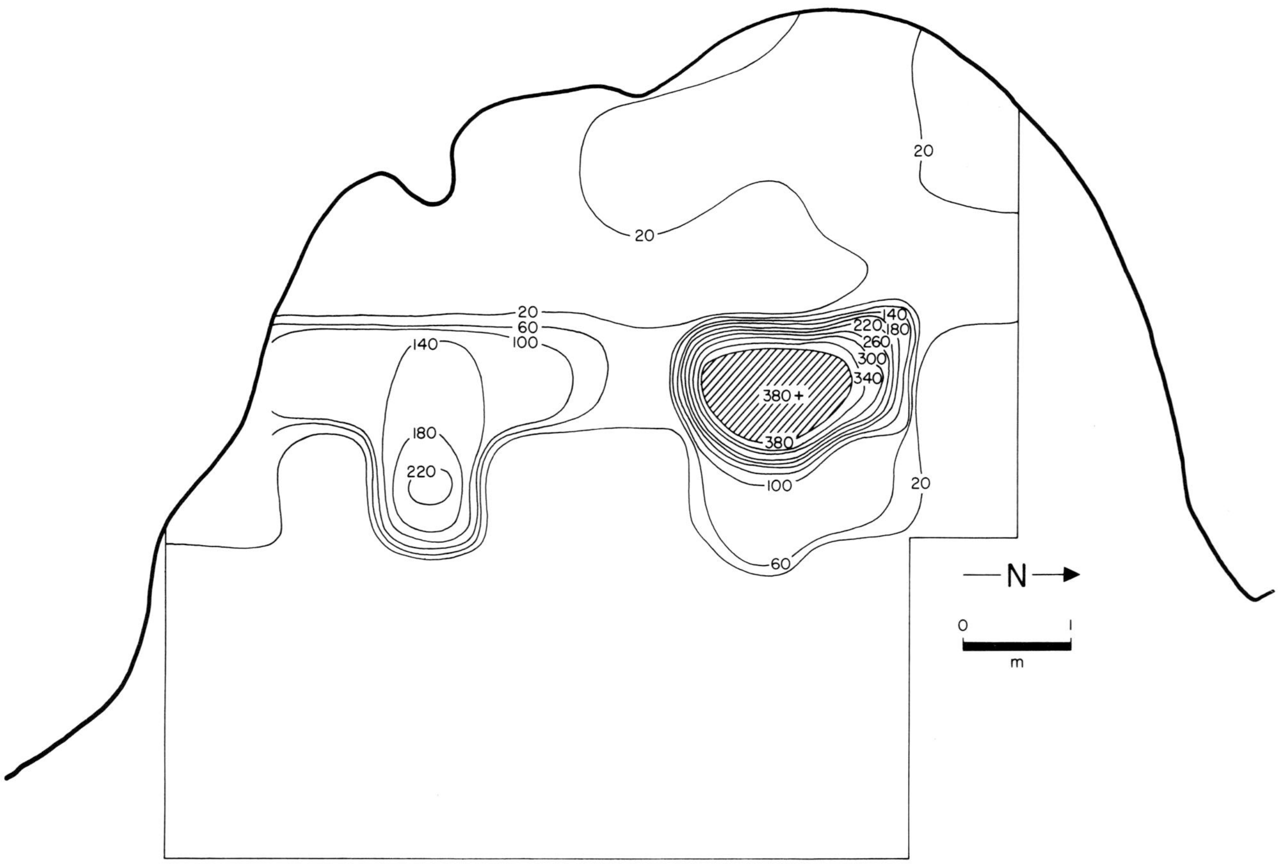

Fig. 27.15. Density contours of nanche fruits, *susí* hulls, piñon nuts, acorns, hackberry seeds, and bean pod valves, taken together as a group over the area subjected to spatial analysis at Guilá Naquitz, Zone C.

These plants represent a mixture of products from the Thorn Forest and the Mesquite Grassland, as well as a domesticated genus. All most probably would have been gathered or cultivated by women, but all the different possible collection or harvesting areas from which these plant products must have come are represented together in this group. Taken together, the distributional pattern of these plant remains exhibits a tight, dense, spatial concentration from the middle to the front of the occupation area on the north side (Fig. 27.16).

Opuntia seeds are rather abundant in this zone at Guilá Naquitz, but they are aggregated spatially quite independently of any of the other materials on this floor, being lightly clustered in the middle of the site on the north side and heavily concentrated on the south side from the middle to the front of the cave (Fig. 26.20).

Zone B1

There were 16 kinds of material present in sufficient quantities to allow analysis of their patterns of spatial distribution on the occupation floor of Zone B1 (Table 27.4). Maximal spatial patterning tended to be exhibited by these materials at Block Size 16, meaning aggregation of counts in units of 16 original 1 × 1-m squares, or quarters of the grid used for this analysis. This is the same scale of patterning as found on the occupation floor of Zone E.

The ordered correlation matrix, based on item counts at Block Size 16, shows two clear clusters of spatially associated materials (Fig. 27.17). The smaller of these appears to consist of rabbit and deer bones, flint flakes, and nanche fruits. Nanche fruits, however, are sparsely distributed over this floor in small, discrete concentrations, each centered on a specific 1 × 1-m square (Fig. 26.43). In contrast, the distribution of the deer and rabbit bones together, combined with the flint flakes, shows a broad distribution over the cave floor, with substantial areas of concentration (Fig. 27.18). The largest and most dense of these concentrations covers a broad area on the south side of the living floor, ranging from the back to the middle of the cave. A smaller, less dense aggregation occurs on the north in the middle of the cave, and there are isolated concentrations within single 1-m squares at the rear of the cave on the north side. Since the statistical association of nanches with this grouping could have resulted from men eating nanche fruits as they worked and spitting out the seeds in and around their work space, the larger, clear-cut group of materials in the south might well be interpreted as representative of the remains of men's animal processing activities. Reynolds (Chapter 28) also considers this a men's work area.

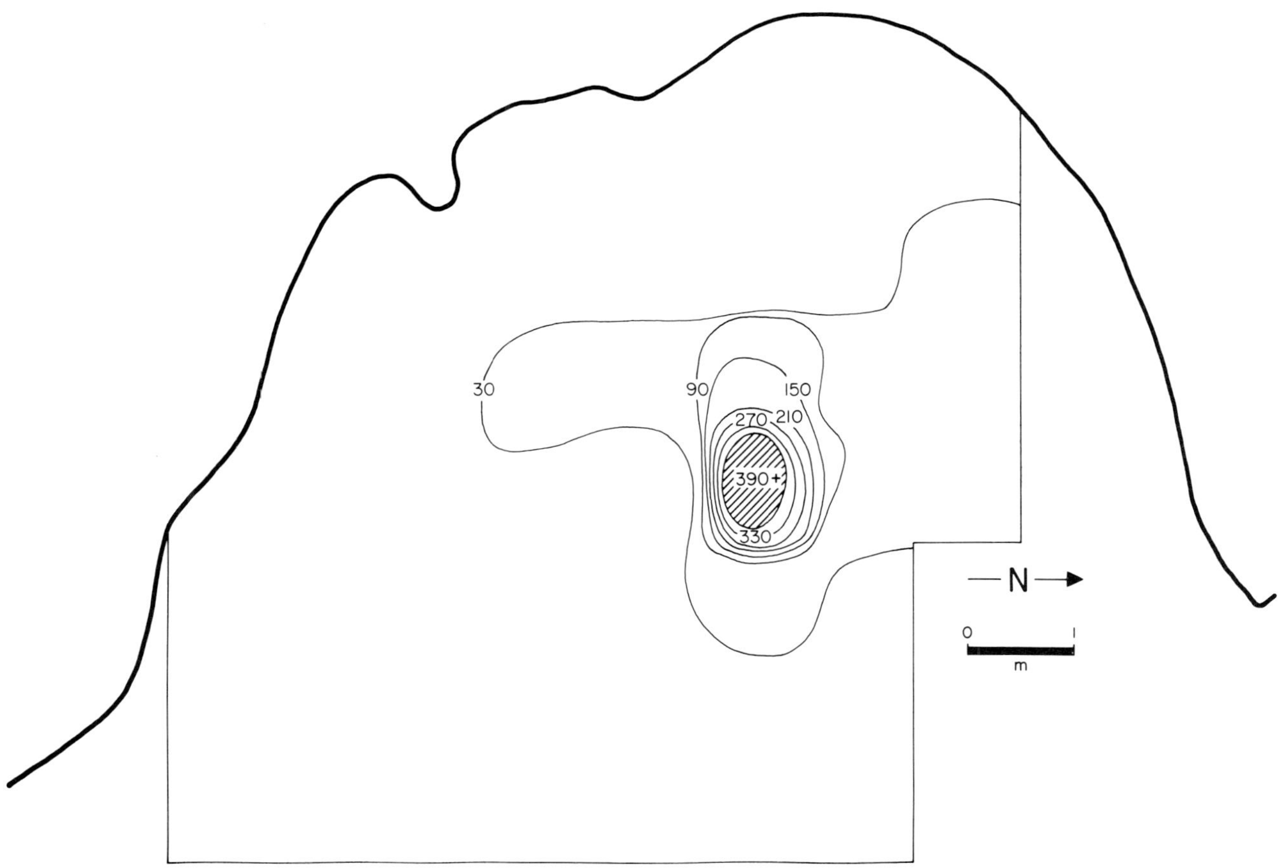

Fig. 27.16. Density contours of nopales, *guaje* pods, *Agave* leaves, *Agave* quids, cucurbit remains, and mesquite seeds, taken together as a group over the area subjected to spatial analysis at Guilá Naquitz, Zone C.

TABLE 27.4
Materials Used in the Spatial Analysis of Zone B1 at Guilá Naquitz

Item no.	Material
1	*Agave* quids
2	Hackberry seeds
3	Cucurbit remains
4	*Susí* hulls
5	*Guaje* pods
6	Nanche fruits
7	*Opuntia* seeds
8	*Opuntia* nopales
9	Bean pod valves
10	Piñon nuts
11	Mesquite seeds
12	Acorns
13	Turtle bones
14	Deer bones
15	Rabbit bones
16	Flint flakes

The second grouping of materials in this matrix comprises most of the other plant remains from this zone. This group includes cucurbits, mesquite seeds, *susí* nuts, hackberry seeds, acorns, *Agave* quids, piñon nuts, nopales, wild beans, and *guajes*. Representatives of all vegetational facies in the area, as well as cultivated plants, are part of this group. All together, it seems to represent the products of women's gathering trips plus their probable cultivation of cucurbits. The spatial distribution of all of these materials, combined, is made up of two distinct areas of concentration (Fig. 27.19). The less dense of these two areas occurs on the south side of the living floor, ranging from the cave wall out to the focus of the concentration in the middle of the site. The major area of distribution of these plant materials, however, is on the north side of the site, largely in the middle-to-rear parts of the cave. Densities in this area are very high, occasionally extremely so.

Turtle bones are localized discretely as a small concentration on this floor, in a pattern not significantly parallel to that of any other material (Fig. 26.42). *Opuntia* seeds, also, exhibit a series of small, localized clusters that are not distributed like any other material on this floor (Fig. 26.42).

In general, the remains resulting from men's and women's activities are separated neatly into two groups by the spatial analysis of this occupation floor. The distributions of these groups overlap substantially, but the group of materials associated with men's activities clearly lies preponderantly on the south side of the cave, while that associated with women's activities clearly is most heavily concentrated on the north side. The clear distinction of the products of men's and women's activities in this zone appears also in the results of Reynolds' analysis (Chapter 28).

CONCLUSION

The approach to spatial analysis taken in this chapter is only one of many. As all approaches so far developed, it has a number of problems that render its results less clear and less definitively interpretable than one ideally would desire. Nonetheless, it has proven useful here, in analyzing the larger occupation floors at Guilá Naquitz and in enlarging our possibilities for observing and interpreting the distributional patterning of the archaeological materials on these floors. For each of the four levels analyzed, such interpretations have been phrased in terms of the organization of subsistence activities of the prehistoric inhabitants of this site. Two dimensions to this organization can be seen to greater or lesser extent in all of these spatial analyses. First, there is usually evidence for a sexual division of labor and occasionally for a clear separation of men's and women's activities. Second, there is often evidence for separate episodes of exploitation of different vegetational communities—Thorn Forest or Mesquite Grassland—in the vicinity of Guilá Naquitz. This

MATERIAL	7	5	9	8	10	1	12	2	4	11	3	15	14	6	16	13
7. OPUNTIA SEEDS	1.00	.60	.27	.50	.37	.68	.43	.33	.36	.30	.15	.69	.49	.44	.48	-.45
5. GUAJE PODS	.60	1.00	.93	.99	.96	.99	.97	.95	.95	.93	.86	.92	.88	.81	.63	-.18
9. BEAN POD VALVES	.27	.93	1.00	.96	.99	.90	.98	.99	.98	.98	.97	.79	.83	.77	.52	-.02
8. NOPALES	.50	.99	.96	1.00	.98	.97	.99	.96	.97	.96	.91	.88	.88	.80	.60	-.14
10. PINON NUTS	.37	.96	.99	.98	1.00	.94	.99	.99	.98	.98	.95	.83	.86	.79	.56	-.07
1. AGAVE QUIDS	.63	.99	.90	.97	.94	1.00	.96	.93	.95	.92	.85	.96	.93	.87	.72	-.10
12. ACORNS	.43	.97	.98	.99	.99	.96	1.00	.99	.99	.98	.94	.88	.90	.83	.62	-.05
2. HACKBERRY SEEDS	.33	.95	.99	.96	.99	.93	.99	1.00	.99	.99	.97	.83	.87	.81	.58	-.01
4. SUSI HULLS	.36	.95	.98	.97	.98	.95	.99	.99	1.00	.99	.97	.88	.93	.88	.68	.07
11. MESQUITE SEEDS	.30	.93	.98	.96	.98	.92	.98	.99	.99	1.00	.98	.85	.90	.85	.64	.08
3. CUCURBITS	.15	.86	.97	.91	.95	.85	.94	.97	.97	.98	1.00	.78	.87	.84	.62	.20
15. RABBIT BONES	.69	.92	.79	.88	.83	.96	.88	.83	.88	.85	.78	1.00	.96	.94	.87	.06
14. DEER BONES	.49	.88	.83	.88	.86	.93	.90	.87	.93	.90	.87	.96	1.00	.99	.90	.26
6. NANCHE FRUITS	.44	.81	.77	.80	.79	.87	.83	.81	.88	.85	.84	.94	.99	1.00	.94	.38
16. FLINT FLAKES	.48	.63	.52	.60	.56	.72	.62	.58	.68	.64	.62	.87	.90	.94	1.00	.49
13. TURTLE BONES	-.45	-.18	-.02	-.14	-.07	-.10	-.05	-.01	.07	.08	.20	.06	.26	.38	.49	1.00

Fig. 27.17. Ordered matrix of correlation coefficients among the 16 classes of material analyzed from the occupation floor of Zone B1, Guilá Naquitz. Coefficients of .90 or greater have been underlined to make clusters of mutually associated materials stand out more clearly in this matrix. (The range of coefficients underlined is arbitrary and does not indicate significant, as opposed to nonsignificant, values.) The associations and clusters evident in this ordered matrix are described and discussed in the text.

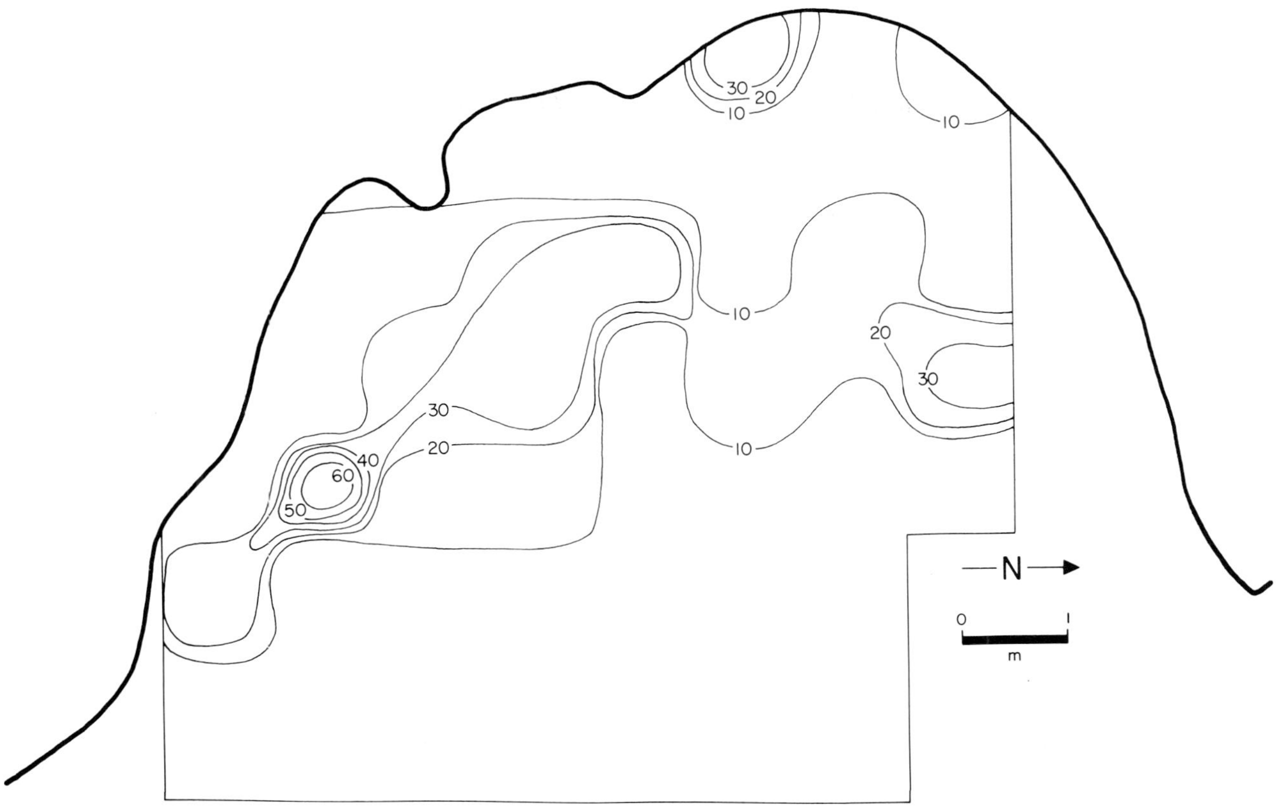

Fig. 27.18. Density contours of rabbit bones, deer bones, and flint flakes, taken together as a group over the area subjected to spatial analysis at Guilá Naquitz, Zone B1.

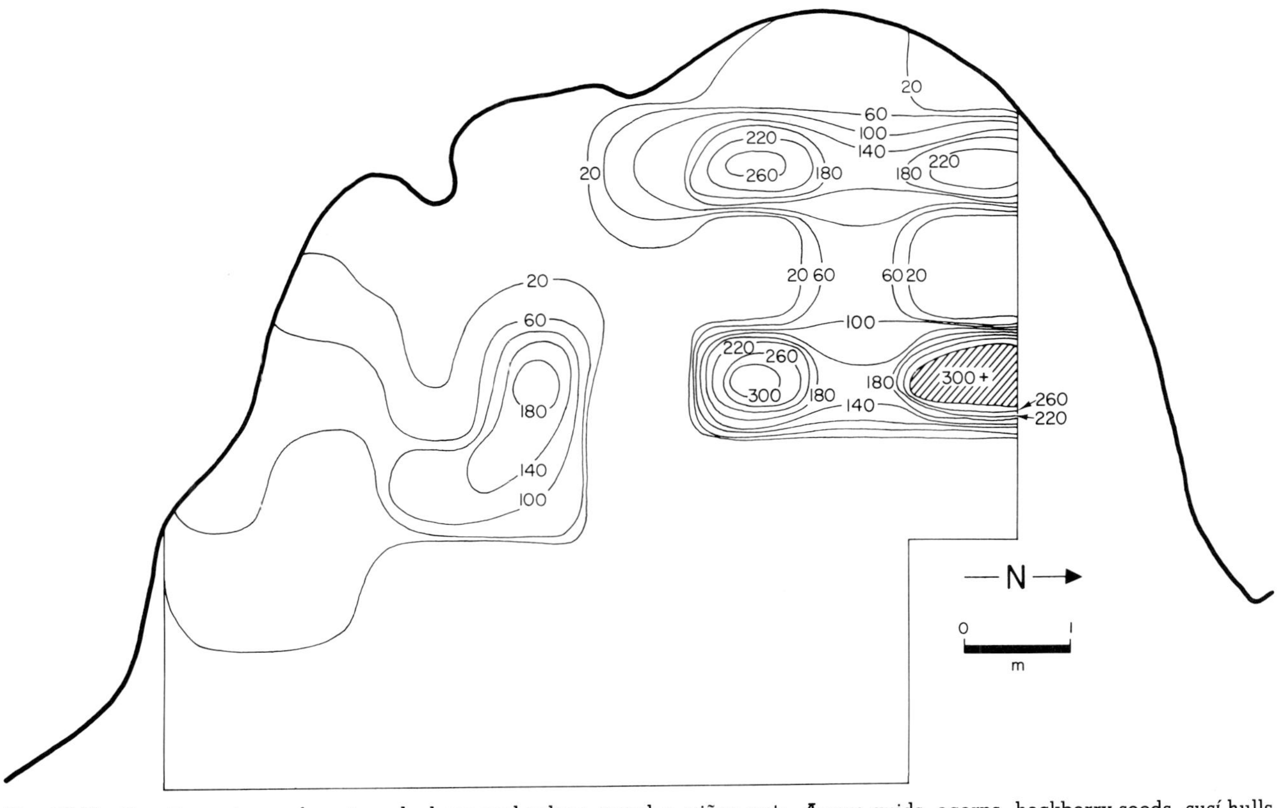

Fig. 27.19. Density contours of *guaje* pods, bean pod valves, nopales, piñon nuts, *Agave* quids, acorns, hackberry seeds, *susí* hulls, mesquite seeds, and cucurbit remains, taken together as a group over the area subjected to spatial analysis at Guilá Naquitz, Zone B1.

evidence, however, is not so strong, or so regularly present, in the analytical results as the indications for the clear sexual division of labor. Thus, although relatively simple to apply, and certainly not as refined as some might demand, the approach to spatial analysis taken in this chapter seems to have produced a series of generally reliable, useful, and behaviorally significant results.

ACKNOWLEDGMENT

I would like to thank Kent Flannery both for allowing and encouraging me to work with the Guilá Naquitz materials in experimenting with the archaeological application of dimensional analysis of variance and for substantial help and advice in the preparation of this chapter.

28

Multidimensional Scaling of Four Guilá Naquitz Living Floors

Robert G. Reynolds

She took her off the table as she spoke, and shook her backward and forward with all her might.

The Red Queen made no resistance whatever; only her face grew very small, and her eyes got large and green: and still, as Alice went on shaking her, she kept on growing shorter—and fatter—and softer—and rounder—and—and it really *was* a kitten after all.

(Lewis Carroll, *Through the Looking Glass*)

INTRODUCTION

Archaeologists are continually manipulating their data, much like Alice, in the hope of discovering the underlying nature of a particular cultural system from its artifactual representation. Such manipulations can vary from a gentle massage to an uncontrolled beating, where the exact strokes used depend on the questions that need to be answered. The data, however, often get a beating when requiring only a massage. Under such duress the data will sometimes tell the interrogator fictions that he or she, in turn, interprets as truth. In fact, the data frequently never recover from this beating; they are then buried quietly and without ceremony in cardboard coffins along with field notes and other reminders of their previous existence.

As it happens, the witnesses to these cases of assault seldom speak up, either out of ignorance or perhaps out of fear. Those readers who feel that they fall into either category might be tempted to ask the following questions, "What do I look for?" and "What can I do about it?" It is suggested here that there are three basic sets of relationships which must be considered by anyone wishing to treat the data fairly during the course of his or her analysis:

1. Ensuring that the structure of the problem is consistent with the structure of the data
2. Checking to see if the assumptions made by the analytical techniques employed are consistent with the problem's structure
3. Making sure that the structure of the data corresponds with the assumptions made by the analytical methods

Of these three basic types of considerations, the third has recently been the focus of attention for several papers dealing with techniques for the spatial analysis of occupation, including dimensional analysis of variance (Whallon 1973), nearest-neighbor analysis (Whallon 1974a), cluster analysis (Cowgill 1968), discriminant analysis (Bettinger 1979), and correlational analysis (Speth and Johnson 1976). Less has been said in print, however, about an equally important consideration—relating the techniques to the structure of the initial problem. Additional analytical insights await the archaeologist when the structure of the analysis is adjusted to that of *both* the data and the problem.

ACTIVITY AREAS: A PROBLEM IN THE PARTITIONING OF SPACE

It has been suggested that many aspects of the archaeological record cannot be adequately understood without some knowledge of the role that information exchanges have played in their formation (Flannery and Marcus 1976; Johnson 1978). The need for analytical techniques to aid in the development of models that deal directly with the processing of information in human ecosystems has also been acknowledged (Reynolds 1978; Reynolds and Zeigler 1979).

While most authors have considered information exchange on a regional level, Jochim (1976) has pointed out that much can be learned from modeling the relationships obtaining within a single occupational unit. One of the most common forms of analysis at this level is the spatial analysis of occupation floors, where "the aim . . . is generally to define 'tool kits' or clusters of artifacts and other items which occur together on occupation floors as a consequence of having been used together" (Whallon 1973:266). Usually, explicit emphasis is placed on the properties of the material distributions, with only implicit emphasis on the conceptual partitioning of space that produced them. While recent studies in ethnoarchaeology have attempted to alleviate this problem (e.g., Yellen 1977), relatively few attempts have been made to incorporate formally both behavioral and conceptual postulates into the procedures used to analyze occupation floors.

To be sure, the material scatters on occupational floors can be statistically demonstrated, while the conceptual partitions that produced them can only be inferred. However, in this chapter I attempt to take such conceptual partitioning into account in the course of setting up an analysis of the four most extensive living floors at Guilá Naquitz, Zones E, D, C, and B1. This attempt will be based partly on ethnographic analogy and partly on the results of earlier analyses of these same living floors by Spencer and Flannery (Chapter 26) and Whallon (Chapter 27). It is hoped that by setting up the analysis this way, we will make it easier to infer the conceptual bases for statistically demonstrable patterns of debris in the cave.

We begin by specifying a set of assumptions that are presumed to characterize the shared conceptual structures that lie behind spatially differentiated activity areas. This shared conceptual structuring of space serves to generate what we shall call *cognized activity areas.* Utilized space is then partitioned in this way into areas associated with particular sets of tasks. Material goods and waste products associated with these tasks *may* then be distributed over space in ways that reflect this spatial segregation of activities. It is this spatial patterning of artifacts that archaeologists commonly refer to as "activity areas." Here the latter may be called "archaeologically inferred activity areas" in order to differentiate them from their conceptual counterparts.

The problem faced in the analysis of occupation floors is not just the recovery of spatially associated groupings of artifacts but the recovery of those groupings that best reflect the underlying conceptual structuring of activity areas. Not all spatial associations will reflect this conceptual structure. Therefore, we want to use analytical techniques that distinguish associations generated by conceptual partitioning from those generated by other mechanisms. Analytical techniques that fail to be selective in terms of the types of associations extracted from an occupational matrix run the risk of producing archaeologically inferred activity areas that bear little resemblance to the cognized activity areas that preceded them and to which they owe part of their existence.

Since our assumptions specify the types of associations likely to be produced by this conceptual partitioning, we then construct a sequence of analytical techniques designed to extract those particular associations. Using these methods we attempt to derive for each zone an *occupational configuration*, composed of both the activity areas and the paths by which they were interconnected. Finally, changes in the spatial structure of these activity areas and activity paths between zones are used to infer corresponding changes in the behavior of the group(s) occupying the cave through time.

THE STRUCTURE OF COGNIZED ACTIVITY AREAS

One of the fundamental operations that people use to model their environment conceptually is that of clustering objects into groups based on the presence or absence of certain common properties. Certainly, hunter–gatherers have been observed to group plants and animals in this way (see Lee 1968). By establishing such representative groupings, one can expedite decision making; for example, when a decision needs to be repeated frequently, the presence of categories that can be easily associated with a particular decision would seem highly advantageous. If, in addition, an individual is able to link these categories together by another set of similarity measures, we have the framework for a hierarchy of levels of knowledge. Each of these levels represents a further set of abstractions produced by the pooling of previously existing categories. In Fig. 28.1, a graphic representation of such a hierarchically based knowledge structure is given.

We can assume that this tendency to partition objects into groups based on observed similarities or differences with respect to certain variables was as typical of the residents of Guilá Naquitz as it is of other hunter–gatherers. In particular, we are interested in how the inhabitants of Guilá Naquitz might have conceptually partitioned the cave floor into distinct areas for performing distinct tasks. Which activities were allowed to take place together within each area, and which were not? How were these areas distributed spatially within the cave? In order to answer these questions, it is important that we first understand what variables might have served as a basis for such partitioning.

To do this, let us briefly return to the 10 possible sources of spatial co-occurrence or separation listed in Chapter 25. These are

1. Sexual division of labor
2. Differences among plant and animal genera in the amount of processing they need
3. Differences among plant and animal genera in the amount of debris they produce
4. Seasonal differences in availability of plants
5. Differences in harvest and processing between seasonally restricted plants and those available over a greater part of the year

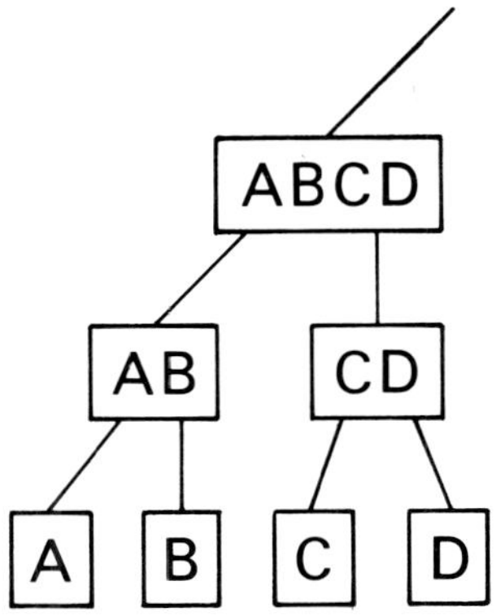

Fig. 28.1. Prototypical hierarchy of conceptual categories that can be produced by basic clustering operations. On the lowest tier are the individual items recognized; on the second tier are larger categories created by pairing those items conceived of as most similar. In the upper tier is a still larger category created by grouping those pairs in the second tier that are seen as most similar.

6. Differences in the vegetational zones where plants are available
7. Differences in the products of daily collecting trips
8. The variety of tasks to which chipped stone might be put
9. Areas of joint activity by men and women
10. The blurring of activity areas by individuals moving around in the cave

Obviously, there are several ways one could group the variables used in our analysis of Guilá Naquitz. Our goal, however, is to cluster the variables in ways that might approximate the categories made by hunters and gatherers. Based on ethnographic analogy, an obvious place to start would be to separate plants from animals. Of all the points listed above, we suspected that sexual division of labor, with women collecting and processing plants and men hunting and/or butchering animals, might be one of the most basic partitions. We also had observed that this partition was one that had emerged in Whallon's matrix analysis (Chapter 27).

Also based on Whallon's matrix analysis, we suspected it might be reasonable to divide the plants into (1) those genera that are seasonally restricted in their availability and produce only modest amounts of processing refuse (such as *susí*, nanches, and hackberries) and (2) those genera that are less seasonally restricted and produce a great deal of processing refuse (such as agaves and prickly pear). We might refer to one category as the "seasonally restricted plants" and the other as the "cactus–*Agave* group".

In Fig. 28.2, we have produced a tiered hierarchy of groupings based on these distinctions. Higher up, the distinction made is between plant processing and animal processing. Lower down, a second distinction is made between the cactus–*Agave* group (with its long-term availability and heavy refuse) and the seasonally restricted plants. Obviously, we could have subdivided these groups still farther, but we felt

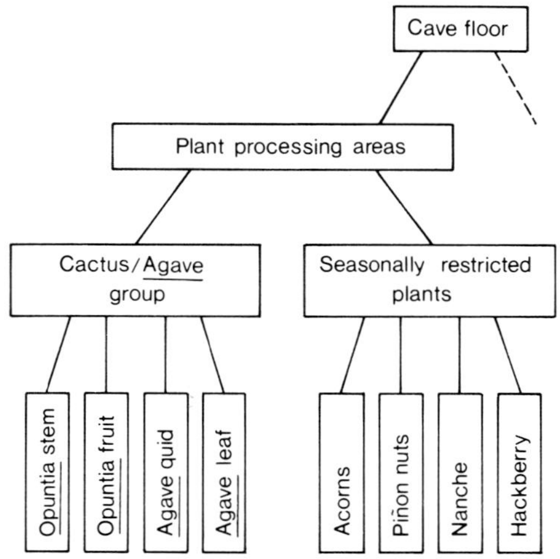

Fig. 28.2. Plant remains from Guilá Naquitz Cave arranged in a hierarchy similar to that given in Fig. 28.1. On the lowest tier are individual genera; on the second tier, these are grouped into two categories, the cactus–*Agave* group and seasonally restricted plants. On the third tier, both groups are joined through the presence of their remains in a plant processing area; this processing area, in turn, is a subdivision of the cave living floor (highest tier).

this would be presuming a greater knowledge of the Naquitz phase Indians' conceptual categories than we really had. We therefore simply allowed the 17 best-represented variables at Guilá Naquitz to serve as the lowest level of the hierarchy, as shown in Fig. 28.2.

A few words should be said about how we expected these groupings to help us get at the conceptual partitioning of the cave floor by its occupants. First, we knew from the previous work of Spencer and Flannery (Chapter 26) and Whallon (Chapter 27) that our 17 variables were not evenly distributed over the cave floor. Certain items tended to show strong association with, or strong disassociation from, other items. We hoped by the use of these groupings to be able to infer the reasons why certain items were, or were not, processed or discarded together.

We recognized the role that might be played by factors such as seasonality—two plant genera might be processed in different seasons. On the other hand, we also suspected that the cave floor might have been conceptually divided into different areas thought to have been appropriate for different tasks. The persons responsible for those tasks might therefore have performed them in the same place throughout the occupation, regardless of season. Moreover, we suspected some tasks might have been *spatially incompatible.* Two tasks can be so defined if they cannot be performed in the same activity area either because the persons responsible for performing them are different or because the two activities are considered by the persons performing them to interfere with each other.

We had already seen in Chapters 26 and 27 that various genera were unlikely to be processed and/or discarded in the same area of the living floor. We hoped our larger groupings would help answer questions such as: Were plants and animals processed in different areas? Were plants of the cactus –*Agave* group, with their large amounts of refuse and their long-term availability, processed or discarded in areas apart from the more seasonally restricted plants? What factors seem to have played the most important role in determining how the occupants of the cave partitioned their work space? In the back of our mind was the hope that we might even be able to go beyond this, to isolate specific areas of the cave floor that were devoted to specific sets of tasks. As we will see, this hope was at least partly realized.

PARTITION DISTANCE TECHNIQUES

In order to measure the amount of spatial partitioning of food processing and discard activities in the cave we first required a means of encoding the excavated data in a form that facilitated this measurement. Therefore, a partitioning was developed for each excavated 1 × 1-m square in the cave.

The 1 × 1-m squares excavated have already been shown in Fig. 5.3. However, not all of the excavated squares contained debris, and only those with debris were used in our analysis. In Table 28.1 the 1 × 1-m squares (referred to in this study as "cells") used to analyze each of the occupation floors are given. If a given cell is used for a stratigraphic zone, it is indexed by an integer number; otherwise, an *X* is shown.

Within each utilized cell, we used only those plant and animal genera that were present with sufficient frequency to be statistically significant. Depending on the living floor, this varied between 15 and 17 variables. The 17 most commonly used variables are given in Table 28.2, grouped according to our notions about their membership in the larger groups discussed earlier.

Next, the combinations of variables that were found together (or not found together) in each cell were calculated. These were first divided into four groups: (1) seasonally restricted plants, (2) the cactus–*Agave* group, (3) animal bones, and (4) flint debitage (unretouched flakes). Unfortunately, flint debitage posed a problem, since it constituted a single-variable group that could not be subdivided during subsequent analysis (see below).[1] After giving it some thought, and noting the results of Whallon's matrix analysis (Chapter 27), we eventually decided we could combine flint debitage with our three animal genera to form a larger group, "animals plus flint flakes." While we were aware that flint flakes surely had resulted from the manufacture of tools for a variety of tasks, they had in fact shown a significant tendency to associate with faunal remains in Whallon's matrices. We regarded this as evidence that we would be merely massaging, not pummeling, our data if we (1) kept flint separate on the *variable level* of our study but (2) combined it with animal remains at the co-occurring group level purely for the purposes of satisfying the program's need for subdivisible groups.

TABLE 28.1
Correspondence between 1 × 1-m Square Designation (from Excavation Grid) and Cell Number (for Q-Mode Analysis)[a]

	Cell Numbers			
Square	Zone E	Zone D	Zone C	Zone B1
B8	X	X	X	1
B9	X	1	1	2
B10	X	X	2	3
B11	X	X	X	4
C7	X	2	X	5
C8	1	3	3	6
C9	X	4	4	7
C10	2	5	5	8
C11	3	X	6	9
C12	X	X	X	10
D5	4	6	X	11
D6	X	X	X	12
D7	X	7	X	13
D8	X	8	7	14
D9	5	9	X	X
D10	6	10	8	15
D11	7	X	9	16
D12	X	X	X	X
E5	X	11	10	17
E6	8	12	11	18
E7	9	13	12	19
E8	10	14	13	X
E9	11	X	14	20
E10	12	X	15	21
E11	13	X	X	22
E12	14	X	X	X
F4	X	15	16	23
F5	X	16	17	24
F6	X	17	18	25
F7	15	18	X	26
F8	16	19	X	X
F9	17	X	19	X
F10	18	X	20	X
F11	19	X	X	X
G4	20	X	X	27
G5	21	20	X	28
G6	X	21	X	X
G7	X	22	X	X
G8	22	23	X	X
G9	X	X	21	X
G10	23	X	X	X
H4	X	X	X	X
H5	X	X	X	X
H6	24	X	X	X
H7	X	X	X	X
H8	X	X	X	X
H9	25	X	X	X
H10	X	X	X	X

[a]An X designates cells not included in analysis because they contained none of the variables used.

Our next step, made possible by the fact that each group now contained at least four variables, was the subdivision of each group into two subgroups. The first represents those variables from each group that are found together in the cell; the second represents those variables that are not. Thus, for any 1 × 1-m square there are associated at most six subgroups of variables. In Table 28.3, we give a purely hypothetical example of how a pattern might develop in a given cell. In our

[1]Unfortunately, no category of *retouched* flint tool was numerous enough to constitute a statistically significant category for each living floor. Had any of the tool types been numerous enough to include in this study, the results might have been even more interesting.

TABLE 28.2
The 17 Most Commonly Used Variables in the Multidimensional Scaling Program

Variable	First grouping	Larger grouping
Agave leaves, *Agave* quids, *Opuntia* stems, *Opuntia* seeds	Cactus–*Agave* group	Plants
Acorns, Piñon nuts, *Susí* nuts, *Guaje* pods, Nanche seeds, Bean pod valves, Cucurbit fragments, Hackberry seeds, Mesquite seeds	Seasonally restricted plants	Plants
Flint debitage, Deer bones, Cottontail bones, Mud turtle bones	Animals plus flint flakes	Animals plus flint flakes

First, a partition (P) of a finite set (S) is a division of that set into a collection of nonempty, pairwise disjoint (nonoverlapping) subsets that taken together contain all the original elements in P (exhaustive). The *cardinality* of the partition P is the number of distinct groups contained in it. For example, given a set of four elements $\{1, 2, 3, 4\}$, P is a partitioning of these elements into two groups $P = \{\{1, 3\}, \{2, 4\}\}$. The cardinality, or number of subsets, is two. For convenience we index each of the n subgroups in a partition P, uniquely from P_1 to P_n. In the example above, P_1 references (1, 3) and P_2 references (2, 4). We denote the size or cardinality of each subgroup by $|P_1|$ and $|P_2|$, respectively. So $|P_1|$ has the value 2 here.

Given two partitions of a set S, labeled respectively P and Q, we define their *intersection* $P \cap Q$ as follows: Two elements of S, a and b, are in the same subgroup of $P \cap Q$ if and only if they are in the same cell of P and in the same cell of Q. To illustrate, if $P = \{\{1, 2\}, \{3,4\}\}$ and $Q = \{\{1\}, \{2, 3, 4\}\}$ then $P \cap Q = \{\{1\}, \{2\}, \{3, 4\}\}$. Thus 3 and 4 occur together in $P \cap Q$ because they are found together in P_2 and Q_2. Since 1 and 2 consistently co-occur with no other elements, each forms a singleton group.

TABLE 28.3
A Hypothetical Example of the Kind of Partitioning of Plant and Animal Subgroups that Could Occur in a Given 1 × 1-m Square

Group	Variables	Presence/absence
Cactus–*Agave*	*Opuntia* seeds, *Opuntia* stems	Present
	Agave quids, *Agave* leaves	Absent
Seasonally restricted plants	Hackberry seeds, cucurbit fragments *susí* nuts, *guaje* pods, nanche seeds	Present
	Bean pod valves, piñon nuts, mesquite seeds, acorns	Absent
Animals plus flint	Deer bones, flint flakes	Present
	Cottontail bones, turtle bones	Absent

hypothetical case, *Opuntia* seeds and stems (from the cactus–*Agave* group) co-occurred in the cell; two other members of the same group, *Agave* quids and leaves, were not found in the cell. Seasonally restricted plants have also been grouped according to whether they did or did not occur in the cell. Animals plus flint flakes are similarly divided in our hypothetical case. By dividing our variable set into those that co-occur and do not co-occur for each group, we are encoding the data for the cell so as to bring out possible complementarities and incompatibilities in processing among variable groups, thereby pursuing our stated desire to search for the conceptual partitioning behind the archaeological debris as well as our intention to relate our technique to the structure of the initial problem.

It is now important to measure the degree of similarity in variable partitioning between pairs of cells. To do this, we must look at the techniques available for generating distance measures for partitions. Arabie and Boorman (1972), among others, have studied the ability of various distance measures to describe the structural dissimilarity between different groupings of objects. In preparation for this type of analysis, we now formally define a partition and some of its properties. This allows us to describe more precisely how such partition distance measures work.

We now have enough information to define an appropriate distance measure. While there are many possible ones from which to choose, we have selected the chi-square measure of association for use here. The chi-square measure generates a distance between 0 and 1 for two partitions, based on the extent to which objects co-occur in the same subgroup in each partition. This is exactly the kind of information we are looking for, since similarity between occupation cells in the cave is based on those activities that are jointly *included* and *excluded*.

If we look back at the definition of $P \cap Q$, for two partitions of a set S, we see that the cardinality or number of groups contained in $P \cap Q$ will decrease as the number of elements that co-occur in distinct subgroups of each increases. This is the type of information used in the chi-square metric to calculate distances between partitions. Formally defined, the chi-square measure is

$$\text{chi-square} = 1 - \frac{\left(\sum_{i,j} \frac{|Z_{ij}|^2}{|P_i| \times |Q_j|} - 1\right)}{\min(|P| - 1, |Q| - 1)}.$$

Then the term

$$\frac{\left(\sum_{i,j} \frac{|Z_{ij}|^2}{|P_i| \times |Q_j|} - 1\right)}{\min(|P| - 1, |Q| - 1)}.$$

measures the extent to which the two partitions P and Q differ. Looking at the numerator first, $\sum_{i,j}$ tells us that for each pair of subgroups, P_i from P and Q_j from Q, we compute $\frac{|Z_{ij}|^2}{|P_i| \times |Q_j|}$.

These values are then summed together. For two subgroups P_i and Q_j, Z_{ij} represents the partition produced by $P_i \cap Q_j$. By taking $|Z_{ij}|$ we get an estimate of the level of disagreement between the two partitions, since, as we mentioned above, $|P_i \cap Q_j|$ will increase as the number of co-occurring elements in the two groups decrease. The denominator $|P_i| \times |Q_j|$ in each case represents an adjustment for the cardinality of the two groups currently compared, P_i and Q_j.

At this point a calculation might be illustrative. Let us say $P = \{\{1, 2, 3, 4\}\}$ and $Q = \{\{1, 2, 3, 4\}\}$. If this is the case, the numerator involves the comparison of only one pair of subgroups. Since Z_{11} is $\{\{1, 2, 3, 4\}\}$ here for the two subgroups, $|Z_{11}|$ equals 4. Thus the numerator is

$$\frac{|Z_{11}|^2}{|P_1| \times |Q_1|} = \left(\frac{|4|^2}{|4| \times |4|} - 1\right),$$

which works out to 0. Therefore the term subtracted from 1 in the chi-square calculation will be 0, and the similarity between the two will equal 1. This is what we would expect, since the chi-square measure is a linear transformation of the standard chi-square statistic.

Q-MODE AND R-MODE ANALYSES

Our study of partitioning used two approaches, a Q-mode and an R-mode analysis. Briefly, in this context an R-mode analysis means we are looking at the individual variables and how they are distributed; a Q-mode analysis means we are looking at the basic units or cells (squares).

We begin with the Q-mode analysis. In order to calculate the associations between the partitions for all relevant cells of each living floor selected for analysis (Zones E, D, C, and B1), I have modified a program originally written by Arabie and Boorman (1972) to calculate this and other measures, so that it could be used in the Guttman–Lingoes nonmetric program series at the University of Michigan (Lingoes 1973).

For each of the four living floors considered here, a matrix of chi-square partition distances between all relevant pairs of cells was computed. This matrix was then used as input to the Guttman–Lingoes Program, SSA-I, which performed a multidimensional scaling of the cells with a maximum of four dimensions. Multidimensional scaling techniques have been well described in the archaeological literature (see Drennan 1976a, 1976b; Doran 1970) and will not be elaborated on here. Nearest-neighbor clustering analysis (Sneath and Sokal 1973:216–222) was then used to cluster the cells based on their scores over the four dimensions. The resulting groups of cells were then considered to represent distinct areas of processing and/or discard within the living floor.

As a check on this analysis, an R-mode approach to analyzing the data was performed as well. This approach represents the more traditional approach to the investigation of activity areas and was used to evaluate the patterns discovered by Q-mode analysis. To perform the R-mode analysis, a matrix of Pearson's r correlations among the variables was produced, using for each zone only those cells that actually had debris. Then a multidimensional scaling of each matrix to two dimensions was performed. Since we were dealing with basic contrasts such as seasonally restricted plants versus the cactus–*Agave* group and plants versus animals, it was felt that two (or at most three) dimensions would be sufficient to detect the patterns seen in the Q-mode analysis. The scores for each cell, based on each of these dimensions, were then computed. Finally, a nearest-neighbor cluster analysis was performed by grouping cells into clusters based on similar scores over these dimensions. It was anticipated that the activity areas defined here should bear some resemblance to those defined by the Q-mode approach based on partition distances. However, we suspected the Q-mode analysis would provide more detail about activity areas, because in this approach information on lack of co-occurrence for the different variables is measured directly as well. Indeed, as we shall see in the next section, this was the case.

Our research design diagram (Fig. 28.3) gives an overall picture of how the Q- and R-mode approaches were combined to contribute to the overall reliability of the study. This dual approach to the problem helped to ensure the consistency of the results, as well as to evaluate the utility of each methodology. As indicated in the diagram, a further check was made by comparing our results with those of Whallon (Chapter 27), who had done a preliminary study of variable correlations using larger blocks of squares (4 × 4- and 8 × 8-m).

RESULTS AND DISCUSSION

In this section we describe and compare the results of both the R-mode and Q-mode analyses for each of the four living floors under consideration. We open with Zone E, the earliest.

Zone E

We began with an R-mode analysis. Figure 28.4 gives a matrix of Pearson's r correlation coefficients for the 15 best-represented variables from Zone E. This matrix was then used as input to a multidimensional scaling program, SSA-I. Figure 28.5 shows a plot of variable scores along each of the two vectors extracted by the program. As we hoped might happen, these two vectors were sufficient to separate different sets of genera associated with different activities. Vector 1 separates

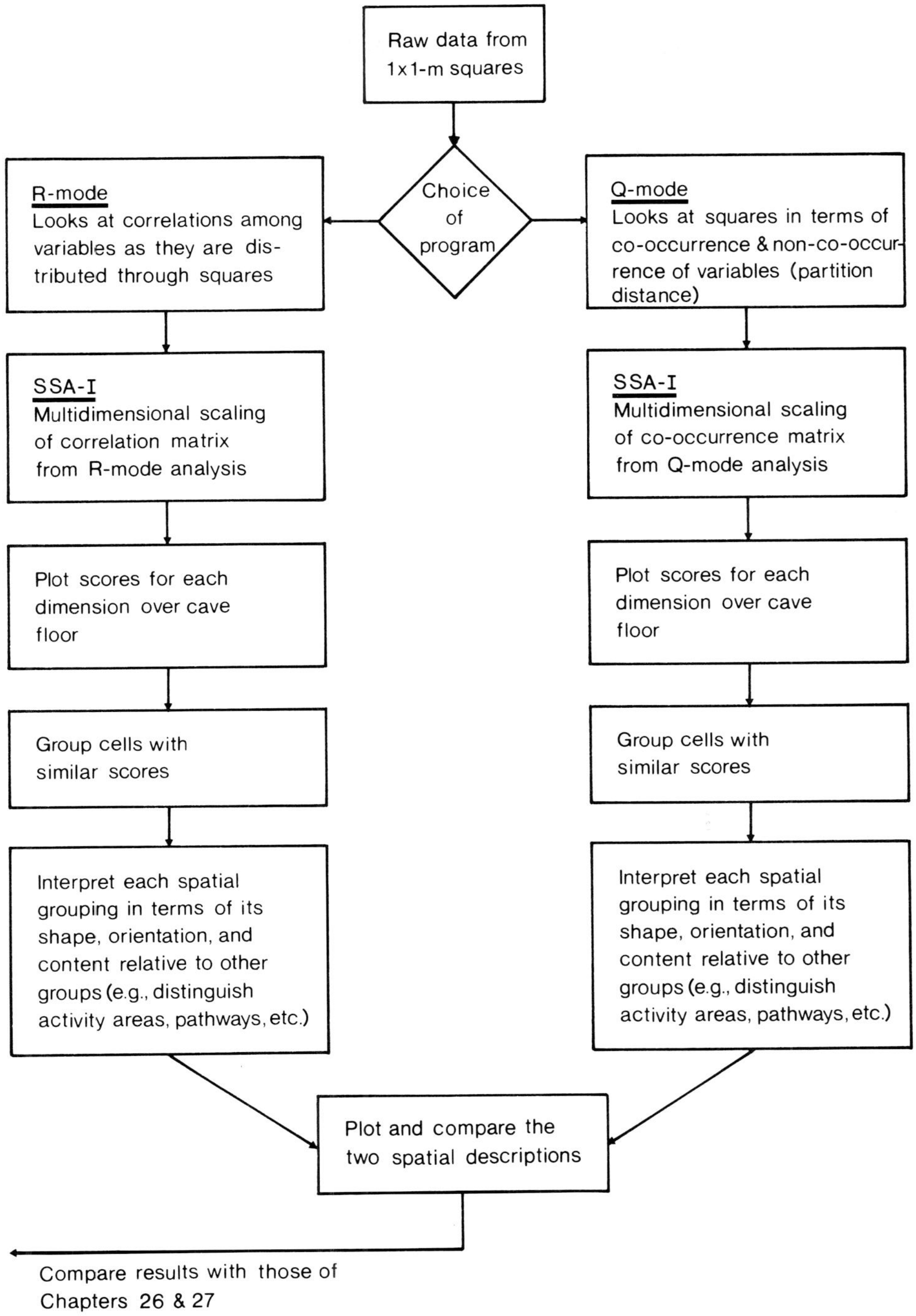

Fig. 28.3. Diagram showing the research design for the multidimensional scaling program described in Chapter 28.

the more seasonally restricted plants from the animal remains. Vector 2, on the other hand, differentiates the animal debris from the cactus–*Agave* group. Thus flint flakes and animal bones all display positive scores for Vector 1 and negative scores on Vector 2 (see Fig. 28.6).

Note also that in Fig. 28.5, flint flakes remain near the center of the graph, presumably because they were associated with processing resources of many different genera. The contiguity of flint flakes, deer, turtle, rabbit, and hackberry may reflect areas in the cave where certain individuals (possibly males) engaged in some casual eating of hackberry fruits while they carried out some activity (possibly butchering or eating) that resulted in animal bones and hackberry seeds being discarded together.

The two vectors generated a configuration of variables in two-dimensional space, such that the distances among the variables preserve 81% of the rank ordering of correlations among the variables in the original matrix (see legend of Fig.

ORIGINAL COEFFICIENTS

COLUMN =	1	2	3	4	5	6	7	8	9	10	11	12	13	14	15
ROW = 1	0.0														
ROW = 2	-0.13	0.0													
ROW = 3	-0.12	-0.09	0.0												
ROW = 4	0.71	-0.10	-0.10	0.0											
ROW = 5	0.44	-0.08	0.16	0.84	0.0										
ROW = 6	-0.10	-0.11	-0.07	-0.07	-0.11	0.0									
ROW = 7	0.06	-0.08	-0.07	0.30	0.38	0.65	0.0								
ROW = 8	-0.06	0.65	-0.04	-0.04	-0.08	-0.07	-0.09	0.0							
ROW = 9	0.03	-0.10	0.00	0.16	0.17	0.54	0.05	-0.04	0.0						
ROW = 10	0.06	-0.17	0.68	0.06	0.15	0.29	-0.03	-0.07	0.44	0.0					
ROW = 11	0.25	-0.09	-0.07	0.40	0.28	-0.04	0.07	0.01	0.09	0.02	0.0				
ROW = 12	0.02	0.08	0.17	0.12	0.23	0.06	0.01	0.23	0.28	0.17	0.00	0.0			
ROW = 13	-0.16	0.27	0.33	-0.14	-0.10	0.14	0.00	0.54	0.14	0.27	-0.04	0.74	0.0		
ROW = 14	-0.13	-0.08	0.46	-0.03	0.10	0.04	-0.06	-0.04	0.35	0.58	-0.06	0.08	0.27	0.0	
ROW = 15	0.01	-0.12	-0.03	0.12	0.23	-0.05	0.03	-0.07	0.17	0.03	-0.09	0.83	0.46	0.23	0.0

Fig. 28.4. R-mode analysis of Zone E, Step 1: matrix of correlation coefficients for 15 selected variables, using Pearson's r. The matrix shows the degree of linear association between all possible pairs of variables: 1, *Agave* quids; 2, *Agave* leaves; 3, hackberry seeds; 4, *susí* nuts; 5, nanche fruits; 6, *Opuntia* seeds; 7, *Opuntia* nopales; 8, bean pod valves; 9, piñon nuts; 10, mesquite seeds; 11, acorns; 12, turtle bones; 13, deer bones; 14, cottontail bones; and 15, unretouched flint flakes.

28.6). Therefore, these two vectors explain a substantial number of the structural relationships among the variables in the original matrix. This seemed consistent with our notion that the cave might be conceptually divided into a set of differing work spaces that would be analytically recoverable even after they had been blurred by depositional processes, and gave us confidence that we were on the right track.

Next, a Q-mode analysis of the data was performed for Zone E. Figure 28.7 gives the matrix of chi-square partition distances calculated between each pair of relevant 1 × 1-m squares in Zone E based on the partitioning of the same 15 variables used in the R-mode analysis. This partitioning was based on co-occurrence and non-co-occurrence of variables, as discussed in the previous section.

A multidimensional scaling of the matrix was performed, and the first four vectors extracted. The scores for each cell on the first two vectors are plotted in Fig. 28.8. As before, the degree of proximity between each cell and all of the others reflects the strength of their association in the original matrix. However, instead of the associations being based on correlations as in the R-mode analysis, in the Q-mode analysis they represent differences in combinations of variables among cells.

Cells scoring high on Vector 1 mainly exhibited co-occurrence of animal bones and flint flakes, with only a few plants co-occurring. This vector may represent squares where animal processing was a strong focus.

Those cells scoring high on Vector 2 possessed many more co-occurring plants of both the seasonally restricted and cactus–*Agave* groups, while animal remains were largely nonco-occurring. The remaining 2 vectors isolated cells in which very few large groups of variables seemed to show strong co-occurrence (or nonco-occurrence). This, one might suggest, could reflect those cells where activities were less

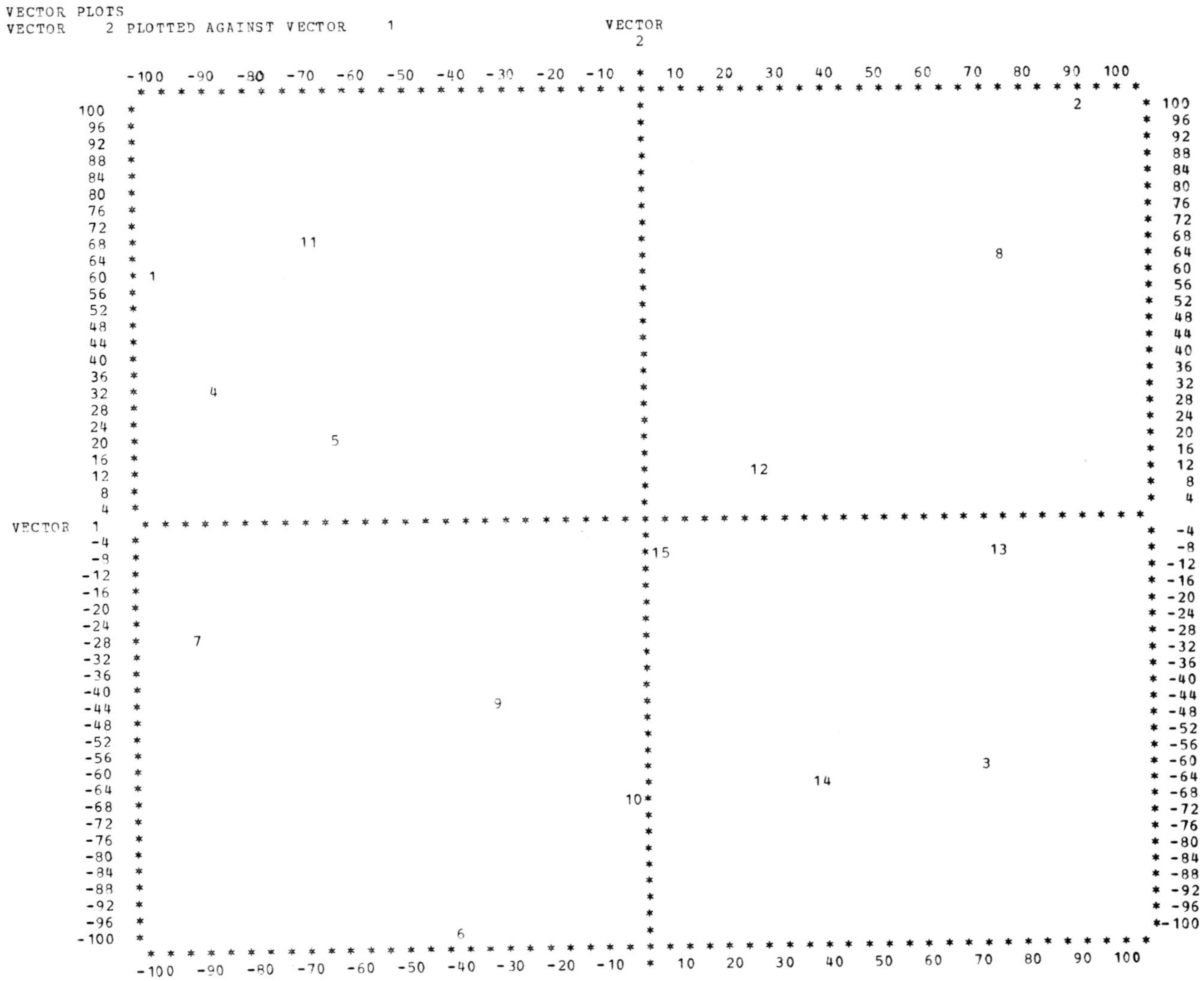

Fig. 28.5. R-mode analysis of Zone E, Step 2: plot of Vector 1 versus Vector 2, where degree of proximity of one variable to another reflects the degree of association as determined by Step 1. (Variables are as in Fig. 28.4.) Note that Vector 1 separates the seasonally restricted plants from the animals and the cactus–*Agave* group; Vector 2, in turn, separates the animals from the cactus–*Agave* group. Flint flakes (Variable 15) remain near the center because they are associated with all three categories. The association of flint flakes, deer, turtle, rabbit, and hackberry may reflect an area where certain individuals (presumably male) butchered animals while eating fresh hackberries. The association of *susí* nuts, nanches, nopales, acorns, and agave quids may reflect an area where certain individuals (presumably female) processed seasonally restricted plants while chewing cooked maguey. Note that plants of the cactus–*Agave* group frequently plot out near the margins, indicating a different processing and/or discard pattern from seasonally restricted plants.

structured and debris from those activities more random in nature. Cells scoring high on these vectors were thought to constitute areas peripheral to the main work areas.

Scores for each cell, on each of the four vectors, are given in Fig. 28.9. These four vectors have a combined Kruskal's Stress of 0.15. What this means is that only 15% of the distances among cells (each plotted as a point in four-dimensional space) do *not* reflect the relative rank ordering of their associations in the original matrix. (Or stated conversely, 85% of the distances *do* reflect their original rank ordering.)

Having completed our Q-mode and R-mode analyses, we can now use the results to map the occupational configuration of Zone E and infer the conceptual partitioning of the living floor by its occupants. To do this, we must return to the grid of 1 × 1-m squares ("cells" in our analysis) into which the cave floor was divided. Using this grid, we draw heavy lines around groups of cells that share very similar patterns of debris (Fig. 28.10). We refer to these groups of cells as "areas," and we describe each area in terms of the plant, animal, and/or artifact debris that make it distinctive.

As we have stated, however, we consider these areas to represent the physical manifestation of a conceptual division of the cave floor into work spaces of different kinds. We will therefore assign each area two designations: a Roman numeral that refers only to its physical manifestation and a verbal description that reflects our interpretation of the conceptual division it represents. On the most "neutral" level, therefore,

ZONE E SCALING

GUTTMAN-LINGOES' SMALLEST SPACE COORDINATES FOR M = 2 (SEMI-STRONG MONOTONICITY).

VARIABLE	CENTRALITY INDEX	DIMENSION 1	DIMENSION 2
1	112.068	-100.000	58.149
2	138.194	88.299	100.000
3	95.712	71.210	-59.273
4	83.926	-83.102	29.656
5	58.397	-60.143	18.232
6	102.240	-33.399	-100.000
7	86.590	-88.280	-26.304
8	103.323	74.716	63.712
9	47.536	-27.965	-43.471
10	65.064	-0.295	-66.584
11	88.050	-61.283	66.216
12	31.565	24.116	9.847
13	78.312	72.795	-8.085
14	73.857	37.052	-62.247
15	13.249	7.445	-5.479

GUTTMAN-LINGOES' COEFFICIENT OF ALIENATION = 0.22049 IN 15 ITERATIONS.
KRUSKAL'S STRESS = 0.19030

Fig. 28.6. R-mode analysis of Zone E: scores of 15 variables for each of the two vectors used in Step 2. Scores reflect the distance from the ordinate (Vector 2) or abscissa (Vector 1) of the graph shown in Fig. 28.5. Note that flint flakes and animal bones display positive values for Vector 1, whereas seasonally restricted plants tend to have negative scores for the same vector. Kruskal's Stress for this stage of the analysis is 0.19, which means that 81% of the relative order of the associations between variable pairs is preserved by the configuration shown in Fig. 28.5.

we refer to Areas I, II, III, and so forth. On the interpretive level, we refer to "men's work areas," "women's work areas," and "pathways" within the cave.

One additional comment on these areas should be made. Some readers may wonder why, when our problem is a locational one, we did not use a statistical procedure that takes location into account. There are good reasons for this. First, there are relatively few ways of assessing the significance of a multidimensional scaling. Purely by chance, two unrelated squares in two different parts of the cave could turn out to have the same pattern of debris. The Q-mode analysis would show them to be similar, but later examination of their spatial relationship would show them to be located far apart. On the other hand, some squares ranked as similar by the multidimensional scaling will turn out to cluster together in the same part of the cave because they really are all part of one activity area. In this way, location can be used as a check on the multidimensional scaling—a truly independent check because spacing was not taken into account by the Q-mode and R-mode analyses. Thus, to qualify as an area, the groups of cells in our analysis must satisfy two criteria: (1) The multidimensional scaling must show they are similar, and (2) they must be contiguous squares.

Area I (Fig. 28.10) consists of Squares D10–D12, E10–E12, F9–F11, and G10. All these contiguous cells score high on vectors related to seasonally restricted plants. Due to its size, Area I could be a locality where one or two individuals processed and/or discarded acorns, *susí* nuts, nanches, and nopales, while chewing cooked agave. On the basis of ethnographic analogy, these were probably women's tasks. Our interpretation would therefore be that in the conceptual partitioning of the cave, this was considered a women's work area.

Area V consists of Squares G4–G5 and would appear to be a locality where at least one person chipped flint, processed and/or discarded deer, rabbit, and mud turtle, chewed cooked agave, and ate hackberries and discarded the seeds. On the basis of ethnographic analogy, such an individual was probably a male.

Both Areas I and V represent cell clusters suggested by the R-mode analysis. However, one of the features of the Q-mode approach is to differentiate between cells on the basis of the number of variables of different types that were present or absent together. Areas where particular processing activities were conducted are characterized by the *presence* of large collections of variables of similar type that co-occurred, as well as by the *absence* of large groups of variables whose processing was evidently not considered compatible with the former. This is particularly true of Areas I and V.

Areas II and IV have a distinct pattern of refuse but lack the *large* groups of variables that are either present or absent in Areas I and V. This suggests that fewer conceptual restrictions were placed on the activities in those areas, at least as far as plant and animal processing go. Indeed, the shape and location of Areas II and IV strongly suggest that they were used as pathways either for entering the cave or moving between work areas. Area II (Squares E7, F7, E9, F8, G8, and H9) appears to be an entrance path for the cave

ORIGINAL COEFFICIENTS

COLUMN =	1	2	3	4	5	6	7	8	9	10	11	12	13	14	15	16	17	18	19	20	21	22	23	24	25	26
ROW = 1	0.0																									
ROW = 2	0.22	0.0																								
ROW = 3	0.0	0.0	0.0																							
ROW = 4	0.36	0.36	0.19	0.0																						
ROW = 5	0.47	0.47	0.33	0.21	0.0																					
ROW = 6	0.31	0.31	0.31	0.32	0.28	0.0																				
ROW = 7	0.45	0.22	0.30	0.46	0.40	0.14	0.0																			
ROW = 8	0.30	0.30	0.33	0.0	0.0	0.33	0.30	0.0																		
ROW = 9	0.32	0.32	0.32	0.25	0.29	0.30	0.27	0.33	0.0																	
ROW = 10	0.14	0.14	0.14	0.25	0.33	0.30	0.27	0.33	0.33	0.0																
ROW = 11	0.32	0.32	0.32	0.33	0.31	0.32	0.32	0.33	0.32	0.25	0.0															
ROW = 12	0.19	0.19	0.19	0.33	0.31	0.32	0.32	0.33	0.32	0.25	0.33	0.0														
ROW = 13	0.45	0.45	0.30	0.46	0.40	0.14	0.22	0.30	0.27	0.27	0.32	0.32	0.0													
ROW = 14	0.30	0.30	0.30	0.32	0.24	0.14	0.0	0.33	0.27	0.27	0.32	0.32	0.0	0.0												
ROW = 15	0.22	0.22	0.33	0.22	0.22	0.33	0.22	0.22	0.33	0.33	0.33	0.33	0.22	0.33	0.0											
ROW = 16	0.44	0.44	0.25	0.47	0.50	0.25	0.49	0.33	0.32	0.19	0.30	0.30	0.24	0.32	0.33	0.0										
ROW = 17	0.39	0.39	0.19	0.50	0.48	0.32	0.49	0.33	0.32	0.25	0.33	0.0	0.24	0.32	0.33	0.22	0.0									
ROW = 18	0.41	0.41	0.33	0.38	0.40	0.21	0.26	0.22	0.33	0.33	0.31	0.31	0.26	0.12	0.0	0.50	0.48	0.0								
ROW = 19	0.44	0.44	0.25	0.47	0.46	0.25	0.39	0.33	0.32	0.19	0.30	0.30	0.33	0.19	0.33	0.43	0.41	0.46	0.0							
ROW = 20	0.38	0.38	0.28	0.40	0.31	0.21	0.26	0.22	0.33	0.24	0.31	0.31	0.20	0.12	0.33	0.21	0.23	0.41	0.27	0.0						
ROW = 21	0.21	0.43	0.28	0.45	0.46	0.21	0.44	0.30	0.33	0.24	0.31	0.31	0.50	0.29	0.22	0.27	0.42	0.41	0.21	0.42	0.0					
ROW = 22	0.21	0.43	0.28	0.45	0.46	0.21	0.31	0.30	0.24	0.24	0.31	0.31	0.40	0.12	0.22	0.40	0.42	0.33	0.40	0.42	0.28	0.0				
ROW = 23	0.29	0.29	0.29	0.32	0.28	0.31	0.30	0.33	0.30	0.14	0.19	0.19	0.30	0.30	0.33	0.25	0.19	0.28	0.25	0.28	0.28	0.28	0.0			
ROW = 24	0.21	0.21	0.21	0.28	0.32	0.28	0.24	0.33	0.33	0.12	0.28	0.28	0.24	0.24	0.33	0.28	0.28	0.31	0.11	0.19	0.31	0.31	0.21	0.0		
ROW = 25	0.19	0.19	0.19	0.33	0.31	0.32	0.32	0.33	0.32	0.25	0.33	0.0	0.32	0.32	0.33	0.30	0.0	0.31	0.30	0.31	0.31	0.31	0.19	0.28	0.0	
ROW = 26	0.39	0.39	0.19	0.50	0.48	0.32	0.49	0.33	0.32	0.25	0.33	0.0	0.43	0.32	0.33	0.41	0.19	0.48	0.22	0.42	0.23	0.42	0.19	0.28	0.0	0.0

Fig. 28.7. Q-mode analysis of Zone E, Step 1: matrix of chi-square partition distances between pairs of cells (1 × 1-m squares) based on relative co-occurrence of the same 15 variables already used in R-mode analysis (see legend for Fig. 28.4).

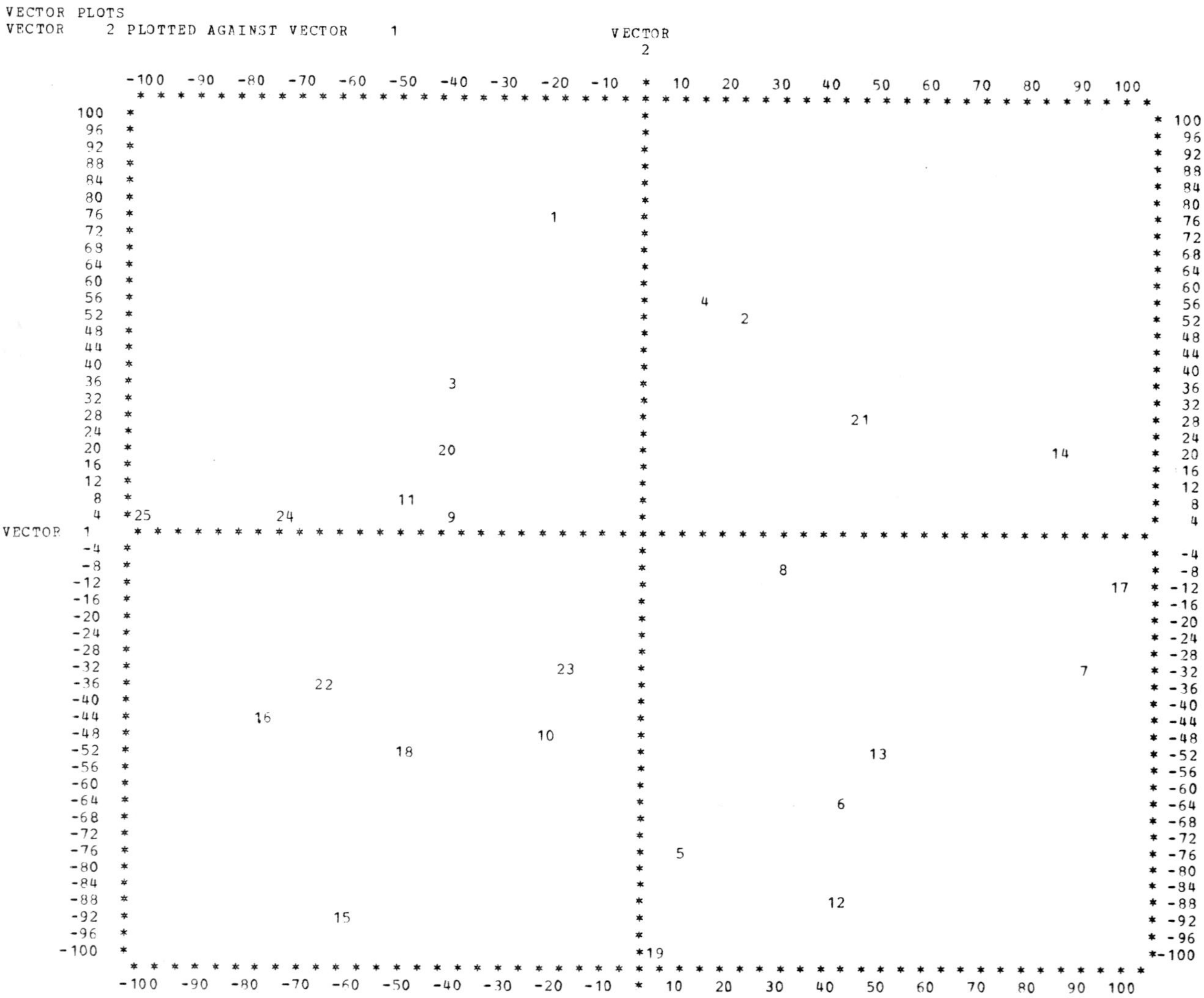

Fig. 28.8. Q-mode analysis of Zone E, Step 2: Plot of Vector 1 versus Vector 2, where the degree of proximity of one numbered cell (1 × 1-m square) to another reflects the degree of association between all possible pairs of cells as determined by Step 1. (Numbered cells 1–25 are as in Table 28.1.) The clusters of associated cells shown here constitute the basis for the activity areas and pathways reconstructed in Fig. 28.10. Vector 1 separates a probable men's work area from a probable women's work area, while Vector 2 separates out a string of peripheral cells elsewhere in the cave. Vectors 3 and 4 (not shown) divide this string of peripheral cells into three different units.

that, once inside, splits into three branches. One branch ends adjacent to Area I, the other adjacent to Area V, and the third adjacent to Area IV near the rear of the cave. Our interpretation is that movement into and out of the cave was principally along this path. Movement from one work area to another within the cave seems to have required crossing the Area II intersection as well. Area IV (Squares D5 and E6) appears to be an extension of the middle branch of Area II but contains even less debris.

Area III (Squares C10–C11) is located near the back of the cave in an area where the roof is low, and its unspecialized nature (when compared with Area I and V) suggests it may be a discard area into which the persons in Area I swept the finer fraction of their debris (those items too small to be thrown out on the talus).[2]

While the Q-mode analysis suggested these five basic areas in the cave, based on the extent to which particular variables were excluded or included in contiguous cells, the R-mode analysis defined only two areas. The first of these was related to animal processing and consisted of a group of cells centered around Area V as defined by the Q-mode analysis. The second area reflected plant processing and was found to

[2]As discussed in previous chapters, we assume that the bulkier refuse was simply thrown out on the cave talus or beyond. When we refer to "discard areas" within the cave, we are therefore referring only to the finer fraction of the debris, such as hackberry seeds and tiny bone fragments.

CHI SQUARE DISTANCE SCALING FOR ZONE E

GUTTMAN-LINGOES' SMALLEST SPACE COORDINATES FOR M = 4 (SEMI-STRONG MONOTONICITY).

VARIABLE	CENTRALITY INDEX	DIMENSION 1	2	3	4
1	93.503	-13.812	73.664	-25.527	-22.580
2	90.488	21.841	51.461	-79.857	-10.409
3	65.606	-33.105	33.735	-51.262	4.411
4	87.882	15.111	53.464	-27.095	44.411
5	93.977	-8.022	-76.318	8.791	58.611
6	80.528	43.501	-61.837	21.090	-19.574
7	91.242	88.600	-30.369	-27.122	-5.011
8	102.059	30.102	-6.440	9.077	84.784
9	55.094	-33.566	3.369	-30.150	-44.559
10	103.003	-13.984	-48.627	-5.621	-100.000
11	86.907	-42.441	4.080	-100.000	-5.830
12	88.931	41.717	-85.595	-55.640	17.567
13	66.326	50.010	-52.417	-6.428	5.147
14	105.961	86.853	19.245	-60.615	-34.335
15	105.881	-53.119	-91.394	-74.933	-31.619
16	100.246	-70.376	-42.445	-80.864	36.667
17	109.534	100.000	-11.404	-66.142	-10.403
18	95.511	-44.713	-50.756	51.006	-18.327
19	86.392	5.529	-100.000	-49.535	1.286
20	99.673	-35.577	19.625	48.099	-48.303
21	98.062	46.972	27.377	45.393	-15.327
22	62.364	-59.702	-36.233	-27.575	-17.759
23	55.743	-11.641	-29.158	-52.337	-51.654
24	83.834	-67.473	3.476	-9.707	39.800
25	104.467	-100.000	0.432	-10.695	21.454

GUTTMAN-LINGOES' COEFFICIENT OF ALIENATION = 0.16630 IN 71 ITERATIONS.
KRUSKAL'S STRESS = 0.14599

Fig. 28.9. Q-mode analysis of Zone E: scores of the 25 numbered cells (1 × 1-m squares) for four vectors, including the two shown in Fig. 28.8. Scores reflect the relative position of the cells on all four axes. Kruskal's Stress for this stage of the analysis is 0.15, which means that 85% of the relative order of the associations between pairs of numbered cells is preserved by the configuration based on these four vectors.

consist of the same cells seen in Area I as defined by the Q-mode analysis. The paths and marginal discard areas defined by the Q-mode analysis do not show up in the R-mode analysis.

In fact, for all of the zones analyzed, the R-mode analysis produced less fine-grained results than the Q-mode approach. As a result, we have not included the diagrams for the R-mode approach and focus instead on the diagrams for the Q-mode approach in each of the remaining zones.

The Q-mode analysis for Zone E suggested the presence of several functionally different activity areas within the cave. We are interested in whether these activity areas, as well as their relative locations, carry over from one living floor to the next. If there is some degree of organizational similarity between zones, then we might have cause to suggest some degree of cultural or behavioral continuity in terms of the occupying group. Looking next at the occupational remains for Zone D, we do in fact discern some similarities with Zone E.

Zone D

In our analysis of Zone D, we used 17 variables. These included all 15 variables used in our Zone E analysis as well as two seasonally restricted plants, *guajes* and cucurbits, that had not been present in Zone E in sufficient quantities to analyze. Figure 28.11 presents the calculated matrix of Pearson's *r* correlations for the Zone D variables over all of the 23 occupied cells. Our multidimensional scaling of the matrix was carried out to three vectors as part of the R-mode analysis, with animal remains and flint flakes again scoring high on Vector 1 (see Figs. 28.12 and 28.13). Interestingly enough, the cactus–*Agave* group scored lower on Vector 1 than did seasonally restricted plants. As in the previous zone, this may be because some seasonally restricted plants (such as hackberries) were eaten by individuals engaged in animal processing. The three vectors, taken together, gave us a Kruskal's Stress of only .11.

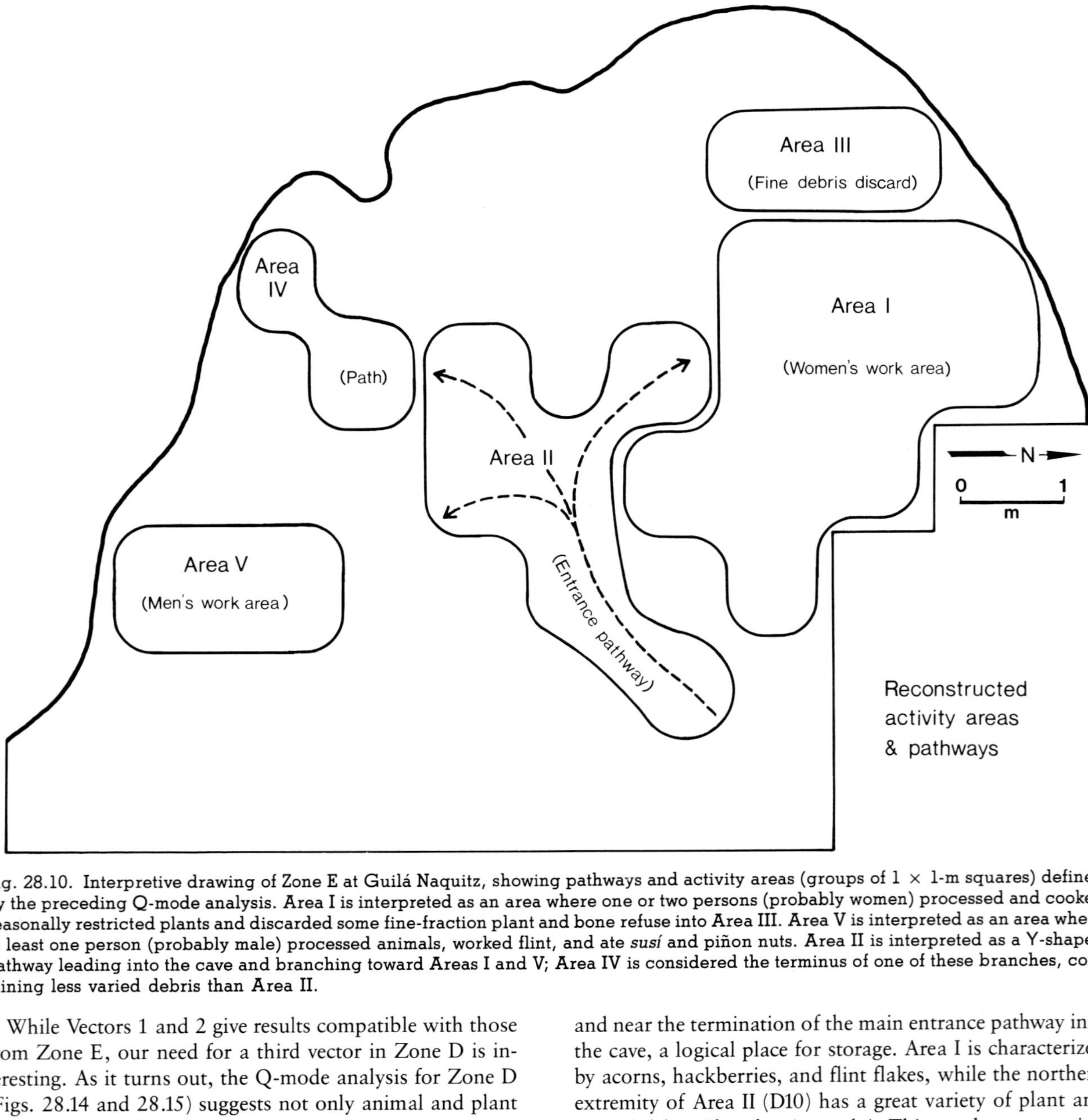

Fig. 28.10. Interpretive drawing of Zone E at Guilá Naquitz, showing pathways and activity areas (groups of 1 × 1-m squares) defined by the preceding Q-mode analysis. Area I is interpreted as an area where one or two persons (probably women) processed and cooked seasonally restricted plants and discarded some fine-fraction plant and bone refuse into Area III. Area V is interpreted as an area where at least one person (probably male) processed animals, worked flint, and ate *susí* and piñon nuts. Area II is interpreted as a Y-shaped pathway leading into the cave and branching toward Areas I and V; Area IV is considered the terminus of one of these branches, containing less varied debris than Area II.

While Vectors 1 and 2 give results compatible with those from Zone E, our need for a third vector in Zone D is interesting. As it turns out, the Q-mode analysis for Zone D (Figs. 28.14 and 28.15) suggests not only animal and plant processing areas but also an apparent acorn storage area as well. It is the presence of this latter area that may produce the need for three vectors to explain 89% of the Zone D matrix ordering.

In Fig. 28.14 we give the matrix of chi-square partition distances for the 23 occupied cells. The multidimensional scaling of this matrix as part of the Q-mode analysis was carried to three vectors as well. Kruskal's Stress for these vectors was .13, again a good fit. Figure 28.16 gives the scores for the 23 cells on each of the three vectors. The cells were then clustered on the basis of their scores for these three vectors to form the groupings shown in Fig. 28.17.

Area I (Squares B9, C7, C8, C10, and D8) and Area II (C9, D9, D10, E7, E8) are interpreted as pathways associated with several storage pits (Features 2, 3, and 10) in the northwest part of the cave. This area is well back under the overhang and near the termination of the main entrance pathway into the cave, a logical place for storage. Area I is characterized by acorns, hackberries, and flint flakes, while the northern extremity of Area II (D10) has a great variety of plant and animal debris (fine fraction only). This northern extension may include fine food preparation debris left near Feature 10.

While no comparable acorn storage area was detected in Zone E, the plant and animal processing areas observed in Zone E do reoccur here; even their general location within the cave is consistent with that seen previously. As in Zone E, the plant processing area is to the north of the animal processing area. Area III (E5, F4–F6, and G5), interpreted as the animal processing area, was characterized by a lack of plant debris, except for items that one might eat while working. Flint, animal bones, hackberry seeds, and agave quids were the items most commonly observed in these cells.

Area IV (E6, F7–F8, and G7–G8), to the north of the proposed animal processing area, consists of cells characterized by the co-occurrence of many seasonally restricted and cactus–*Agave* group plants. This set of cells is associated with

ORIGINAL COEFFICIENTS

COLUMN =	1	2	3	4	5	6	7	8	9	10	11	12	13	14	15	16	17
ROW = 1	0.0																
ROW = 2	0.09	0.0															
ROW = 3	0.12	-0.09	0.0														
ROW = 4	0.40	0.20	-0.18	0.0													
ROW = 5	0.79	0.46	-0.02	0.56	0.0												
ROW = 6	0.24	-0.08	-0.02	0.16	0.14	0.0											
ROW = 7	0.32	-0.08	0.01	0.02	0.23	0.95	0.0										
ROW = 8	0.31	-0.05	0.76	-0.16	0.04	-0.07	-0.05	0.0									
ROW = 9	0.08	0.15	0.64	-0.03	0.07	-0.08	-0.00	0.51	0.0								
ROW = 10	0.47	-0.15	0.25	-0.07	0.21	0.65	0.63	0.21	0.07	0.0							
ROW = 11	0.37	0.54	0.46	0.05	0.48	-0.15	-0.15	0.53	0.23	0.11	0.0						
ROW = 12	-0.04	-0.10	0.10	-0.02	-0.07	0.14	0.12	0.15	0.23	0.26	0.01	0.0					
ROW = 13	0.47	0.13	0.40	0.10	0.52	0.11	0.13	0.55	0.27	0.42	0.70	0.15	0.0				
ROW = 14	-0.13	-0.06	-0.11	-0.08	-0.11	-0.06	-0.06	-0.10	-0.08	0.04	0.06	-0.06	0.18	0.0			
ROW = 15	-0.03	0.64	0.09	0.05	0.34	0.03	0.02	0.03	-0.10	-0.00	0.49	0.20	0.19	-0.04	0.0		
ROW = 16	0.01	-0.04	0.09	-0.13	0.05	0.44	0.47	-0.16	-0.14	0.40	0.02	0.25	0.01	0.17	0.32	0.0	
ROW = 17	0.02	-0.01	-0.15	-0.13	0.02	0.07	0.08	-0.13	-0.13	0.33	0.10	0.23	0.28	0.84	0.08	0.31	0.0

Fig. 28.11. R-mode analysis of Zone D, Step 1: matrix of correlation coefficients for 17 selected variables, using Pearson's r. The matrix shows the degree of linear association between all possible pairs of variables: 1, agave quids; 2, agave leaves; 3, hackberry seeds; 4, cucurbit remains; 5, *susí* hulls; 6, *guaje* pods; 7, nanche fruits; 8, *Opuntia* seeds; 9, *Opuntia* nopales; 10, bean pod valves; 11, piñon nuts; 12, mesquite seeds; 13, acorns; 14, turtle bones; 15, deer bones; 16, cottontail bones; and 17, unretouched flint flakes.

a hearth (Feature 22) and may have been an area for women's food processing and cooking activity. Area IV becomes elongated toward the rear of the cave, suggesting movement between it and Features 2 and 3, as well as some possible movement between Areas III and IV.

Proposed directions for movement within the cave are given in Fig. 28.17. Some plants, such as acorns, were evidently carried to the area of Features 2 and 3 in the rear of the cave, others to Feature 10. Of course, these features may have been used sequentially (rather than simultaneously) during the course of the occupation. Other plants (especially those producing a lot of debris, such as cactus–*Agave*) were carried to Area IV, from whence they could easily be tossed outside. Animals were processed off to the south where such activity would not interfere with plant processing and cooking. Traffic between Areas II and III and between Areas III and IV is indicated by the shape of the areas.

While Zone D has its own pattern, there seem to be several continuities in the orientation and structure of work areas between Zones E and D. In particular, the placement of the main plant and animal processing areas relative to each other is a common feature of both living floors.

Zone C

We may now look at the configuration of areas associated with Zone C. Figure 28.18 shows the matrix of Pearson's r correlations that served as input to the SSA-I multidimensional scaling program for Zone C, using the same 17 variables seen in Zone D. The scaling produced two vectors (Fig. 28.19), and the values for the relevant cells are plotted in Fig. 28.20. The relative distances between the points in this two-dimensional configuration preserve 88% of the rank ordering of associations in the original matrix.

These two vectors show similarities to those produced in Zone E. Vector 1, for example, separates the seasonally restricted plants from the animals-plus-flint-flakes group and the plants of the cactus–*Agave* group. Vector 2 serves to separate the animal remains and flint flakes from cactus–

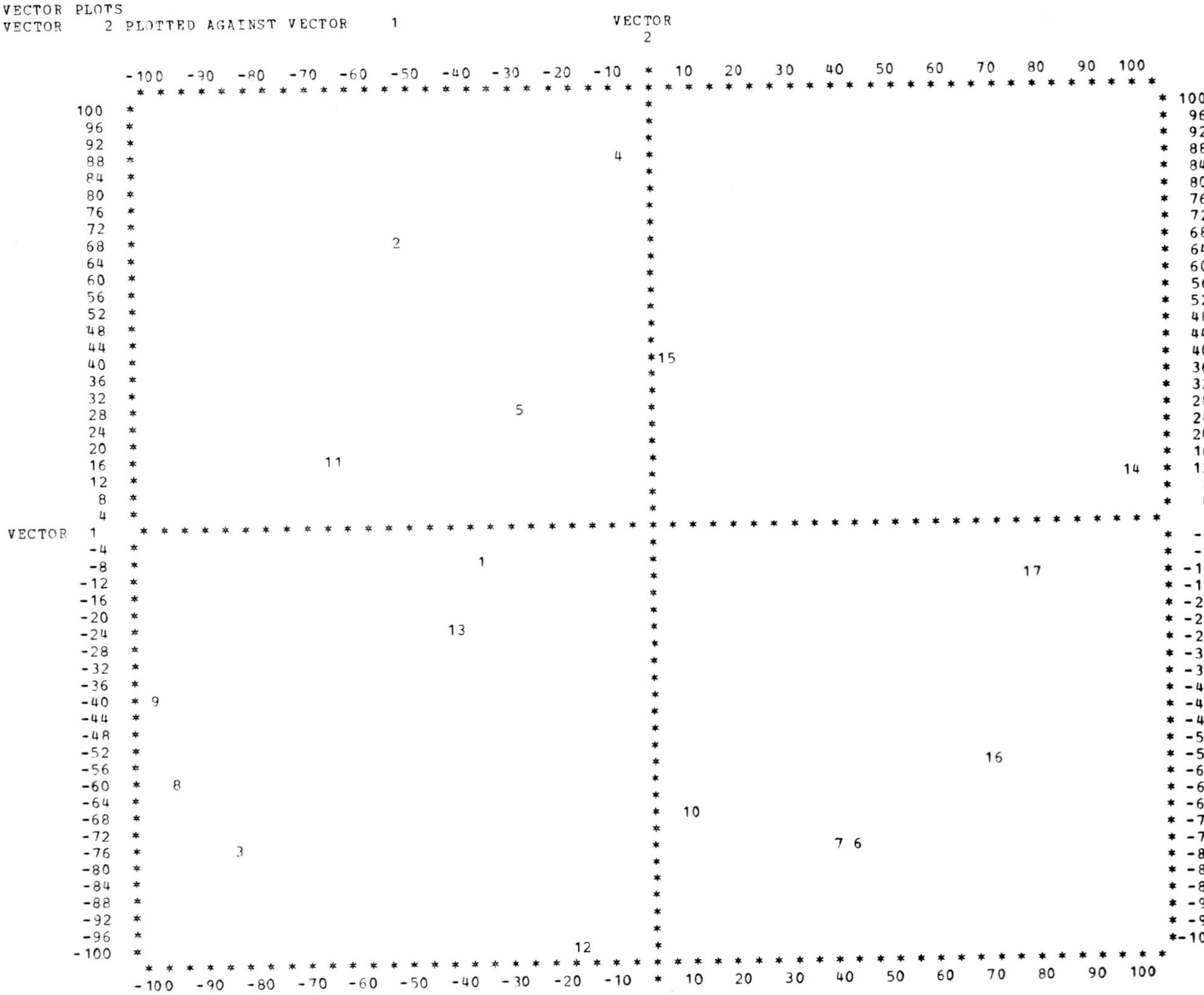

Fig. 28.12. R-mode analysis of Zone D, Step 2: plot of Vector 1 versus Vector 2, where degree of proximity of one variable to another reflects the degree of association as determined by Step 1. (Variables are as in Fig. 28.11.) Note that animal remains and flint flakes score relatively high on Vector 1, while most plant remains score relatively low. Plants of the cactus-*Agave* group tend to score lower on Vector 1 than seasonally restricted plants, suggesting they may have been processed in different squares.

Agave remains. Animal remains therefore score low on Vector 1 and high on Vector 2. It is also interesting to note that cucurbits are associated with plants that would be found in areas downslope from the cave, such as agave and nopales. Runner beans, on the other hand, were more strongly associated with plants which are collected upslope from the cave, such as acorns and piñon nuts. These associations were discovered independently by Whallon in his earlier analysis of Zone C (Whallon 1973).

In Fig. 28.20, the scores of our 17 variables on each of the vectors are given. Notice that only two vectors are important here, as in the case of Zone E; there does not seem to be any evidence of a storage area, as there was in Zone D.

Next, the Q-mode analysis for Zone C was conducted. The matrix of chi-square distances used as input to the multidimensional scaling program is shown in Fig. 28.21. Twenty-one cells were used to produce these figures. Three vectors were extracted, producing a three-dimensional configuration in which the distance between the cells preserved 84% of the original rank ordering of cell associations (see Fig. 28.22). The scores for all 21 cells on each of the three vectors are given in Fig. 28.23.

Cells were then clustered according to their scores on Vectors 1 and 2. These clusters are mapped onto the cave floor in Fig. 28.24. Area V (Squares E8 and F9–F10) is interpreted as a pathway into the middle of the cave, similar to that for Zone E, and there are other similarities to that earlier living floor. As in Zone E, to the north of the central path is a plant processing area. Areas I and II (Squares B9–B10, C9–C11, D10–D11, and E9–E10) make up the plant processing area for Zone C. These areas resemble parallel longitudinal bands stretching from the front to the back of the cave, each

ZONE D SCALING

GUTTMAN-LINGOES' SMALLEST SPACE COORDINATES FOR M = 3 (SEMI-STRONG MONOTONICITY).

VARIABLE	CENTRALITY INDEX	DIMENSION 1	2	3
1	74.272	-43.361	-16.239	-87.350
2	110.210	-64.795	-58.356	34.575
3	100.402	-69.548	94.535	-43.719
4	120.345	-26.707	-100.000	-61.370
5	72.907	-38.995	-49.742	-47.851
6	95.413	55.264	-3.322	-95.881
7	95.173	50.749	5.413	-100.000
8	100.776	-86.610	75.224	-55.334
9	107.383	-100.000	73.610	-17.816
10	67.423	24.313	44.031	-77.443
11	72.771	-71.284	8.714	12.116
12	93.891	28.025	100.000	-28.871
13	32.051	-39.958	27.024	-22.834
14	124.004	61.324	37.076	70.184
15	85.246	-15.351	-38.945	38.143
16	95.725	85.800	16.478	-29.277
17	101.245	65.897	32.017	35.555

GUTTMAN-LINGOES' COEFFICIENT OF ALIENATION = 0.12959 IN 39 ITERATIONS.
KRUSKAL'S STRESS = 0.11451

Fig. 28.13. R-mode analysis of Zone D: scores of the 17 variables for three vectors, including the two plotted in Fig. 28.12. Scores reflect the relative position of the variables on all three axes. Kruskal's Stress for this stage of the analysis is 0.11, which means that 89% of the relative order of the associations between variable pairs is preserved by the configuration based on these three vectors.

culminating in a discard area for fine debris. This suggests there was gradual movement of waste material from a plant processing area near the center of the cave to some fine-fraction discard areas near the rear wall, perhaps because Areas I and II also served as paths. The cells of Areas I and II that are closest to Area V contain large amounts of plant remains. Further, the crescentic shape of Areas I and II suggests the presence of a missing feature in Square D9 that had to be avoided. It should be remembered that during excavation, concentrations of ash and charcoal were found in D9, suggesting a surface campfire that had been kicked around until it became indistinct (Chapter 5). It is suggested that the main area of plant processing was in the squares contiguous to this "erased" feature.

From our analyses of Zones D and E, we expected to find an animal processing area located to the south of the central path. Indeed, a high co-occurrence of animal bones and flint flakes appeared in Area III (Squares E5–E6). This animal processing area, however, seems smaller than those of previous occupations. Area IV (Squares F5–F6), immediately next to Area III, consisted of flint flakes and plant remains, most notably acorns and *Opuntia* seeds.

So far, all three living floors examined have shown a tendency to have plant processing or discard areas located in the more sheltered north or northwest part of the cave, while animal processing or discard areas appear in the south or southeast. Sometimes a central entrance pathway separates these areas, as in Zones E and C. When hearths or storage features are present, they tend to be in the plant processing (presumably women's work area) part of the cave. The fine fraction of debris was often swept against the cave wall.

Zone B1

Given the trends seen in the previous three zones, we now look to Zone B1 to see which of them continue. In Fig. 28.25, we give the matrix of Pearson's r correlations among the same 17 variables used in previous zones. In Fig. 28.26, we present the scaling of these 17 variables along the first two extracted vectors. Again, as with all living floors except Zone D, only two vectors are required to distinguish between our major categories. Vector 1 separates the animal bones and flint flakes from all the plants except for agave leaves. In our discussion of previous zones we have suggested that men might have chewed cooked agave as they worked, and these leaf fragments might have resulted from the chewing of the leaf bases, which are utilized when the main agave heart is finished. If anything, Vector 2 separates these two categories even more strongly, having high scores for plant components and uniformly low scores for animal components. In other words, animal processing seems to have been segregated more sharply from plant processing here than in Zone C. Scores for each of the 17 variables on the two vectors are given in Fig. 28.27. Altogether, 77% of the relative ordering of Pearson's r correlations between variable pairs is preserved by this R-mode configuration.

A Q-mode analysis of the Zone B1 remains was then conducted for the 28 relevant 1 × 1-m squares. The original matrix of associations between all pairs of cells is given in

ORIGINAL COEFFICIENTS

COLUMN =	1	2	3	4	5	6	7	8	9	10	11	12	13	14	15	16	17	18	19	20	21	22	23
ROW = 1	0.0																						
ROW = 2	0.14	0.0																					
ROW = 3	0.0	0.14	0.0																				
ROW = 4	0.22	0.36	0.22	0.0																			
ROW = 5	0.22	0.36	0.22	0.22	0.0																		
ROW = 6	0.31	0.17	0.31	0.31	0.31	0.0																	
ROW = 7	0.28	0.31	0.28	0.28	0.28	0.31	0.0																
ROW = 8	0.25	0.21	0.25	0.47	0.47	0.38	0.19	0.0															
ROW = 9	0.43	0.45	0.43	0.21	0.43	0.40	0.0	0.36	0.0														
ROW = 10	0.41	0.44	0.41	0.19	0.41	0.39	0.28	0.47	0.21	0.0													
ROW = 11	0.32	0.38	0.32	0.32	0.32	0.46	0.19	0.41	0.31	0.25	0.0												
ROW = 12	0.22	0.24	0.22	0.45	0.45	0.40	0.29	0.25	0.44	0.46	0.48	0.0											
ROW = 13	0.41	0.44	0.41	0.19	0.41	0.39	0.28	0.47	0.21	0.16	0.20	0.52	0.0										
ROW = 14	0.47	0.47	0.47	0.24	0.47	0.41	0.28	0.45	0.21	0.22	0.32	0.48	0.18	0.0									
ROW = 15	0.27	0.35	0.27	0.27	0.27	0.19	0.24	0.38	0.35	0.31	0.34	0.42	0.31	0.40	0.0								
ROW = 16	0.45	0.46	0.45	0.22	0.45	0.40	0.29	0.47	0.22	0.14	0.35	0.47	0.14	0.24	0.37	0.0							
ROW = 17	0.39	0.40	0.39	0.39	0.39	0.24	0.12	0.35	0.26	0.31	0.34	0.43	0.32	0.28	0.20	0.35	0.0						
ROW = 18	0.32	0.33	0.32	0.32	0.32	0.33	0.31	0.31	0.31	0.30	0.28	0.32	0.30	0.33	0.32	0.32	0.32	0.0					
ROW = 19	0.41	0.40	0.41	0.41	0.41	0.48	0.31	0.40	0.40	0.38	0.34	0.28	0.48	0.47	0.50	0.41	0.49	0.28	0.0				
ROW = 20	0.38	0.40	0.38	0.38	0.38	0.23	0.22	0.40	0.33	0.25	0.31	0.46	0.30	0.30	0.18	0.26	0.07	0.31	0.54	0.0			
ROW = 21	0.43	0.45	0.43	0.21	0.43	0.40	0.32	0.46	0.24	0.21	0.47	0.35	0.35	0.32	0.35	0.25	0.46	0.28	0.32	0.48	0.0		
ROW = 22	0.24	0.25	0.24	0.47	0.47	0.41	0.31	0.23	0.45	0.44	0.38	0.24	0.44	0.47	0.40	0.46	0.40	0.0	0.38	0.40	0.43	0.0	
ROW = 23	0.23	0.24	0.23	0.46	0.46	0.41	0.33	0.21	0.47	0.41	0.33	0.33	0.38	0.37	0.39	0.45	0.31	0.19	0.41	0.30	0.53	0.14	0.0

Fig. 28.14. Q-mode analysis of Zone D, Step 1: matrix of chi-square partition distances between pairs of cells (1 × 1-m squares) based on relative co-occurrence of the same 17 variables already used in R-mode analysis (see legend for Fig. 28.11).

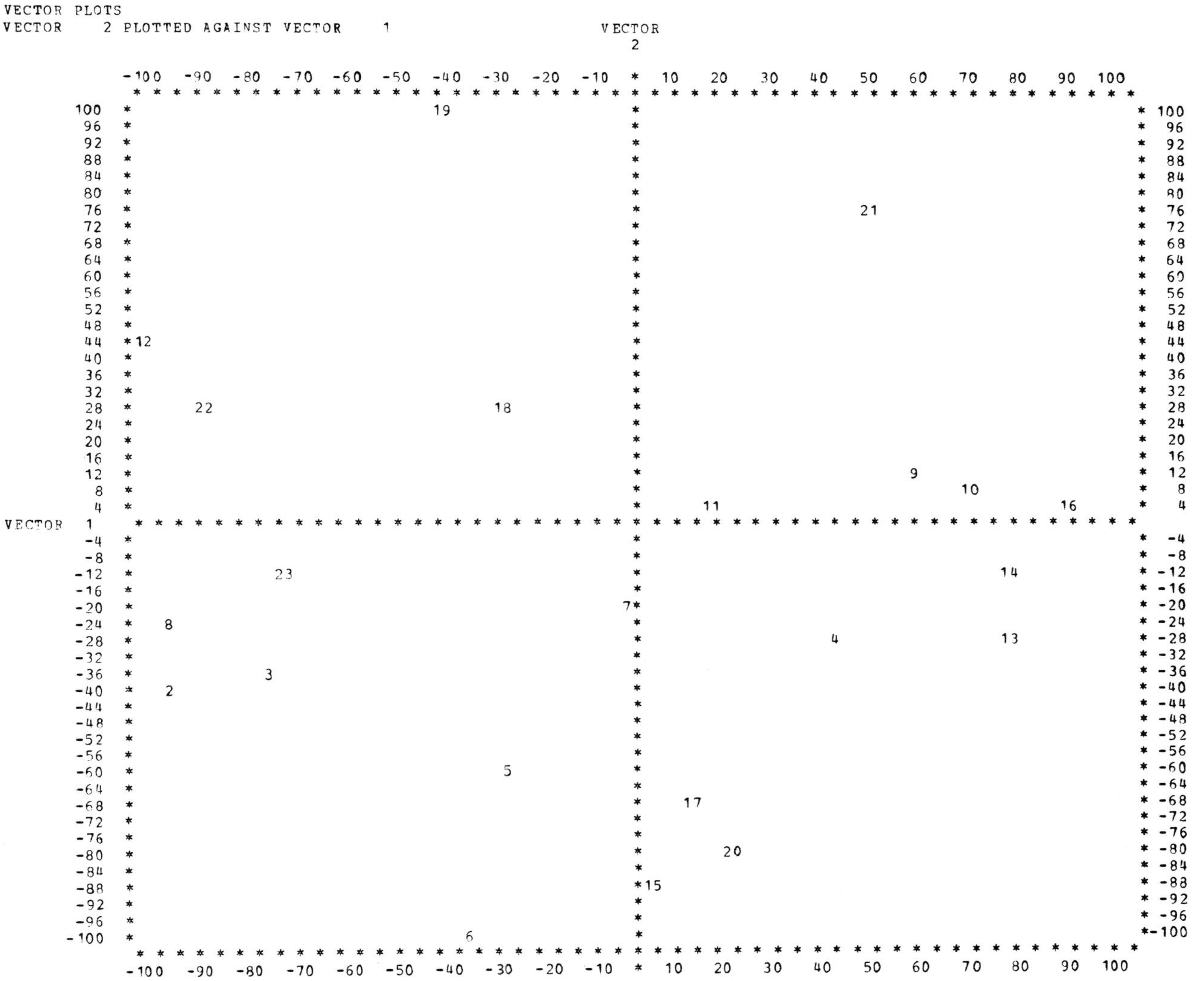

Fig. 28.15. Q-mode analysis of Zone D, Step 2: plot of Vector 1 versus Vector 2, where the degree of proximity of one numbered cell (1 × 1-m square) to another reflects the degree of association between all possible pairs of cells as determined by Step 1. (Numbered cells 1–23 are as in Table 28.1.) The clusters of associated cells shown here constitute the basis for the activity areas and pathways reconstructed in Fig. 28.17.

Fig. 28.28. This matrix of chi-square associations was then input to the multidimensional scaling program. Two vectors were produced, and the relative positions of each cell on these two dimensions are plotted in Fig. 28.29. The goodness of fit for this analysis was 74%, as compared to 77% for the R-mode analysis. The scores of the 28 cells are then given in Fig. 28.30.

Next, the cells were clustered into groups showing similarity based on scores for these two dimensions, producing the interpretive drawing given in Fig. 28.31. As in the case of several previous zones, we see a central pathway (Area V, Squares D6, E7, and F7) dividing the cave into two unequal sections. To the south we find a group of cells (D5, E4, F4, F6, and G5) linked by the co-occurrence of animal bone and flint flakes; this is Area IV, presumably a men's work area. The R-mode analysis for Zone B1 had suggested a larger animal processing area on this living floor than was found in Zone C, and such certainly appears to be the case.

Running north–south, almost perpendicular to Area V, is another pathway designated Area III-south. This pathway (Squares F5, E6–E7, and D8) appears to connect Area IV, our proposed men's work area, with a possible women's work area on the other side of the cave. This latter area (Area II, Squares B9–B10, C9–C10, and D9–D10) consists of cells characterized by the co-occurrence of many seasonally restricted plants, with smaller amounts of co-occurring animal remains. It is significant that Area II, which was probably an area for plant processing and food preparation, surrounds one hearth (Feature 14) and lies adjacent to another (Feature 11). Two areas apparently used for the discard of fine-fraction debris (both plant and animal) occur in Squares B8 and B11, between Area II and the cave wall. Nearby is Area

CHI SQUARE DISTANCE SCALING FOR ZONE D WITH 17 VARIABLES

GUTTMAN-LINGOES' SMALLEST SPACE COORDINATES FOR M = 3 (SEMI-STRONG MONOTONICITY).

VARIABLE	CENTRALITY INDEX	DIMENSION 1	2	3
1	76.387	-72.401	-33.579	-44.360
2	88.370	-90.514	-40.574	-22.859
3	75.815	-71.638	-33.346	-44.656
4	70.948	42.839	-27.597	-58.125
5	102.474	-21.743	-60.314	-100.000
6	89.104	-30.322	-100.000	-18.260
7	14.569	0.067	-18.476	2.418
8	99.817	-92.629	-23.024	41.287
9	76.831	57.996	11.568	21.794
10	79.601	69.039	4.801	-22.227
11	68.164	17.219	2.943	-70.800
12	109.570	-100.000	42.823	-20.180
13	86.479	78.089	-28.169	-14.121
14	100.095	78.268	-11.133	42.499
15	72.415	4.837	-85.106	-19.576
16	97.775	89.470	0.466	-10.328
17	79.179	14.124	-68.352	44.346
18	51.226	-22.829	25.345	18.713
19	122.355	-36.388	100.000	-41.640
20	80.865	21.394	-80.484	27.275
21	105.433	50.140	73.659	-18.381
22	94.990	-81.185	26.584	33.625
23	88.498	-68.693	-9.138	53.836

GUTTMAN-LINGOES' COEFFICIENT OF ALIENATION = 0.14438 IN 92 ITERATIONS.
KRUSKAL'S STRESS = 0.12932

Fig. 28.16. Q-mode analysis of Zone D: scores of the 23 numbered cells (1 × 1-m squares) for 3 vectors, including the two shown in Fig. 28.15. Scores reflect the relative positions of the cells on all three axes. Kruskal's Stress for this stage of the analysis is 0.13, which means that 87% of the relative order of the associations between pairs of numbered cells is preserved by the configuration based on these three vectors.

I (Squares C8 and D7), whose function is ambiguous although it may be related to discard from Area II.

Area III-north, the northernmost group of cells (C11–C12, D12, and E10–E11), seems to represent a pathway linking the two hearths. Its crescentic shape suggests that it detoured around an obstacle in Square D11—perhaps a seated person, since no feature was found in this square. The content of D11 suggests the discard of fine refuse, which would not rule out the presence of a seated worker (perhaps a woman). Note also that Areas III-north and III-south, which share a number of characteristics, are separated only by the area of a hearth (Feature 11). Were it not for this hearth, the two areas would probably constitute one long, interconnected pathway between men's and women's work areas.

It should be stressed that our separation of Area III into two sections is purely the result of the Feature 11 gap. In analytical terms, all the cells in Areas III-north and III-south were present in the same Q-mode-generated cluster. It is presumed that the similarity of the variable structure in all of these cells relates to their common function as components of pathways. The results for Zone B1 indicate that such pathways were very important areas within a living floor. They served to demarcate the limits of each work area and to channel the movement of people and resources through the cave.

Three basic patterns of movement can be discerned from the four living floors we analyzed. First, in virtually all zones there is evidence for a central pathway leading into the cave from the talus slope outside. Second, we usually also see evidence for pathways between men's and women's work areas in the cave. Third, we frequently also see some evidence for movement *within* a work area, such as between hearths in a women's work area. Clearly, the placement and orientation of such paths seems as crucial to our overall understanding of the group's activity as do the placement of the more frequently discussed plant and animal processing areas.

CHANGES THROUGH TIME

One final set of tests was run on our four living floors at Guilá Naquitz. In order to see what changes might have taken place in the partitioning of the cave through time, we performed an additional set of multidimensional scaling runs in which each living floor was compared with every other living floor in terms of the results of its Q-mode analysis. In this section we discuss the overall results of these comparisons but have selected only four cases to illustrate graphically (Figs. 28.32–28.35). Since the legends for Figs.

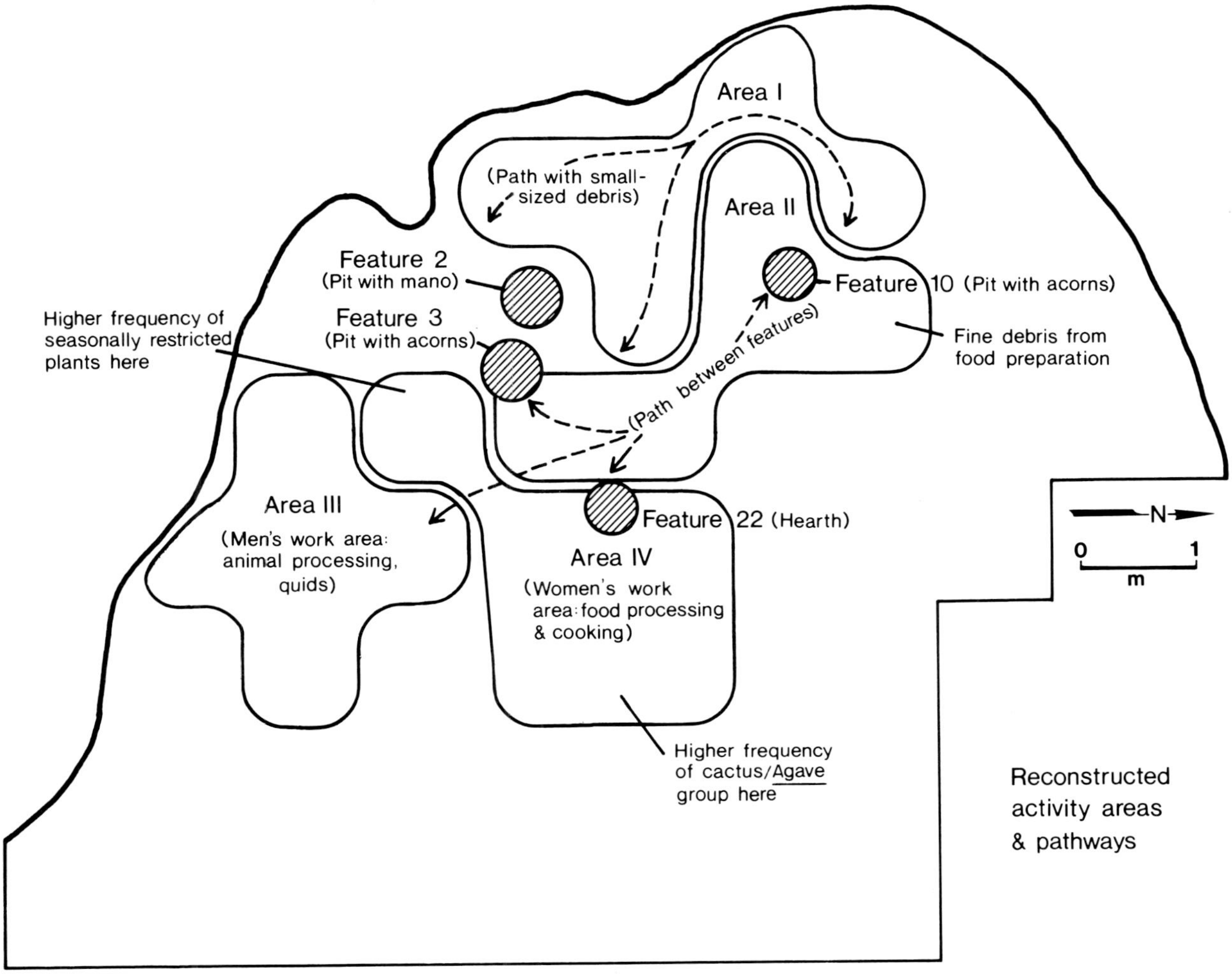

Fig. 28.17. Interpretive drawing of Zone D at Guilá Naquitz, showing pathways and activity areas (groups of 1 × 1-m squares) defined by the preceding Q-mode analysis. Area I is interpreted as a curving pathway with acorn, hackberry, and flint debris that linked Feature 2 with the open area north of Area II. Area II is interpreted as a pathway running between acorn storage and food preparation areas. Area III is interpreted as an area where one or two persons (probably men) engaged in tasks related to animal processing. Area IV is interpreted as an area where one or two persons (probably women) processed and cooked both seasonally restricted and cactus-*Agave* group plants.

28.32–28.35 are largely self-explanatory, we simply summarize the two most important results of the comparisons here.

1. First, there is a tendency for men's and women's activity areas to become increasingly distinct through time. Both men's and women's work areas in Zone D were more distinct than in Zone E; both areas were also more distinct in Zone C than in Zone E. In Zone B1, both men's and women's work areas were quite distinct, as well as different from their Zone C counterparts. What this might mean (tentatively speaking) is that sexual division of labor became more pronounced or more specialized during the Naquitz phase.

2. Second, even if we have correctly identified work areas as being used by men or women, it appears that the activities within them were far from uniform. For example, the men's work areas in Zones E and D are more alike than are the women's work areas; on the other hand, when Zones D and C are compared, the women's work areas are more alike than are the men's work areas. In the case of Zone B1, both men's and women's work areas differ somewhat from their Zone C counterparts. What this may mean is that, although the cave floor was indeed conceptually partitioned into men's and women's work space, the specific tasks performed in each could vary somewhat from occupation to occupation.

Finally, we are left with an unanswered question: Were the variations in (and increasing distinctness of) sexual division of labor simply natural trends during the course of the Archaic period, or were they accelerated by the adoption of primitive agriculture?

SUMMARY AND CONCLUSIONS

As we stated at the outset, our analysis had several goals. One was to see what kinds of archaeologically inferred activity areas would emerge from a study of the 1 × 1-m squares

ORIGINAL COEFFICIENTS																	
COLUMN =	1	2	3	4	5	6	7	8	9	10	11	12	13	14	15	16	17
ROW = 1	0.0																
ROW = 2	0.98	0.0															
ROW = 3	-0.06	0.01	0.0														
ROW = 4	0.98	0.99	0.05	0.0													
ROW = 5	0.27	0.26	0.80	0.32	0.0												
ROW = 6	0.98	1.00	-0.03	0.99	0.24	0.0											
ROW = 7	0.11	0.15	0.64	0.20	0.62	0.14	0.0										
ROW = 8	-0.08	-0.09	-0.03	-0.07	-0.00	-0.10	0.17	0.0									
ROW = 9	0.98	0.93	-0.08	0.94	0.27	0.94	0.10	-0.13	0.0								
ROW = 10	-0.09	-0.02	0.95	0.02	0.81	-0.06	0.68	0.12	-0.12	0.0							
ROW = 11	0.01	0.10	0.65	0.11	0.54	0.07	0.45	-0.07	-0.01	0.67	0.0						
ROW = 12	0.81	0.85	0.48	0.88	0.63	0.83	0.49	-0.11	0.78	0.44	0.42	0.0					
ROW = 13	-0.03	0.04	0.88	0.09	0.86	0.01	0.70	0.11	-0.06	0.95	0.75	0.46	0.0				
ROW = 14	-0.14	-0.09	-0.05	-0.12	-0.20	-0.10	-0.20	-0.07	-0.14	-0.04	0.21	-0.04	-0.04	0.0			
ROW = 15	0.11	0.17	0.18	0.15	0.04	0.15	0.0	-0.29	0.12	0.10	0.09	0.19	0.09	-0.18	0.0		
ROW = 16	-0.24	-0.17	-0.01	-0.21	-0.16	-0.18	-0.28	-0.19	-0.22	0.00	0.31	-0.14	0.04	0.66	0.17	0.0	
ROW = 17	0.40	0.45	0.41	0.48	0.48	0.45	0.35	-0.18	0.34	0.42	0.39	0.69	0.45	0.15	0.26	0.12	0.0

Fig. 28.18. R-mode analysis of Zone C, Step 1: matrix of correlation coefficients for 17 selected variables, using Pearson's r. The matrix shows the degree of linear association between all possible pairs of variables. (Numbered variables are as given in the legend for Fig. 28.11.)

for each living floor. A second was to see whether we could, by grouping our data in certain ways, infer something about the conceptual partitioning of the cave by its occupants. Our first goal resulted in the areas we have designated by Roman numerals. The second goal has led us to an interpretation of those areas as "men's work space," "women's work space," "entrance pathways," and so on.

An additional by-product of the analysis, however, has been a glimpse at the impact that different statistical techniques can have on the result of the analysis. Studying Zones E, D, C, and B1 by means of an R-mode analysis based on the multidimensional scaling of a matrix of Pearson's r correlations allowed us to isolate basic plant and animal processing zones in all cases. However, the second analytical technique we tried, a Q-mode analysis, showed us that more information on the spatial structuring of activities (including some data on conceptual partitioning) was available if only we knew how to look.

Previous analyses, as well as ethnographic analogy, had suggested reasons why certain plant and animal products might be conceptually "grouped" by the cave occupants and hence processed or discarded together; other products whose collecting or processing might be considered incompatible with the first group might be found elsewhere. This led us to employ a measure of association using information about the co-occurrence or nonco-occurrence of resources from different groups. This measure, the chi-square partition distance, derives from the basic chi-square statistic. Cells (1 × 1-m squares) were then compared, not on the basis of the *amount* of a resource present (as in the R-mode analysis), but on the basis of *presence or absence* of groups and subgroups.

Utilization of this Q-mode approach, in which a matrix of chi-square associations served as input for a multidimensional scaling, produced an activity configuration (or "map" of the living floor) that included not only those activities already suggested by the R-mode analysis but also a new set of linear areas relating to the movement of people and materials into, through, and out of the cave. These linear areas, designated activity paths, were interpreted as corresponding to cells through which movement of materials

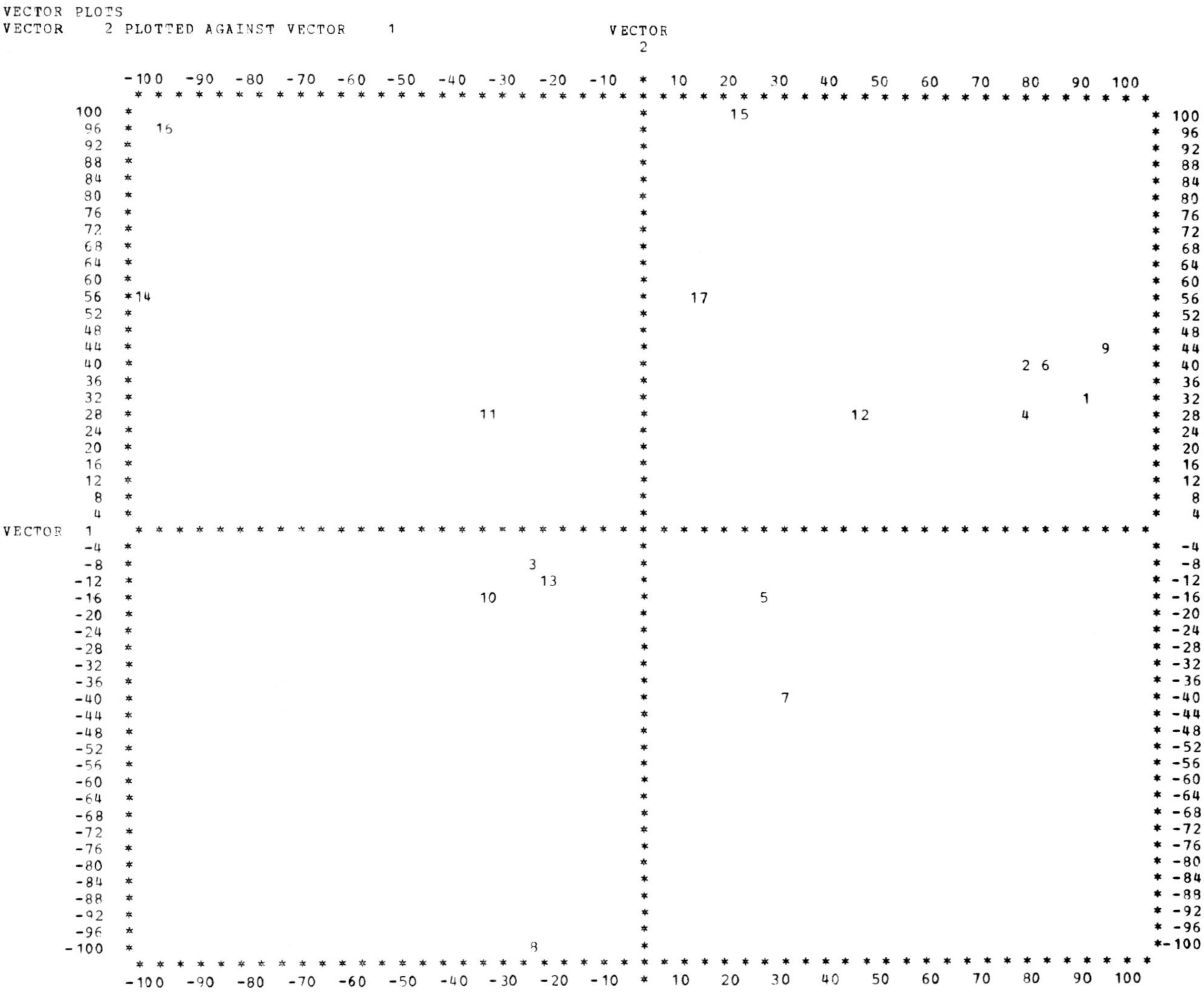

Fig. 28.19. R-mode analysis of Zone C, Step 2: Plot of Vector 1 versus Vector 2, where degree of proximity of one variable to another reflects the degree of association as determined by Step 1. (Variables are as in Fig. 28.11.) Note that Vector 1 separates the seasonally restricted plants from the animals-plus-flint group and the plants of the cactus–*Agave* group; Vector 2, in turn, separates the animals and flint from the plants of the cactus–*Agave* group. Note that cucurbits show an association with plants that would be collected downslope from the cave, while wild runner beans are more strongly associated with plants that would be collected upslope; these associations were independently detected by Whallon (1973).

occurred and that consequently developed their own distinct pattern of debris. The pathways could be isolated using our chi-square partition distance measure precisely because debris within them had originated from at least two different activity areas. Thus, pathway cells had a mixture of variables that co-occurred, some from each of the areas that the path connected. These groups of cells were therefore easily isolated in the analysis from the more straightforward activity areas.

The isolation of pathways within a living floor is one step toward the recreation of the cycle of processing activities (*activity cycle*) that occurred there. It provides us with information about the flow of materials through the living space, as well as the ways in which the processing areas are situated relative to that flow. It is clear, however, that not all measures of association are selective enough to extract the information needed to discern these areas. This is one more illustration of the fact that a choice of statistic should be based in large part on the type of information needed to answer a specific question.

The results of our analysis also suggest some overall patterns in the occupation of the cave. Activity areas and paths for Zones E, C, and B1 are quite similar in terms of positioning, and the placement of features (where present) reinforces the pattern. This implies some continuity in the behavior of the group(s) that occupied the cave through time. Zone D shows some differences, but these may be partly related to the placement of pits for acorn storage.

In view of the long-term similarities in spatial configuration exhibited by these four zones, it would be of interest to know which of these properties were the least likely to be

ZONE C SCALING

GUTTMAN-LINGOES' SMALLEST SPACE COORDINATES FOR M = 2 (SEMI-STRONG MONOTONICITY).

DIMENSION		1	2
VARIABLE	CENTRALITY INDEX		
1	75.388	91.045	29.685
2	63.170	76.211	36.438
3	44.204	-20.955	-5.649
4	63.001	78.036	25.640
5	35.736	24.778	-14.860
6	67.777	80.667	37.609
7	60.914	31.179	-39.211
8	124.562	-19.470	-100.000
9	81.916	94.292	41.389
10	55.245	-26.871	-16.053
11	43.578	-27.895	25.168
12	32.684	47.497	25.372
13	43.146	-13.539	-12.465
14	120.615	-100.000	54.931
15	80.822	23.614	100.000
16	128.691	-90.446	92.910
17	32.917	12.362	52.387

GUTTMAN-LINGOES' COEFFICIENT OF ALIENATION = 0.13823 IN 43 ITERATIONS.
KRUSKAL'S STRESS = 0.11631

Fig. 28.20. R-mode analysis of Zone C: scores of 17 variables for each of the two vectors used in Step 2. Scores reflect the distance from the ordinate (Vector 2) or abscissa (Vector 1) of the graph shown in Fig. 28.19. Note that flint flakes and animal bones display high positive values for Vector 2 and low values for Vector 1; that tendency is reversed for plants. Kruskal's Stress for this stage of the analysis is 0.12, which means that 88% of the relative order of the associations between variable pairs is preserved by the configuration shown in Fig. 28.19.

changed in the presence of new activities. The configuration of Zone D allows us to make a few suggestions. Although in Zone D the plant processing area is in an atypical location, it is still placed to the north of the animal processing area, a characteristic of all other zones. On the other hand, the animal processing region is invariant over all four living floors. Its consistent placement off to the south may suggest either (1) that plant collecting requires more movement into and out of the cave during the day, and as such is more likely to be situated closer to the main entrance for convenience, or (2) that women preferred to locate their work space well back in under the overhang, in the most suitable part of the cave for placing hearths and pits. Thus, there seem to have been certain constants in the way work space in the cave was conceptualized, and these constants carried over from one group of occupants to the next. Other spatial properties seem to be more flexible, which could reflect either lesser priorities or gradual changes in the conceptualization of work space.

Finally, we should make two additional observations: (1) Sexual division of labor seems to have been a much stronger determinant of the use of space than were seasonality or the vagaries of vegetational zonation, and (2) this division of labor becomes increasingly more clearly displayed as time goes on. Men's and women's work areas are more distinct, more separate, and more easily delimited in Zone B1 than they ever were in Zone E. Possibly, this reflects an increasing specialization in tasks that accompanied increasing familiarity with the Guilá Naquitz environment as well as increasing efficiency at using it.

Obviously, there is much more we would like to know about what went on in Guilá Naquitz, but we are at the limits of what the data will yield to a gentle massage. To convert our Red Queen into a true kitten would take a more vigorous shaking than we are prepared to apply.

ORIGINAL COEFFICIENTS

COLUMN =	1	2	3	4	5	6	7	8	9	10	11	12	13	14	15	16	17	18	19	20	21
ROW = 1	0.0																				
ROW = 2	0.19	0.0																			
ROW = 3	0.33	0.25	0.0																		
ROW = 4	0.10	0.25	0.40	0.0																	
ROW = 5	0.16	0.28	0.44	0.22	0.0																
ROW = 6	0.19	0.33	0.25	0.25	0.28	0.0															
ROW = 7	0.40	0.32	0.42	0.35	0.38	0.25	0.0														
ROW = 8	0.21	0.31	0.40	0.18	0.23	0.31	0.41	0.0													
ROW = 9	0.28	0.31	0.24	0.24	0.19	0.31	0.24	0.22	0.0												
ROW = 10	0.19	0.30	0.37	0.24	0.21	0.30	0.40	0.25	0.28	0.0											
ROW = 11	0.24	0.33	0.46	0.24	0.23	0.33	0.47	0.21	0.28	0.22	0.0										
ROW = 12	0.22	0.32	0.47	0.25	0.25	0.32	0.33	0.18	0.24	0.24	0.34	0.0									
ROW = 13	0.47	0.28	0.44	0.44	0.47	0.31	0.48	0.36	0.19	0.47	0.36	0.40	0.0								
ROW = 14	0.25	0.33	0.47	0.25	0.25	0.33	0.35	0.25	0.33	0.25	0.19	0.19	0.41	0.0							
ROW = 15	0.0	0.0	0.0	0.0	0.0	0.0	0.0	0.0	0.0	0.0	0.0	0.0	0.0	0.0	0.0						
ROW = 16	0.28	0.31	0.24	0.33	0.32	0.31	0.29	0.32	0.31	0.28	0.28	0.29	0.32	0.33	0.0	0.0					
ROW = 17	0.47	0.30	0.46	0.46	0.43	0.30	0.32	0.47	0.33	0.46	0.51	0.37	0.35	0.41	0.0	0.28	0.0				
ROW = 18	0.46	0.32	0.46	0.45	0.43	0.32	0.31	0.47	0.33	0.47	0.52	0.37	0.35	0.41	0.0	0.28	0.15	0.0			
ROW = 19	0.40	0.32	0.39	0.40	0.41	0.32	0.24	0.38	0.28	0.35	0.41	0.27	0.32	0.42	0.0	0.21	0.42	0.40	0.0		
ROW = 20	0.39	0.32	0.37	0.42	0.41	0.32	0.35	0.41	0.33	0.40	0.32	0.43	0.31	0.35	0.0	0.29	0.47	0.46	0.10	0.0	
ROW = 21	0.25	0.30	0.46	0.24	0.25	0.30	0.47	0.08	0.28	0.24	0.18	0.26	0.24	0.19	0.0	0.33	0.52	0.53	0.35	0.32	0.0

Fig. 28.21. Q-mode analysis of Zone C, Step 1: matrix of chi-square partition distances between pairs of cells (1 × 1-m squares) based on relative co-occurrence of the same 17 variables already used in the R-mode analysis. The numbers 1–21 refer to 1 × 1-m squares, as shown in Table 28.1.

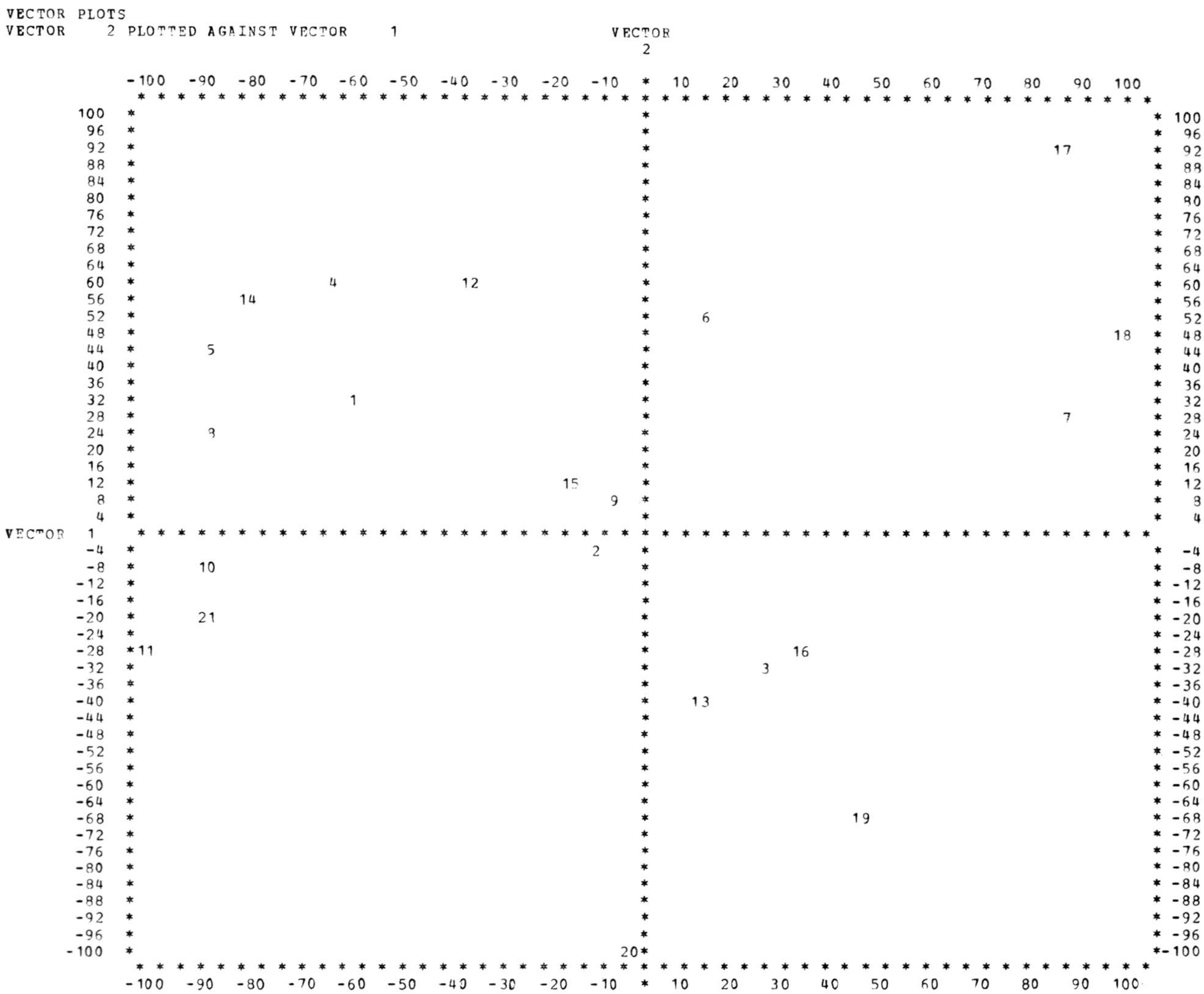

Fig. 28.22. Q-mode analysis of Zone C, Step 2: plot of Vector 1 versus Vector 2, where the degree of proximity of one numbered cell (1 × 1-m square) to another reflects the degree of association between all possible pairs of cells as determined by Step 1. (Numbered cells 1–21 are as in Table 28.1.) The clusters of associated cells shown here constitute the basis for the activity areas and pathways reconstructed in Fig. 28.24.

ZONE C CHI SQUARE DISTANCE SCALING FOR 17 VARIABLES

GUTTMAN-LINGOES' SMALLEST SPACE COORDINATES FOR M = 3 (SEMI-STRONG MONOTONICITY).

VARIABLE	CENTRALITY INDEX	DIMENSION 1	DIMENSION 2	DIMENSION 3
1	68.755	-56.156	29.040	-45.958
2	59.677	-7.068	-3.415	-55.283
3	115.812	27.390	-29.727	-100.000
4	72.491	-58.702	59.712	-18.756
5	81.705	-83.526	42.797	-18.209
6	58.172	13.961	49.398	-28.202
7	102.711	85.745	24.198	-21.727
8	73.546	-81.810	20.012	24.996
9	33.293	-3.575	4.220	34.794
10	81.751	-82.255	-5.240	-37.725
11	94.320	-100.000	-27.246	6.642
12	65.764	-31.766	59.351	41.322
13	109.036	14.139	-37.066	98.543
14	93.687	-75.996	53.142	55.324
15	2.862	-9.827	8.609	3.201
16	61.769	32.143	-28.767	-18.167
17	123.246	85.069	90.738	11.728
18	127.099	100.000	45.150	48.778
19	94.928	45.312	-65.085	20.072
20	109.933	2.821	-100.000	15.371
21	85.347	-81.070	-18.851	46.284

GUTTMAN-LINGOES' COEFFICIENT OF ALIENATION = 0.18611 IN 33 ITERATIONS.
KRUSKAL'S STRESS = 0.16032

Fig. 28.23. Q-mode analysis of Zone C: scores of the 21 numbered cells (1 × 1-m squares) for three vectors, including the two shown in Fig. 28.22. Scores reflect the distance from the ordinate (Vector 2) or abscissa (Vector 1) of the latter graph. Kruskal's Stress for this stage of the analysis is 0.16, which means that 84% of the relative order of the associations between pairs of numbered cells is preserved by the configuration based on these three vectors.

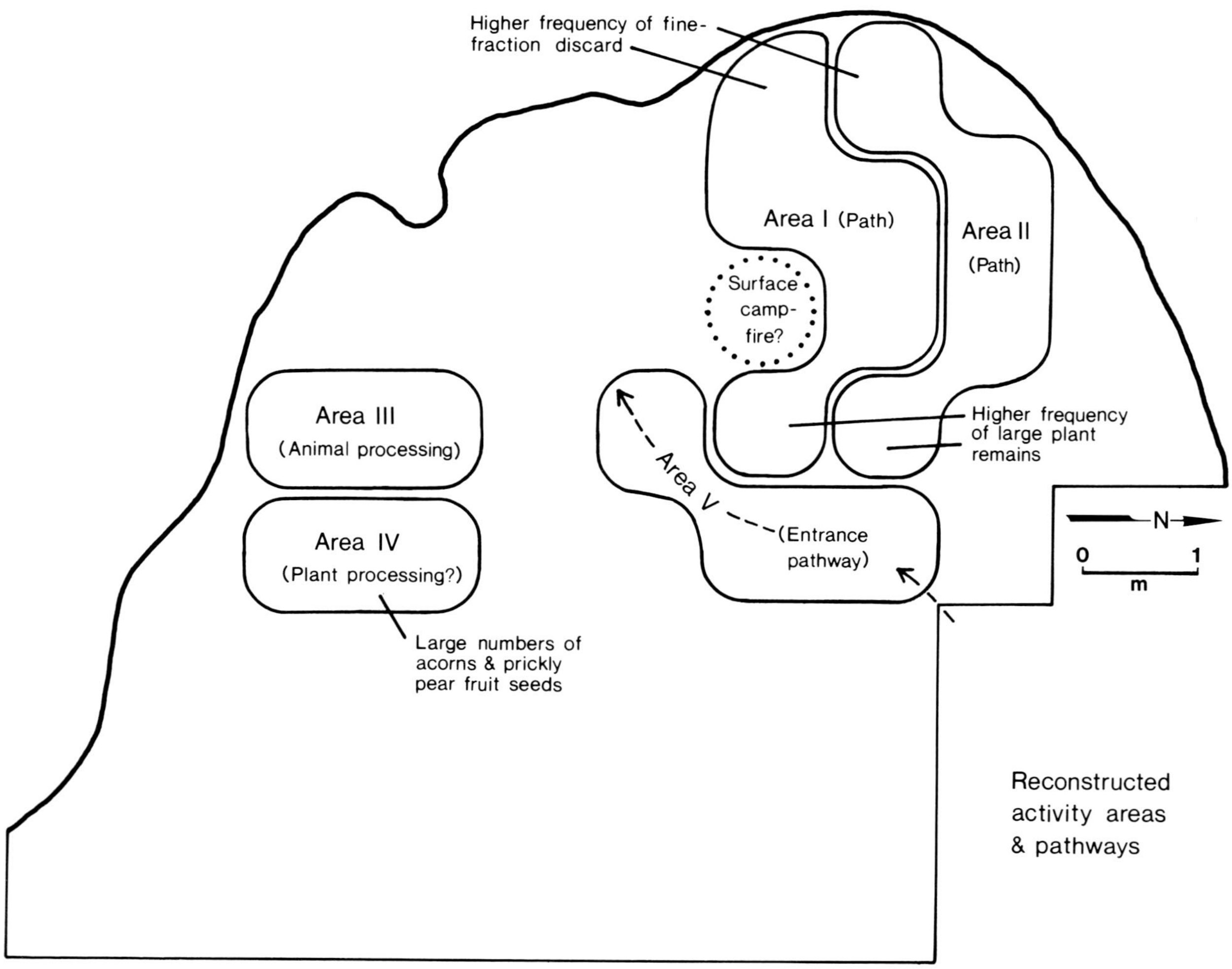

Fig. 28.24. Interpretive drawing of Zone C at Guilá Naquitz, showing pathways and activity areas (groups of 1 × 1-m squares) defined by the preceding Q-mode analysis. Areas I and II are each interpreted as pathways leading east-west between a plant processing area and a discard area near the wall of the cave; there are differences in the diversity of debris between them, with Area I being more complex. Area V is interpreted as the debris area associated with the main pathway entering the cave from the northeast; it includes several plants of the cactus–*Agave* group. Area III is interpreted as an area where at least one person (probably male) processed animals. Area IV is interpreted as an area where at least one person (possibly the same person responsible for Area III) processed some plants.

ORIGINAL COEFFICIENTS

COLUMN =	1	2	3	4	5	6	7	8	9	10	11	12	13	14	15	16	17
ROW = 1	0.0																
ROW = 2	-0.11	0.0															
ROW = 3	0.14	-0.14	0.0														
ROW = 4	0.20	0.03	0.33	0.0													
ROW = 5	0.46	-0.10	0.44	-0.01	0.0												
ROW = 6	0.61	-0.06	0.02	-0.11	-0.02	0.0											
ROW = 7	0.12	-0.10	-0.10	0.03	0.06	0.15	0.0										
ROW = 8	0.51	-0.07	0.04	0.20	-0.06	0.16	-0.00	0.0									
ROW = 9	0.59	-0.05	0.01	-0.10	-0.03	0.97	0.24	0.09	0.0								
ROW = 10	0.26	-0.05	0.14	0.77	-0.04	-0.00	-0.04	0.36	-0.04	0.0							
ROW = 11	0.56	-0.04	0.16	0.14	0.75	-0.01	0.31	-0.05	0.07	-0.01	0.0						
ROW = 12	0.63	0.11	0.10	0.21	0.69	0.07	0.25	0.00	0.14	0.08	0.96	0.0					
ROW = 13	0.38	-0.12	0.62	0.17	0.43	0.38	0.06	-0.08	0.37	-0.05	0.30	0.27	0.0				
ROW = 14	0.02	-0.08	-0.05	-0.13	0.09	-0.11	-0.15	-0.13	-0.10	-0.08	-0.08	-0.11	-0.05	0.0			
ROW = 15	0.40	-0.03	0.06	0.48	0.25	-0.08	-0.11	0.26	-0.08	0.62	0.35	0.40	-0.00	-0.05	0.0		
ROW = 16	0.02	0.12	-0.06	0.02	0.15	-0.11	-0.01	-0.08	-0.08	0.02	0.25	0.22	-0.07	-0.08	0.42	0.0	
ROW = 17	-0.05	0.25	-0.36	-0.11	-0.08	-0.01	0.27	-0.18	0.04	-0.24	0.12	0.13	-0.28	0.17	0.14	0.31	0.0

Fig. 28.25. R-mode analysis of Zone B1, Step 1: matrix of correlation coefficients for 17 selected variables, using Pearson's r. The matrix shows the degree of linear association between all possible pairs of variables. (Numbered variables are as given in the caption for Fig. 28.11.)

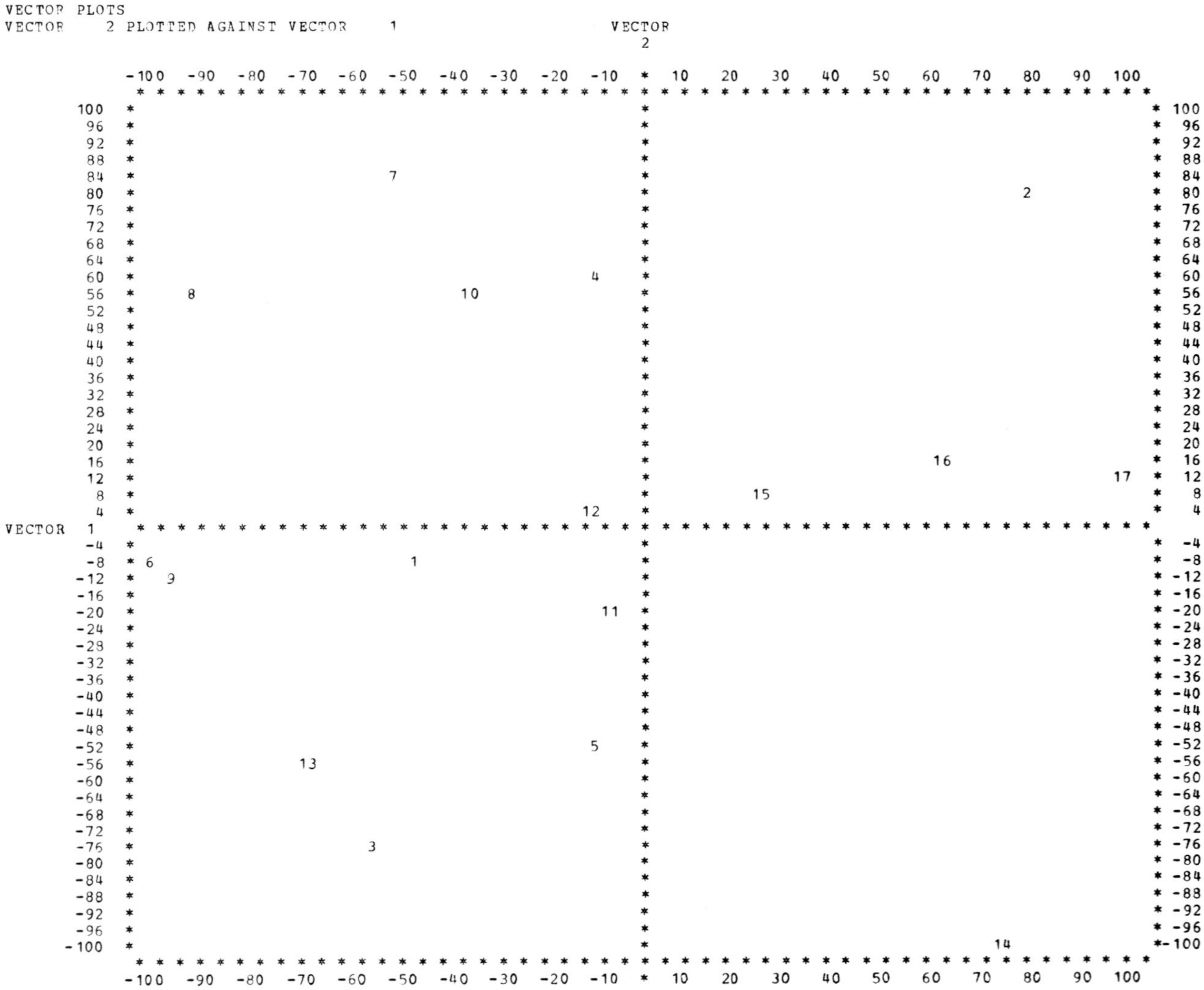

Fig. 28.26. R-mode analysis of Zone B1, Step 2: plot of Vector 1 versus Vector 2, where degree of proximity of one variable to another reflects the degree of association as determined by Step 1. (Variables are as in Fig. 28.11.) Note that Vector 1 separates the animal bones and flint waste flakes from all plants except agave leaves (Variable 2). *Opuntia* seeds, nopales, and *guaje* pods all score very low on Vector 1, suggesting that *guaje* and prickly pear may have been processed and/or discarded together.

ZONE B1 SCALING

GUTTMAN-LINGOES' SMALLEST SPACE COORDINATES FOR M = 2 (SEMI-STRONG MONOTONICITY).

VARIABLE	CENTRALITY INDEX	DIMENSION 1	DIMENSION 2
1	34.417	-43.900	-8.910
2	117.066	78.277	77.462
3	84.308	-49.300	-73.177
4	57.040	-6.251	58.890
5	53.165	-7.688	-50.971
6	89.082	-100.000	-5.980
7	87.157	-46.816	81.657
8	90.215	-85.133	53.887
9	79.442	-89.895	-9.378
10	54.036	-31.640	52.127
11	22.383	-1.619	-18.118
12	4.165	-7.423	0.507
13	76.471	-63.752	-53.557
14	132.931	73.876	-100.000
15	37.534	25.932	6.960
16	75.790	63.648	13.454
17	111.598	100.000	10.454

GUTTMAN-LINGOES' COEFFICIENT OF ALIENATION = 0.27288 IN 38 ITERATIONS.
KRUSKAL'S STRESS = 0.22761

Fig. 28.27. R-mode analysis of Zone B1: scores of 17 variables for each of the two vectors used in Step 2. Scores reflect the distance from the ordinate (Vector 2) or abscissa (Vector 1) of the graph shown in Fig. 28.26. Note that all types of animals display high values for Vector 1 and low values for Vector 2. Kruskal's Stress for this stage of the analysis is 0.23, which means that 77% of the relative order of the associations between variable pairs is preserved by the configuration shown in Fig. 28.26.

ORIGINAL COEFFICIENTS

COLUMN =	1	2	3	4	5	6	7	8	9	10	11	12	13	14	15	16	17	18	19	20	21	22	23	24	25	26	27	28
ROW = 1	0.0																											
ROW = 2	0.22	0.0																										
ROW = 3	0.22	0.16	0.0																									
ROW = 4	0.0	0.22	0.22	0.0																								
ROW = 5	0.0	0.22	0.22	0.0	0.0																							
ROW = 6	0.22	0.50	0.50	0.22	0.22	0.0																						
ROW = 7	0.22	0.14	0.23	0.22	0.22	0.40	0.0																					
ROW = 8	0.22	0.21	0.08	0.22	0.22	0.40	0.17	0.0																				
ROW = 9	0.22	0.46	0.50	0.22	0.22	0.39	0.32	0.40	0.0																			
ROW = 10	0.30	0.38	0.38	0.30	0.30	0.41	0.40	0.39	0.41	0.0																		
ROW = 11	0.22	0.41	0.50	0.22	0.22	0.23	0.46	0.50	0.48	0.41	0.0																	
ROW = 12	0.30	0.35	0.41	0.30	0.30	0.40	0.26	0.40	0.40	0.46	0.39	0.0																
ROW = 13	0.33	0.50	0.41	0.33	0.33	0.23	0.25	0.19	0.21	0.49	0.46	0.49	0.0															
ROW = 14	0.22	0.47	0.47	0.22	0.22	0.49	0.50	0.49	0.22	0.35	0.49	0.42	0.49	0.0														
ROW = 15	0.33	0.33	0.28	0.33	0.33	0.25	0.33	0.32	0.33	0.30	0.33	0.32	0.25	0.32	0.0													
ROW = 16	0.0	0.33	0.33	0.0	0.0	0.33	0.24	0.20	0.30	0.43	0.41	0.40	0.25	0.38	0.33	0.0												
ROW = 17	0.0	0.25	0.38	0.0	0.0	0.35	0.38	0.41	0.35	0.44	0.35	0.46	0.50	0.31	0.32	0.21	0.0											
ROW = 18	0.22	0.46	0.46	0.22	0.22	0.31	0.39	0.38	0.19	0.41	0.49	0.40	0.24	0.24	0.30	0.30	0.39	0.0										
ROW = 19	0.30	0.32	0.41	0.30	0.30	0.46	0.45	0.48	0.47	0.24	0.38	0.45	0.42	0.39	0.25	0.53	0.41	0.39	0.0									
ROW = 20	0.0	0.40	0.40	0.0	0.0	0.45	0.48	0.48	0.41	0.45	0.41	0.47	0.43	0.35	0.28	0.34	0.21	0.47	0.38	0.0								
ROW = 21	0.22	0.46	0.46	0.22	0.22	0.39	0.48	0.47	0.24	0.41	0.49	0.40	0.49	0.24	0.30	0.38	0.39	0.0	0.31	0.40	0.0							
ROW = 22	0.30	0.40	0.40	0.30	0.30	0.48	0.47	0.46	0.48	0.25	0.41	0.46	0.43	0.40	0.30	0.49	0.44	0.49	0.19	0.34	0.41	0.0						
ROW = 23	0.33	0.41	0.41	0.33	0.33	0.34	0.42	0.40	0.40	0.46	0.50	0.50	0.43	0.34	0.33	0.35	0.33	0.40	0.25	0.14	0.46	0.23	0.0					
ROW = 24	0.30	0.32	0.32	0.30	0.30	0.38	0.38	0.35	0.40	0.14	0.40	0.45	0.48	0.40	0.33	0.38	0.41	0.31	0.21	0.43	0.31	0.24	0.41	0.0				
ROW = 25	0.22	0.46	0.46	0.22	0.22	0.35	0.52	0.48	0.50	0.39	0.25	0.40	0.44	0.49	0.32	0.35	0.41	0.53	0.33	0.30	0.47	0.31	0.21	0.35	0.0			
ROW = 26	0.22	0.25	0.22	0.22	0.22	0.38	0.17	0.19	0.39	0.40	0.49	0.35	0.10	0.50	0.19	0.33	0.40	0.39	0.46	0.48	0.49	0.47	0.44	0.39	0.54	0.0		
ROW = 27	0.22	0.25	0.25	0.22	0.22	0.50	0.35	0.34	0.55	0.39	0.50	0.40	0.44	0.49	0.21	0.45	0.41	0.53	0.48	0.48	0.47	0.46	0.40	0.42	0.54	0.34	0.0	
ROW = 28	0.0	0.38	0.38	0.0	0.0	0.42	0.38	0.33	0.35	0.44	0.42	0.46	0.50	0.40	0.32	0.08	0.24	0.39	0.47	0.21	0.39	0.44	0.46	0.41	0.17	0.40	0.41	0.0

Fig. 28.28. Q-mode analysis of Zone B1, Step 1: matrix of chi-square partition distances between pairs of cells (1 × 1-m squares) based on relative co-occurrence of the same 17 variables already used in R-mode analysis. The numbers 1–28 refer to 1 × 1-m squares as shown in Table 28.1.

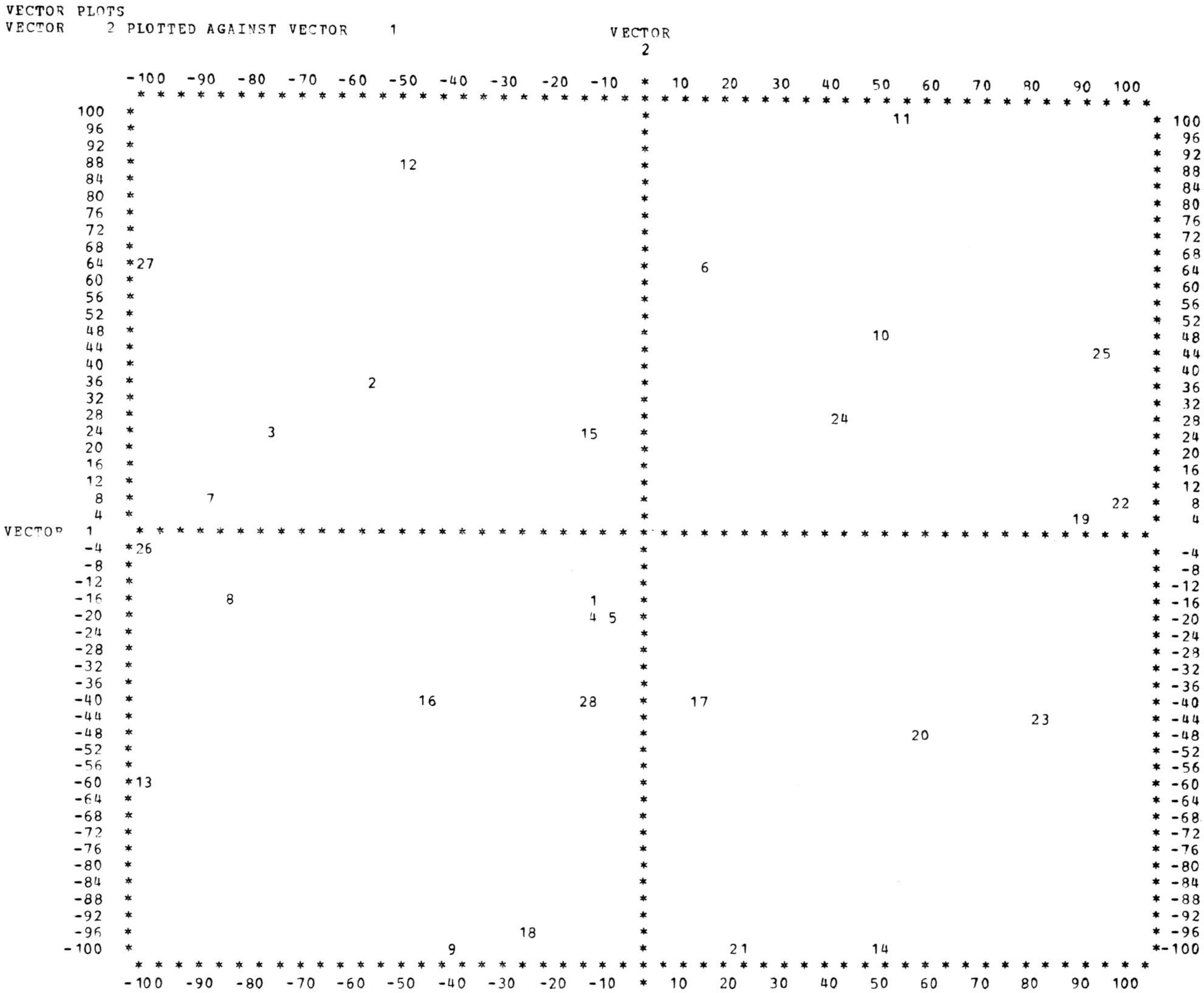

Fig. 28.29. Q-mode analysis of Zone B1, Step 2: plot of Vector 1 versus Vector 2, where the degree of proximity of one numbered cell (1 × 1-m square) to another reflects the degree of association between all possible pairs of cells as determined by Step 1. (Numbered cells 1–28 are as in Table 28.1.) The clusters of associated cells shown here constitute the basis for the activity areas and pathways reconstructed in Fig. 28.31.

CHI SQUARE DISTANCE SCALING FOR ZONE B1 WITH 17 VARIABLES

GUTTMAN-LINGOES' SMALLEST SPACE COORDINATES FOR M = 2 (SEMI-STRONG MONOTONICITY).

VARIABLE	CENTRALITY INDEX	DIMENSION 1	DIMENSION 2
1	9.219	-5.103	-16.573
2	63.171	-49.917	33.675
3	73.554	-69.325	22.428
4	10.201	-5.515	-17.471
5	9.704	-4.548	-17.233
6	72.606	13.467	63.080
7	79.703	-81.076	4.122
8	77.818	-79.869	-13.620
9	96.064	-33.881	-98.514
10	74.552	49.613	45.737
11	119.009	52.567	97.823
12	104.121	-43.172	87.950
13	109.804	-100.000	-57.857
14	106.731	51.621	-99.928
15	30.726	-7.320	22.507
16	48.173	-38.300	-39.775
17	37.297	15.388	-40.652
18	88.443	-19.334	-94.581
19	91.371	88.546	2.262
20	72.197	57.631	-48.104
21	95.407	22.261	-100.000
22	102.989	100.000	4.354
23	91.158	81.318	-44.171
24	56.348	43.990	24.373
25	108.054	93.520	42.200
26	91.825	-93.952	-2.693
27	115.884	-94.128	62.848
28	33.302	-8.043	-40.598

GUTTMAN-LINGOES' COEFFICIENT OF ALIENATION = 0.30146 IN 32 ITERATIONS.
KRUSKAL'S STRESS = 0.26373

Fig. 28.30. Q-mode analysis of Zone B1: scores of 28 numbered cells (1 × 1-m squares) for the two vectors shown in Fig. 28.29. Scores reflect the distance from the ordinate (Vector 2) or abscissa (Vector 1) of the latter graph. Kruskal's Stress for this stage of the analysis is 0.26, which means that 74% of the relative order of the associations between pairs of numbered cells is preserved by the configuration based on these two vectors.

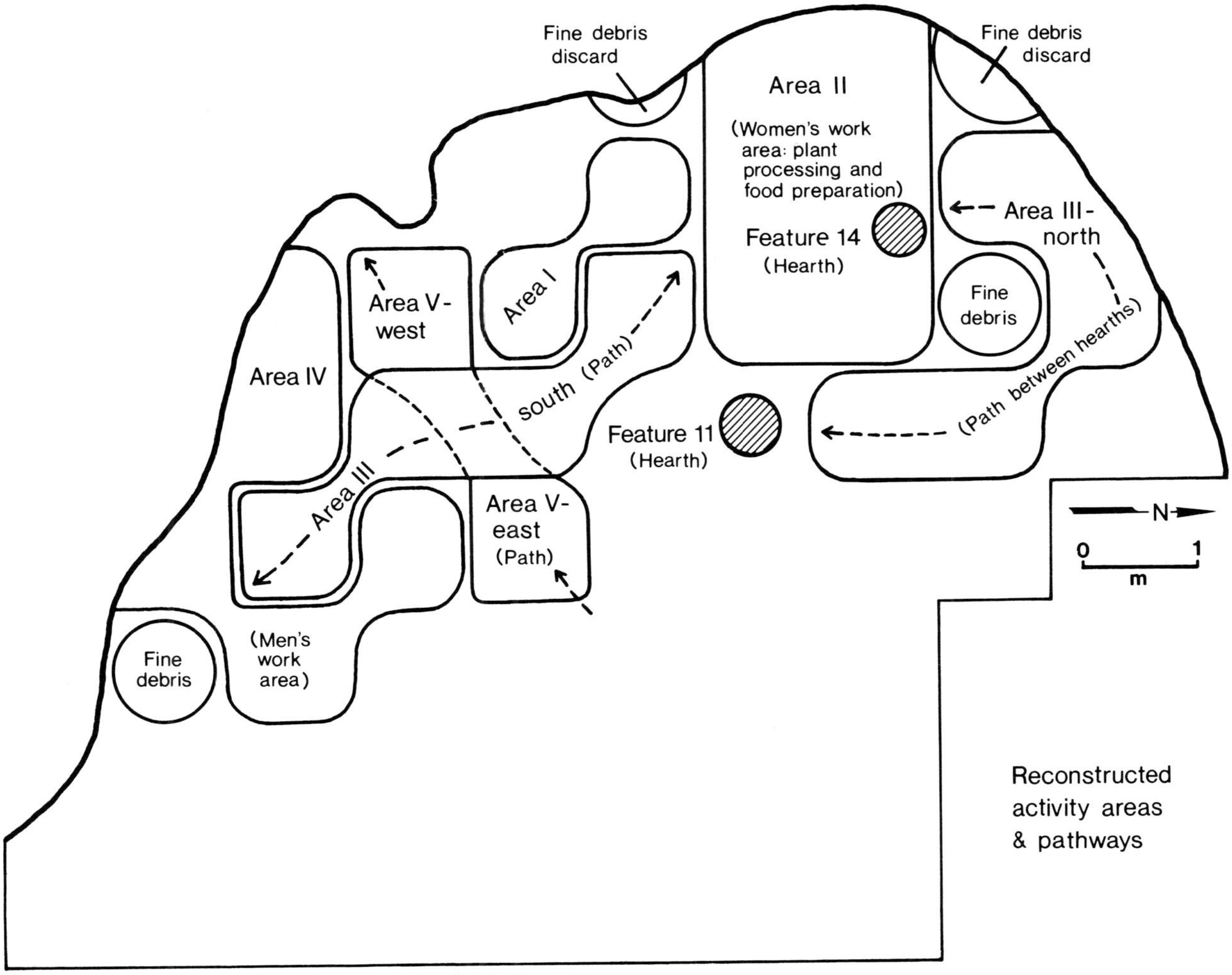

Fig. 28.31. Interpretive drawing of Zone B1 at Guilá Naquitz, showing pathways and activity areas (groups of 1 × 1-m squares) defined by the preceding Q-mode analysis. Area II is interpreted as a work area where one or two persons (probably women) processed seasonally restricted plants, cooked, and discarded debris against the cave wall. Area I and the two small unnumbered areas to the north of II are all considered to be discard areas related to Area II activities. Area III-north is interpreted as a pathway between two hearths (Features 11 and 14), perhaps used primarily by women. Area IV is interpreted as an area where one or two persons (possibly men) processed animals, chipped flint, and chewed agave. A small, unnumbered discard area to the southeast is considered related to Area IV. Area III-south is interpreted as a pathway connecting Areas II and IV, while Area V is considered part of the main pathway entering the cave from the northeast (and crossing Area III).

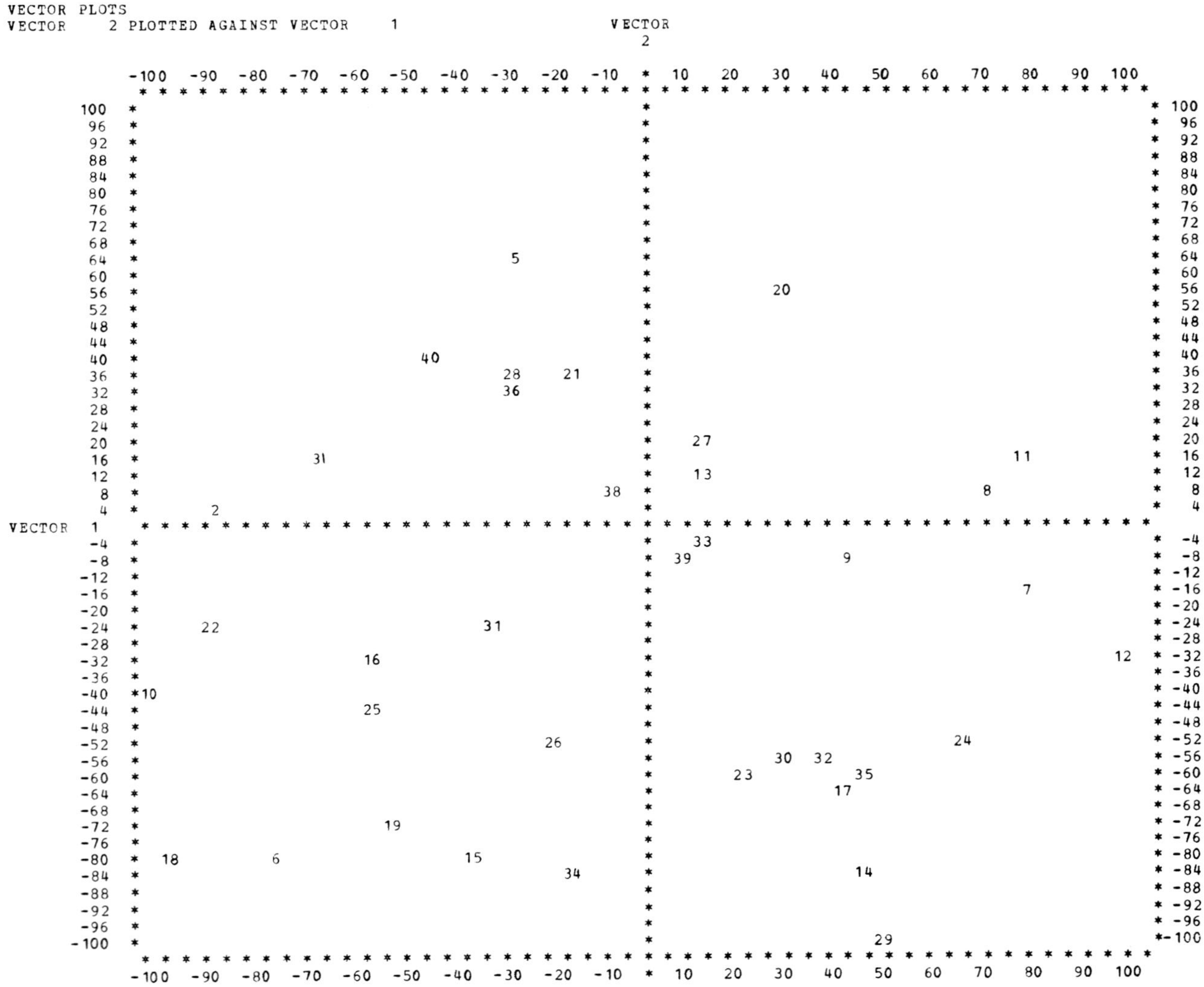

Fig. 28.32. A comparison of the Q-mode analyses for Zones E and D. The numbers refer to cells (1 × 1-m squares) that fell in activity areas; cells that fell outside activity areas have been omitted. Numbers 1–19 refer to cells in Zone D; numbers 20–40 refer to cells in Zone E. Note that presumed men's and women's activity areas in Zone D are more distinct (i.e., more widely separated) than in Zone E; in addition, the men's activity areas in E and D are more alike than are the women's activity areas from the same two levels. Kruskal's Stress for this stage of the analysis is 0.15, which means that 85% of the relative order of the associations between pairs of numbered cells is preserved by the configuration based on four vectors.

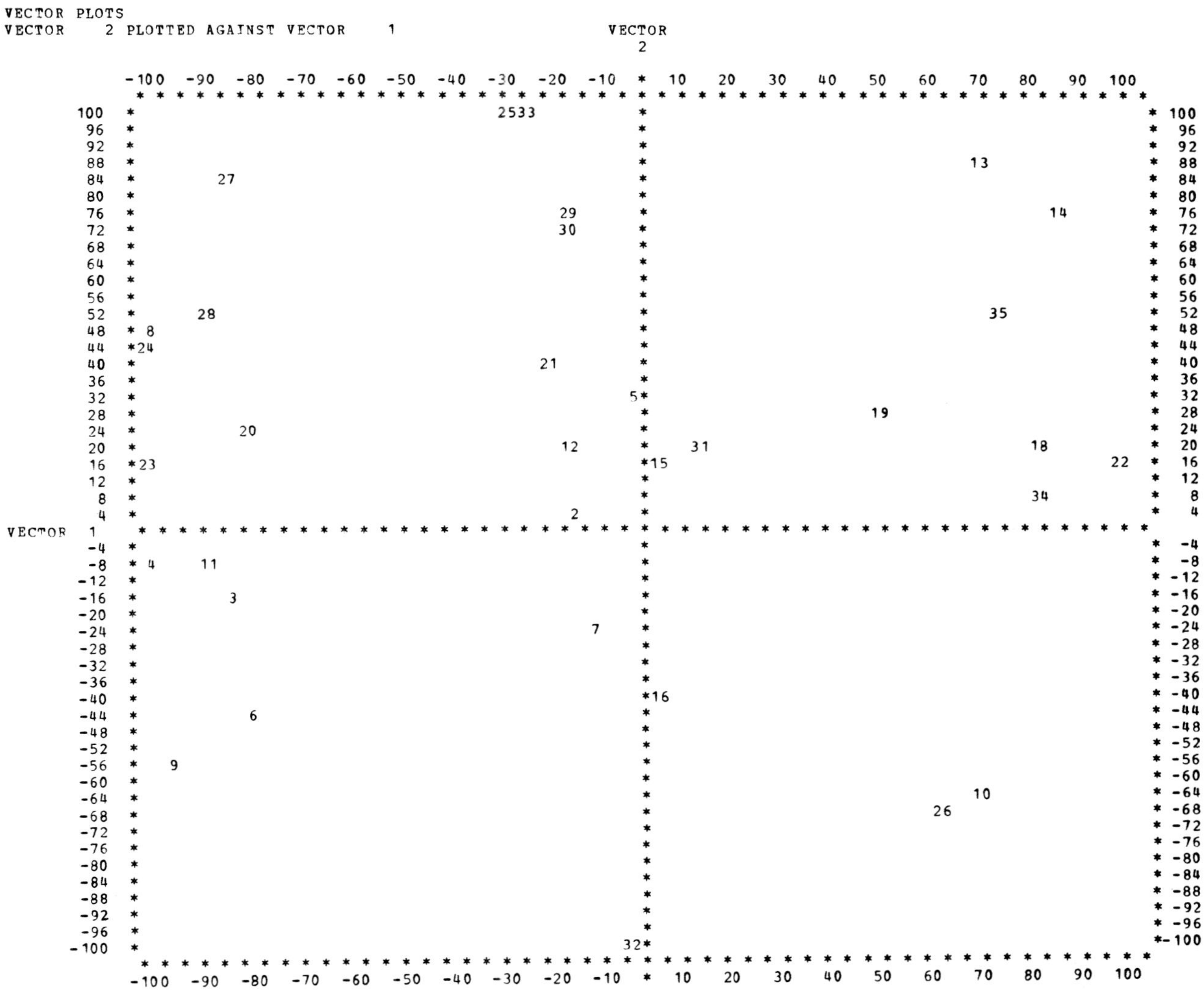

Fig. 28.33. A comparison of the Q-mode analyses for Zones D and C. The numbers refer to cells (1 × 1-m squares) that fell in activity areas; cells that fell outside activity areas have been omitted. Numbers 1–16 refer to cells in Zone C; numbers 17–35 refer to cells in Zone D. Note that the presumed women's activity areas in Zones D and C are more similar than are the presumed men's activity areas in the same two levels. Kruskal's Stress for this stage of the analysis is 0.11, which means that 89% of the relative order of the associations between pairs of numbered cells is preserved by the configuration based on four vectors.

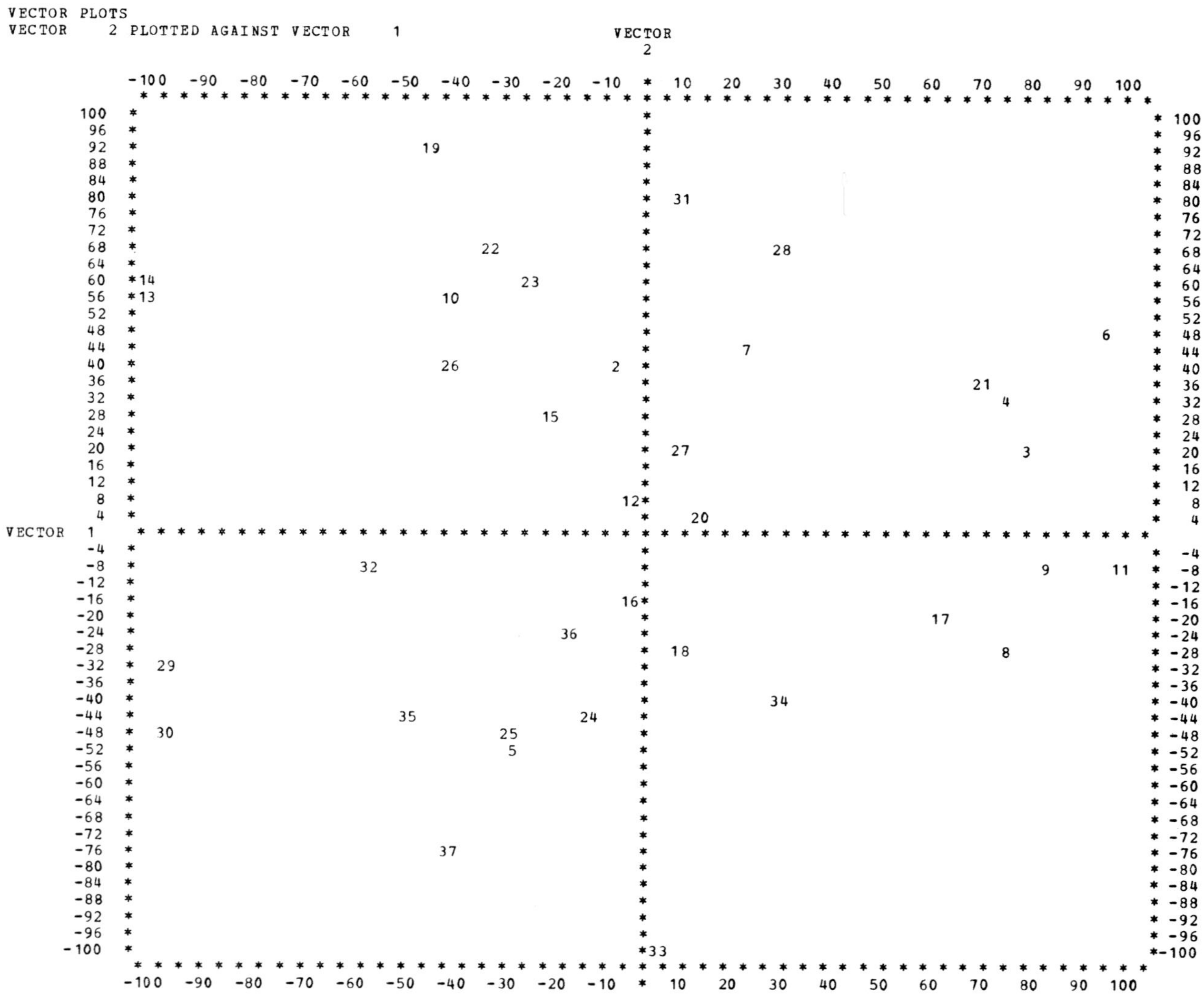

Fig. 28.34. A comparison of the Q-mode analyses for Zones E and C. The numbers refer to cells (1 × 1-m squares) that fell in activity areas; cells that fell outside activity areas have been omitted. Numbers 1–16 refer to cells in Zone C; numbers 17–37 refer to cells in Zone E. Note that presumed men's and women's activity areas in Zone C are more distinct (i.e., more widely separated) than in Zone E. Kruskal's Stress for this stage of the analysis is 0.16, which means that 84% of the relative order of the associations between pairs of numbered cells is preserved by the configuration based on four vectors.

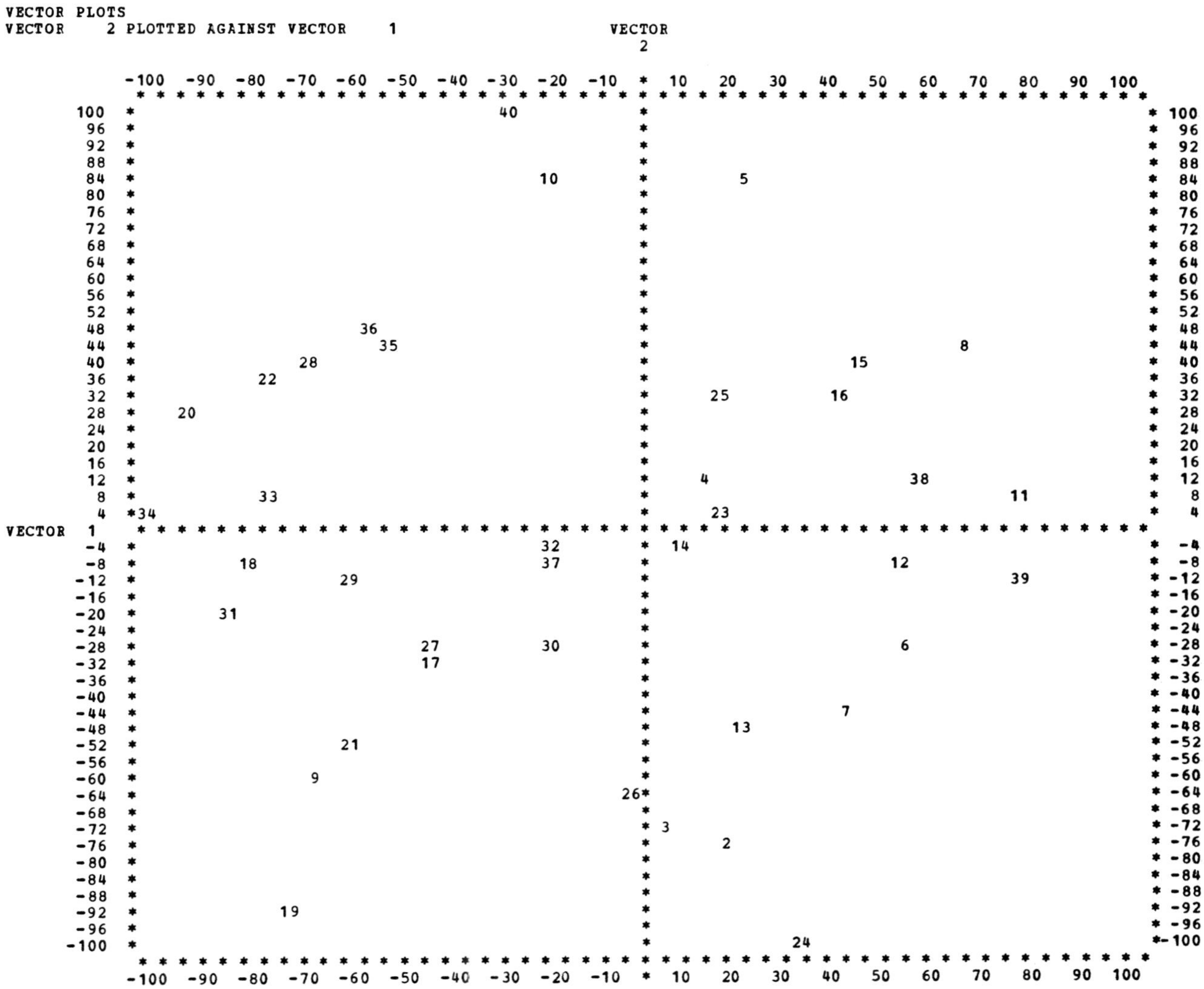

Fig. 28.35. A comparison of the Q-mode analyses for Zones C and B1. The numbers refer to cells (1 × 1-m squares) that fell in activity areas; cells that fell outside activity areas have been omitted. Numbers 1–26 refer to cells in Zone B1; numbers 27–40 refer to cells in Zone C. Note that in the case of *both* presumed men's and women's activity areas, the Zone B1 areas differ from their Zone C counterparts. Kruskal's Stress for this stage of the analysis is 0.23, which means that 77% of the relative order of the associations between pairs of numbered cells is preserved by the configuration based on three vectors.

29

Episodal Analysis of Guilá Naquitz: A Synthesis of Spencer's, Whallon's, and Reynolds' Results

Kent V. Flannery

Chapters 26, 27, and 28 have presented three different approaches to the spatial analysis of Guilá Naquitz. Although there were major points of agreement, each approach brought out features not found by the others. In this chapter we attempt to synthesize the results of all three studies and to add a temporal dimension to their spatial one. The latter is possible because we know the seasonal order in which various plants and animals become available (Chapter 18) and the logical order in which various steps in their processing must be carried out (Chapter 17).

This chapter also gives us a chance to add to our study those artifacts that were not sufficiently numerous to include in the analyses of Chapters 26–28. By superimposing the activity areas and pathways proposed by Reynolds in Chapter 28 (Figs. 28.10, 28.17, 28.24, and 28.31) over the maps of artifact distributions given in Chapter 5 (Figs. 5.32–5.38), we can see which artifacts were left behind in which areas. We can also compare the locations of the retouched flint tools defined by Hole (Chapter 6) with the contour maps of flint debitage provided in Chapter 26. Reading of this chapter should therefore be accompanied by an examination of all the figures mentioned in this paragraph, and because those figures contain so much of the data we can keep our syntheses brief.

Since neither Whallon nor Reynolds dealt with Zones B3 or B2, no syntheses of those living floors are necessary.

EPISODAL ANALYSIS

A convenient format for our synthesis is *episodal analysis,* which was applied by Ascher and Clune (1960) during their study of Waterfall Cave, Chihuahua. Ascher and Clune describe an *episode* as a series of events that defines the temporal and contextual limits of the use of the site. Each of these events is further subdivided into *incidents*; each incident further defines the content of the event (Ascher and Clune 1960:272–273). The authors go on to describe the formation of the archaeological deposits in Waterfall Cave as a sequence of events, each made up of two or more incidents. Like the paper by Heizer (1960) mentioned in Chapter 17, Ascher and Clune's study reminds us that it did not really take archaeologists until the 1970s to develop a concern for "site formation processes." We need only read the older archaeological literature to learn that it has been a topic of discussion for many decades.

In our version of Ascher and Clune's format, the formation of each living floor at Guilá Naquitz is considered an episode. Except for Zone E, whose matrix was predominantly weathered ignimbrite that had accumulated prior to the first Naquitz phase occupation, all living floors in the cave were composed mainly of material brought in by humans (see Chapter 5). Thus our incidents include the bringing in and processing of plant foods, the butchering and eating of animals, the chipping of flint, the digging of pits, the use of hearths, the laying-in of oak-leaf beds, and so on. While we cannot specify the exact order in which each incident took place, we can group them into events according to the seasonal and logical order in which they are believed to have occurred.

It goes without saying that our episodal analysis is a reconstruction, based on a number of the same assumptions given in Chapter 17. We cannot prove that everything happened the way we suggest, but our suggestions are at least informed by the preceding 28 chapters of this volume.

EPISODE 1: THE DEPOSITION OF ZONE E

Late Summer Events

A microband that has been foraging in the mesquite forest (C. E. Smith's [1978] reconstructed "ancestor" of today's Mesquite Grassland B) moves into Guilá Naquitz in anticipation of the fall harvest season in the upper piedmont. During late summer, they continue to harvest the plants and animals of the mesquite forest. *Incidents:* Flint knapping is carried out in two different areas, with debitage discarded against the south wall (Area V) and the northwest wall (Area III). Pods are collected from mesquite groves and processed in Area I, presumably by women. Hackberries are collected from the same area and eaten in large numbers in Area I, perhaps mainly by women. The men hunt animals—cottontails, a young deer, a mud turtle and possibly a barn owl that was already living in the cave. Some rabbits are processed and eaten in Area I. Areas II and IV are becoming pathways between men's and women's work areas.

Early Fall Events

The microband begins collecting prickly pear fruits (tunas), whose season of maturity spans the transition from summer to fall. *Incidents:* Both men and women, sitting in two different areas of the cave, despine and eat tunas. However, at least two piles of leftover fruits, one in Area I and the other in Area V, are allowed to turn mushy on the cave floor and leave their seeds behind.

Late Fall Events

The microband begins harvesting the products of the lower oak–pine forest and thorn–scrub–cactus forest (C. E. Smith's [1978] reconstructed "ancestor" of today's Thorn Forest A), while continuing to hunt and trap animals. *Incidents:* Percussion flaking of flint continues, as does the discarding of debitage and hammerstones in Areas III and V. Area III is developing as a fine-debris discard area for the activities of the women working in Area I, and the great amounts of debitage included in Area III suggest that they did a lot of flint knapping in the course of their work.

Foragers return from the oak groves with basketloads of acorns; at least one of these is dumped in Area I and another near the mouth of the cave, just outside Area V. This may indicate that several different persons participated in acorn processing. Persons in both Areas I and V hull and eat piñon nuts. *Susí* nuts, another upper piedmont product, are hulled and eaten in Area V (presumably by men), while other hulls are tossed into the Area II pathway. Whallon's analysis suggests that nanche fruits and maguey quids were associated with this debris and may therefore have been eaten and discarded during the same period.

Small black runner beans are processed in Area I, with their pods discarded in a spot not typical of other plant debris. Some 812 pods of *Dalea,* a possibly medicinal plant not known to have served as food, are collected and left on the Area II pathway near the entrance to the cave, not far from a slab metate fragment.

The men hunt deer, killing at least one doe; some of its bones show up in Areas I and V, perhaps where two different persons ate parts of it. Notched flakes occur with both deer bone scatters. Cottontails, a pigeon, a dove, and other animals may have been taken during this period.

Final Events

The microband abandons the cave, leaving behind a well-worn entrance pathway (Areas II and IV), a large area of food preparation and women's work (Area I) from which fine-fraction debris was tossed toward the northwest cave wall (Area III), and an area of debris from men's work near the south wall (Area V). However, sexual division of labor is still not as statistically clear in Zone E as it will be on later living floors. An unfinished projectile point is left near the cave mouth; fragments of cordage, perhaps from tied bundles of food or fuel brought into the cave, are left on the entrance pathway and in the women's work area. Significant artifacts left behind in the activity areas proposed by Reynolds are:

Area I (women's work)
1 crude blade
1 notched flake
1 biface, Variety C
cordage

Area V (men's work)
1 notched flake
1 utilized flake
1 hammerstone

Area III (fine discard from I)
1 steep denticulate scraper
1 notched flake
1 hammerstone
1 antler artifact

Area II (pathway)
1 slab metate fragment
1 knot in cordage

Area IV (pathway)
nothing

Postoccupation Events

Peromyscus and other rodents enter the cave, gnawing bones and plant remains left on the surface. This marks the end of Episode 1.

EPISODE 2: THE DEPOSITION OF ZONE D

Late Summer Events

As in the case of Zone E, a microband that has been harvesting in the mesquite forest moves into Guilá Naquitz, continuing their summer foraging activities while they wait for the fall harvest season. *Incidents:* The group brings quantities of oak leaves into the cave for use as bedding, excavates the hearth basin that is to become Feature 22, and excavates at least one storage pit (Feature 10). The establishment of these features sets in motion the processes by which a women's food processing and cooking area (Area IV) will develop near the hearth, while a path (Area II) will develop between the hearth and the storage pit.

Women collect mesquite pods, storing them in Feature 10 prior to processing. Hackberries are eaten and their seeds discarded by at least two persons, one sitting near Feature 10 and the other near the southwest extreme of Area IV. This latter hackberry concentration is accompanied by a "higher frequency of seasonally restricted plants" (Fig. 28.17) lying in an area that may indicate that Feature 3, a second storage pit, was already in use by that time. (If this is the case, it would be our second hint that women ate most of the hackberries brought to the cave. Alternatively, however, some specimens could be seeds from hackberries eaten by men working in Area III, along the south wall of the cave.)

Men kill at least one juvenile deer, some cottontails, and perhaps a mud turtle. Bones from these animals are left behind in Area III (men's work area, with some animal processing and/or eating) and in Area II (pathway between hearth and Feature 10), the latter debris perhaps resulting from women's meals. In addition, deer bones are discarded in an area of "fine debris from food preparation" in northernmost Area II (Fig. 28.17). Three of the four notched flakes found on the living floor occur near deer remains.

Early Fall Events

Women collect prickly pear fruits (tunas) and tender stem sections (nopales). *Incidents:* Removal of the spines and roasting of the nopales on wooden skewers takes place over the Feature 22 hearth; discarding of debris takes place near the eastern limits of Area IV (Fig. 28.17). Tunas are eaten by both men and women, with leftover fruits turning mushy and leaving their seeds in both Areas III and IV. *Guaje* seeds and nanche fruits are eaten raw by people sitting around the Feature 22 hearth, perhaps suggesting that the women ate a lot of nanches as well as hackberries.

Late Fall Events

The group begins its autumn foraging in the thorn–scrub–cactus forest and lower oak–pine forest. *Incidents:* Storage capacity of the cave is increased by the digging of the Feature 2 pit, and perhaps Feature 3 as well (assuming it had not been dug earlier; see above). Most of the unretouched flakes from flint knapping are discarded in the neighborhood of these two storage pits, although hammerstones, cores, and core fragments are scattered more widely around the cave. Men kill at least one deer, a buck with fully hardened antlers, some of whose bones are discarded in Areas III (a men's work area) and II (a depository for "fine debris from food preparation"). A collared peccary, whose jaw was left near the hearth, may also have been killed in this season. Cottontails are cooked and eaten, with bones discarded in Areas III and II.

Domestic *Cucurbita pepo*, perhaps planted somewhere near the cave(?) during the previous spring, is harvested and processed in Area IV. While these tasks are presumably carried out by women, men working in Area III evidently eat some of the roasted squash seeds, since one is left behind there (Fig. 5.33). Runner beans are hulled near the hearth in Area IV, their distribution partly overlapping that of squash.

At least one maguey heart is dug up, cooked, and eaten, with quids left in two major piles about 2 m apart—one in Area III (the quids presumably chewed by men) and another near the hearth in Area IV (the quids presumably chewed by women). Two flakes with sheen, discarded in Area III, may suggest the men did most of the heavy cutting that removed the tough leaves from the maguey heart. The same two people (or groups of people) who chewed these quids seem also to have eaten *susí* nuts, discarding their hulls in the same two areas.

Masses of acorns are harvested and stored in Features 2, 3, and 10. One of the last major basketloads is dumped out along the border between Areas III and IV, with many unused acorns left both here and in the bottom of Feature 3. The women evidently use one-hand manos to process these into flour; two such manos are left on the eastern limits of the last basketload dump mentioned above, while two more are left in or near Feature 2 (Fig. 5.33). Movement between Feature 10 and Features 2 and 3 creates the Area I pathway along the west wall of the cave and extends and reinforces the Area II pathway to the east. Piñon nuts, probably collected with the acorns, are shelled and eaten in Areas III (by men) and IV (by women).

Final Events

At some point late in the occupation, many of the now-dry oak leaves brought into the cave as bedding catch fire and smoulder. The microband abandons the cave, leaving behind a hearth and three storage pits; two well-worn paths (Areas I and II) that connected those features; an area of debris from women's plant processing and food preparation (Area IV); and an area of debris from men's animal processing, quid chewing, and eating. Fine-fraction debris has gravitated toward the west and north walls of the cave. For the second living floor in a row, a fragment of cordage from a tied bundle of food or fuel is left behind on a pathway (Area II). Men's and women's work areas are more well-defined in a statistical sense than they were in Zone E. Significant artifacts left behind in the activity areas proposed by Reynolds are

Area IV (women's work)
- 1 steep denticulate scraper
- 2 burins
- 2 notched flakes
- 1 hammerstone
- 2 core faces
- 1 burned wooden slat

Area III (men's work)
- 1 crude blade
- 1 notched flake
- 2 flakes with sheen
- 1 biface, Variety C
- 1 core face
- 1 core edge

Area I (path with small debris)
- 1 drill
- 1 pointed piece
- 1 notched flake
- 1 utilized flake
- 1 biface, Variety C
- 1 core face

Area I (path between features)
- 1 steep denticulate scraper
- 1 burin
- 1 notched flake
- 1 utilized flake
- 1 antler artifact
- cordage

Postoccupation Events

Peromyscus and other rodents enter the cave, gnawing bones and plant remains left on the surface. This marks the end of Episode 2.

EPISODE 3: THE DEPOSITION OF ZONE C

Late Summer Events

As in previous zones, a microband that has been harvesting in the mesquite forest moves into Guilá Naquitz, continuing their summer foraging activities while they wait for the fall harvest season. *Incidents:* The group brings quantities of oak leaves and grass into the cave to use as bedding. Using a fire drill, they make a surface campfire on the spot that will later become our Square D9. (At some time during the course of the occupation, some of the leaves and grass will catch fire and smoulder; this partly burned vegetal layer will become the gritty compost of lower Zone C. Still later, accumulated white ash from the surface campfire will be kicked around over the whole living floor, becoming upper Zone C.)

This surface campfire serves as the pivot around which three curving traffic pathways begin to develop (see Fig. 28.24). Area V, the entrance pathway, enters from the northeast and ends in the center of the cave. Area I curves around the north side of the campfire, connecting Area V with the west wall of the cave; Area II runs parallel to I, a meter farther to the north. Both Areas I and II display a higher frequency of large plant remains on their eastern ends (nearest the cave mouth) and a higher frequency of fine-fraction debris on their western end (nearest the cave wall). This is in keeping with other living floors, where large debris seems to have been thrown toward the talus while fine refuse was swept toward the wall.

In addition, the surface campfire in D9 seems to become a central locality around which evenly spaced areas of flint knapping develop. Three separate concentrations of debitage (perhaps left by three different knappers) appear at 2-m intervals—one in Area II, one in Area V, and the third in an open area between Areas II and III. Since Area III (near the south wall of the cave) is interpreted by Reynolds as an area for animal processing, this third debitage concentration may have resulted from men's work; the concentration north of the campfire may well have resulted from women's work.

Women collect mesquite pods, processing them and discarding the seeds in one major pile east of the campfire (where Fig. 28.24 shows a "higher frequency of large plant remains" for Area I). Whoever did this processing seems also to have eaten at least 100 g of hackberries, discarding the seeds in the same area as the mesquite. This is yet another hint that women ate most of the hackberries brought back to the cave. Meanwhile, the men killed a barn owl (perhaps one already living in the cave when they arrived), at least one juvenile deer, several cottontails, and a mud turtle. Whallon's analysis shows that deer, rabbit, and turtle are associated strongly with each other, and perhaps with some of the flint debitage as well; in addition, some five to seven notched flakes are left near deer remains. Deer bones are discarded both in Area III ("animal processing") and in Areas I and V near the campfire, where meat may have been cooked and eaten. Cottontail bones occur in those same areas, as well as in the "fine-fraction refuse" where Areas I and II reach the cave wall.

This is the first living floor in which there is a superficial disagreement between Spencer's and Reynolds' interpretations of the remains. Reynolds considers Area V a pathway, while Spencer's contour maps show considerable plant processing refuse left behind in those same squares. What we may be dealing with is an area where plant debris is being tossed from the campfire toward the cave mouth, but a true midden cannot develop because foot traffic into and out of the cave keeps extending the refuse in linear fashion along the long axis of Area V.

Early Fall Events

The season of *guajes,* nanches, and prickly pear fruits begins to divert foragers' attention away from the mesquite forest. *Incidents:* Women eat *guaje* seeds, discarding the pods in Area V not far from where they had earlier left their mesquite seeds. They eat nanche fruits and discard the seeds in Area V, not far from where they had earlier left their hackberry seeds. However, a second pile of nanche seeds, tossed toward the cave wall south of Reynolds' Area IV, suggests a second person or persons (perhaps a man) eating nanches. A large pile of uneaten prickly pear fruit also seems to have turned mushy in Area IV, releasing its seeds into the floor debris.

Tender young nopales are skewered on sticks and cooked over the campfire, with the resulting debris tossed some 2 m away into Area V; one nopal on a skewer and three burned sticks are discarded there.

Late Fall Events

The group begins its autumn foraging in the thorn–scrub–cactus forest and lower oak–pine forest. *Incidents:* Although no storage pits are dug to receive them, acorns are harvested in large numbers. At least two piles are left incompletely processed into flour. One of these is dumped just east of the campfire, in the "higher frequency of large plant remains" sections of Areas I and II. This may be an area where women grind up acorns; two one-hand manos are discarded about a meter to the east, in Area V (Fig. 28.24). A second pile of acorns is dumped in Area IV near the south wall of the cave, with a mortar fragment in apparent association (Fig. 28.24). For this reason, Area IV is considered a possible plant processing area by Reynolds. It is not clear whether women occasionally worked in this south part of the cave or whether Zone C presents us with an instance where the men (who most often seem to have worked near the south wall) also assisted in acorn processing. The hulls from piñon nuts, which would almost certainly have been harvested in the same season and general area as acorns, are concentrated in the same two major piles.

At least 1 maguey heart is dug up and cooked, with more than 100 quids expectorated in Area V; this may mean a lot of quid chewing was done by persons sitting near the campfire. A flint flake with sheen, perhaps used in trimming the *Agave* heart, eventually winds up on the Area II pathway. *Susí* nuts, another upper piedmont product, are eaten raw, with their hulls tossed just to the east of the campfire in a part of Area I where there are already a lot of discarded acorns and hackberry seeds. When runner beans are harvested, their pods are also discarded in this same scatter of debris.

Domestic squash, perhaps planted somewhere near the cave during the previous spring, is harvested and its seeds are roasted over the campfire, with unwanted rinds discarded in Area V. This may also be the season when *Lagenaria* fruits were harvested, to be converted into gourd containers.

Men kill at least one deer (a buck with fully hardened antlers) and several cottontails, whose remains are scattered through Areas III (men's work?); V (women's food preparation?); I; and II. This may also be the season in which they hunted down a peccary and a raccoon, most of whose remains were evidently tossed out of the cave.

Final Events

The group abandons the cave, leaving behind several piles of unprocessed acorns and a great deal of plant debris near the center of the living floor. The white ash from their surface campfire has now been spread over the whole occupied area, and the traffic patterns within the cave have left some of their garbage strung out along three crescentic pathways that partially encircle the remains of the campfire. While the division of the cave into a southern work area (Areas III and IV) and a northern work area (Areas I, II, and V) is still relatively clear, constant circular traffic around the campfire has blurred the details of what was almost certainly the main focus of women's plant processing and food preparation. As on earlier living floors, bits of cordage from bundles of food or fuel tend to be left behind on pathways. Underneath it all, the oak leaves and grass originally brought into the cave as bedding are turning to compost. Significant artifacts left behind in the activity areas proposed by Reynolds are

Area III (animal processing)
nothing

Area V (entrance pathway)
1 sidescraper/knife
2 notched flakes
1 utilized flake
1 flake core
2 one-hand manos
1 pebble pigment grinder
cordage
1 nopal on a stick
3 burned sticks

Area II (path)
2 steep denticulate scrapers
2 notched flakes
1 flake with sheen
1 hammerstone
1 core edge
1 fire drill
cordage

Area IV (plant processing?)
1 mortar fragment

Area I (path)
3 notched flakes
2 utilized flakes
1 flake core
1 core edge
1 antler artifact

Postoccupational Events

Peromyscus and other rodents enter the cave, gnawing bones and plant remains left on the surface. This marks the end of Episode 3.

EPISODES 4 AND 5: THE DEPOSITION OF ZONES B3 AND B2

As indicated earlier, no extensive syntheses of Zones B3 and B2 are necessary, since Whallon and Reynolds did not consider those units in their analyses. Additional insight into Zone B3 can be obtained by superimposing our map of significant artifacts (Fig. 5.36) over Spencer's plant and animal distribution maps (Figs. 26.26–26.31). Similarly, for Zone B2, the artifact map (Fig. 5.37) can be superimposed over the plant and animal distributions (Figs. 26.32–26.38).

EPISODE 6: THE DEPOSITION OF ZONE B1

Late Summer Events

As in previous zones, a microband that has been harvesting in the mesquite forest moves into Guilá Naquitz, continuing their summer foraging activities while they wait for the fall harvest season. For reasons given in Chapter 26, we believe this microband consists of at least two men and two to three women. *Incidents*: Two hearths, Features 11 and 14, are excavated into the previous deposits. Feature 11, a large hearth lined with stones, is created near the center of the living floor and remains independent of any of Reynolds' activity areas (Fig. 28.31). Feature 14, a less elaborate hearth, becomes an integral part of Reynolds' Area II, a women's work area for plant processing and food preparation in the northwest quadrant of the cave.

The establishment of Area II and Features 11 and 14 will have a profound influence on traffic flow and refuse discard in the course of the Zone B1 occupation. Area III-north will gradually develop as a crescentic pathway linking the two hearths, a path probably used mainly by women; it will be crescentic because it detours around an area of fine debris just outside Area II. Three more areas of fine debris will develop near Area II, two of them against the back wall of the cave and one dribbling off toward the south. The latter, called Area I by Reynolds (Fig. 28.31), will probably become elongated as the result of foot traffic between men's and women's work areas.

Widespread flint knapping takes place in Zone B1, with four peak areas of debitage developing. The three smallest peaks are in the northwest quadrant and suggest that two or more individuals (probably women) sitting near the hearths engage in a modest amount of tool preparation, discarding their debitage toward the cave wall. The fourth and largest peak accumulates in Reynolds' Area III-south, where that traffic pathway penetrates into and is surrounded by Area IV, a men's work area. Considerable numbers of retouched tools will eventually be discarded by the men in this part of the cave, including a hammerstone, cores and core fragments, a biface and projectile point, scrapers, and retouched flakes. Of course, many of those tools will pass through multiple contexts and uses before they become part of the cave refuse.

Harvesting and processing mesquite pods are two of the major activities of this microband, and discarded mesquite seeds are the most common single plant left on the living floor. This suggests the microband may be spending more time in the mesquite vegetation belt than in previous years, perhaps because they are gradually shifting incipient agriculture to the riverine alluvium. One huge pile accumulates in Area III-north, presumably as the result of women's work. The women also collect hackberries in the same mesquite forest belt, and evidently eat them as they work. Two women—one discarding her seeds in Area II near the Feature 14 hearth, the other discarding hers in nearby Area I—eat more than 100 hackberries between them. Once again, our evidence suggests that women eat most of the hackberries brought to the cave (although the men may have eaten their share outdoors, while hunting or trapping in the mesquite forest).

As for the men, they kill at least one deer (a pregnant doe whose foetus was discarded in Zone B1) and several cottontails. The bones of these animals are discarded all around the women's food preparation area, especially in Area III-north or in the fine-debris piles associated with Area II; they also show up in Area IV and nearby Area III-south, either as the result of butchering or eating by the men in the cave. A barn owl (perhaps a former resident of the cave) is killed, along with at least two mud turtles. The turtles are roasted in the shell over hot coals (probably in Feature 14) and their carbonized shells discarded in Area I. A few lizards, typically active during the summer rainy season, are also eaten.

Early Fall Events

The season of prickly pear, *guaje,* and nanche begins to divert the foragers' attention away from the mesquite forest. *Incidents: Guaje* seeds are eaten raw by both men and women, the men discarding their pods in Area IV and the women discarding theirs around the Feature 11 hearth. Perhaps as many as two men and two women eat nanches, leaving behind four piles of seeds; two piles on the edge of Area IV are presumably left by men, while the other two (in Area III-north and in Area III-south near Feature 11) are presumably left by women. Both men and women eat prickly pear fruit, each leaving behind piles of uneaten fruits that turn to mush on the living floor. The men's pile leaves its seeds in Area IV, the women's leaves its near the Feature 14 hearth.

Late Fall Events

The group begins its autumn foraging in the thorn–scrub—cactus forest and lower oak–pine forest. *Incidents:* Several trips are made for the purpose of collecting acorns, but not all acorns are ground into flour. At least four basketloads, dumped on the living floor, are left partially unprocessed. In the northwest quadrant of the cave, the area where women do most of their work, three piles are left—one in Area II with a mano nearby, a second near the Feature 11 hearth, and a third on the Area III-north pathway. A fourth pile is left in Area V-east, a pathway in the south half of the cave.

The women eat piñon nuts, discarding the hulls in Area III-north. Both men and women eat *susí* nuts, with the men discarding their hulls in Area V-east, just outside their work area. The women discard their *susí* hulls in two piles, one in Area II near the Feature 11 hearth, the other in Area III-north.

At least one maguey heart is roasted, and everyone in the cave participates in chewing the cooked leaf bases. In Area IV, it appears that one to two men produce two piles of quids; in the northwest quadrant of the cave, a woman near Feature 14 and one to two women near Feature 11 leave three similar piles of quids.

Runner beans and domestic cucurbits are processed together in the women's work area of the cave, with empty pods and discarded rind fragments tossed toward the cave wall in Area III-north, not far from the Feature 14 hearth. The squashes and bottle gourds are definitely domestic, perhaps harvested from a crop planted on the alluvium the previous spring. The beans are phenotypically "wild," but their unexpectedly high numbers, plus the fact that they seem to have been processed along with known domesticates, provide circumstantial evidence that there may have been an effort to cultivate them (see Chapter 26).

Meanwhile, the men hunt deer, killing at least one buck with fully hardened antlers. They also bring back several rabbits and birds, possibly including the hawk and ground dove found in Zone B1. These animals are partially processed by men working in the south part of the cave, eventually cooked over the Feature 14 hearth, and later eaten by both men and women, with bone fragments discarded both inside and outside the cave.

Final Events

The last fires are kindled in Features 11 and 14, perhaps with the wooden fire drill hearth discarded in Area II. By this stage of the occupation, gray ash from both hearths has been kicked around over the whole living floor, providing much of the matrix for Zone B1. As the acorn harvest season draws to a close, the group abandons Guilá Naquitz. They leave behind a mass of debris from women's plant processing and food preparation (Area II), areas of fine-debris discard nearby, two hearths (Features 11 and 14) connected by a curving pathway (Area III-north), masses of debris from men's work (Area IV), areas of fine debris nearby, a path connecting the men's and women's work areas (Area III-south), and a pathway leading from the cave center to the back wall (Area V). Significant artifacts left behind in the activity areas proposed by Reynolds are

Area II (women's work)

- 2 notched flakes
- 2 utilized flakes
- 1 core face
- 1 one-hand mano
- 1 ball of cactus gum
- 1 stick, cut on both ends

Fine debris associated with Area II

- 1 notched flake

Area I (discard associated with Area II)

- 1 crude blade
- 1 antler artifact

Area III-north (path between hearths, probably used mainly by women)

- 1 sidescraper/knife
- 2 crude blades
- 3 notched flakes
- 1 utilized flake
- 2 cores
- 1 core face
- 1 core edge
- 1 mortar fragment

Area IV (men's work)

- 1 burin
- 1 notched flake
- 1 Trinidad(?) point fragment
- 1 core face
- 1 slab paint palette

Fine debris associated with Area IV

- 1 hammerstone
- 1 core face

Area V (path, probably used mainly by men)

- 1 steep denticulate scraper
- 5 notched flakes
- 2 slab metate fragments
- cordage

Area III-south (path between men's and women's work areas)

- 1 steep denticulate scraper
- 1 crude blade
- 3 notched flakes
- 2 utilized flakes
- 1 biface, Variety B
- 2 cores
- 1 core face

Postoccupation Events

Peromyscus and other rodents enter the cave, gnawing bones and plant remains left on the surface. This marks the end of Episode 6.

ARTIFACTS AND ACTIVITY AREAS

One of the frustrating aspects of analyzing a cave like Guilá Naquitz is that we cannot specify the tasks for which many of the tools were used. It is not hard to imagine the uses for projectile points, mortars, and fire drills; but the uses for such things as notched flakes, utilized flakes, and Variety C bifaces are not so easy to specify.

In a few cases, however, we may be able to suggest whether certain artifacts were more likely women's tools, men's tools, or both. We can do this by looking to see which tools tend to be left behind in paths or activity areas used mainly by women, or in paths and activity areas used mainly by men. Tables 29.1 and 29.2 list all the artifacts that occurred in the activity areas or pathways proposed by Reynolds in Chapter 28. Examination of these tables should be accompanied by the usual warnings: (1) Artifacts do not necessarily get discarded in the very areas where they were used, and (2) not all the artifacts at Guilá Naquitz occurred in one of Reynolds' areas. In spite of these disclaimers, we feel the following observations, based on Tables 29.1 and 29.2, are worth thinking about:

1. In general, hammerstones, core fragments, and debitage could occur in either men's or women's work areas, suggesting that primary flint knapping was done by both sexes.
2. Pathways, especially those passing near the cave wall, were evidently good places to toss discarded tools. Paths contained more flake cores, more notched flakes, and more utilized flakes than any other area.
3. Five of seven steep denticulate scrapers were left in paths or work areas used mainly by women. This suggests that many of our steep denticulate scrapers were used by women, without specifying the task(s) for which they were used.
4. Three of four burins were left in paths or work areas used mainly by women.
5. Twice as many notched flakes were left in paths or work areas used mainly by women as in paths or work areas used mainly by men. This suggests that many of our notched flakes were used by women, without specifying the task(s) for which they were used.
6. Ground-stone tools, perhaps because they are somewhat bulky, were more frequently tossed into paths than left in work areas.

7. All four antler artifacts were left in paths or work areas used mainly by women. We do not know, however, what task(s) most of these antlers were used for.

8. Five of six cordage fragments occurred on pathways, perhaps because bundles of food or fuel were often untied there.

TABLE 29.1
All Chipped-Stone Tools that Can Be Associated with One of Reynolds' Activity Areas in Guilá Naquitz Zones E, D, C, and B1

Categories	Women's work and associated discard areas	Path used mainly by women	Path (neutral)	Path used mainly by men	Men's work and associated discard areas
Choppers/knives	—	—	—	—	—
Steep denticulate scrapers	2	3	1	1	—
Sidescrapers/knives	—	1	1	—	—
End scrapers	—	—	—	—	—
Burins	2	1	—	—	1
Drills	—	1	—	—	—
Pointed pieces	—	1	—	—	—
Crude blades	2	2	1	—	1
Notched flakes	7	10	5	5	3
Flakes with sheen	—	1	—	—	2
Utilized flakes	2	5	3	—	1
Projectile points	—	—	—	—	1
Bifaces, Variety B	—	—	1	—	—
Bifaces, Variety C	1	1	—	—	1
Tanged picks	—	—	—	—	—
Hammerstones	2	1	—	—	2
Flake cores	—	3	3	—	—
Core faces	3	2	1	—	3
Core edges	—	3	—	—	1

TABLE 29.2
All Tools (Other than Chipped Stone) that Can Be Associated with One of Reynolds' Activity Areas in Guilá Naquitz Zones E, D, C, and B1

Categories	Women's work and associated discard areas	Plant processing	Path used mainly by women	Path (neutral)	Path used mainly by men	Men's work and associated discard areas
Metate fragments	—	—	—	1	2	—
Mortar fragments	—	1	1	—	—	—
Manos and mano fragments	1	—	—	2	—	—
Pebble pigment grinders	—	—	—	1	—	—
Slab paint palettes	—	—	—	—	—	1
Cordage	1	—	2	2	1	—
Fire drills	—	—	1	—	—	—
Antler artifacts	2	—	2	—	—	—
Balls, cactus gum	1	—	—	—	—	—

VII

SIMULATING FORAGING AND EARLY AGRICULTURE IN OAXACA

30

The Modeling of Foraging Strategy: An Introduction to Part VII

Kent V. Flannery

> Adaptation is the process of evolutionary change by which the organism provides a better "solution" to the "problem", and the end result is the state of being adapted.
>
> (Lewontin 1978:213)

In Part VI of this volume we dealt with division of labor and partitioning of work space by the group occupying Guilá Naquitz. In Part VII we deal with a pair of questions that are even more difficult: (1) What strategy led the occupants of the cave to select the mix of plants we see on the preagricultural living floors? and (2) How did this strategy change during the period of incipient agriculture?

We stated in Chapter 2 that we intended to give "equal time" to the role of information in the origins of agriculture, and the adaptive computer simulation model designed by Reynolds (Chapter 31) is consistent with that approach. Reynolds rejected the notion of a mechanistic-deterministic simulation, opting instead for a stochastic-probabilistic model in which thousands of alternative decisions could be made on the basis of remembered information.

The approach taken in Chapter 31 owes a lot to our University of Michigan colleague John Holland, whose book *Adaptation in Natural and Artificial Systems* (Holland 1975) explores the mathematical complexities of the process of adaptation. Holland argues that adaptation is a nonlinear phenomenon that is not amenable to the mathematical formulae that are easiest to use. For example, he demonstrates that if adaptation by animals were a simple matter of trying out mutations one by one, taking advantage of the occasional more advantageous allele, there would be insufficient time in the life of the universe to accommodate the progress that has been made. Rather, adaptation deals with "chunks" of information or cooperating sets of genes. Holland's suggestions for modeling such adaptation are incorporated into Reynolds' simulation, along with some of Zeigler's (1976) procedures for simplifying models to increase their tractability, while preserving their essential features with respect to the questions being asked.

In Reynolds' model, a hypothetical microband of 5 foragers, starting from a position of ignorance, "learns" how to schedule its collection of the 11 major plant foods of the Guilá Naquitz environment by trial and error over a very long period of time. They do this by attempting to improve the efficiency of their recovery of calories and protein per area searched during each cycle, or time step, of the simulation. They are confronted with an unpredictable sequence of wet, dry, and average years that change the productivity of the plants; and information on their past performance gradually builds up in the memory of the system, informing their decisions on whether or not to modify their strategy when a similar year comes up again. Each strategy considers the vegetational zone searched, the rank order in which the plants are searched for in each zone, and the size of the harvest area for each. By evaluating *sets* of changes (rather than single modifications, as would be the case if the group were testing "mutations" one by one), the group arrives at what would be called "coadapted sets of decisions" in Holland's framework. As time goes on, increasing amounts of experience are available to inform the microband's decisions, until their performance is so efficient that no new modification has much probability of being an improvement (although thousands of potential modifications are available should the parameters of the group's environment change). At this point, Reynolds compares the model group's strategy to the remains from Zone D of Guilá Naquitz, our last largely preagricultural level.

Agricultural plants (cucurbits, beans, and primitive maize) are introduced into the simulation at this point, and the whole process begins again. Slowly but surely, the microband shuffles its priorities until it develops a new coadapted set of strategies that are hard to improve upon. Here Reynolds can compare the mix of plants generated by this set of collecting strategies with that from Zone B1 of Guilá Naquitz. But the analysis does not stop there, for one of the advantages of a simulation is the possibility of varying the parameters of the model to see what might have happened under different conditions. Reynolds therefore makes several changes in the model:

1. The frequency of wet years is gradually increased to see what effect a climatic improvement might have.
2. The frequency of dry years is gradually increased to see what effect a climatic deterioration might have.
3. The frequency of average years is set at 100% to see what effect reduced climatic variation might have.
4. The population of the group is allowed to (a) increase steadily or (b) fluctuate unpredictably around a mean to see what effect such demographic change might have.

Rather than anticipate Reynolds' results, we withhold comment on his conclusions until Chapter 32.

REYNOLDS' MODEL IN THE CONTEXT OF OTHER COMPUTER SIMULATIONS

Computer simulations have a considerable time depth in archaeology. One of the first archaeologists to simulate a hunting–gathering strategy was David H. Thomas in his dissertation on the Shoshone of Nevada (Thomas 1971). Not long after, Wright and Zeder (1973, 1977) simulated a tribal-level exchange system and Wobst (1974) simulated boundary conditions for Paleolithic social systems. By the late 1970s, whole symposium volumes had been dedicated to simulation (e.g., Hodder 1978; Sabloff 1981).

There are many kinds of simulations, and to put Reynolds' in perspective we may briefly consider some of these. A simulation is an imitation of reality, and the more realistic the data put into it, the closer the imitation should be. At the head of the list we might put such simulations as Paulik and Greenough's (1966) study of Pacific Northwest salmon populations or McCullough's (1979) study of the George Reserve deer herd. After collecting the necessary quantified data on salmon recruitment and mortality, Paulik and Greenough generated 250-year runs on the computer to see what would happen to salmon fisheries under different conditions. After collecting similar data on deer recruitment and mortality over a 12-year period, McCullough generated a series of herd histories to see what would happen under different hunting strategies. Such studies imitate reality well because (1) all variables contain values empirically derived from study of the animals involved, (2) all formulae used are based on recurrence relationships among variables that have been generated by empirical observation, and (3) the number of values "made up" by the investigator are few to none.

More common in archaeology are cases where some of the values are empirically based but others amount to "missing data" and have to be made up by the investigator. In still other cases, the values are derived from the real world but the recurrence relationships among variables are not yet empirically generated in the way the formulae used by McCullough and Paulik and Greenough have been. In Wright and Zeder's (1973, 1977) simulation, for example, the villages participating in the exchange system are fictional but their rates of population gain or loss are taken from actual quantitative accounts in the ethnographic literature on New Guinea.

Finally, one can find simulations in which *all* values have to be made up by the investigator because no one has yet done the basic research necessary to provide reliable quantitative values. For example, a simulation of the Classic Maya collapse by Hosler *et al.* (1977) uses a set of variables not one of whose values could possibly be known. These include the birthrate and death rate for Classic Maya commoners, their average lifetimes, the number of kilograms of crops eaten per person per year, the number of commoners involved in monument construction, the desired number of monuments per commoner, and so on. It will certainly be a long time before archaeologists possess values for those variables that are as empirically generated as the figures one can collect for Pacific Coast salmon or George Reserve deer; and since we still have no idea what the "reality" of the Classic Maya collapse was, we cannot yet evaluate the extent to which it is imitated by the model.

Reynolds' simulation in Chapter 31, like that of Thomas (1971), lies somewhere between McCullough's and Wright and Zeder's. By far the majority of the values used come from our decade-long study of the natural vegetation near Guilá Naquitz (described in Chapters 4, 18, 23, and 24) and are therefore empirically generated. Occasionally we have to deal with missing data, such as the piñon nuts that no longer occur in the Guilá Naquitz environment; in such cases, values have been taken from the literature. Because it is impossible for us to specify how many calories preceramic foragers expended while harvesting the foods of the eastern Valley of Oaxaca, we have used square meters of area harvested—a variable for which we *do* have quantified data—as our measure of work effort. To make the model more realistic, Reynolds has presented his hypothetical band of foragers with a succession of wet, dry, and average years that occur in the frequency displayed by Oaxaca's rainfall records, but in random and therefore unpredictable order. These years change the plant resources of the model in the ways we have observed them to change the Mitla area vegetation in real life. In other words, while hampered by our inability to observe and interview the preceramic occupants of the cave, we have done everything we could think of to make the model realistic.

ALTERNATIVE WAYS OF MODELING HUNTER-GATHERER SUBSISTENCE STRATEGIES

There are many alternative ways in which we might have modeled the use of plants by the preceramic inhabitants of Guilá Naquitz. Two that have been used and discussed in the recent literature are optimal foraging theory (Winterhalder and Smith 1981) and linear programming (Keene 1981a, 1981b; Reidhead 1979, 1981). While we have no objection to either of these approaches, we chose to use an alternative method and therefore feel that we should explain that choice. Our reasons had to do with (1) the diachronic nature of our data, which included strategies changing through time; (2) our desire to give information equal time with energy and matter; and (3) our suspicion that resiliency or risk reduction might have been more of a goal than optimization in the Guilá Naquitz case.

Optimal foraging theory is derived from evolutionary ecology, which we regard as a point in its favor. As pointed out by Durham (1981), however, it was largely developed for animals where competition occurs not only with other species but also between individuals of the same species—even between mated males and females, and between parents and their offspring. It is grounded in the notion that where organisms compete for resources, selective advantage is conferred on those organisms having the more efficient techniques for acquiring energy. Hence, foragers should attempt to maximize the number of calories (energy) they acquire per unit of time. Reynolds' model overlaps to the extent of measuring efficiency in terms of calories and protein acquired per hectare searched. However, Cashdan (1982) points out that while there are good ecological reasons for focusing on energy, anthropologists can easily think of situations where some other optimization goal might be more appropriate, such as "reliability of energy capture."

Winterhalder and Smith measure the extent to which Cree Indians and Inuit Eskimos meet the expectations of an optimal foraging model, but since their studies are ethnographic they do not face the diachronic problems we face. Zones E, D, C, and B1 of Guilá Naquitz present us with a history of changes in foraging strategy, and while we could presumably have tested each living floor against an optimization model, we strongly suspected we should try to simulate the changes as a long-term learning process.

In contrast to optimal foraging theory, linear programming analysis was developed in economics. It uses a computer program to arrive at the one single combination of available resources that both satisfies the subsistence needs of a group and does so at the lowest possible cost of time and risk. Its major advantage over optimal foraging theory lies in the fact that it considers more variables than caloric needs and time constraints. Reynolds' simulation is similar in the sense that it considers both calories and protein, and indeed we wish we could have included even more variables—for example, deer hunting, for which we have "missing data" because of a lack of deer in the Guilá Naquitz region today.

Keene's (1981a, 1981b) linear program for the prehistoric Saginaw Valley suggests that energy was a binding constraint only during the low-sunlight months of March, April, September, and November; during other months there were different limiting factors such as calcium, ascorbic acid, thiamine, and hides. Keene's results remind us that we are prevented from modeling a whole calendar year at Guilá Naquitz because the preceramic occupants moved on to other campsites from January to August.

Among the problems posed by linear programming is the fact that it, like optimal foraging theory, presents us with one, static, optimal solution to what is for us a diachronic problem. Moreover, that solution requires what Moore (1981) calls the "all-knowing, computationally perfect decision maker." In other words, linear programming can tell us the best way to exploit an environment, but it does not tell us how to get there from a position of ignorance; and there is some reason to believe you might not get there without a computer, or else a high level of competition combined with intense selection pressure.

How often did prehistoric people arrive at the optimum solution? In his linear programming study of prehistoric Indiana, Reidhead (1981) found that behavior observed in the archaeological record "departed from a pure optimizing strategy in a few notable ways" (p. 95) and that "strict effort minimization. . . does not seem to have been the end nor the goal of food procurement" (1981:103). Keene (1981b:237) goes so far as to ask, "Are optimization models realistic?" He concludes that "a perfect correspondence between an optimal solution and real behavior should not be expected," but that "optimizational modeling provides a baseline against which observed behavior can be compared with a theoretical optimum" (1981b:237). Significantly, he adds that "in benevolent environments, when resources are abundant and when risk is minimal, there is greater leeway to deviate from the optimal pattern of behavior without risk to security and survival" (Keene 1981b:237–238). Without giving away the results of Reynolds' analysis (Chapter 31), I suggest the reader make note of the way our hypothetical microband behaved during wet years when risk was reduced.

THE RATIONALE FOR OUR APPROACH

It should be clear by now that our approach to the foraging strategy of the Naquitz phase is one that focuses on adaptation through learning by trial and error, relies on memory to inform decisions, and concerns itself more with resiliency in the face of unpredictable variation than with optimizing energy capture. We might summarize our rationale for this approach as follows:

1. In linear programming and optimal foraging theory, time is not a factor. The programs yield brilliant solutions to foraging problems; but linear programming does not tell you exactly how to get there, while optimal foraging theory tells you to get there by competition. If you are an ethnologist dealing with contemporary peoples, this lack of time depth is not a problem, but if you are an archaeologist faced with a sequence of changes, it is. Reynolds' simulation tells us how

to get to the adaptations seen in Zones D or B1 of Guilá Naquitz, starting from a position of ignorance and without assuming competition from another group. Looking back over the output, one can see just what decisions were made to get to a certain point because the program is process oriented.

2. Both optimal foraging theory and linear programming descriptions of foraging lean heavily on energy and matter, while Reynolds' hypothetical microband also relies on information and through experience gets better and better at making decisions over time.

3. In Reynolds' model, not all changes are improvements; some make things worse. Slowly changing by means of "chunks," or "coadapted sets of decisions," the model group increases its efficiency to the point where there is a low probability that any new change will be an improvement; yet even at this stage, there is a tremendous reservoir of potential variation available to the group in case of unpredictable stress. This is what we mean by "preserving resiliency," and it seems to us to be at least as realistic an imitation of the prehistoric world as the solution of an "all-knowing, computationally perfect decision maker."

4. We wanted to be able to vary the parameters of the model to see what would happen if population grew or fluctuated or if the climate got wetter, drier, or more uniform. Reynolds' simulation allows us to do this.

5. We wanted our scheduling model to be a systems model, with feedback relationships within it, and we wanted to be able to test the importance of those feedback relationships. In Chapter 31, Reynolds experimentally disconnects the feedback loop between the memory bank of past performances and the present-day decision makers, and we learn something about the effects of such feedback within the system.

6. No deer herd or school of salmon has the kind of multigenerational memory constructed by Reynolds for his adaptive model; hence the latter's responses should be more like those of a human group, which is, after all, what we are trying to model.

Finally, the approach taken in Chapter 31 is only one alternative—one that we felt met our goals and the nature of our data but far from the only alternative. While we have given our reasons for not using linear programming or optimal foraging theory here, we would be curious to see what the results of such analyses would be. Indeed, we hope that we have included enough information in the earlier chapters of this volume so that some of our colleagues with expertise in optimization models will be tempted to apply them to our data. Only in this way can the Guilá Naquitz data transcend the parochiality of our analyses and contribute to the general discussions of foraging theory carried on by our colleagues.

31

An Adaptive Computer Model for the Evolution of Plant Collecting and Early Agriculture in the Eastern Valley of Oaxaca

Robert G. Reynolds

PART 1: INTRODUCTION

One aspect of all interesting social systems would seem to be the need to synchronize reliably sets of concurrent or parallel social processes. While many observers will readily admit to the existence of mechanisms to meet this need, few have attempted to model them. Some years ago, Flannery (1968) suggested that such processes of synchronization were instrumental in the gradual evolution of primitive agricultural systems in central Mexico, drawing most heavily on MacNeish's work in the Tehuacán Valley of Puebla.

While many of the crucial processes involved in agricultural evolution involve exchanges of *matter* and *energy*, Flannery argued that they were synchronized by exchanges of *information*. In a later paper, he and Marcus argued that most paleoecological studies in archaeology had dealt only with exchanges of matter and energy and that this was "distressing because it is the information exchanges which regulate many of the matter–energy transactions" (Flannery and Marcus 1976:374).

The widespread bias toward the modeling of human ecosystems in terms of matter and energy reflects a corresponding bias in the nature of the available data. That is, aspects of a prehistoric society's material culture, such as tools, animal bones, and plant remains, are more likely to be preserved in the archaeological record than evidence concerning its information-processing and decision-making structure. The archaeologist is, therefore, in a situation somewhat akin to that of the computer scientist who is given a detailed description of the hardware for a particular computing system and then asked to make inferences about the structure of the programs that are most often run on it. Making inferences in such situations often involves "leaps of faith" that most researchers are hesitant to make.

Flannery's 1968 model for agricultural evolution can be briefly described as follows. Given the seasonal aspects of the semiarid Mexican highlands, many edible plants and animals are available only at certain times of the year. For example, some species bear fruit at the end of the dry season, which lasts from November to May, so that their seeds can germinate during the rainy season to follow. A large number of other species fruit at the end of the rainy season so that their seeds lie dormant during the following dry season and sprout when the rains begin in June. As a result, a number of plant and animal resources are available simultaneously in large amounts at certain times of the year and not at others. This situation produces scheduling conflicts such that a small group of hunter–gatherers with limited time and labor capabilities must decide what to collect, when, and how. Since they are effectively competing with other animals, what is not collected immediately is often lost to the group.

Incipient agriculture, Flannery suggested, began as an extension of efforts to increase the local density of desirable plants. While the scope of these new activities was initially quite restricted, even then scheduling conflicts began to arise. These conflicts were especially important during the months from August to November. This is the peak time for the collection of most of the seasonal wild plants recovered at Guilá

Naquitz. Any harvesting of newly cultivated plants would have to compete with these other collecting activities for the group's time. Therefore, the achievement of a sedentary agricultural system from one based principally on the collection of wild plants with some incipient agriculture involves the gradual rescheduling of a group's activities over time. In the Valley of Oaxaca, this process would have taken place between 9000 and 1500 B.C., by which time the first permanent village settlements had appeared.

One driving force behind this rescheduling process was the fact that repeated planting of maize and cucurbits increased the plants' availability, first by increasing their density and later by selecting for favorable genetic changes. Therefore, while the yields associated with the hunting and collecting sector of the economy remained approximately the same, those associated with certain domesticates increased. If rescheduling by the group was based at least in part on the schedule's observed performance, one would then expect over time a gradual but directed shift toward scheduling strategies that emphasized the agricultural component.

While the overall model deals with the set of adaptations that led to sedentary agriculture from a system based entirely on the collection of wild plants, it can be divided into several different stages: (1) collection of exclusively wild plants, (2) an early form of agriculture that simply involved altering the distribution of available species, (3) a gradual production of higher-yielding strains due to genetic change and selection, and (4) sedentary agriculture. In this chapter I am concerned mainly with modeling the first two stages, which took place during the period 9000–6500 B.C. Extension of our basic model to cover the remaining stages should be a subject for future research.

With regard to these two early stages, Flannery argues that some of the most important pressures involved in producing the initial series of adaptive changes were *internal* ones based on decision-making problems. This provides a contrast with some other models for the origins of agriculture, in which *external* pressures such as climatic change were seen as crucial. Our intent here is to operationalize those portions of the 1968 model dealing with agricultural incipience and to test them against the data presented in earlier sections of this volume. In the process, we test two basic claims: (1) that scheduling decisions based on a number of different criteria served to structure the group's preagricultural collecting behavior and (2) that these scheduling considerations regulated the manner in which the group adjusted its existing collecting behavior to accommodate new activities dealing with incipient agriculture. Since Flannery's 1968 model was based largely on MacNeish's 1960–1964 work in Tehuacán, and the data in this volume result largely from Oaxacan analyses completed subsequent to 1968, no tautology is involved.

Another of my purposes in undertaking this analysis is to challenge the notion that systems models cannot be quantified or operationalized. For example, Doran (1970:289) has already questioned whether Flannery's 1968 model could be made operational:

> Flannery can, for example, point to an example of positive feedback in the ecosystem of Mesoamerica in the period he considers. But there is little he can do to quantify that observation. And if he could, by some miracle, attach meaningful figures to the various food-procurement systems he studies, what would happen then? Although there certainly is a solid body of mathematical theory and techniques labelled "system theory", it forms a branch of control engineering and is most unlikely to be of practical use in such a context.

Doran thus sees two obstacles that stand in the way of operationalizing a model such as Flannery's: (1) the lack of mathematical techniques to aid in the modeling and eventual computer simulation of such a system and (2) the lack of archaeological data against which the model could be tested.

In the decade since Doran made these comments, a number of advances in both of the above areas have been made. For example, in the realm of control theory, Holland (1975) has developed a sophisticated framework in which to model adaptation in both natural and artificial systems. Holland's framework is employed here as a formal basis in which to embed Flannery's model. However, this initial formal model still needed to be simplified somewhat before it could be simulated on a computer. Zeigler (1976) has dealt extensively with techniques for translating a formal "base" model into a simpler "lumped" model whose behavior preserves certain desirable properties of the original. We employ Zeigler's approach here to generate a simpler version of our initial formal model, one that is amenable to computer simulation and produces results that can be compared with current archaeological data.

While these mathematical techniques were being worked out, new quantitative data were being collected on the seasonality, productivity, and nutritional makeup of the eastern Valley of Oaxaca. These data, which were presented and analyzed in Chapters 18–24 of this volume, were not available in 1968 and thus effectively prevented Flannery from quantifying his model. It is these newly collected Oaxaca data that can now be combined with Holland's and Zeigler's mathematical approaches to operationalize a model originally generated from Tehuacán data.

With the above background in mind, the remainder of this chapter is structured as follows. The first portion, consisting of Parts 2–6, deals with the developing and testing of the scheduling model for the preagricultural system. We begin in Part 2 by formally demonstrating that a group of hunter–gatherers in a semiarid environment like that of the Valley of Oaxaca are faced with a number of complex problems in information processing. These problems are felt to have generated internal pressures that helped to orient the group's scheduling of resource acquisition activities.

In Part 3 we begin our modeling of preagricultural scheduling activities by looking first at the available archaeological and environmental data. The nature of these data motivates us to make certain initial assumptions about the structure of our model. In Part 4 we derive an informal characterization of our scheduling model. In Part 5 this informal model description is given formal interpretation as a network of adaptive systems. This basic formal model is then simplified to allow for computer simulation. It is shown that the behavior of this simplified system formally preserves certain desired aspects of the original model. As a result, simulation of this simpler model can be expected to produce behavior like that of the original, more complex one. The results of this simulation are then presented in Part 6. Our intent there is to characterize the general set of resource scheduling adaptations acquired by the model group in the process of attaining an equilibrium set of scheduling strategies. The properties of these strategies are then compared with the archaeological data.

Having developed our adaptive computer model of preagricultural scheduling activities and characterized its steady state behavior, we proceed to make a number of experimental structural changes to the system in Part 7. These changes correspond to the acquisition of incipient agriculture in the manner suggested earlier in this chapter. In Part 8, this new system is simulated. Starting with the system in its preagricultural steady state, we observe what happens as it begins to experiment with incipient agriculture. It will be interesting to see whether the basic preagricultural scheduling adaptations previously acquired by the group are retained in the wake of these changes. Also, we want to see what forms of rescheduling are produced by the simulated system and how well these changes conform to the corresponding changes in the archaeological record over time. In Part 9, we alter the parameters of the model in order to see what effect different regimes of climate and population might have had on the rate of acquisition of incipient agriculture.

Finally, in Part 10 we summarize our results, draw conclusions about the behavior of our model group in both preagricultural and early agricultural times, and set forth what we see as the wider implications of our study.

PART 2: MODELING HUNTER–GATHERER DECISION MAKING

Introduction

In this section, we examine a formal mathematical model that suggests that consensus-based egalitarian hunter–gatherers (especially those living in mosaic environments such as those in the Near East, highland Mesoamerica, and the Andes) were faced with a number of significant problems dealing with acquisition of reliable information about the location of necessary resources. By concentrating on problems of information acquisition, we also suggest that agriculture could have arisen in some areas as a solution not to climatic change or population problems but to the predictable problems of resource search and scheduling encountered by any group lacking a decision-making hierarchy.

We are concerned with modeling the decisions made by a group of hunter–gatherers about how to best utilize a two-dimensional distribution of resources. In particular, we are concerned with the following problems:

1. the ability of each member to collect and process information about the resource distribution,
2. the extent to which information is shared among members,
3. the specific sets of decisions available to each member, and
4. the way in which the individual decisions are integrated to produce a group decision.

In this context, it is formally demonstrated that constraints on the communication between individuals impose limitations on the utilization that any egalitarian group can make of the spatially distributed resources accessible to it.

Our basic model of group decision making is that of the "linear threshold" or "voting" type that demands only minimal communication facility by its members. By relating our model to the perceptrons studied by Minsky and Papert (1969), we are able to employ some of their fairly deep results on the limitations of perceptron recognition in our own context.

Our specific aim here is to characterize formally certain aspects of egalitarian decision-making systems and to demonstrate within the context of the model that it is hard for such a group to answer certain questions about its environment. In particular, we demonstrate that the group's ability to decide on the direction to take for maximum resource exploitation is strongly limited by the information gathering of its individual members and *not* by the number of individuals in the group.

More concretely, let the area searchable by an individual be bounded by some constant M. Then no matter how large the group, or how complex the information-processing power of the individual (limited of course to the search area), there is a maximum region size N, determined by M, in which the group can decide the following question: In which region is the largest supply of resources concentrated? We stress that this is not a trivial limitation due to limited accessibility (since with a sufficient number of individuals an area of any size can be fully searched) or to individual processing capacity but a genuine limitation imposed by the structure of the decision-making process.

We show that, by augmenting the basic group model with a centralized decision maker[1] so that the structure is now of the pandemonium form (see Minsky and Papert 1969), the limitation on maximum region disappears. In fact, very little in the way of information collection capacity is required of the individual members to solve the "best-direction" problem, though mental capacity by the decider for perception and comparison of directional proposals is now necessary.

[1]This term refers to a simple pattern-recognizing scheme in which the model's components vote on the presence or absence of a particular pattern. The final decision is a weighted linear sum to the components' individual decisions.

Some Preliminaries: The Modeled Environment

We begin by providing a formal characterization of the environment with which our several decision-making models must deal.

Definition: Let R represent an arbitrary set of cells in the standard two-dimensional cellular space (i.e., a planar region divided into R discrete subregions [cells] of unit area).

Here it will be convenient for us to think of R as the catchment area, or current set of locations about which the group can acquire information. Each location i will refer to a particular cell in the two-dimensional space. With each cell in this space is associated a finite set of properties or attributes. At this point it is best to think of these properties as being resources of interest to the group. For our purposes, all we need to know is whether a certain resource is present (coded here as 1) or absent (coded as 0) in a specific location at a certain time t. No assumption is made about the nature of these properties except that they can be recognized by an individual scanning the location. Therefore, if resource Z is present at location i, then an individual looking for resource Z at location i will observe it.

The subset of all locations in which a resource Z is found at time t exhibits a certain distribution over the space R, as does the set of all locations at which it is not found. We denote by Z_t the subset of cells in R at which resource Z is found at the time t. Since time plays no role in our current discussion we drop the t subscript and refer directly to distributions as subsets Z or R.

In order to exploit a resource distribution Z, a group must be able to categorize it in various ways. The most fundamental categorization is of the binary kind in which the subset Z is determined to have some property P (such as being empty) or not.

Definition: A configurational predicate is a function that assigns to each subset of the catchment set R one of two possible values.

If the distribution of cells in R that possess resource Z is characterized by a property P, then $\Psi_P(Z)$ takes the value of 1; otherwise it has the value of -1.

There are essentially two general classes of predicates:

1. Position-dependent predicates are defined precisely with reference to specific points in R. For example, the query, "Are there nut crops in the canyon on the south side of the river?" is defined relative to a particular subarea of R.
2. Position-independent predicates are concerned with the recognition of general classes of patterns independent of where they occur within the region R. One example would be the predicate Ψ_{connect}, which serves to identify the class of all connected regional distributions. Other position-independent predicates that might be of concern are whether any grove of trees within its territory R has ripe fruit or whether there are any potential predators within the region R.

We now specify what it means for a group potentially to recognize a distributional predicate.

Definition: A distributional predicate is potentially recognizable by a group if there exists a decision-making algorithm or procedure executable by the group whereby it can decide with absolute certainty whether or not the predicate is true for any given distribution.

The Group Theoretic Framework

In the preceding sections we developed a formalism in which to phrase both the distribution of resources and the properties of these distributions over the space R. Now we develop our basic model in which each individual is able only to vote for or against a particular question that is posed to the group. He or she is not, however, able to express to the rest of the group the specific information used to form his or her opinion.

The following assumptions are made about the behavior of any member:

1. The region of interest R is larger than the area over which the member can collect data. This is certainly a reasonable assumption when one is concerned with hunter–gatherers, where walking is the principal form of transportation.
2. If an observable resource is present within a cell, then an individual looking there will observe it.

Although the latter assumption is particularly optimistic, since more realistically some erroneous data might be collected, the presence of this assumption in the model serves a good purpose. We show that even these hypothetical groups composed of error-free individuals cannot recognize certain types of distributions without some additional processing. More realistic models with the same decision-making structure but employing error-prone individuals can at best only hope to do as well.

With the above in mind, we express the data collection activities of any one individual X as δ_X, a predicate whose domain of reference is a subset of points in R. This subset of points is termed the *support* of the local predicate δ_X. For example, suppose an individual X searches a subregion S_X at time t in order to check whether there are any cells containing a resource Z. Formally, if $S_X \cap Z$ is not empty, $\delta_X(Z) = 1$, otherwise $\delta_X(Z) = -1$. The support of δ_X is then the subset of locations within R that are checked by an individual.

Local predicates of this kind, called *masks*, are basic to perception studies, but the theory is not limited to their use.[2] Supports of local predicates can overlap, so a location may be checked by more than one individual. Moreover the support of a local predicate need not consist of contiguous cells. The requirement of contiguity is a natural one in our context, but since we are interested in showing limitations

[2]The restriction to masks turns out not to limit perceptron capabilities according to a theorem of Minsky and Papert (1969).

of the primate group decision model under even the most optimistic circumstances, we do not restrict supports to contiguous sets or local processing to the mask type.

It now remains for us to discuss how the several individual decisions are woven together by the group to produce an overall decision.

The Decision Function

As already indicated, we initially consider the case in which an individual can only communicate his or her opinion, either favorable or not, with respect to a question (predicate) posed to the group.

First we take the localized predicate δ_X, which symbolizes the individual's decision based on his or her particular experience. With each individual X is associated a weight W_X that reflects the relative influence his or her opinion has on the group's decision about the particular class of problems. Every member is able then to vote on the matter, with some opinions counting more than others and some counting not at all ($W_X = 0$).

Now the group decision Ψ_{GROUP} can be expressed as a weighted linear function of the individual ones, where

$$\Psi_{\text{GROUP}}(Z) = \begin{cases} 1 \text{ if } \sum_{X \in \text{GROUP}} W_X \, \delta_X(Z) > \theta \text{ and} \\ -1 \text{ otherwise.} \end{cases}$$

The constant θ represents a certain level of confidence that must be attained before consensus is reached. If the sum of the weighted individual opinions taken together exceeds this level, a positive consensus is attained. Otherwise the consensus is a negative one. Note that the threshold θ is problem specific. It can therefore take small values for unimportant questions and larger values for more important ones.

We refer to the above models as *voting groups*. Such models belong to the class of linear threshold devices and are formally isomorphic to the one-level perceptrons of Minsky and Papert (1969).

A predicate Ψ is potentially recognizable by a group if there are weights W_X, $X \in$ GROUP, and a threshold θ such that for each subset Z in R, $\Psi_{\text{GROUP}}(Z) = \Psi(Z)$.

Individual Decision-making Capabilities

An individual can recognize a class of position-dependent predicates, namely those whose reference cells fall within his or her support. If his or her domain is relatively small, this class of predicates is correspondingly restricted. Moreover, in such a case, the individual is incapable of recognizing any nontrivial translation-invariant predicates. A *translation-invariant predicate* is a special case of position-independent predicate in which the defining property is preserved under all spatial translations.

Theorem: An individual searching a maximum area of size L within a sufficiently large region R cannot recognize any nonconstant translation-invariant regional predicates.

Proof: Suppose Ψ is a translation-invariant predicate on R that is recognizable by X. Then any two distributions that agree on L are Ψ equivalent (i.e., assigned the same Ψ value). Let R be large enough so that a translation L' of L can be found disjoint from L. Let d_{empty} be the distribution with all zeros in R. Let $d_{L'}$ consist of any arbitrary assignment of zeros and ones to L' and zeros everywhere else. Translate d_{empty} and $d_{L'}$ so that L' is placed over L. Since d_{empty} and d_L agree on L, they are Ψ equivalent; and since Ψ is translation invariant, their translates are also Ψ equivalent. Thus Ψ assigns the same value to all distributions over L and hence to all distributions over R, that is, Ψ is constant predicate (Ψ_{TRUE} or Ψ_{FALSE}) Q.E.D. With our intuition concerning the limits of individual capability thus reinforced we turn to group decision capabilities.

Recognition Capabilities of Voting Groups

Minksy and Papert (1969) have established certain capabilities and limitations of perceptron models. In our context, the most relevant strength of perceptrons is their ability to do certain kinds of counting and their ability to perform certain kinds of numerical comparison. To make this precise, we need to introduce the notion of "order." The order of a local predicate δ_X is the size of its support (i.e., the number of cells it looks at). The order of a perceptron or voting group employing a set of predicates $\{\delta_X | X \in \text{GROUP}\}$ is the maximum of the orders of its local predicates. Finally, the order of a predicate Ψ_R on region R is the smallest order of all the perceptrons that can recognize Ψ_R.

We have just now purposefully emphasized that a predicate always refers to a particular region. The reason is that even though this is true, we often think of the same predicate being applied to any arbitrary region. Technically, such a predicate is really a *predicate scheme* (i.e., a rule that associates to each region R a particular predicate Ψ_R). For example, the predicate scheme $\Psi_{\text{everywhere}}$ assigns to any region R the predicate $\Psi_{\text{everywhere}}$, where $\Psi_{\text{everywhere}}(Z) = 1$ if and only if Z is a distribution on R and $Z = R$.

With this in mind we can define the *order of a predicate scheme* to be the smallest of the orders of all the predicates defined by the scheme. If this order is finite, all the predicates, no matter what the size of the underlying region, are recognizable by perceptrons of this order or less; if the order is infinite, then the order of perceptrons needed to recognize the predicate Ψ_R grows without bound as the size of the region increases.

We summarize some results of Minsky and Papert (1969) in Table 31.1. Let us interpret the table in terms of the voting group model. Our hunter–gatherer group can decide whether or not a region has any resource. To do this, it need only have enough individuals to cover the entire region, with each individual examining a subset of cells for presence of the resource. The weights can be all -1 for Ψ_{empty}, and all $+1$ for $\Psi_{\text{somewhere}}$, with thresholds $|R|$ and $-|R|$ respectively ($|R|$ is the size of R). More usefully it can decide whether one region is more profitable to utilize than a second region. It can recognize $\Psi_{A|>|B}$ by weighting individuals sampling A by +1 and those sampling B by −1. All the results above the dashed line in the table are variations on this theme except for $\Psi_{=N}$. This last result has an interesting implication in our context:

Observation: The predicate $\Psi_{A=B}$ (regions A and B contain the same number of resource cells) is of Order 2. Moreover, at least one local predicate must have support in both A and B.

Proof: $\Psi_{A=B}$ is true if and only if $\Psi_{A \geq B}$ and $\Psi_{B \geq A}$ are both true or both false (i.e., $\Psi_{A=B} = \Psi_{A \geq B} \cdot \Psi_{B \geq A}$). By substitution of the linear forms, or by Theorem 1.5.4 of Minsky and Papert, the order of $\Psi_{A=B}$ is seen to be no greater than 2. If the order were 1, then the order of $\Psi_{=N}$ would also be 1, since we could set up a distribution of N cells in A and recognize when any distribution in B had this same number of nonempty cells using $\Psi_{A=B}$.

Now suppose $\Psi_{A=B}$ is recognizable with predicates having supports restricted to either A or B but not both. Then the sum in the linear form splits into two disjoint parts. By a familiar argument, we can show that the following four situations lead to a contradiction:

1. $|A \cap Z| = |B \cap Z| = 0$,
2. $|Z \cap Z| = 1, |B \cap Z| = 0$,
3. $|A \cap Z| = 0, |B \cap Z| = 1$,
4. $|Z \cap Z| = |B \cap Z| = 1$.

Q.E.D.

Thus, to check the equality of resources in two disjoint regions at least one member of the group must sample both regions, in effect doing a comparison. In fact, in the realization with Order 2, many individuals make elementary comparisons of pairs of cells.

From Table 31.1, we see that counting Modulo 2 (Ψ_{parity}) is not easy for perceptrons. In fact, Minsky shows that *all* cells must be examined by at least one individual in the group (meaning that the individual must be smart enough to do the calculation!). Also noteworthy is the fact that $\Psi_{\text{connected}}$ is not of finite order, so that for any group of individuals there is a largest region in which the group (voting model) can handle the concept of connectivity. A more meaningful limitation is that of the infinite order of the minimization predicate $\Psi_{A > C, P > C}$. This implies that deciding which of more than two alternative regions is best is hard for our voting model. But to see this requires that we formulate the notion of choice of direction for our model.

TABLE 31.1
Predicate Schemes and Their Orders

Predicate	Definition	Order
Ψ_{empty}	Resource absent in region.	1
$\Psi_{\text{somewhere}}$	Resource present somewhere in region.	1
$\Psi_{\text{everywhere}}$	Resource is found everywhere.	1
$\Psi_{>N}$	At least N cells have the resource.	1
$\Psi_{<N}$	No more than N cells have the resource.	1
$\Psi_{=N}$	Exactly N cells have the resource.	2
$\Psi_{A>B}$	Subregion A contains at least as much resource as subregion B.	1
$\Psi_{A>C, B>C}$	Subregion C has less resource than both A and B each do.	∞
Ψ_{parity}	An odd number of cells have the resource.	∞
$\Psi_{\text{connected}}$	The resource distribution is connected.	∞

Group Directional Decision Making

While we have so far formulated our decision problem in binary form, much field observation suggests that hunter–gatherer groups are very concerned with directional choices. Finding itself located at some point in a large region, with which its members have some experience, the group must decide among the diverse proposals which direction to move. The essence of this problem, we feel, is capturable by our voting model extended as follows to enable directional decisions:

Let a point in a region R be designated as the origin of a rectangular co-ordinate system (this is to represent the group's current position). Let QUAD designate the set of quadrants represented counterclockwise by the points (1,1), (−1,1), (−1,−1), and (1,−1), respectively.

Definition: A directional function Ψ is a mapping from the subsets of R to QUAD. Thus Ψ assigns to each distribution Z an element $\Psi(Z)$ in QUAD that represents a direction in which to move.

Our primary example is the maximization function $\Psi_{\max}$ defined as $\Psi_{\max}(Z) = (i^*, j^*)$, where (i^*, j^*) represents the quadrant having the largest number of resource cells. Formally, $|Q_{(i^*, j^*)} \cap Z| = \max_{(i,j) \in \text{QUAD}} |Q_{(i,j)} \cap Z|$, where $Q_{(i,j)}$ is the quandrant represented by (i,j). (If there is a tie for the most profitable quadrant, we allow an arbitrary choice.)

While we work with four directions, our results are easily extendable to an arbitrary number of directions.

We now extend our voting model to enable directional decision making.

Definition: A vector voting model specifies a linear threshold directional function: $\Psi_{\text{GROUP}} = \text{QUAD}[\sum_{X \in \text{GROUP}} W_X \, \delta_X(Z)]$, where each δ_X is a local directional function mapping Z into QUAD ∪ (0, 0) and QUAD maps the region R into quadrants—namely $\text{QUAD}(x,y) = [\text{bin}(x), \text{bin}(y)]$, where

$$\text{bin}(x) = \begin{cases} 1 \text{ if } x > 0 \text{ and} \\ -1 \text{ otherwise.} \end{cases}$$

Our vector voting model is truly an extension of the basic voting model. In the vector model each individual can propose a direction based on the region he has explored. Each direction is treated as a unit vector, the associated weight gives a magnitude to the vector and then the vectoral sum is taken. The resultant vector is then categorized according to the quadrant in which it lies. No generality is lost by not including a threshold explicitly since this may be included as one of the local functions. In these terms the basic voting model is a one-dimensional version of the extended model of the form $\Psi_{GROUP} = \text{bin}[\sum_{X \in GROUP} W_X \delta_X(Z)]$.

Definition: A vector voting model computes a directional function Ψ if $\Psi_{GROUP} = \Psi$.

We now establish the promised limitation on the vector model's directional decision capabilities.

Theorem: The directional function Ψ_{max} is not of finite order in the class of all vector voting models.

Proof: Suppose to the contrary that the directional function is of finite order M. Let R be an arbitrary region with designated origin and let a vector model of order M compute Ψ_{max}.

$$\begin{aligned}\Psi_{max}(Z) &= \text{QUAD}\ [\textstyle\sum_{X \in GROUP} W_X\delta_X(Z)]\\ &= \text{QUAD}\ [\textstyle\sum_{X \in GROUP} W_X(\delta^1_X(Z), \delta^2_X(Z))]\end{aligned}$$

(where δ^1 and δ^2 are the projections of δ on the first and second co-ordinates, respectively)

$$\begin{aligned}&= \text{QUAD}\ [\textstyle\sum_{X \in GROUP} W_X\delta^1_X(Z), \sum_{X \in GROUP} W_X\delta^2_X(Z)]\\ &= \text{bin}[\textstyle\sum_{X \in GROUP} W_X\delta^1_X(Z)], \text{bin}[\sum_{X \in GROUP} W_X\delta^2_X(Z)]\\ &= [\Psi^1_{GROUP}(Z)\Psi^2_{GROUP}(Z)],\end{aligned}$$

where Ψ^1_{GROUP} and Ψ^2_{GROUP} are of Order M. (The projections have the same order as the original functions.)

Thus $\Psi^1_{GROUP}(Z) = 1 \Leftrightarrow \Psi_{max}(Z) = (1,1)$ or $\Psi_{max}(Z) = (1,-1)$. Now restrict Ψ^1_{GROUP} to quadrants (1,1), (−1,1), and (−1,−1). Then

$$\begin{aligned}&\Psi^1_{GROUP}(Z) = 1 \Leftrightarrow \Psi_{max}(Z) = (1,1)\\ &\Leftrightarrow |Q_{(1,1)} \cap Z| = \max\{|Q_{(1,1)} \cap Z|, |Q_{(-1,1)} \cap Z|, |Q_{(-1,-1)} \cap Z|\}\\ &\Leftrightarrow |Q_{(1,1)} \cap Z| > |Q_{(-1,1)} \cap Z|\\ &\quad \text{and } |Q_{(1,1)} \cap Z| > |Q_{(-1,-1)} \cap Z|.\end{aligned}$$

Thus it is possible to decide which of three quadrants has the largest concentration of resources. But by defining $\overline{\Psi}(Z) = \Psi^1_{GROUP}(\overline{Z})$, where $\overline{Z}$ is the complementary distribution to Z, we then have

$$\begin{aligned}\overline{\Psi}(Z) = 1 \Leftrightarrow\ &|Q_{(1,1)} \cap Z| < |Q_{(-1,1)} \cap Z| \text{ and}\\ &|Q_{(1,1)} \cap Z| < |Q_{(-1,-1)} \cap Z|;\end{aligned}$$

and thus the minimization predicate $\Psi_{A > C, B > C}$ is of finite order M, a contradiction. Q.E.D.

We conclude that for every vector voting group there is an upper bound on the size of the region in which the group can decide in which direction to go to find the largest concentration of resources. This upper bound is determined by the area sizes accessible to group members.

Decision Capabilities of Groups with a Central Decision Maker

Consider now the vector model augmented with a maximizer D. The direction function computed by such a pandemonium structure is

$$\Phi(Z) = (i^*, j^*)$$

where

$$\text{MAG}_{(i^*, j^*)}(Z) = \max_{(i,j) \in QUAD} \text{MAG}_{(i,j)}(Z),$$

where

$$\text{MAG}_{(i,j)}(Z) = |\delta_X(Z) = \textstyle\sum_{(i,j)}(i,j)\, W_X|.$$

In other words, the output direction is the one in which the magnitude of the associated vector is largest. Note that this requires that proposals of each of the members be classified into directional categories, the strength in each category totaled, and the results perceived by the demon D who selects the category with greatest strength. Clearly much more in the way of intragroup communication is required to realize this structure.

It is easy to see, however, that pandemonium models can compute the maximization function with finite order local functions. Indeed, let each cell (x,y) in R be scanned by a first-order local function

$$\delta_{(x,y)}(Z) = Z(x,y) \bullet \text{QUAD}\ (x,y);$$

that is, if resource is present at cell (x,y), then a proposal is made for motion toward the quadrant in which it lies. Using unit weights the resultant proposals in the four alternative directions are equal to their respective concentrations of resources. The maximizer thus selects the most promising direction.

While we have shown how a group making decisions in the manner described by our model can theoretically compute a number of spatial predicates and functions presented to them by the environment, there is no guarantee that it will. In terms of the model, the following criteria must be met in order for a model computable function to be actualized:

1. There must be a set of expected values for the pertinent variables corresponding to locations within the environment.
2. The search areas must be coordinated so that all cells in the space necessary to the recognition of the pattern are scanned.
3. The weights for individual opinions regarding a problem must be coordinated with their experience and the level of consensus needed to obtain a group decision.

Note that Criterion 2 requires there be a sufficient number of members in the group so that the region is covered. Our limitation results show that even if this is the case, there are limitations on the group's decision-making ability. It is conjectured here that the coordination of weights is fostered by the presence of a relatively fixed set of social niches or roles that group participants might play. Such roles are often structured with respect to age and sex in egalitarian societies. Therefore, if a conflict between members arises and the one is able to displace the other, the principal effect is merely to change roles. If each role is characterized by an associated set of weights in the decision-making process, then the group by insuring the presence of a certain set of roles is maintaining an associated set of weights as well. This increases the likelihood of a good decision being duplicated in the future, even though the individuals who made the initial decision may no longer be present.

Predictions and Open Questions

Our results suggest that the size of the region in which resources may be maximally exploitable by hunter–gatherer groups that do not possess centralized decision making is strictly limited by the individual's abilities to gather information. In semiarid environments where the distribution of resources is mosaic, with these patches scattered over a large area, this limitation could be an important one. In this situation, a group in order to collect sufficient resources is often forced to forage over an area so large that members can only make educated guesses about the locations of available resources. Since we have shown that the group cannot always make decisions with certainty, it is important that they be able to increase their probability of making good decisions.

The remainder of this chapter is concerned with modeling the way individuals acquire and adjust their resource collecting strategies based on experience. In particular, we want to see how this process affects the way in which the group members incorporate agricultural innovations into an economy based principally on the collection of wild plants. Here we are concerned only with modeling this process within the Valley of Oaxaca, although aspects of the model may be applicable to other situations.

PART 3: THE EASTERN VALLEY OF OAXACA DATA

Introduction

In Part 2 it was suggested that good resource collecting schemes are not easy to generate since many decisions within them are based on collected information that is inherently unreliable. As a result, not all the strategies employed by the group will necessarily be good ones. It is therefore important that the group be able to find out, over time, which strategies are less successful than others. Accordingly, the model developed here deals with how a hunter–gatherer group acquires and uses performance information to restructure its repertoire of strategies.

Although the task of generating such a model is made difficult by the fact that the real system no longer exists, some archaeological data concerning the nature of the Guilá Naquitz group's behavior and its environment have been presented in the earlier chapters of this volume. Here we use that information to make certain suggestions about the model's design.

The Environment

The environment of the Guilá Naquitz area has been described in Chapters 3 and 4. In Chapters 15 and 16, Schoenwetter and Smith and Flannery and Wheeler have suggested on the basis of pollen and microfauna data that the present-day environment is a useful guide to the past. In that present-day environment, there is both (1) *predictable* variation in rainfall between the May–November wet season and the December–April dry season and (2) *unpredictable* variation in rainfall from one year to the next (Kirkby 1973:Figs. 9, 10, 58). Available precipitation data suggest that over a 40-year period about half the years will have "average" rainfall (between 420 and 600 mm), while a quarter of the years will be significantly drier than average and a quarter will be significantly wetter than average. In Chapter 18, the effects of "good" and "bad" years on the productivity of wild plants was documented. In our model, a hypothetical band of hunter–gatherers are presented with an unpredictable (randomly patterned) sequence of years 50% of which are "average," 25% of which are "rainy," and 25% of which are "dry." As suggested in Chapter 18, the effect of such years may be that there is a 25% decline in productivity in dry years and a 25% increase in productivity in wet years, which may be augmented if two wet or two dry years occur in succession (see Part 5). Since comparable data on animal resources are not available, our model is confined to the use of plants.

Vegetational zones in the vicinity of Guilá Naquitz have been presented in Chapter 4 (see especially Fig. 4.6) and their productivity estimated in Chapter 18. Virtually all the plant foods used at Guilá Naquitz were available within 4 km of the cave, and we possess data on (1) how far one would have to walk to each vegetational zone today and (2) how many square meters of that zone would have to be harvested to produce a specified amount of a given vegetal food. In Chapter 23 we have been given data on the nutritional makeup of the most important plants, and in both Chapters 23 and 24 we have reconstructions of hypothetical daily intakes. These are the raw data on the basis of which we attempt to model which zones our foragers used, the priorities they assigned to certain plants, and the extent of their success in terms of calories and protein recovered.

Even if we restrict the problem to one of scheduling a sequence of autumn plant collection activities, the task facing our group is not an easy one. This is evident from Fig. 18.1 in Chapter 18, which depicts the large number of plants available from August through November. Not only are there a number of potentially available plants but they also exhibit differing densities and distributions, especially as these parameters vary from year to year as a function of available rainfall.

The inhabitants of Guilá Naquitz were faced then with a rather complex problem, but one that had to be effectively dealt with if the group were to survive. The plant remains deposited on the living floors of the cave reflect the extent to which the real-life group solved this problem. It is the ultimate goal of our model to help us determine how. In order to do this we need to compare our hypothetical group's behavior with the archaeological data that we presently have about the real Guilá Naquitz group's solution.

The Archaeological Data

The excavation of Guilá Naquitz has already been described in Chapter 5, and the raw counts of major plant foods have been given in Chapter 25. These are the data against which the performance of our model must be evaluated. To what extent will the hypothetical group of hunter–gatherers in our simulation come up with the same plant frequencies displayed by living floors E, D, C, and B1 at Guilá Naquitz? To what extent will they display, over time, the changes in plant collecting documented for the Zone E–D–C–B1 sequence in Chapter 24?

Basic Assumptions

On the basis of the data presented in early chapters of this volume, we make the following assumptions for the purposes of our model.

1. The environment of the valley today is sufficiently similar to that during the period of study so as to allow us to use current environmental data in the model.
2. The distribution and density of wild plants in sample transects near the cave are sufficiently similar to what was present between 9000 and 6500 B.C. (There are a few exceptions to this, such as piñon nuts.)
3. During the period in question the cave was occupied principally between August and November by a group of at most four to five persons.
4. At this time, the group's focus was mainly on the collection of available wild resources. Due to the difficulty of evaluating the relative amounts of time spent hunting versus plant collecting, our model considers only the scheduling of plant collecting activities by the group.
5. Archaeological data from the cave support the assumption of behavioral continuity between the groups that successively occupied the cave. In other words, the data do not suggest the intrusion of another group with different patterns of resource exploitation.
6. There was no increase in the size of the group that occupied Guilá Naquitz from Zone E (9000 B.C.) through Zone B1 (6700 B.C.).

With these basic assumptions in mind, we begin the task of generating the preagricultural resource acquisition model in Part 4.

PART 4: THE INFORMAL PREAGRICULTURAL MODEL

Introduction

We are now ready to construct a model for the preagricultural foraging behavior of our hypothetical group of four to five hunter–gatherers in the eastern Valley of Oaxaca. In Part 4 that model is presented informally, in plain English. In Part 5, we present it formally in order to facilitate its implementation as a simulation program.

However, before proceeding, it may be beneficial to look at the experience of others who have used a systems perspective in modeling hunter–gatherer subsistence. Thomas (1971), in particular, has employed this approach in order to develop a model of subsistence for the precontact Shoshone in the Great Basin of Nevada. There he used ethnographic data collected by Steward (1938) to characterize the group's resource procurement and decision-making subsystems. Simulation of this model on the computer produced results that were then compared with the available archaeological data. This was done in order to test the ability of the model based on current ethnographic information to explain the group's behavior prior to first contact with Europeans. In the next section, we describe this excellent study in some detail. This will not only help us to characterize our own system but also provide us with some benchmarks against which our model can be compared.

Thomas's Model

The Great Basin of Nevada is a vast area of interior drainage that lies mainly between the Wasatch Mountains of Utah and the Sierra Nevada range of California. The landscape is characterized by large arid valleys nestled among mountain ranges that run from north to south (Steward 1955:103). These valleys lie between 4000 and 6000 feet above sea level and receive from 5 to 20 inches of rainfall in a year. Such low rainfall, coupled with high evaporation, supports largely drought-resistant vegetation. Much of the vegetation has limited nutritional value to animals or men. The largest concentrations of roots and edible seeds are found along the banks of streams that etch the landscape. However, in the extensive sandy areas between these streams, the quantity of edible plants is low and their distributions vary from year to year and place to place, depending on rainfall. This paucity of vegetation severely restricted the numbers of available game. Hunting was therefore an intermittent activity conducted on a communal basis. Fish were an additional source of food, but their runs were seasonal.

The second major vegetational zone was characterized by stands of piñon and juniper trees. This zone occurs at elevations from 6000 to 9000 feet above sea level, largely along the sides of the mountain chains. Increased rainfall at these altitudes resulted in more available seeds, roots, grasses, and game (especially deer) than along the valley floor. However, the most important available resource was the nut of the piñon pine. These were available seasonally in large numbers and were intensively collected. Large quantities of the nuts were cached and used for food during the winter months.

These biotic communities served to structure the spatial and temporal activities of the area's hunter–gatherers. Foragers, therefore, moved seasonally from one zone to the next depending on the availability of local resources. Steward described the typical movements of a Shoshone family in the following way:

> The typical Shoshoni family living in the piñon area of Nevada traveled alone or with one or two related families during the spring and summer, seeking seeds, roots, various small mammals, rodents, insects, larvae, and other edible items. In the late summer when a family heard reports that the pine nuts seemed very promising in a certain portion of a mountain range, it arranged its travels so as to arrive in that locality in late October or early November, when the first frosts would have opened the cones and made the nuts ready to harvest. Other families who had also been foraging for food within a radius of perhaps twenty to thirty miles of that locality came there for the same reason.
>
> In gathering the pine nuts, each family restricted itself by common understanding to a limited area, because there were so many pine nuts in the locality as a whole that no one could gather them all before they dropped and because each family could harvest more if it worked alone. The different families remained from several hundred yards to a mile or more apart. Each gathered pine nuts as rapidly as it could and stored them in earth caches. If the harvest was good, it might support the family throughout most of the winter.
>
> The winter encampment consisted of perhaps twenty or thirty families within easy visiting distance of one another. Early spring generally found the people suffering more or less acutely from hunger. The families then had to go their separate ways to forage for greens, game, and any other foods they could find. Throughout spring and summer, the migrations of a particular family, although limited in general to the terrain it knew well, were determined almost week to week by its knowledge of available foods. It might learn that sand grass seeds were promising in one place, rabbits numerous elsewhere, fly larvae abundant in a certain lake, and antelope ready for a communal hunt under a shaman or medicine man over in the next valley. (Steward 1955:105–106; reprinted by permission of the University of Illinois Press)

Steward's observations were taken by many anthropologists to be representative of Shoshonean lifestyles prior to their first contact with the white man. Since the environment has changed considerably as a result of this contact, one can only hypothesize that such a correspondence exists. Thomas undertook to test this hypothesis in the following manner. He first developed a computer program that simulated the seasonal round of activities for a "typical" family in an "average" year based on Steward's observations. The purpose of this simulation was to estimate the relative portions of artifacts deposited in an average year by a group that behaves in the manner suggested by Steward.

> The concern is not to chart change over a cultural trajectory, but rather to establish the "equilibrium basin" for the Great Basin Shoshonean technoeconomic system. The simulation model does not employ time in the conventional sense; that is, a 1000 year simulation model does not attempt to array systemic behavior over 1000 continuous years. Rather, a computer run simulating 1000 years simply repeats the artifactual deposition for the *same* year, 1000 times. (Thomas 1971:12)

In order to do this, Thomas embedded Steward's theory in a systems structure (Thomas 1971: Fig. 2.1). Each box in his flowchart represented a particular resource-acquisition strategy with respect to a certain resource. The decision-making scheme dealt only with how these different subsystems are selected during a year; in fact, Thomas constrained the sequence of selected activities to be that suggested by Steward. He was not, therefore, concerned with modeling change in resource procurement strategies over time, but rather with modeling the result of a particular seasonal scheduling strategy at one point in time (specifically A.D. 1840).

Significantly, in Thomas's Fig. 2.1 there are no connections between the different resource procurement subsystems. That is, each set of activities is seen to be done independently of all others:

> *BASIN I* obviously ignores interactions between subsystems. In more advanced simulation models, the interaction effects may be more significant than any of the main effects (Watt 1968:151). But *BASIN I* is not yet in such complex form. The attempt here has been to simplify an intricate extractive system into component subsystems, so that artifactual deposition can be simulated. (Thomas 1971:38–39)

As Thomas indicates, this assumption was quite consistent with the goal of his simulation. However, in developing our model of daily decision making and resource scheduling, these interaction effects have to be included.

Thomas was therefore primarily concerned with modeling a particular schedule of seasonal activities, the one proposed by Steward. As a result, the decision-making subsystem consists of the single seasonal sequence of activities. Associated with each of these activities is a set of artifacts deposited on the ground as a result of that activity. The model therefore generates a spatial pattern of artifacts that is then compared with what is known archaeologically about the distribution of these artifacts. Data were then collected from the Reese River in order to test the model. As it turns out, 75% of the predicted frequencies were verified by the archaeological data. From this, Thomas was led to conclude the postcontact Shoshone behavior, as described by Steward, well represented subsistence activities in the prehistoric Reese River valley.

Thomas's systems model was designed to model the spatial distribution of artifacts produced by one set of annual procurement strategies at a particular point in time. He suggests, however, that such a basic model could be extended to describe changes in resource procurement strategies over time. "Goal-directed systemic change (cultural trajectories) could perhaps also be studied if key variables could be integrated" (Thomas 1971:21).

In order to test our Oaxaca model, we have to simulate the types of changes suggested above by Thomas. This is what we begin to do in the next section.

Operationalizing the Oaxaca Model

The work of Thomas described in the previous section suggested that the seasonal scheduling of activities observed by Steward was also employed by prehistoric populations in the Reese River valley. While such patterned behavior is most readily observed in terms of seasonal shifts in resource use, local scheduling of collecting activities over a shorter period (say, several days) should be important as well. For example, Leakey and Lewin point out the importance of local decision making in describing the foraging activities of the !Kung bushmen:

> The real skill of food gathering is knowing where to go and when to go there. With a wide range of fruits, nuts, roots, and shoots coming into season at different times of the year and in different places, food gatherers must balance up the probability of success in traveling, say, three miles in one direction to a potentially good source of food, against going four miles in the opposite direction to an even richer source, but perhaps with a lower probability that it is ready to collect just yet.
>
> To make a success of a food-gathering economy you need highly efficient mental maps, not just of space but of time; you have to know where to go and at what time. . . . So, the key to this type of economy lies in the information and analytical skills inside the head, rather than in fancy technology wielded in the hand. (Leakey and Lewin 1978:109; reprinted by permission of Anchor Press)

As a result we might easily conceive of a resource collection schedule as a sequence of activities, where each activity consists of searching a number of locations in a particular vegetational zone for a certain resource. The principal resource of interest is termed the *focus* of the activity. While looking for this resource at the specified locations, an individual may collect small amounts of other associated resources found at the same location or on the way to that location. Such practices, however, contribute only small amounts to the overall yield for an activity and therefore are not explicitly dealt with here.

In our model, an activity is defined as occurring within one of the four vegetation zones (Thorn Forest A and B and Mesquite Grassland A and B) described in Chapter 4. This is because each vegetational zone in the vicinity of Guilá Naquitz has a distinct set of component species and associated densities. Therefore, the set of species associated with a focal resource changes from zone to zone, as do the densities of that resource. This implies that the nature of collecting activities will also vary from zone to zone. In fact, this variation is well documented by Steward for the Shoshone. By restricting the location of an activity to points within a vegetational zone, these distinctions can be taken into account.

A resource collecting schedule or strategy is a sequence of these activities that represent the behavior of the group over a certain period of time. In our model we look at collecting activities over a period of 10 days. During this 10-day period, group members execute each of the proposed activities in turn. At the end of this period, it may happen that not all of the planned-for activities have been carried out. These "intended activities" are still associated with that schedule, and together with those completed activities they describe the whole collecting strategy. Later, we describe the mechanisms that allow the schedule to be restructured. One possible result of this restructuring is to allow the use of activities that the group previously did not have time for.

In terms of our present approach, it is not important how the schedule is represented in the minds of the group members. By describing an activity on the basis of the resources collected and locations searched, we are really characterizing a strategy in terms of the observable results it produces when used. It is, therefore, not important whether it is specified by a single individual or jointly specified by several members; what counts is the fact that it is specified and used.

To this point we have talked only in terms of one generalized scheduling strategy. However, Flannery has suggested that there are probably a number of possible strategies that are available to the group: "The outlines of a schedule, albeit with conflicts, were present. . . . 'scouting reports' helped resolve conflicts and give precision to the dates of each kind of resource exploitation, depending on individual variations in growing season from year to year" (Flannery 1968:90). That is, at any point in time a number of alternative, possibly conflicting, schedules were available to be used. Exactly which one was selected depended on the current state of the environment. Collected information about the environment was then used to select a strategy to employ.

In our model, we presume that there were no more than 10 basic alternative strategies available to our group during a 4-month period between late August and early December, which seems reasonable for a group of five persons. From this set of 10, a strategy is selected for use based on environmental data. Since each strategy spans only a 10-day interval, a number of strategies could each be used several times during the period from late August to early December. In terms of our model, the plant materials found on the living floors of Guilá Naquitz cave were the result of applying a number of small-scale scheduling strategies over time. The more frequently a strategy is used, the more likely it is to influence the relative amounts of plant material deposited. While certain strategies may be used more often than others, our model assumes that it is the interplay of a number of strategies that contributes to the group's subsistence in any given year. It would, therefore, be very interesting to see what the equilibrium mix of strategies is for our model system. If our model is a good one, then the relative proportions of plant materials collected by this equilibrium mix should resemble the proportions found in the cave. For this reason we need to transform our present verbal description into a formal model. The equilibrium behavior of this formal model can then be simulated on the computer and the results compared with the archaeological data from the cave.

Associated with each strategy is information about its performance that is used when deciding whether or not to use the strategy. We now discuss, in turn, the types of information used to describe a schedule's performance in our present model.

Starting sometime in August and continuing into early December, the group selects a number of the currently available schedules to structure its collecting activities. Every time a strategy is used it allows the group to acquire a certain amount of food material, measured here in terms of protein and calorie yield. These measures represent the extent to which the energetic needs of the population are satisfied. While hunter–gatherers obviously do not evaluate a strategy directly in terms of the acquired grams of protein or number of calories, there is a distinct correspondence between the strategies' ability to satisfy the group's energetic needs and the group's physical well-being over the course of a season. During the season, the group members produce an estimate of the ability of each strategy to acquire food resources, based on the number of times that it is used. In our model, this estimate is based on the total amount of protein and calories acquired over the season. These estimates are then used to evaluate the relative performance of each strategy for the season.

Along with estimates of total yield, group members also form an impression about the amount of effort it took to collect that much food. Such considerations have been observed among present day hunter–gatherers. For example, the !Kung bushmen of the Kalahari "tend to cut only the most attractive foods available at a given season and bypass the less desirable ones in terms of taste and/or ease of collection" (Lee 1972:343).

Also, Silberbauer argues that conservation of collecting effort is important among the G/Wi bushmen. "In choosing from among a number of available species . . . [the G/Wi's] criteria of preference are, in order of importance, the thirst and hunger alloying properties of the plant food, the ease with which it may be exploited, and lastly its flavor" (Silberbauer 1972:283).

Therefore, the effort expended by group members in the course of employing a particular resource collecting schedule should be taken into account when evaluating its performance. In our model, estimates of the total amount of protein and calories acquired per unit area searched are used as a reflection of the effort needed to acquire them. The data needed to derive these estimates are drawn from Chapters 18 and 24 of this volume.

In addition to the specific impressions of each strategy's performance given above, an overall impression of the scheduling strategy's performance, relative to the other strategies used for the season, is developed. The factors that enter into the development of this impression are the following:

1. The strategy must satisfy certain minimal energetic needs for the group. The fact that hunter–gatherers seem to be well aware of this has been noted by Sahlins (1968), among others. Jochim expresses this principle quite nicely when he says that:

> The minimum number of calories necessary for the biological viability of the population provides a minimum aspiration level. The lack of large surplus accumulation and of large-scale redistribution systems in most hunter–gatherer societies, and the presence of conflicting demands on time and energy indicate that the actual aspiration level is not far above this minimum. (1976:16)

2. The strategy must continually be able to provide yields above the minimum even in the worst possible years. This is obviously important for hunter–gatherers in semiarid environments, as mentioned by Lee with reference to the !Kung: "During the dry season the [!Kung] diet becomes much more eclectic. . . . It is this broad base that provides an essential margin of safety during the end of the dry season when the mongongo nut forests are difficult to reach (1968:35)." Flannery (1968:90) also notes that consistent performance above the starvation level must have been important for hunter–gatherers in Oaxaca as well. Here the ability of a strategy to perform better than others in terms of both yield and required effort is taken to be a good indicator of relative reliability.

3. A strategy must take into account the state of the environment in which it was last employed. As described in Chapter 4, the state of the environment is primarily a function of rainfall, which will determine the density and distribution of relevant plants. Associated with each available collecting schedule is information that determines the "state" of the environment in which the strategy was last used. The extent to which this state agrees with the current year is undoubtedly one factor in determining whether to use a given strategy.

All of the above factors play a part in determining whether a schedule will be used and whether it will be modified. In our Oaxaca model, the group must select between 10 available strategies. This involves comparing the strategies on the basis of their relative merits. For our purposes, two comparisons are made in order to select a strategy: (1) What is the degree to which the present "state" of the environment matches the "state" in which the strategy was last used? (2) What is the group's impression of the strategy's performance relative to those employed the last season in which it was used? Taken together, these two criteria determine the likelihood of a particular strategy's being used at least once during a season.

A second major function of the decision making system is to "reschedule" a strategy, based on its performance. Once a schedule has been selected and its performance observed, the group members may elect to alter its structure. If the schedule performs well there is little need to alter its structure; indeed, tampering with it may do more harm than good. Therefore, it is more likely that a poorly performing schedule would be altered (since it is doing poorly already, the group has little to lose by making an adjustment).

Given that the likelihood of a schedule's being changed is associated with its performance, what manner of adjustments can be made to it? We program into our model a number of alternative ways in which changes can be made, based on suggestions by Flannery (1968):

1. Change the focal resource for an activity.
2. Change the vegetational zone in which the activity is carried out.
3. Change the set of locations over which to search for the resource.
4. Given two activities in a schedule, change the order in which they are executed.
5. Exchange activities between two schedules.
6. Replace an activity on one schedule with an activity from a more productive schedule. (This is equivalent

to imitating a successful activity associated with another strategy.)

7. Given two activities in a schedule that occur in the same vegetational zone, exchange their focal resources. (This means that the locations of the two resource collecting activities are shifted.)
8. Given two activities in a schedule that are in the same zone, exchange subsets of locations between them.
9. For two activities in a schedule, exchange their zones of execution and location sets, on the proviso that the new activities are defined. (In order for an activity to be defined, the resources must be present in the vegetational zone to be searched.)

Notice that each of these decisions involves at most two activities in two schedules. However, by applying a number of these decisions at the same time, one can represent more complicated decisions dealing with multiple activities over a number of schedules. In this way, the model group can create "its own decisions" out of these simple parts.

A rescheduling strategy assigns to each of these basic decisions a certain likelihood that it will be used to alter a given collecting strategy. This set of likelihoods develops as the result of past decision-making experience. In particular, those combinations of local decisions that have produced improvements in the relative performance of a schedule from one year to the next will have an increased likelihood of being used with the strategy again. Combinations of decisions that reduce the observed relative performance for a schedule will be less likely to be used again to modify the schedule. Therefore, the group uses observed changes in relative performance for a strategy to adjust its rescheduling policy for that strategy. As the structure of the group's collecting strategies changes with experience over time, so will the structure of its rescheduling decisions associated with each of them. That is, there will be a tendency to use only those decisions that are most likely to improve the performance of the current strategies.

In this section we have taken a first step toward operationalizing our model for the eastern Oaxaca Valley by providing a structural framework on which to build our formal model in Part 5. This structure contains a number of basic rules by which the group is able to alter its resource collecting strategies based on its experience. As a result of the operation of these decision-making rules over a period of time, the system should attain a relatively stable structure with some basic properties. What we attempt to do in Part 6 is to simulate the long-term behavior of the model and compare its structure with that of the real world. If the model adequately represents the real system, the simulation results should bear some similarity to the data from the real world, that is, the living floors in Guilá Naquitz.

There is still another way to verify our model. We can introduce a change in the model that corresponds with a known change in the real system and see if the two systems produce similar responses. If they do, we will have constructed a strong argument for the model's ability to represent the real system. Since the real system is known to have acquired incipient agriculture sometime between 8750 and 6800 B.C., we allow the model group to experiment with incipient agriculture and see if the model undergoes the same changes in resource use that the real system did. This is our task in later sections of this chapter.

Recapitulation

In Fig. 31.1, we present a simplified diagram that suggests how our model operates over time. Our model begins with the entry of a hypothetical family of five hunter–gatherers into the eastern Valley of Oaxaca. The area they enter consists of four vegetation zones that resemble Thorn Forest A and B and Mesquite Grassland A and B as defined earlier in this volume. Those zones are characterized by the differing frequencies of piñon nuts, acorns, *susí* nuts, agaves, nanches, wild beans, mesquite, hackberry, prickly pear, *guaje*, and wild cucurbits already described for them in earlier chapters; each of these foods is assigned the protein and calories it was found to contain in Chapters 23 and 24.

It is not assumed that our hypothetical group has as yet worked out any efficient way of using these resources. To begin with, they are assigned a 10-day strategy in which the order in which they utilize these resources, the vegetational zones they go to, and the amount of time spent in each zone are selected randomly. Then, over the course of a 4-month collecting season, they are allowed to modify their strategies at 10-day intervals; this procedure is followed year after year as the simulation continues to run.

Our hypothetical group is presented with an unpredictable (randomized) succession of years in which about 50% are average, 25% are wet, and 25% are dry. Plant yields are adjusted to year type as suggested earlier in this chapter. The system has a memory, so that the group can remember how a particular strategy worked in each type of year. The "efficiency" of a strategy is measured in terms of calories and protein recovered versus square meters of vegetation searched; the smaller the effort required to produce 2000 cal and 40 g of protein per person per day, the more efficient the strategy. We feel that considerable "realism" is introduced into the model through the fact that the group cannot predict in advance what kind of a year to expect.

Over the course of a season, the group uses a subset of plant collecting schedules to structure their resource acquisition activities. A schedule is selected to be used based on its past performance relative to other schedules, and the current state of the environment. Associated with the use of a schedule is a notion of its relative performance that is derived from a number of different considerations, such as reliability, yield, and expended effort. For those schedules that had been changed since they were last used, their new relative performances are contrasted with their old. If the schedule's performance increased as a result of the past rescheduling decision, then the rescheduling policy for the strategy is adjusted to make this decision more likely in the future. On the other hand, if the change produced an observed reduction in performance, then the rescheduling policy is adjusted

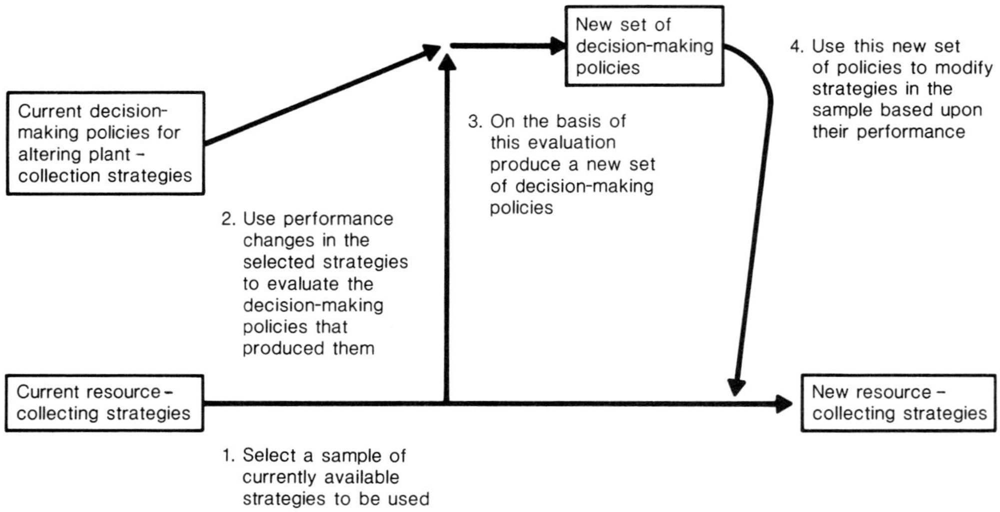

Fig. 31.1. Interaction between the two adaptive systems.

in order to make it less likely that the same type of change will be made again.

These adjusted rescheduling policies are then used to alter the structure of the utilized schedules on the basis of their relative performance. This serves to generate a new set of schedules that will then be available for the next year. It is this general sequence of processes that form the basis for the formal model described in Part 5.

While no simulation could exactly imitate a system that existed 8000–10,000 years ago, we feel that ours includes a number of elements that add to its realism:

1. The environment is based on quantified data that are believed, on the basis of paleoenvironmental studies, to be like those of the past.
2. With the exception of piñon nuts, the nutritional data are based on studies of plants collected within 5 km of Guilá Naquitz.
3. Our group must adjust to year-to-year variations in rainfall based on the best available precipitation data for the area.
4. The group can choose from a wide variety of strategies during any 4-month period (more than 1,000,000 possible sequences of collecting activities), making it impossible to "stack the deck" in favor of any one strategy.

PART 5: DEVELOPING THE FORMAL MODEL

Introduction

In Part 5, the informal base model presented in the preceding pages is given a formal interpretation. Since the informal model is concerned with the decision-making adaptations that served to structure the collecting activities of preceramic hunter–gatherers in the Valley of Oaxaca, any formal version should necessarily highlight the model's adaptive aspects. Along these lines, Holland (1975) has developed a formal mathematical framework in which to describe complex systems exhibiting adaptation. This framework has been used in a number of situations to describe adaptation in both natural and artificial systems (Cavicchio 1970; De Jong 1975; Holland and Reitman 1978). It is, therefore, well suited to our present purposes.

We begin the following section by providing a general overview of how the informal model is represented in terms of Holland's basic framework. It is convenient to view our model as comprising two interacting adaptive systems. The first deals with the specification and use of the various existing strategies for plant resource collection by the group. The second deals with the formulation of policies whereby the group can alter these strategies based on experience. In the subsequent 5 sections the formal specification for both subsystems is presented.

Given that we can rephrase our informal base model in terms of a formal adaptive system, the fact remains that our base model, as pointed out previously, makes predictions that cannot be verified using presently collectible data. We would, therefore, like to be able to translate the formalized base model into yet another simplified model, one that utilizes currently available data and can be tested via computer simulation. Recent work on the theory of modeling and simulation (Zeigler 1976) suggests that in certain situations precise formal inference can be made about the properties of a model even when the collected data are insufficient to test the model in all its aspects. This is done by generating a simplified or *lumped* version of the original model that preserves selected structural and behavioral properties.

The formal connection between the base and lumped model is termed a *system homomorphism*, and its existence guarantees that (1) there is a precise relationship between the structure of both models and (2) there is a similarity between the behavior of both models when subjected to corresponding sequences of inputs. In other words, the existence of a system homomorphism guarantees that the lumped model, designed to make predictions comparable to known data, responds to input in the same fashion as does the original.

The key to insuring such a correspondence is to ascertain that each change made in producing the simplified model preserves the particular behavioral and structural correspondences sought by the modeler. In the section titled, "Simplifying the Base Model," we state the structural properties we wish to preserve in the simplification process, as well as the type of simplifications we want to make. Next, a sequence of lumped models is produced by the successive application of these model simplifications. The end result is a model that uses only currently collectible data and makes predictions easily comparable with existing data. In addition, we are able to prove formally that this lumped model preserves the desired set of structural and behavioral properties found in the original. Once this can be substantiated we need only to simulate *this lumped model* and observe its behavior with regard to the properties of interest.

The Formal Base Model: An Overview

As suggested by Lewontin (1978), "the modern view of adaptation is that the external world sets certain 'problems' that organisms need to 'solve' . . . . [a]daptation is the process of evolutionary change by which the organism provides a better and better 'solution' to the 'problem'." Holland (1975) has extended this biological notion of adaptation in order to develop a formal framework in which to pose a problem in adaptation for any arbitrary system. Within this framework a problem is considered to be well posed if a formal specification can be given of the system undergoing adaptation, the specific problems posed by its environment to which the system can adapt, the set of available policies that can generate new adaptations, and some criterion by which to evaluate the system's adaptive success.

The two major problems facing our group of hunter–gatherers are what collecting strategies to use and how to change them based on their performance. It is possible, however, that the group may adapt to each of these problems concurrently. Also, the two problems are interdependent in the sense that adaptations made in one problem domain may affect the nature of subsequent adaptations in the other. In most interesting problems in evolution, parallel adaptations will play an important part in determining the way in which a system develops. It is, therefore, quite important that we be able adequately to model such parallelism formally. In the remainder of this section we examine Holland's framework for adaptation in some detail, as well as suggest how our problem in concurrent adaptation might fit into such a framework.

The basic paradigm around which Holland's formalism is developed, concerns a system (S) that is able to alter its structure and/or behavior based on experience in some set of performance environments (E). Such a system can be specified as a collection of mathematical objects in the following way:

> by the set of objects $(\mathscr{A},\Omega,I,\tau)$ where
> $\mathscr{A} = \{A_1,A_2, \ldots\}$ is the set of attainable structures, the domain of action of the adaptive plan,
> $\Omega = \{\omega_1,\omega_2, \ldots\}$ is the set of operators for modifying structures with $\omega \in \Omega$ being a function of $\omega:\mathscr{A} \rightarrow \mathscr{P}$, where $\mathscr{P}$ is some set of probability distributions over $\mathscr{A}$,
> I is the set of possible inputs to the system from the environment, and
> $\tau:I \times \mathscr{A} \rightarrow \Omega$ is the adaptive plan which, on the basis of the input and structure at time t, determines what operator is to be applied at time t.
> Under the intended interpretation
>
> $$\tau(I(t), \mathscr{A}(t)) = \omega_t \in \Omega \text{ and } \omega_t(\mathscr{A}(t)) = \mathscr{P}(t + 1)$$
>
> where $\mathscr{P}(t + 1)$ is a particular distribution over $\mathscr{A}$. $\mathscr{A}(t + 1)$ is determined by drawing a random sample from $\mathscr{A}$ according to the distribution $\mathscr{P}(t + 1)$. Given the input sequence $<I(1), I(2), \ldots>$, τ completely determines the stochastic process. (Occasionally, when the adaptive system is to be deterministic with $\mathscr{A}(t + 1)$ being uniquely determined once $I(t)$ and $\mathscr{A}(t)$ are given, τ will be defined without the use of operators so that $\tau:I \times \mathscr{A} \rightarrow \mathscr{A}$. The structure of the adaptive system at time t, $\mathscr{A}(t)$, will be required to summarize whatever aspects of the input history are to be available to the plan. Hence it will often be useful to represent $\mathscr{A}$ as $\mathscr{A}_1 \times \mathscr{M}$, where $\mathscr{A}_1$ is the set of structures to be directly tested and $\mathscr{M}$ is the set of possible memory configurations, for retaining past history not directly incorporated in the tested structures. (Holland 1975:28, reprinted by permission of the University of Michigan Press)

Now that we have a general notion of what it takes to describe formally an adaptive system, just how can we express our informal model in this framework? It was mentioned earlier that the group of hunter–gatherers must be concerned not only with the problem of when to use a collecting strategy but also with the problem of when and how to change it. Each of these problems characterizes a different set of adaptations, and this suggests that we construct a specific adaptive system for each. These problems are not independent in the sense that the system's responses to one affects its responses to the other. So it is necessary that we model the interaction between these systems as well. A schematic diagram of the proposed system and interactions is given in Fig. 31.1. The interactions are of the following form:

1. At the beginning of each model cycle or iteration the group selects a subset of available strategies to use between August and November. The relative performances of each in terms of bulk plant material acquired and the effort taken to acquire it is gauged.
2. Changes in the relative performances of strategies that have been modified since they were last used are noted. These changes are then used to evaluate the effectiveness of the decision-making policy that produced them.

Those policies that produced improvements are more likely to be preserved than those that do not.

3. This information about strategy performance is then used to generate a new set of adjusted decision-making policies.
4. These new decision-making policies are used to alter selected members of the current sample of strategies.

With this general scheme in mind, let us now see how much of these two systems might be defined in terms of the informal model given in Part 4. The most natural subsystem to start with is the one dealing with the specification and selection of collecting strategies. This system, (S^C), is represented formally as a collection of mathematical objects $\{\mathcal{A}^C, \Omega^C, I^C, \tau^C\}$, where each formal object characterizes portions of the informal model in the following ways.

The set of possible collecting strategies available to the group is represented by $\mathcal{A}^C$. Each strategy specifies a sequence of collecting tasks over a 10-day period from August through November. At any one point in time only a subset of those possible strategies are actually in use.

The set of mechanisms by which the system alters the structure of its collecting strategies based on experience is represented by Ω^C. Here the mechanisms are effectively decisions to alter certain aspects of the current set of strategies.

The information available to members of the group at each model time step is specified by I^C. In our model, this information concerns the relative performance of strategies that are selected for use by group members, as well as some information about the state of the environment when the strategy was last used.

The transition function τ^C describes two basic processes. The first, τ_1^C, models the selection of a set of strategies to be used based on their past relative performances and how well the climate in which they were last used matches the current climate. The second, τ_2^C, describes the changes made to each of the sampled strategies based on their performance. The exact nature of these changes depends on the current decision-making policies held by group members with regard to each of these strategies. How the group members are able to generate these policies based on experience is the concern of the next subsystem we discuss.

The decision-making subsystem, S^D, can be specified by the set of mathematical objects $\{\mathcal{A}^D, I^D, \Omega^D, \tau^D\}$. Here $\mathcal{A}^D$ represents the set of possible decision-making policies that could be held by group members regarding a strategy. At any point in time, only a subset of these are actually used. Every policy is expressed in terms of the probability of applying each one of the basic decisions.

The set I^D represents the possible information that could be used by members of the group to evaluate the effectiveness of their current policies. At each time step the group acquires information about the effects of past decisions on the current performance of the sampled strategies. The policies associated with each of these may then be adjusted depending on whether the policy has produced an increase or decrease in the performance of the strategies since last used.

The basic adjustments that can be made to a decision-making policy based on its peformance is described as Ω^D. The nature of these adjustments was discussed in Part 4.

The transition function for the decision-making subsystem, τ^D, details how the above operators are used based on their performance. In general, the better the schedule's relative performance, the less likely it will be subject to change by the group.

Given the two subsystems, it now remains for us to specify the set of environments, E, in which they can perform. As in the informal model, the current plant densities and distributions are seen to be a function of annual rainfall. Three general classes of performance environments are then defined to be wet, dry, and average on the basis of total rainfall. The basis for this classification was given earlier in this chapter. Associated with each type of year is a characteristic distribution of plant densities that is conditional on the preceding year type. Since there are three basic year types, it follows that there will be 3^2, or 9, possible performance environments. The exact nature of the differences between each are discussed later.

We have now given a general overview of how our informal model can be rephrased formally as a set of problems in parallel adaptation. Our new perception of the basic model can be seen in Fig. 31.2. Note that our entire system can be described in terms of two interacting adaptive subsystems. Taken together their associated transition functions represent the transition function for the whole system.

In addition we have indexed each aspect of the model by the variable referred to as "model time." That is, $\mathcal{A}^C(t)$ can be referred to as the current set of collecting strategies at time t. Applying the transition function τ^C to $\mathcal{A}^C(t)$ produces a new set of strategies that is now designated as $\mathcal{A}^C(t + 1)$. Note that the model time is just a method of indexing the succession of group structures generated by the model. It does not necessarily have to have any direct correspondence to real time. In actuality, the length of time between each evaluation cycle may vary, and there is no information that presently allows us to describe such variation. However, since the main purpose of the model is to discern the equilibrium mix of strategies, the time between evaluations is not important here. Thus a model time of 1 between each successive transition is used for convenience.

Even though the model time does not directly correspond to real time, if the model is a valid representation of the decision-making adaptations of the group, then the changes in equilibrium structures produced by the model when incipient agriculture is introduced should correspond in some reasonable fashion to what we know from the archaeological record. This is a topic of interest later in this chapter.

Formal Model Development: Strategy Specification

Now that we have a general impression of what the model's structure will look like, it is time to construct a more specific realization of the base model. First we must have a way to

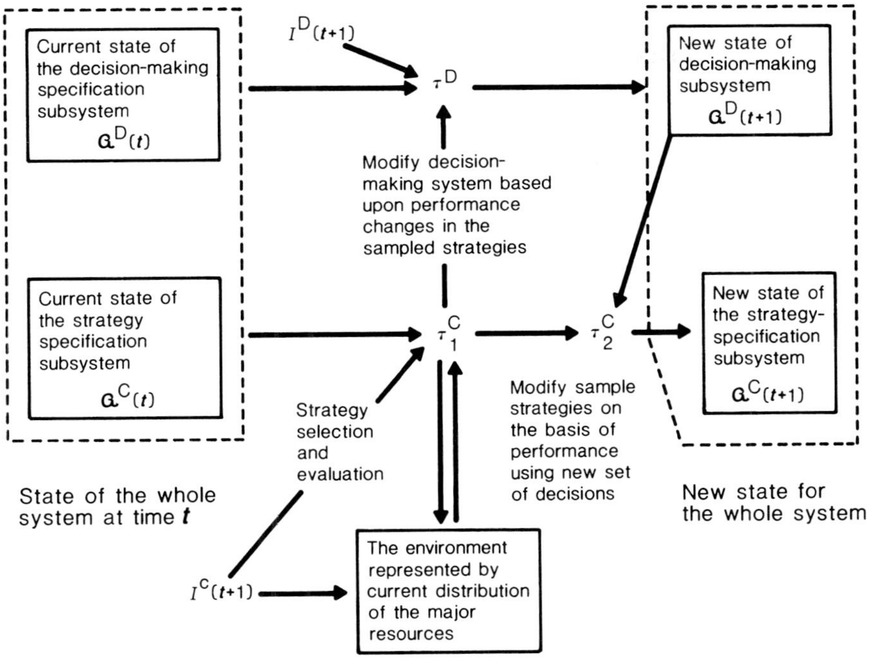

Fig. 31.2. Revised specification of the resource scheduling system.

represent the structure of our adaptive system. The representation should be sufficient to describe all relevant changes to the system during the period of interest. To construct such a representation the system can be thought of as comprising subsystems, where each of these is described via a corresponding set of descriptive variables. Thus, if we subdivided an arbitrary system into N subsystems, each resulting system S_i (where $i \in \{1, \ldots N\}$) would have an associated set of M_i descriptive variables. The number of descriptive variables for subsystem S_i is represented by M_i, since not all components will necessarily have the same variable set. The sum of these terms

$$\sum_{i=1}^{N} M_i,$$

is the total number of variables used to describe the entire system. We refer to this number as K. Each of these M_i variables will be indexed so that the jth descriptive variable in the ith subsystem is referenced as D_{ij}. Associated with this variable is its *range set*, which is the range of measured values that the variable can have for the system. The range set for an arbitrary variable D_{ij} is represented as R_{ij} and corresponds then to a set of allowable values for the variable.

An arbitrary system that was first decomposed into an indexed set of n interconnected systems, can now be respecified by replacing each subsystem with its associated descriptive variables as shown below.

$$\begin{gathered} S \\ \Downarrow \\ S_1, S_2, \ldots, S_n \\ \Downarrow \\ D_{11}, D_{12}, \ldots, D_{1m_1}, D_{21}, \ldots, D_{nm_n} \end{gathered}$$

The Cartesian product, or cross product of this collection of sets, denoted by $D_{11} \times D_{12} \times \ldots D_{nm_n}$, is the set of all K-tuples $\{<d_{11}, d_{12}, \ldots>|d_{ij} \in R_{ij}\}$ formed by selecting a particular measured value d_{ij} for each descriptive variable. This stands for the set of all states that our system might have during the course of our observation.

In the previous section we initially decomposed our system into two subsystems, each representing a certain type of adaptation. Now we develop the descriptive variables for both systems, beginning with the strategy-specification subsystem. These descriptive variables together characterize the elements of $A^C(t)$, the current set of strategies.

As mentioned previously our descriptive variables should detail the *observable* results of using a strategy or making a decision. They describe to us the current behavior of the system and, therefore, can be expressed via our choice of measurements. For example, we can describe the area searched by a strategy in terms of hectares even though the group members probably represented this in a much different manner. We do presume, however, that there exists a correspondence between the internal structuring of this information and the external behavior.

First, recall that in our base model a scheduling strategy is represented by a set of activities, or *tasks*. Associated with each task is information about the vegetational zone in which it occurs, what the focal resource is, which locations are to be searched, and how much of the resource is present at each location. In addition, this set of tasks is ordered, where the ordering reflects their relative sequence of execution. These tasks when taken together specify a sequence of collecting activities for the group over a 10-day period from August into November.

The model, therefore, presupposes that this collection of variables is effectively specified for each task. It means that the information used to specify a collecting activity effectively associates with that activity certain properties. These are properties that characterize the task's execution and may not be directly represented in the mind of an individual member of the group. They are, however, the observable results of the mentally specified task and therefore constitute the descriptive variables for our strategy-specification subsystem.

Since we are dealing with a segment of the behavioral repertoire of a small group, it is to be expected that certain limits will exist in terms of the total number of tasks and associated strategies that can be specified by the group. The model assumes that at any one point in time there is an upper bound on the number of strategy specifications retained by the group; we designate this MAX_STRAT. Here the maximum number of schedules is set equal to 10. We might expect, also, that an individual strategy will consist of no more than some maximum number of tasks since the strategy's specification is over a very limited period of time—10 days. This maximum is designed as MAX_TASKS in the model.

We are now in a position to list the descriptive variables that represent the strategy-specification subsystem for the model. The strategy-specification component descriptive variables are as follows:

1. MAX_STRAT is the maximum number of currently active strategies. Here the maximum number of currently active strategies is taken to be 10.
2. MAX_TASKS is the maximum number of tasks associated with a given strategy. The maximum number of tasks for this model is taken to be 25.

For every STRAT(i) • TASK(j) where $i \in \{1, \ldots$ MAX_STRAT$\}$ and $j \in \{1, \ldots$ MAX_TASKS$\}$ the following variables are included in the model:

3. STRAT(i)•TASK(j)•ZONE specifies the vegetation zone in which the activity is carried out. It can take values from the set {Mesquite Grassland B, Thorn Forest A, Thorn Forest B, Mesquite Grassland A}.
4. STRAT(i)•TASK(j)•RESOURCE represents the focal resource for the collecting activity. Depending on the zone, this variable can take values from the set of major plant resources {piñon nuts, hackberries, *susí* nuts, nanche fruit, *Opuntia* nopales, *Opuntia* fruits, acorns, *guaje* pods, mesquite pods, *Agave*, beans}.
5. STRAT(i)•TASK(j)•LOCATIONS_TO_SEARCH represents the set of locations in a particular zone that are searched during the collecting activity. This takes values from the set of all allowable subsets of locations with respect to the cave for each zone. The total area encompassed by these locations cannot refer to more than 0.33 ha. Since 0.33 ha is the maximum area that the group is allowed to search in a day, this means that one task can specify no more than one day's activities. Also we presume that the locations searched for each task in a strategy are not included in tasks found either in that strategy or in other current strategies. Since the minimum size of a location is taken to be 0.001 ha, this condition seems quite reasonable.

In addition to the set of variables that describe the sequence of plant collecting activities, a strategy can also be specified in terms of its performance. Recalling the informal base model, two pieces of information were found to be important in determining which strategies to employ during any one year. In making such a decision, one would first need to know how well a strategy performed the last time it was used. Since the performance for a strategy is a function of the environment, one also needs information about the type of environment in which it was last used in order to compare its performance with other strategies. Therefore, for each strategy i there is a variable STRAT(i)•LAST_USED that stores the last year type in which the strategy was used. Accordingly, there will be information about how the strategy performed the last time it was used, STRAT(i)•LAST_PERF. A discussion of the exact nature of this performance index is deferred until later when we discuss the performance function.

Since a strategy has been selected for use at time t, we need to record its current performance. This variable is labeled STRAT(i)•PERF and takes on a value for those strategies used at time t. As mentioned in the previous chapter, this represents a derived impression of the schedule's relative performance in terms of a number of criteria. While we have no information as to how these impressions were derived, it is quite conceivable that they were generalized descriptions of each strategy's performance. Over the course of the season, a general set of expectations regarding the performance of a strategy is developed. Here we represent such expectations by the strategy's average performance. For each utilized strategy i, the following expectations are produced:

1. STRAT(i)•CALORIES is the average yield in terms of calories
2. STRAT(i)•PROTEIN is the average yield in terms of protein
3. STRAT(i)•CALORIES_EFFORT is the average effort expended relative to the amount of calories acquired. Here this is represented by dividing the number of acquired calories by the area searched, STRAT(i)•CALORIES/STRAT(i)•AREA_SEARCHED. STRAT(i)•AREA_SEARCHED is computed by adding together all the individual locations searched
4. STRAT(i)•PROTEIN_EFFORT is the average effort expended relative to the amount of protein acquired. This is represented by dividing the amount of acquired protein (in grams) by the area searched, STRAT(i)•PROTEIN/STRAT(i)•AREA_SEARCHED.

Next, expectations for each strategy are then compared with those for the other sampled strategies as well as with the basic energetic needs of the group. This produces a generalized impression of the strategy's overall performance for the year, STRAT(i)•PERF. These comparisons are made in the following way for each strategy i:

1. Is the strategy's caloric yield, STRAT(i)•CALORIES, less than the average for the sample? This information is obtained by averaging over all the sampled performances for this variable. If the answer to the above question is yes, then add 1 to the performance index STRAT(i)•PERF. The initial value for this variable is set to 0.
2. Is the strategy's protein yield, STRAT(i)•PROTEIN, less than the average for the sample? If so, then increment its performance index by 1.
3. Is the effort expended to acquire the current number of calories greater than the average for the sample? If it is, then increment the performance index by 1.
4. Is the effort expended to acquire the current amount of protein in excess of the average for the sample? If it is, then increment STRAT(i)•PERF by 1.
5. Does the yield in calories fall below the group's minimal requirements in terms of calorie yield? If it is not able to meet these minimal energetic requirements, then increment STRAT(i)•PERF by 1.
6. Does the yield in protein fall below the group's minimal protein requirements? The performance index is again incremented by 1 if it does.

The performance index therefore stands for an integrated impression of a strategy's expected performance with respect to both the performance of the sample as a whole and the group's minimal performance requirements. It takes a range of integer values from 0 to 6, where 0 represents the best possible performance for a strategy in a sample and 6 denotes the worst.

Our strategy specification subsystem, therefore, includes several additional variables that describe the group members' impression of the performances for the current set of scheduling strategies. For each STRATEGY(i) where $i \in \{1, \ldots \text{MAX\_STRAT}\}$ the following descriptive variables are included in the model:

1. STRAT(i)•LAST_USED stores the type of year in which the schedule was last used. It will take values from the set of possible year types {dry, wet, average} given in Part 4.
2. STRAT(i)•PERF stores the current impression of the strategies' performance. It takes values from the set of integers $\{0, \ldots, 6\}$.
3. STRAT(i)•LAST_PERF records the impression of the strategies' relative performance the last time it was used. It also takes integer values from 0 to 6.
4. STRAT(i)•PROTEIN represents the total protein yield for the strategy. It takes a range of values from R^+ (the positive reals).
5. STRAT(i)•CALORIE represents the total caloric yield expected for the strategy. It takes values from R^+.
6. STRAT(i)•PROTEIN_EFFORT stands for the effort expended relative to the amount of protein acquired. It can take values from R^+.
7. STRAT(i)•CALORIE_EFFORT stands for the effort expended relative to the number of calories acquired. The range of this variable is over R^+.
8. STRAT(i)•AREA_SEARCHED represents the sum over all the locations searched by the strategy. It takes values over the positive reals.
9. STRAT(i)•USED takes values from the set {yes, no} and indicates whether or not the strategy was used during the current time step.

We have completed our description of the variables associated with the specification of the collecting strategies and their associated performances. However, it remains for us to specify the group's decision-making subsystem. Remember that a decision-making policy assigns to each of the available decision-making operators a certain probability of being applied to each current strategy. Given a particular set of strategies, certain policies will be more successful than others in generating improvements to this set. Therefore, one critical aspect of the adaptation process is the ability of the group to adjust its current policies to fit its present set of collecting structures using information from its environment. The *observable* result of this adjustment process is to alter the frequency with which certain rescheduling decisions are made. In this model a rescheduling policy for a specific schedule is expressed in terms of the frequency with which certain rescheduling decisions are made. Altering these frequencies of application represents changing the associated rescheduling policy. In this way we are able to concern ourselves only with the observable results of a policy rather than how that policy was represented in the minds of certain individuals. However, before we can discuss how these policies are described in model terms, we first need to define the basic set of decisions to which a policy can refer.

Basic Rescheduling Operators

The set of operators by which an adaptive system can change its structure is represented by Ω^D. Here it constitutes an elementary set of strategy modification decisions as described in Part 4. This set can be subdivided into two broad classes. The first group consists of those operators that deal with recombining existing specifications (analogous to crossover-type operations in genetics). One type of recombination is to exchange two activities between different scheduling strategies. Another way is to exchange the sequence of execution for two activities within a strategy. At an even more detailed level, specifications for two activities (such as location) can be exchanged between two existing activities in the same schedule.

The second group is composed of decisions that can generate new values for the descriptive variables associated with an activity. While the first set of operators reflected the exchange of *existing* ideas, these operators reflect the production of *new* ones. These operators in conjunction with the recombination-type operations are enough to describe the basic decisions given in Part 4.

In this model we are not concerned with specifying the precise set of circumstances that produce a decision, only its likelihood of occurrence given certain key variables. Here this likelihood is expressed in terms of probabilities. Each rescheduling decision is conceived of as a stochastic process with a certain probability of occurring. Associated with each process is an ideal random number generator with a state set in the interval [0,1] and having a transition function of the form RAND:[0,1] → [0,1]. Given an initial seed r_0, this function will produce a sequence of numbers $< r_0, r_1, r_2, r_3, \ldots >$ that appear to be uniformly distributed throughout the interval [0,1]. Every number in this sequence appears to be unrelated to those previously sampled. This number is translated into an associated model event by a random variable that takes the form of a function mapping the numbers in [0,1] into a set of outcomes. Here the set will be {yes, no}. Once a decision to reschedule a strategy has been made, a number of other random variables may come into play depending on the exact nature of the decision. We now specify each of these decisions in detail, beginning with the recombination operators.

ω_1 = SWAP_EXTERNAL: This operator represents the exchange of tasks between strategies. This exchange is a stochastic process mediated by three random variables in the following way:

1. Given a pseudorandom number generator with the transition function RAND:[0,1] → [0,1], apply this mapping to the current value for the SWAP_EXTERNAL_SEED to produce a real number between [0,1], r_1. In addition, it will generate a new value for EXTERNAL_SEED.
2. Next apply the SWAP_EXTERNAL_DECISION map to r_1 to obtain the sampled SWAP_EXTERNAL_DECISION. SWAP_EXTERNAL_DECISION is, therefore, a random variable that takes real numbers in [0,1] into the set {yes, no} in the following way:

$$\text{STRAT}(i) \cdot \text{OPER(1)} \cdot \text{USED} = \begin{cases} \text{“yes” if } r_1 \leq \text{STRAT}(i) \cdot \text{OPER(1)\_PROB and} \\ \text{“no” if otherwise.} \end{cases}$$

   STRAT(i)•OPER(1)_PROB is the current probability that the swap external operation will be performed on this strategy.
3. If the decision is yes then the following processes occur. First a new random number, r_2 is produced, RAND:(PARTNER_SEED) → $r_2 \in [0,1]$. This value is used to determine which other member of the current sample is to participate in the exchange by applying the SWAP_PARTNER map to r_2. SWAP_PARTNER:[0, 1] → set of sampled strategies excluding STRAT(i). This map corresponds to dividing the unit interval into K subintervals of equal length, where K is the cardinality of the range set. The index of the subinterval into which r_2 falls is the index of the strategy to be selected.
4. Next a third random number, r_3, is generated. The ORDER map is applied to r_3 in the same manner as above. ORDER[0,1] → {1, . . . MAX TASKS}. This determines the execution order of the tasks to be exchanged.
5. Finally, the selected tasks with a given execution ORDER are exchanged between STRAT(i) and STRAT(PARTNER) in a manner that depends on their performance indices. If the two strategies have similar performance indices, |STRAT(i) • PERF_STRAT (PARTNER)•PERF| ≤ 2, then the selected tasks are exchanged between them. However, if their difference is 3 or more, it means that one performed better in at least two of the three basic performance categories (effort, yield, and minimal yield). As such, this represents more than just a slight difference in performance. In this situation the poorer performer replaces its current task with a copy of a task from the other strategy.

$$\text{SWAP\_EXTERNAL} = \begin{cases} \text{If } |\text{STRAT}(i)\cdot\text{PERF}-\text{STRAT (PARTNER)}\cdot\text{PERF}| \leq 2|, \\ \text{then} \\ \quad (\text{STRAT}(i)\cdot\text{TASK(ORDER), STRAT (PARTNER)}\cdot\text{TASK(ORDER)}) \rightarrow \\ \quad (\text{STRAT}(i)\cdot\text{STRAT(PARTNER)}\cdot\text{TASK(ORDER), STRAT(PARTNER)}\cdot(\text{STRAT}(i)\cdot\text{TASK(ORDER)})). \\ \text{else if} \\ \quad \text{STRAT}(i)\cdot\text{PERF} > \text{STRAT(PARTNER)}\cdot\text{PERF}, \\ \text{then} \\ \quad (\text{STRAT}(i)\cdot\text{TASK(ORDER), STRAT(PARTNER)}\cdot\text{TASK(ORDER)}) \rightarrow \\ \quad (\text{STRAT}(i)\cdot\text{STRAT(PARTNER)}\cdot\text{TASK(ORDER), STRAT(PARTNER)}\cdot\text{TASK(ORDER)}); \\ \text{else} \\ \quad (\text{STRAT}(i)\cdot\text{TASK(ORDER), STRAT(PARTNER)}\cdot\text{TASK(ORDER)}) \rightarrow \\ \quad (\text{STRAT}(i)\cdot\text{TASK(ORDER), STRAT(PARTNER)}\cdot(\text{STRAT}(i)\cdot\text{TASK (ORDER)})). \end{cases}$$

This operator is described schematically for two arbitrary strategies in Fig. 31.3.

One might suggest that our operator is overly restrictive since, realistically, exchanges can presumably be made between TASKS at different levels of priority. We show later that our system, by combining this operator with other elementary operations, can represent these types of decisions as well as more complicated ones. Thus we are able to generate a hierarchy of increasingly complex decision-making rules based on our elementary operators in Ω^D. This provides us with a very powerful tool in the analysis of the model's decision-making behavior.

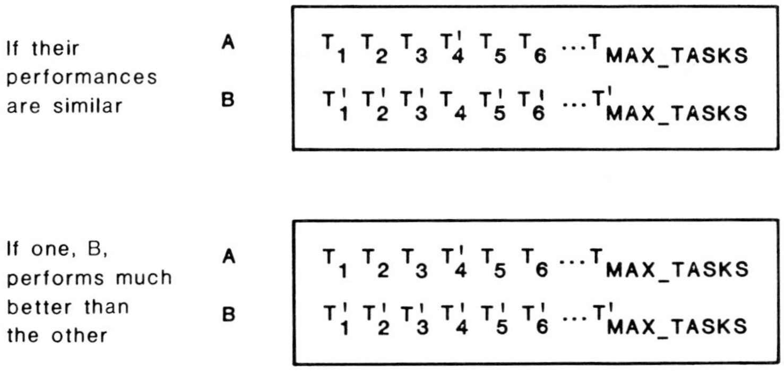

Fig. 31.3. The SWAP_EXTERNAL operator exchanges TASKS with the same order of execution between two strategies with similar performances. If one strategy is a much better performer, then a copy of its TASK is made.

ω_2 = SWAP_INTERNAL: This stochastic operator performs an exchange of tasks within a strategy. Intuitively, it reflects a decision to exchange the order of execution between two tasks. One reason for doing this is to reorder the schedule so that several activities in the same zone are executed in sequence. This could lead to a reduction in the distance traveled by the group members involved.

The SWAP_INTERNAL operator is mediated by three random variables, each of which specifies an aspect of the overall process in the following way:

1. Input an initial seed, SWAP_INTERNAL_SEED, into the pseudorandom number generator, which maps this into $r_4 \in [0,1]$.
2. Next, apply the SWAP_INTERNAL_DECISION map to r_4 to obtain the sampled decision SWAP_INTERNAL_DECISION. This map represents a random variable that takes numbers in [0,1] into the set {yes, no} in the following way for a sample strategy i:

$$\text{STRAT}(i) \bullet \text{OPER(2)\_USED} = \begin{cases} \text{“yes” if } r_4 \leq \text{STRAT}(i)\bullet \\ \text{OPER(2)\_PROB and} \\ \text{“no” otherwise.} \end{cases}$$

   STRAT(i) • OPER(2)_PROB is the current probability of performing this rescheduling operation on strategy i.
3. If SWAP_INTERNAL•(STRAT(i) = yes, the process continues; otherwise no exchange is made. Assuming that the decision is in the affirmative, we now must select the two tasks to exchange. Using another seed, TASK_1_SEED, we derive a new random number r_5. Next, we define TASK_1 to be a random variable that maps r_5 into the set of available tasks for Strategy i: TASK_1: $[0,1] \rightarrow \{T_1, T_2, \ldots T_{\text{MAX\_TASKS}}\}$.
   This is done by subdividing the unit interval [0,1] into 10 subintervals of equal length, where the jth subinterval corresponds to the probability of selecting the jth task. The index of the subinterval into which r_5 falls is then the index of the task to be selected.

   The second task to be exchanged is selected in the same manner as above using the variables TASK_2_SEED, r_6, and TASK_2.
4. The SWAP_INTERNAL operator then exchanges the ORDER of execution for the two selected tasks in STRAT(i) as shown schematically in Fig. 31.4.

ω_3 = SHUFFLE_LOCATION: This operator symbolizes the decision to exchange locational specifications between two activities in a strategy. Associated with each activity is a set of locations where the activity is carried out. This operator exchanges subsets of locations between activities on the same schedule, providing that these activities are located in the same zone. It is a stochastic operation that is specified in the following way for a given strategy i:

1. First, SHUFFLE_LOCATION_SEED is used to generate a random number r_7. SHUFFLE_LOCATION_DECISION is defined to be a random variable that maps r_7 into the set {yes, no} for a given strategy i.

$$\text{STRAT}(i)\bullet\text{OPER(3)\_USED} = \begin{cases} \text{“yes” if } r_7 \leq \text{STRAT}(i) \bullet \\ \text{OPER(3)\_PROB and} \\ \text{“no” otherwise.} \end{cases}$$

   STRAT(i) • OPER(3)_PROB represents the current probability of making this decision for Strategy i.
2. Next the two tasks that are to be involved are determined stochastically in the same manner as given for the SWAP_INTERNAL operator defined previously. Here SHUFFLE_TASK_1_SEED and SHUFFLE_TASK_2_SEED are used to generate random numbers r_8 and r_9, respectively. These are then mapped by their respective random variables, SHUFFLE_TASK_1 and SHUFFLE _TASK_2, into the set of available tasks for Strategy i as before.
3. Now the SHUFFLE_LOCATION operator is applied to exchange the LOCATION sets associated with the two selected tasks for Strategy i.

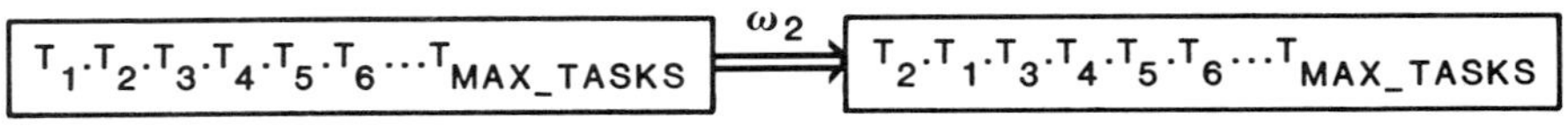

Fig. 31.4. The SWAP_INTERNAL operator describes the exchange of execution priorities between TASKS. Here the execution sequence for the first two TASKS is reversed.

$$\text{SHUFFLE\_LOCATION:}\begin{cases}\text{STRAT}(i)\cdot\text{TASK(SHUFFLE\_TASK\_1)}\\ \cdot\text{LOCATIONS} \rightarrow \text{STRAT}(i)\cdot\text{TASK}\\ \text{(SHUFFLE\_TASK\_2)}\cdot\text{LOCATIONS}\\ \text{and}\\ \text{STRAT}(i)\cdot\text{TASK(SHUFFLE\_TASK\_2}\\ \cdot\text{LOCATIONS} \rightarrow \text{STRAT}(i)\cdot\text{TASK}\\ \text{(SHUFFLE\_TASK\_1)}\cdot\text{LOCATIONS.}\end{cases}$$

In order to perform the mapping, it is presumed that the resulting tasks are well defined. In other words, the operator is constrained to exchange locations that make sense for the currently specified TASKS, since one does not want to allow exchanges that associate a plant species with a location where it does not grow. Therefore, this operator changes only allowable combinations of TASK variables into other allowable combinations. Such constraints apply not only to this operator but also to all that follow.

While SHUFFLE_LOCATION does serve to change the nature of an activity, the change is somewhat conservative in that it only involves information already used in some form to specify the schedule. For example, the shifts in locations produced by SHUFFLE_LOCATION for a schedule will only be in terms of locations already associated with the schedule. This is the type of change represented by each of the shuffle operations to be described next.

ω_4 = SHUFFLE_ZONE: This stochastic operator represents the exchanging of zone specifications between compatible activities in the same schedule. Since the location set is zone specific, this operator involves more than just exchanging zones. For each participating task, if the new zone differs from the old then a new set of locations in the new zone must be produced. The total area encompassed by these new locations is constrained to be the same as before for each strategy. The process for some strategy i goes as follows:

1. Initially the SHUFFLE_ZONE_SEED is used to produce a random number r_{10} that is then mapped by SHUFFLE_ZONE_DECISION into the set {yes, no}.

$$\text{STRAT}(i)\cdot\text{OPER(4)\_USED} = \begin{cases}\text{“yes” if } r_{10} \le \text{STRAT}(i)\cdot\\ \text{OPER}(j)\text{\_PROB and}\\ \text{“no” otherwise.}\end{cases}$$

   As before, STRAT(i)•OPER(j)_PROB represents the probability that the group will decide to make the change in light of its current policy for Strategy i.
2. If the decision is to reschedule the strategy in this way, then the tasks involved are determined stochastically. This is done via two seeds, SHUFFLE_ZONE_TASK_1_SEED and SHUFFLE_ZONE_TASK_2_SEED. These seeds generate random numbers r_{11} and r_{12} which are then mapped into the sets of available tasks by the random variables SHUFFLE_ZONE_TASK_1 and SHUFFLE_ZONE_TASK_2 in the same manner as for SHUFFLE_LOCATION.
3. This is followed by the application of the SHUFFLE_ZONE operator that exchanges the zones where these tasks are carried out.

$$\text{SHUFFLE\_ZONE:}\begin{cases}\text{STRAT}(i)\cdot\text{TASK(SHUFFLE\_ZONE\_}\\ \text{TASK\_1)}\cdot\text{ZONE} \rightarrow \text{STRAT}(i)\cdot\text{TASK}\\ \text{(SHUFFLE\_ZONE\_TASK\_2)}\cdot\text{ZONE}\\ \text{and}\\ \text{STRAT}(i)\cdot\text{TASK(SHUFFLE\_ZONE\_}\\ \text{TASK\_2)}\cdot\text{ZONE} \rightarrow \text{STRAT}(i)\cdot\text{TASK}\\ \text{(SHUFFLE\_ZONE\_TASK\_1)}\cdot\text{ZONE.}\end{cases}$$

4. In addition, if the two zones that are exchanged are different, then two random numbers r_{13} and r_{14} are generated from AREA_SEED_1 and AREA_SEED_2. Each of these is mapped into a subset of points for the new zone that encompasses the same amount of area as the previous set. Each subset of available points is assigned a subinterval of equal length over the real line between 0 and 1. The new subsets are the ones whose intervals overlap with the numbers r_{13} and r_{14}, respectively.

ω_5 = SHUFFLE_RESOURCE: This operator exchanges the focal resources between two compatible activities on the same schedule. The process is mediated by three random variables that determine the basic characteristics of this decision for a strategy i:

1. The current SHUFFLE_RESOURCE_SEED is used to produce a random number r_{15}. The random variable STRAT(i)•OPER(5)_USED maps r_{15} into the set {yes, no} and is the same as for the previous operator.
2. If the value for STRAT(i)•OPER(5)_USED is yes, then SHUFFLE_RES_SEED_1 and SHUFFLE_RES_ SEED_2 are used to generate two random numbers r_{16} and r_{17}. These are mapped into the set of available tasks for Strategy i by the random variables SHUFFLE_RES_TASK_1 and SHUFFLE_RES_TASK_2.
3. Finally, the SHUFFLE_RESOURCE operator is applied to these two selected tasks to produce the required scheduling change:

SHUFFLE_RESOURCE:
- STRAT(i)•TASK(SHUFFLE_RES_TASK_1)•ZONE → STRAT(i)•TASK(SHUFFLE_RES_TASK_2)•ZONE
- and
- STRAT(i)•TASK(SHUFFLE_RES_TASK_2)•ZONE → STRAT(i)•TASK(SHUFFLE_RES_TASK_1)•ZONE.

This closes out our discussion of those rescheduling decisions that involve recombining existing information within and between strategies. Now we turn our attention to a set of operators that represent the outright change of a descriptive variable associated with a strategy.

ω_6 = ALTER_RESOURCE: This operator models those rescheduling decisions that consist of replacing the focal resource for an activity with another that is compatible with the current vegetation zone for the activity. Unlike previous operators, this allows the introduction of resources that are previously not employed within any current strategy into a schedule. Formally, it can be described as follows for some strategy i:

1. ALTER_RES_SEED is used to generate r_{18}. STRAT(i)•OPER(6)_USED then maps r_{18} into the set {yes, no} in the same fashion as with the previous operators.
2. If STRAT(i)•OPER(6)_USED = yes, then we proceed. Another seed ALTER_RES_TASK_SEED is used to produce r_{19}, which is mapped into the set of available tasks for Strategy i by ALTER_RES_TASK. This designates the task to be changed.
3. Then RES_VALUE_SEED is used to derive r_{20}. This value is mapped by NEW_RES_VALUE into a resource compatible with the present location specification by assigning each compatible resource to a subinterval in [0,1] of equal length. The index of subinterval into which r_{20} falls is, therefore, the index of the new resource. This NEW_RES_VALUE is then assigned to STRAT(i)•TASK(ALTER_RES_TASK)•RESOURCE.

ω_7 = ALTER_ZONE: The function of this rescheduling operator is to represent a change in the vegetation zone for the task. This is only possible if the present resource is found in another zone. The operator is able, therefore, to describe decisions that shift resource collecting activities from one zone to another.

1. For a given strategy (i), employ ALTER_ZONE_SEED to get r_{21}, which is then mapped by STRAT(i)•OPER(7)_USED into the set {yes, no} in the same manner as the other operators.
2. If the decision is yes, then we proceed to select a task to be changed. This is done using the ALTER_ZONE_TASK_SEED to produce r_{22}, which is then mapped by ALTER_ZONE_TASK into one of the available tasks. If the focal resource for that acitivity can be found in another zone, we continue by generating r_{23} via NEW_ZONE_SEED. This number is then mapped into the set of zones compatible with the focal resource in our standard manner.
3. Once this is done we use NEW_ZONE_LOCATION_SEED to get r_{24}. The random variable NEW_ZONE_LOCATION maps this number into a set of locations with the same amount of area in the new zone. The details of this mapping need not concern us now.

ω_8 = ALTER_LOCATION: This represents rescheduling changes that involve the relocation of a collecting activity within a zone. Given a current set of locations the operator can modify a subset of these to produce a new set in the following way.

1. First, the decision is made in light of current policy for the strategy regarding whether or not to shift locations. Here this decision is represented stochastically by the generation of a random number, r_{25}, from ALTER_LOC_DECISION_SEED. If the value of r_{25} is less than or equal to the current probability of such a decision being made for this strategy, then the random variable STRAT(i)•OPER(8)_USED is assigned the value "yes"; otherwise it is given the value "no."
2. Provided that the decision is in the affirmative, the task to be changed is selected at random. ALTER_LOC_TASK_SEED produces r_{26}, mapped into the set of 25 available tasks by the random variable ALTER_LOC_TASK.
3. Next ALTER_LOC_VALUE_SEED is input to the pseudorandom number generator and r_{27} is produced. This random number is then mapped into a new subset of available locations where each subset is assigned a subinterval in [0,1]. The new subset is the one whose interval overlaps with the value for r_{27}.

The above set of stochastic operators represents the basic set of decisions that we can expect a group to make with respect to modifying its collecting strategies. However, these are not the only decisions that *can* be represented in the model. For example, by applying these decisions in sequence, we can represent more-complex operations as demonstrated in Fig. 31.5. These two operators are applied in sequence to a strategy, and the resulting strategy can be thought of as being produced by $\omega_1 \bullet \omega_2$. We signify this sequential application of operators by a "•", which signifies the successive composition of the basic operations. By this means we can represent more-complex modification decisions in terms of our original set.

Therefore, rather than specify the types of complex decisions available to the group, we have defined a simple set of basic decisions and allow the group to combine them in any manner. The group through its own experience is able to generate complex decisions that prove to be advantageous. As illustrated in Fig. 31.6, these complex operators can be

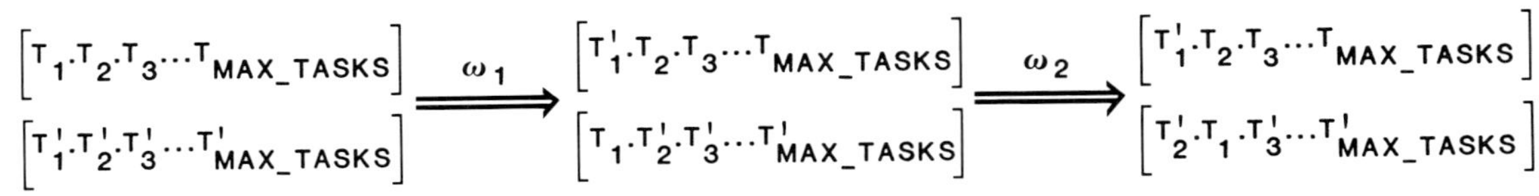

Fig. 31.5. The composition of two operators $\omega_1 \cdot \omega_2$ for two strategies.

n+1 order	. . .	. . .	. . .
Third order	$\omega_8 \cdot \omega_6 \cdot \omega_4$	$\omega_1 \cdot \omega_2 \cdot \omega_3$	$\omega_6 \cdot \omega_1 \cdot \omega_2$
Second order	$\omega_6 \cdot \omega_8$	$\omega_1 \cdot \omega_1$	$\omega_1 \cdot \omega_6$
First order	ω_6	ω_1	-none
Type	**Mutation**	**Recombination**	**Hybrid**

Fig. 31.6. Hierarchy of higher-order decisions achieved by the successive application of operators. The order of a decision represents the number of basic operations required to produce it. While only three orders are illustrated here, many more are possible.

of three types. In the first class all decisions in the sequence are recombination-type operators. Their composition produces a more complex recombination decision. Similarly the second group stands for more-complicated mutation operations. The third class, however, is new. It is a hybrid of the first two. In it one finds decisions that contain both mutation- and recombination-style operators. The *order* of a decision reflects in model terms the minimum number of basic decisions required to represent it. Later, in analyzing the simulation results we are able to observe the exact nature of the hierarchy generated by the group during each phase of its adaptation.

Model Group Rescheduling Policies

Now having described the set of operators, Ω^D, we again turn our attention to the specification of a decision-making policy. Implementing such a policy means that certain operators will be applied more freqently than others. In the model, a policy is specified in terms of the relative probabilities with which each of the basic set of operators is to be used to modify a particular strategy. Therefore, with every current strategy we associate a set of descriptive variables that reflect the probability that each operator will be applied to it under the current policy.

In addition, the group members must recall the operators involved in the last rescheduling decisions for each strategy. This information is used to apportion credit for an increase or decrease in performance among the operators that participated in a decision. For example, if a decision to modify a particular strategy was third order (e.g., $\omega_1 \cdot \omega_2 \cdot \omega_6$) and it resulted in an improvement in the strategy's performance, the group would want to modify the operator probabilities for each of the three operators so that this favorable decision can be repeated more often.

We can now list the descriptive variables associated with the model's decision-making policy component. The decision-making policy-specification descriptive variables are as follows:

1. For every strategy, STRAT(i), where $i \in \{1, \ldots$ MAX_STRAT$\}$, and OPERATOR (j), where $j \in \{\omega_1, \ldots \omega_8\}$, we have STRAT($i$) • OPERATOR($j$) • PROB.

   This represents the probability that the current decision-making policy will apply Operator $\omega(j)$ to Strategy i. The range of values taken by this variable are from the real numbers in [0,1].
2. The variable STRAT(i)•OPERATOR(j)•USED reflects the participation of operator j in the last decision to reschedule strategy i. This variable will take values from the set {yes, no}.

One important point worth noting here is that the effects of these rescheduling operators discussed above are not necessarily additive. For example, a successful higher-order decision involving several operators may fail to produce improvements altogether if one of them is not used in conjunction with the rest. The presence of such nonlinearities means that adaptation for the model group becomes in part a search for coadapted sets of operators—operators that when applied together produce improvements in a schedule's performance. This search is complicated by the large number of potential combinations of available operators and the ability of the group to store information about the performance or only a subset of these combinations.

Holland, however, has investigated certain means by which a system can cope with these problems. To illustrate his approach, let us take the set of all third-order decisions. Any one of the eight available operators can be used in each of the three decisions. This means that there are 8^3 basic types of decisions that can be made. We can denote subsets of these decisions that have attributes in common using the following notation: $\omega_1 \cdot \square \cdot \square$, for example, represents all those third-order decisions beginning with ω_1, where a $\square$ stands for a "don't care." These representations are called schema, and 9^3 such schemata exist for third-order decisions. A schema is said to be defined on those positions not specified by "don't care" symbols.

Any given third-order decision (e.g., $\omega_1 \cdot \omega_3 \cdot \omega_8$) is an instance of 2^3 possible schema that can be defined by substituting $\square$'s for one or more of three attributes. Therefore, when we look at the performance for $\omega_1 \cdot \omega_3 \cdot \omega_8$ we are really acquiring information about the performance of 2^3 constituent schemata. This situation is called intrinsic parallelism. Algorithms that are able to exploit such parallelism have been shown to exist by Holland and Reitman (1978:313). These algorithms are able to test many schemata with a single trial and implicitly store the results in the current population of available structures. We now proceed to develop the transition function for the model that determines how the group is able to alter both its resource collecting schedules and their associated decision-making policies over time. In this context the current set of available schedules and their associated rescheduling policies can be thought of as the data base in which information acquired from experience is stored. Holland refers to this set of existing information as $B(t)$ for some time t (Holland 1975:91). The transition function for the model group describes how this data base is updated by the group members over time based on input from the environment. As we see in the following, certain portions of this transition function possess the intrinsic parallelism discussed above and facilitate the search for compatible resource schedules and rescheduling policies.

The Basic Transition Function

Having specified the structure of both the resource collecting schedules and their associated decision-making policies, it now remains for us to describe the transition function that allows the group to change these structures based on its experience in the environment. The transition function for a typical cycle can be viewed as the composition of a number of basic processes.

1. Initially the state of the environment is determined stochastically.
2. The next process deals with the selection of the subset of strategies that group members intend to use. Schedules are selected for use on the basis of current environmental information as well as past performance.
3. This process models the day-to-day foraging activities for the group members.
4. As the season ends, an impression of the expected total performance for each selected schedule is developed.
5. The performance of each strategy is compared with the others used, as well as with certain minimal standards of performance. The result of these comparisons is a relative impression of the schedule's performance.
6. If the strategy has been rescheduled since it was last used, the two relative performances are compared. If the change was associated with an improvement in relative performance, then the probabilities of all the decision-making operators that were combined to produce the decision are increased. If, on the other hand, there was a reduction in performance, the probability of applying each of the associated operators is reduced.
7. Next, a decision to apply each of these updated rescheduling policies to its associated schedule is made based on the schedule's current relative performance.

Together, these processes characterize the changes made to the system during one cycle in the model. We now describe each of the constituent processes in turn.

1. The previous state of the environment is mapped by δ_1 into a new state. Here this is represented by a random variable, CURRENT_YEAR. From data on current rainfall patterns in the Valley of Oaxaca (Kirkby 1973), the following environmental states and their associated probabilities were derived:

   a. A "wet year" has rainfall in excess of 600 mm. The probability of this is approximately .25.
   b. Years with rainfall less than 420 mm are designated "dry years." The possibility of this occurring is .25.
   c. A year that is neither wet nor dry with rainfall between 420 and 600 mm is termed "average" and has a .50 probability of occurrence.

   CURRENT_YEAR, therefore, assigns a random number generated by YEAR_SEED into the set {wet, dry, average} in the following fashion:

$$\text{CURRENT\_YEAR} = \begin{cases} \text{dry if } r_{28} \le .25, \\ \text{average if } .25 < r_{28} \le .75, \text{ and} \\ \text{wet otherwise.} \end{cases}$$

   Note that the current rainfall does not depend on the rainfall for previous years. This assumption was made on the basis of Kirkby's analysis of current rainfall data in the Valley of Oaxaca.

2. After having determined the climatic conditions for the current year, the next phase represents the process of selecting a set of strategies that the group intends to use. This set is generated on the basis of the group's current information and needs. The function δ_2 maps STRAT(i) • USED into the set {yes, no} for each $i \in \{1, \ldots 10\}$. It stochastically represents the selection of a subset of currently available strategies to be used in the following manner:

$$\text{STRAT}(i)\text{\_USED} = \begin{cases} \text{yes if RAND(EVAL\_SEED)} \ge \\ \quad \text{(STRAT}(i) \cdot \text{PERF*.14} + .08 \text{ and} \\ \text{if RAND(CLIM\_SEED)} \le \\ \quad \text{COMPATIBILITY (CURRENT\_YEAR,} \\ \quad \text{STRAT}(i) \cdot \text{LAST\_USED), and} \\ \text{no otherwise.} \end{cases}$$

TABLE 31.2
Probabilities for Selecting a Strategy Given the Current Year Type and the Type in Which It Was Last Used

Current year type	Strategy (i) last used in year type	Probability of selection on basis of climate compatibility
Dry	Dry	1.0
Dry	Average	.5
Dry	Wet	.25
Average	Dry	.5
Average	Average	1.0
Average	Wet	.5
Wet	Dry	.25
Wet	Average	.5
Wet	Wet	1.0

RAND(EVAL_SEED) represents the random number obtained by applying the EVAL_SEED to RAND, the pseudorandom number generator. Notice that the probability of selecting a strategy is based on two factors. First, the better the performance impression associated with a strategy, the more likely it is to be used. The best performance has a .92 probability of use in terms of performance, while the worst has only a .08 probability. However, the type of year in which the strategy was last used is important too. The more compatible the present environment is with that in which the strategy last performed, the more likely that its past performance reflects its current potential. The index of compatibility employed here assigns the probabilities listed in Table 31.2 for selecting a strategy, given the current year type and the type in which it was last used. If the two year types agree, then the strategy's performance is the critical factor in making the decision. The more they disagree, the more influence these environmental considerations have on the decision not to use a strategy.

3. The third part of the transition function, δ_3, deals with the use of these selected strategies throughout the period from August to November. Initially, each strategy can be used a number of times within the 120-day period when it is initiated. For the next 10 days after each initiation, the collecting activity of the group is governed by this strategy. The exact times at which the strategy will be used depends stochastically on the distribution and availability of the focal resources associated with the schedule. Based on scouting reports, the group is able to get some idea of the current status of its evironment so that it can employ the collecting strategy it feels is best suited to the situation. In other words, the group has a fixed probability of knowing the amount of plant materials currently found at a location x at time t. For each accessible location x, there is a random variable KNOWLEDGE_OF(x) that maps into the set {yes, no} with fixed probabilities at time t. The group's perception of its environment at time t then consists of information for all those locations x with the value KNOWLEDGE_OF(x) = yes. If it is time to select a new strategy, the strategies that have been intended for use this year are compared with the group's current information. Each strategy then has a certain probability of being selected based on the results of this comparison. The strategy that contains the most locations about which the group has current knowledge is the one selected. If there is a tie, they pick the one with the highest relative performance. Since the group intends to use each of the strategies selected by δ_2, we presume that over the course of the season subsets of location will be searched in such a manner as to allow each intended strategy to be used at least once. This is accomplished by altering the KNOWLEDGE_OF(x) probabilities over the season. In the base model, the probability of gaining information about location x is a function of the number of times that it is referenced in the set of intended strategies. At the onset of the season, the probability that the group will have knowledge about location x is the number of times it is referenced by the strategies the group intends to use divided by LOCATION_TOTAL. LOCATION_TOTAL represents the maximum number of references that can be made to the location in terms of the intended strategies. Since a location can be referenced once by each of the 25 tasks in a strategy, LOCATION_TOTAL = 25 × number of intended strategies. Once a strategy has been used, the probabilities for searching the available locations are adjusted. Such an adjustment is due to the depletion of resources at those locations as a result of the activity. This means that the probability of searching other locations is higher. In the model, the adjustment is accomplished for each location x by removing the total number of references made to it by the strategy and dividing by a new value for LOCATION_TOTAL, where LOCATION_TOTAL = (old)LOCATION_TOTAL − 25. This does not rule out the possibility of the strategy being used again. It only says that the strategy is less influential in governing the information collecting activities of the group in the near future. This readjustment of the probabilities occurs only after the initial use of each intended strategy. Subsequent use of a strategy does not serve to readjust the probabilities until each of the intended strategies has been used at least once. At this point, the original probabilities that prevailed at the beginning of the season are reinstated and the process begins anew, selecting a strategy every 10 days until the 120-day duration of the season is reached. These probabilities are reinstated because now most of the referenced locations have been visited at least once and locations that have been searched previously may again begin to look attractive.

Each of the intended strategies is used, therefore, at least once by the group. We now discuss the foraging activities of the group based on a particular strategy *i* at time *t* as described in the model.

Starting at Guilá Naquitz in Thorn Forest A, the collectors proceed to the zone specified by the first activity, STRAT(*i*)•TASK(1) • ZONE. There they search each of the specified locations, STRAT(*i*) • TASK(1) • LOCATIONS, for the focal resource STRAT(*i*) • TASK(1) • RESOURCE. Associated with each location at time *t* is the amount of the resource available there, which is expressed in terms of proteins and calories. This amount is added to the variables STRAT(*i*)_PROTEIN and STRAT(*i*)_CALORIES. The area encompassed by the locations searched is added to STRAT(*i*)_AREA as well as to DAY_AREA, a variable that keeps track of the area searched in a day. Each of these variables was set to 0 before the strategy was used. In addition, the distance traveled to get to the zone from the cave is represented by a generalized index, DISTANCE, in the following way. At the start of each collecting day DIST is set to 0. Each time the collectors move from one zone to an adjacent zone, the index is incremented by 1. This represents not only the movement between zones but also the distance traveled within the specified zone searching for the focal resource. If the collectors must first move through other zones to get to the one specified by the task, then the index is incremented by 1 for each zone crossed into. Thus, if the group moved from Thorn Forest A to Mesquite Grassland B in order to collect mesquite, the index would be increased by 3 since the collectors would travel through Thorn Forest B, Mesquite Grassland A, and Mesquite Grassland B. A collecting day is measured in terms of both the area searched by the group and the distance traveled. It was mentioned earlier that the maximum area the group can search in a day (DAY_AREA) is 0.33 ha; the maximum distance that it can travel in a day is here set to a value of 6. To provide an idea of just how far this is, movement by the group down from the cave to the river in Mesquite Grassland B and back would be of Order 6. While moving directly to and from Mesquite Grassland A will take less than a day by itself, we must remember that the distance travled within a zone in order to check the specified locations also contributes to the total distance traveled.

Therefore, beginning with the first activity in a schedule, the group collects a certain amount of resources, as well as having to travel a certain DISTANCE and search a certain AREA. If the potential increments to both DISTANCE and AREA associated with the next task in the sequence will not cause either total to exceed its daily limit, then the group proceeds with the next task. Otherwise, the next task will begin on a new day, DAY+1, while the day indices DAY_AREA and DISTANCE are reset to 0.

This cycle of activities continues until the addition of a new task would cause the DAY index to be in excess of 10 (DAY was initially set to 0 when the group began to use the strategy). At that point, STRAT(*i*)_PROTEIN and STRAT(*i*)_CALORIES will represent the total plant resources collected, while STRAT(*i*)_AREA stands for the total area searched. Every time a strategy is used during the season the acquired yield is added to STRAT(*i*) • PROTEIN and STRAT(*i*) • CALORIES, respectively. By the end of the season both of these variables represent the total yield for the strategy. STRAT(*i*)•AREA represents the area encompassed by the tasks employed over the 10-day period. This value is constant for the strategy over the season.

4. At the end of the season a general expectation of the performance for each utilized strategy is produced by δ_4. Each estimate of total yield, STRAT(*i*) • PROTEIN and STRAT(*i*) • CALORIES, is divided by STRAT(*i*) • AREA to produce STRAT(*i*) • PROTEIN_EFFORT and STRAT(*i*) • CALORIES_EFFORT. These variables measure the effort taken to acquire the total yield for the strategy.
5. The subsequent formation of an expectation about the overall performance of the sample of strategies used is symbolized by δ_5. In the model, this is produced by taking an average over the expectations for each of the individual strategies used. Therefore, SAMPLE_PROTEIN stands for the average expected protein yield over the set of utilitized strategies. The same holds true for SAMPLE_ CALORIES, SAMPLE_CALORIE_EFFORT, and SAMPLE_PROTEIN_EFFORT.

   Next, the expected performance for each strategy is compared with that for the sample in order to generate an impression of the strategy's relative performance, STRAT(*i*)_PERF. The basis of this comparison has been described previously and is not repeated here.
6. Those utilized strategies that had been altered since they were last used are now scrutinized. If the state of the environment is the same now as it was the last time the strategy was tested, CURRENT YEAR = STRAT(*i*) • LAST_USED, and STRAT(*i*) • PERF−STRAT(*i*) • LAST_PERF $\neq$ 0, then the rescheduling decision has definitely affected the strategy's performance. The corresponding adjustment of the probabilities for the decision-making operators that contributed to the decision is represented by δ_6. The nature of the adjustment depends on whether an improvement in performance was achieved. For a particular strategy *i* the process works as follows:
   a. Generate a random number RAND(MODIFY_SEED). The random variable MODIFY_DECISION maps this number into the set {yes, no} such that

MODIFY_DECISION = "yes" if RAND(MODIFY_SEED) ≤ (|STRAT(*i*)•PERF−STRAT(*i*)•LAST_PERF|*.14 + .08 and CURRENT_YEAR=STRAT(*i*)•LAST_USED, and "no" otherwise.

Note that even if no change was produced, the strategy's resource scheduling policy still has some chance of being altered, although that chance is rather small.

b. If the decision is yes, then the following procedures are carried out for each of the eight operators associated with the strategy's policy: For each $j \in \{1, \ldots 8\}$ if STRAT(i) • OPER(j) • USED = yes, stochastically adjust the probability of using that operator upward if the change resulted in an improvement and downward if the change produced a decrease in performance.

$$\delta_6\text{: (STRAT}(i)\text{•OPER}(j)\text{•PROB)} = \begin{cases} \text{If STRAT}(i)\text{•PERF\_STRAT}(i)\text{–} \\ \text{LAST\_PERF} < 0 \\ \quad \text{then GENERATE} \\ \text{RAND(OPER\_SEED)}\in[0, \\ \text{STRAT}(i)\text{•OPER}(j)\text{•PROB]} \\ \\ \text{else} \\ \text{if STRAT}(i)\text{•PERF\_STRAT}(i)\text{•} \\ \text{LAST\_PERF} \geq 0 \\ \quad \text{then GENERATE} \\ \text{RAND (OPER\_SEED)}\in \\ \text{[STRAT}(i)\text{•OPER}(j)\text{•PROB,1]} \end{cases}$$

LAST_PERF ≥ 0.

Here RAND(OPER_SEED) is mapped into the uniformly distributed set of numbers between the designated end points. Next the STRAT(i) • OPER(j) • USED is reset to no for each of the operators in Strategy i.

Notice that by incrementing the probability for an operator that participated in a decision to improve a schedule, we are increasing the probability of every schema for which that operator is a defining element. For example, if the probability of ω_1 is increased, so is the probability $\omega_1\square$, $\square\omega_1$, $\omega_1\square\square$, $\square\omega_1\square$, $\square\square\omega_1$, etc. Therefore, by increasing the probability of an operator, one also increases the expected number of instances for its associated schema. This property was referred to earlier as intrinsic parallelism and allows the model group to test implicitly and store information about a number of related schemata simultaneously. Those schema that are associated with improvements will become more common while those that produce reductions will become less frequent.

b. If the decision to reschedule the strategy is in the affirmative, the following procedures are performed on each of the eight decision-making operators: Generate a random number RAND(OPERATOR(j) _SEED). STRAT(i) •OPER(j)•USED is then the random variable that maps this number into the set {yes, no} such that

$$\delta_7\text{: RESCHEDULE} = \begin{cases} \text{"yes" if} \\ \text{RAND(RESCHEDULE\_SEED)} \leq \\ \text{STRAT}(i)\text{•PERF* .14 + .08) and} \\ \text{"no" otherwise.} \end{cases}$$

Therefore, the probability of a strategy being rescheduled is a linear function of its performance. Notice that even the best performance, STRAT(i) • PERF = 0, results in some possibility of change.

b. If the decision to reschedule the strategy is in the affirmative, the following procedures are performed on each of the eight decision-making operators: Generate a random number RAND(OPERATOR(j) _SEED). STRAT(i) •OPER(j)•USED is then the random variable that maps this number into the set {yes, no} such that

$$\text{STRAT}(i)\text{•OPER}(j)\text{USED} = \begin{cases} \text{"yes" if} \\ \text{RAND(OPER}(j)\text{\_SEED} \leq \\ \text{STRAT}(i)\text{•OPER}(j)\text{•PROB) and} \\ \text{"no" otherwise.} \end{cases}$$

Therefore, on the basis of the current rescheduling policy for strategy i, a decision is made regarding use of the operator. If the decision is in the affirmative, then STRAT(i)•OPER(j)•USED is set to "yes" and the appropriate stochastic operator is applied to the system according to the specifications given earlier.

The Modeled Environment

This completes our discussion of the basic adaptive subsystems that comprise the model. Now we turn our attention to the environmental subsystem with which they interact. The initial state of this subsystem, at the beginning of each season, is a function of both the present and past year types. Associated with each location is the average total collectible yield for the current combinations of year types—PROTEIN_YIELD•RES(i)•LOC(j) and CALORIE_YIELD•RES(i)•LOC(j). In general, the yield at every location relative to sequences with two "average" years in a row is as shown in Table 31.3. This relative yield is assumed to apply to every resource in that location.

The yield for a location in two average years is defined to be the average measured yield for that location observed in the field.

Every time the group chooses to search a specific location j for a resource i, a random variable YIELD is assigned a value between MIN_COLLECT and REMAIN_PERCENT _RES(i)•LOC(j). That is, the group can potentially collect no less than some minimum amount, MIN_COLLECT, and no more than the total remaining yield. If REMAIN_PERCENT_RES(i)•LOC(j) is less than MIN_COLLECT, then YIELD is assigned the value for REMAIN_PERCENT_RES(i)•LOC(j). YIELD represents the fraction collected, and REMAIN_PERCENT_RES(i)•LOC(j) stands for the percentage remaining. This latter variable is initially set equal to 1. Every time the location is searched, the group collects some fraction, YIELD, of this total. The value for REMAIN_YIELD_RES(i)•LOC(j) is then decremented by the fraction collected. The number of proteins and calories associated with the collected fraction is added to STRAT(i) •PROTEIN and STRAT(i)•CALORIES.

TABLE 31.3
The Yield at Every Location Relative to Sequences of Two "Average" Years in a Row

Previous year	Present year	Relative yield
Average	Wet	1.25 as much
Avg.	Dry	.75 as much
Wet	Dry	1.0 as much
Wet	Wet	2.0 as much
Wet	Avg.	1.25 as much
Dry	Dry	.5 as much
Dry	Wet	1.0 as much
Dry	Avg.	.75 as much

The value for MIN_COLLECT is determined by the group from experience and is large enough to insure that the group acquires all of the resource available at the location.

We have now finished our description of the basic model. The model is quite detailed, and would be too expensive to simulate on the computer. Therefore, we developed a simplified version in the next section that preserves this model's behavior but is easier to simulate on the computer.

Simplifying the Base Model

While the base model presented in the previous section well characterizes those aspects of the system we are interested in, testing the model requires data about the total yield of each major plant resource in every location during the season. The only data that are presently available is the total yield for each major plant species per hectare. What we must do, therefore, is simplify our base model so that it uses these available data. However, we must do this in such a way so as to produce a new model whose behavior corresponds with the original. We begin by describing the changes made to the base model and then demonstrate that this new model behaves in much the same manner as the original.

Turning first to the resource scheduling component, the following changes are made: If we are going to use the average yield per 0.001 ha for each resource in each zone, we no longer need specific information about the locations. Instead, we replace the location specificiation with a variable that represents the total area encompassed by these locations. Therefore, for every STRAT(*i*)•TASK(*j*)•LOCATIONS_TO_SEARCH, we substitute a new variable STRAT(*i*)•TASK(*j*)•AREA that represents the area encompassed by the locations searched.

The corresponding changes are now made for the rescheduling adaptive system in terms of each operator:

1. With ω_1 = SWAP_EXTERNAL, there is no real change except that the tasks that are exchanged use STRAT(*i*)• TASK(*j*)•AREA instead of STRAT(*i*)•TASK(*j*)• LOCATIONS_TO_SEARCH.
2. With ω_2 = SWAP_INTERNAL there is no change aside from that mentioned with ω_1.
3. Here ω_3 = SHUFFLE_LOCATION symbolizes not the exchange of location sets between tasks on the same strategy, as before, but the exchange of area specifications between tasks. However, the mechanism of the process is the same as before.
4. With ω_4 = SHUFFLE_ZONE there is no real change except that it is no longer necessary to specify a new set of locations for each participating task. As a result, Step 4 in the specifications can now be omitted.
5. With ω_5 = SHUFFLE_RESOURCE there is no change.
6. With ω_6 = ALTER_RESOURCE there is no change.
7. The modification of ω_7 = ALTER_ZONE is the same as that of ω_4. Since it is no longer necessary to regenerate a new set of points with equal area for a new zone, Step 4 can be omitted.
8. With ω_8 = ALTER_LOCATION, rather than selecting a subset of new locations, we need only to select a different amount of area to search. Perhaps the easiest way to visualize this is to select a subset of locations as done previously and assign to STRAT(*i*)•TASK(*j*)•AREA the area covered by this subset. In our lumped model it is therefore not the specific locations that are important but the area that they cover.

With these basic structural alterations in mind, let us now look at changes made in the transition function for the system:

1. The determiner of the new environmental state, δ_1, is not changed.
2. The manner in which the strategies to be used during the season are selected, δ_2, is also not changed.
3. The only process that is changed significantly is δ_3. Previously, δ_3 modeled the day-to-day foraging activities for the group. The result of these activities was the production of a performance estimate based on total yield. However, this requires estimates of the yield for every major resource in each location in an "average" year. Since our present data consist of estimates of the total yield for an average location in each zone during an average year, we restructure δ_3 so that these data can be used.

Given that the total yield for an arbitrary collecting task is the sum of the total yields for each location that comprises the strategy, this sum can be rephrased as in the following:

1. The average total yield for Resource *i* at *n* locations in Zone *j*, where *n* is the number of locations searched for the task, is represented as n^*. As *n* increases, the average yield for the *n* locations approaches the average yield over all locations in the zone. In add' `n, since the tasks are defined to be locationally independent (no one location is referenced more than once over all currently used tasks), the average total yield for each task

that uses a particular resource, i, in a certain zone j is m^*.

2. Given that m^* is the average total yield for resource i in n locations for zone j where n is the total number of such locations in the strategy, then if m is large, the average yield for the n locations will be very close to the average yield over all locations in the zone. It is shown elsewhere that as long as n is of a certain minimum size, using average yields to represent the performance of each unit area rather than using the yield for each individual unit area will produce the same average performance values for a strategy (Reynolds 1979). Since the remaining portions of the transition function are only influenced by the relative performance index, this means that the substitution of the one set of calculations for the other will not affect the system's overall behavior.

   So our new version of δ_3 emphasizes the average total yield per strategy and does not attempt to deal with the variability of foraging activities. The process therefore takes the following form: For each strategy selected to be used by the group, the total yield for each is calculated. This is done by finding out which tasks will actually be used over the 10-day period in the same way as the base model. Once this set of tasks has been delimited, then the total yield for each task is computed. This is done by multiplying the average total yield for Resource i per 0.001 ha in Zone j by the number of 0.001-ha units searched for the task. The yields for each of these tasks are then added together to get the total yield for the strategy.
3. The process of deriving a generalized performance expectation, δ_4, for each strategy is equivalent to that in the base model.
4. The formation of expectations about the overall performance of the entire sample of strategies (δ_5) is unchanged as well.
5. The process of changing the rescheduling policy for a strategy based on its performance is the same as before, although the operators' function, as mentioned previously, has changed slightly.
6. The basic process of rescheduling each utilized strategy based on its performance is the same as before, although the specific rescheduling decisions have been altered slightly, as mentioned earlier.

It can be shown in fairly straightforward fashion that the above model exhibits the same average behavior as our base model, when both are started in corresponding states and are exposed to the same sequence of inputs. Given that we have a formal description for each, it remains only to construct a correspondence between the two. Zeigler (1976:267) has specified certain relations that must hold between two systems in order for them to exhibit the same behavior. It is shown in detail elsewhere that such relations can be constructed between our present models (Reynolds 1979). As a result, we can be assured that the results obtained by simulating our simplified model in the next chapter also characterize the more detailed base model as well.

PART 6: THE PREAGRICULTURAL SIMULATION

Introduction

The early stages of agricultural evolution in the southern Mexican highlands have been characterized by Flannery (1968) as featuring a series of positive feedback loops that allowed successively greater divergence from the preagricultural pattern. These loops were seen as taking a series of small (and perhaps initially accidental) deviations in the preexisting hunter–gathering subsistence system and magnifying them to the point where they began to effect widespread changes in some of their fellow subsystems.

In this section, we suggest that these later cycles of deviation amplification were made possible by a preexisting pattern of systemic interaction that had been "primed" by certain adaptations that encourage the system to search for and exploit sources of positive system change. For example, we see in Fig. 31.7 the basic set of interactions between the adaptive subsystems. Notice that there is a potential for both positive and negative feedback cycles. The exact nature of these interactions depends on the current state of the model and its environment. What we do in the next section is to begin with a random set of resource scheduling strategies that will allow the rescheduling subsystem plenty of room in which to make improvements. Initially, we would expect the group to generate and maintain a directed sequence of performance improvements over time. If the group can do this, it will be interesting to characterize the pattern of behavior associated with this positive feedback loop, as well as any adaptations that the group might acquire to sustain such a loop. When the schedules have been improved to the point where the current information is not sufficient to allow for continued success in rescheduling the available strategies, the interaction between the two systems changes gradually so that a negative feedback loop comes to dominate. In this way, the group's rescheduling policy becomes more conservative and avoids the risk of losing the performance improvement that it already has made.

However, the decision-making adaptations acquired during the positive feedback cycle are retained by the group. These adaptations "prime" the system so that its search for improvements can still go on as an undercurrent within an otherwise conservative decision-making environment. In the event that an improvement is found, these adaptations can serve to restart the positive feedback cycle. This is, in fact, exactly what happens with the introduction of incipient agriculture in Part 8.

We then compare the performance of our preagricultural model against data from the same period, Zone E and Zone D at Guilá Naquitz. Such a comparison allows us to see how well the behavior of our model group matches that of the real world. If they correspond, we will have good reason to believe that it is the result of corresponding rescheduling behavior over time.

Finally, we alter the network of interactions between the subsytems in the model in order to see how the system's per-

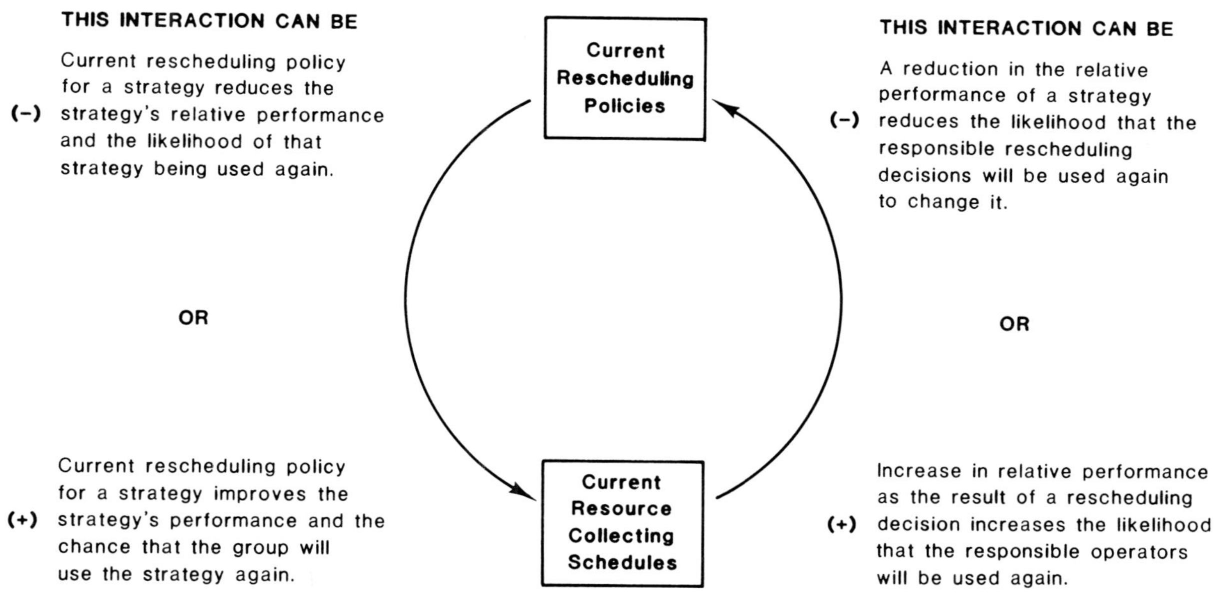

Fig. 31.7. Basic interaction between the system's components.

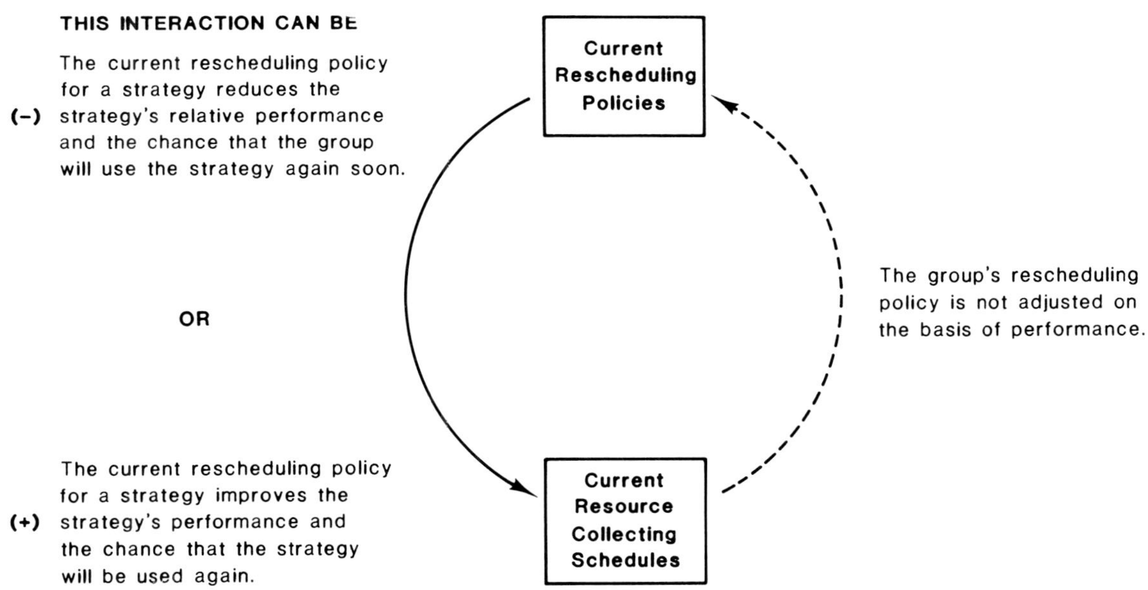

Fig. 31.8. Removing the feedback cycle. (Those interactions that have been disabled are given as dashed lines.)

formance is affected. In particular, we no longer allow the rescheduling system to adjust its decision-making policies based on their performance. This disables the feedback loop as shown in Fig. 31.8.

The extent to which this adjusted system is able to mirror the archaeological data provides insight into the importance of these feedback mechanisms in the functioning of the real system.

Behavior of the Model Group: Learning To Adapt within a Challenging Environment

Since our interest is in seeing how well our model group is able to assimilate information from its environment into its decision-making structure over time, we start off the basic set of structures in an undifferentiated, random fashion. The same sequence of random activities is assigned to each of the

TABLE 31.4
The 25 Tasks that Were Each Used Once in Each Initial Strategy

Resource	Zone	Area (ha)
Acorns	Thorn Forest A	.075
Piñon nuts	Thorn Forest A	.075
Nanches	Thorn Forest A	.075
Nanches	Thorn Forest B	.075
Wild beans	Thorn Forest A	.075
Wild beans	Thorn Forest B	.075
Susí nuts	Thorn Forest A	.075
Susí nuts	Thorn Forest B	.075
Susí nuts	Mesquite Grassland A	.075
Agave hearts	Thorn Forest A	.075
Agave hearts	Thorn Forest B	.075
Agave hearts	Mesquite Grassland A	.075
Opuntia nopales	Thorn Forest A	.075
Opuntia nopales	Thorn Forest B	.075
Opuntia nopales	Mesquite Grassland A	.075
Opuntia nopales	Mesquite Grassland B	.075
Opuntia fruits	Thorn Forest A	.075
Opuntia fruits	Thorn Forest B	.075
Opuntia fruits	Mesquite Grassland A	.075
Opuntia fruits	Mesquite Grassland B	.075
Mesquite pods	Thorn Forest B	.075
Mesquite pods	Mesquite Grassland A	.075
Mesquite pods	Mesquite Grassland B	.075
Hackberries	Mesquite Grassland A	.075
Hackberries	Mesquite Grassland B	.075

10 initial strategies, where the basic tasks used are as described in Table 31.4. A fixed search area of 0.075 ha is associated with each of the tasks in every strategy.

This assignment of strategies is taken to represent the knowledge that the group initially has of its environment when exposed to it for the first time. Corresponding to this lack of scheduling experience is a lack of decision-making experience as well. In this situation, each of the basic rescheduling decisions has a 50–50 chance of being used. If our model at all characterizes the adaptive aspects of the group's scheduling system, the structure of the model group's behavior should change over time based on its experience.

If the model group can indeed learn from its experiences, its performance over time should become better. In order to gauge changes in the group's performance, an index was formed by adding together the following variables:

3*SAMPLE_PROT_AVE+SAMPLE_CAL_AVE+3*
SAMPLE_PROT_EFFORT+SAMPLE_CAL_EFFORT

These represent the four variables used by the group to judge each strategy's performance in a given season. Thus, for a given season, the index represents the overall performance of the sample strategies, based on the average amount of protein and calories acquired along with the effort expended to acquire these amounts. The variables associated with the acquisition of protein were both multiplied by a factor of three, so that the contribution of all four variables to the index would be of the same order of magnitude.

In Figs. 31.9–31.11, the changes in the above index are displayed over 500 time steps for each of the 3 year types. Notice that the group improves its performance in all 3 situations over time, although the greatest rate of improvement is associated with both dry and average year types. In these years, the reduced yield in many species put a certain amount of pressure on the group to survive. This selective pressure apparently leads to rapid improvements in the group's collecting activities in these year types.

It is also interesting that the group seems to treat both dry and average year types alike in terms of performance. Notice, for example, that the group is able to effect few performance improvements in either year type after about 300 iterations. Perhaps this is because there is still too much selective pressure on groups in an average year to allow them to treat such a year any differently from a dry year.

Wet years, on the other hand, are associated with relatively high yields, and survival is much easier in such a context. This reduction in performance constraints means that a wider variety of collecting strategies can be utilized in such an evironment. As a result, the group's performance in wet years takes a much longer time, almost twice as long, to level off. This is because there is less concern with improving performance in a situation that is favorable already. Also, the lack of long-term storage technology meant that there was little incentive to collect more than the group could consume within a reasonable length of time.

While our hypothetical group was given three year types to work with, they reduced this to a binary classification (wet versus dry or average) based on the relative selection pressures in each. This is interesting, since many Zapotec speakers in the Valley of Oaxaca today seem to perceive of all years within recent memory as belonging to one of two classes, "wet" or "dry" (see for example Kirkby 1973:Appendix I). Thus Zapotec farmers, given a wide variety of actual annual rainfall patterns, behave as if there were only two kinds of years. The fact that our model group does the same may suggest that there is a functional basis for such environmental perception.

The model group is able to produce a number of modifications in its resource collecting activities that, at first, lead to marked increases in performance. However, over time, it becomes much harder for the group to effect new improvements with the information provided. As a result, the group reaches a stable performance state by 500 iterations. The use that the group makes of the plant resources at this stage is evident from Table 31.5. In it, the average area searched per task for each major resource is given. Notice that both acorns and hackberries are given the most emphasis. This is a good sign, since we have already seen in Chapter 19 that these were the two most abundant resources in Zones E and D of Guilá Naquitz. In the following section we discuss correspondence between these data and those from the cave in more detail.

The majority of activities carried out by the group at this

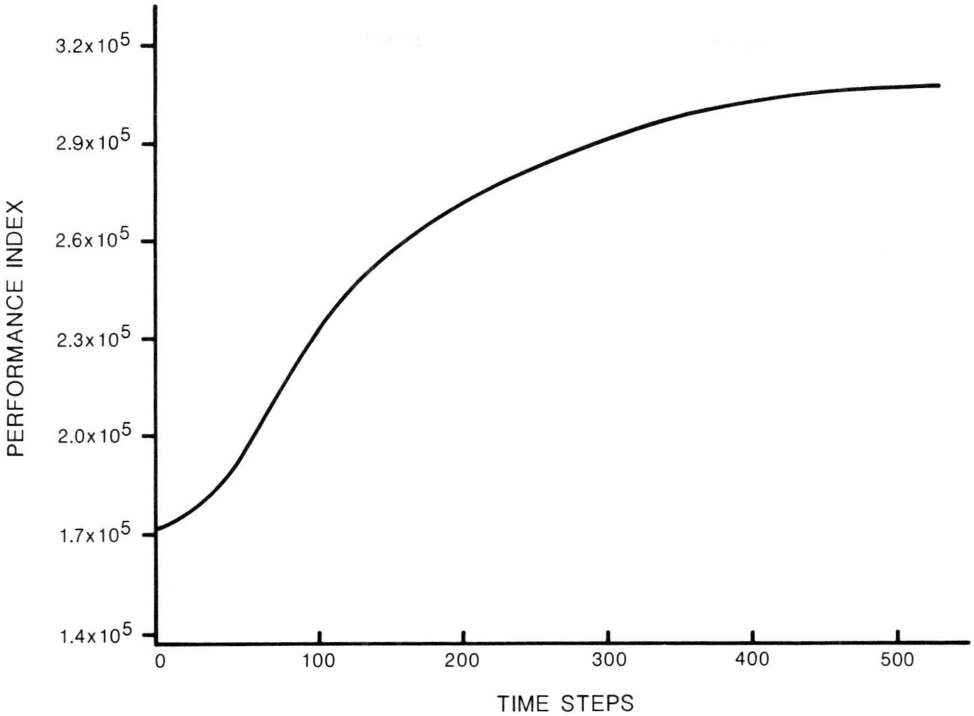

Fig. 31.9. Changes in the model group's performance in "dry" year types over time.

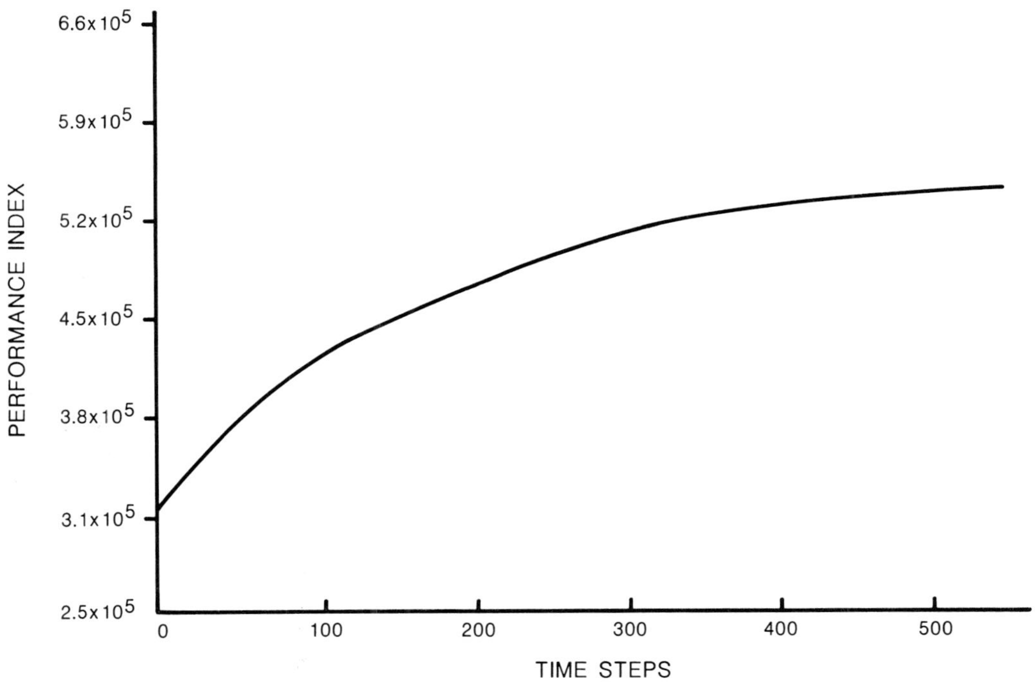

Fig. 31.10. Changes in the model group's performance in "average" year types over time.

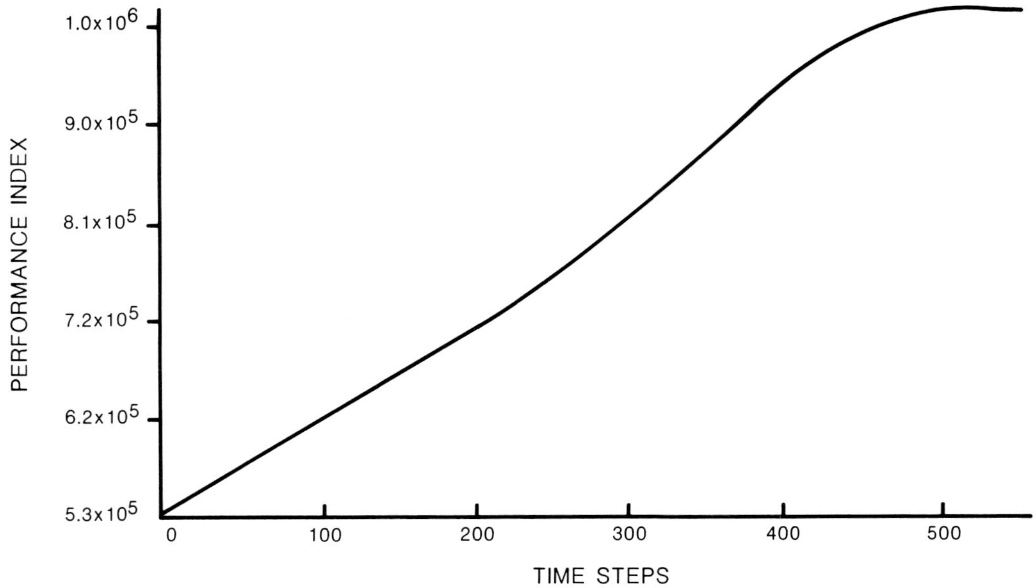

Fig. 31.11. Changes in the model group's performance in "wet" year types over time.

TABLE 31.5
Average Area Searched per Activity Involving Each Major Resource after 500 Time Cycles

Resource	Average area used (ha)	Rank based on average area
Piñon nuts	.133	3
Acorns	.164	2
Guaje	.104	10
Susí nuts	.115	9
Agave	.1175	8
Nanche fruits	.123	6
Wild beans	.080	12
Mesquite	.1275	5
Cucurbits	.129	4
Hackberries	.171	1
Opuntia fruits	.121	7
Opuntia nopales	.106	11

time centered around the Thorn Forest zones near the cave. Of the 10 strategies, only 1 had more activities outside of these zones than within them. That strategy emphasized tasks in the Mesquite Grassland. While there were not as many activities carried out in the Mesquite Grassland zones, those that were carried out involved a much larger share of the day's activities than most tasks from other environments. This is due, perhaps, to the time that the group would have to spend to get to the Mesquite Grassland from the cave in the first place.

The rescheduling policies for the group also changed as its experience with the environment increased. These trends are shown in Fig. 31.12. Notice that all the operators exhibit an initial decrease in use. At this point, the group has no experience on which to base its decision. It therefore takes a number of cycles before the policies become adjusted to their associated schedules. Once this occurs, there is then a chance for improvement to take place.

The first operators to exhibit increased use are those that recombine existing specifications rather than creating new ones. The use of these operators alone explains the slow-growth phase of the performance curve for dry years. However, after about 150 cycles, the mutation-type operators come into play as well. Together, the two account for the fast-growth phase of the performance curves for dry years. Next, certain recombination operators begin to be used less; and by 300 cycles, the alter operators exhibit a similar change. From here on the group's rescheduling policy becomes more and more conservative, and with this the performance curves start to flatten out.

In other words, schema consisting of recombination-style operators were the first to exhibit major increases. As the model group's experience with rescheduling decisions increased, the group began to use high-order schema more often. By 350 model time steps, all the mutation-style operators have been brought into play along with the recombination operators. However, it becomes harder for any of these schema to effect improvement in the performance of the majority of available strategies sampled. As a result, the overall complexity of available schema for the set of sampled strategies begins to decrease by 400 iterations. This does not mean, however, that the policies for particular strategies are necessarily reduced in complexity.

While on the average the rescheduling policies of the group tend to become more conservative over time, there are exceptions. These exceptions are strikingly obvious by 500 time

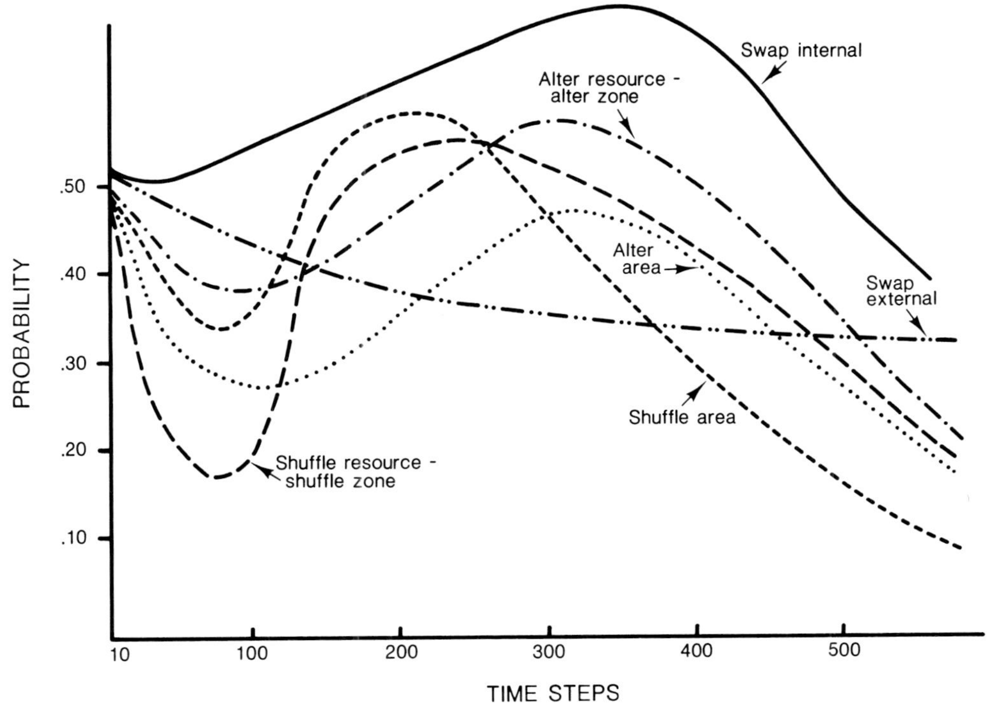

Fig. 31.12. Changes in average probability for each operator based on the sampled strategies. "Alter resource" and "alter zone" are shown together, as are "shuffle resource" and "shuffle zone".

cycles. In particular, a small number of strategies develop rescheduling policies that encourage experimentation. One strategy, after 500 model cycles, has a .75 average probability for each rescheduling operator. This means that it is probably changed every time that it is used.

At first glance, this does not appear to be consistent with our earlier statement that the group's rescheduling behavior had become more conservative. It turns out, however, that this strategy is a relatively poor performer and is used very infrequently. Unlike most of the available strategies, the majority of its tasks (18 out of 25) occur in the 2 least-productive zones, Thorn Forest B and Mesquite Grassland A. Accordingly, it tends to use less-productive species in less-productive zones.

What is interesting here is really not the schedule's productivity but when it is used. This below-average strategy, with a rather unusual collection of tasks and a high probability of being modified, is used primarily in wet years. This tactic makes a good deal of sense. In doing this, the system is able to continue its experimentation even while it becomes more conservative. Changes are made, not to those strategies that have a lot to lose if they are unsuccessful, but to poor performers such as the one described, which have little to lose and more to gain. In addition, the group hedges its bets by making such changes predominantly in wet years, when plant densities are high and rescheduling mistakes would not be felt as much.

Therefore, the group, in associating specialized decision-making policies with poorer performing strategies, can introduce changes into the system via years in which selective pressure is reduced. This consistent process of change generates a schedule whose structure is unique to wet years. These wet-year strategies, therefore, provide for an undercurrent of change within a system that is becoming more conservative.

This notion of experimenting with schedules in wet years is an interesting one, and examples of it can still be found among present day agriculturalists in the valley. Kirkby (1973:94), in describing planting strategies among Zapotec farmers, state that "marginal land provides an area in which opportunities for gambling on wet years takes place." It may even be that this modern behavior includes an extension of resource collecting policies developed in preagricultural times.

Comparing the Model's Performance with the Archaeological Data

We have seen in the previous section that the group acquired a number of decision-making adaptations before achieving a stable performance level. In this section, we want to see how well the strategies generated by the group using these adaptations characterize the resource collecting behavior of the real group. In order to do this, we compare the relative emphasis placed on each of the major plant species in the model group with the relative frequency of each plant in Zones E and D of Guilá Naquitz. Both of these zones, as mentioned in earlier chapters, are the results of human

TABLE 31.6
Raw Counts, Relative Frequency, Rank Order of Relative Frequency, and Corresponding Rank Order Predicted by the Model for Each Major Plant in Zone E

Resource	Raw frequency counts	Relative frequency	Rank order based on relative frequency	Predicted rank order in model
Piñon nut hulls	155	.042	4	3
Acorns	1846	.511	1	2
Susí seed coats	98	.027	6	8
Agave quids and leaves	8	.002	8	7
Nanche seeds	6	.001	9	5
Bean pod valves	161	.044	3	10
Mesquite seeds	32	.008	7	4
Hackberry seeds	1166	.322	2	1
Opuntia seeds	135	.037	5	6
Opuntia nopales	3	.0008	10	9

occupation prior to 7500 B.C. and represent the preagricultural activities of the hunter–gatherers of the valley.[31]

We begin with Zone E, the earliest occupational zone found in the cave. In Table 31.6 the raw counts for each of the major species found in the zone are given, as well as their relative frequency. A rank ordering based on their relative frequency was computed and given as well. It is with this rank ordering that we compare the model results.

For each of the major resources in the model that correspond to those in Zone E, a ranking was given, on the basis of the average area allocated to it by each task in which it occurred, for all the existing schedules at Time Step 500. This time slice was chosen because by then our model group had achieved a stable performance state, as discussed earlier. This rank ordering, therefore, reflects the relative emphasis placed on each resource by the model group at this time.

In comparing the two sets of rankings, we notice a great deal of correspondence between them. Indeed, 7 of the 10 plant resources have rankings that differ by 2 or less. Only mesquite, nanche fruits, and wild beans exhibit any greater discrepancy.

Although it seems quite apparent that the two sets of trends correspond, can we quantify this correspondence statistically? This can be done using a nonparametric statistic that tests the agreement between K sets of observations. In our case, the rank correlation between the predicted and observed changes is computed. The *concordance statistic* is a linear function of these paired comparisons and represents the goodness of fit between them (Gibbons 1971). This statistic is used to test the hypothesis that the model makes predictions that are independent of the observed rank ordering of plant use based on the archaeological data from Zone E. If the value for the concordance statistic differs enough from the value that would be expected, given that the two sets were not related (i.e., they are independent), then we would have good reason to reject this hypothesis of independence.

The concordance statistic was computed for this case and found to be 0.75, where 1.00 denotes a perfect correspondence. The probability that the two orderings were generated by independent processes, given the present level of agreement, is .07. This is sufficient grounds for us to reject our original hypothesis in favor of one that postulates a correspondence between the model and the real system.

While the similarities between the two systems are important, their differences can be instructive as well. The fact that mesquite and nanche fruit are found less often in the cave than predicted, while wild beans are found more often, is interesting. From Fig. 18.1 we learn that both mesquite and nanche are available earlier in the season (by August), while beans become available slightly later (September). It is possible, therefore, that the difference between the two systems lies in the fact that the real group arrived at the cave slightly later than presumed in the model (see also discussion in Chapter 24).

Looking next at Table 31.7, we see the frequency data for Zone D. At this time, there are 12 major resources present, rather than just the 10 found in Zone E. The 2 new plants include cucurbits as well as *guajes*. The corresponding ranks for each of the 12 resources generated by the model are given as well. By inspection, we observe that 9 out of the 12 pairs of rankings differ by at most 2 positions. The concordance statistic for these 2 sets of rankings was .78 and the associated level of significance was .03. Again, according to our criteria, this leads to a rejection of our original hypothesis that the two rankings were independent.

Therefore, the relative frequency of the plant resources collected by the model group does not differ much from the rank ordering of materials actually found in the cave for both

[31]One possible *Cucurbita pepo* seed occurred in Zone D, but it was so heavily outnumbered by wild plant remains that Zone D can be considered preagricultural for our purposes.

TABLE 31.7
Raw Counts, Relative Frequency, Rank Order and Corresponding Rank Order Predicted by the Model for Each Major Plant in Zone D Based on Relative Frequency

Resource	Raw frequency counts	Relative frequency	Rank order based on relative frequency	Predicted rank order in model
Piñon nut hulls	94	.020	5	3
Acorns	3182	.707	1	2
Susí seed coats	237	.052	4	9
Agave quids and leaves	53	.011	7	8
Nanche seeds	49	.010	8	6
Bean pod valves	17	.003	10	12
Mesquite seeds	59	.013	6	5
Hackberry seeds	416	.092	2	1
Opuntia seeds	348	.077	3	7
Opuntia nopales	28	.006	9	11
Guaje pods	11	.002	11	10
Cucurbit remains	6	.001	12	4

Zones D and E. On this basis, it seems quite likely that rescheduling processes of the kind modeled here are of fundamental importance in the formation of the preagricultural group's resource collecting activities over time.

Testing the Feedback Cycle Assumption

In order to substantiate our assumption that feedback cycles played a necessary part in the development of the real group's collecting activities, we can disable the basic cycle in our model as shown in Fig. 31.8. We then start our simulation using the same set of strategies and rescheduling policies as before. However, this time the group is not able to alter its basic rescheduling policy based on experience. This situation is somewhat akin to what happens within more complex societies when the decision-making system either is not able to keep tabs on the very subsystems it is supposed to be controlling or has become too inflexible in its policy making. What happens is that the modified system exhibits a sequence of initial improvements in somewhat the same manner as the original system did. This is because almost any change in this initial set will represent an improvement. Not only does the group's performance peak earlier than before, but its performance in each of the three year types also peaks concurrently. This suggests that the distinction made between wet and dry climatic types in our original model is not made here. This makes sense, because the group now cannot adjust its rescheduling policies to take advantage of the differences in selection pressure in the different year types. As a result, it cannot generate either the decision-making or resource collecting specializations found in the other model.

In addition, our group is not able to adjust its rescheduling policy to fit the structure of the existing schedules. This means that it will make unnecessary and perhaps unwise adjustments in the current schedules. It is because of this and the other problems mentioned above that the group's performance begins to exhibit wild oscillatory behavior soon after experiencing its initial increases. Such unstable behavior continues on for over 2000 iterations, although only part of that is graphed here (Fig. 31.13).

Not only did the lack of articulation between the two component subsystems result in intrinsic performance instability, but also, the overall performance got progressively worse. For example, Table 31.8 gives the relative emphasis placed on each of the major plant resources by this model compared with the relative amounts of collected resources found in Zone D of the cave. A number of high-yielding plants, such as acorns and mesquite, are deemphasized and replaced instead by less-productive genera.

Additional runs were made not only with this set of fixed policies but with other fixed policies as well. In every case, the system exhibited inherent instability after a certain period of time. Not only is there no evidence for such instability archaeologically, but the mix of resources produced does not correspond with existing data either. For example, when the ranked preferences produced by this model were compared with the actual rank observed for the plants in Zone D, the concordance statistic took the value of .35. This degree of association was not significant statistically. These results strongly suggest that feedback cycles within the group's decision-making system had to be present in the preagricultural period in order for the group to exhibit the observed scheduling behavior.

Conclusions

Analysis of the simulation results described in this part of the chapter has revealed a number of interesting points. First, it seems that the basic feedback cycle between the two decision-making subsystems articulated in our model is needed to insure performance comparable with the real

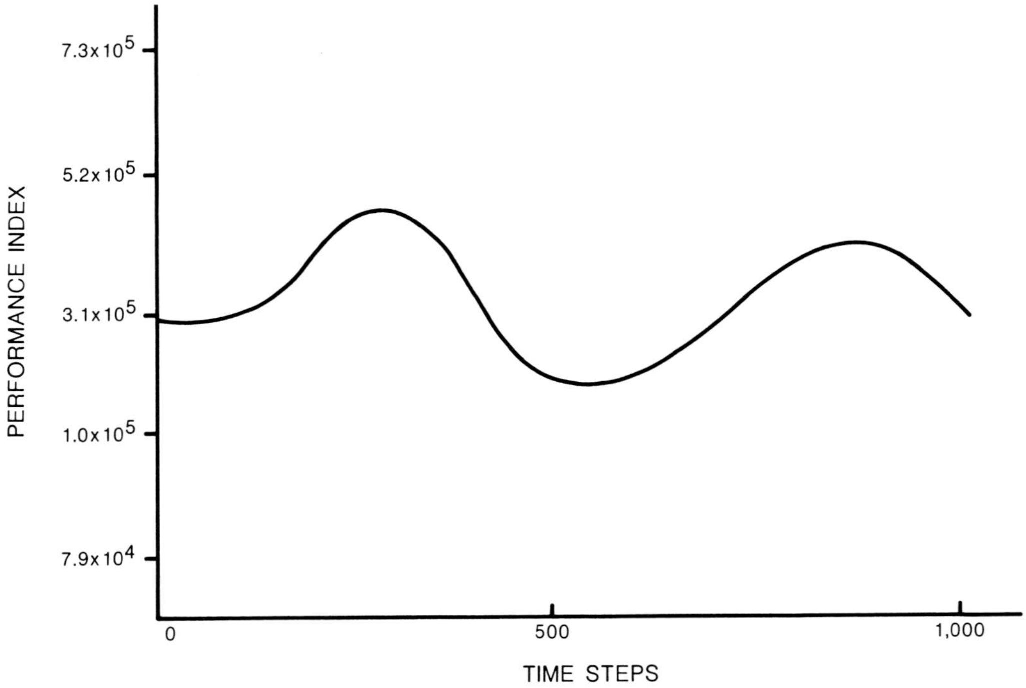

Fig. 31.13. Performance trends for the modified group model in "average" year types.

TABLE 31.8
Comparative Rank Ordering of Collected Plants between the Adjusted Model and Actual Remains from Zone D[a]

Resource	Rank order based on relative amounts in adjusted model	Rank order based on relative amounts in Zone D
Piñon nut hulls	7	5
Acorns	9	1
Guaje pods	3	11
Susí seed coats	12	4
Agave quids and leaves	8	7
Nanche seeds	2	8
Bean pod valves	4	10
Mesquite seeds	10	6
Cucurbit remains	11	12
Hackberry seeds	5	2
Opuntia seeds	6	3
Opuntia nopales	1	9

[a]The value 1 denotes the most important and 12 the least important.

system. This feedback loop allows both experimentation and system stability at the same time. This is accomplished through the aid of a number of adaptations acquired by the group over time. Such adaptations appear designed to deal with differences in selective pressure exhibited by the different environmental states encountered by the group. Among these adaptations we find the following:

1. The group tends to make rescheduling decisions more frequently in wet years, when the overall yields are greater than normal. These decisions are associated with strategies that possess relatively poor performance. By doing this, the group is able to make rescheduling decisions under favorable circumstances. That is, since the target strategy is not as good as the others, there is less to lose by changing it; and even

if the change produces a reduction in performance, it is least likely to be noticed in a wet year, when yields are increased for most species.

2. This differential decision-making produces a set of wet-year strategies that exhibit more variability in collecting behavior than those used in other year types. For example, the more successful strategies could be termed "general purpose" and focus their activities on the most-productive species in the most-accessible zones. Wet-year strategies, on the other hand, are more likely to use less-productive species as well.

3. Although the group is given information regarding three different year types, it behaves as if there are only two. This binary classification corresponds to the traditional Zapotec dichotomy between wet and dry years. It turns out that the average year type does not possess enough selective advantage over dry years to permit specialized decision-making as with wet years. As a result, dry and average years are effectively lumped together.

Employing the above adaptations, the group is able to produce resource collecting behavior that corresponds significantly with data from Zones E and D of Guilá Naquitz.

The entire basis for the above behavior is performance information acquired from the environment. It may be that the rate of system change is in some sense limited by the amount of information provided to the group by its environment. The more information available to the group, the more adjustments it is able to make if needed.

The climate of the valley is such that the sequence of year types is unpredictable over time. It is perhaps this unpredictable succession of climatic types with different selective pressures that provides the context in which the group can acquire a good deal of information about its existing schedules through experimentation. It will be our goal in Part 9 of this chapter to see just how important this climatic sequence is to the rate of system change. By varying the associated probabilities for each year type, we can hypothetically change the climate to which the group is exposed and measure the corresponding achievement of the group.

PART 7: MODELING INCIPIENT AGRICULTURE

Introduction

In Part 6, we demonstrated that considerations of foraging efficiency in the face of annual variation are by themselves sufficient to generate a pattern of collecting behavior that corresponds well with the archaeological data on preagricultural plant use. The relatively stable pattern of collecting seen after Time Step 500 did not, however, last throughout the preceramic era. Part way through the Naquitz phase came hints that incipient agriculture was now part of the subsistence pattern.

Perhaps the first group of plants to come under domestication were the cucurbits. The bottle gourd was represented by rind fragments as early as Zone C of Guilá Naquitz; 1 possible *Cucurbita pepo* seed was present in Zone D, and this squash was even better represented in Zones C–B1, where 14 seeds and/or peduncles were present (Chapter 20). Small black runner beans were also present in large numbers at Guilá Naquitz. While the species involved is one that has left no domestic descendants, the beans were present in such unexpectedly high quantities relative to more common plants that it is possible the occupants of the cave were experimenting with their cultivation (Chapter 24). Finally, pollen grains of the genus *Zea* were present in Zones C–B1 of Guilá Naquitz (Chapter 15). While it cannot be determined whether teosinte or early domestic maize was the source of the pollen, tiny but indisputable cobs of maize have been recovered from sixth-millennium B.C. levels near Tehuacán, not far to the north of Guilá Naquitz.

For all these reasons, we decided it would be interesting to model incipient agriculture in the eastern Valley of Oaxaca by adding maize, beans, and squash to the group of plants already present in the preagricultural simulation and observing how the preexisting collecting schedule was modified to accommodate them.

The Mechanisms of Incipient Agriculture

This chapter is not the place to speculate on why the first seed was planted in the Valley of Oaxaca. In our model, incipient agriculture is viewed as a process by which (1) the density of certain useful plants was artificially increased and (2) the location of these atypically dense stands within the overall Guilá Naquitz environment was made more predictable. Judging from the archaeological record, any increases in productivity associated with these early agricultural activities were small compared to later periods (after 5000 B.C.), when genetic changes in some of the plants produced substantial increases in productivity. However, the initial changes probably produced certain shifts not only in the group's collecting activities but also in the type of rescheduling decisions that were likely to be made. These adjustments favored continued experimentation[4] with incipient agriculture, which ultimately led to the "fixing" of genetically modified strains.

Our purpose here is to introduce the potential capability for incipient agriculture into the preagricultural group model at equilibrium in order to discern the nature of these initial shifts in scheduling. In terms of our preagricultural lumped model, this entails the following adjustments. First, associated with each task in every strategy is a new descriptive variable STRAT(i)•TASK(j)•DENSITY. This variable denotes the extent to which the density of the focal resource has been altered by the group for the associated locations. It effectively represents the results of effort spent in the preceding months

[4]We use the term *experimentation* here in the vernacular sense only, as in the "trying out" of various subsistence strategies over a long period of time.

TABLE 31.9
The Distribution and Maximum Density for Maize, Beans, and Squash as the Result of Incipient Agriculture[a]

Plant	Mesquite Grassland B	Mesquite Grassland A	Thorn Forest B	Thorn Forest A
Maize	.33	1	1	.16
Beans	5.0	1	3	4.0
Squash	5.0	2	2	4.0

[a]Each figure represents the maximum relative increase in density that can be attained by incipient agriculture. A 1 represents a zone in which the density is not appreciably affected by incipient agriculture.

to adjust the distribution of the focal resource at the prescribed locations. These distributional shifts will have the same relative effect on the yield of the focal resource throughout the season.

While it might be expected that a number of resources could have their distributions adjusted, only three—cucurbits, beans, and maize—are dealt with here. These are the plants that were to become the backbone of the later sedentary agricultural systems in the valley. Although the densities of other plants may have been affected as well, the adjustments made to them are relatively minor in comparison and therefore are not considered.

Each of the three plants—maize, beans, and squash—has a characteristic range of possible densities for each of the four vegetational zones. In Table 31.9, the maximum potential increase in yield as the result of incipient agriculture is given for the three species. The greatest improvements in yield are to be made in Thorn Forest A and in Mesquite Grassland B, as the result of increased moisture availability. Of these two, Mesquite Grassland B has the higher potential yields because of the availability of good alluvial soils.

Therefore, in a given year, the distribution of certain plants may be altered during the spring and summer by the group. The degree to which this is done will affect the character of the strategies that the group intends to use in the fall, since their knowledge of and interest in these areas will be increased. The STRAT(i) • TASK(j) • DENSITY variable serves to associate with the collecting task the amount of prior effort taken to produce the observed yield. If no prior effort was given to altering the associated distribution, then the variable takes a value of 1. Otherwise, it can have a value up to the maximum potential increase for the particular resource in the particular zone. These values represent, in model terms, the increase in yield as the result of the group's activities, where yield is said to be directly related to prior effort.

The only additional structural change that needs to be made in the model is with regard to the basic set of rescheduling operators. Now that the group can potentially alter the density of the three plants, new sets of decisions must be made by the group members. These new decisions are represented by two new stochastic operators in the following way:

ω_9 = ALTER_DENSITY: This represents a decision to alter the density of the focal resource for a task. For an arbitrary strategy i and task j, the process is modeled stochastically in two stages:

1. Generate a random number RAND(ALTER_DENSITY_SEED) and apply this number to the random variable STRAT(i) • OPER(9) • USED that maps it into the set {yes, no}, where

$$\text{STRAT}(i)\cdot\text{OPER}(9)\cdot\text{USED} = \begin{cases} \text{“yes” if} \\ \text{RAND(ALTER\_DENSITY\_} \\ \quad \text{SEED} \le \text{STRAT}(i)\cdot\text{OPER}(9) \\ \quad \text{PROB} \\ \qquad\qquad \text{and} \\ \text{“no” otherwise.} \end{cases}$$

2. If the decision is in the affirmative, then RAND (ALTER_DENS_TASK_SEED) is produced and mapped by ALTER_DENS_TASK into the set {1, . . . 25} of indices for the available tasks.
3. Next ALTER_DENS_VALUE_SEED is used to provide a new random number that is then mapped into the set of $R^+ \in$ [1, MAX_DENS_ZONE]. MAX_ DENS_ZONE is taken from Table 31.9 and represents the maximum potential increase in yield as the result of incipient agriculture for the focal resource in the designated zone.

ω_{10} = SHUFFLE_DENSITY: This operator stands for the exchange of density specifications between two compatible tasks in a strategy i. It is determined stochastically by the following sequence of processes:

1. First a random number, RAND(SHUFFLE_DENS_SEED), is mapped into the set {yes, no} by the random variable STRAT(i) • OPER(10) • USED such that

$$\text{STRAT}(i)\cdot\text{OPER}(10)\cdot\text{USED} = \begin{cases} \text{“yes” if} \\ \text{RAND(SHUFFLE\_DENS\_} \\ \quad \text{SEED)} \le \text{STRAT}(i)\cdot \\ \quad \text{OPER}(10)\cdot\text{PROB} \\ \qquad\qquad \text{and} \\ \text{“no” otherwise.} \end{cases}$$

   STRAT(i)•OPER(10)•PROB stands for the probability that a favorable decision will be made to modify the strategy in this way, given the current rescheduling policy for the strategy.
2. If the decision is yes, then the seeds SHUFFLE_DENS_SEED_1 and SHUFFLE_DENS_SEED_2 are used to generate two random numbers. These are mapped into the set of available tasks for the strategy by SHUFFLE_DENS_TASK_1 and SHUFFLE_DENS_TASK_2 in the same manner as described for the previous shuffle operators.

3. If the two tasks are performed in the same vegetational zone and involve the same resource—either maize, beans, or squash—then the density specifications are exchanged such that

$$\text{SHUFFLE\_DENS:}\begin{cases} \text{STRAT}(i)\cdot\text{TASK(SHUFFLE\_DENS\_TASK\_1)}\cdot\text{DENS} \rightarrow \\ \text{STRAT}(i)\cdot\text{TASK(SHUFFLE\_DENS\_TASK\_2)}\cdot\text{DENS} \\ \text{and} \\ \text{STRAT}(i)\cdot\text{TASK(SHUFFLE\_DENS\_TASK\_2)}\cdot\text{DENS} \rightarrow \text{STRAT}(i)\cdot\text{TASK} \\ \text{(SHUFFLE\_DENS\_TASK\_1)}\cdot\text{DENS} \end{cases}$$

Aside from the above changes, the rest of the model for incipient agriculture is the same as with the preagricultural model.

Conclusions

After having made the above structural adjustments to our preagricultural model, we are now in the position to simulate the development of incipient agriculture by the group. In Part 8, we describe the results of running the agricultural model, where initially all the values for the resource collecting schedules defined in the preagricultural model are the same as they were at equilibrium for that model. The new density specification for each task is initialized to 1. This means that no distributional change has yet been made to any of the focal resources. In addition, since the group has had no experience with making decisions that affect the density specification, the initial probability of making each of the new decisions is initially set to 50-50.

The group, while initially displaying the scheduling behavior described in Part 6, is now in a position to employ these new agricultural procedures as well. It is our goal in Part 8 to see whether the group is able to develop incipient agriculture in this context. If it does, it will be interesting to see the manner in which these rescheduling processes will take place and the new mix of resource acquisition strategies produced.

PART 8: SIMULATING THE ACQUISITION OF INCIPIENT AGRICULTURE

Introduction

In Part 6 we observed that our preagricultural model, which was based on a very simple set of decision rules, achieved a stable level of performance. The relative mix of plants collected by the model group compared well with the existing archaeological data for the last essentially preagricultural level (Zone D) in Guilá Naquitz. This correspondence suggests that the occupational remains found in the cave represent the results of collecting activities for a group that had attained a stable performance relationship with its environment. Since the model group achieved this relationship via a series of gradual adjustments in its resource collecting and policy-making behavior, it is possible that the real inhabitants could have achieved it in this way as well.

This stable performance level was maintained by the group through a specific set of decision-making and resource collecting adaptations. These adaptations focus on the sequence of different performance environments observed by the group. Since critical resources are more available in wet years than in dry ones, it is to be expected that the group members should adjust their resource collecting and decision-making strategies in order to reflect these differences. In particular, the group developed specialized collecting strategy for both wet and dry climatic types. While a third category (average years) was included, the group effectively behaved as if these were dry years. Thus, given three possible categories, the group only used two. As it turns out, a binary classification scheme not only is consistent with local ethnological data, but also reduces the probability that the group will employ a specialized collecting strategy in the wrong year type. Indeed, if the group was not able to reduce the probability of misapplication in this way, it would be hard to develop and maintain these specialized collecting strategies.

Therefore, in addition to a set of general purpose strategies that were used by the group in any type of year, there were two classes of special purpose strategies. *Dry-year strategies* used relatively fewer available species and emphasized plants with high yields. In *wet-year strategies*, a greater variety of plant species were used and less emphasis was placed on high-yield resources. This increased flexibility exhibited by wet-year collecting strategies was mirrored by their associated policy-making schemes. Wet-year strategies were more likely to be altered if they performed poorly than were dry-year strategies. The decision-making policies associated with wet-year types were much more complex than those of dry-year types. At the stable performance level, there is little turnover between wet-year strategies and general-purpose strategies. That is, special-purpose strategies rarely become general-purpose strategies, and vice versa. This is because few changes will lead to improvements in performance. While increased performance will allow a specific wet-year strategy to be employed in other types of years, there is little likelihood that this will occur once the system has achieved a stable level of performance. Such specialization effectively allowed the group to take advantage of good years through increased experimentation with certain resource collecting strategies.

Specialized strategies constituted 30% of the total number of currently available strategies. The remainder were general purpose in the sense that they had a good probability of being used, regardless of the type of year, because of their high performance. Since general-purpose strategies are relatively successful, their associated decision-making policies are quite conservative. Their focus is on resources with high yields in the vicinity of the cave. The mix of resources collected with these strategies compares very well with the cave remains.

These are the strategies that seem to determine the basic character of the cave remains and that represent in performance terms the best results of the group's experiences. It is important to note, however, that the special-purpose strategies that are archaeologically less visible still play an important part in the development of incipient agriculture.

In this chapter, we see if a group possessing the above specializations is able to incorporate incipient agriculture into its structure. Specifically, we want to know how the preexisting adaptations affect the way in which the group acquires these new agriculturally based tasks. Also, in the wake of these changes, we would like to know if the group is able to retain the decision-making and resource collecting specializations acquired earlier.

Acquiring Incipient Agriculture

Earlier we characterized the first stages of incipient agriculture as attempts to alter explicitly the densities of particular plants. Now we begin with a group in the stable performance state described above and simulate their acquisition of this new activity. We first observe how the group's current resource collecting and decision-making structures will affect this process. It might be expected that the initial experimentation will take place in wet years, since only wet-year decision-making policies are flexible enough to allow for such changes. It was mentioned previously that at the stable performance level there was little chance of a wet-year strategy improving its performance and displacing a current general-purpose strategy. However, wet-year strategies that incorporate useful innovations into their structure will have a decided advantage. Their increased performance should allow them to displace existing general-purpose strategies, and they will therefore be used more often in drier years. With increased use, these strategies will become more visible and certain aspects of their structure are likely to be copied by their competitors. This is in fact what the model group does.

As the simulation progresses, the frequency of use for each of the strategies begins to change over time (Fig. 31.14). These changes are shown in Table 31.10. Strategy 1, which characterized the group's basic preagricultural routine, was still used most often. However, after 100 time steps a number of changes had occurred. In particular, Strategy 2, which was a wet-year strategy in the preagricultural phase, was used much more frequently. Its policy was one of experimentation. Therefore, it was modified almost every time it was used.

These early modifications to Strategy 2 introduced a number of tasks associated with the incipient cultivation of beans in Thorn Forest A. Since this is the vegetation zone in which the cave is located today, it meant a refocusing of collecting strategies nearer the cave. This allowed the individuals to travel less but collect more by increasing the density of beans through cultivation in the zone near the cave. Such an adaptation would seem to be advantageous, in terms of both increasing the total yield and reducing the amount of effort spent collecting those resources. Indeed, judging from the associated performance curves, this is exactly what happened (Figs. 31.15–31.17). In each figure, the performance gradually improves from the onset of the simulation until tapering off after 90 time steps for the sample strategies in each year type.

Since Strategy 6 was also initially a wet-year strategy, we would expect to see a similar increase in its performance. This happened between Time Steps 100 and 150. During this period, Strategy 6 was used 27 times and was second only to Strategy 1 in usage (see Table 31.10). The increased performance of both wet-year strategies allowed them to be used more frequently. It also meant that they would displace other strategies in terms of frequency of use. This happened after approximately 150 time steps. By that time, the relative performance advantages of Strategies 3 and 9 had eroded to the point that they were used predominantly in wet years. Again, as their relative performances decreased, their corresponding probability of being modified increased. Since they were in a performance environment that was slightly different from that of the original wet-year strategies, one might expect differences in their modification policies as well. Now that other strategies had the innovations, the exchange of tasks between and within strategies was quite advantageous. As a result, the schema involving the recombination operations were employed more often in Strategy 3 to produce improved scheduling changes.

These changes produce an increase in the performances of both Strategies 3 and 9. This makes it possible for them to displace other strategies, and the cycle continues. The presence of this cycle of gradual improvement can be seen in Fig. 31.18. The figure compares the probability of modification of strategies selected to be used in wet years with those selected to be used in dry years. In the latter year type, this probability remains around .32. However, that for wet years is consistently higher. This reaffirms our earlier observations that increased experimentation takes place in wet-year types. In addition, the probability of modification for wet years seems to oscillate in a fairly regular way. This oscillation is due to the replacement of the current wet-year strategies (with their high probabilities of modification) by new ones. As they begin to produce successful changes, their probabilities of use increase as well. Once they become good enough, they displace other strategies, their decision policies become more conservative and the cycle continues. Through this cycle of improvement and displacement, the group gains experience in the scheduling of these new agricultural tasks.

After 450 time steps, all strategies in the population have become involved in this cycle, including Strategy 1. Strategy 1 was the most frequently used during the preagricultural phase and the last to become involved in the displacement cycle. As long as it performed well, there was no need to make substantial adjustments. Although agricultural tasks were initially associated with this strategy, they were assigned a low priority of use. In fact, their priority was so low they were never actually used. This is shown in Table 31.11. From the table we can also see that they were not actively employed by the strategy until about 250 iterations. At this point in time, other agriculturally based strategies were beginning to

```
STRAT                1  DAY              2  STARTED AT           2  DAY             3  STARTED AT          3  DAY         4  ST
ARTED AT                4  DAY              5  STARTED AT           5  DAY             6  STARTED AT          6  DAY      7
STARTED AT                 7  DAY              8  STARTED AT          10  DAY             9  STARTED AT         11  DAY
            10  STARTED AT           12
STRAT                2  DAY              2  STARTED AT           3  DAY             3  STARTED AT          5  DAY         4  ST
ARTED AT                7  DAY              5  STARTED AT           8  DAY             6  STARTED AT         10  DAY      7
STARTED AT                13  DAY              8  STARTED AT          15  DAY             9  STARTED AT         16  DAY
            10  STARTED AT           18
STRAT                4  DAY              2  STARTED AT           3  DAY             3  STARTED AT          4  DAY         4  ST
ARTED AT                6  DAY              5  STARTED AT           7  DAY             6  STARTED AT          8  DAY      7
STARTED AT                 9  DAY              8  STARTED AT          10  DAY             9  STARTED AT         11  DAY
            10  STARTED AT           12
STRAT                8  DAY              2  STARTED AT           2  DAY             3  STARTED AT          3  DAY         4  ST
ARTED AT                4  DAY              5  STARTED AT           5  DAY             6  STARTED AT          6  DAY      7
STARTED AT                 8  DAY              8  STARTED AT           9  DAY             9  STARTED AT         10  DAY
            10  STARTED AT           11
PROT_EFF   5488.3590000   SAM_CAL_EFF238668.1000000   CAL_AVE584058.0000000   PROT_AVE 13430.8700000   AREA_AVE      2.4471550
CULTIV_AREA AVE. TFA       0.1780000
CULTIV AREA AVE TFB       0.1741436
CULTIV AREA AVE MGA       0.0000000
CULTIV AREA AVE. MGB        0.0000000
STRAT                2  MODIFIED
ALTER_RES               2              8
ALT_AREA               2             10
SWAP_EXT_1               2              1              2
SWAP_INT               2              8              19
SHUFFLE_AREA               2              2              6
SHUFFLE_RES               2              16              1
CURRENT_YR_TYPE               3
SAMPLE_SIZE               4
THE STRATEGIES IN THE SAMPLE ARE              1              2              4              8
TOTAL CULTIVATED AREA FOR TFA IS        0.7119999   TOTAL CULTIV AREA TFB      0.6965745
TOTAL CULTIV AREA MGA        0.0000000   TOTAL CULTIVATED AREA FOR MGB IS      0.0000000   TFBB      0.3010000   TFBC      0.3955745   MGAC
     0.0000000   TFAM       0.4340000   TFAB       0.2779999   TFAC       0.0000000   MGBM      0.0000000   MGBB      0.0000000   MGBC
     0.0000000
STRAT_NO                1   STRAT_CAI680202.2000000   STRAT_PROT  17037.0300000
STRAT_PROT_EFFORT   7066.3750000   STRAT_CAL_EFFORT282124.7000000
STRAT_LAST_USED               3   STRAT_PERF       0.0000000   STRAT_TRIP_DIST      22.0000000   TASKS      12.0000000
STRAT_OPERATORS FOR               1       0.5000000       0.0110000       0.2010000       0.0620000       0.4480000       0.0090000

     0.5000000       0.1280000
STRAT_NO                2   STRAT_CAI643794.0000000   STRAT_PROT   9891.0080000
STRAT_PROT_EFFORT   3814.6240000   STRAT_CAL_EFFORT248289.3000000
STRAT_LAST_USED               3   STRAT_PERF       2.0000000   STRAT_TRIP_DIST      26.0000000   TASKS      19.0000000
STRAT_OPERATORS FOR               2       0.8200000       1.0000000       0.9899999       1.0000000       0.9000000       0.8300000
     0.7800000       0.9000000
STRAT_NO                3   STRAT_CAI118991.2000000   STRAT_PROT   6719.2500000
STRAT_PROT_EFFORT   3070.8900000   STRAT_CAL_EFFORT 54382.3800000
STRAT_LAST_USED               2   STRAT_PERF       4.0000000   STRAT_TRIP_DIST      32.0000000   TASKS      17.0000000
STRAT_OPERATORS FOR               3       1.0000000       0.9000000       0.9899999       1.0000000       0.9800000       0.5100000
     0.9400000       0.8300000
```

Fig. 31.14A. One time step from a sample computer run for the incipient agricultural simulation. This was Time Step 66 in the simulation, which falls early in the process of adding incipient agricultural tasks to the model group's repertoire. "CURRENT_YR_TYPE:3" indicates that this was a "wet" year. In this time step, Strategies 1, 2, 4, and 8 were selected to be used.

```
STRAT_NO                  4  STRAT_CAI662460.3000000   STRAT_PROT 10383.2000000
STRAT_PROT_EFFORT  4090.6170000   STRAT_CAL_EFFORT260986.2000000
STRAT_LAST_USED               3  STRAT_PERF      2.0000000  STRAT_TRIP_DIST     20.0000000  TASKS     12.0000000
STRAT_OPERATORS FOR           4         0.5000000        0.1250000        0.3600000        0.0400000        0.1140000        0.3900000
    0.5000000       0.1500000
STRAT_NO                  5  STRAT_CAI242694.2000000   STRAT_PROT  3202.6500000
STRAT_PROT_EFFORT  1451.1750000   STRAT_CAL_EFFORT109968.8000000
STRAT_LAST_USED               2  STRAT_PERF      4.0000000  STRAT_TRIP_DIST     30.0000000  TASKS     17.0000000
STRAT_OPERATORS FOR           5         0.8300000        0.7100000        0.5500000        0.9899999        0.9200000        0.5200000
    0.5000000       0.8300000
STRAT_NO                  6  STRAT_CAL 76158.1200000   STRAT_PROT  1644.0970000
STRAT_PROT_EFFORT   671.8040000   STRAT_CAL_EFFORT 31119.4100000
STRAT_LAST_USED               1  STRAT_PERF      5.0000000  STRAT_TRIP_DIST     29.0000000  TASKS     20.0000000
STRAT_OPERATORS FOR           6         0.5599999        0.5300000        0.7000000        0.9200000        0.8699999        0.5300000
    0.6000000       0.5300000
STRAT_NO                  7  STRAT_CAI246940.6000000   STRAT_PROT  2715.9860000
STRAT_PROT_EFFORT  1005.8930000   STRAT_CAL_EFFORT 91456.9400000
STRAT_LAST_USED               2  STRAT_PERF      4.0000000  STRAT_TRIP_DIST     26.0000000  TASKS     16.0000000
STRAT_OPERATORS FOR           7         0.3370000        0.0030000        0.4000000        0.4710000        0.1760000        0.0010000
    0.5000000       0.2620000
STRAT_NO                  8  STRAT_CAL349776.5000000   STRAT_PROT 16412.2700000
STRAT_PROT_EFFORT  7305.9960000   STRAT_CAL_EFFORT155704.6000000
STRAT_LAST_USED               3  STRAT_PERF      2.0000000  STRAT_TRIP_DIST     21.0000000  TASKS     11.0000000
STRAT_OPERATORS FOR           8         0.5360000        0.1100000        0.1350000        0.3090000        0.0170000        0.6100000
    0.5000000       0.5450000
STRAT_NO                  9  STRAT_CAL106438.3000000   STRAT_PROT  6058.1450000
STRAT_PROT_EFFORT  2458.0090000   STRAT_CAL_EFFORT 43185.8600000
STRAT_LAST_USED               2  STRAT_PERF      4.0000000  STRAT_TRIP_DIST     30.0000000  TASKS     18.0000000
STRAT_OPERATORS FOR           9         0.0570000        0.5000000        0.0680000        0.1479999        0.1370000        0.0010000
    0.5000000       0.0110000
STRAT_NO                 10  STRAT_CAI239865.9000000   STRAT_PROT  8091.3400000
STRAT_PROT_EFFORT  3648.3750000   STRAT_CAL_EFFORT108155.2000000
STRAT_LAST_USED               2  STRAT_PERF      2.0000000  STRAT_TRIP_DIST     31.0000000  TASKS     15.0000000
STRAT_OPERATORS FOR          10         0.0160000        0.6799999        1.0000000        0.3830000        0.2399999        0.0030000
    0.5000000       0.9400000
CURR_TIME                66
SAMPLE OPER. PROB             1         0.5889997
SAMPLE OPER. PROB             2         0.3114998
SAMPLE OPER. PROB             3         0.4214995
SAMPLE OPER. PROB             4         0.3527496
SAMPLE OPER. PROB             5         0.3697495
SAMPLE OPER. PROB             6         0.4597497
SAMPLE OPER. PROB             7         0.5699999
SAMPLE OPER. PROB             8         0.4307497
THE SAMPLE OPERATOR AVE. IS       0.4381247
```

Fig. 31.14B. In this, the second half of the same time step shown in Fig. 31.14A, the current performance averages for the strategies are printed out so that the analyst can observe any changes in performance.

TABLE 31.10
Changes in the Frequency of Use for Each of the Ten Strategies (1–10) over 450 Time Steps[a]

50 time steps	100 time steps	150 time steps	200 time steps	250 time steps	300 time steps	350 time steps	400 time steps	450 time steps
1	1	1	1	1	1	1	3	1
10	8	6	10	7	10	10	1	10
8	2	10	3	9	9	9	6	8
3	4	8	7	10	3	8	10	2
4	10	4	5	5	4	3	7	6
9	3	2	8	3	7	7	2, 4, 8, 9	9
5	5	5	4	8	5	2	—	7
7	7	7	6	4	8	6	—	3
6	9	9	2	2	6	4	—	5
2	6	3	9	6	2	5	5	4

[a]Strategies are listed in order of decreasing frequency of use.

TABLE 31.11
The Position of Agricultural Tasks in Strategy 1 over Time[a]

Time step	Strategy (tasks used from left to right)
50	[(0 0 0 0 0 0 0 0 0 0 0 0) 0 0 0 0 0 0 0 0 0 C 0 0] used
250	[(0 0 0 0 0 0 0 0 0 0 0 0) 0 0 0 0 0 C 0 0 0 C C 0] used
550	[(0 B 0 0 0 C 0 0 0 0 0 B B) 0 0 M 0 0 0 0 0 0 0 C B] used

[a]An agricultural task is specified by either B (beans), C (cucurbits), or M (early maize); all other tasks are represented by 0.

erode the performance advantage of Strategy 1. As its relative performance decreased, it became more likely to be changed. These early changes were successful and led to policy changes that encouraged experimentation. By Time Step 550, Strategy 1 had again become the most frequently used strategy. Now, however, it included an agricultural component.[5]

Table 31.11 also provides a good example of how a particular strategy incorporates agricultural tasks into its structure. Beginning at a position of low priority, they are seldom used. As time goes on, they increase in frequency and priority of use. At Time Step 550, incipient agriculture is employed through the strategy. It would be interesting to know what policy changes were involved in producing these shifts. More generally, is there a difference in the way that the innovation is accepted by a strategy at this stage as compared to earlier ones?

We have already seen that the policy at 200 iterations favored both the generation of new tasks and the rescheduling of existing ones. The mutation operators produce new tasks, and the recombination operators attempt to reschedule the existing sequence of tasks to accommodate these changes. However, by the time Strategy 1 is in a position to be modified (about 300 iterations), the situation is somewhat different. A number of agricultural tasks are already available and in regular use by other strategies at that time. Strategy 1 also has associated with it a number of agricultural tasks that it just has not used. As a result, one would expect that the strategy's policy would be modified to favor the recombination-type operators. Such operators can function to alter the priority of execution for a task, as well as allowing the "copying" of tasks from other, more successful strategies. Looking at Table 31.12, this possibility seems confirmed. Both "swap external" and "swap internal" (exchanging the priorities of tasks in the same strategy) increase in probability by 550 iterations. This increase is preceded by a corresponding increase in the tendency to change the zone or resource associated with a task. While this pattern is similar to the succession of operators at the start of the

[5]In Strategy 1 (preagricultural version), the group spent nearly 80% of its time in Thorn Forest A, not even leaving it until Day 6; the second most frequently visited vegetational area was Mesquite Grassland B, where they spent nearly 20% of their time. About 8 days out of every 10, they concentrated on only 1 major resource per day, typically harvesting an area of 1700–2900 m^2/day. Less frequently, they might devote a day to harvesting 2–3 minor resources. When incipient agriculture was added to Strategy 1, the harvest of 900–1400 m^2 of domestic squash became an additional task. This is about half the area per day over which they had to range for wild products, illustrating the way agriculture further reduces search area.

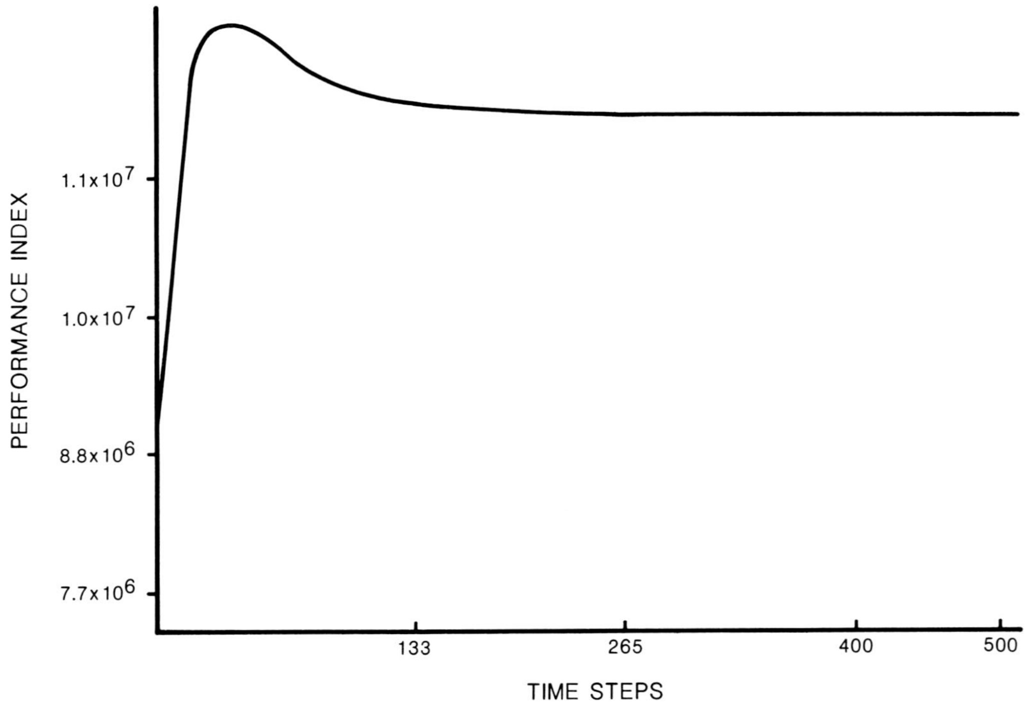

Fig. 31.15. Changes in the performance index for "wet" years brought about by experimentation with incipient agriculture.

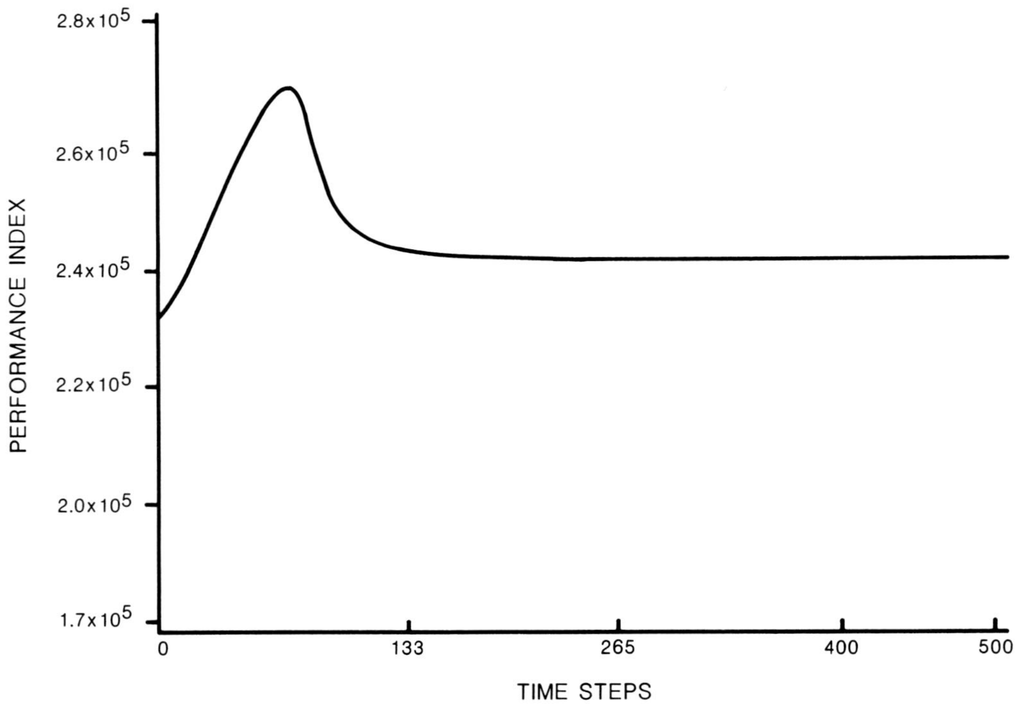

Fig. 31.16. Changes in the performance index for "dry" years brought about by experimentation with incipient agriculture.

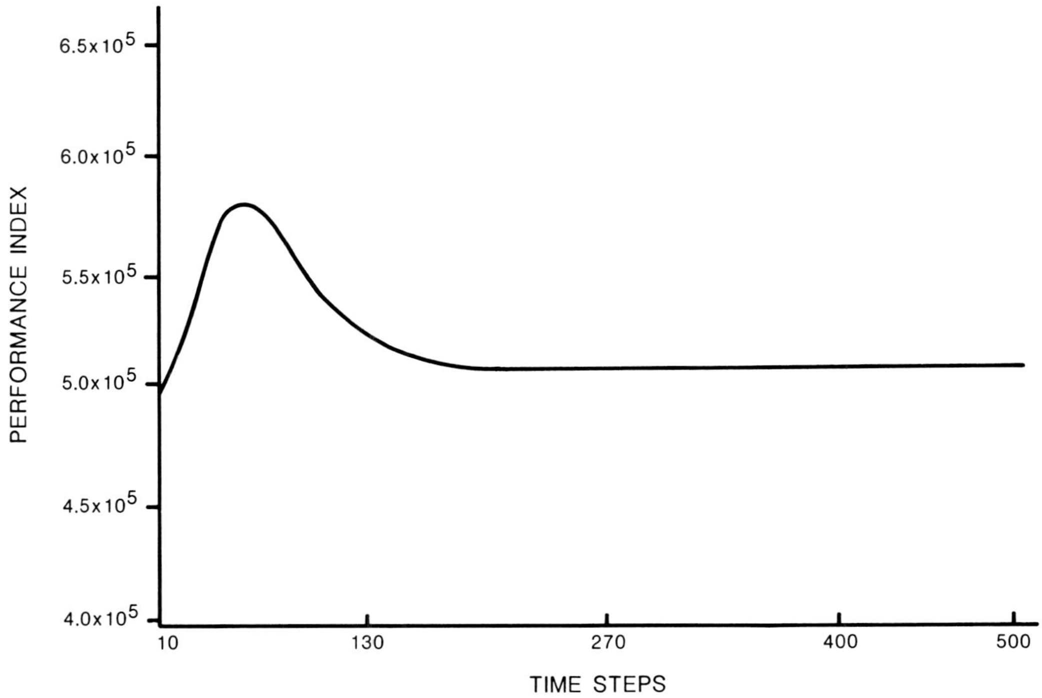

Fig. 31.17. Changes in the performance index for "average" years brought about by experimentation with incipient agriculture.

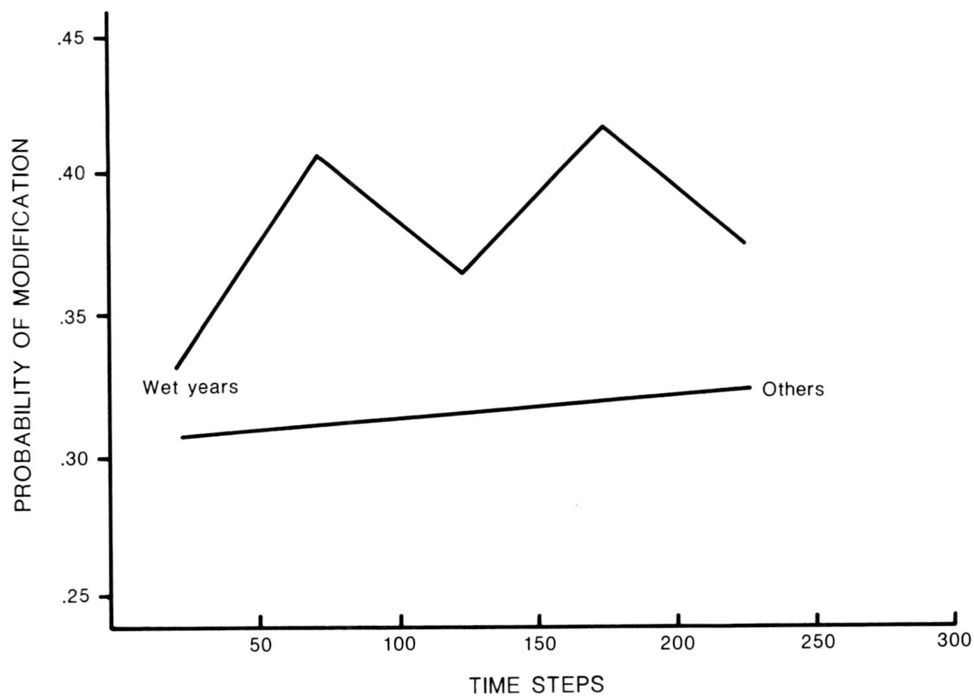

Fig. 31.18. The probability of modifying sample strategies in "wet" years as opposed to other year types.

TABLE 31.12
Changes in Operator Probabilities for Strategy 1 over Time

Time step	Alter resource	Alter zone	Alter area	Swap internal	Swap external	Shuffle resource	Shuffle zone	Shuffle area	Shuffle density	Alter density
50	.5	.5	.01	.20	.06	.44	.44	.009	.12	.5
250	.5	.5	.01	.20	.06	.44	.44	.009	.12	.5
550	.003	.003	.01	.86	.86	.065	.065	.062	.94	.14

simulation, the tendency to change resources is not as high now as it was. This is to be expected, since there are already a number of agricultural tasks available to the strategy. The problem here is not so much to generate new tasks but to schedule efficiently the ones already available.

Once the above operators have supplied the strategy with several viable agricultural tasks, then the shuffle operators come into play. These recombination operators then "fine tune" the strategy by exchanging resource, zone, and density specifications among the group of agricultural tasks for the strategy. Therefore, the policy shifts required to change strategies after the group has had some experience with the innovation are significantly different from those initially used. The emphasis in the latter stages is on rescheduling the original sequence to accommodate the new tasks. This rescheduling takes the form of large-scale shifts of tasks between and within strategies, as well as fine tuning the sequence of agricultural tasks.

We have, therefore, been able to observe a gradual shift in wet-year policies over the course of the simulation. Initial emphasis is on those operators that can generate new tasks. As these new tasks become more common, increased emphasis is on operators that efficiently reschedule the strategy in order to accommodate the new tasks. It soon becomes harder to make successful "coarse-grained" adjustments, so the emphasis shifts to operators that produce "fine-grained" changes within the group of agricultural tasks for a strategy. After 500 iterations, the system again achieves a stable performance state, with Strategy 1 again being the most frequently used strategy. At this point, few changes are made to the set of scheduling strategies (less than 12% of the operators have more than a 50% chance of being applied).

It is interesting to note that the incipient agricultural system takes around 550 iterations to reach a stable level of performance, while it took only 300 iterations to reach stability in the preagricultural phase. This slowdown has to do with the decision-making adaptations acquired by the group during the preagricultural phase. Such adaptations seem designed to maintain the majority of structures intact, while deferring experimentation to wet years. This made the system much more conservative. Since the current structures represent the results of a long period of experience with the environment, it seems "logical" that the system would be reluctant to change. Such logic is represented in the model by a low probability of making successful changes, and therefore a low probability of making adjustments. As a result, major policy shifts were not made by group members until after they had been exposed to the innovation for over 50 iterations. Even then, the changes diffused through the population quite slowly. It turns out, however, that this acceptance process has a number of advantages. In order to appreciate what these are, we need to know a bit more about the external results of the processes just described. In particular, we want to find out how these processes change the relative emphasis placed on each of the available resources by the group. This allows a direct comparison between the sequence of model changes and the corresponding changes in the archaeological record. If they match up, we will then have reason to believe that the model's behavior is a good characterization of the real system.

Spatial and Temporal Changes in Resource Use

In the previous section, we focused on the general adjustments made by each component in our network of adaptive systems to the presence of the new innovation. Here we look in more detail at the specific changes in resource use exhibited by the model. These changes are then compared with the related archaeological data.

We begin by tracing changes in the total number of tasks associated with the 3 species involved in incipient agriculture. These changes are displayed graphically in Fig. 31.20. Notice that initially, 32 out of the 250 available tasks are ascribed to these 3 species. Of these, only 13 are actually of high enough priority to be used in their respective strategies. In general, it would seem that initially the 3 species played a rather minor role in the group's resource collecting activities. However, after approximately 50 iterations, the number of tasks involving maize, beans, and squash began to increase. At this point, the group began to increase its experimentation during wet years as the result of initial successes. The current wet-year strategies improve their performance and temporarily displace other strategies whose relative performance is now worse. These strategies are relegated to infrequent use during wet years and are subject to increased experimentation. Such changes produce new incipient agricultural tasks that improve their relative performance. These, in turn, displace others and the cycle continues. These developments produce an almost linear increase in the number of tasks for the 3 species through 350 time steps. Over this interval, a new task is generated every 8 time steps.

After 350 iterations, agricultural tasks are present in every strategy, and the slope of the curve approaches 0. At this

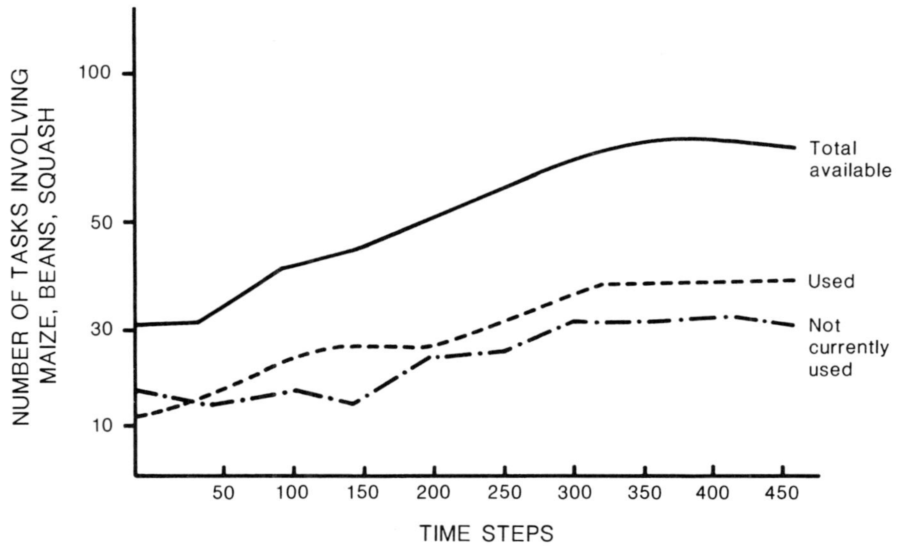

Fig. 31.19. Changes in the total number of tasks involving maize, beans, or squash through time.

point, the probability of generating a new task that is better than the one it must replace also is reduced. Thus, as the system approaches a stable performance state, almost 30% of all the available tasks relate to incipient agriculture.

Since the individuals only have a limited amount of time to apply a strategy, they may only be able to complete a portion of the tasks they have scheduled. In Fig. 31.19, the total number of tasks for the three species is broken down into two categories. The first represents the number of these tasks that are important enough to be used. In the second, the number of tasks that are specified but not used is given. Notice that after the first 50 iterations, the number of agricultural tasks used is always greater than the number of those tasks that are not. This relationship continues to hold as the system attains a stable state. There, 60% of all available tasks for these three species are used. Also, they represent about 30% of the number of tasks used by the group.

While we have seen that the use of agricultural tasks increases over time, it is important that we discern the relative contribution made by each of the three species. In Figs. 31.20–31.22, the changes in the number of tasks for each species is displayed. It is apparent that both beans and cucurbits account for a major share of the observed increase. Maize is used to a significantly lesser extent. Since the yield per unit area of this early maize is quite low, it is interesting that the group uses it at all. In fact, because of its low original productivity, the group uses it, for the most part, in wet-year strategies. In this situation, its yields are high enough to make it competitive. This result suggests that early experimentation with maize was performed in wet years and that this fostered its continued use for a long period of time. This would guarantee its availability for repeated use and set the stage for the emergence of new, more-productive forms in the future via genetic modification.

It is now quite apparent that group members are slowly incorporating agricultural tasks into their repertoire. But in doing so, what changes are they making in their relative use of other resources? The answer to this is given in Table 31.13. In this table, the change in the total number of tasks for each seasonally available resource used in the model is given. Notice first that the total number of tasks has increased. Since the maximum distance that individuals can travel in the 10-day period is constant, the reason for this must lie in the group's increased ability to consolidate resource collecting tasks over time. This corresponds well with earlier observations that particular strategies maintained incipient agriculture primarily within Thorn Forest A, the vegetation zone that is nearest to the cave in our model (just as it is under today's climatic conditions in the eastern Valley of Oaxaca). This trend served to refocus collecting activities in Thorn Forest A; in fact, it seems that one of the advantages for this earliest agriculture rests in its ability to encourage such shifts.

More evidence for the subsequent refocusing of collecting activities is provided by Fig. 31.23. There we see that in the preagricultural phase, collecting was done primarily within the three vegetation zones nearest the cave, with only slight preference given to collecting resources in Thorn Forest A. With the onset of incipient agriculture, there is a shift in emphasis towards Thorn Forest A and away from the other two zones, which are farther from the cave. Thorn Forest A is also the principal zone in which the group practices incipient agriculture. Mesquite Grassland B, which is the farthest zone from the cave, exhibits little change as far as the number of nonagricultural tasks executed there is concerned. Since this zone was furthest from the cave, activities that caused the individuals to travel this distance had to be of relatively high preference. For example, the collection of mesquite was quite important to the survival of the group and represents a sizable

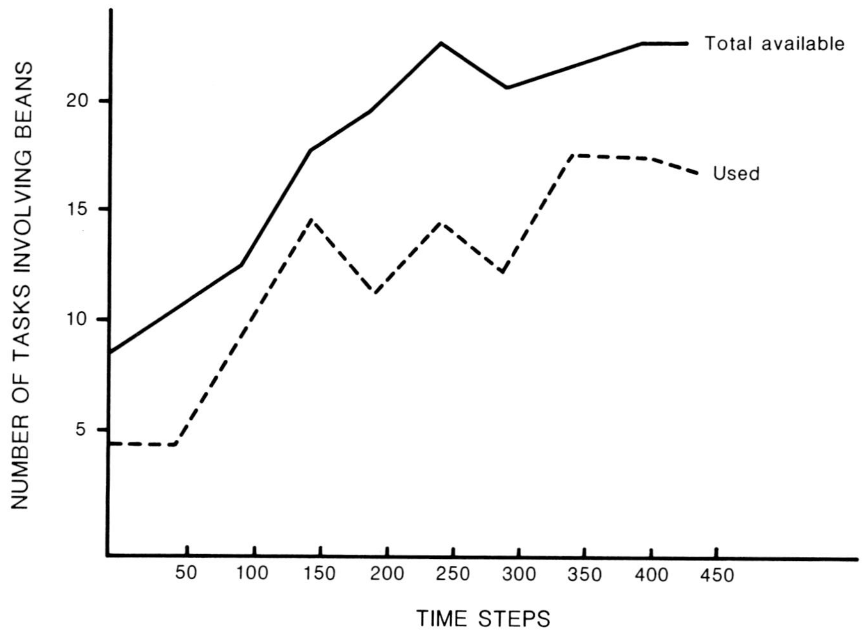

Fig. 31.20. The number of tasks allocated to beans through time.

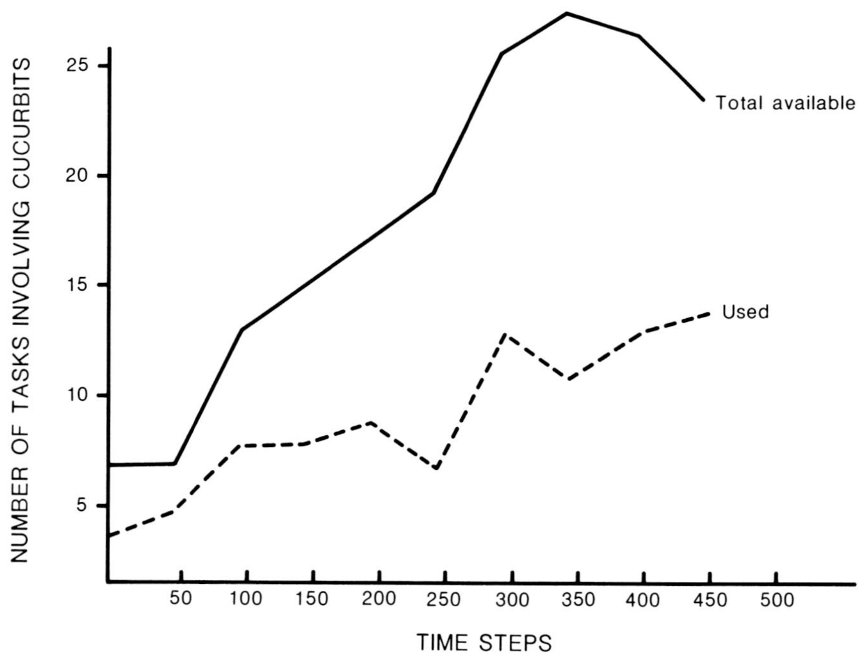

Fig. 31.21. The number of tasks allocated to cucurbits through time.

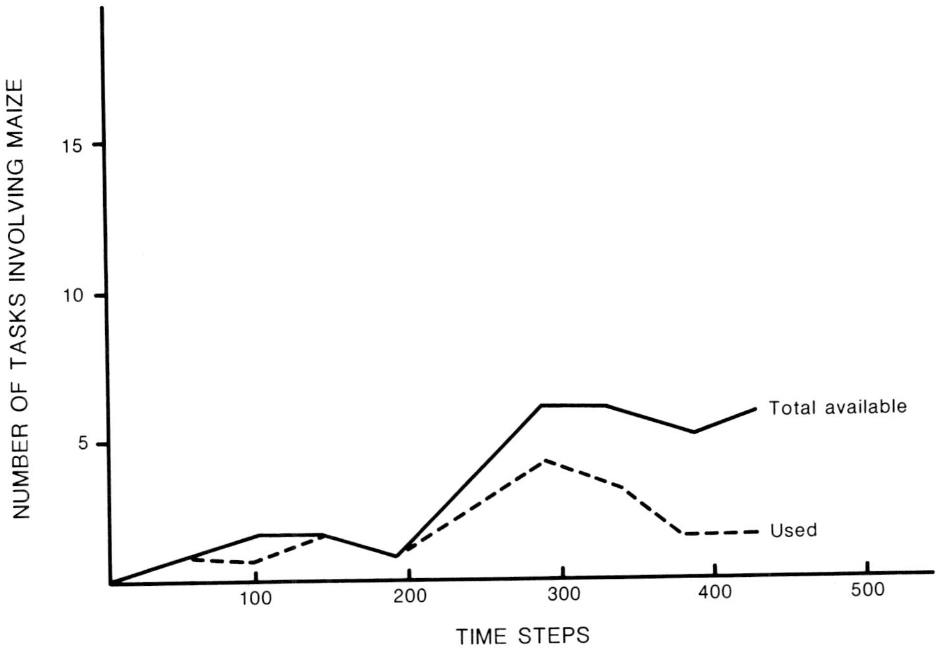

Fig. 31.22. The number of tasks allocated to maize through time.

TABLE 31.13
Changes in the Number of Tasks Used for Each of the Seasonally Available Resources in the Model

Plant	Number in the pre-agricultural phase (0 time steps)	Number at stable performance state for incipient agriculture (450 time steps)	Change in the total number over the period 0–450
Piñon nuts	6	3	−3
Hackberries	17	11	−6
Susí nuts	24	16	−8
Nanches	13	5	−8
Opuntia fruit	46	37	−9
Opuntia nopales	41	31	−10
Acorns	5	9	+4
Guajes	5	7	+2
Mesquite	17	21	+4
Agave	29	32	+3
Beans	9	23	+14
Cucurbits	7	23	+16
Maize	0	7	+7

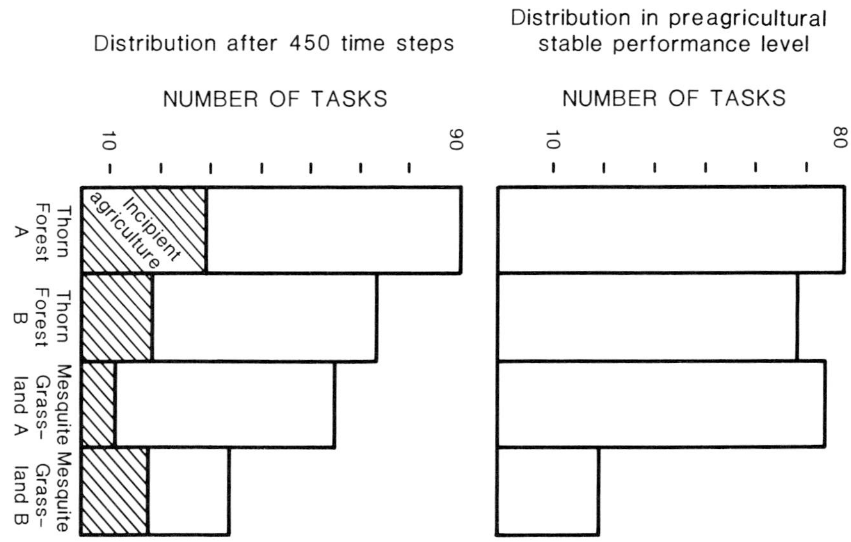

Fig. 31.23. The number of tasks per vegetation zone in all current strategies.

fraction of the available tasks for this zone. In this version of the model, the yields associated with incipient agriculture were originally not enough to produce substantial changes in activities in Mesquite Grassland B. However, in Part 9, when additional variability is added to the model, such changes are observed.

While the number of nonagricultural tasks remains about the same for Mesquite Grassland B, the number of agricultural tasks represents a large fraction of the activities carried out there. It appears that even now, the group is beginning to realize the advantages of agriculture in this zone nearest the river. (This zone would in later periods become the site of the first sedentary villages in the valley, as the group's activities shifted away from Thorn Forest A.)

Incipient agriculture, by allowing higher densities of plants nearer the cave, surely allows the individual to spend more time collecting and less time traveling. As a result, the group is able to increase the number of collecting activities. These increases are most marked in those species found closest to the cave in our model. For example, acorns—which are found principally in Thorn Forest A—are increasingly emphasized; this is true as well for *guaje*, nanche, and agave. Mesquite also increases, presumably because the group is now spending more time in Mesquite Grassland B. Whether this is because they are now trying out agriculture on the riverine alluvium or for other reasons, it is significant.

Another effect of adding incipient agriculture to this system is that the individuals will now spend less time on marginal (low density, low-yield) resources in return for increased emphasis on beans, squash, and maize. Since the yields associated with those early methods of cultivation are relatively low, only four wild species exhibit a reduction in use. These are piñon, hackberry, nanche, and *susí*. The number of tasks involving *susí* decrease by 34%, those involving hackberry by 36%, those involving piñon by 50%, and those involving nanche by 62%. *Susí*, while very productive in Thorn Forest A, is much less so in Thorn Forest B and Mesquite Grassland A. Therefore, we see a decrease in tasks for *susí* collection outside of Thorn Forest A, perhaps due to reduced yields and increased travel distance.

While the use of highly productive resources such as *susí* was cut back along the margins of their ranges, less-productive species experienced a reduction throughout their ranges; this was true of nanches, piñon, and hackberries. (Piñon and nanches are today found primarily in Thorn Forest A.) Of the resources employed in the model, only wild beans were less productive than piñon and nanche; and when the yields associated with incipient agriculture surpassed both of these, their relative use decreased.

Hackberries are in a similar situation in Mesquite Grassland B. Of the four major resources available there, hackberries are the least productive. It is therefore understandable that they are the only species in this zone to exhibit a decrease in use.

In other words, the main impact of adding incipient agriculture to the system was to refocus collecting activities to vegetation zones nearer the cave. Secondarily, it caused a decrease in the use of both productive species along the margins to their ranges and marginally productive species near the center of their range.

In terms of the model, adding agriculture meant that the group could increase the densities of selected plants near the cave. This increase in efficiency allowed the group more time to collect plants nearby. Also, the yields associated with incipient agriculture ultimately made it more productive than collecting certain other species. Since it would take time for the group to achieve such yields, one might expect that the first changes in use observed in the model resulted from the increases in efficiency. Because the individuals do not have to travel quite as far to collect what they did before, they have additional time in which to accomplish more activities. This results in an increase in the number of tasks allocated to certain species. As the group's experience increases, the yield associated with incipient agriculture continues to increase as well. In addition, certain activities are no longer as productive relative to incipient agriculture, and they are done less often. Therefore, the model predicts that decreases in the use of piñon, hackberries, *susí*, and nanches should follow initial increases in other species such as acorns, *guaje*, and cucurbits.

The model predicts not only the expected shifts in the use of the major resources but the order in which these shifts will take place as well. Now it remains for us to see how well these predictions correspond to the archaeological data from the cave. In Table 31.14, the raw frequency counts for each of the major plant resources found in the cave are given for Zones D, C, and B1. The relative frequency of every plant in terms of the total in each of the zones is given in Table 31.15, as well as the net change between Zones D and B1. Table 31.16 compares the net changes in the observed total counts and relative frequencies with the predicted changes.

In Table 31.16, notice that for 10 of the 12 resources, the predicted changes match those changes in total frequency observed between Zones D and and B1. Only acorns and nopales exhibited trends not consistent with the model. This represents an 83% agreement between the model and the data for these resources. It also suggests that the process that produced the changes in the model group may also be the principal factor affecting the changes in the behavior of the real group. Therefore, what appears on the surface to be a very complicated pattern of resource shifts can be characterized quite parsimoniously in terms of the few basic adaptations discussed above.

The concordance statistic between the two sets of predictions was computed and found to have a value of .83, where 1.00 denotes complete correspondence. The probability that the two sequences were generated by independent processes, given the present level of agreement, is .009. This is sufficient to reject our original hypothesis in favor of one that postulates dependence between the model and the real system.

Up to this point we have dealt only with changes in raw frequencies. What happens to our correspondence if we compare the observed changes in *relative frequency* with our model's predictions? From Table 31.16 we note that 9 out of the 12 changes in relative frequency were predicted by the model. Acorns and nopales again do not correspond with the model's predictions. Also, nanches exhibited an increase in relative frequency, contrary to what the model would predict.

Still, 75% correspondence is quite good, especially since relative frequencies are a much more conservative index of change than raw counts. With this in mind, the null hypothesis that the observed changes in relative frequencies shown by the real system do not correspond with those for the model was tested. This time the concordance statistic had a value of .75 and the probability that this would be produced by two independent processes is .04. Given that we stated that any probability less than .10 would be sufficient ground for rejection, the null hypothesis that the two sets of values are unrelated is rejected.

While the general shifts in resource use for the model compare well with the changes in the archaeological record between Zones D through B1, what about the *order* in which these shifts occur? The model suggests that the initial effects of incipient agriculture would be to increase the use of a number of species, particularly in the vegetational zone nearest the cave. As the yields associated with early agriculture improve, there will be a decrease in the use of some plants whose yields are less than those associated with incipient agriculture. Does the sequence of changes hold true for the archaeological data as well? Looking again at our tables, one notices that 9 of the 12 changes in relative frequency from Zone D (8750–7840 B.C.) to Zone C (7450–7280 B.C.) were increases. *All* of the 6 species predicted to exhibit increases do just that. In the later transition fron Zone C (7450–7280 B.C.) to B1 (6910–6670 B.C.), the opposite trend occurs. During this period, 9 of the 12 species exhibited decreases. In addition, 5 out of the 6 species that were predicted to increase do increase.

From this it is quite apparent that the shifts in the archaeological record bear a remarkable correspondence to those predicted by the model. To illustrate just how well they match up, let us now rank each of the 12 species in ascending order of predicted increase, with the species with the largest predicted increase ranked Number 12. The species can next be ranked in ascending order based on the observed increase in relative frequency between Zones D and B1. The two rankings are then compared in Table 31.17. The question now is, How well do they correspond?

Again we use the concordance statistic to test the null hypothesis that the two data sets were generated by unrelated processes. The observed concordance was computed to be .70. The probability that two statistically unrelated processes could have generated sequences that match this well is .08. This is low enough for us to again reject the null hypothesis. As a result we conclude that the processes governing the behavior of the real group and the model must be quite similar. Not only does the model predict the general behavior of the group in terms of changes in resource use, but it emulates the relative magnitude of these changes as well.

This correspondence between the changes in the archaeological data and the real system can be used to verify another aspect of the model. It was assumed originally that the basic cycle of evaluation and modification was performed annually by the group. The fact that the sequence of predicted changes over a particular number of time steps matches the observed changes in the dated archaeological record allows us to estimate, in a very general way, the average length of time required by one cycle in the model. In other words, we can estimate what the length of time between cycles would be if the model were to generate its changes over the same period of time as the real system. It turns out that the length

TABLE 31.14
Raw Frequency Counts for Each of the Major Plant Species in Zones D through B1

Plant	Zone D	Zone C	Zone B1	Net change from D to B1
Agave quids and leaves	53	216	56	+3
Hackberry seeds	416	969	240	−176
Cucurbit remains	6	50	21	+15
Susí seed coats	237	68	22	−215
Guaje pods	11	247	105	+94
Nanche seeds	49	196	46	−3
Opuntia seeds	348	751	143	−205
Opuntia nopales	28	49	45	+17
Bean pod valves	17	43	111	+94
Piñon nut hulls	94	80	31	−63
Mesquite seeds	59	554	1833	+1774
Acorns	3181	1570	892	−2290
	4500	4793	3545	

TABLE 31.15
Relative Frequency of Occurrence of Each Major Species in Zones D through B1

Plant	Zone D	Zone C	Zone B1	Change in relative percentage from D to B1
Agave quids and leaves	.0117	.0450	.0157	+.0040
Hackberry seeds	.0924	.2021	.0677	−.025
Cucurbit remains	.0013	.0104	.0059	+.0046
Susí seed coats	.0526	.0141	.0062	−.0462
Guaje pods	.0024	.0515	.0296	+.0272
Nanche seeds	.0108	.0408	.0129	+.0021
Opuntia seeds	.0773	.15669	.0403	−.0370
Opuntia nopales	.0062	.0102	.0126	+.0064
Bean pod valves	.0037	.0089	.0313	+.0276
Piñon nut hulls	.0208	.0166	.0087	−.0121
Mesquite seeds	.0131	.1155	.5170	+.5039
Acorns	.7071	.3275	.2516	−.4555

TABLE 31.16
Comparison of Predicted Changes in Relative Frequency with Actual Changes between Zones D and B1[a]

Plant	Predicted change in frequency	Observed change in total frequency	Observed change in relative frequency
Piñon nut hulls	Decrease	Decrease	Decrease
Hackberry seeds	Decrease	Decrease	Decrease
Susí seed coats	Decrease	Decrease	Decrease
Opuntia seeds	Decrease	Decrease	Decrease
Opuntia nopales	Decrease	*Increase*	*Increase*
Nanche seeds	Decrease	Decrease	*Increase*
Acorns	Increase	*Decrease*	*Decrease*
Guaje pods	Increase	Increase	Increase
Mesquite seeds	Increase	Increase	Increase
Agave quids and leaves	Increase	Increase	Increase
Bean pod valves	Increase	Increase	Increase
Cucurbit remains	Increase	Increase	Increase

[a]Italicized words represent changes that do not match the predictions of the model.

TABLE 31.17
A Comparison of the Predicted Order of Relative Increase for Each Species Relative to the Others with the Observed Order of Increase Between Zones D and B1

Plant	Predicted ranking in ascending order of increase	Observed ranking in ascending order of increase
Piñon nut hulls	2	5
Hackberry seeds	3	4
Susí nut hulls	4	2
Opuntia seeds	5	3
Opuntia nopales	6	9
Nanche seeds	1	6
Acorns	10	1
Guaje pods	9	10
Mesquite seeds	8	12
Agave quids and leaves	7	7
Bean pod valves	11	11
Cucurbit remains	12	8

of time falls generally between 1 and 1.5 years, depending on one's asssumptions. Given the number of assumptions that have to be made in order to do these calculations, the order of magnitude is strikingly close to what would be expected if the real results were actually produced by an annual sequence of local decisions like those in the model. This only serves to confirm our other findings, all of which indicate a detailed correspondence between the observed behavior of the real system and that of the model.

One would expect that this behavioral correspondence stems from a basic similarity between the decision-making structures of the two systems. As we have seen, the behavior of the model is strongly determined by the structure of its acquired adaptations. This suggests a more detailed examination of the performance advantages offered by these adaptations. In particular, it has already been suggested on theoretical grounds in Part 2, and experimentally in Part 6, that changes that reduce the impact of decision-making "mistakes" on the system's performance would be favored. Such mechanisms certainly would reduce the variation associated with the system's performance. This is particularly important in the present situation, where the environment is inherently unpredictable to start with.

Conclusions

While factors such as population pressure may be important at later periods in the Valley of Oaxaca, it was hypothesized in Part 1 that the initial structuring of resource use in the valley was principally the result of information-processing considerations. These considerations would seem to be of special importance in Oaxaca, where the distribution and density of resources fluctuate markedly from year to year. With this in mind, the main goal of the simulation was to see how well a basic model of hunter–gatherer information processing can explain the archaeological data associated with the acquisition of incipient agriculture.

The beginnings of incipient cultivation generated two basic types of changes within the model. First, there was an initial refocusing of activities nearer the cave. The ability to increase the densities of certain plants meant that less effort had to be spent collecting them. This left the group with additional time in which to collect other resources nearby. Also, with experience the yields associated with incipient agriculture began to increase. This made incipient agriculture more productive than certain other resource collecting activities. As a result, the group began to shift its effort away from collecting and toward incipient agriculture.

The main effects of incipient agriculture were, then, to produce a spatial refocusing of collecting activities about the cave as well as a displacement of less-productive strategies. The shifts in resource use resulting from these changes corresponded quite well with the observed changes in relative frequency for the major plants found in the cave. This strong correspondence suggests that the archaeological data may have been produced in a similar fashion.

Information-processing considerations are quite helpful in describing the behavior of the model group in an uncertain performance environment. Decision-making mistakes in such a context tend to produce large-scale system instability, as shown in Part 6. Adaptations that reduced the likelihood of these errors would be quite advantageous. The model group in fact employed a number of adaptations that do just that. It seems highly unlikely that the real system would be able to display such a detailed correspondence with the stable model unless it, too, was able to reduce the possibility of error. In addition, the types of adaptations used by the model correspond with known hunter–gatherer behavior. This suggests that the adaptations used by the real system may not have been too far removed from the ones in the model.

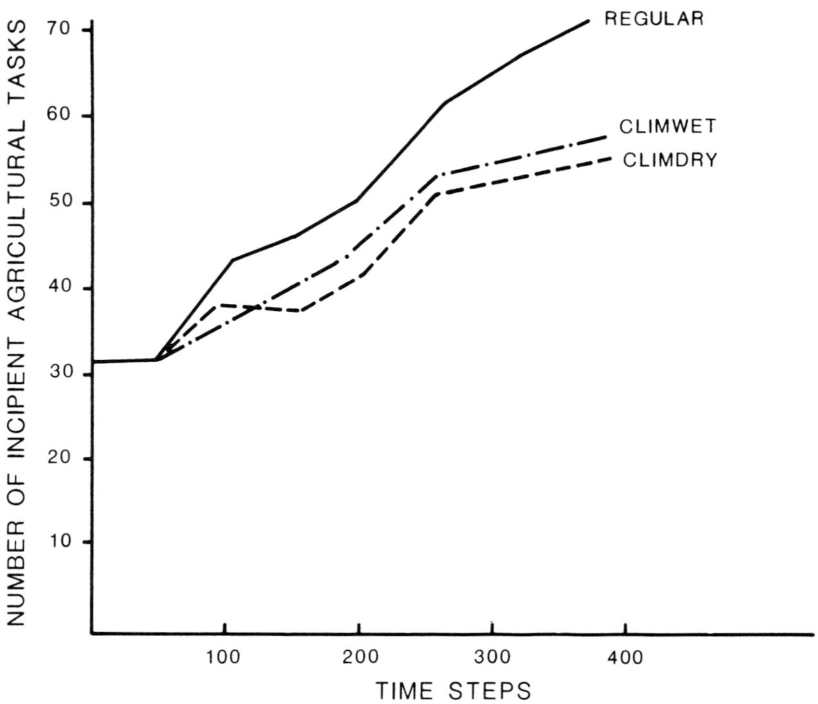

Fig. 31.24. The number of incipient agricultural tasks acquired through time for the three experiments **REGULAR**, **CLIMDRY**, and CLIMWET.

PART 9: SIMULATING CLIMATIC CHANGE AND POPULATION GROWTH

Introduction

Having now demonstrated a resonably good fit between the performance of our model and the evidence from the archaeological record, we are in a position to "experiment" by changing one variable of the model and holding all others constant. In Part 9, we simulate changes in two aspects of the model, rainfall and population. There are several reasons for selecting those variables. One reason is that both climatic change and population growth have been suggested as causal factors in the acquisition of incipient agriculture. A second reason is that either, or both, variables might have affected the movement of agriculture from the piedmont barrancas out onto the riverine alluvium. A third reason is that the climate of the Naquitz phase is thought to have been somewhat drier than today's, and we need to see how our model would respond to drier conditions. The latter observations should help make up for the fact that our environmental data were not collected under conditions identical to those of the Naquitz phase.

Prior to Part 9, our simulations were carried out using a fixed group size (five persons) as well as a fixed probability for the occurrence of each year type (wet, dry, average). A year with average rainfall had a .5 probability of occurrence, whereas our wet and dry year types each had a probability of .25. About 550 time steps after the first signs of incipient cultivation, our model group had achieved a stable adaptation, characterized by a mix of resource collecting strategies that correspond well with the plant remains from Zone B1 of Guilá Naquitz. In the model, these strategies emphasized incipient agriculture primarily within Thorn Forest A, the vegetation zone in which the cave is located under today's climatic regime. However, through increased trial and error (and genetic improvement in domesticates), one would eventually find that agriculture is more productive in Mesquite Grassland B, the vegetation belt associated with riverine alluvium. By 1500 B.C., almost all agricultural communities were located so as to take advantage of this alluvium. One of our goals in Part 9, therefore, is to ascertain what variation in model parameters would encourage increased planting in Mesquite Grassland B.

We therefore allow some of the model's basic demographic and climatic variables to change over time, observing their effects on the rate and extent to which the model group develops incipient agriculture. Three fundamental measures are used to estimate the extent to which agriculture is acquired. The first is the number of distinct tasks devoted to incipient cultivation by the group over time. Second, we observe the relative amounts of use of the four vegetation zones (Thorn Forest A and B and Mesquite Grassland A and B) by our model group. Third, we examine the extent to which the group has brought about changes in the density of maize, beans, and squash.

Making the Climate Wetter

In our first experiment, we decided to simulate the gradual onset of a wetter climatic phase. The probability that a wet year (>600 mm) would occur was increased by .0005 each model time step, while the probability that a dry year (<420 mm) would occur was decreased by .0005 each time step. This meant that the climate slowly became wetter and wetter over an extended period of time.

Some results are shown in Fig. 31.24, where REGULAR indicates the acquisition of incipient agricultural tasks under today's "regular" climate (50% average, 25% wet, and 25% dry years) and CLIMWET indicates the same acquisition under our simulated wetter climatic conditions. Note that there is a marked reduction in the rate at which the model group increased its number of incipient agricultural tasks under wetter conditions. There are probably several reasons for this, perhaps including some difficulty on the group's part in recognizing and adjusting to the continual .0005 change. However, one important possibility that emerges from an examination of the output is that pressure for efficiency is lowered. The increased number of wet years allows a lot of trying out of new strategies (many of them representing no improvement), and the decreased number of dry years reduces the rate at which poor strategies are eliminated.

Making the Climate Drier

Our second experiment was the reverse of the first: We increased the probability that a dry year would occur by .0005 each time step while decreasing the probability that a wet year would occur by .0005. In this way we simulated the onset of a drier climatic phase that would gradually become more xeric as time went on. Once again, some results are given in Fig. 31.24, where CLIMDRY indicates the acquisition of incipient agricultural tasks under our simulated drier conditions. Note that the model group did even more poorly than under CLIMWET conditions, apparently because they became more conservative; the reduced number of wet years allowed them fewer opportunities to try out new strategies, while the increased aridity put more pressure on them to be efficient without taking risks. Once again, the difficulty of detecting and adapting to the steady decrease in moisture may have been a contributing factor.

Making the Climate Uniform

Having modeled wetter and drier climatic phases, we found ourselves growing increasingly curious about the tradeoffs between climatic uniformity and variation. From the beginning, our model had featured an unpredictable succession of wet, dry, and average years, and even our simulated wet and dry climates had retained this strong annual variation. This was in keeping with one of the general themes of the whole Guilá Naquitz research project, which was an investigation of the way hunter–gatherers deal with variation. But what if variation had been substantially less?

Our third experiment, therefore, consisted of observing the behavior of the model group in a situation where the two extreme year types—wet and dry—were not present. For each of the computer runs during this experiment, therefore, every time step was programmed to be an average year. Figure 31.25 displays the performance of our model group during 500 time steps of this uniform climate and compares it with the maximum efficiency reached by the group under our original (REGULAR) climatic conditions. Two features of this new graph are particularly interesting. First, the performance of the group under uniform climatic conditions was not as good, on the average, as when all three year types were present. Second, the group's performance oscillated up and down over time, rather than reaching a plateau as it had under our original (varied) climatic conditions. Both these features suggest that lack of climatic variation tends to reduce the group's ability to isolate effective new collecting strategies.

We had not anticipated the results of the uniform climate experiment, but in retrospect they probably should not have been surprising. We had seen earlier that, under "regular" Oaxaca conditions, wet years were used to try out new strategies under low-risk conditions while dry years were used to highlight the more efficient strategies and delete the less efficient ones. In our uniform climate simulation, the loss of wet years reduces the opportunity for new strategies to be tried out and picked up; the loss of dry years reduces the pressure for efficiency, hence the "peaks" in the graph are lower. Without sufficient climatic variation, the group simply oscillates at a level that is well below the performance standards achieved under regular conditions.

There is also another way to view this phenomenon, one that is consistent with the overall theoretical approach of this volume. Variable environments provide a good context in which to develop and test new strategies because they contain so much information. Our new, uniform climate provides much less information about stress, plant productivity, efficiency, and all the other factors the group needs to have in its long-term memory in order to make good decisions. Deprived of this information, the group cannot adapt as well, or as rapidly, as it did under a regime of low-stress–average-stress–high-stress conditions. This discovery does more than confirm our belief that information must be a part of the ecological–evolutionary equation. It also shows us that unpredictable environmental variation, far from being the curse some have considered it, may in fact have been providing early man with a good context for long-term adaptive improvement.

Letting Population Grow

For the fourth experiment, we decided to let our model group begin to grow at a rate that would be reasonable for a foraging population. After looking through the literature for estimates of hunter–gatherer population growth (see, for example, Hassan 1981), we finally decided to let our group increase at a rate of .002% every 100 model time steps. We reasoned that even this small rate of increase, if allowed to continue long enough, should increase human demands on the plant foods of the region. Among other things, we wanted to see if our growing population would speed up its adoption of incipient agricultural tasks.

Some results are shown in Fig. 31.26, which compares our increasing forager population (POPINCR) with the stable group of five persons used in earlier simulations (REGULAR). Notice that the performance of the system in the wake of this gradual, long-term population growth is reduced relative to our original simulation. While there was an initial increase in the adoption of incipient agricultural tasks between Time Steps (or iterations) 50 and 100, the curve soon leveled off and fell some 20 tasks short of the REGULAR curve by Time Step 450. This was true of all computer runs made during this experiment. Evidently, our growing model population met its increasing subsistence needs as much by intensifying its use of certain wild plants as by intensifying agriculture.

This experiment shows that, for our model at least, the addition of population growth does not necessarily speed up the adoption of incipient agricultural tasks. This is significant, but we hasten to add that it should not necessarily be construed as a test of Cohen's (1977) population-pressure model. Cohen argues that population pressure is what caused foragers to turn to agriculture in the first place. Our simulation asks whether foragers, having already been given incipient agriculture as an option, would adopt it more quickly if their population were growing. The two questions are related, but they are not identical.

Letting Population Fluctuate

In our fifth experiment, we took a different approach to demographic change. Instead of letting population grow steadily, we decided to let our model group fluctuate randomly between four and six persons (with a mean of five) from one time step to the next. The underlying distribution of this new random variable was uniform. That is, the probability of a group having four, five, or six members was equally likely. In effect, this meant that an unpredictably variable population had to adapt to an unpredictably variable climatic regime. Although the source of such variability does not concern us here, fluctuations of this magnitude are not uncommon among local groups of hunter–gatherers. The results are shown in Fig. 31.26, where the curve for our varying microband (POPVAR) is compared with that of the REGULAR population.

It is clear that the POPVAR group adopted incipient agricultural tasks at a greater rate than the steadily increasing group (POPINCR), and superficially POPVAR shows a performance profile like that of REGULAR. However, striking differences appear when the output for all POPVAR runs are compared closely with those of REGULAR.

While both the REGULAR and POPVAR simulations accumulated comparable numbers of incipient agricultural tasks, the nature of these tasks differed extensively. In particular, increased short-term variation in population size put additional pressure on the system to generate more productive strategies, strategies able to satisfy a larger-than-average population in a drier-than-average year. As a result, within 150 time steps, the POPVAR group has assigned more than 20 tasks to incipient agriculture in Mesquite Grassland B; it had taken the REGULAR group approximately 500 time steps to achieve the same level of agricultural use of the alluvium. After 150 time steps, the POPVAR group was spending over 50% of its time in Mesquite Grassland B; the REGULAR group had spent less than 25% of its time there, even after 500 model time steps. The POPVAR group also increased crop densities for maize, beans, and squash in Mesquite Grassland B, while the REGULAR group did not.

Looking at the output in detail, it appears that the need to compensate for unpredictable short-term population shifts led to increased wet-year experimentation and dry-year testing, with the result being the production of more efficient strategies. This is not to say that these improved strategies were always necessary, but their existence provided the group with additional access to resources in those stressful times when "six persons" and "dry year" both come up in the same time step. Most significantly, although POPVAR did not adopt quite as many incipient agricultural tasks as REGULAR did overall, it did a far better job of moving agriculture out of the piedmont barrancas and down to the riverine alluvium of Mesquite Grassland B. It also did this better than POPINCR, which suggests that unpredictable annual variation in microband size might select more strongly for valley-floor agriculture than would gradual, long-term population increase. While this is obviously determined in part by the design of our simulation, it could also have implications for Binford's (1968) "density equilibrium model."

Summary

In the course of five separate experiments, we changed one of our model's variables while holding all others constant. First, the climatic regime was altered in three ways: (1) to become wetter, (2) to become drier, and (3) to remain uniformly average. Next, the size of the human population was altered in two ways: (1) to increase at a small, steady rate, and (2) to vary randomly between four and six with a mean of five.

Long-term climatic change, whether toward a wetter or drier climate, did nothing but slow the rate at which agricultural tasks were added to the model's repertoire. Even worse was the model's performance under conditions of climatic uniformity, which reduced the information in the system. Uniformity prevented the development of the binary strategy seen in our original model, where wet years allowed the trying out of new strategies and dry years weeded out the ones that were inefficient.

Our two kinds of demographic change had very different results. Our model group adapted to long-term population growth without increasing its commitment to agriculture; indeed, it failed to acquire as many agricultural tasks as the original model. The small but random fluctuation of the

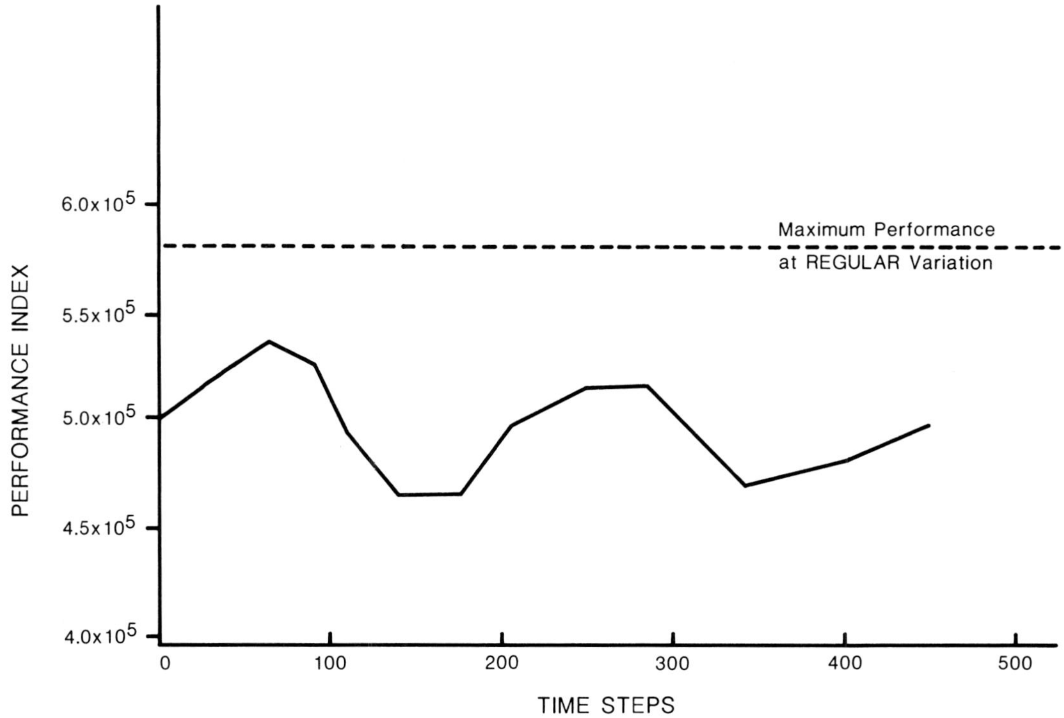

Fig. 31.25. Changes in the performance index brought about by experimentation with incipient agriculture in a situation with greatly reduced climatic variation.

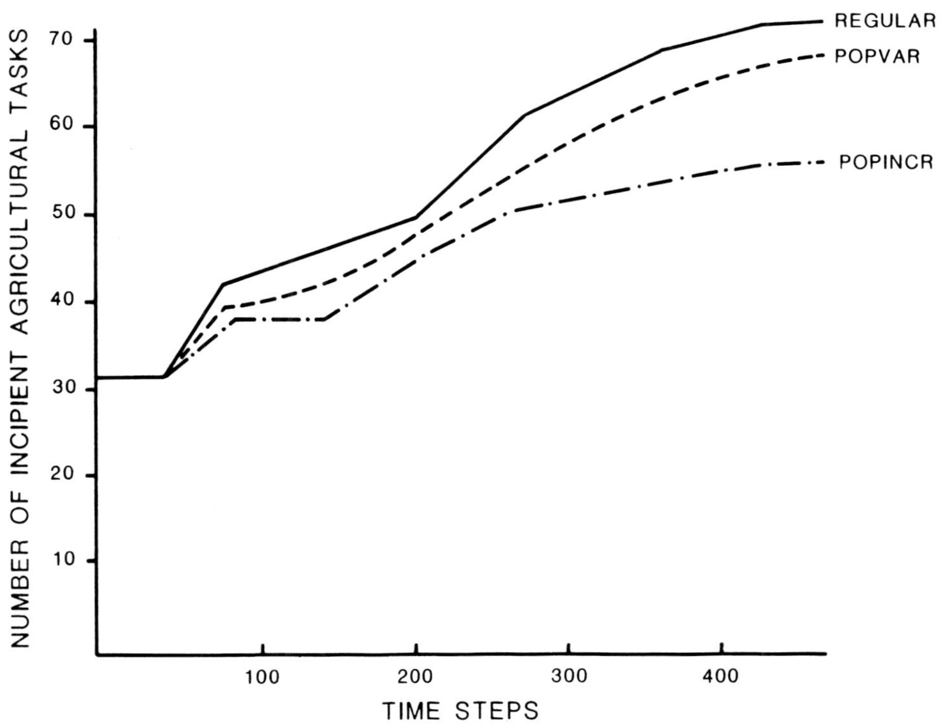

Fig. 31.26. The number of incipient agricultural tasks acquired through time for the three experiments REGULAR, POPVAR, and POPINCR.

population around a mean, while it led to an overall performance profile similar to that of the original model, shifted agriculture to the alluvium of Mesquite Grassland B more quickly. This is significant, because the increase of maize, bean, and squash densities on the alluvium is thought to have been an important step in the direction of Formative village life.

This allows us a partial answer to our earlier questions about what kind of variation in model parameters would encourage increased planting in Mesquite Grassland B. Given *two* sources of uncertainty—unpredictable climatic fluctuation and unpredictable population fluctuation—the model moved more quickly to increase the density of domestic plants in the environmental zone of their greatest potential productivity. In our experiments, at least, this double dose of uncertainty had a greater effect on the system than steady, predictable population growth. We should perhaps add that two sources of variation also mean two sources of information.

Obviously, there are dozens of other experiments that could be carried out now that we have the model set up, and we hope to explore other possibilities in the future. We content ourselves here with these five additional simulations, which can serve as an example of the way the model can be used to attack some of the questions raised in other sections of this volume.

PART 10: SUMMARY AND CONCLUSIONS

The Preagricultural Simulation

We began by developing an adaptive computer model of hunter–gatherer scheduling and resource collecting activities prior to incipient agriculture. In the model, information on the relative performance of a schedule with respect to other schedules used during the season is "fed back" to the system and used to perform the following functions:

1. Adjust the probability of using the schedule again, based on its performance.
2. Adjust the group's rescheduling policy for a strategy, based on the policy's observed ability to affect the schedule's performance.
3. Adjust the schedule's structure, using the current rescheduling policy associated with it.

Starting with a random sequence of resource acquisition schedules and an arbitrary decision-making policy, our preagricultural group, through this network of interactions, is able to construct a stable set of resource collecting schedules that demonstrate a highly significant correspondence with the results from Guilá Naquitz Cave. In the stable performance state, the relative emphasis placed on the available plant resources by the model schedules demonstrates a marked correspondence (80%) with the relative amounts of these same resources found in the cave for both Zones E and D. Therefore, on the basis of performance information alone, the system is able to develop a quite sophisticated sequence of collecting activities that produce results similar to those produced by the real occupants of the cave.

The key to the continued operation of this feedback cycle is that the measured performance of one strategy is always relative to the performance of the others used during that season. As a result, there are always strategies that do not work as well as others. They are more likely to be modified and therefore provide the grist for our evolutionary mill. If these new modifications improve a schedule's performance, others begin to emulate its successful aspects. Thus, over time, favorable combinations of the basic rescheduling operators are developed and employed more frequently. This acquisition of coadapted sets of decisions is facilitated by the intrinsically parallel nature of the data acquisition process, as described in Part 5. In this way, more and more information is mapped into the group's decision-making structure over time, until the resulting strategies are difficult to improve. Once this happens, rescheduling decisions become more likely to produce decreases in performance and the model group becomes more conservative with regard to change.

At this stable performance level for the preagricultural model, those high-yielding schedules that represent the best results of the rescheduling process are associated with very conservative rescheduling policies. Such "general-purpose" strategies are used often and seem to determine the basic character of the cave remains. While these strategies are used in each possible type of year, another set of "special-purpose" strategies is developed to be used in specific year types. The most interesting of these are the schedules used most often during wet years. They generally incorporate less-productive genera into their collecting repertoire and as a result are relatively poor performers. However, during wet years, the increased environment productivity makes them more productive.

Although wet-year strategies contribute less to the overall productivity than the more visible, general-purpose strategies, they perform an important function nonetheless. Since a wet-year strategy does not perform as well as others, one would expect that its associated rescheduling policy would be less conservative. The model group will make adjustments to such a strategy even when system-wide improvements become less probable. This has several basic implications:

1. There is more variability in resource scheduling behavior in periods of relatively high productivity than in drier years, when the system falls back on tried and true methods.
2. Although most of the group's behavior is determined by relatively high-performing schedules that it is reluctant to change, there is a less visible ongoing undercurrent of change generated within wet years.
3. It is quite possible that the cumulative success of this overall adaptation process is due to the unpredictable variability in rainfall. Wet years, while infrequent, allow more variability in behavior and offer a chance for

trying out new strategies. Dry years, on the other hand, place more selective pressure on the model group, and as a result it employs schedules with relatively high reliability and yield.

The selective pressure placed on the group can vary unpredictably, and it is this variation that may be an important factor in determining the rate of change within the system. If, for example, the group was exposed only to a sequence of dry years that constantly put selective pressure on the group, the wet-year strategies that introduce most of the variation into the system would disappear or never be used. The resource scheduling system as a whole would be extremely conservative. This is, in fact, how most ethnologists describe the scheduling behavior of a number of hunter–gatherer groups that have been pushed into extreme desert environments.

On the other hand, an overbalance of wet years would introduce a good deal of variability into the system, but the group would seldom have an opportunity to test the worth of these adaptations in a more strenuous environment. As a result, a number of adaptations might be acquired that would not stand up well within an environment that is more selective. Between these two extremes stands the environment of Oaxaca. There, wet years occur only about 25% of the time, but this is enough regularity to allow a steady number of new strategies to be tried out over time. On occasion, there is even opportunity to try some of the more successful wet-year schedules in more strenuous environments. If they fail to be competitive, they may well be changed again. This allows the group to maintain only those scheduling changes that do not erode the system's performance.

The Incipient Agriculture Simulation

The same basic adaptations described above continued to hold as the model group was given the opportunity to acquire incipient agriculture. The principal changes made to the existing set of schedules were as follows:

1. Reduce the use of some high-density species along the margins of their range, where the density gets relatively low.
2. Reduce the use of species with characteristically low density in all vegetation zones in which they occur.
3. Concentrate incipient agriculture in the vegetation zone nearest the cave—in our model, Thorn Forest A—with some activity occurring in Mesquite Grassland B near the river.
4. Replace less-productive species with more-productive species that could be planted closer to the cave so that the group had more time to collect additional wild plants in the same vegetation zone.

As a result, the model predicted a sequence of shifts in resource use over time as a result of this gradual acquisition of incipient agriculture. The shifts predicted by the model, based on the above principles, correspond with 80% of the observed shifts in the relative use of the same species from Zone D to Zone B1 of Guilá Naquitz Cave, where only one of the unpredicted shifts involved a species that exhibited a marked change. This correspondence not only is statistically significant but also provides a fairly straightforward explanation of what on the surface appears to be a rather complex unpatterned shift in resource use.

Thus, our predictions concerning wild plant use rescheduling during the initial phases of incipient agriculture in the Valley of Oaxaca correspond remarkably well with the archaeological data. The model's ability to predict the group's resource collecting behavior prior to the onset of incipient agriculture, as well as a significant proportion of the major shifts in resource use that did occur with incipient agriculture, lends strong support to our belief that day-to-day resource scheduling decisions play an important role in the formation of a group's subsistence activities. In this way we are able to construct a sequence of performance curves and associated rescheduling policies that, in turn, can be associated with the different occupational levels within the cave. The occupational levels and their associated remains, therefore, gain a new interpretation as part of a long-term process of cultural adaptation within the valley. Therefore, our model has allowed us to make the necessary connection between the information-processing structure of a group and the archaeological record.

In addition, we have suggested that certain feedback cycles were necessarily present within the real group's decision-making network prior to the advent of incipient agriculture. The presence of such cycles was necessary to generate the basic preagricultural scheduling behavior. They also provided the means by which the group was able to incorporate selected incipient agriculture tasks into its schedule of activities. In fact, it seems quite likely that these cycles formed the nucleus about which the later, more complex feedback cycles, based on genetic changes in maize, came about. In order to test this, we will need to extend our scheduling model to encompass the group's entire seasonal round of activities. This proposed extension will require data on the group's hunting activities, since hunting was an important activity at other times during the year. Data from Cueva Blanca, a slightly later site at which hunting was pursued more intensely than it was at Guilá Naquitz, should provide us with additional information on these activities. This extended decision-making model will form the basis for a more detailed simulation that will ultimately trace both the evolution of sedentary agriculture in the valley and the development of Formative decision-making systems.

Altering the System's Parameters

Finally, we ran a series of five experiments in which certain variables of the incipient agricultural model were changed while all other variables were held constant. In the first, the climate was allowed to become gradually wetter; in the second, it was allowed to become gradually drier. In the third, variation was reduced by making all years average ones. In the fourth experiment, population was allowed to grow slowly but steadily; in the fifth, it was allowed to oscillate around a fixed mean.

Although much work remains to be done, our basic findings were as follows:

1. Gradual, long-term climatic change, whether toward a wetter or a drier climate, did not speed up the rate at which incipient agricultural tasks were adopted.
2. The reduction of annual climatic variation slowed the adoption of agriculture significantly, since the model group no longer had (a) wet years in which to try out new strategies under low-risk conditions, (b) dry years which put pressures on them to be efficient, or (c) as much information available to inform their decisions.
3. Slow, long-term population growth, at a rate reasonable for hunter–gatherers, did not speed up the rate at which incipient agricultural tasks were adopted; nor did the resulting "mix" of resources compare well with that in Zone B1 of Guilá Naquitz.
4. Short-term fluctuation in population around a fixed mean *did* speed up the shift of agriculture from the piedmont (Thorn Forest A) to the river alluvium (Mesquite Grassland B).
5. The latter finding suggests that the prehistoric evolution of agriculture in the Valley of Oaxaca can be more realistically simulated using (a) a model with several sources of short-term uncertainty, even if the sources' ranges of variation are low, than (b) a single long-term "prime mover" process, such as climatic change or population growth, even if that process is allowed to continue for hundreds of time steps.
6. Finally, we do not want to give the impression that we have ignored the importance of long-term unidirectional trends in either climate or population. We are well aware that many parts of the ancient world experienced long-term unidirectional climatic changes such as the retreat of glaciers from Europe at the end of the Pleistocene or demographic trends such as the permanent and dramatic population increases of Formative Mexico. We also know from our simulations that such unidirectional long-term changes are difficult for our model group to adapt to, since the stored data from past situations become increasingly inappropriate for evaluating the group's performance under the permanently changed conditions of the future. Such unidirectional changes may require significant modifications in the group's performance evaluation process, and we hope in the future to investigate how this happened in Formative Mexico. All we are saying here is that (a) no such trends are apparent in the archaeological record for Preceramic Oaxaca, and (b) our computer modeling suggests that it is not necessary to postulate them in order to simulate the adoption of incipient agriculture in the Valley of Oaxaca.

32

Adaptation, Evolution, and Archaeological Phases: Some Implications of Reynolds' Simulation

Kent V. Flannery

One reason for undertaking a computer simulation is that it sometimes yields insights that are unexpected, or even counterintuitive. That happens because humans tend to assume the existence of causal relationships between variables that occur close together in time and space; in fact, cause and effect in complex systems may be temporally and chronologically separated. In this chapter we look at some possible implications of Reynolds' adaptive computer model, both expected and unexpected.

THE PREAGRICULTURAL STAGE

Reynolds' model microband starts from a position of ignorance (or better said, no memory of past use of the eastern Valley of Oaxaca), and within 500 time steps achieves a strategy not unlike that of Guilá Naquitz Zone D. It should be stressed that in this model, time steps are *not* years, however analogous they may seem; hence we do not know how long it might have taken a real-life group to achieve the same strategy.

The group is not instructed to "optimize" or to "maximize" any particular variable. They are merely told to make small changes in their foraging strategy each time step, to remember how well each strategy did, and to improve through time by repeating more successful strategies and disdaining less successful ones. The hypothetical group comes up with a two-part strategy, one for dry and average years, another for wet years; the former is more conservative, the latter more experimental, yet both show resiliency when the model parameters are changed. In dry years the group works harder, in wet years less, thereby reducing the difference in productivity between the two year types. They concentrate their efforts in Thorn Forest A and Mesquite Grassland B, suggesting that they are more interested in reducing search area than travel time. Their strategies seem to emphasize calories at the expense of plant protein, but without an optimal foraging analysis we would not want to guess how far short of optimizing either of these nutrients they might have fallen.

IMPLICATIONS

1. Reynolds' model group did best when generating its own decision-making policies, rather than responding to rates set by the computer programmer. Here we may see one of the many evolutionary advantages of a creature with multigenerational memory, logic, detailed perception, and decision-making ability.

2. Further, when Reynolds disconnected the feedback loop between the multigenerational memory and the decision-making apparatus, strategy changes began to wander from gradual improvement toward random fluctuation. This suggests that the use of a systems approach is a perfectly reasonable way to model cultural adaptation, despite the reservations thoughtfully expressed by Doran (1970) and

Salmon (1978). Indeed, when in the future it becomes possible for archaeologists to model human decision-making on a truly serious scale within the context of prehistoric ecosystems, it is hard to imagine how it will be done without some kind of systems approach (see, for example, Thomas 1971).

3. It is noteworthy that Reynolds' simulation gave his hypothetical group more classifications, more options, more tasks, and more possible strategies than they needed to achieve the level of efficiency they eventually displayed. It thus appears that an even simpler model, while less "realistic," might have produced a pattern similar in some of its most basic aspects.

4. One of the most interesting and unexpected results of Reynolds' analysis was the fact that his hypothetical foragers consistently acted as if there were only two kinds of years. Dry years and average years were treated as one type, which required a conservative strategy; wet years, perhaps because of the increased resources alluded to by Keene (1981b:237–238), featured a more experimental strategy. In other words, the group took Reynolds' tripartite year classification and reduced it to the binary opposition, wet–dry. As Reynolds points out, this is just what Mitla Zapotec speakers do today. The Zapotec tend to remember peak rainy or peak drought years, and believe (erroneously) that they can detect cycles in them (Kirkby 1974). If farmers perceive late spring rainfall as "high" (> 80 mm in our terms), they predict a wet year and act accordingly. If they perceive late spring rainfall as "low" (< 40 mm in our terms), they predict a dry year and act accordingly. Since Zapotec farmers have no rain gauges, they cannot possibly make all the fine distinctions we might make; they have focused on one aspect of local rainfall that allows them to reduce a great deal of ambiguity to a binary decision.

While structuralists and ethnoscientists might well conclude that wet–dry was a cognitive classification based on Zapotec perceptions of the world, the fact that Reynolds' computer did the same thing suggests that there may be perfectly good adaptive reasons for reduction to two strategies, one conservative and one experimental. I like to think that the implications of this discovery are that (a) cultural ecologists should not simply dismiss all native binary oppositions as mentalist epiphenomena and (b) structuralists should not be too quick to argue that the "code" in their informants' minds works independently of ecological adaptation.

5. Improvement of foraging efficiency proceeded more rapidly in dry years, presumably because of tighter selection pressure; it proceeded more slowly in wet years, presumably because of relaxed selection pressure. These findings will not startle those who believe in punctuated equilibrium or stepwise evolution, accelerated by periods of stress. But Reynolds' model also shows that the fastest improvement may take place in the context of an unpredictable stream of wet and dry years, suggesting that the role of annual variation in environmental conditions in setting evolutionary rates may have been underestimated. In Reynolds' simulation, adaptation in dry years benefited from the adoption of improvements that had been tried out first under the relaxed pressure of wet years. This may mean that adaptation moves fastest not simply under conditions of stress, but when its setting combines (a) stressful periods during which conservative and resilient strategies are rigorously selected with (b) more relaxed periods during which the pool of potential innovations can be increased without risk of elimination.

6. As if to underscore this point, one of the most interesting insights came about when Reynolds experimentally instructed the model to present his group with a sequence of exclusively "average" years. Intuitively, one might have expected the group to settle on a uniform and long-term stable adaptation under these conditions. Instead, the group began to wander aimlessly from strategy to strategy rather than progressing toward greater efficiency. Examination of the printout suggested to Reynolds that removing the variation in annual rainfall and vegetational productivity deprived the group of much of the information it needed to make good decisions. If this is so, there can be no clearer evidence for the importance of information in evolving systems.

7. The paragraph above raises one additional question. Could a group presented with an unchanging environment, as is often the case in applications of linear programming or optimal foraging theory, ever reach the point of optimizing its adaptation? Optimal foraging theorists might say "Yes, under conditions of competition, which are lacking in Reynolds' model." Perhaps under those conditions, the necessary stress would come from competition rather than annual variation. But what Reynolds' model suggests is that annual variation can speed adaptation even in the *absence* of competition; and this finding has implications for many human groups that, like the microband at Guilá Naquitz, do not seem to have had much in the way of competition from other humans.

8. As Reynolds' group approaches a stable adaptation similar to that of Guilá Naquitz D, they seem to be sacrificing plant protein in order to gain more calories and reduce their search area. In real life, of course, the group had access to protein from deer, cottontails, birds, and turtles, none of which are included in our simulation. But their sacrifice of plant protein should not be ignored, because it sets up an interesting situation—one in which the cultivation of a high-protein plant that could be grown in a small search area might be selected for.

A glance at Robson and Elias' Fig. 23.1 reveals that the Guilá Naquitz plants highest in protein were cucurbit seeds, followed at a distance by *susí* nuts, *guaje* seeds, and mesquite. When one considers that *susí, guaje,* and mesquite are all perennial trees or woody shrubs that take years to grow to maturity, the case for cucurbit cultivation becomes very strong. Cucurbits are annuals; they are "weedy camp followers" that do well on disturbed soils, such as the talus slope of an occupied cave (Cutler and Whitaker 1967); and they are 33.5% protein. There was probably no way the occupants of Guilá Naquitz could more easily have increased their plant protein, while continuing to lower search area, than by raising cucurbits for the seeds.[1]

[1]In general, wild cucurbits have a distasteful-to-nonexistent flesh; abundant and good tasting flesh is a product of centuries of genetic change following domestication.

9. One could almost phrase the above situation as a testable hypothesis: In cases where increased plant collecting efficiency sacrifices protein in the course of reducing search area, the group may either (a) increase hunting, (b) domesticate a high-protein plant species, or (c) both. In the case of the eastern Valley of Oaxaca, the evidence is still ambiguous. At both Gheo-Shih and Cueva Blanca, two later preceramic sites near Guilá Naquitz, projectile points and other hunting implements are more abundant than at Guilá Naquitz (Flannery *et al.* 1981:56–63); thus we cannot rule out the possibility that deer hunting was intensified in the Jícaras and Blanca phases. On the other hand, given our low estimated densities for deer in the Guilá Naquitz area (Chapter 24), deer hunting would certainly have worked *against* the reduction of search area seen in Reynolds' model. Squash cultivation, by producing localized high-protein patches near Guilá Naquitz, would have facilitated sedentism if this happened to be one of the group's goals; intensive deer hunting might have necessitated frequent moves that prevented sedentism. This may have been one of many factors involved in the long delay between the origins of agriculture and the origins of sedentary life in Oaxaca.

10. One way of accommodating the centrifugal pull of deer hunting and the centripetal pull of horticulture, of course, would be a shift from "foraging" to "collecting" in Binford's (1980) terms—or to a system like that of the San Bushmen, who forage for plants but organize their big-game hunting logistically (Binford 1982). The first alternative could account for 1.5-ha base camps (such as Gheo-Shih) in suitable farming localities, coupled with small hunting camps (such as Cueva Blanca) in the piedmont canyons. It could also be phrased as a testable hypothesis: When there is a conflict between highly localized plant resources and extremely dispersed animal resources, logistically organized collecting is a very likely solution.

11. While we would not want to push the implications of the Oaxaca case too far, we are struck by an interesting contrast between Mesoamerica and the Near East. In Oaxaca, where high-calorie plants are abundant and ungulate species few, one of the first plants domesticated was a protein source. In the Near East, where ungulate species are far more abundant and diversified, two of the first plants domesticated—wheat and barley—were carbohydrate sources.

THE INCIPIENT AGRICULTURAL STAGE

In an earlier section of this book, we discussed the possibility that the first plant domesticated in the New World may have been the bottle gourd (Heiser 1979:81–82; Lathrap 1977). I am intrigued with this hypothesis for two reasons: (1) It could mean that the later domestication of cucurbits (and other plants) in Mesoamerica was partly a long-term consequence of prior gourd domestication, and (2) it could mean that the earliest Mexican agriculture featured a plant used as an artifact, rather than as a food source freeing man from dietary stress or population pressure. This, in turn, would spare us from having to search for phantom population increases or nutritional deficiencies in an archaeological record that so far does not seem to display them.

Because of this uncertainty over the ultimate *causes* of domestication, Reynolds wisely did not attempt to have his model somehow "trigger" cultivation. What he did was to add cucurbits, beans, and primitive maize to the adaptation achieved in late preagricultural times—as if incipient agriculture were introduced into the eastern Valley of Oaxaca from a neighboring region, which could in fact be what happened. He then watched to see (1) if agriculture was adopted, (2) what form that adoption took, (3) how it modified the use of other plants, and (4) how it gradually changed the collecting strategy of his theoretical microband over time.

What Reynolds found is that acceptance of domesticates began as all innovations in his model began: They were used first in wet years, and only later, after they had proved reliable, were they introduced into dry and average years. Little by little, as the use of cultivars increased in importance, the "mix" of wild plants diverged from that seen in Zone D of Guilá Naquitz. In Chapter 24 we made note of the fact that during the transition from Zone E to Zone B1 of the cave, "the use of mesquite pods increased through time, while the use of acorns, piñon nuts, *susí*, and hackberry declined." Reynolds' hypothetical group made virtually the same changes and arrived at a pattern of wild plant use not unlike that seen in Guilá Naquitz B1. The group also gradually began to shift its main focus of activity away from Thorn Forest A (a vegetational zone crucial to preagricultural subsistence) and toward Mesquite Grassland B (a zone crucial to later agricultural adaptation). This supports MacNeish's (1967) Tehuacán model, in which agriculture began in the piedmont barrancas and only later spread to the alluvial valley floor.

IMPLICATIONS

1. Reynolds experimentally altered his model to observe the effects of (a) population increase, (b) a wetter climate, and (c) a drier climate. The results were interesting and have implications for several models of incipient agriculture. First, neither increased humidity nor increased aridity sped up the rate at which cultivars were adopted by the system. The latter increased stress so much that the system became overly conservative, while the first reduced it so much that there was less pressure for efficiency; in both cases, adaptation moved more slowly than under the original rainfall regime.

Second, steady population growth at a rate considered normal for hunters and gatherers did not speed the adoption of cultivars either. There was enough resiliency in the system to accommodate it—even if there were good evidence for population growth in the Oaxaca preceramic, which there is not. Causing the population to fluctuate unpredictably around a mean, while it moved agriculture out of the piedmont and onto the alluvium more effectively, also slowed the overall rate at which incipient agricultural tasks were adopted. Parenthetically, if such local fluctuations (which would be almost invisible archaeologically) did take place, they might be one more factor that would help to explain the incredibly long time it took agriculture to evolve in the Mexican highlands.

Because no trigger to set off agriculture is present in Reynolds' model, we cannot claim to have proven that population pressure or climatic change had nothing to do with the planting of the first squash seed. However, we doubt they had much to do with the planting of the first bottle gourd seed. And on the basis of our discussion in the two preceding paragraphs, we would say this: It is not necessary to invoke either population growth or climatic change to explain the adoption of cultivars in the Valley of Oaxaca. Quite simply, the system does not need them, and they only slow the process down when they are added.

2. Despite the importance of annual variation in giving our hypothetical group both the stress and information they need to reach a new adaptation, the fact that they work harder in dry years and less in wet years suggests a "satisfising" strategy that actually reduces the effect of the differences between years. In Chapter 1 we discussed Richard Ford's suggestion that one of the reasons agriculture was adopted was because it helped to "even out" the effects of annual variation. If anything, Ford's suggestion is strengthened by Reynolds' results. It would make the origins of agriculture not a startling innovation or a response to some new stress, but the extension of a strategy already displayed by the group in preagricultural times.

In Chapter 18 we indicated that the Valley of Oaxaca climate varies along two axes: *predictably* from dry season to rainy season and *unpredictably* from year to year. Preceramic foragers handled seasonal variation by moving from one environment to another, a considerable (but unavoidable) investment in travel time. They handled annual variation by reducing all year types to the binary pair wet–dry, for which they worked out two strategies—one conservative and more labor intensive and one experimental and less labor intensive. In addition to reducing the differences between years, this strategy provided the combination of risk reduction and innovation necessary to achieve an efficient yet resilient adaptation through time.

Squash cultivation increased the efficiency of this adaptation by further reducing the search area for a high-protein plant food. When beans and primitive maize were added, the search area was reduced even further—for Oaxaca Indians, unlike U.S. farmers, did not segregate crops by plot. They grew maize, beans, and squash in the same field, thus concentrating three species in the same patch of vegetation. This was not an Indian invention, since nature had provided the model (see Fig. 1.3, as well as Flannery 1973:291; Flannery and Ford 1972).

However, if nature provided the model, the Indians provided the disturbance. For example, in some thorn forest areas near Chilpancingo, Guerrero, and Valle de Bravo, Estado de México, any patch cleared for the cultivation of squash would ultimately be invaded by teosinte, wild runner beans, and other pioneer weeds of the local succession. It is interesting to speculate on the ways this might have influenced the Indians' choice of plants to cultivate. In such an area, one would not even have to plant to produce a field of *Zea, Cucurbita,* and *Phaseolus*; just clear a hectare of thorn forest, and the next year when you return, nature will already have done the job for you. Of course, if you wanted beans with nonshattering pods and squash with good-tasting flesh, you would have to select and plant.

AGRICULTURE AS AN EXTENSION OF THE PREAGRICULTURAL PATTERN

Let us briefly consider agriculture, then, in terms of the kinds of strategies and ecological relationships we saw in the preagricultural adaptation of the Naquitz phase.

1. As we have seen, agriculture lowers search time; the cultivator knows exactly where the densest stand is, since he created it.
2. It can convert a zonal environment into a patchy environment, where the milpa is the patch.
3. It can produce a patch that is "coarse grained" for maize, beans, and squash in the midst of an environment that is otherwise "fine grained" in MacArthur and Wilson's (1967) terms.
4. It can create a patch in which prehistoric people are likely to (a) spend more time, (b) travel less, and (c) deplete the vegetation more before they move on.
5. If Rindos (1984) is correct, it can also make an *r*-selected plant more *K*-selected, since humans are now investing more "parental care" in the seeds.
6. Rindos has also suggested that agriculture made humans more *r*-selected, but as we have seen, any significant increase in *r* in the Valley of Oaxaca took place millennia after the origins of agriculture and was probably associated with sedentary village life.
7. Agriculture can encourage human groups to move their basic settlement strategy from foraging, or "mapping on to resources" in Binford's (1980) terms, farther along the continuum toward collecting, or "logistically-based food procurement."
8. Agriculture can convert a negative-feedback loop into a positive-feedback loop, thereby setting in motion a series of changes that greatly modify the role of humans within the local ecosystem. One way to illustrate this is to return to one of the simple systems diagrams we used in Chapter 1. That diagram, the one showing the relationship between humans, primitive maize, and mesquite trees, is repeated as Fig. 32.1A of this chapter.

To recapitulate: groves of mature mesquite along the Mitla River yield an average of 183.6 kg of edible portion per ha (Chapter 18), much more than the 60–80 kg/ha estimated for the earliest corncobs of the Tehuacán sequence by Kirkby (1973:Fig. 48). Under such conditions, the best strategy was to continue to cultivate maize in the piedmont barrancas of Thorn Forest A while harvesting mesquite pods on the alluvium of Mesquite Grassland B. The two plants did not compete directly. In May or June, foragers would note the appearance of green mesquite pods and perhaps camp nearby to collect them in July or August. These harvests probably aided in the dispersal of mesquite seeds, since evidence from Guilá Naquitz shows many were never eaten. By September or October, foragers had moved to the piedmont to harvest maize and await the acorns, piñon nuts, and *susí*.

However, as Fig. 32.1 shows, maize steadily increased in productivity as cob length increased. Eventually it reached 250 kg/ha, the lower limit of productivity that Zapotec farmers in the eastern Valley of Oaxaca today require before they will make the effort to clear mesquite forest for agriculture. By the time that point was reached—a point that presumably varied from valley to valley—the Indians were probably well aware of corn's soil and water requirements. The Río Mitla alluvium gets less rainfall than Thorn Forest A, but its soil is considerably richer and more moisture retentive.

Fig. 32.1B attempts to model what happened then. With maize crossing the 250 kg/ha threshold, a new loop in the system was established; mesquite trees were cut down to make way for maize, and although preceramic peoples continued to collect mesquite pods, they were harvesting from ever-dwindling *Prosopis* groves. Instead of propagating mesquite seeds by their collecting and threshing behavior, they were propagating maize kernels by eliminating mature mesquite.

It is particularly interesting to note the way information flow changed between Figs. 32.1A and 32.1B. With primitive maize producing only 80 kg/ha, the most important information foragers could obtain by "reading" mesquite was *an estimate of when its pods would be ready to harvest.* With maize at 250 kg/ha, the most important information obtained from reading mesquite was *the presence of mature trees,* which are considered a sure indication of good corn land. Today's Zapotec say that when fallow land has mesquite trees on it "as thick as a man's arm," it is a sign that that land has recovered enough fertility to be farmed again. Thus, changes in information combined with changes in matter and energy to establish the new feedback loop seen in Fig. 32.1B.

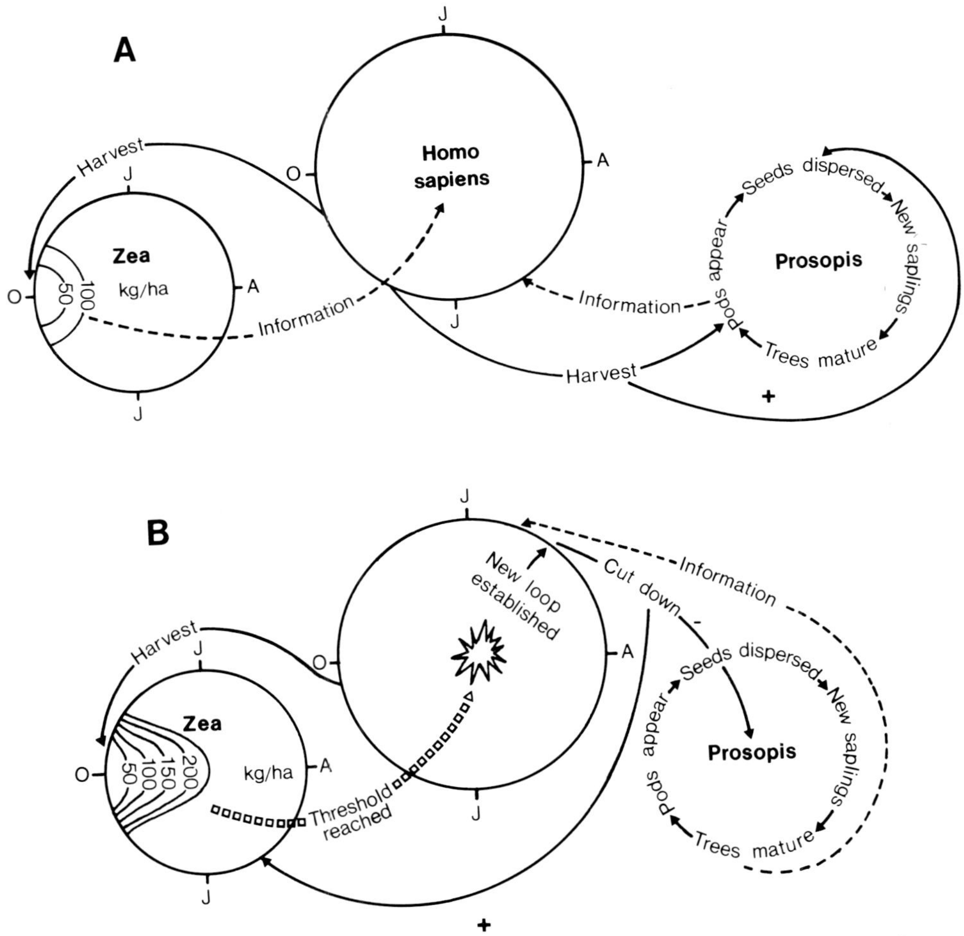

Fig. 32.1. Changes in the systemic relationship of humans and mesquite trees following significant increases in the yield of preceramic maize. A, At 4000 B.C., with maize (*Zea*) yielding only 50–100 kg/ha, mesquite (*Prosopis*) pods are used as food; the appearance of pods is the information triggering a July–August harvest that aided the mesquite's seed dispersal. B, At 2000–1500 B.C., with maize reaching a threshold arbitrarily set by the Indians (perhaps 250 kg/ha in the Oaxaca case), mesquite is now cut down so that corn can be planted on the riverine alluvium; here the appearance of mature *Prosopis* is the information triggering an April–May land clearance that reduces mesquite and increases maize populations (see text).

One other point that should be made is that this systemic change was not forced on our incipient cultivators by the environment, the plants involved, or the increase in productivity. The Zapotec threshold of 250 kg/ha is an arbitrary one, set by the farmers themselves; they could just as easily set a threshold of 100 or 500 kg, depending on how hard they want to work. The notion that a man's arm is an appropriate measure of mesquite thickness, and hence soil fertility, is arbitrary as well. Factors such as human perception and environmental interpretation, a satisfising ethic, and a whole set of notions about how hard one should work are just as important in this model as kilograms and calories. Thus the model illustrates what Harris (1979) might call "feedback between the infrastructure and superstructure of a culture," and it should not be ingenuously applied to some other part of the world where the superstructure was quite different.

9. The long-term consequences of agriculture on Oaxaca peoples' population and protein supply were quite different from the short-term consequences.

This will come as no surprise to those who have watched the long-term evolution of systems, but the details in this case are worth noting. While squash cultivation may have been adaptive in terms of plant protein, it seems to have had little or no effect on population size for thousands of years. If "fitness" is measured by the leaving behind of more offspring, one could legitimately ask how much "fitter" Oaxacan peoples were at 3000 B.C. than at 8000 B.C. Unfortunately, the archaeological record does not tell us whether incipient cultivators in Oaxaca left more offspring than foragers in Coahuila or Durango who remained plant collectors. It tells us only that population densities remained very low until the establishment of village life between 2000 and 1500 B.C.

Ironically, it was not a plant high in protein that changed population rates for the Oaxaca peoples but yet another carbohydrate source—one that could produce 250 kg/ha by the late preceramic and went on to produce 1000 or more kg in later periods. By doing this, maize in fact returned the system to the strategies of (1) increasing calories, (2) sacrificing protein, and (3) reducing search area, which we already observed in the preagricultural era. Indeed, so drastic was the reduction of search area that by Formative times a family of four to five persons could get almost all the calories they needed from a single hectare of land whose location was 100% predictable.

While maize agriculture raised the carrying capacity of the eastern Valley of Oaxaca, that may not even have been the major variable leading to population growth. The decision to become sedentary in hamlets of permanent houses and storage features on the edge of the alluvium in Mesquite Grassland B could have been an equally important factor. For example, Binford and Chasko (1976) have shown that the transition from nomadism to sedentism can have an effect on population growth rates even when agriculture is not involved. It may therefore have been sedentism, rather than agriculture per se, that dramatically increased Oaxacan peoples' fitness.

That population increase, of course, merely worsened the protein sacrifice whose first hints emerged from Reynolds' simulation. A population expanding on the calorie base of maize, beans, squash, and avocados reached levels of 50–700 persons/community by 900 B.C. (Flannery *et al.* 1981:69). There were not enough deer in the Valley of Oaxaca to supply such a population with meat, and the villagers' only domestic animal was the dog. It is therefore no surprise that the system of hereditary ranking that arose in Oaxaca included a restriction of venison to the elite (Whitecotton 1977:143), ameliorating the ruler's protein sacrifice but not that of lower-ranking individuals. In fact, various signs of protein deficiency can be seen in the skeletons of Formative Oaxaca peoples (Richard G. Wilkinson, unpublished data, 1977), who nonetheless continued to display population growth rates (and hence "fitness") that are archaeologically impressive.

ADAPTATION AND TIME

Given the above, we cannot leave Reynolds' simulation without a word about time. If adaptation is the solution of a problem, then time was the greatest ally prehistoric man had on his side. Given a long enough period—thousands of years in the case of the Oaxaca Archaic—even the tiniest incremental changes and growth rates could ultimately be transformed into major cultural changes. This fact makes it unnecessary for us to search for dramatic "revolutions" in the archaeological record: all we need are minor adaptive improvements, and lots of time.

ADAPTATION AND ARCHAEOLOGICAL CHRONOLOGY

Finally, let us consider a major problem with the way we archaeologists deal with time. We can begin by looking again at Reynolds' Figs. 31.9–31.18, which describe the group's gradually improving efficiency both before and after the introduction of cultivated plants.

In Fig. 31.9, for example, improvement in the model group's performance resembles a sigmoid curve. Early Naquitz phase foragers, presented with many alternatives and little experience, improve slowly for a while; then comes a period when their efficiency curve rises steeply for perhaps a hundred time steps; next comes a leveling off at a new adaptive plateau. When our group is presented with a few early cultivated plants, a new curve—less sigmoid but equally striking—arises from the previous plateau and eventually, after several hundred more time steps, levels off at a new plateau of its own (see Fig. 31.16).

As anthropologists, what we claim we most want to know about are the processes underlying these curves. In our grant proposals we talk about "preagricultural adaptation," about agriculture "reaching the takeoff point," about ancient cultures "achieving a new adaptive plateau." One would therefore expect that our preceramic chronologies would be based on the major landmarks of these sigmoid curves. We might expect to hear statements such as, "Guilá Naquitz E lies near the top of the curve's upswing, just before it levels off, while Guilá Naquitz D lies on the stable plateau formed after the upswing levels off." We might expect to hear complaints about how hard it is to find living floors dating to the upswings (which are shorter), and how much easier it is to find occupations from the curve's plateaus (which are longer).

Do we hear such things? We do not. Instead, we have living floors assigned to archaeological phases that are based on projectile point styles. We learn that squash cultivation in the Tehuacán Valley may have begun during the El Riego phase, a period dating from 6500 to 5000 B.C. and defined by the survival of Plainview and Abasolo points, the presence of El Riego, Flacco, Tortugas, Agate Basin, La Mina, Hidalgo, Trinidad, and Nogales points, and the appearance "toward the end of the phase" of San Nicolás and Tilapa points (MacNeish *et al.* 1967:55). Maize cultivation, on the other hand, began in the Coxcatlán phase, which ran from 5000 to 3500 B.C. and was characterized by Abasolo, Trinidad, Nogales, Tilapa, San Nicolás, Abejas, Almagre, and Coxcatlán points (1967:55). We have done things no differently in Oaxaca, since one of the bases for our preceramic chronology is typological overlap with the Tehuacán sequence. In this volume we have learned that squash cultivation in Oaxaca began during the Naquitz phase, which ran from 8900 to 6700 B.C. and was characterized by Lerma(?), Pedernales, and possibly Almagre and Trinidad points.

Thus, the whole sequence of plant collecting, incipient cultivation, and gradually developing preceramic agriculture in the valleys of Oaxaca and Tehuacan has been broken down into time segments based on stylistic changes in deer-hunting equipment. This situation makes no sense in terms of the performance improvement curves we have just looked at, but we are temporarily stuck with it. Radiocarbon dates are too gross to resolve our problems, and most ground-stone artifacts and preceramic flake tools changed too slowly to be useful chronological indicators. We are therefore confronted with a paradox: the processes we wish to document proceed as a series of logistic curves, while our chronologies are composed of linear phases based on stylistic changes in artifacts that may have little or nothing to do with those processes. In Fig. 32.2, I have tentatively tied the living floors of Guilá Naquitz to a series of points along the curve of increasing plant-use efficiency in preceramic Oaxaca. But plant use is only one aspect of human adaptation, and Fig. 32.2 is only a temporary measure; it serves mainly to remind us that so long as our evolutionary sequences are tied to stylistic phases, we have an unresolved dilemma.

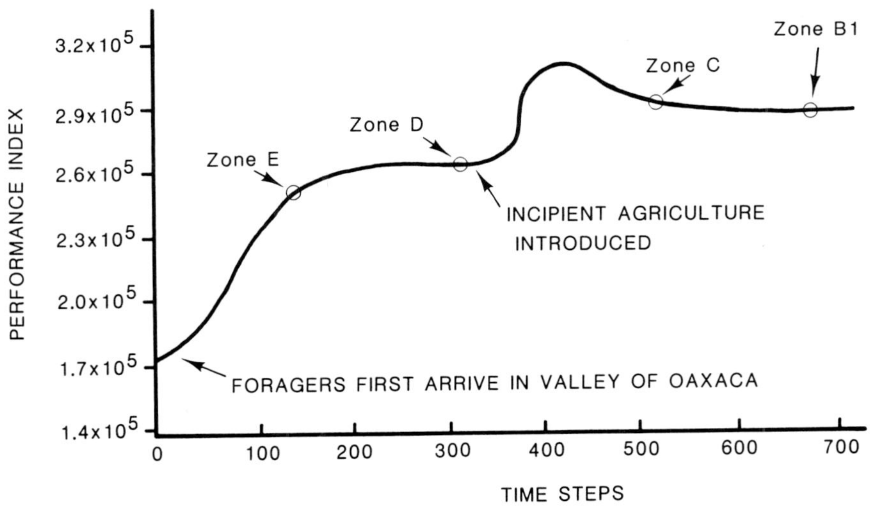

Fig. 32.2. Four living floors from Guilá Naquitz, tentatively fitted to a curve of improving foraging performance like those used by Reynolds in Chapter 31. The purpose is to show what archaeological chronologies might look like if they were based on adaptive or evolutionary processes rather than artifact styles. (While the curve is modeled on Reynolds' Figs. 31.9–31.18, it is not intended to match a specific one of those graphs.)

VIII

SUMMARY AND CONCLUSIONS

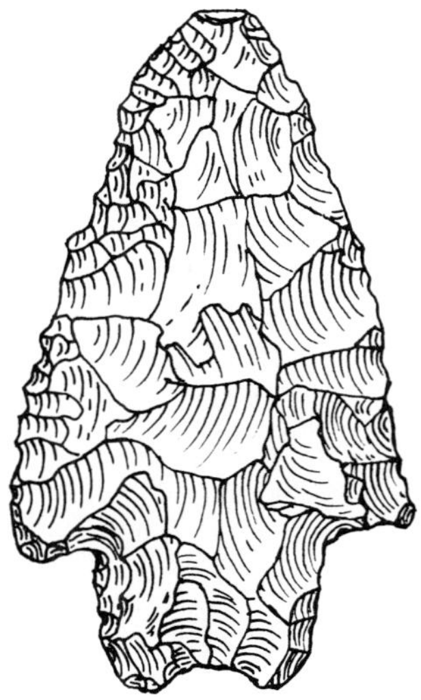

33

A Visit to the Master

Kent V. Flannery

Guilá Naquitz was only a small site, the encampment of a microband, and we have probably carried its analysis as far as we should. In a very real sense, it now needs to be understood not in its own right but in the context of sites such as Cueva Blanca and Gheo-Shih, on which we hope to report later.

This chapter does not provide a lengthy summary and conclusions, because such features have largely been presented in earlier chapters. Chapter 1 describes the theories on agricultural origins that competed for our attention at the time we began this volume. Chapter 2 outlines the ecosystem model we have used here, while Chapter 31 makes it operational. The major results of our analyses of subsistence are given in Chapter 24. Chapter 29 lists the major results of our spatial analyses of the cave, while Chapter 32 lists the major results of our simulation of the transition from foraging to incipient agriculture. The usual summary and conclusions expected in a volume of this type can therefore be obtained from a reading of those chapters.

This chapter, in contrast, concentrates on two topics we have incompletely explored: explanation and causality. After all we have said and done, what caused the Indians of Oaxaca to turn to cultivation? Have our studies contributed to an explanation of the origins of agriculture or merely described it?

The issues of causality and explanation are heated topics in archaeology, ones that have carried some archaeologists to the threshold of philosophy. A whole series of logical positivists—many of them named Karl—have raised nagging questions about the way we conduct our profession. Must explanations in archaeology involve covering laws, as some of my best friends have suggested? Is the notion of multivariate causality a cop out, or can crucial events really have multiple causes? Can systems models, with their circular causality, really provide explanations?

Quite frankly, I do not feel comfortable at the threshold of philosophy, and when it came time to write this chapter I felt I should seek the aid of an expert. Someone who had spent a lifetime considering the organization, methods, and principles of knowledge, and the mysteries of a universe where cause and effect seem often to confound human logic.

As a Mesoamericanist, my first choice would have been a visit to Don Juan Matus, the Yaqui shaman immortalized by the writings of Carlos Castañeda. For a variety of reasons, I had to rule this out. First, the visit would have involved a difficult trip across the blistering north Mexican desert, mostly in 4-wheel drive. Second, I was advised that Don Juan had been so thoroughly "discovered" by the universities in the Southwest that his appointment calendar was usually full. "The archaeologists from Arizona, New Mexico, and U.C.L.A. have him pretty well tied up," I was assured by a friend from El Paso. "Some weekends you even see Colorado and Texas license plates in the line outside his place. Last time I went down there, I waited 4 hours before I got in, and even then all he did was tell me my Hypothesis was really just an Assumption."

"I kind of had my heart set on Don Juan," I said.

"Too crowded. Not worth it," said my friend from El Paso. "Besides, I've heard people say there's a new guy around now who may be even better. Not too many people know about him yet."

"Where can I find him?" I wondered.

"Somewhere up north; somewhere around the Oregon–Nevada border. A place called Antelope Springs—at least, that's what I heard. And he's not a shaman; he's supposed to be a real Holy Man from the Orient."

"Does he know anything about scientific explanation?"

"Well," said my friend from El Paso, "they say his all-time specialty is New World archaeology."

THE MASTER

At first glance, Antelope Springs is like any one of a hundred valleys in the Great Basin of southeast Oregon, southwest Idaho, and northern Nevada. The gray-green mountains are speckled with piñon and juniper, almost as thick as a forest on the upper ridges, then thinning along the flanks of the dissected piedmont where dry gullies carve their way down to the alkali stain of a long-dead playa lake. On the floor of the valley, sagebrush and shadscale and Mormon tea stretch interminably across a carpet of yellow-gray dust. Some 200 m above the valley, a high cliff of Tertiary volcanic rock provides almost permanent shadow for an extensive series of aboriginal petroglyphs.

What makes Antelope Springs unique is that it was the place selected by a genuine spiritual master, an Eastern wise man whose full name I omit here simply because I have no intention of sharing him with other archaeologists. The Master, as all his followers call him, arrived in Antelope Springs in 1974 accompanied by a handful of disciples. By 1976 he had a following of 50, including the resident postmaster, 2 volunteer firemen, and the manager of the local laundromat. By 1977 the town's only restaurant was exclusively vegetarian, and 40% of the registered voters were disciples of The Master. In 1978, his holiness was announced as a candidate for mayor, and it was revealed that following his election it would be a felony to kill any animal for food. This appears to have been a miscalculation, since for more than 70 years Antelope Springs' only source of income had been cattle ranching.

One week after his candidacy had been filed, The Master was visited by the sheriff, 2 deputies, the president of the Cattleman's Association, and a group of some 100 ranchers in pickup trucks with bumper stickers reading "Register Criminals, not Guns." His holiness was offered two choices: leave town, or be processed into silage through a belt-driven International Harvester shredder.

A compromise was struck. Forsaking the town and all aspirations to earthly political power, The Master withdrew to a cave in the volcanic cliff high above Antelope Springs. Here he is free to read, teach, meditate, study petroglyphs, and do some occasional contract archaeology. His awakened interest in Great Basin prehistory and the other vanished cultures of the New World make him the ideal authority for an archaeologist to consult—much more so than Don Juan, or anyone named Karl.

A visit to The Master, such as I made in 1983, begins with a long, hot climb up Antelope Mountain. One is met at the shore of the playa by an acolyte in a saffron robe who examines your credentials and guides your trip up the winding trail. Stunted junipers cling to the steep slopes, and the air is dry and dusty until you enter the shadow of the cliff. There are three resting places along the way, where you can catch your breath and reflect on the fact that you are ascending to a calmer and less materialistic world, one where the theoretical issues of archaeology become sharper and clearer, unencumbered by fads and grant proposals and publishing or perishing.

At the top, between two areas of human hand prints pecked into the living rock, lies the large triangular mouth of the cave. The acolyte enters first to announce your arrival, and after a brief ritual of purification you are allowed to enter the chamber. It took me some time for my eyes to become adjusted to the darkness, which was relieved only by guttering candles set against the pictograph-covered walls.

"Come in," said The Master, "and sit down."

I sat uncomfortably on a small carpet, through whose worn fabric I could feel the grainy material of which the cave floor was formed—pack rat droppings, I suspected, but at least they were good and desiccated.

The Master seemed unexpectedly petite, with piercing black eyes set in a face like tanned leather. I could not place him ethnically, sensing that he could have been from any one of a number of Asian countries stretching from the Indian subcontinent to Tibet or Indochina. The loose-fitting saffron robe hid a body that could have been any shape but seemed to be lean and muscular wherever it was exposed. The Master sat hunched over a wooden bowl into which he periodically dipped his fingers, each time forming a small ball of rice which he transferred casually to his mouth. I say "casually" because not every grain of rice made it to its destination, and this may have been the one flaw in an otherwise imposing presence. Visitors to The Master were supposed to be gripped by the candlelight reflections in his penetrating eyes, but this was hard to achieve if your attention was riveted on the stray grains of rice in his beard.

"Tell me the reason for your visit," he suggested, in a voice that put me at ease without being condescending or overly familiar.

As best I could, I explained that I was an archaeologist working on the origins of agriculture in Mexico. After reviewing a number of current theories on the topic, I had designed a multivariate model based on the best data available to me from Tehuacán. One of my colleagues had converted it to a mathematical model that could be run on a computer, and with that we had been able to simulate the pattern of subsistence activities seen in a whole new batch of data collected more recently in Oaxaca. The results allowed us to reject some current opinions on the origins of agriculture, defend others, and contribute some new opinions of our own.

"But you are still not satisfied," he said knowingly.

"No," I said. "In today's archaeology, people want you to come up with clear-cut *causes* for prehistoric phenomena. They want those causes to be explained in terms of universal covering laws of human behavior. I think I can suggest *how* the transition from foraging to agriculture in Oaxaca was accomplished, but if you asked me, 'Why did agriculture begin?' I'm not sure what I'd give you as a cause. And if you asked me, 'What law did you come up with?' I'm not sure what I'd say."

An understanding smile, marred only by an errant grain of rice, illuminated The Master's face. "Tell me something," he said. "Of all the other sciences, which one do you feel most resembles archaeology: mathematics, physics, chemistry, or biology?"

"In my opinion, biology," I replied. "Specifically, within biology, I feel that archaeology most resembles paleontology. We both attempt to reconstruct evolutionary sequences on the basis of an incomplete fossil record."

"In that case," said The Master, "why don't we look at what a distinguished senior biologist has to say about covering laws?"

By the flickering light of the candles, his hand moved along a wooden shelf behind his seat, and for the first time I saw that it was stacked with books of every shape and size. He stopped at a thick brown volume; his fingertips moved gently over the gold printing on the spine, and withdrawing the book he opened it on the carpet before me.

"Ernst Mayr is one of the architects of the modern evolutionary synthesis," said The Master. "This is his new book, *The Growth of Biological Thought* (Mayr 1982). I have opened it to his second chapter, 'The Place of Biology in the Sciences and its Conceptual Structure,' so that we can read it together."

AFTERNOON

Until that day in the cave I had not read Mayr's book; by the time I had reached his fortieth page it was clear that many of his views would have been condemned as heresy by a generation raised on sages named Karl. "Philosophers of science, most of them with a background in physics," he began, "have unfortunately based their treatment of the philosophy and methodology of science almost exclusively on the physical sciences" (Mayr 1982:33). Yet the physical sciences are not an appropriate yardstick, since biology cannot be reduced to the laws of physics. In the biological sciences "we deal with phenomena unknown for inanimate objects" (p. 35). On the average, "systems in the world of organisms are infinitely more complex than those of inanimate objects. . . . Living systems are invariably characterized by elaborate feedback mechanisms unknown in their precision and complexity in any inanimate system" (p. 53).

Perhaps the most striking characterization of biology by Mayr, however, was that it proceeds virtually without laws such as those in physics. "Instead of formulating laws, biologists usually organize their generalizations into a framework of concepts," yet philosophers of science have paid little attention to concepts (p. 43). While Smart has denied the existence of universal laws in biology, and Ruse and Hull have argued for them, "biologists have paid virtually no attention to the argument, implying that this question is of little relevance for the working biologist" (p. 37). Virtually every law proposed for biology has too many exceptions to be a covering law; "they are explanatory as far as past events are concerned, but not predictive, except in a statistical (probabilistic) sense" (p. 37). That is why generalizations in biology are almost invariably of a probabilistic nature, and predictions about evolution are impossible. "No one would have predicted at the beginning of the Cretaceous that the flourishing group of the dinosaurs would become extinct by the end of this era" (p. 58).

There are several reasons, Mayr argues, why evolution is not predictable. One reason is its systemic nature, with multiple recurrence relationships and feedback loops. Absolute predictions "are impossible owing to the complexity of the hierarchical systems, the high number of possible options at each step, and the numerous interactions of simultaneously occurring processes. . . . The number of potentially possible interactions in [such] highly complex systems is far too large to permit prediction as to which one will actually take place" (p. 42). Another reason is that systems "almost always have the peculiarity that the characteristics of the whole cannot (even in theory) be deduced from the most complete knowledge of the components, taken separately or in other partial combinations. This appearance of new characteristics in wholes has been designated as *emergence*" (p. 63). This concept, which Popper also calls *emergent novelty,* is a characteristic of higher systemic levels that cannot be predicted from knowledge of lower systemic levels.

"Then how can biologists prove anything?" I asked The Master.

"By ridding themselves of the notion that proof in biology is the same thing as proof in a predictive science such as physics," he replied. "See here: 'stochastic processes, even though making predictions probabilistic (or impossible) rather than absolute, are just as causal as deterministic processes' (Mayr 1982:42)."

Modern biology, Mayr suggests, does not demand the absolute proof of physics. Biologists are satisfied to consider true "that which appears most probable on the basis of the available evidence" or "that which is consistent with more, or more compelling, facts than competing hypotheses" (p. 26). This is especially true in fields such as paleontology, where "experiments" are impossible and "future predictions" meaningless. A paleontologist "would insist that the former existence of dinosaurs and trilobites was as certain as any mathematical theorem" (p. 40). Biology "should not focus most of its attention on laws, considering what small role laws actually play in much of biological theory" (p. 76). Rather, biology should focus on *concepts* such as mutation, natural selection, predation, competition, and symbiosis. It is unlikely that these concepts will ever be reduced to physics or chemistry, Mayr argues. For example, predation is a chemical and physical process, but it is only a biological concept and cannot be reduced to a physicochemical concept or a law (p. 62).

"Now," said The Master, "if there aren't any covering laws in biology that work, what do you think are the chances that any will be found in anthropology or archaeology?"

"Slim enough," I replied, "so that I don't think I'll lose much sleep if I don't come up with one."

"That is good," said The Master, slowly forming a new ball of rice between his fingers. "I recently had a visit from a young man who told me he needed to discover at least one law before he came up for tenure."

"What did you advise him?"

"I provided him with a law," said The Master, pausing while he transferred the rice to his mouth. "It reads, 'The most important burial in your site will be found at 4 P.M. on the last day of the field season, and most of it will extend into the profile.' "

"Did he like it?"

A few additional grains of rice quivered in The Master's beard. "He asked me not to publish it before he did."

We sat quietly in the subdued light for a minute or two, sipping tea and reflecting on Mayr's vision of a new philosophy of biology. One that would "include and combine the cybernetic–functional–organizational ideas of functional biology with the populational–historical program-uniqueness-adaptedness concepts of evolutionary biology" (Mayr 1982:73). One that would use concepts and recurrence relationships, but also recognize that some events in evolution may have been virtually unique.

"So," said The Master at length, "without laws and without reduction to mathematical formulae, what form should an archaeological explanation take?"

"According to Mayr, in biology, and particularly in evolutionary biology, explanations take the form of what he calls 'historical narratives' (p. 58). Narrative explanations use concepts but are constructed without mentioning any general laws (p. 71). For some sciences, such as paleontology, historical narratives play an important role, and Mayr actually uses the phrase 'historical narrative theory'. He goes on to say that historical narratives 'have explanatory value because earlier events in a historical sequence usually make a causal contribution to later events' (p. 72). At the same time, he acknowledges that 'philosophers trained in the axioms of essentialistic logic seem to have great difficulty in understanding the peculiar nature of uniqueness and of historical sequences of events'. They often try to deny the importance of historical narratives, or axiomatize them in terms of covering laws, but Mayr is unconvinced."

The Master's fingers strayed momentarily to his beard, and I was relieved to see at least a couple of rice grains disappear into the cave floor. "All right," he said. "Why don't you start by giving me a historical narrative explanation for the origins of agriculture in Oaxaca? When you have finished, we'll talk a little about causality."

"That could take all night," I said, somewhat apologetically.

"Quite literally," said The Master with a knowing smile, "I have all the time in the world."

EVENING

The sun was beginning to set over Antelope Springs as I started my historical narrative on preceramic Oaxaca. I laid the groundwork by describing the prior contributions of MacNeish, Binford, Cohen, Wright, Hassan, and Rindos, all summarized in Chapter 1. Each of these contributions had its strong points, I explained, yet each author also made points that were contradicted by the archaeological record. What I had done was to synthesize their strong points as a preliminary stage of our analysis.

In this synthesis, the end-Pleistocene climatic changes described by Wright and the growth of world population described by Cohen and Hassan combined between 10,000 and 5000 B.C. to bring about a density-dependent shift in human cultural behavior over much of the world. Emigration and high mobility, which had been major solutions to man–land imbalances in the past, were reduced in importance; humans now had to improve their strategies for dealing with predictable (seasonal) and unpredictable (annual) environmental variation on a more localized basis. Regionalism in tool kits, a broadening of the diet to include more lower-trophic-level foods, an increase in storage facilities, and greater diversity in the "foraging" and "logistically organized collecting" described by Binford become more apparent in the archaeological record. Given this pattern, it would seem that agriculture arose in the context of efforts to deal locally with long-term environmental variation. If this was the case, we could not fully understand incipient cultivation by subsuming it under a covering law, or a biological concept such as coevolution. We needed to understand the system of late preagricultural strategies out of which incipient cultivation arose.

Humans are well suited to the working out of resilient strategies for dealing with long-term environmental variation, since they possess multigenerational memory, can modify their behavior without waiting for genetic change, and have unique methods for exchanging information and establishing cooperative relationships over long periods. In Chapter 2, therefore, we presented an ecosystem model for man–plant relationships that includes information as one of its components. Indeed, as extended in Chapter 32, the model can even accommodate cognitive or ideological variables, such as arbitrary beliefs about how large a yield makes land clearance worthwhile.

The model consists of a number of interacting subsystems by which matter and energy are exchanged between plant and animal populations, with information serving to regulate the order, timing, and nature of such exchanges. There are a number of "recurrence relationships" in the model, which we asked the reader not to confuse with laws. "Of course," I said, "now that I've read Mayr I realize we didn't need to concern ourselves with laws."

I didn't have to look at The Master to know that somewhere in the darkness, he was smiling.

Reynolds' conversion of my crude verbal model into a more polished mathematical model allowed us to run it on the computer, I explained, following a hypothetical microband through centuries of environmental variation. When we add this to our studies of nutrition and use of living space, an outline of the preagricultural subsistence strategy begins to emerge.

Part of that strategy was division of labor, which (along with the digging of storage pits and hearth basins) did more to determine the partitioning of cave space than any other factor. Women seem to have processed most plant foods, stored acorns and mesquite, and eaten much of the fruit brought back to the cave. Men hunted, trapped, butchered animals, chewed agave, and probably did much of their fruit eating while out on the trail. The Guilá Naquitz people were largely foragers who "mapped onto" resources, and our studies suggest they were likely to exhaust the deer in the area long before they exhausted the edible plants. What limited their plant food supply was not acreage but the changing of the seasons; storage facilities and seasonal settlement shifts were designed to combat this problem.

Another part of the strategy was to reduce search area. Our simulations show the foragers' efficiency increased as they traveled directly to the groves of densest plant productivity, reducing the number of square meters harvested rather than the distance traveled from cave to grove. However, those same simulations also showed that the "mix" of plants collected by our hypothetical foragers emphasized calories at the expense of protein. While this strategy was "efficient" in terms of kilocalories recovered per hectare searched, it revealed an area of vulnerability already mentioned in Chapter 24: The Naquitz phase plant diet needed more protein, such as 100 g of meat daily could provide. Given the fact that they were more likely to run out of deer than wild plants, it would seem that the Naquitz adaptation could have been substantially improved by the domestication of a high-protein plant food, such as cucurbit seeds.

One of the most interesting aspects of the adaptation displayed by our simulated foragers was the way they treated unpredictable annual variation. The group settled on two basic strategies, one for wet years and one for dry or average years. Under the greater stress of dry or average years, they were more conservative; under the lessened stress of wet years, they experimented more with new combinations of plants. Just as interesting as their two responses was the fact that they reduced three year types to a binary opposition, something done by today's Zapotec as well. Almost all innovation in preagricultural strategy took place in wet years, and was added to dry year strategies only after it had proved itself over a period of time.

"Now tell me," suggested The Master, "how agriculture began."

"One intriguing possibility," I said, "is that the earliest New World domesticate may have been the bottle gourd *Lagenaria*. This is a plant that is used as an artifact, not as food. It would be useful to foragers no matter how low their population density, and the most obvious advantage provided by its domestication would be to insure a supply of water bottles as foragers traveled widely over the continent. It might even have been domestic *before* the climatic and population changes suggested by Wright, Cohen, and Hassan."

"Interesting," said The Master.

"However," I went on, "the bottle gourd is an enigma because its wild ancestor may not even occur in the New World; some botanists have suggested it floated to the Americas, where it became adapted to areas of human disturbance. If so, it is the only one of our early domesticates that comes close to fitting Rindos' model of man–plant interdependence in preagricultural times. If the early 'wild' bottle gourds were limited to habitats disturbed by man and if their primary mechanism of seed dispersal was human intervention, they meet Rindos' expectations. The other wild plants don't, because they occurred by the millions, while people occurred only by the hundreds. You see, ant–acacia coevolution works because there are millions of ants—enough to have an impact on millions of acacias. Twenty-five or 50 people in the Valley of Oaxaca aren't going to have that kind of impact on millions of wild runner beans. Ninety-nine percent of the beans will still disperse their seeds with the wild-type corkscrew pod. Only people who have *already begun to cultivate beans* can take the occasional limp-pod mutant and, by planting it and continuing to select for it, eventually produce domestic fields of limp-pod beans. Even today, after thousands of years of nearby bean domestication, the wild beans of Oaxaca still have corkscrew pods."

Night had fallen on Antelope Springs, and from the mouth of the cave we had a good view of a cold, white moon suspended above the darkened junipers. Choosing my words carefully, I tried to explain to The Master that it might be no accident that *Cucurbita pepo* was one of Oaxaca's first domesticates. It belonged to the same family as the bottle gourd and would probably have been recognized as just as simple to cultivate. It was a weedy annual whose seeds could be roasted, stored, and carried and eaten on trips. Unknown to its early cultivators was the fact that its seeds could provide them with the extra dose of protein their diet needed, but this unsuspected nutritional benefit could have conferred a selective advantage on early squash cultivators. *Susí* nuts could have done the same, but the *susí* is a perennial shrub and therefore not nearly as amenable to casual cultivation as a weedy annual.

Why would Naquitz phase foragers begin to cultivate squash? Surely not because their population was so high that they had exhausted all sources of wild food. No, I explained, I preferred to see early agriculture as a logical extension of the preagricultural pattern. Our simulated foragers had shown a concern with reducing search area; cultivation did that by concentrating lots of squash in a small area. It also made the location of the densest squash stands totally predictable from year to year; and it increased a storable seed crop, which extended the harvest season.

Our simulated foragers had shown a preagricultural tendency to try out new strategies in wet years. Presented with domestic squash, they had also tried growing it first in wet years, expanding its cultivation to dry and average years only after it had proved itself. Selecting for greater numbers of seeds per fruit, we suspect that preceramic cultivators unwittingly selected for a fruit with edible flesh, positively reinforcing the advantages of cultivation and eventually producing a phenotype that could not survive without human intervention. As agriculture increased the efficiency of their search for calories and protein, they increased the number of agricultural tasks in virtually all their strategies and eventually moved agriculture out of the piedmont and onto the alluvium where its greatest strides would be made. Thus a very small initial change in behavior, amounting to no more than an extension of the preagricultural strategy, was amplified into a major change by time and positive feedback.

Perhaps most interestingly, we were able to simulate the pattern seen at Guilá Naquitz without resorting to population pressure or climatic change. In fact, when Reynolds experimentally altered those variables, he found that the rate at which agriculture was adopted did not increase with (1) increasing rainfall, (2) increasing aridity, (3) steady population growth, or (4) fluctuation of population around a mean. Rather, it would appear that our original succession of wet, dry, and average years provided the best adaptive context: times of stress that encourage efficiency plus times of abundance that allow experimentation.

"And in the end," said The Master, "you came to prefer a probabilistic ecosystem model to a mechanistic-deterministic theory?"

"For several reasons," I replied, "For one thing, such a model leaves open the options for several different responses to systemic change. For example, if a group's strategy of reducing search area leads to a protein deficit, there are several alternative changes that might be selected for. They could move in the direction the Naquitz phase people moved, domesticating a high-protein plant while further reducing search area. Or they could move in the direction of more intensive hunting or fishing; this may have happened in parts of the world where agriculture *didn't* arise between 10,000 and 5000 B.C. If we make agriculture the *only* possible outcome of our model, we fail to account for those areas where it *didn't* appear independently."

"And what are the other reasons?" he asked.

"I like being able to incorporate information into the model along with matter and energy," I said, "because without it you deny one of the most significantly human aspects of human adaptation. What happened when Reynolds experimentally 'disconnected' the feedback loop between the multigeneration memory and the decision-making apparatus? Strategies began to slide from gradual improvement toward random fluctuation. You see, humans don't have to wait for natural selection or coevolution to make their decisions for them if they can remember the advice of their grandparents. And what happened when Reynolds made all years in the model 'average' years? The group began to wander aimlessly from strategy to strategy; they had been deprived of much of the environmental information they needed to make good decisions. It's almost as if the group were saying, 'Go ahead, hit us with more variation; it'll just make us wiser decision-makers.'"

"I gather," said The Master, "that for you the word *intentional* may still have a place in human adaptation,"

"I think it depends on whether you mean short-term or long-term intentionality," I said. "If a forager spends 2 hours chipping and hafting projectile points, I would say he is about to go deer hunting intentionally. If he spends 2 days clearing thorn forest off the cave talus and planting beans from a limp-pod mutant, I would say he is about to cultivate intentionally. Those are short-term intentions, and we don't know of a human group that doesn't have them. On the other hand, I do *not* believe that our hunter would say, 'I'm going out today to contribute to the long-term survival of the deer population by eliminating weak, elderly individuals.' Nor do I believe our bean planter would say, 'Today I'm initiating an agricultural revolution that will one day change the course of Mesoamerican history.' Those are long-term intentions, and on *that* level I agree with Rindos that the origins of agriculture were 'unintentional.'"

The Master signaled to the acolyte to bring him another bowl of rice. "And are there still further reasons for your choice of model?" he wondered.

"Yes," I replied. "In addition to information, it even allows me to include such features as world view or ideology: for example, the Zapotec notions that mesquite 'as thick as your arm' is an indicator of good corn land, or that land is not worth clearing unless the corn yield will be around 250 kilos per hectare. These notions may have some underlying basis in ecology or economics, but essentially they are arbitrary cultural beliefs. Yet we need to take them into account because they help determine when land is cleared or taken out of fallow. If you lower your arbitrary threshold to 50 kilos per hectare, or raise it to 1000 kilos, it significantly affects the rate at which maize gets moved to the alluvium."

"Would it be fair to say," The Master suggested, "that while you regard coevolution as a powerful 'concept' in Mayr's sense—profoundly heuristic in the case of creatures without culture—you consider it to provide only part of the story in the case of humans?"

"That's right," I said. "If there is such a thing as cultural evolution, then whatever the biological processes involved, we *also* need to understand how the cultural pattern 'incipient cultivation' developed out of the cultural pattern 'foraging'."

"And I sense," said The Master, "that you regard culture as so flexible that it confounds most mechanistic-deterministic models."

"Exactly," I said. "If the biological world is made up of complex feedback loops and recurrence relationships, as biologists like Watt and Mayr evidently believe—and if even paleontologists depend on historical narrative explanations—can we really expect to reduce cultural behavior to a set of laws like those of physics?"

Patiently, The Master formed a new ball of rice between his fingertips. I swear, I didn't know where he was putting it all, but I decided not to worry about it.

"Did anything strike you as familiar about your historical narrative explanation?" he asked.

"It struck me that it was not unlike the kind of summary and conclusions that many archaeologists have offered in the past," I admitted. "Does that mean that it's been all right to do that all along?"

"Would that surprise you?" he asked.

"It would certainly be reassuring to a lot of archaeologists over 35," I told him.

The Master struggled to suppress a laugh. He suppressed it partly because it would have spoiled his dignified image and partly because it would have caused him to spit a lot of rice back into his bowl.

"I have another question," I said. "Does 'historical narrative theory,' to use Mayr's term, qualify as 'middle-range theory'?"

"Based on his Chapter 2," said The Master, "it should qualify as 'theory' if it employs 'concepts': coevolution, symbiosis, intentionality, recurrence relationships, foraging efficiency, logistically organized collecting, reduction of search area, density dependence, and division of labor are some of the concepts that come to mind. But what do you mean by 'middle range theory'?"

"I'm not completely sure," I admitted, "but from what I've read, I suppose it's the generation of the concepts we later use in historical narratives."

"So, ironically," said The Master, "middle-range theory is the highest-range theory any archaeologist has so far been involved in."

Now it was my turn to suppress a laugh. "Yes, so far," I said, "but the 'high-level theory' category has been left open in case anyone comes up with a truly Newtonian law."

A faraway look came over The Master's face. "The archaeologist who does that," he said, "will ascend directly, through seven cosmic levels, to Archaeological Heaven."

The acolyte had remained silent for so long a time that I was now startled to hear his voice behind me. "Master," he said emotionally, "what *is* Archaeological Heaven?"

The Master looked intently at the new sphere of rice in his fingers, as if it were a miniature crystal ball in which he could see the cloudy outlines of paradise. "Heaven," he said softly, "is a giant archaeological site, stretching on for mile after mile, to the limits of our visible universe. On it you excavate for all of eternity, and everywhere you put your pits you find fabulous material—Clovis points, Olmec jades, Mimbres vessels, Luristan bronzes, unbelievably pornographic Mochica pots. You never have to take a note, you never have to count a body sherd, and best of all, you never have to write the material up."

"Master, I must already be in Heaven," said the acolyte with genuine relief. "I didn't take a note or count a sherd on our last salvage project, and I have no intention of writing it up."

"Unworthy one," said The Master, "that is why your Maker has a very different destination planned for you."

MIDNIGHT

It was the dead of night in Antelope Springs, and a slight chill had descended over the cave. After a meal of yogurt The Master, his acolyte, and I sat huddled around a small kerosene space heater, wrestling with the last of my theoretical problems.

"All right," I said, "we've proposed that archaeology finds a better model in paleontology than in physics. This allows us to use historical narratives and to recognize that unique past events, as well as universal long-term processes, may be causally important. It allows us to focus on concepts rather than laws and on probabilistic rather than mechanistic-deterministic explanations. It justifies the use of an ecosystem framework, with its ability to integrate the kind of complex feedback relationships that distinguish living from nonliving matter. It also recognizes that emergent novelty, although typical of higher-order systems, cannot be predicted."

The Master smiled. "But you are not yet satisfied."

"No," I replied, "not until I have a clearer image of what 'causality' means in a historical narrative explanation. What do I say when someone asks me, 'What caused the origins of agriculture in Oaxaca?' "

"Begin by asking them which cause they are referring to," he answered. "The material cause, the efficient cause, the formal cause, or the final cause."

The Master reached toward his darkened shelves again and brought down an ancient, leather-bound volume. Hand lettered across the worn cover were the words *Analytica Posteriora.*

"In Book II, Chapter 10," he explained as he opened the yellowed pages, "Aristotle proposes that every event has four causes. The material cause is that *out of which* something is made: 'leather is the material cause of the shoe.' The efficient cause is that *by which* something is made: 'the shoemaker is the efficient cause of the shoe.' The formal cause is that *into which* something is made: Thus, the shoemaker forms the leather according to a model, a concept of 'shoeness.' The final cause is that *for the sake of which* something is made: 'to cover a man's foot is the final cause of the shoe.' Now put your Oaxaca narrative in those terms."

"You don't want much, do you?" I protested.

"Take all the time you need," he smiled.

"All right," I began. "A bunch of weedy, genetically plastic annual plants were the material cause, the thing 'out of which' early agriculture was made. Woody perennials couldn't be; they were the wrong 'material'."

"And the efficient cause?"

"It would depend on whom you talk to. If 'the shoemaker is the efficient cause of the shoe', then humans were the efficient cause of agriculture because they cleared the land, did the planting, and selected the mutant strains. I'm sure you could find people who would argue that population pressure or climatic change might have been the efficient cause, but our simulations don't make it seem necessary to claim this."

"How about positive feedback?" he suggested.

"You have a point there," I admitted. "If positive feedback could transform minor and occasional planting into permanent cultivation, it sounds like an efficient cause. But feedback isn't an 'actor', and as I understand it, the efficient cause is supposed to be an actor."

"And the formal cause?"

"That's a hard one," I said. "But as I remember, Aristotle used the example of an acorn growing into an oak tree—taking on the 'form' of 'oakness'. If that's the case, then maybe the genetic code in the plants would qualify, with the domestic phenotype being 'that into which something is made'."

"The botanists would like that," he smiled. "But how about the anthropologists?"

"They might say that the jícara, or Precolumbian water bottle, was the formal cause for gourd cultivation. Or that a dense field of limp-pod beans was the 'archetype' the incipient cultivators had in their minds."

"And the final cause?" he asked.

"I find the final cause one of the most interesting," I replied, "because the phrase 'for the sake of which something is made' shows me that Aristotle believed in intentionality. If that's the case, I have several candidates. 'To further reduce search area' would be one. 'To further reduce the differences in productivity between wet and dry years' would be another. 'To further extend the season of plant availability' would be a third. 'To have water bottles wherever we go' is a fourth. Notice that none of these implies an awareness by humans that they were doing anything new, which I gather to be the kind of intentionality that Rindos objects to. All would be 'final causes' shared with foragers and thus seem reasonable in light of our results."

"But surely," said The Master, "there must be a final cause that is not shared by foragers, or agriculture would not have gone on to replace foraging."

I took a deep breath. "If we make it clear that we are speaking in an *evolutionary* sense," I said, "and rejecting both human intentionality and teleology, I can suggest at least one. Reynolds' simulation suggests that Oaxaca's incipient cultivators had a higher performance index than their forager ancestors; under conditions of competition, the former would have beaten out the latter, given enough time. So an evolutionary final cause might be 'for the sake of having a selective advantage over neighboring foragers.' It's also the case that agriculture had the potential for supporting more human offspring, and hence 'fitness', although sedentism seems to have been an intervening variable in the Oaxaca case."

The Master gestured to the acolyte to turn up the space heater a few degrees. "You are clearly bothered by the notion of teleology," he said.

"Archaeologists are frequently criticized for it."

"Yes," he acknowledged, "but mainly by philosophers for whom physics is the prototype. Let's see what Mayr has to say about it in the context of biology."

He reopened *The Growth of Biological Thought* and thumbed slowly forward to page 72, folding down one corner of the page and handing the book to me.

"All biological processes," wrote Mayr (1982:72–73), "have both a proximate cause and an evolutionary cause. Much confusion in the history of biology has resulted when authors concentrated exclusively either on proximate or on evolutionary causation." He went on to explain that the biology of proximate causes concentrates on the questions, "What?" and "How?", much as physics might. On the other hand, the biology of evolutionary causes asks the question, "What for?", a query that is meaningless in the world of inanimate objects. Far from being a teleology, "in the living world the question 'what for?' has a powerful heuristic value" (p. 72).

Why does a warbler migrate? Mayr points out that the proximate cause is a change in photoperiodicity which sends the bird south. But other species of birds, exposed to the same day length, stay put. The evolutionary cause is that over thousands of years, there was natural selection for an insect-eating bird that could venture north when summer brought out the insects, then retreat to the south when the insects disappeared in winter. The warbler is the product of a genetic program that took millennia to develop and that depended on previous evolutionary events such as the rise of insects and the evolution of the wing.

"You see," said The Master, his knees pressed so close to the heater that his robe seemed ready to smoke, "although we do not often hear Aristotle's four causes mentioned today, there is continued awareness of the fact that there is more than one kind of causality. Multivariate causality is not simply a device used to avoid having to specify a cause; it is a recognition of the fact that causality in living systems may be much more complex than causality in nonliving systems. Essentially, what Mayr is saying is that biology needs a philosophy that takes into account the unique nature of living matter, rather than being based on physics."

DAWN

The first rays of sunlight had just cleared Antelope Mountain, and in the background I could hear the acolyte making three cups of tea.

"Wouldn't it be nice," I suggested, "if archaeology had a philosophy that took into account the unique nature of human adaptation?"

The Master accepted his cup of tea from the acolyte and stood sipping it in the warm column of air from the space heater. "All it would take," he said, "is the addition of one more concept."

I looked at him expectantly.

"Mayr points out that biological phenomena can only be understood when both the proximate and evolutionary causes are understood," he said. "But neither attacks the unique nature of human adaptation, which relies more heavily on cultural behavior than on genetic programming. Archaeological explanation will therefore have to take into account all three: proximate causes, evolutionary processes, and human decision making."

"That would explain a lot of the arguments over the origins of agriculture," I said. "For example, when archaeologists talk about population pressure or density-equilibrium imbalance, they're really only talking about proximate causes. That won't be satisfying to a botanist, who's more concerned with evolutionary processes such as symbiosis and coevolution; at the same time, the botanist's apparent satisfaction with coevolution will puzzle the archaeologist, who wants to know which proximate cause triggered agriculture."

"And neither will satisfy you," said The Master, "given your view that human behavior is at least partly the result of human decisions."

"That's right," I said. "Human hunting and gathering is, after all, economic behavior. Foragers make decisions about which vegetational facies to harvest in, how far to travel, how large an area to search, which plants to concentrate on, and when to move on because they've reached the point of diminishing returns. Their shift to agriculture wasn't a change in genetic programming, such as happened when ants adopted acacia trees. It represents a change in their pattern of decisions, a reversible change, possibly only one of several alternative reactions to a new situation they perceived. It has to be understood in terms of their previous decision-making pattern, the options open to them, and that new situation."

"And in turn," said The Master, "the advocates of proximate causes and evolutionary processes will find your culturally based explanation inadequate to stand alone. That is why it will be important for you to give the former equal time. Your proximate causes must emerge from the worldwide end-Pleistocene phenomena you spoke of, and they must create the situation during which your changes in human decision making might be selected for. Your evolutionary explanations must deal with the selective advantage conferred on early cultivators, and the fact that they became the means of seed dispersal for a group of plants whose genetic program they had altered, in defiance of natural selection."

We stood together at the mouth of the cave, enjoying the slight rise in temperature and the view of dawn spreading over the valley below.

"I don't know how to thank you," I said at last. "I had no idea how I was going to deal with causality. Now I think I can write my last chapter."

The Master smiled. "Do not thank me too hastily," he said. "I really only tried to direct you away from the laws of physics and toward the concepts of biology. To fully understand the origins of agriculture, you will have to eschew science completely and ascend to a higher spiritual level where there is no essential difference between plants and man. You are not yet ready for that level. Even this unworthy student," he gestured at the acolyte, "is not yet ready for that."

"No, Master," admitted the wistful acolyte.

"You see," said The Master, "now that we have finished talking I can admit to you that I was present at the origins of agriculture. It was in an earlier life, of course, and in another land and another cave. And it did not happen just the way you think. At that time I did not expect that it would become as important as it has. I was too young then to understand."

"Could you give me just a hint?" I asked.

"No," said The Master. "That would take away the whole reason for your search. In some fields, such as archaeology and paleontology, the search is half the fun."

Outside, on the dusty talus of the cave, I tried to find the words for an appropriate goodbye.

"In the Orient," he said affectionately, "we believe that the most important human attribute is humility. I try to leave every visitor with a thought that will make him more humble, and therefore more worthy of respect."

I waited apprehensively.

"I want you to remind yourself," he said, "that you have now spent 1 field season of excavation, 15 years of analyses, and thousands of dollars of taxpayers' money to find out no more than this: what a family of five did during the autumn, on six different occasions, scattered over a 2000-year period. I earnestly hope that the thought will keep you humble."

"Would you believe 'deeply depressed'?"

"You Americans," he smiled, "have such an amusing way of expressing yourselves."

Not a word was spoken as the acolyte led me down the trail to the second of three resting places. We paused for a moment to enjoy the smell of the junipers, and I told him I could find my way down from there.

Only at the last resting place, where the sagebrush and grazing Herefords on the valley floor came into focus, did I really begin to feel I had returned to the secular world. The cave half-hidden in the shadow of the cliffs above me seemed somehow unreal, but for a moment I felt almost overwhelmed with a desire to be back there again; there was so much still to ask, so much I hardly knew where to start asking.

Of course, that's the trouble with oriental wisdom. An hour later, you're hungry for knowledge again.

34

Resumen en Español

David J. Wilson

Este trabajo presenta los resultados de un estudio interdisciplinario sobre los orígenes de la agricultura en el Valle de Oaxaca, México, en base a los datos recogidos en Guilá Naquitz, una cueva pequeña ubicada sobre una región montañosa unos 6 km al noroeste del pueblo de Mitla. Las excavaciones permitieron aislar 6 pisos de ocupación precerámica, identificados con las letras E (el más antiguo), D, C, B3, B2 y B1 (el más reciente). Al parecer, cada piso se trata del campamento de una "microbanda" de 3–5 personas en la terminología de MacNeish (1964, 1972), y todos corresponden a la fase Naquitz, ubicada según el radiocarbono entre 8900 y 6700 años a. de C.

La ocupación más profunda (nivel E) no proporcionó plantas domesticadas. La planta cultivada más temprana, o sea la *Cucurbita pepo* (calabaza), apareció en el nivel D (8750–7840) a. de C.), y luego en el nivel C (7450–7280 a. de C.) se registró *Lagenaria siceraria* (jícara). Los frijoles negros pequeños (*Phaseolus* sp.) se encuentran en toda la secuencia, pero sobre todo en nivel B1 (6900–6700 a. de C.), lo cual sugiere que estaba experimentándose con su cultivo. Sin embargo, esta especie de *Phaseolus* es una que nunca ha llegado a producir una raza domesticada, así que nuestra evidencia no es absoluta. En los niveles C–B1 de la cueva se encontró polen del género *Zea* (sea maíz o teosinte), pero no se halló ningún espécimen de *Zea*.

Este volumen se divide en 8 secciones, como sigue.

PARTE I

En el Capítulo 1 se discuten los orígenes de la agricultura como un argumento teórico mundial en la antropología, considerando los puntos de vista de Binford (1968, 1980), Cohen (1977), Hassan (1981), Rindos (1980, 1984), y otros. Al respeto, cada autor presenta planteamientos válidos e inválidos, pero podemos sintetizarlos argumentando que los cambios climáticos del Pleistoceno Final descritos por Wright, y el crecimiento de la población mundial descrito por Cohen y Hassan, se unieron entre 10,000 y 5,000 anos a. de C. para producir un cambio densidad-dependiente en la conducta cultural humana en muchas partes del mundo. La emigración y la alta movilidad declinaron en cuanto a su importancia, y las estrategias concernientes con la variación *predecible* (estacional) y *no predecible* (anual) en una base local empezaban a emerger. El regionalismo en los artefactos líticos, la ampliación de la dieta para incluir más alimentos de nivel trófico más bajo, un aumento en el almacenaje, y mayor diversidad en el forraje y la colección logísticamente organizada descritos por Binford, se hacían más visibles en el testimonio arqueológico. En tal situación la agricultura temprana quizá se inició sencillamente como una extensión de los métodos que ya venían siendo utilizados por los recolectores y cazadores para enfrentarse a las variaciones ambientales de largo plazo. Por medio del aumento artificial de la densidad de ciertas plantas anuales en lugares específicos, cuya ubicación se podía predecir con certeza, podrían extender los patrones ya establecidos de reducir las diferencias de productividad entre buenos años y malos años. La evidencia arqueológica mundial sugiere que los humanos empezaron a experimentar con el cultivo *antes de* que los cambios genéticos favorables se manifiesten en las plantas, y no *después de* ellos como en el modelo de Rindos.

Los humanos se hallan en capacidad de producir estrategias duraderas para enfrentar la variación ambiental prolongada, pues no sólo poseen una memoria multigeneracional, sino que también pueden modificar su conducta sin esperar los cambios genéticos, y tienen métodos únicos para el intercambio de la información y el establecimiento de vínculos de reciprocidad a través del tiempo. Por eso, en el Capítulo 2 presentamos un modelo ecosistemático para la relación hombre–planta en la cual la información es uno de sus componentes. El modelo consiste en una serie de subsistemas entrelazados, por los cuales la materia y la energía se intercambian entre las poblaciones de plantas y animales, y la información sirve para regular el orden, la periodicidad, y la naturaleza de tales intercambios. Hay gran cantidad de "relaciones recurrentes" en el modelo, pero nos esforzamos por señalar que estas relaciones—que algunos arqueólogos suelen llamar "leyes"—en realidad no son de ningún modo leyes. La aplicación de los datos de Guilá Naquitz a este modelo de sistemas es uno de los tópicos de este volumen.

PARTE II

En el Capítulo 3 se ubica a Guilá Naquitz en su contexto espacial, temporal, y cultural. En el Capítulo 4 se examina el contexto ambiental de la cueva. Según el botánico C. Earle Smith, la vegetación original del Valle de Oaxaca incluía (1) un bosque mesofítico a lo largo del Río Atoyac y sus tributarios principales; (2) un bosque de mezquite–huizache en el terreno aluvial del río; (3) un bosque espinoso en las faldas del piedemonte; y (4) un bosque de robles y pinos en las montañas. Hoy en día, se puede dividir la región de Guilá Naquitz en 4 zonas de vegetación, llamadas: Bosque Espinoso A y B (cerca de la cueva) y Mezquital–Pastizal A y B (cerca del río que corre abajo). Los habitantes de la cueva explotaron los recursos de estas 4 zonas.

PARTE III

En el Capítulo 5 se describe la excavación de Guilá Naquitz, considerando las unidades verticales (niveles estratigráficos o pisos de ocupación) y las horizontales (cuadrados de 1 × 1 m, y ubicación tridimensional de los artefactos). En el Capítulo 6 Hole describe los artefactos líticos, que incluyen puntas de proyectil Lerma(?), Pedernales, Almagre, y Trinidad(?); bifaces; buriles; varias formas de raspadores; y varios tipos de lascas retocadas o deliberadamente o por el uso. Las canteras de sílex y toba volcánica silicificada que se utilizaron para producir los artefactos líticos se describen por Whalen en el Capítulo 7. Las manos, los metates, los morteros y otros artefactos de piedra para moler se describen en el Capítulo 8; los artefactos de madera, hueso, y asta de venado se dan en los Capítulos 10 y 12. En el Capítulo 9, King describe los especimenes de red, mecate, y canasta. En los otros capítulos de esta sección se describen los combustibles, los coprolitos, y las fechas radiocarbónicas de Guilá Naquitz.

PARTE IV

Uno de los objetivos de la Parte IV es de establecer si la vegetación actual del área de Guilá Naquitz es suficientemente semejante a la de la fase Naquitz para permitirnos sugerir un modelo sobre las relaciones ambientales prehistóricas. En el Capítulo 15, Schoenwetter y L. D. Smith dan un informe sobre la evidencia del polen para toda la secuencia precerámica oaxaqueña, de la cual Guilá Naquitz es sólo una parte. En el Capítulo 16, Flannery y Wheeler comparan la microfauna precerámica y moderna. Ambos estudios sugieren que aunque el ambiente de la fase Naquitz fue un poco más templado y seco que el de hoy, fue sin embargo suficientemente semejante como para permitir la elaboración de los modelos que deseábamos presentar.

PARTE V

En esta sección, varios autores analizan los datos sobre la subsistencia de Guilá Naquitz. En el Capítulo 17 se presenta un informe sobre la manera por la cual las plantas y los animales llegaron a la cueva, además de las suposiciones que tienen que hacerse para poder cuantificar sus restos. En el Capítulo 18 se examinan unos 7 censos anuales, que abarcan un período de 10 años (1966–1976), y que suministran densidades de las plantas silvestres (y de algunos animales) cerca de Guilá Naquitz; estos censos incluyen años "lluviosos", "secos," y "típicos", y también cálculos sobre las diferencias de productividad entre ellos. Utilizando los datos de los capítulos siguientes, estos datos se convierten a unidades como kilos de porción comestible, kilocalorías, y gramos de proteína por hectárea para las especies principales de plantas y animales.

En el Capítulo 19, Smith describe las plantas silvestres excavadas en Guilá Naquitz, que incluyen bellota (*Quercus*), piñón (*Pinus*), guaje (*Leucaena*), susí (*Jatropha*), maguey (*Agave* spp.), nanche (*Malpighia*), tuna y nopal (*Opuntia*), cebolla silvestre (*Allium*), mezquite (*Prosopis*), rompecapa (*Celtis*), y otras. En el Capítulo 20, Whitaker y Cutler describen las calabazas de la cueva; en el Capítulo 21, Kaplan describe los frijoles negros. En el Capítulo 22, Flannery y Wheeler identifican los animales de caza, de los cuales los más comunes fueron venado cola blanca (*Odocoileus*), conejo (*Sylvilagus* spp.), y tortuga de charco (*Kinosternon*). También capturaron jabalí (*Dicotyles*), mapache (*Procyon*), y varias aves como codorniz (*Colinus*), paloma (*Columba*), huilota (*Zenaidura*), tortolita (*Columbigallina*), y tecolote (*Tyto*).

En el Capítulo 23, Robson y Elias presentan un análisis nutricional sobre las plantas que consumían en la época prehistórica del área de Guilá Naquitz. Muestras modernas de estas plantas fueron recogidas y enviadas en forma congelada por vía aérea a la Universidad de Michigan para su estudio respectivo. El análisis de los mencionados autores los lleva a la reconstrucción de la dieta de la fase Naquitz en términos nutricionales. En el Capítulo 24, utilizamos los datos compilados en capítulos anteriores para reconstruir el área del territorio que habrían necesitado los cazadores–recolectores de Guilá Naquitz para sobrevivir a base de los alimentos encontrados en la cueva.

PARTE VI

En la Parte VI se presenta un análisis espacial de la flora, la fauna, y los artefactos más numerosos en los pisos de ocupación de Guilá Naquitz. En el Capítulo 25, se establece la base de este análisis al demostrar la cantidad variable de las 17 categorías mayores de desechos en cada uno de los cuadrados de 1 × 1 m, y para cada piso de ocupación. En los siguientes capítulos, se utilizan tres métodos para estudiar la distribución de estas 17 variables de un cuadrado a otro en los pisos de ocupación. Spencer y Flannery (Capítulo 26) presentan un programa de computadora que dibuja los contornos de densidad para cada tipo de los desechos sobre cada piso de ocupación. Whallon (Capítulo 27) proporciona una matriz de la *r* de Pearson para cada nivel, demostrando que ciertas categorías de desechos tienden a asociarse en forma muy marcada en grupos compuestos de 4–8 cuadrados vecinos. Reynolds (Capítulo 28) presenta dos programas para el escalamiento multidimensional, uno de ellos para las categorías de los desechos y el otro para los cuadrados de 1 × 1 m; basándose en estos resultados, propone para cada piso de ocupación cuáles son los cuadrados que pueden combinarse para señalar las áreas de trabajo de las mujeres, las áreas de trabajo de los hombres, las áreas de desecho, y las rutas de tráfico dentro de la cueva. En el Capítulo 29 se sintetizan estos tres métodos, obteniéndose un "análisis episódico" (Ascher y Clune 1960), o sea una relación narrativa de los episodios mediante los cuales se formó cada piso de ocupación.

PARTE VII

En la Parte VII del estudio se presenta una simulación sobre la evolución de la recolección de plantas y la agricultura temprana en la región de Guilá Naquitz, Oaxaca. En el Capítulo 31 Reynolds pone en operación, en forma matemática, el modelo del ecosistema que habíamos presentado en forma verbal en el Capítulo 2, al convertirlo primero a un modelo algebraico y luego a un programa de computadora. En el programa de Reynolds, se coloca a una microbanda de 4 personas en un ambiente compuesto de Bosque Espinoso A y B, y de Mezquital–Pastizal A y B, basado en nuestros estudios cuantitativos de estas zonas de vegetación. A la microbanda se presenta una sucesión de años lluviosos, secos, y típicos en un orden al azar (o sea no predecible), pero en las proporciones en que sí se ocurren según los archivos de lluvias pasadas de Oaxaca. Después de pasar por una etapa de inexperiencia, este grupo desarrolla una estrategia para la recolección de las plantas silvestres a través de centenares de "unidades de tiempo." Después de cada unidad de tiempo, ellos evalúan su obra previa en cuanto a su habilidad de conseguir 2000 kilocalorías y 40 gramos de proteína por persona al día, mientras que tratan de reducir el área en hectáreas sobre el cual hacen su recolección de plantas. Se acuerdan bien de cómo salieron en cada tipo do año, y mientras que se va aumentando la información en su memoria, poco a poco modifican su estrategia de recolección hasta llegar a tal punto que no hay muchos cambios que representan mejoramientos. En este punto, utilizan una combinación de plantas que es muy semejante a la del nivel D de Guilá Naquitz.

Algunos de los resultados interesantes son los siguientes: (1) El grupo decide en dos estrategias básicas, la una para los años lluviosos y la otra para los años secos y típicos. Durante su estrategia del año seco/típico, son más conservadores; durante los años lluviosos, suelen experimentar más. (2) El grupo parece concentrarse más en las calorías que en la proteína. (3) El grupo parece interesarse más por reducir el área de la cosecha *dentro de* una zona de vegetación, que por reducir la distancia que se viaja de la cueva *a* una zona de vegetación.

Luego, se le permite al grupo la opción de incorporar la calabaza, los frijoles, y el maíz primitivo a sus estrategias de subsistencia. El grupo pone a prueba estas plantas, primero durante los años lluviosos, pero finalmente también en los años secos/típicos. Su eficiencia de forrajear se mejora con el cultivo incipiente, pero el aumento en el tiempo que se invierte en la calabaza, los frijoles y el maíz, a la vez, reduce paralelamente el tiempo que se invierte en ciertas plantas silvestres. Cuando se estabiliza una nueva adaptación, después de centenares de unidades de tiempo, la combinación de plantas utilizadas es marcadamente semejante a la del nivel B1 de Guilá Naquitz.

Finalmente, Reynolds mantiene constante a todas las otras variables y hace 5 cambios experimentales en los parámetros del ecosistema. En estos experimentos, (1) el clima llega a ser más seco; (2) el clima llega a ser más lluvioso; (3) los extremos del clima desaparecen, y quedan solamente años "típicos" uniformes; (4) la población crece lentamente a través de centenares de unidades de tiempo; (5) la población del grupo fluctúa en forma no predecible pero mantiene un promedio de 4 personas. Ninguno de estos experimentos acelera el proceso de la adopción de la agricultura, y algunos hasta resultan en un proceso de adopción más lento. En cuanto al clima, parece que una combinación de años secos (que fomentan la eficiencia) y años lluviosos (que permiten la experimentación) provee el contexto de adaptación más favorable. Nuestros resultados no apoyan la teoría que el crecimiento de la población acelera la adopción de la agricultura; sin embargo, la fluctuación no predecible de la población sí aceleró el proceso por el cual se concentraba la agricultura en la zona Mezquital–Pastizal B, donde los suelos aluviales son mejores.

En el Capítulo 32 se examinan algunas de las implicaciones de la simulación de Reynolds para los métodos y la teoría arqueológicos.

PARTE VIII

En el Capítulo 33 se presenta un resumen de los resultados más importantes de este estudio, se evalúan varias teorías recientes acerca de los orígenes de la agricultura, y se reflexiona sobre el significado de la "causalidad" en el contexto de los cambios culturales prehistóricos. Además se examinan las opiniones del distinguido biólogo Ernst Mayr sobre el papel muy reducido que las "leyes" juegan en la explicación de los fenómenos biológicos.

References

Ahrens, E. H., Jr., and C. A. Boucher
1978 The composition of a simulated American diet. *Journal of the American Dietetic Association* 73:613–620.

Álvarez del Toro, Miguel
1952 *Los Animales Silvestres de Chiapas.* Tuxtla Gutiérrez: Ediciones del Gobierno del Estado.

Arabie, Phipps, and Scott A. Boorman
1972 Multidimensional scaling of measures of distance between partitions, *Psychology and Education Series, Institute for Mathematical Studies in the Social Sciences, Stanford University, Technical Reports* 189.

Ascher, Robert, and Francis J. Clune, Jr.
1960 Waterfall Cave, southern Chihuahua, Mexico. *American Antiquity* 26(2):270–274.

Aschmann, Homer
1980 Comment on Rindos' "Symbiosis, Instability, and the Origins and Spread of Agriculture." *Current Anthropology* 21(6):765.

Aveleyra Arroyo de Anda, L.
1956 The second mammoth and associated artifacts at Santa Isabel Iztapan, Mexico. *American Antiquity* 22(1):12–28.

Bailey, K. V.
1963 Nutrition in New Guinea, *Food and Nutrition Notes and Reviews* 20(7 and 8):89–112.

Banerjee, V. C., and E. S. Barghoorn
1972 Fine structure of pollen grain ektexine of maize, teosinte, and tripsacum. *Thirtieth Annual Proceedings of the Electron Microscopy Society of America*: 226–227.

Barash, David P.
1982 *Sociobiology and Behavior* (second ed.). New York: Elsevier Press.

Barkley, Fred A.
1934 The statistical theory of pollen analysis. *Ecology* 15:283–289.

Beadle, George W.
1972 The mystery of maize. *Field Museum of Natural History Bulletin* 43(10):2–11.
1977 The origin of *Zea mays.* In *Origins of Agriculture*, edited by Charles A. Reed. The Hague: Mouton Press. Pp. 615–635.

Berlin, E. A., and E. K. Markell
1977 An assessment of the nutritional and health status of an Aguaruna Jívaro community, Amazonas, Peru. *Ecology of Food and Nutrition* 6(2): 69–81.

Bernabo, J. Christopher, and Thompson Webb, III
1977 Changing patterns in the Holocene pollen record of northeastern North America. *Quaternary Research* 8:64–96.

Bertalanffy, Ludwig von
1962 General system theory: a critical review, *General Systems, Yearbook of the Society for General Systems Research* 7:1–20.

Bettinger, Robert L.
1979 Multivariate statistical analysis of a regional subsistence–settlement model for Owens Valley. *American Antiquity* 44(3):455–470.

Binford, Lewis R.
1968 Post-Pleistocene Adaptations. In *New Perspectives in Archeology,* edited by Sally R. Binford and Lewis R. Binford. Chicago: Aldine Publishing Co. Pp. 313–341.
1980 Willow smoke and dogs' tails: Hunter–gatherer settlement systems and archaeological site formation. *American Antiquity* 45(1):4–20.
1981 Behavioral archaeology and the "Pompeii Premise." *Journal of Anthropological Research* 37:195–208.
1982 The archaeology of place. *Journal of Anthropological Archaeology* 1(1):5–31.

Binford, Lewis R., and W. J. Chasko, Jr.
1976 Nunamiut demographic history: A provocative case. In *Demographic Anthropology: Quantitative Approaches*, edited by Ezra B. W. Zubrow. Albuquerque: University of New Mexico Press. Pp. 63–143.

Birdsell, Joseph B.
1968 Some predictions for the Pleistocene based on equilibrium systems among recent hunter–gatherers. In *Man the Hunter,* edited by Richard B. Lee and Irven DeVore. Chicago: Aldine Publishing Co. Pp. 229–240.

Birks, H.J.B., and Hillary H. Birks
1981 *Quaternary Paleoecology.* Baltimore: University Park Press.

Birks, H.J.B., T. Webb III, and A. A. Berti
1975 Numerical analysis of pollen samples from central Canada: A comparison of methods. *Review of Paleobotany and Palynology* 20:228–269.

Blake, Emmet Reid
1953 *Birds of Mexico.* Chicago: University of Chicago Press.

Bradbury, John Platt
1971 Paleolimnology of Lake Texcoco, Mexico: Evidence from diatoms. *Limnology and Oceanography* 16:180–200.

Braidwood, Robert J., and Bruce Howe
1960 Prehistoric investigations in Iraqi Kurdistan. *Studies in Ancient Oriental Civilization* 31. Chicago: University of Chicago Press.

Bressani, R., and N. S. Scrimshaw
1958 Effect of lime treatment on *in vitro* availability of essential amino acids and solubility of protein fractions in corn.

Journal of Agricultural and Food Chemistry 6:774–778.

Bryant, Vaughn M., Jr.
1975 Pollen analysis of 10 soil samples from Cache River basin. In The Cache River Archaeological Project: An experiment in contract archaeology, edited by M. B. Schiffer and J. H. House. *Arkansas Archaeological Survey Research Series* 8:311–312.

Bryant, Vaughn M., Jr., and Richard G. Holloway
1983 The role of palynology in archaeology. In *Advances in Archaeological Method and Theory*, Volume 6, edited by M. B. Schiffer. Pp. 191–224.

Bryson, Reid, D. A. Baerreis, and W. M. Wendland
1970 The character of late-Glacial and post-Glacial climatic changes. In Pleistocene and Recent environments of the central Great Plains, edited by W. Dort and J. K. Jones. *University of Kansas Department of Geology Special Publication* 3. Lawrence, Kansas. Pp. 53–74.

Burton, Robert J.
1973a Progress report on the statistical analysis of Koster surface pollen. Manuscript submitted to J. Schoenwetter, Department of Anthropology, Arizona State University, Tempe.
1973b Discriminant analysis of Koster pollen samples. Manuscript submitted to J. Schoenwetter, Department of Anthropology, Arizona State University, Tempe.

Byers, Douglas S. (editor)
1967 *The Prehistory of the Tehuacán Valley*. Volume 1: *Environment and Subsistence.* Austin: University of Texas Press.

Caldwell, M. J.
1972 Ascorbic acid content of Malaysian leaf vegetables. *Ecology of Food and Nutrition* 1(4):313–317.

Callen, Eric O.
1967a Analysis of the Tehuacán coprolites. In *The Prehistory of the Tehuacán Valley* (Vol. 1): *Environment and Subsistence*, edited by Douglas S. Byers. Austin: University of Texas Press. Pp. 261–289.
1967b The first New World cereal. *American Antiquity* 32(4):535–538.

Calloway, D. H., R. D. Giague, and F. M. Costa
1974 The superior mineral content of some American Indian foods in comparison to federally donated counterpart commodities. *Ecology of Food and Nutrition* 3(3):203–211.

Canseco, Alonso de
1580 Relación de Tlacolula y Mitla hecha en los Días 12 y 23 de Agosto respectivamente. *Papeles de Nueva España: Segunda Serie, Geografía y Estadística* (Vol. 4), edited by Francisco del Paso y Troncoso. Madrid 1905. Pp. 144–154.

Carr, Christopher
1984 The nature of organization of intrasite archaeological records and spatial analytic approaches to their investigation. In *Advances in Archaeological Method and Theory*, Volume 7, edited by M. B. Schiffer. Orlando: Academic Press.

Cashdan, Elizabeth
1982 Review of *Hunter–Gatherer Foraging Strategies: Ethnographic and Archaeological Analyses*, edited by Bruce Winterhalder and Eric Alden Smith. *Science* 216:1308–1309.

Cavicchio, Donald J.
1970 Adaptive search using simulated evolution. Ph.D. dissertation, Department of Computer and Communication Sciences, University of Michigan.

Ceci, Lynn
1980 Comment on Rindos' "Symbiosis, Instability, and the Origins and Spread of Agriculture." *Current Anthropology* 21(6):766.

Cohen, Mark Nathan
1977 *The Food Crisis in Prehistory: Overpopulation and the Origins of Agriculture.* New Haven: Yale University Press.
1980 Comment on Rindos' "Symbiosis, Instability, and the Origins and Spread of Agriculture." *Current Anthropology* 21(6):766–767.

Cooley, William W., and Paul R. Lohnes
1971 *Multivariate Data Analysis.* New York: John Wiley and Sons.

Córdova, Fray Juan de
1578 *Vocabulario en Lengua Zapoteca.* México: Pedro Charte y Antonio Ricardo. (Reprinted 1942.)

Cowgill, George L.
1968 Archaeological applications of factor, cluster, and proximity analysis. *American Antiquity* 33:367–375.

Cravioto, R. B., E. E. Lockhart, R. K. Anderson, F. de P. Miranda, and R. S. Harris
1945 Composition of typical Mexican foods. *Journal of Nutrition* 29:317–329.

Craytor, W. B., and L. Johnson, Jr.
1968 Refinements in computerized item seriation. *University of Oregon Museum of Natural History Bulletin* 10. Eugene.

Curwen, E. Cecil
1930 Prehistoric flint sickles. *Antiquity* 4(14):182–186.

Cutler, Hugh C., and Thomas W. Whitaker
1961 History and distribution of the cultivated cucurbits in the Americas. *American Antiquity* 26:469–485.
1967 Cucurbits from the Tehuacán Caves. In *The Prehistory of the Tehuacán Valley* (Vol. 1): *Environment and Subsistence*, edited by Douglas S. Byers. Austin: University of Texas Press. Pp. 212–219.

Dacey, M. F.
1973 Statistical tests of spatial association in the locations of tool types. *American Antiquity* 38:320–328.

Dawkins, R.
1976 *The Selfish Gene*. London: Oxford University Press.

De Jong, Kenneth A.
1975 Analysis of the behavior of a class of genetic adaptive systems. Ph.D. dissertation, Department of Computer and Communication Sciences, University of Michigan.

Delcourt, Paul A., Hazel R. Delcourt, and Thompson Webb, III
1984 Atlas of mapped distributions of dominance and modern pollen percentages for important tree taxa of Eastern North America. *American Association of Stratigraphic Palynologists, Contribution* 14.

Department of Geography, University of Michigan
1974 Interactive contouring program. *Cartographic Laboratory Report* 7. Ann Arbor.

De Wet, J. M., and Jack R. Harlan
1972 Origin of maize: The tripartite hypothesis. *Euphytica* 21:271–279.

Dixon, Wilfrid J., and Frank J. Massey
1969 *Introduction to Statistical Analysis.* New York: McGraw-Hill.

Doran, James E.
1970 Systems theory, computer simulations and archaeology. *World Archaeology* 1:289–298.

Doran, James E., and F. R. Hodson
1975 *Mathematics and Computers in Archaeology*. Cambridge, Mass.: Harvard University Press.

Drennan, Robert D.
1976a A refinement of chronological seriation using nonmetric multidimensional scaling. *American Antiquity* 41(3):290–302.

1976b Fábrica San José and Middle Formative society in the Valley of Oaxaca. Prehistory and human ecology of the Valley of Oaxaca (Vol. 4). *Memoirs of the University of Michigan Museum of Anthropology* 8. Ann Arbor.

Durham, William H.
1981 Overview: Optimal foraging analysis in human ecology. In *Hunter–Gatherer Foraging Strategies: Ethnographic and Archaeological Analyses,* edited by Bruce Winterhalder and Eric Alden Smith. Chicago: University of Chicago Press. Pp. 218–231.

Food and Agriculture Association (FAA)
1972 *Food Composition Table for Use in East Asia.* New York: Food and Agriculture Association.

Faegri, Knut, and J. Iverson
1975 *Textbook of Pollen Analysis* (third ed.). New York: Hofner Press.

Fish, Suzanne K.
1969 A pollen record from the Valley of Oaxaca. Paper presented at the thirty-fourth annual meeting of the Society for American Archaeology, Milwaukee.
1977 Cultigens, cultivation, and chronology: Palynology at Edzná. Paper presented at the forty-second annual meeting of the Society for American Archaeology, New Orleans.
1978 Palynology of Edzná and Aguacatal: Environment and economy. Paper presented at the forty-third annual meeting of the Society for American Archaeology, Tucson.

Flannery, Kent V.
1966 The Postglacial 'readaptation' as viewed from Mesoamerica. *American Antiquity* 31(6):800–805.
1967 Vertebrate fauna and hunting patterns. In *The Prehistory of the Tehuacán Valley* (Vol. 1): *Environment and Subsistence,* edited by Douglas S. Byers. Austin: University of Texas Press. Pp. 132–177.
1968 Archeological systems theory and early Mesoamerica. In *Anthropological Archeology in the Americas,* edited by Betty J. Meggers. Washington, D.C.: Anthropological Society of Washington. Pp. 67–87.
1969 Origins and ecological effects of early domestication in Iran and the Near East. In *The Domestication and Exploitation of Plants and Animals*, edited by P. J. Ucko and G. W. Dimbleby. London: Gerald Duckworth and Co. Pp. 73–100.
1972 The cultural evolution of civilizations. *Annual Review of Ecology and Systematics* 3:399–426.
1973 The origins of agriculture. *Annual Review of Anthropology* 2:271–310.
1976 Empirical determination of site catchments in Oaxaca and Tehuacán. In *The Early Mesoamerican Village,* edited by Kent V. Flannery. New York: Academic Press. Pp. 103–117.
1983 Settlement, subsistence, and social organization of the proto-Otomangueans. In *The Cloud People: Divergent Evolution of the Zapotec and Mixtec Civilizations,* edited by Kent V. Flannery and Joyce Marcus. New York: Academic Press. Pp. 32–36.

Flannery, Kent V. (editor)
1970 Preliminary archaeological investigations in the Valley of Oaxaca, Mexico, 1966 through 1969: Report to the Instituto Nacional de Antropología e Historia and the National Science Foundation. Manuscript on file, Museum of Anthropology, University of Michigan, Ann Arbor.

Flannery, Kent V., and Richard I. Ford
1972 A productivity study of teosinte (*Zea mexicana*), Nov. 22–25, 1971. Manuscript on file, Museum of Anthropology, University of Michigan, Ann Arbor.

Flannery, Kent V., Anne V. T. Kirkby, Michael J. Kirkby, and Aubrey W. Williams, Jr.
1967 Farming systems and political growth in ancient Oaxaca. *Science* 158:445–454.

Flannery, Kent V., and Joyce Marcus
1976 Formative Oaxaca and the Zapotec cosmos. *American Scientist* 64(4):374–383. New Haven, Conn.: Sigma Xi, The Scientific Research Society of North America, Inc.

Flannery, Kent V., Joyce Marcus, and Stephen A. Kowalewski
1981 The preceramic and Formative of the Valley of Oaxaca. In *Supplement to the Handbook of Middle American Indians* (Vol. 1): *Archaeology*, edited by Jeremy A. Sabloff (general editor, Victoria R. Bricker). Austin: University of Texas Press. Pp. 48–93.

Flannery, Kent V., and James Schoenwetter
1970 Climate and man in Formative Oaxaca. *Archaeology* 23:144–152.

Flannery, Kent V., and C. Earle Smith, Jr.
1983 Monte Albán IV foodstuffs in Guilá Naquitz Cave. In *The Cloud People: Divergent Evolution of the Zapotec and Mixtec Civilizations,* edited by Kent V. Flannery and Joyce Marcus. New York: Academic Press. P. 206.

Ford, Richard I.
1968 An ecological analysis involving the population of San Juan Pueblo, New Mexico. Ph.D. dissertation, Department of Anthropology, University of Michigan.
1976 Carbonized plant remains. In Fábrica San José and Middle Formative Society in the Valley of Oaxaca, edited by Robert D. Drennan. *Prehistory and Human Ecology of the Valley of Oaxaca* (Vol. 4), *Memoirs of the University of Michigan Museum of Anthropology* 8. Ann Arbor. Pp. 261–268.

Fowler, Melvin L., and Richard S. MacNeish
1972 Excavations in the Coxcatlán locality in the alluvial slopes. In *The Prehistory of the Tehuacán Valley* (Vol. 5): *Excavations and Reconnaissance,* by Richard S. MacNeish, Melvin L. Fowler, Angel García Cook, Frederick A. Peterson, Antoinette Nelken-Terner, and James A. Neely. Austin: University of Texas Press. Pp. 219–340.

Fox, Dennis, and Kenneth Guire
1976 Documentation for MIDAS. Manuscript on file, University of Michigan Statistical Laboratory, Ann Arbor.

Fried, Morton H.
1967 *The evolution of political society.* New York: Random House.

Galinat, Walton C.
1970 The cupule and its role in the origin and evolution of maize. *Massachusetts Agricultural Experiment Station Bulletin* 585:1–18.
1971 The origin of maize. *Annual Review of Genetics* 5:447–478.

García Moll, Roberto
1977 Análisis de los materiales arqueológicos de la Cueva del Texcal, Puebla. *Colección Científica 56 Arqueología*. México, D.F.: Instituto Nacional de Antropología e Historia, Departmento de Prehistoria.

Gibbons, Jean Dickinson
1971 *Non-parametric Statistical Inference.* New York: McGraw-Hill.

Gilbert, Lawrence E., and Peter H. Raven (editors)
1975 *Coevolution of Animals and Plants.* Austin: University of Texas Press.

Gish, Jannifer W.
1976 Palynology of the Robinson site, north-central Wisconsin. *Arizona State University Anthropological Research Paper*

5. Tempe.

Greig-Smith, P.

1952 The use of random and contiguous quadrats in the study of the structure of plant communities. *Annals of Botany* (n.s.) 16:293–316.

1961 Data on pattern within plant communities I: The analysis of pattern. *Journal of Ecology* 49:695–702.

1964 *Quantitative Plant Ecology* (second ed.). New York: Plenum Publishing Co.

Greig-Smith, P., K. A. Kershaw, and D. J. Anderson

1963 The analysis of pattern in vegetation: A comment on a paper by D. W. Goodall. *Journal of Ecology* 51:223–229.

Hall, E. Raymond, and Keith R. Kelson

1959 *The Mammals of North America* (2 vols). New York: The Ronald Press.

Harris, Marvin

1979 *Cultural Materialism: The Struggle for a Science of Culture.* New York: Random House.

Hassan, Fekri A.

1977 The dynamics of agricultural origins in Palestine: A theoretical model. In *Origins of Agriculture,* edited by Charles A. Reed. The Hague: Mouton Publishers. Pp. 589–609.

1981 *Demographic Archaeology.* New York: Academic Press.

Heiser, Charles B.

1979 *The Gourd Book.* Norman: University of Oklahoma Press.

Heizer, Robert F.

1960 Physical analysis of habitation residues. In *The application of quantitative methods in archaeology,* edited by Robert F. Heizer and Sherburne F. Cook. *Viking Fund Publications in Anthropology* 28. New York: Wenner-Gren Foundation. Pp. 93–142.

Heizer, Robert F., and Sherburne F. Cook, (editor)

1960 The application of quantitative methods in archaeology. *Viking Fund Publications in Anthropology* 28. New York: Wenner-Gren Foundation.

Helbaek, Hans

1969 Plant collecting, dry-farming, and irrigation agriculture in prehistoric Deh Luran. In *Prehistory and Human Ecology of the Deh Luran Plain,* by Frank Hole, Kent V. Flannery, and James A. Neely. *Memoirs of the University of Michigan Museum of Anthropology* 1. Ann Arbor. Pp. 383–426.

Hevly, Richard H.

1968 Studies of modern pollen rain in northern Arizona. *Journal of the Arizona Academy of Sciences* 5:111–115.

Hipsley, E. H., and N. E. Kirk

1965 Studies of dietary intake and the expenditure of energy by the New Guineans. *South Pacific Commission Technical Papers* 147. Noumea.

Hodder, Ian (editor)

1978 *Simulation studies in archaeology.* Cambridge: Cambridge University Press.

Hodder, Ian, and Clive Orton

1976 *Spatial Analysis in Archaeology.* Cambridge: Cambridge University Press.

Hole, Frank, Kent V. Flannery, and James A. Neely

1969 Prehistory and human ecology of the Deh Luran plain. *Memoirs of the University of Michigan Museum of Anthropology* 1. Ann Arbor.

Holland, John H.

1975 *Adaptation in Natural and Artificial Systems.* Ann Arbor: University of Michigan Press.

Holland, John H., and Judith R. Reitman

1978 Cognitive systems based on adaptive algorithms. In *Pattern Directed Information Systems,* edited by Roth F. Hayes and D. Waterman. New York: Academic Press. Pp. 313–329.

Holmes, W. H.

1897 Archaeological studies among the ancient cities of Mexico (Part II): Monuments of Chiapas, Oaxaca and the Valley of Mexico. *Field Columbian Museum Anthropological Series* (1)1. Chicago.

Hosler, Dorothy, Jeremy A. Sabloff, and Dale Runge

1977 Simulation model development: A case study of the Classic Maya collapse. In *Social Process in Maya Prehistory,* edited by Norman Hammond. New York: Academic Press. Pp. 553–590.

Iltis, Hugh H.

1983 From teosinte to maize: The catastrophic sexual transmutation. *Science* 222:886–894.

Iltis, Hugh H., J. F. Doebley, R. Guzmán M., and B. Pazy

1979 *Zea diploperennis* (Gramineae): A new teosinte from Mexico. *Science* 203:186–188.

Jacobs, A.

1969 Availability of food iron. *British Medical Journal* 1:673–676.

Jennings, Jesse D.

1957 Danger Cave. *Memoirs of the Society for American Archaeology* 14.

Jochim, Michael A.

1976 *Hunter–Gatherer Subsistence and Settlement: A Predictive Model.* New York: Academic Press.

Johnson, Frederick, and Richard S. MacNeish

1972 Chronometric dating. *In The Prehistory of the Tehuacán Valley* (Vol. 4): *Chronology and Irrigation,* edited by Frederick Johnson. Austin: University of Texas Press. Pp. 3–55.

Johnson, Gregory A.

1978 Information sources and the development of decision-making organizations. In *Social Archaeology: Beyond Subsistence and Dating,* edited by C. L. Redman, M. J. Berman, E. V. Curtin, W. T. Langhorne, Jr., N. M. Versaggi, and J. C. Wanser. New York: Academic Press. Pp. 87–112.

Johnson, Irmgard W.

1967 Textiles. In *The Prehistory of the Tehuacán Valley* (Vol. 2). *The Non-ceramic Artifacts,* by Richard S. MacNeish, Antoinette Nelken-Terner, and Irmgard W. Johnson. Austin: University of Texas Press. Pp. 191–226.

Kaplan, Lawrence

1965 Archaeology and domestication in American *Phaseolus* (beans). *Economic Botany* 19(4):358–368.

1981 What is the origin of the common bean? *Economic Botany* 35(2):240–254.

Kaplan, Lawrence, and Richard S. MacNeish

1960 Prehistoric bean remains from caves in the Ocampo Region of Tamaulipas, Mexico. *Harvard University Botanical Museum Leaflets* 19(2):33–56.

Katz, S. H., M. L. Hediger, and L. A. Valleroy

1974 Traditional maize processing techniques in the New World. *Science* 184:765–773.

Keene, Arthur S.

1981a Optimal foraging in a nonmarginal environment: A model of prehistoric subsistence strategies in Michigan. In *Hunter–Gatherer Foraging Strategies: Ethnographic and*

Archaeological Analyses, edited by Bruce Winterhalder and Eric Alden Smith. Chicago: University of Chicago Press. Pp. 171–193.

1981b *Prehistoric Foraging in a Temperate Forest: A Linear Programming Model.* New York: Academic Press.

Kershaw, K. A.

1957 The use of cover and frequency in the detection of pattern in plant communities. *Ecology* 38:291–299.

1964 *Quantitative and Dynamic Ecology.* New York: Elsevier Press.

King, James E., W. E. Klippel, and R. Duffield

1975 Pollen preservation and archaeology in eastern North America. *American Antiquity* 40:180–190.

Kirkby, Anne V. T.

1973 The use of land and water resources in the past and present Valley of Oaxaca, Mexico. Prehistory and human ecology of the Valley of Oaxaca. (Vol. 1) *Memoirs of the University of Michigan Museum of Anthropology.* 5. Ann Arbor.

1974 Individual and community responses to rainfall variability in Oaxaca, Mexico. In *Natural Hazards: Local, Regional, and Global,* edited by Gilbert F. White. New York: Oxford University Press. Pp. 119–128.

Klecka, William R.

1975 Discriminant analysis. In *Statistical Package for the Social Sciences (second ed.),* by Norman H. Nie, C. Hadla Hull, Jean G. Jenkins, Karin Steinbrenner, and Dale H. Brent. New York: McGraw-Hill. Pp. 434–467.

1980 Discriminant analysis. In *Quantitative Applications in the Social Sciences* 19, edited by J. L. Sullivan. Beverly Hills, Calif.: Sage Publications Inc. Pp. 1–71.

Konlande, J. E., and J.R.K. Robson

1972 The nutritive value of cooked camas as consumed by Flathead Indians. *Ecology of Food and Nutrition* 1(3):193–195.

Kuhnlein, H. V., and D. H. Calloway

1979 Adventitious mineral elements in Hopi Indian diets. *Journal of Food Science* 44(1):282–285.

Lamprecht, Herbert

1968 Wesen and Wedegang Naturbedingter Arten. *Phyton* 13(1–2):1–14.

Lathrap, Donald W.

1968 The "hunting" economies of the tropical forest zone of South America: An attempt at historical perspective. In *Man the Hunter,* edited by Richard B. Lee and Irven DeVore. Chicago: Aldine Publishing Co. Pp. 23–29.

1977 Our father the cayman, our mother the gourd: Spinden revisited, or a unitary model for the emergence of agriculture in the New World. In *Origins of Agriculture,* edited by Charles A. Reed. The Hague: Mouton Publishers. Pp. 713–752.

Leakey, Richard E., and Roger Lewin

1978 *People of the Lake.* New York: Anchor Press.

Lee, Richard B.

1968 What hunters do for a living, or how to make out on scarce resources. In *Man the Hunter,* edited by Richard B. Lee and Irven DeVore. Chicago: Aldine Publishing Co. Pp. 30–48.

1972 The !Kung Bushmen of Botswana. In *Hunters and Gatherers Today,* edited by M. G. Bicchieri. New York: Holt, Rinehart, and Winston. Pp. 327–368.

Lee, Richard B., and Irven DeVore, (editors)

1968 *Man the Hunter.* Chicago: Aldine Publishing Co.

Leopold, A. Starker

1959 *Wildlife of Mexico: The Game Birds and Mammals.* Berkeley: University of California Press.

1959 *Wildlife of Mexico: The Game Birds and Mammals.* Berkeley: University of California Press.

Lewin, Roger

1983 How did vertebrates take to the air? *Science* 221:38–39.

Lewontin, Richard C.

1978 Adaptation. *Scientific American* 239(3):213–230.

Lindsay, Alexander J., Jr.

1968 Current research: Western Mesoamerica. *American Antiquity* 33:418.

Lingoes, James C.

1973 *The Guttman–Lingoes Nonmetric Program Series.* Ann Arbor: Mathesis Press.

Lorenzo, José Luis

1958 Un sitio precerámico en Yanhuitlán, Oaxaca. *Instituto Nacional de Antropología e Historia, Dirección Prehistoria, Pub.* 6. México, D.F.

1960 Aspectos físicos del Valle de Oaxaca. *Revista Mexicana de Estudios Antropológicos* 16:49–63.

Lorenzo, José Luis, and Lauro Gonzales Q.

1970 El más antiguo teosinte. *Instituto Nacional de Antropología e Historia, Boletín* 42:41–43. México, D.F.

Lorenzo, José Luis, and Miguel Messmacher

1963 Hallazgo de horizontes culturales precerámicos en el Valle de Oaxaca. In *Homenaje a Pedro Bosch-Gimpera,* edited by Santiago Genovés. México D.F.: Instituto Nacional de Antropología e Historia and Universidad Nacional Autónoma de México. Pp. 289–301.

McArthur, M.

1960 Food consumption and dietary levels of groups of Aborigines living on naturally occurring foods. In *Records of the American-Australian Scientific Expedition to Arnhem Land* (Part 2): *Anthropology and Nutrition,* edited by C. P. Mountford. Melbourne: Melbourne University Press. Pp. 90–135.

MacArthur, Robert H., and Edward O. Wilson

1967 The Theory of Island Biogeography. *Monographs in Population Biology* 1. Princeton: Princeton University Press.

McCullough, Dale R.

1979 *The George Reserve Deer Herd: Population Ecology of a K-Selected Species.* Ann Arbor: University of Michigan Press.

MacNeish, Richard S.

1958 Preliminary archaeological investigations in the Sierra de Tamaulipas, Mexico. *Transactions of the American Philosophical Society.* (Vol. 48, Part 6). Philadelphia.

1961 *First Annual Report of the Tehuacán Archaeological–Botanical Project, R.S. Peabody Foundation for Archaeology.* Andover, Mass.: Phillips Academy.

1964 Ancient Mesoamerican civilization. *Science* 143:531–537.

1967 A summary of the subsistence. In *The Prehistory of the Tehuacán Valley.* (Vol. 1): *Environment and Subsistence,* edited by Douglas S. Byers. Austin: University of Texas Press. Pp. 290–309.

1972 The evolution of community patterns in the Tehuacán Valley of Mexico and speculations about the cultural processes. In *Man, Settlement, and Urbanism,* edited by P. J. Ucko, R. Tringham, and G. W. Dimbleby. London: Gerald Duckworth and Co. Pp. 67–93.

1978 *The Science of Archaeology?* N. Scituate, Mass.: Duxbury Press.

MacNeish, Richard S., Melvin L. Fowler, Angel García Cook, Frederick A. Peterson, Antoinette Nelken-Terner, and James A. Neely
1972 *The Prehistory of the Tehuacán Valley* (Vol. 5): *Excavations and Reconnaissance.* Austin: University of Texas Press.

MacNeish, Richard S., and Antoinette Nelken-Terner
1972 Introduction. In *The Prehistory of the Tehuacán Valley* (Vol. 5): *Excavations and Reconnaissance,* by Richard S. MacNeish, Melvin L. Fowler, Angel García Cook, Frederick A. Peterson, Antoinette Nelken-Terner, and James A. Neely. Austin: University of Texas Press. Pp. 3–13.

MacNeish, Richard S., Antoinette Nelken-Terner, and Irmgard W. Johnson
1967 *The Prehistory of the Tehuacán Valley* (Vol. 2): *Nonceramic Artifacts.* Austin: University of Texas Press.

MacNeish, Richard S., and Frederick A. Peterson
1962 The Santa Marta Rockshelter, Ocozocoautla, Chiapas, Mexico. *Papers of the New World Archaeological Foundation* 14. Provo, Utah.

Mangelsdorf, Paul C.
1947 The origin and evolution of maize. *Advances in Genetics* 1:161–207.
1958 Reconstructing the ancestor of corn. *Proceedings of the American Philosophical Society* 102:454–463.
1974 *Corn: Its Origin, Evolution, and Improvement.* Cambridge, Mass: Harvard University Press.
1983 The search for wild corn. *Maydica* 28:89–96.
n.d. Preliminary report on archaeological remains of corn from caves in Oaxaca, Mexico. Unpublished manuscript to appear in a future Memoir of the University of Michigan Museum of Anthropology, Ann Arbor.

Mangelsdorf, Paul C., Richard S. MacNeish, and Walton C. Galinat
1967 Prehistoric wild and cultivated maize. In *Prehistory of the Tehuacán Valley.* (Vol. 1): *Environment and Subsistence,* edited by Douglas S. Byers. Austin: University of Texas Press. Pp. 178–200.

Mangelsdorf, Paul C., Richard S. MacNeish, and Gordon R. Willey
1964 Origins of agriculture in Middle America. In *Handbook of Middle American Indians* (Vol. 1): *Natural Environment and Early Cultures,* edited by Robert Wauchope and Robert C. West. Austin: University of Texas Press. Pp. 427–445.

Mangelsdorf, Paul C., and R. G. Reeves
1938 The origin of maize. *Proceedings of the National Academy of Sciences* 24:303–312.
1939 The origin of Indian corn and its relatives. *Texas Agricultural Experiment Station Bulletin* 574:1–315.

Marks, Anthony
1971 Settlement patterns and intrasite variability in the Central Negev, Israel. *American Anthropologist* 73:1237–1244.

Martin, Paul Schultz
1963 *The Last 10,000 Years: A Fossil Pollen Record of the American Southwest.* Tucson: University of Arizona Press.

Martin, Paul S., J. R. Rinaldo, E. Bluhm, H.C. Cutler, and R. Grange
1952 Mogollon cultural continuity and change: The stratigraphic analysis of Tularosa and Cordova caves. *Fieldiana: Anthropology 40.*

Maruyama, Magoroh
1963 The second cybernetics: deviation-amplifying mutual causal processes. *American Scientist* 51(2):164–179.

Mayr, Ernst
1982 *The Growth of Biological Thought.* Cambridge, Mass.: Harvard University Press.

Mehringer, Peter J., Jr.
1967 Pollen analysis of the Tule Springs Site. In Pleistocene studies in southern Nevada, edited by M. Wormington and D. Ellis. *Nevada State Museum Anthropological Papers* 13:130–200.

Messer, Ellen
1978 Zapotec plant knowledge: Classification, uses, and communication about plants in Mitla, Oaxaca, Mexico. In Prehistory and human ecology of the Valley of Oaxaca (Vol. 5, Part 2), edited by Kent V. Flannery and Richard E. Blanton. *Memoirs of the University of Michigan Museum of Anthropology* 10. Ann Arbor.

Metzger, Duane, and Gerald Williams
1966 Some procedures and results in the study of native categories: Tzeltal firewood. *American Anthropologist* 68:389–407.

Miller, James G.
1965 Living systems: Basic concepts; structure and process; cross-level hypotheses. *Behavioral Science* 10:193–237, 337–379, 380–411.

Minsky, M., and P. Papert
1969 *Perceptrons.* New York: Princeton Hall.

Moore, James A.
1981 The effects of information networks in hunter–gatherer societies. In *Hunter–Gatherer Foraging Strategies: Ethnographic and Archaeological Analyses,* edited by Bruce Winterhalder and Eric Alden Smith. Chicago: University of Chicago Press. Pp. 194–217.

National Academy of Sciences (NAS)
1974 Food and nutrition board: Recommended dietary allowances (eighth ed.). Washington, D.C.: National Academy of Sciences.

Neter, J., and W. Waserman
1974 *Applied Linear Statistical Models.* Homewood, Ill.: Richard D. Irwin Inc.

Niederberger, Christine
1976 Cinco milenios de ocupación humana en un sitio lacustre de la Cuenca de Mexico. *Instituto Nacional de Antropología e Historia, Colección Científica* 30. México, D.F.
1979 Early sedentary economy in the Basin of Mexico. *Science* 203:131–142.

Niklewski, J. and W. van Zeist
1970 A Late Quaternary pollen diagram from northwestern Syria. *Acta Botanica Neerlandica* 19:737–754.

Ohngemach, Dieter
1973 Análisis polínico de las sedimentas del Pleistoceno Reciente y de Holoceno en la región Puebla–Tlaxcala. *Comunicaciones, Proyecto Puebla-Tlaxcala* 7. Puebla: Fundación Alemana Para la Investigación Científica. Pp. 47–49.

Oomen, H.A.P. C.
1961 The Papuan child as a survivor. *Journal of Tropical Pediatrics* 6:103–121.

Orton, Clive
1980 *Mathematics in Archaeology.* London: Collins Publishers.

Parsons, Elsie Clews
1936 *Mitla, Town of the Souls.* Chicago: University of Chicago Press.

Paulik, G. J., and J. W. Greenough, Jr.
1966 Management analysis for a salmon resource system. In *Systems Analysis in Ecology,* edited by Kenneth E. F. Watt. New York: Academic Press. Pp. 215–252.

Pielou, E. C.
1969 *An Introduction to Mathematical Ecology.* New York: Wiley-Interscience.

Piper, C. V.

1926 Studies in American phaseolinae. *Contributions from the U.S. National Herbarium* 22:663–701. Washington, D.C.

Platt, B. S.

1962 Tables of representative values of foods commonly used in tropical countries. *Medical Research Council, Special Report Series* 302. London: Her Majesty's Stationery Office.

Rankin, Adrianne

1980 Pollen analytical studies in Corduroy Creek. In Studies in the Prehistory of the Forestdale Region, Arizona, edited by C. Russel Stafford and Glen E. Rice. *Anthropological Field Studies,* No. 1. Office of Cultural Resource Management, Arizona State University.

Reidhead, Van A.

1979 Linear programming models in archaeology. *Annual Review of Anthropology* 8:543–578.

1981 A Linear Programming Model of a Prehistoric Subsistence Optimization: A Southeastern Indiana Example. *Indiana Historical Society Prehistory Research Series* (Vol. 6, No. 1). Indianapolis.

Reinhold, J. G.

1972 Phytate concentrations of leavened and unleavened Iranian breads. *Ecology of Food and Nutrition* 1(3):187–192.

Reinhold, J. G., H. Hedayati, A. Lahimgarzader, and K. Nasr

1973 Zinc, calcium, phosphorus and nitrogen balance of Iranian villagers following a change from phytate-rich to phytate-poor diets. *Ecology of Food and Nutrition* 2(2):157–162.

Reynolds, Robert G.

1978 On modeling the evolution of hunter–gatherer decision-making systems. *Geographical Analysis* 10(1):31–46.

1979 Multiformalism modeling of cultural adaptations. Manuscript on file, Department of Computer and Communication Sciences, University of Michigan, Ann Arbor.

Reynolds, Robert G., and Bernard P. Zeigler

1979 A formal mathematical model for the operation of consensus-based hunting–gathering bands. In *Transformations: Mathematical Approaches to Culture Change,* edited by Colin Renfrew and Kenneth L. Cooke. New York: Academic Press. Pp. 405–418.

Riley, Thomas J.

1974 Constraints on dimensions of variance. *American Antiquity* 39:489–490.

Rindos, David

1980 Symbiosis, instability, and the origins and spread of agriculture: a new model. *Current Anthropology* 21(6):751–772.

1984 *The Origins of Agriculture: An Evolutionary Perspective.* New York: Academic Press.

Robson, J.R.K.

1978 Fruit in the diet of prehistoric man and of the hunter–gatherer. *Journal of Human Nutrition* 32:19–26.

Robson, J.R.K., and Douglas E. Yen

1976 Some nutritional aspects of the Philippine Tasaday diet. *Ecology of Food and Nutrition* 5(2):83–89.

Sabloff, Jeremy A. (editor)

1981 *Simulations in Archaeology.* Albuquerque: University of New Mexico Press.

Sahlins, Marshall D.

1968 *Stone Age Economics.* Chicago: Aldine Publishing Co.

Sahlins, Marshall D., and Elman R. Service (editors)

1960 *Evolution and Culture.* Ann Arbor: University of Michigan Press.

Salmon, Merrilee H.

1978 What can systems theory do for archaeology? *American Antiquity* 43:174–183.

Schiffer, Michael B.

1974 On Whallon's use of dimensional analysis of variance at Guilá Naquitz. *American Antiquity* 39:490–492.

Schlegel, S. A., and H. A. Guthrie

1973 Diet and the Tiruray shift from swidden to plow farming. *Ecology of Food and Nutrition* 2(3):181–191.

Schmieder, Oscar

1930 The settlements of the Tzapotec and Mije Indians, State of Oaxaca, Mexico. *University of California Publications in Geography* 4. Berkeley.

Schoenwetter, James

1972 Appendix C: Archaeological pollen studies in western Sonora. In An archaeological reconnaissance in southern Sonora and reconsideration of the Río Sonora culture, by Richard A. Pailes. Ph.D. dissertation, Department of Anthropology, Southern Illinois University.

1974 Pollen records of Guilá Naquitz Cave. *American Antiquity* 39(2):292–303.

1981 Contributions of pollen analysis in investigations of New World Agriculture. *Proceedings of the IV International Palynological Conference, Lucknow, India (1976–1977).* Birbal Sahni Institute of Paleobotany.

Schoenwetter, James, Michael J. Kirkby, and Anne V. Whyte

1967 Environment of the Valley of Oaxaca, past and present. Paper presented at the Thirty-Second Annual Meeting of the Society for American Archaeology, Ann Arbor.

Shaffer, Jim G.

1980 Comment on Rindos' "Symbiosis, Instability, and the Origins and Spread of Agriculture." *Current Anthropology* 21(6):768.

Silberbauer, G. B.

1972 The G/Wi Bushmen. In *Hunters and Gatherers Today,* edited by M. G. Bicchieri. New York: Holt, Rinehart, and Winston. Pp. 271–326.

Smith, C. Earle, Jr.

1965 Flora, Tehuacán Valley. *Fieldiana: Botany* 31(4):101–143.

1980 Plant remains from Guitarrero Cave. In *Guitarrero Cave,* edited by Thomas F. Lynch. New York: Academic Press. Pp. 87–119.

1969 Additional notes on pre-conquest avocados in Mexico. *Economic Botany* 23:135–140.

1978 The vegetational history of the Oaxaca Valley. In *Prehistory and human ecology of the Valley of Oaxaca* (Vol. 5, Part 1), edited by Kent V. Flannery and Richard E. Blanton. *Memoirs of the University of Michigan Museum of Anthropology* 10. Ann Arbor.

1980 Plant remains from Guitarrero Cave. In *Guitarrero Cave,* edited by Thomas F. Lynch. New York: Academic Press.

Smith, Judith E.

1981 Formative botanical remains at Tomaltepec. In Excavations at Santo Domingo Tomaltepec: Evolution of a Formative community in the Valley of Oaxaca, Mexico, by Michael E. Whalen. *Prehistory and human ecology of the Valley of Oaxaca* (Vol. 6), edited by Kent V. Flannery and Richard E. Blanton. *Memoirs of the University of Michigan Museum of Anthropology* 12. Ann Arbor. Pp. 186–194.

Smith, Landon D.

1974 The multidiscriminate analysis of surficial and fossil pollen samples from Oaxaca, Mexico. Manuscript on file,

Department of Anthropology, Arizona State University, Tempe.
1975 Statistically defining the environmental expression of pollen in Oaxaca, Mexico: Present and prehistoric. Paper prepared for presentation at the Conference on Computer Applications in Archaeology, 1976, Birmingham, England.

Sneath, Peter H. A., and Robert R. Sokal
1973 *Numerical Taxonomy.* San Francisco: W.H. Freeman.

Sokal, Robert R., and F. James Rohlf
1969 *Biometry.* San Francisco: W.H. Freeman.

Somogyi, J. C., and K. Schiebe
1966 Der Vitamin-C Gehalt Verscheidener Kartoffelsorten und Seine Abnahme Während der Lagerung. *International Journal of Vitamin and Nutrition Research* 36:337–359.

Speth, John D., and Gregory A. Johnson
1976 Problems in the use of correlation for the investigation of tool kits and activity areas. In *Cultural Change and Continuity,* edited by Charles Cleland. New York: Academic Press. Pp. 35–57.

Steward, Julian
1938 Basin–plateau aboriginal sociopolitical groups. *Bureau of American Ethnology Bulletin of the Smithsonian Institution* 120. Washington, D.C.
1955 *Theory of Culture Change.* Urbana: University of Illinois Press.

Suhm, Dee Ann, and Edward B. Jelks (editors)
1962 Handbook of Texas Archaeology: Type descriptions. *Texas Archeological Society Special Publications* 1, *Texas Memorial Museum Bulletin* 4. Austin.

Tchernov, Eitan
1968 *Succession of Rodent Faunas during the Upper Pleistocene of Israel,* Mammalia Depicta. Hamburg: Parey.

Thomas, David Hurst
1971 Prehistoric subsistence–settlement patterns of the Reese River valley, central Nevada. Ph.D. dissertation, Department of Anthropology, University of California at Davis.
1972 A computer simulation model of Great Basin Shoshonean subsistence and settlement patterns. In *Models in Archaeology,* edited by David L. Clarke. London: Methuen. Pp. 671–704.
1973 An empirical test for Steward's model of Great Basin settlement patterns. *American Antiquity* 38:155–176.
1976 *Figuring Anthropology: First Principles of Probability and Statistics.* New York: Holt, Rinehart and Winston.
1983 The archaeology of Monitor Valley (Part 1): Epistemology. *Anthropological Papers of the American Museum of Natural History* 58(1). New York.

Thompson, H. R.
1958 The statistical study of plant distribution patterns using a grid of quadrats. *Australian Journal of Botany* 6:322–342.

Townsend, P. K., S. C. Liao, and J. E. Konlande
1973 Nutritive contributions of sago ash used as a native salt in Papua, New Guinea. *Ecology of Food and Nutrition* 2(2):91–97.

Tsukada, Matsuo, and Edward S. Deevey, Jr.
1967 Pollen analyses from four lakes in the southern Maya area of Guatemala and El Salvador. In *Quaternary paleoecology,* edited by E. J. Cushing and H. E. Wright, Jr. *Proceedings of the Seventh Congress of the International Association for Quaternary Research* 7:303–333.

Turner, Judith, and Joyce Hodgson
1979 Studies in the vegational history of the Northern Pennines I. Variations in the composition of the early Flandrian forests. *Journal of Ecology* 67:629–646.
1983 Studies in the vegetational history of the Northern Pennines III. Variations in the composition of the mid-Flandrian forests. *Journal of Ecology* 71:95–118.

Van de Geer, John P.
1971 *Introduction to Multivariate Analysis for the Social Sciences.* San Francisco: W.H. Freeman.

Van Loon, Maurits, James H. Skinner, and W. van Zeist
1968 The oriental institute excavations at Mureybit, Syria: Preliminary report on the 1965 campaign (Part 1): Architecture and general finds. *Journal of Near Eastern Studies* 27:265–290.
1970 The oriental institute excavations at Mureybit, Syria: Preliminary report on the 1965 campaign (Part 3): The paleobotany. *Journal of Near Eastern Studies* 29:167–176.

Van Zeist, W., and W. A. Casparie
1968 Wild einkorn wheat and barley from Tell Mureybit in northern Syria. *Acta Botanica Neerlandica* 17(1):44–53.

Veldman, D. J.
1967 *Fortran Programming for the Behavioral Sciences.* New York: Holt, Rinehart and Winston.

Villa R., Bernardo
1954 Contribución al conocimiento de las épocas de caída y nacimiento de la cornamenta y de su terciopelo en venados cola blanca (*Odocoileus virginianus*) de San Cayetano, Estado de México, México. *Anales del Instituto de Biología* 25. México, D.F.: Universidad Nacional Autónoma de México. Pp. 451–461.

Vita-Finzi, Claudio, and Eric S. Higgs
1970 Prehistoric economy in the Mt. Carmel area of Palestine: Site catchment analysis. *Proceedings of the Prehistoric Society* 36:1–37.

Walker, A.R.P.
1951 Cereals, phytic acid and calcification. *The Lancet* 2:244–248.

Walker, A.R.P., B. F. Walker, and M. Wadvalla
1975 An attempt to measure the availability of calcium in edible leaves commonly consumed by South African Negroes. *Ecology of Food and Nutrition* 4(2):125–130.

Warner, Rose Ella, and C. Earle Smith, Jr.
1968 Boll weevil found in pre-Columbian cotton from Mexico. *Science* 162:911–912.

Watt, Bernice K., and Annabel L. Merrill
1963 Composition of foods. *U.S. Department of Agriculture, Agricultural Research Service, Consumer and Food Economics Research Division, Agriculture Handbook* 8. Washington, D.C.

Watt, Kenneth E. F.
1966a *Systems Analysis in Ecology.* New York: Academic Press.
1966b The nature of systems analysis. In *Systems Analysis in Ecology,* edited by Kenneth E. F. Watt. New York: Academic Press. Pp. 1–14.
1968 *Ecology and Resource Management.* New York: McGraw-Hill.

Webb, Thompson, III
1974a Corresponding patterns of pollen and vegetation in lower Michigan: A comparison of quantitative data. *Ecology* 55:17–28.
1974b A vegetational history of northern Wisconsin: Evidence from modern and fossil pollen. *American Midland Naturalist* 92:12–34.

Webb, Thompson, III, and Reid A. Bryson
1972 Late- and Postglacial climatic change in the northern Midwest, U.S.A.: Quantitative estimates derived from fossil

pollen spectra by multivariate statistical analysis. *Quaternary Research* 2:70–115.

Webb, Thompson, III, and D. R. Clark

1977 Calibrating micropaleontological data in climatic terms: A critical review. *Annals of the New York Academy of Sciences* 288:93–118.

Webb, Thompson, III, and John H. McAndrews

1976 Corresponding patterns of contemporary pollen and vegetation in central North America. *Geological Society of America Memoir* 145:267–302.

Wendorf, Fred, Angela E. Close, and Romuald Schild

1983 Letter to Boyce Rensberger, Senior Editor, *Science '83,* June 3, 1983.

Wendorf, Fred, Romuald Schild, N. el Hadidi, Angela E. Close, M. Kobusiewicz, H. Wieckowska, B. Issawi, and H. Haas

1979 The use of barley in the Egyptian Late Paleolithic. *Science* 205:1341–1347.

West, Gerald James

1978 *Recent Palynology of the Cedar Mesa Area, Utah.* Ph.D. dissertation, Department of Anthropology, University of California, Davis.

Whalen, Michael E.

n.d. The Oaxaca Chert Survey Project. Manuscript on file, Phoenix Memorial Laboratory, University of Michigan, Ann Arbor.

Whallon, Robert

1973 Spatial analysis of occupation floors I: Application of dimensional analysis of variance. *American Antiquity* 38:266–278.

1974a Spatial analysis of occupation floors II: Application of nearest neighbor analysis. *American Antiquity* 39:16–34.

1974b Reply to Riley and Schiffer. *American Antiquity* 39:492–494.

Whitaker, Thomas W., and W. P. Bemis

1975 Origin and evolution of the cultivated *Cucurbita. Bulletin of the Torrey Botanical Club* 102:362–368.

Whitaker, Thomas W., and Hugh C. Cutler

1971 Pre-historic cucurbits from the Valley of Oaxaca. *Economic Botany* 25(2):123–127.

Whitaker, Thomas W., Hugh C. Cutler, and Richard S. MacNeish

1957 Cucurbit materials from three caves near Ocampo, Tamaulipas. *American Antiquity* 22:352–358.

White, Theodore

1953 A method of calculating the dietary percentage of various food animals utilized by aboriginal peoples. *American Antiquity* 18:396–398.

Whitecotton, Joseph W.

1977 *The Zapotecs: Princes, Priests, and Peasants.* Norman: University of Oklahoma Press.

Wilkes, H. Garrison

1972 Maize and its wild relatives. *Science* 177:1071–1077.

Willey, Gordon R.

1966 *An Introduction to American Archaeology.* (Vol. 1): *North and Middle America.* Englewood Cliffs, N.J.: Prentice-Hall.

Williams, Howel, and Robert F. Heizer

1965 Geological notes on the ruins of Mitla and other Oaxacan sites, Mexico. *Contributions of the University of California Archaeological Research Facility* 1:41–54.

Wilson, Edward O.

1979 Comparative social theory. *The Tanner Lecture on Human Values.* University of Michigan, Ann Arbor.

Winer, B. J.

1962 *Statistical Principles in Experimental Design.* New York: McGraw-Hill.

Winterhalder, Bruce, and Eric Alden Smith (editors)

1981 *Hunter–Gatherer Foraging Strategies: Ethnographic and Archaeological Analyses.* Chicago: University of Chicago Press.

Witthoft, John

1967 Glazed polish on flint tools. *American Antiquity* 32(3):383–388.

Wobst, H. Martin

1974 Boundary conditions for paleolithic social systems: A simulation approach. *American Antiquity* 39:147–178.

World Health Organization (WHO)

1963 Expert Committee on medical assessment of nutritional status. *World Health Organization Technical Reports Series* 258. Geneva: World Health Organization.

1973 Protein and energy requirements. *World Health Organization Technical Reports Series* 522. Geneva: World Health Organization.

Wright, Henry T., and Melinda Zeder

1973 The simulation of a linear exchange system under equilibrium conditions. Paper presented at the Annual Meeting of the Society for American Archaeology, San Francisco, May 1973.

1977 The simulation of a linear exchange system under equilibrium conditions. In *Exchange Systems in Prehistory,* edited by Timothy K. Earle and Jonathan E. Ericson. New York: Academic Press. Pp. 233–253.

Wright, Herbert E., Jr.

1974 The environment of early man in the Great Lakes area. In Aspects of upper Great Lakes anthropology: Papers in honor of Lloyd A. Wilford, edited by Elden Johnson. *Minnesota Historical Society, Minnesota Prehistoric Archaeology Series* 11:8–14.

1977 Environmental change and the origin of agriculture in the Old and New Worlds. In *Origins of Agriculture,* edited by Charles A. Reed. The Hague: Mouton Publishers. Pp. 281–318.

Yarnell, S. H.

1965 Cytogenetics of the vegetable crops IV: Legumes. *The Botanical Review* 31(3):247–330.

Yellen, John E.

1977 *Archaeological Approaches to the Present: Models for Reconstructing the Past.* New York: Academic Press.

Zeigler, Bernard P.

1976 *Theory of Modeling and Simulation.* New York: John Wiley and Sons.

Index

A

B

C

T

V

W

Z